THE WAY NATURE WORKS

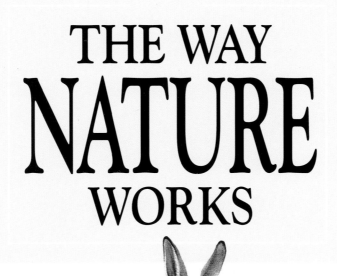

THE WAY NATURE WORKS

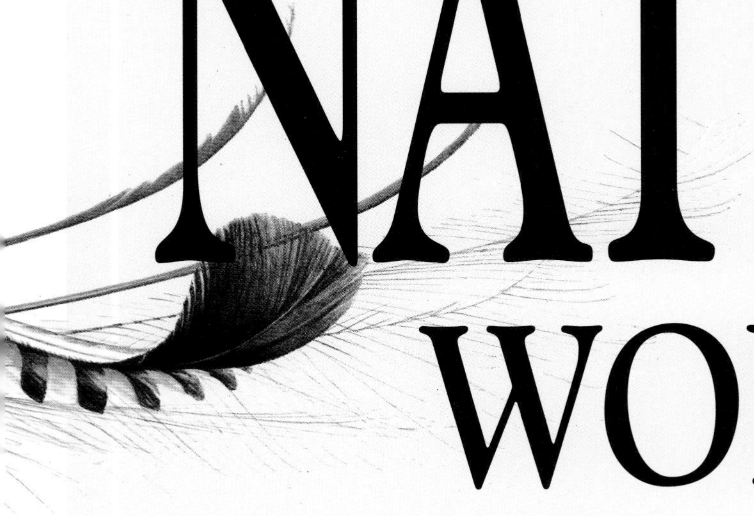

Macmillan Publishing Company
New York

Maxwell Macmillan International
New York Oxford Singapore Sydney

The Way Nature Works
Contributors:

Jill Bailey
Jonathan Elphick
Dr. Derek Elsom
Linda Gamlin
Dr. Tony Hare
Eleanor Lawrence
Dr. Rebecca Renner
Dr. Sara Oldfield
Chris Pellant
Brenda Walpole
Martin Walters

Consultants:

Dr. Gisela Benecke
Dr. Paul Browning
Dr. Mike Cheadle
Dr. Malcolm Coe
Dr. Bill Dolling
Thomas Heppel
Dr. Manfred Hoffmeister
Dr. Charles Jarvis
Dr. Henning Kahle
Professor Franz-Dieter Miotke
Sabine Müller
Ursula Rzepka
Penny Tranter

Senior Executive Editor Robin Rees

Senior Editor Clifford Bishop

Editors Steve Luck
Gavin Sweet
John Morton
Marek Waliesewicz
Mike Darton
Clint Twist

Art Editor Ted McCausland

Senior Designer Iona McGlashan

Designer Jean Jottrand

Picture Research Ann-Marie Erlich

Proofreaders Fred and Kathie Gill

Indexer Marie Lorimer

Typesetter Kerri Hinchon

Production Sarah Schuman

Edited and Designed by
Mitchell Beazley International Ltd,
Michelin House, 81 Fulham Road,
London SW3 6RB

Macmillan Publishing Company
866 Third Avenue
New York, NY 10022

Macmillan Publishing Company is part of the Maxwell Communication Group of Companies.

Library of Congress Cataloging-in-Publication Data

The Way nature works
p. cm.
Includes index.
ISBN 0-02-508110-1
1. Natural history – Enyclopedias. 2. Science – Encyclopedias.
QH13.W38 1992
508–dc20 92-12283
CIP

10 9 8 7 6 5 4 3 2 1
Printed in the United States of America

Introduction

The Way Nature Works has been created to give non-specialists a real understanding of the way our planet and its living organisms function. It provides scientific answers to questions that arise when looking at the world around us, watching nature documentaries, or studying – answers that strip away the mystery and cut through to the fundamental processes.

From our Earth and its atmosphere, through evolution, and on to the extraordinary diversity of form and behavior of today's living species, the encyclopedia covers the broad sweep of both earth and life sciences, employing hundreds of illustrations to translate detailed technical information into terms that everyone can follow.

Some of the easiest questions to ask are the most difficult to answer, but the book does not avoid "difficult" topics – if a process is essential to a true understanding of the dynamics of Earth and life, then it is included, along with the best explanation science has to offer.

How the Book Works

The Way Nature Works is organized in dynamic themes that reflect the approach of modern science, drawing together common processes and activities rather than dividing and classifying. It covers earth sciences in two sections, and life sciences in seven sections, though there are frequent cross-connections between the two disciplines. Within each section, information is arranged in self-contained double-page topics, which have been selected both for their immediate interest and to convey essential components of information, so that when taken as a whole the encyclopedia provides a full and balanced foundation of knowledge. Cross-connections lead on to topics of related interest, weaving an intriguing web throughout the book that enables the reader to obtain a complete overview of a particular subject.

The encyclopedia may be used in several different ways. Looking at the illustrations and reading their captions will provide a concise and rapid understanding of a specific topic. Reading the main text in addition gives the broader context and general principles and leads to a more thorough appreciation of the subject. The book may also be used to look up specific subjects in the extensive index, which includes all the species of animals and plants found in the book. A glossary of common technical terms forms a useful additional reference.

The Earth Sciences

In recent years geology has tended to be merged with many other physical sciences under a broader category, earth sciences, as knowledge of how the Earth works has moved on from mere description to a much better understanding of the underlying processes. Section 1, *Shaping the Planet*, and section 2, *The Power of the Atmosphere*, demonstrate the new, dynamic approach, and the illustrations clearly show the forces that are at work and the effects they cause – within the Earth, at the surface, in the oceans and in the atmosphere.

The Life Sciences

Similarly, the study of life on Earth has become more interdisciplinary, and subjects such as botany and zoology are commonly referred to as life sciences. The processes which underlie evolution, behavior and ecology are now key areas of study, and this is reflected in the seven sections of the encyclopedia covering the life sciences.

Section 3, *Evolution and Adaptation*, examines the rise of life from primitive organisms to present-day species and their extraordinary adaptations. In section 4, *Reproducing to Survive*, the reproductive strategies of plants and animals are explained, and the section includes the elaborate courtship rituals employed by some species. Section 5, *The Search for Food*, deals with the way plants and animals obtain nourishment, and section 6, *Movement and Shelter*, looks at the mechanics of animal movement and how they build their homes, including details of the social structures of species that collaborate to build. *Attack and Defense* (section 7) presents the weaponry and tactics employed in the battle between predator and prey. Section 8, *Senses and Communication*, reveals the secrets of the acute senses that many animals possess – some of which are unknown in humans – and describes the wide range of techniques they use to communicate with each other.

The final section, *The Living Environments*, looks at global habitats and how different species of plants and animals live and interact. The impact of human activities on the environment is an increasing concern for everyone, and an explanation is included of the effects of pollution on land, air and sea.

Contents

4 *Reproducing to Survive*

5 *The Search for Food*

6 *Movement and Shelter*

7 *Attack and Defense*

8 *Senses and Communication*

Units of Measurement

Physical Quantity	International System of Units
time	second (s)
length	meter (m)
area	square meter (m^2)
volume	cubic meter (m^3)
mass	kilogram (kg)
density	kilogram/cubic meter (kg/m^3)
acceleration	meter/second squared (m/s^2)
speed	meter/second (m/s)
force	newton ($N = m\ kg/s^2$)
pressure	pascal (Pa = force/area = kg/ms^2)
energy	joule (J = N m)
electrical potential	volt (V)
temperature	degree Celsius (°C)

Conversions

1μ (micron) = 0.000001 meters
10 millimeters = 1 centimeter = 0.39 inch
100 centimeters = 1 meter = 3.28 feet
1000 meters = 1kilometer = 0.62 miles

12 inches = 1 foot = 30.40 centimeters
3 feet = 1 yard = 0.91 meter
1760 yards = 1 mile = 1.61 kilometers

10,000 sq. centimeters = 1 sq. meter = 1.20 sq. yards
1,000,000 sq. meters = 1 sq. kilometer = 0.39 sq. miles

144 sq. inches = 1 sq. foot = 929 sq. centimeters
9 sq. feet = 1 sq. yard = 0.84 sq. meter
3,097,600 sq. yards = 1 sq. mile = 2.59 sq. kilometers

1,000 cubic millimeters = 1 cubic centimeter = 0.06 cubic inch
1,000,000 cubic centimeters = 1 cubic meter = 1.31 cubic yard

1,728 cubic inches = 1 cubic foot = 28,317 cubic centimeters
27 cubic feet = 1 cubic yard = 0.76 cubic meter

16 ounces = 1 pound = 0.45 kilograms
2,240 pounds = 1 tonne = 1.02 tons
1,000 grams = 1 kilogram = 2.2 pounds
1,000 kilograms = 1 ton = 0.98 tonnes

X°Celsius = (5X/9 + 32)°Fahrenheit

1 mile per hour = 1.61 kilometers per hour

1 atmosphere = 101,325 Pa
1 bar =0 1,000 millibars = 100,000 Pa

1 Shaping the Planet

Earth's Structure
Earth's Magnetic Field
Plate Tectonics
Mountains
Earthquakes
Ice Ages
Glaciers
Polar Regions
Seas and Oceans
Ocean Circulations

Earth's History

Era	Period	Epoch	Millions of years ago
Cenozoic	Quaternary	*Holocene (Recent)*	0.01
		Pleistocene	2
	Tertiary	*Pliocene*	5
		Miocene	25
		Oligocene	38
		Eocene	55
		Paleocene	65
Mesozoic	Cretaceous		144
	Jurassic		213
	Triassic		248
Paleozoic	Permian		286
	Carboniferous	*Pennsylvanian*	320
		Mississippian	360
	Devonian		408
	Silurian		438
	Ordovician		505
	Cambrian		590
Pre-Cambrian			4,600

Volcanoes
Rock Formation
Sedimentary Rocks
Fossilization
Soils
Minerals
Coal, Oil and Gas
Caves
Tides
The Ocean Floor
Island Formation
Deserts
Coastlines
Rivers
Lakes

Journey to the Center of the Earth

How the Earth is structured

Travel down vertically through the Earth's crust and you would find an alien and truly inhospitable world. Even a few miles down there is an inferno of heat and pressure, and below lie molten rock materials that rarely reach the surface. At deeper levels still, the planet seems to be structured in "shells" of different materials separated by abrupt changes in density and composition. The deepest hole ever drilled into the Earth reached less than 8 miles (12 kilometers), so scientists must use more ingenious methods to find out about the Earth's interior.

The Earth probably began in the coming together of dust and gas particles around the newly formed Sun about a billion years after the Big Bang (the probable origin of the universe, some 20 billion years ago). Extraterrestrial materials therefore provide a surprisingly good way of investigating the internal structure of our planet.

Meteorites are thought to be the fragmented remains of small planetlike bodies only a few dozen miles across. The different types of meteorites would then represent different layers in the original planetoid structure. Iron meteorites (*siderites*) are about 97 percent metal, mainly nickel-iron, and being so dense probably represent the planetoid core. *Chondrites* are of silicate composition and may represent the lower mantle. *Achondrites* are also silica based, but with a different composition. They may be fragments of crust and upper mantle.

Volcanic information

Volcanic outpourings also give us an insight into the Earth's hidden depths, but not as much as might be expected because, although lavas are generated well below the surface, the actual depth is shallow when compared with the Earth's radius of 3,950 miles (6,370 kilometers). The basaltic lavas that issue through the ocean ridges are created from the Earth's mantle by differentiation (a number of processes by which different rocks can be produced from a single parent magma). This suggests that the mantle itself is far denser than these basalts, because material that is less dense tends to rise to the surface.

Shocks to the system

Another major source of information about the Earth's structure is the study of the seismic, or shock, waves produced by earthquakes. One of the main measurements determined by the paths of these waves is the extent of the Earth's core, since shear waves do not pass through liquid material and are absorbed by a zone that starts at a depth of 1,800 miles (2,900 kilometers). From here to the center of the Earth is the planet's core, although the core is not liquid throughout.

The boundary that separates the core from the overlying mantle is called the Gutenberg Discontinuity and is the region where, over a few miles, the mantle rocks change to matter

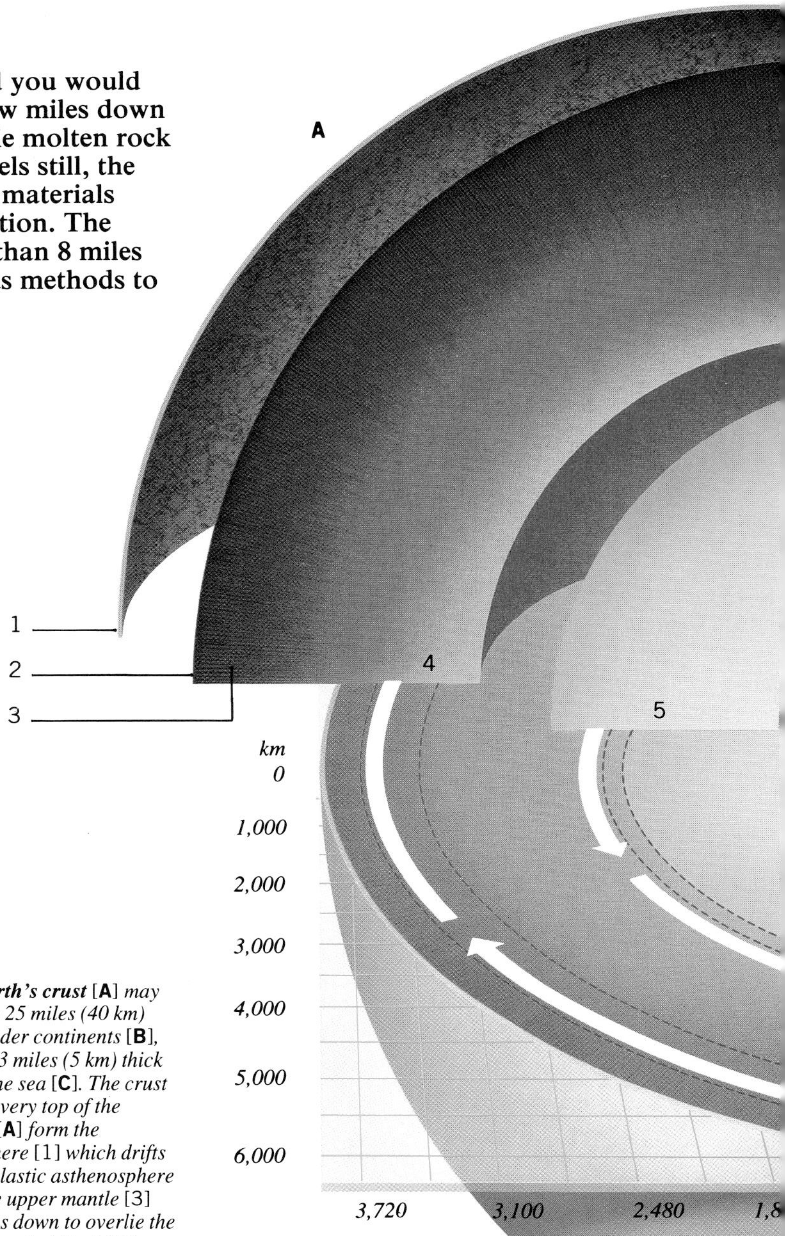

*__The Earth's crust__ [**A**] may be up to 25 miles (40 km) thick under continents [**B**], or only 3 miles (5 km) thick under the sea [**C**]. The crust and the very top of the mantle [**A**] form the lithosphere [1] which drifts on the plastic asthenosphere [2]. The upper mantle [3] stretches down to overlie the lower mantle [4] at 430 miles (700 km) depth. From the surface the temperature inside the Earth increases by 85°F/mile (30°C/km) so that the asthenosphere is close to melting point. After 60 miles (100 km) depth the rate of temperature increase slows dramatically. Because we know so little about what happens deep within the Earth, there is a great uncertainty about the temperature at the core. Pressure also increases with depth: already in the asthenosphere the pressure is equal to 250,000 atmospheres. At the Earth's core it may go as high as 4 million atmospheres. The density near the surface of the Earth is only about 2 ounces/cubic inch, and this does not vary by much more than 30 percent right down through the lower mantle. The increase in density that does occur is due to closer atomic packing under pressure. There is a leap in density to 10 grams/ cubic centimeter in passing to the outer core [5], and a further leap to between 7 and 9 ounces/cubic inch in the inner core [6]. The heat at the core fuels vast convection currents [7] in the material of the mantle.*

*Continental crust is made by many different processes and is therefore difficult to generalize. A typical cross section [**B**] might well consist of deformed and metamorphosed sedimentary rocks at the surface [1]*

Connections: Coal, Oil and Gas 34 Earthquakes 20 Earth's Magnetic Field 14 Island Formation 52 Minerals 32 Mountains 18 Plate Tectonics 16

of much higher density. From the behavior of seismic waves and from meteor-ite and other evidence it is believed that the inner core is solid iron and nickel at a temperature over 7,000°F. The outer core is less dense because it comprises liquid metal.

The mantle has also been investigated seismically. The upper limits of the mantle are where the seismic waves are slowed down at the Mohorovičić Discontinuity (*Moho*). This marks the change from mantle to less dense crustal rock. Within the mantle, density of rock material increases rapidly with depth.

Scanned images

In recent years the mantle has, in addition, been investigated by seismic tomography, a method similar to CAT (computer-aided tomography) used in medical scanning.

The shock waves of more than 10,000 earthquakes are monitored every year. The actual velocity at which the waves travel in the mantle is the most significant factor. "Slow" and "fast" areas in the mantle are determined, and from an intersecting network of seismic waves a three-dimensional model of the mantle can be plotted. Hot parts of the mantle transmit waves more slowly than areas that are cooler. This difference allows the mapping of the heat-transporting convection cells in the mantle, which cause movement in the continental plates.

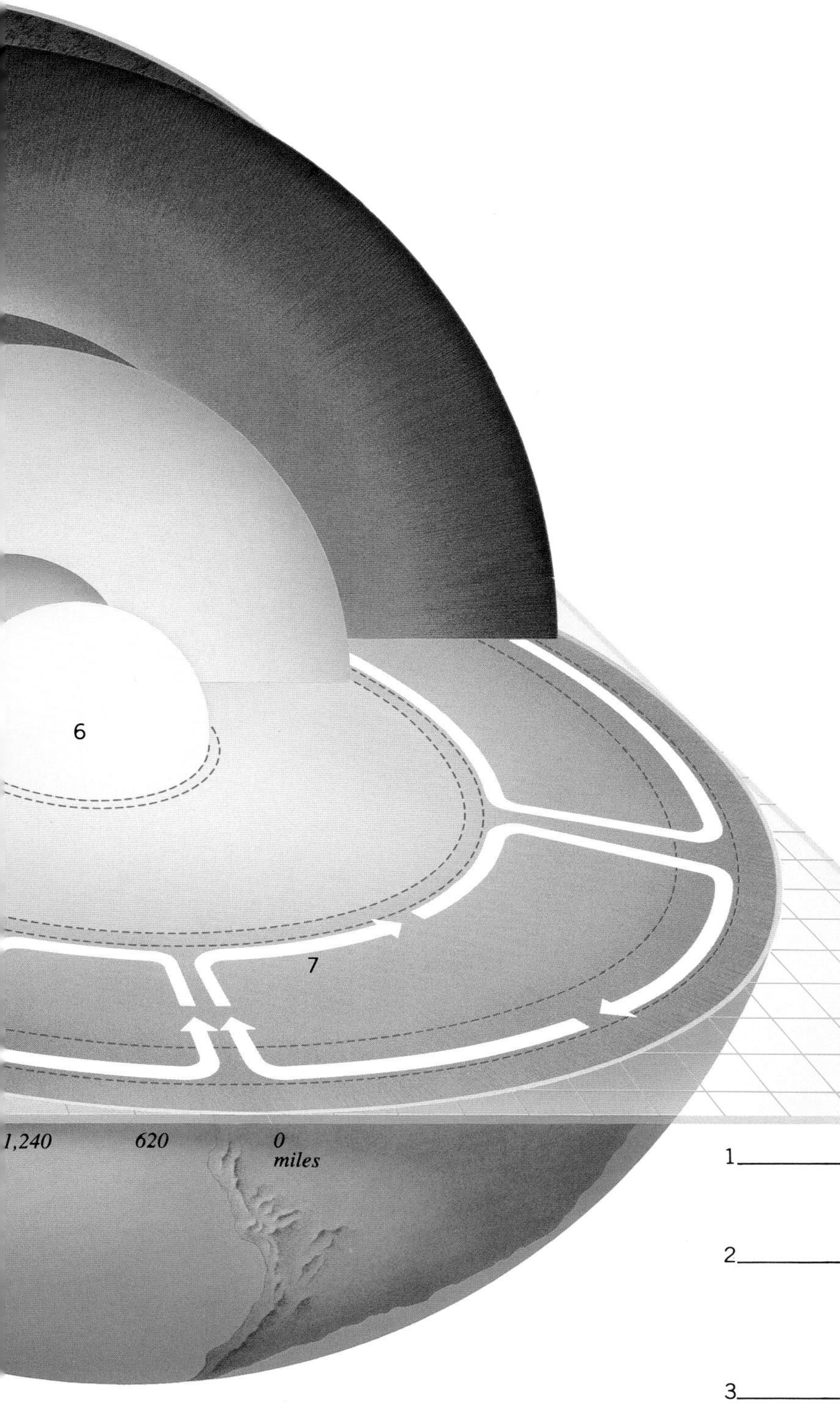

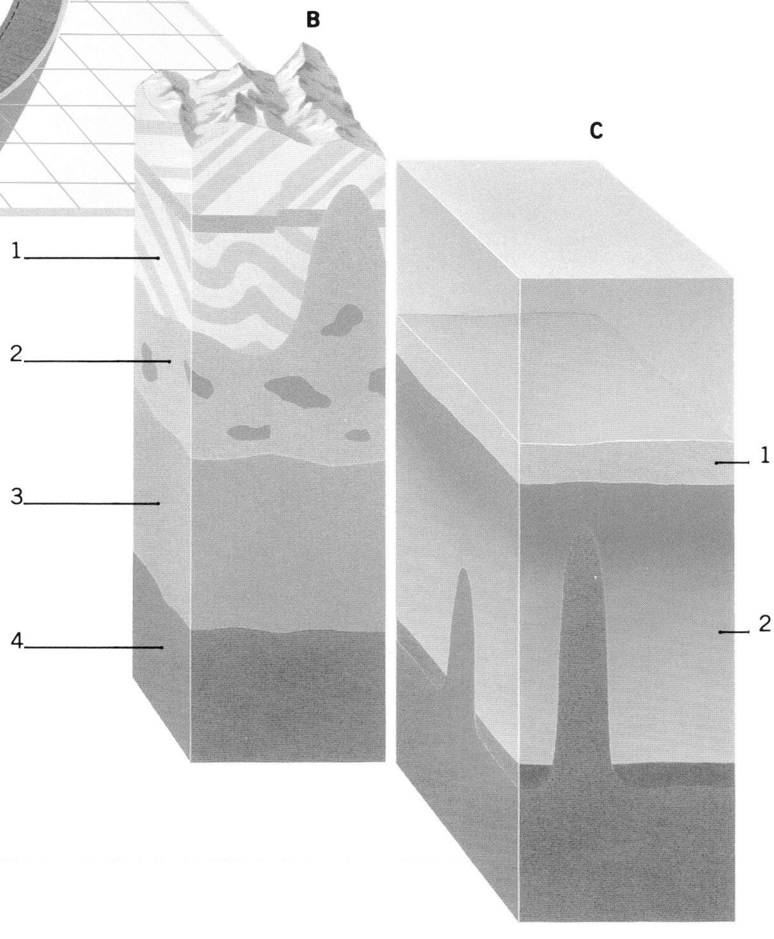

underlain by a granite intrusion [2]. *The remaining crust would consist of metamorphosed sedimentary rock and igneous rock* [3] *reaching to the mantle* [4].

In oceanic crust [**C**], *ocean sediments* [1] *overlie a basalt that is pillowed and also chemically altered by sea water. Below this layer* [2] *the basalt is made up of dikes – small vertical intrusions that are packed so close together they intrude each other.* Gabbros *are coarse-grained igneous rocks with a very similar composition to basalt. They are sometimes layered.*

Rock Formation 24 **Sedimentary Rocks** 26 **Volcanoes** 22

Force of Attraction

How the Earth acts as a magnet

The Earth's magnetic field has been getting weaker for the last 2,000 years and may temporarily disappear altogether within the next 2,000 years. But this is only a part of a cycle that has already been enacted many times in the Earth's history, as the magnetic field slowly reverses itself and north becomes south. Despite the constant variation, people have been using compasses to navigate for millennia. Animals such as pigeons, bees and salmon – which contain grains of magnetic material in their bodies – may have been using magnetism to navigate for much, much longer.

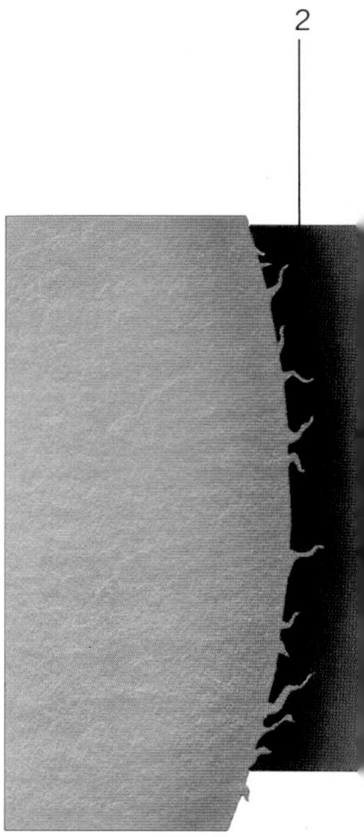

A

The magnetic field *[A] of the Earth extends far out into space as the magnetosphere [1]. Streams of charged particles from the Sun – called the solar wind [2] – bend the magnetosphere out of the normal ball shape of a dipole magnet and into the shape of a teardrop. The magnetic field of the Earth deflects the dangerous solar wind, which is mainly protons, electrons and alpha radiation, and keeps the particles from the planet's surface. Without it life on Earth would be impossible. Some solar wind particles do enter the atmosphere at the polar cusps [3], however, as shown by the* auroras *(below).*

The Earth's magnetic field is similar to that of a simple bar magnet. Because of its two opposing poles – north and south – it is known as a *dipole* field. But the Earth obviously does not have a giant magnet buried in its center. A dipole field can in fact also be created by a coil of wire carrying an electric current. The north pole is at the end of the coil where the current appears to be turning counterclockwise. Therefore scientists speculate that the Earth's field is due to charged particles – like the electrons in an electric current – moving about in the molten center of the planet. The Earth's core is made of iron and nickel, and in fact the inner core is solid. But the temperatures here – some 9,000°F – produce convection currents in the molten metal surrounding the central core. These currents, it is argued, create magnetic fields that will continue for as long as the heat and the Earth's rotation are sustained to cause them.

The origins of magnetism

The Earth's magnetic field was probably initiated by the Sun's magnetic field when the Earth was still a swirling ball of dust and gas, billions of years ago. As electrically conducting material in the dust cloud swept through the solar field, the Sun's magnetism exerted a force on its electrons. The electrons began to move, which created an electric current which in turn created the magnetic field that was the origin of the field of today.

The permanent record

For a number of reasons the Earth's magnetic field varies in strength. There are daily variations in the overall magnetic field that may be caused by disturbances in the atmosphere. Much more locally, changes in the rocks near the surface produce variations called magnetic anomalies, which are measurable with sensitive electronic apparatus.

There are three main types of magnetic materials. *Ferromagnetic* substances like nickel or iron will become magnetized in a magnetic field and will still be magnetized after they are taken from the field. *Paramagnetic* materials like copper – or even oxygen – will be magnetized by a field but will lose their magnetism when taken out of it. When *diamagnetic* materials are placed in a magnetic field they become magnetized in the opposite direction to the field.

The rocks affected most by magnetism are those that contain iron-bearing minerals such as magnetite and titanomagnetite. The latter is the most common magnetic mineral. Dark volcanic rocks such as basalt are over 100 times more likely to acquire magnetism than sedimentary rocks such as sandstone. When basalts are molten, the tiny crystals of magnetite forming in the liquid lava may take on a magnetic polarity aligned to that of the Earth's magnetic field at that precise time.

Geologists can map rock magnetism to show where basalts and related rocks occur in intrusions such as sills and dikes. Iron ore can also be found in this way, and when a mining company wishes to prospect a new region, one of the initial stages is often an aeromagnetic survey. This rapid and inexpensive method can be used over land or sea and was used extensively in the preliminary exploration of the North Sea oilfields.

The curtains of light *that occur in the Arctic and Antarctic skies (above), shimmering in bright neon colors, are auroras. They are caused by the focusing effect of the Earth's magnetic field on the highly charged particles of the solar wind. These particles are funneled down through the rarefied upper atmosphere (ionosphere) above the poles. There they strike atoms or molecules of air, either ionizing them (knocking free an electron) or exciting (energizing) them. As the excited air molecules or atoms return to their normal states they emit electromagnetic energy in the form of visible light.*

Connections: **Atmospheric Phenomena** 66 **Coal, Oil and Gas** 34 **Earth's Structure** 12 **Minerals** 32 **Migration Techniques** 230 **Plate Tectonics** 16

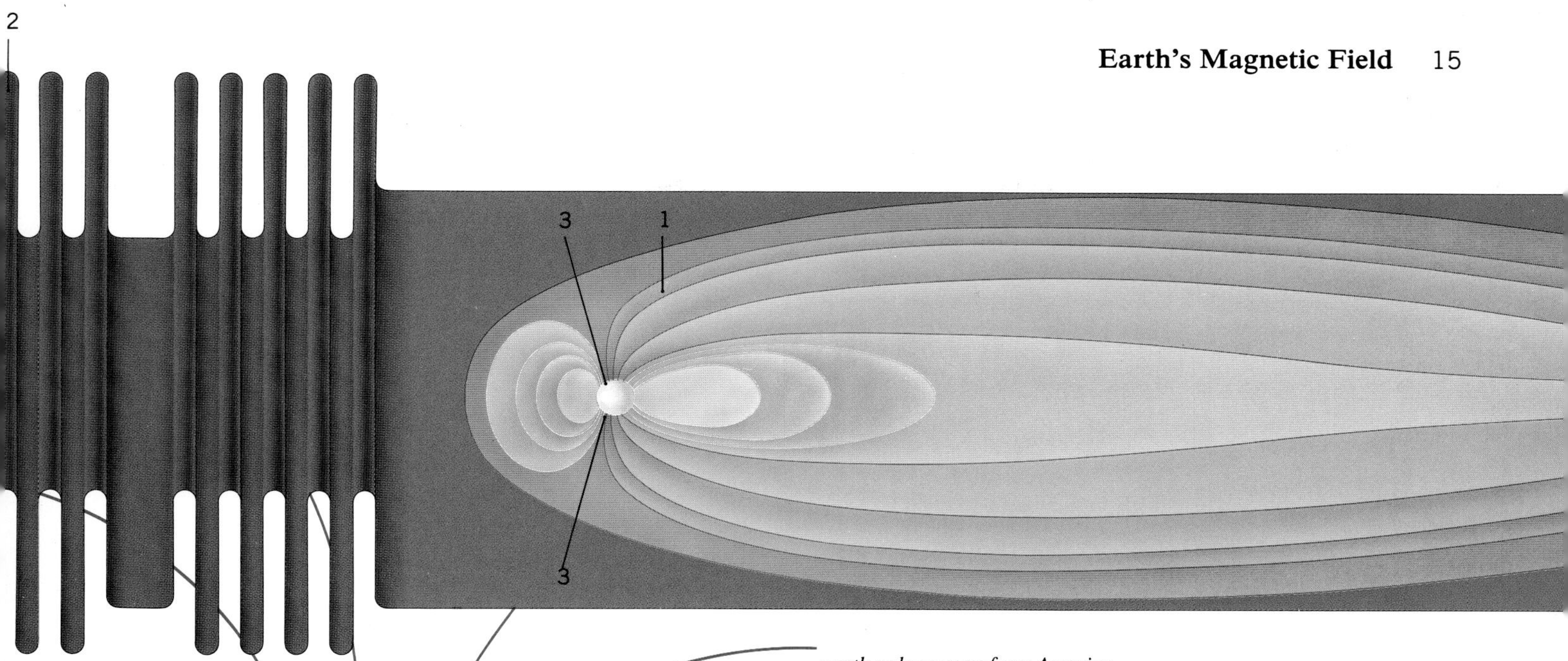

B

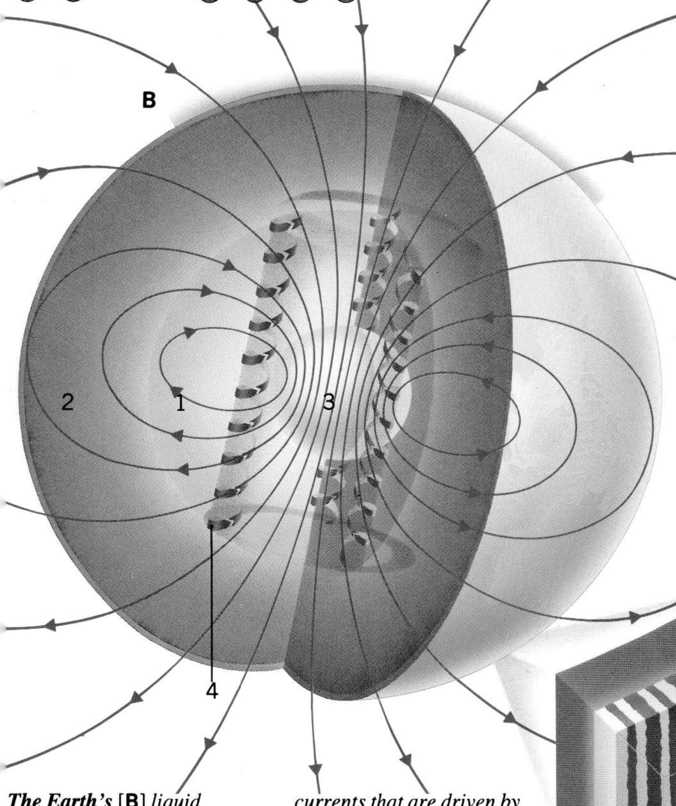

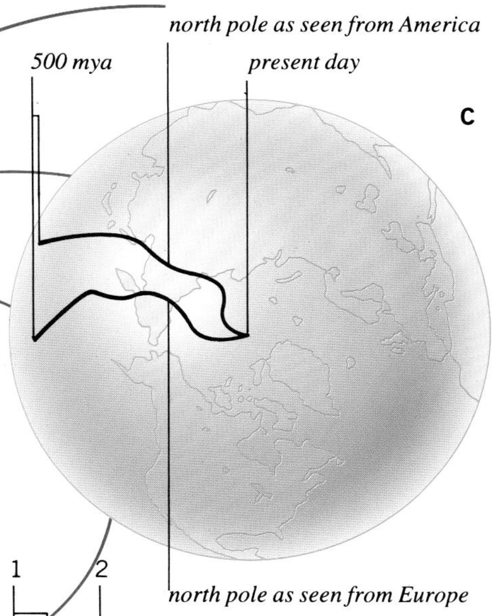

C

The magnetic poles [C] *do not coincide with the geographic poles, and in fact they do wander as the Earth's magnetic field varies. Yet if the position of the poles is plotted back in time by studying the magnetized rock on different continents, it seems that the poles have wandered many thousands of miles more than seems plausible. The magnetic pole positions seem to become increasingly farther away from the geographic poles with the increasing age of the rocks studied. But the apparent polar wandering is a product of continental drift and not actual large-scale movement of the magnetic poles.*

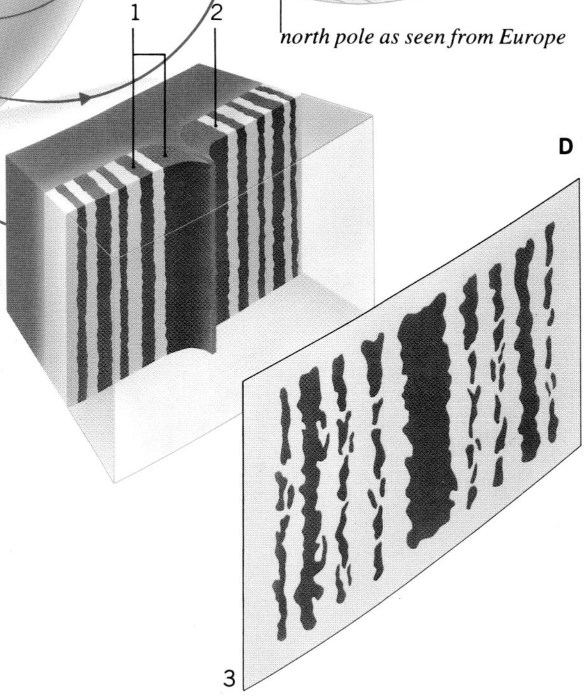

D

As lavas that are formed at mid-ocean ridges [D] *cool, the magnetically susceptible minerals, like magnetite, that form within them get magnetized in the direction of the existing magnetic field. As the polarity of the field reverses with time, "stripes" of normally magnetized* [1] *and reverse magnetized* [2] *rock get laid down successively by the spreading sea bed. Of course the pattern of magnetic stripes is not very neat, and in reality shows many irregularities and breaks* [3]. *Rocks that cool more slowly, deeper down, record the direction of the average magnetic field throughout the time that the magma was solidifying.*

The Earth's [B] *liquid outer core* [1] *is 1,400 miles (2,200 km) thick and flows between the mantle* [2] *and the solid inner core* [3]. *It is a circulating mass of nickel-iron alloy, but is not intrinsically magnetic because no material could retain its magnetism at the temperatures at the Earth's core. Instead the Earth's magnetic field is generated by the circulation of the electric charges in the molten mass. The matter moves because of convection currents that are driven by the difference in heat between the inner and outer core (about 4,530°F). The coriolis forces that arise from the Earth's rotation – and, for example, cause cyclonic storms on the surface – could twist these into "rollers" of fluid* [4]. *The rollers – nobody is sure exactly how many would be aligned along the Earth's axis of rotation – act like the coil of wire in a dynamo to generate a dipole magnetic field.*

Rock Formation 24 **Volcanoes** 22

The Shifting Surface

How plate tectonics shape the Earth

The surface of the Earth is made up of gigantic rocky "plates" of crustal material up to 60 miles (100 kilometers) thick. But these plates are not set fast, immobile; they are moving very slowly – on average no faster than a human fingernail grows – riding on top of more fluid layers beneath, colliding and separating again, and carrying with them the oceans, the continents – and us. When two of these gigantic plates collide, huge energies are released, manifesting themselves in the birth of volcanoes, in earthquakes, and in the formation of mountain ranges.

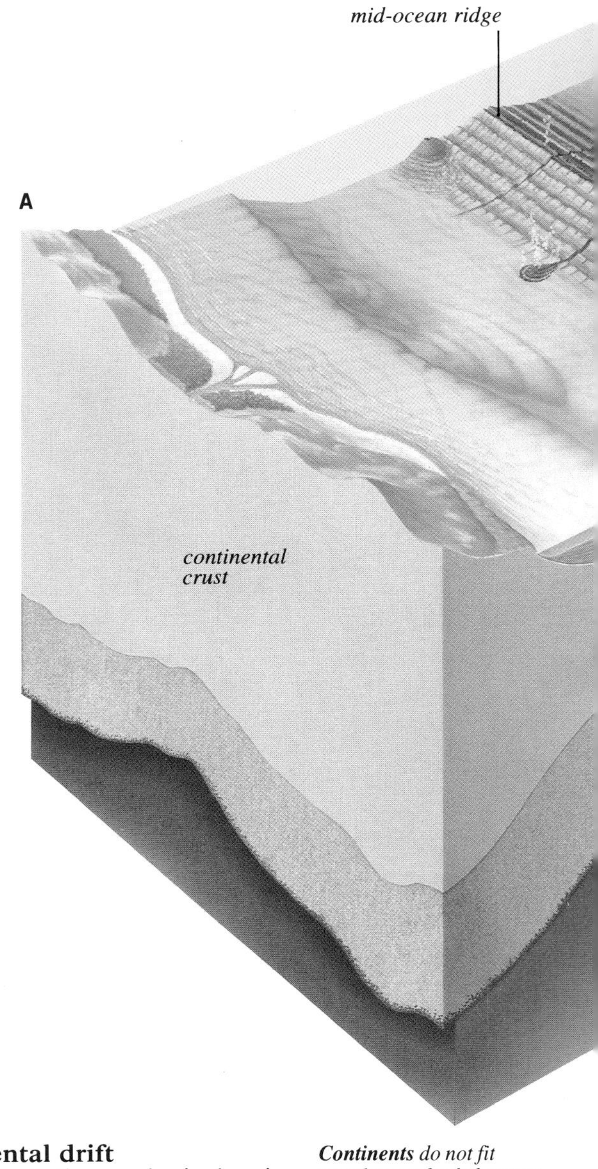

In 1912 the German scientist Alfred Wegener realized that the continents had not always been in the same relative positions and proposed the theory of continental drift. But the mechanism for moving the continents remained a mystery to him. Plate tectonics – the idea that the continents ride on a series of rigid but highly mobile plates – has provided an answer in the last 30 years.

The plates are referred to by geologists as the *lithosphere* and are made up of the rocky outer coating of the Earth – the crust – and the upper part of the less solid region, called the mantle, that is found below the crust. Many different plates have been identified, including a number of very large plates (such as the Pacific plate and the Eurasian plate) as well as many smaller, fragmented ones (like those around the Mediterranean, the Middle East and the Caribbean).

Ways of colliding

When an oceanic plate and continental plate collide (as is occurring along the west coast of South America), the less dense continental plate rides above the oceanic plate, which is forced downward into the mantle in a process known as *subduction*. As the oceanic plate is slowly subducted, it is gradually destroyed, pulling the sea bed with it to form a deep ocean trench; descending further, the plate melts. Some of this molten rock then rises into the continental crust above to form volcanoes and other igneous features.

Such wholesale destruction of oceanic crust in subduction zones may suggest that the Earth's skin is shrinking. Plate tectonics, however, works to maintain the balance. The creation of new basaltic crust along mid-ocean ridges compensates for material that is destroyed. The classic instance is found in the Atlantic. Here, an upwelling convection current in the mantle beneath the middle of the Atlantic Ocean is forcing the sea bed apart as molten material rises to the surface and then cools, in the process extending the oceanic plates on either side of the ridge. The longest volcanic mountain range on Earth is found along this ridge.

Thus the sea floor may be seen as a conveyor belt for material, much of which is continually recycled in the Earth's mantle. This allows the ocean floor to spread without increasing the overall size of the Earth's crust.

Below the Earth's plates *[**A**], powerful temperature differences create massive slow movements of molten and plastic rock [1]. Known as* convection currents, *these streams circulate toward the surface of the mantle. There are weak points at the plate margins – usually the mid-ocean ridges where the crust is much thinner. Here two currents collide and separate and the plates bulge and move apart. Magma – molten rock from the mantle – flows up to the surface and solidifies. It pushes the relatively mobile ocean crust apart and, if there are no intervening subduction zones, pushes the continents [2] with them. The rocks on the sea bed grow progressively younger the nearer they are to the ridge.*

Subduction [3] is the plunging of the heavy oceanic crust beneath the lighter continental crust. It results in chains of magmatic volcanoes. Subduction zones exist beneath such diverse landscapes as the Aleutian Islands and the Andes.

Evidence for continental drift

Biological clues support the geological evidence for continental drift. *Glossopteris*, a deciduous seed-fern, is found fossilized in South America and southern Africa as well as in Antarctica, India and Australia. This indicates that all these land areas once formed a single landmass. It is thought that, between 350 and 250 million years ago, South America and Africa were joined together. Tremendous glacial erosion took place on what is now Africa, leaving scratches and grooves on bare rock.

The erosion of the African surface led to the transport and deposition of huge glacial deposits now found in Brazil. Because glaciations are found only in narrow climatic belts this is strong evidence that Africa and South America were joined. Rock formations that are similar in age can also be found in parts of other continents.

Continents *do not fit together perfectly because of erosion and the buildup of sediments at the coastlines. But the plates they ride on do form a global jigsaw. Within the last 250 million years Europe and North America have collided and separated again. This Atlantic Ocean is the fourteenth. At one time South America, Antarctica, Africa and Australia formed one large landmass, a supercontinent called Gondwanaland, and even earlier eastern Asia, Europe and North America were one continent, Laurasia. The Red Sea and African Rift Valley are examples of areas that will form new oceans.*

Connections: **Earth's Structure** 12 **Earthquakes** 20 **Island Formation** 52 **Mountains** 18 **Ocean Floor** 50 **Rock Formation** 24 **Volcanoes** 22

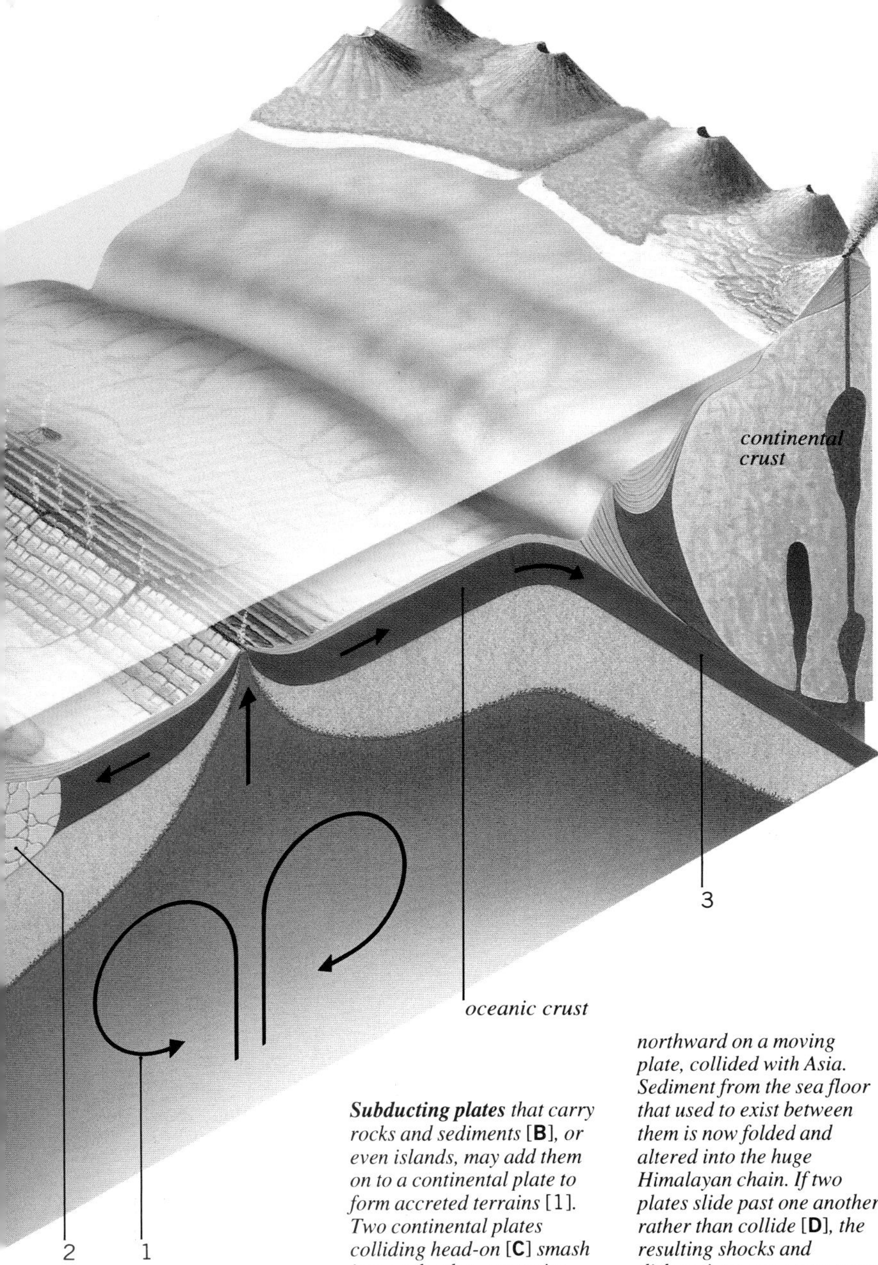

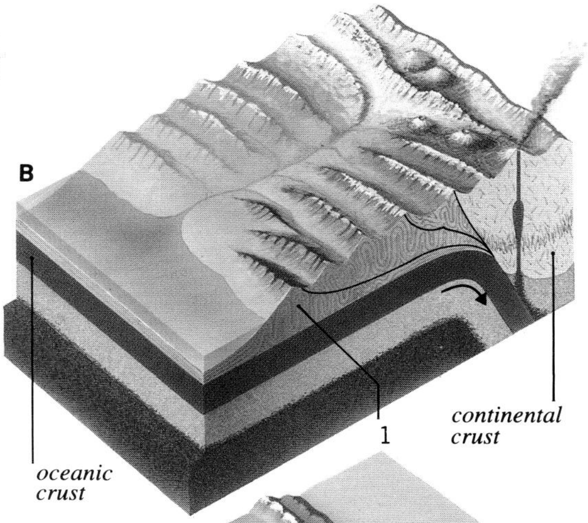

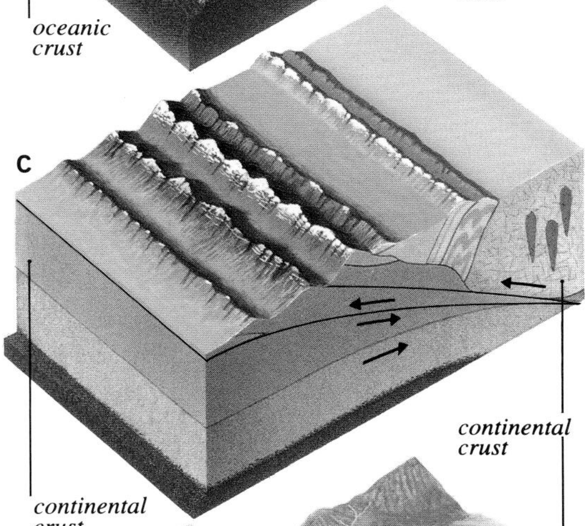

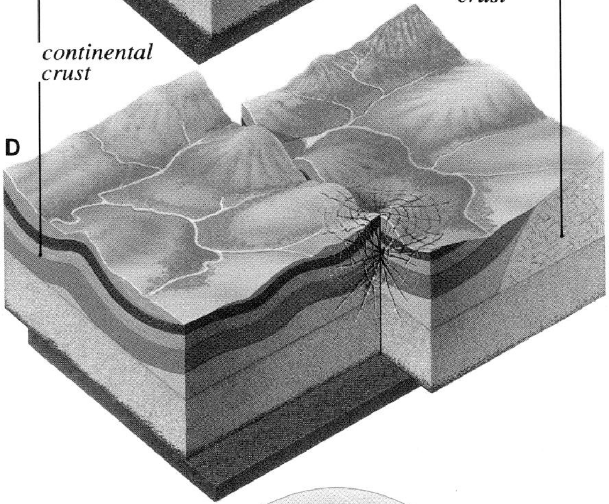

***Subducting plates** that carry rocks and sediments* [**B**]*, or even islands, may add them on to a continental plate to form accreted terrains* [1]*. Two continental plates colliding head-on* [**C**] *smash into each other, rupturing and folding as the energy of the collision is diffused. India, for example, riding northward on a moving plate, collided with Asia. Sediment from the sea floor that used to exist between them is now folded and altered into the huge Himalayan chain. If two plates slide past one another rather than collide* [**D**]*, the resulting shocks and dislocations can cause great surface damage, as in the renowned San Andreas Fault in California.*

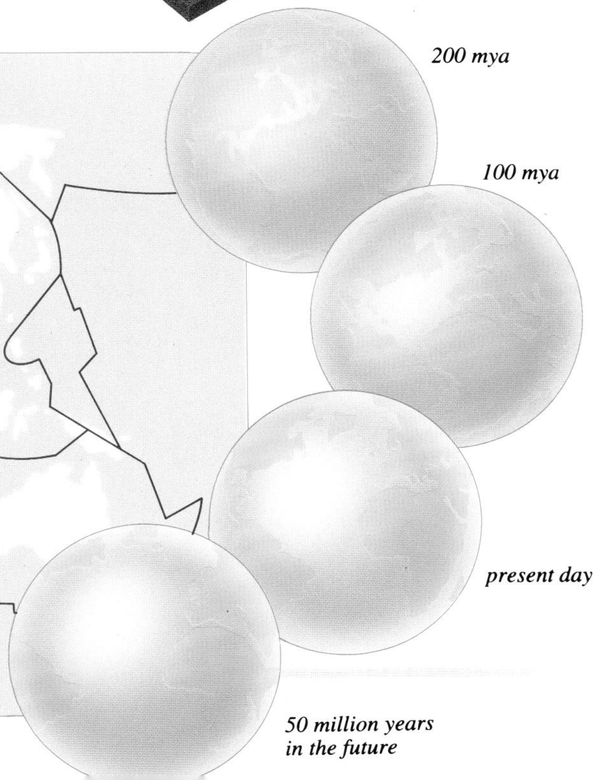

The Roof of the World

How mountains are formed

From the top of Ben Lomond in the Scottish Highlands it is hard to imagine that all the mountains stretching out to either side were once connected to the Appalachian mountains of the eastern United States. But once, before the opening up of the Atlantic Ocean, the Highlands were part of a range of mountains, known as the Caledonides, that formed one huge chain stretching for thousands of miles, from Spitzbergen, north of the Arctic Circle, through Newfoundland to the Blue Ridge and Smoky Mountains of the southern United States.

K

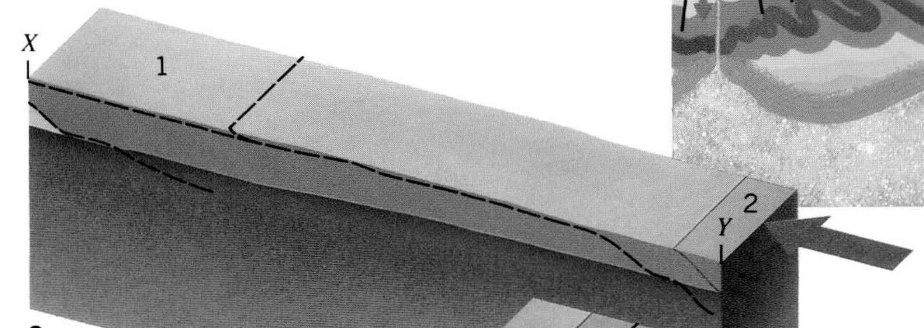

C

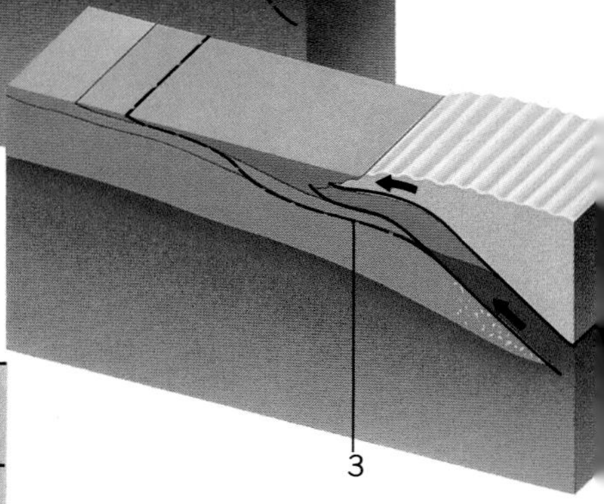

D

A

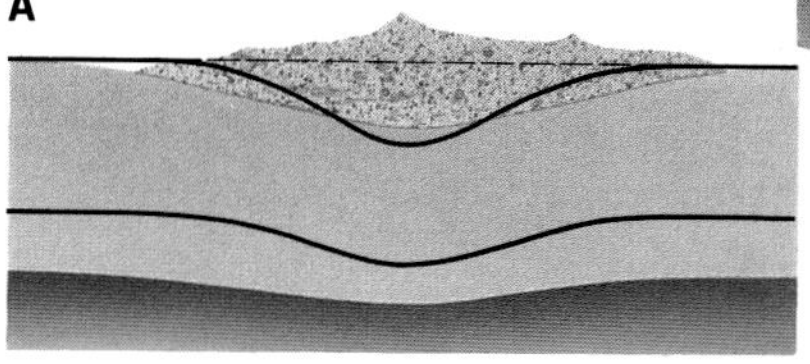

B

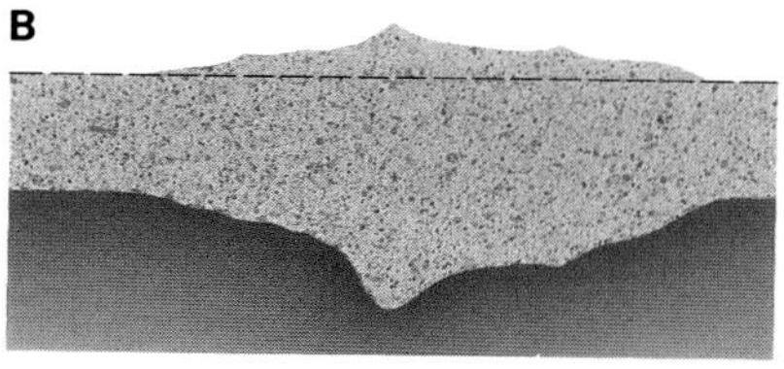

All the great mountain ranges were formed at the edges of the Earth's huge, mobile plates of crust and upper mantle – the lithospheric plates. There, rocks either bend or break under the compressive and extensional forces that result when the plates collide. Folds, which come in all shapes and sizes, mainly form deep within the continental crust. Rocks tend to fault at the shallower levels.

A section through the mountain belt might start with flat-lying sedimentary rocks that gradually, over a distance of a few miles, become more and more folded to create alternating arch-shaped *antiforms* and trough-shaped *synforms*, as in the Jura Mountains of France and Switzerland, or the Valley and Ridge province of the Appalachian Mountains. In other instances, for example the eastern Rocky Mountains in Canada, rocks may be stacked on top of each other by a series of low-angle thrust faults.

In the central part of mountain ranges like the Alps or the Himalayas, where the deformation is most intense, not only are the rocks folded but the folds have been flattened and sheared so that the upper part is moved forward, in some instances by many miles, in structures called nappes. In such mountains, which are formed by the collision of two continental plates, a few very deep major thrust faults slice packets of rock and pile them on top of each other. These faults cut deep into the basement of old, strong, crystalline rocks.

The collapsing crust

But mountains are not static features. As soon as they start to rise, the slow process of erosion starts to wear them down, and sometimes when the horizontal forces that thicken the crust stop – for example, if subduction of one plate under another ceases – then the thickened crust starts to spread. Evidence of this kind of crustal extension is plentiful in the Great Basin and Range area to the west of the Rocky Mountains in the United States, where the alternating basins and upward-tilted ranges are bounded by normal faults.

As the crust spreads apart, blocks drop along the faults to form basins. One – Death Valley – is now below sea level, but at one time it may have been at an altitude of several miles. Some geologists speculate that the Andes and the Tibetan plateau, which lies to the north of the Himalayan range, may also one day collapse in a similar way because these mountains have deep roots of lighter buoyant material that can spread out if it is not squeezed by horizontal forces. Rifts, such as those found in the North Sea, the East African Rift or the Rhine valley, are another type of extensional zone in which a long, narrow block has sunk down between two parallel normal faults.

Erupting giants

But the biggest mountains are formed by volcanic activity, not structural deformation. Two volcanoes in the Hawaiian Islands – Mauna Kea and Mauna Loa – have bases over 88 miles (140 kilometers) across. Although single volcanic mountains occur, they can also be grouped in clusters or chains like the Cascades of the American North-West, which contains Mount St. Helens.

The Himalaya Mountains *are nearly twice as high as the Alps, even though both chains of mountains were formed by the self-same process – the collision of two continents. The reason is partly that the European lithospheric plate is only about half as thick and consequently half as strong as the Indian plate. One theory of how mountains are prevented from sinking into the Earth holds that the weight of the mountain bends the lithosphere and so spreads the load across a broad area* [**A**]. *An older theory assumes that the higher mountain ranges sit on top of areas where the crust is much thicker* [**B**].

Connections: **Earth's Structure** 12 **Earthquakes** 20 **Island Formation** 52 **Minerals** 32 **Mountain Life** 306 **Plate Tectonics** 16 **Rock Formation** 24

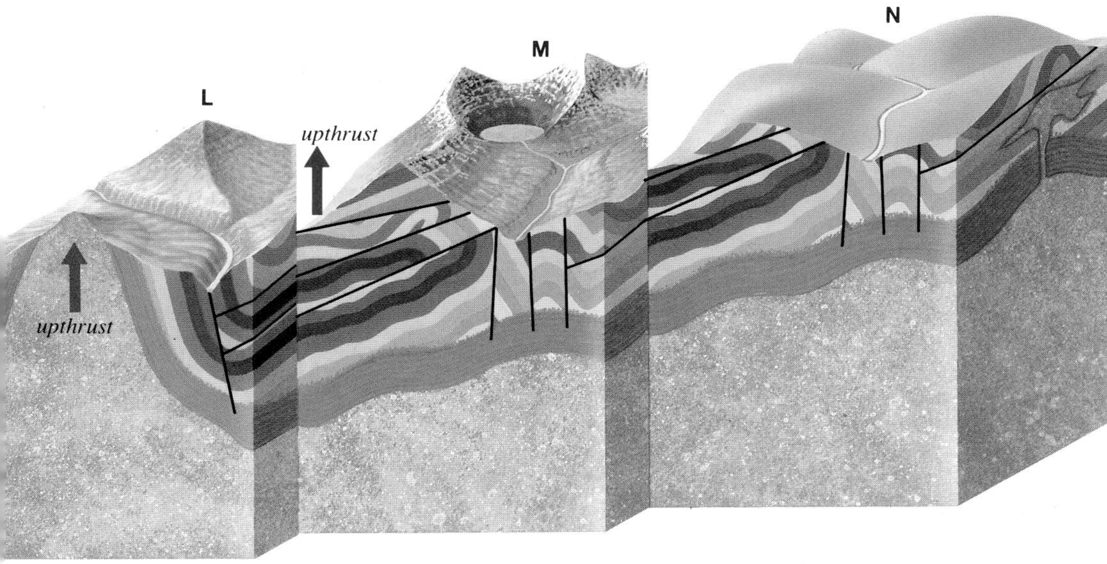

Idealized folds *take various forms, the commonest being the syncline* [**H**] *and the anticline* [**I**]. *When an anticline has one fold forced over the other it is an overfold. Extreme overfolds are called recumbent folds. A nappe* [**J**] *occurs when a recumbent fold is sheared through so that the upper limb can be moved forward. Nappes are especially prominent in the Alps. But folds like these are only the beginning. A present-day mountain profile* [**G**] *can come about only as a result of erosion, by agents such as wind, glaciers and rivers. Sediments that have been horizontally deposited deep underground are folded, fractured and overthrust* [**K**]. *Temperature, pressure and the strain rate – how fast rocks are squeezed – determine whether they fold or fault. Near the surface, or in the upper crust where pressures are low, rocks tend to fault, while more ductile deformation (folding) often occurs lower down. The complex rock mass resulting from all these processes is then lifted up* [**L**]. *As soon as it appears above sea level it is broken up by weathering processes and can then be transported by water, ice and wind. Water erodes the rock and makes deep valley incisions into the uplifted mountain complex* [**M**]. *Erosion will alter the appearance of the surface rock, and will expose lower layers. Resistant layers will form ridges. It is only through this valley formation that a high plateau is transformed into the peaks and troughs of a typical mountain chain. Upthrust is continuing at this stage, so although the mountain chain is being eroded away, its level above the sea is still rising. The debris transported from the middle of the uplifted area gathers at the foot of the mountains (the* piedmont*), or is transported to the sea. The foreland deposits* (molasse) *can even be folded later (making* flysch*) in a process of extension of the mountain chain. The constant erosion transforms the mountain chain into hilly country* [**N**] *and finally into a so-called* peneplain.

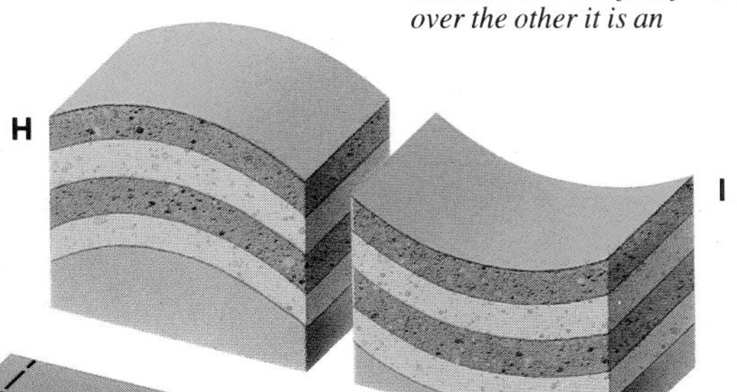

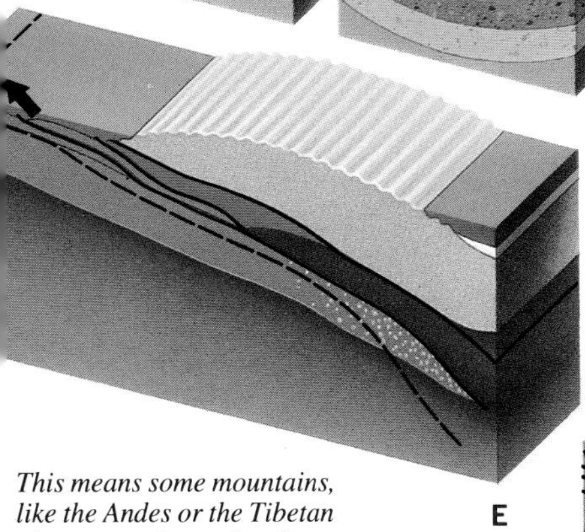

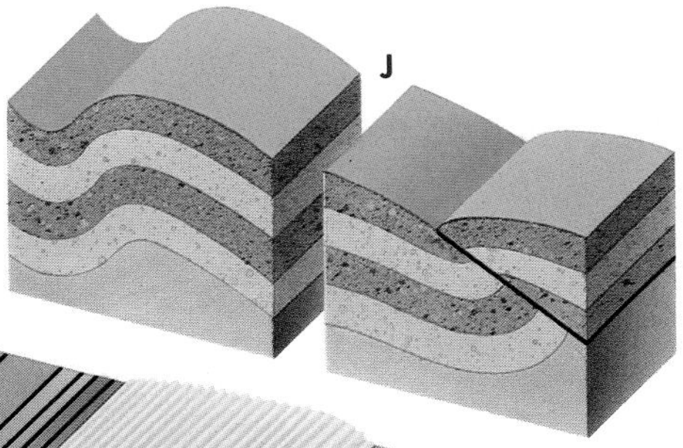

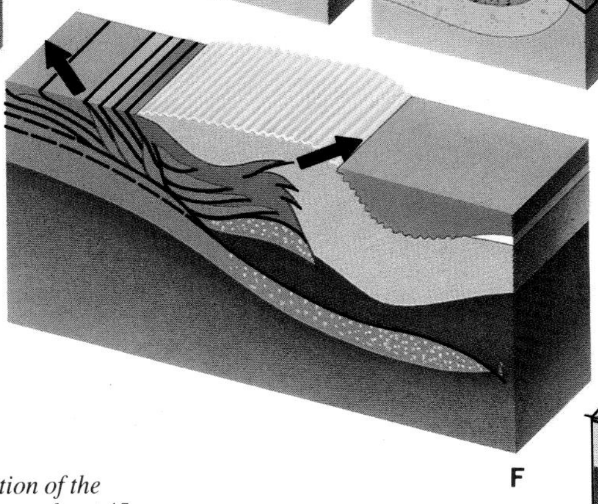

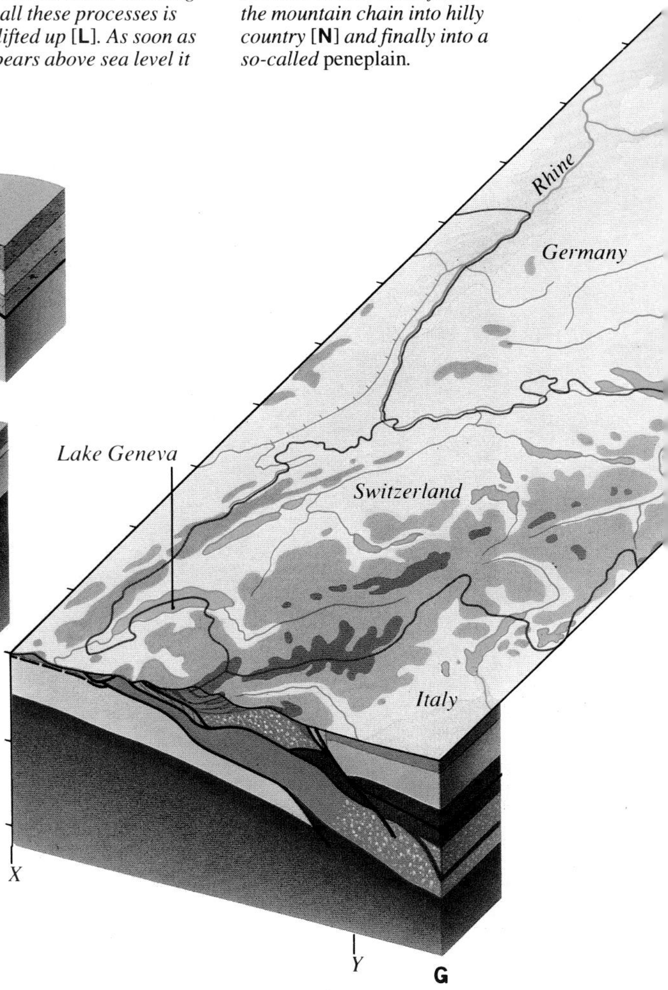

This means some mountains, like the Andes or the Tibetan plateau, are buoyed up by deep low-density rock, so that the mountains float on the crust in much the same way that an iceberg floats in the sea.

The Alps *exhibit most of the common forms of mountain folding. Their geological history extends back over 250 million years, but before the Cenozoic era, which began some 65 million years ago, there was very little evidence* [**C**] *of the extensive crumpling and dislocation that can be observed today* [**G**]. *The Alps arose from the collision of the Eurasian* [1] *and African* [2] *continental plates. Deformation of the western Alps began about 45 million years ago* [**D**]. *The eastern Alps, formed in the Italian prong of the African plate, was already deformed by folding. Firstly, thrusting (small arrows) began near the suture between the plates* [3], *but successive faults* [**E–F**] *lifted up previous thrust slices and brought deep crustal rocks to the surface. One theory holds that the horizontal shortening* [**X–Y**] *that created the Alps was of the order of 250 miles (400 km).*

Earth-Shaking Events

What causes an earthquake

There are 150,000 noticeable earthquakes every year, and over a million can be measured with sensitive apparatus. A million deaths have been caused by earthquakes in the last 100 years. For the Earth's crust is not stable. It is riddled with cracks and faults along which the ground can slide, often unpredictably. The human cost is frequently appalling, but the science of earthquake prediction is not exact and the only resource for millions living in areas of high seismic risk is to design better earthquake-resistant buildings. Sadly, this happens all too rarely.

__The Earth's crust__ is elastic. For a long time it may absorb strains within itself without reacting [**A**]*. Then, abruptly, it will rupture along the nearest fault line or lines* [**B**]*. The focus – the point of rupture* [1] *– may be at the Earth's surface or up to 450 miles (700 km) below. The epicenter of the earthquake* [2] *is at ground level directly above. Most damage is done by earthquakes that occur at a depth of 6 miles (10 km) or less. Normally, the longer the interval between movements on an active fault line the greater the eventual shock. The San Andreas Fault in California is a 750 mile*

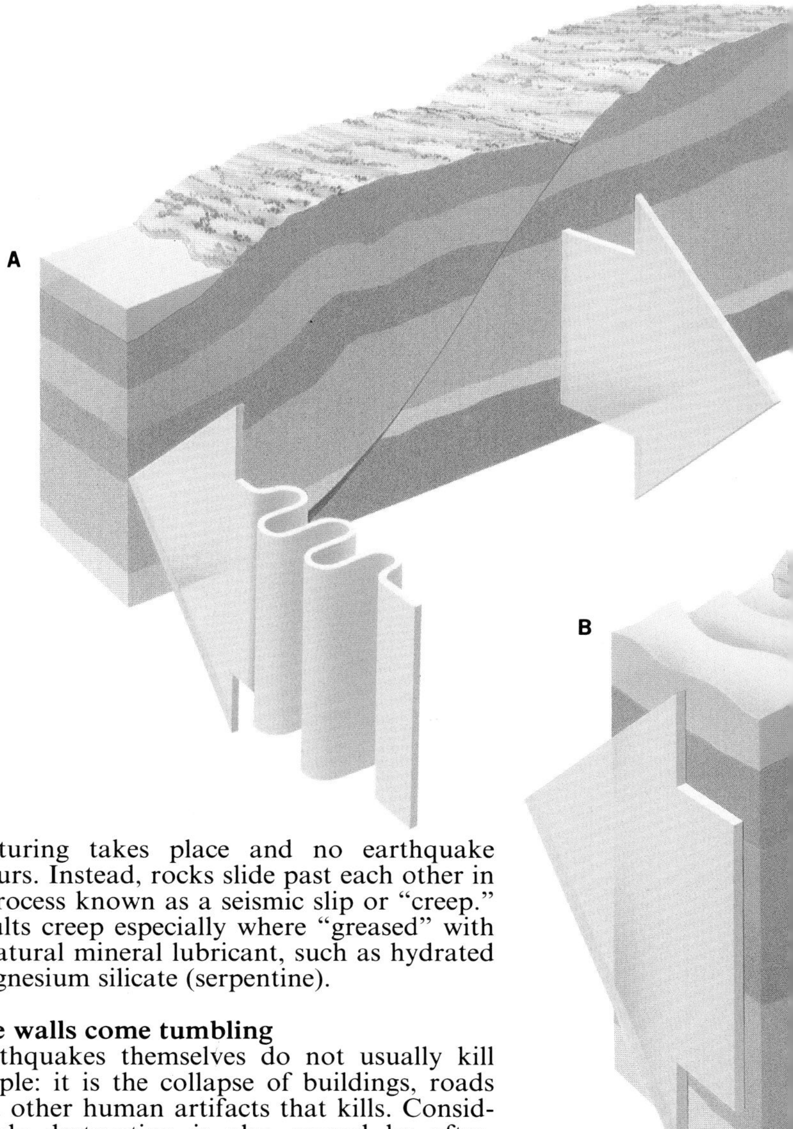

Several kinds of fault in the surface of the Earth commonly lead to earthquakes. The 1985 quake that killed 10,000 people in Mexico City occurred because a subduction zone – where one tectonic plate slips beneath another – generated a thrust fault when the overlying plate became compressed by the plate subducting under it. The San Andreas Fault is the most famous example of a strike-slip fault, where one plate slides laterally past another. But perhaps the most common earthquake-causing fault is exemplified by the devastating Armenian quake of 1988, which occurred on a thrust fault where one block of earth was driven vertically up above another by compressional forces.

Making waves

An earthquake's energy is transmitted in the form of three main types of wave: primary or P waves, secondary or S waves and surface waves. Surface waves are the largest and most slow-moving and, in shallow earthquakes, transmit the bulk of the energy. The surface waves of particularly strong earthquakes may travel right around the Earth several times and still be measured on seismograms days after the event took place.

Two concepts are commonly used to measure earthquakes: magnitude and intensity. Magnitude represents a measure of the total energy generated in an earthquake, and is calculated by measuring the maximum amplitude of seismic waves involved. Most commonly used now is the Richter scale, which is logarithmic: each unit represents a 10-fold increase in the amplitude of the waves (and nearly a 30-fold increase in the energy generated). The largest earthquakes measure around 8.8 on this scale, but there is no upper limit. However, discrepancies can occur depending on whether S, P or surface waves are measured. Surface waves are frequently used because they bear the most direct relationship to ground movement.

Intensity, on the other hand, corresponds to the subjective experience of observers near the earthquake and is a guide to the degree of shaking. The Mercalli Intensity Scale runs from I (indicating that the event is not felt at all) to XII (the total destruction of land-based objects, ripples in the ground, and people thrown into the air). Sometimes the strain energy is released slowly, so that no sudden rupturing takes place and no earthquake occurs. Instead, rocks slide past each other in a process known as a seismic slip or "creep." Faults creep especially where "greased" with a natural mineral lubricant, such as hydrated magnesium silicate (serpentine).

The walls come tumbling

Earthquakes themselves do not usually kill people: it is the collapse of buildings, roads and other human artifacts that kills. Considerable destruction is also caused by after-effects like fire, flood, landslide and tsunamis. A high standard of construction work, carried out to earthquake-resistant design, is vital. Modern skyscraper designs rely on reinforced frames that can absorb and distribute seismic energy. Some skyscrapers in San Francisco sit on bearings made of layers of steel and rubber, which allow the buildings to shift with ground movement.

Connections: Coastlines 56 Earth's Structure 12 Mountains 18 Plate Tectonics 16 Tides 48 Volcanoes 22

(1,200 km) boundary between the Pacific [3] *and the North American* [4] *tectonic plates. Although the Pacific plate is heading northwestward at an average 2.5 in (6 cm) a year, most of its movement consists of sudden jumps. Earthquake science – seismology – is not advanced enough to predict such jumps accurately. In the 1989 Loma Prieta quake, the focus was 11 miles (18 km) below the surface. Although the Pacific plate slipped 6 ft (2 m) northwest and rode a yard upward on the North American plate, most of the energy of the quake was absorbed below ground, so there was only comparatively minor surface cracking* [5].

Tsunamis

Tsunamis – also known as tidal waves – are caused by underwater earthquakes or landslides. Either of these events causes displacement of the sea bed, and a similar displacement of the water immediately above. A drop in the level of the sea surface causes huge amounts of water to flood in from all directions, in turn causing the surface of the sea to bulge at that point. Gravity then causes tsunami waves to spread out from the area. In deep water, these waves can move at very high speeds – some have been noted moving as fast as 500 mph (800 km/h). As the waves approach the shore, the shallower water slows them down, but the crest of the waves piles up higher and higher, until it arrives at the coast as a destructive, towering wall of water up to 160 ft (50 m) high and perhaps traveling at 50–60 mph (80–100 km/h).

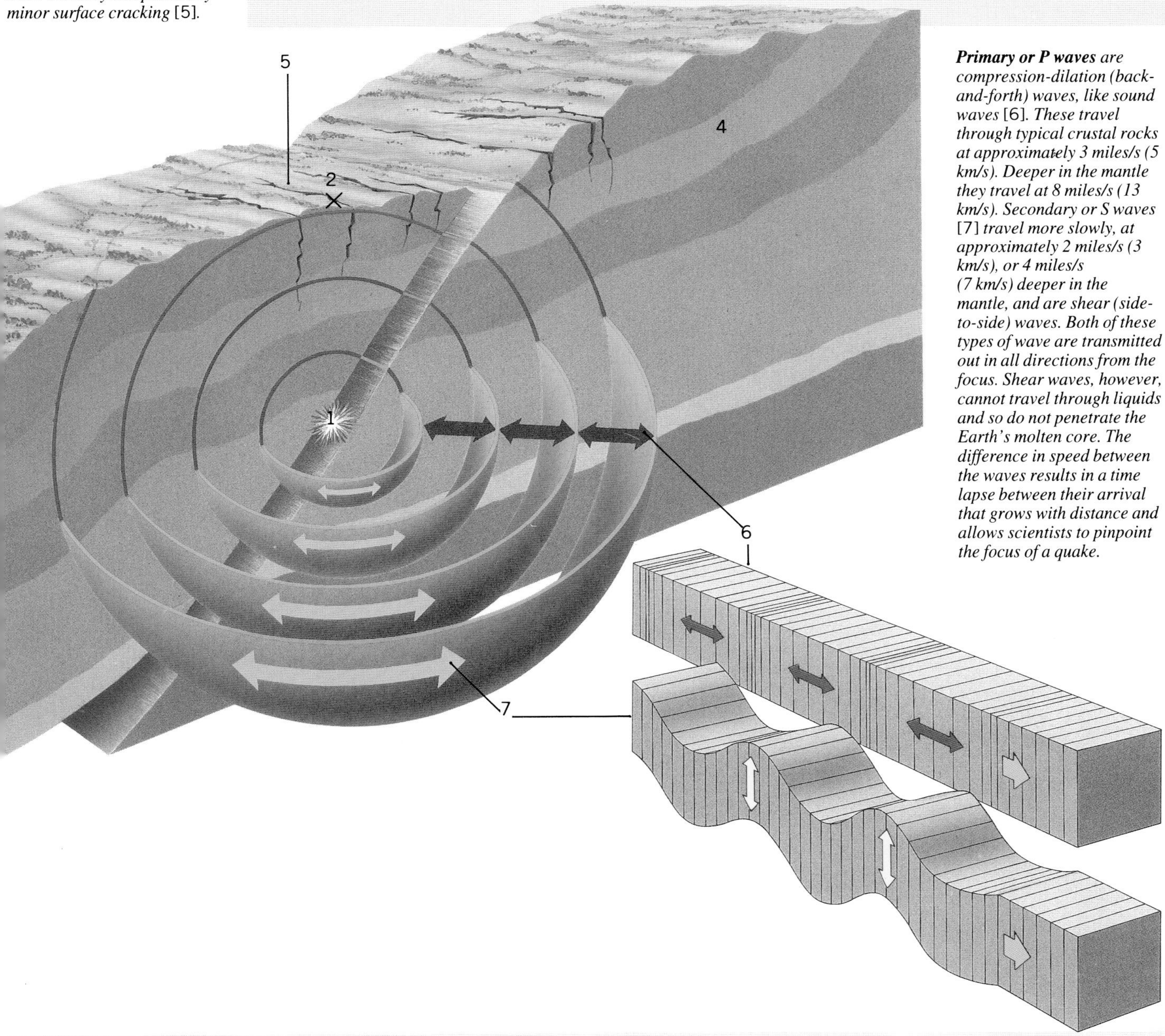

***Primary or P waves** are compression-dilation (back-and-forth) waves, like sound waves* [6]. *These travel through typical crustal rocks at approximately 3 miles/s (5 km/s). Deeper in the mantle they travel at 8 miles/s (13 km/s). Secondary or S waves* [7] *travel more slowly, at approximately 2 miles/s (3 km/s), or 4 miles/s (7 km/s) deeper in the mantle, and are shear (side-to-side) waves. Both of these types of wave are transmitted out in all directions from the focus. Shear waves, however, cannot travel through liquids and so do not penetrate the Earth's molten core. The difference in speed between the waves results in a time lapse between their arrival that grows with distance and allows scientists to pinpoint the focus of a quake.*

Mountains of Fire

What causes volcanoes

When Krakatoa exploded, it blasted out 4.5 cubic miles (18 cubic kilometers) of rock in under a day, creating massive tidal waves and forming a circular depression 4 miles (6 kilometers) across and two-thirds of a mile (a kilometer) deep. Although volcanoes operate on a far faster time scale than most geological phenomena – it took just four days for the volcanic island of Surtsey to grow up out of nothing – their activity varies enormously: from violent explosions that destroy entire mountains and communities to quiet but extensive flows of lava.

Volcanoes occur when the Earth's surface is breached and magma flows forth as lava or explodes into the air as tuff. All volcanic eruptions are driven by rapidly expanding gases within the magma, so the two factors that determine the violence of an eruption are the amount of dissolved gases and how easily they can escape. It is the magma's viscosity or fluidity that controls how the gas escapes, and it is the composition of the magma that determines its viscosity. The relatively easy-flowing basaltic magma that comes out of Hawaii's Kilauea has a low silica content, about 50 percent; less fluid andesite magma, about 60 percent silica, feeds explosive volcanoes like Mount St. Helens. Impending eruptions may be signaled by small earthquakes, or by a volcano swelling and emitting gas.

Fountains of fluid rock

When the magma is low viscosity and basaltic, with a low content of dissolved gas that rises slowly through the crust, the gas escapes slowly and gradually as the magma rises. Near the surface, perhaps a few yards away, the magma may start to foam and this is what causes lava fountains. Quiet, effusive eruptions of lava gradually build up more gently sloping mountains, like Kilauea, called shield volcanoes. They can even form lava plateaus. At the top of Kilauea glowing lava shoots high into the air above a lake of molten rock, and rivers of slow-moving lava ooze down the volcano's sides, traveling so slowly that a walker can outdistance them. But lava can also outpace a family car when pouring down a steep slope.

Gas-powered explosions

A high-silica magma with high viscosity and high dissolved gas content behaves very differently. As the magma rises the gas starts to escape from solution, but the high viscosity of the melt holds the bubbles back, so that they have an internal pressure that can be as high as several hundred atmospheres.

Then, when enough bubbles form, or the external pressure decreases, the gas pressure blows the rock apart and starts off an explosive eruption. Violent eruptions that throw dust and rock fragments into the air are called *pyroclastic* eruptions. These eruptions alternate with lava flows to produce cone-shaped, crater-topped mountains, sloping up at angles of about 30° like Mount Fuji in Japan. Such mountains are *stratovolcanoes.*

The "ring of fire" – the volcanoes surrounding the Pacific Ocean – consists of stratovolcanoes that form above the subduction zones encircling the Pacific.

An embryonic atmosphere

The volcanic gases consist of water vapor – the most important component – carbon dioxide, sulfur dioxide, hydrochloric acid and nitrogen. In fact the ratios of these volcanic gases are extraordinarily close to the ratios of water, carbon, chlorine, and nitrogen in the air, oceans and surface rocks of the Earth. The main difference is that volcanoes produce too much sulfur. Nevertheless, it may be that the volcanoes, expelling the Earth's gases throughout the millennia of geological time, were what created the stuff of atmosphere, oceans and rocks.

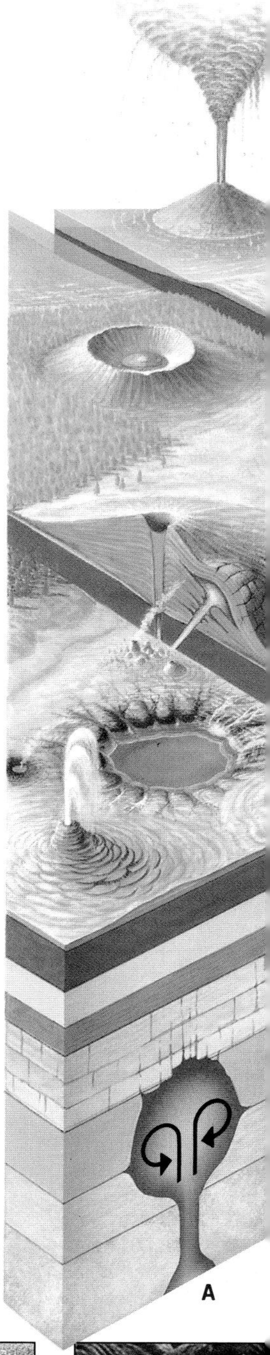

***Small differences** in the viscosity – or fluidity – of lava can make very big differences in its appearance after it cools. One type of lava,* aa *(below left), has a high viscosity and freezes into piles of blocky rubble. Low-viscosity (highly fluid) lava flows can form ropy lava, or* pahoehoe *(below right), when the cooler surface crust is stretched and twisted as it is carried along by its molten interior. Where the lava flow is quick, pahoehoe beds are thin and smooth with wrinkled tops. The same eruption can produce both, with aa usually released somewhat later than pahoehoe.*

Connections: **Earthquakes** 20 **Earth's Structure** 12 **Island Formation** 52 **Minerals** 32 **Mountains** 18 **Plate Tectonics** 16 **Rock Formation** 24

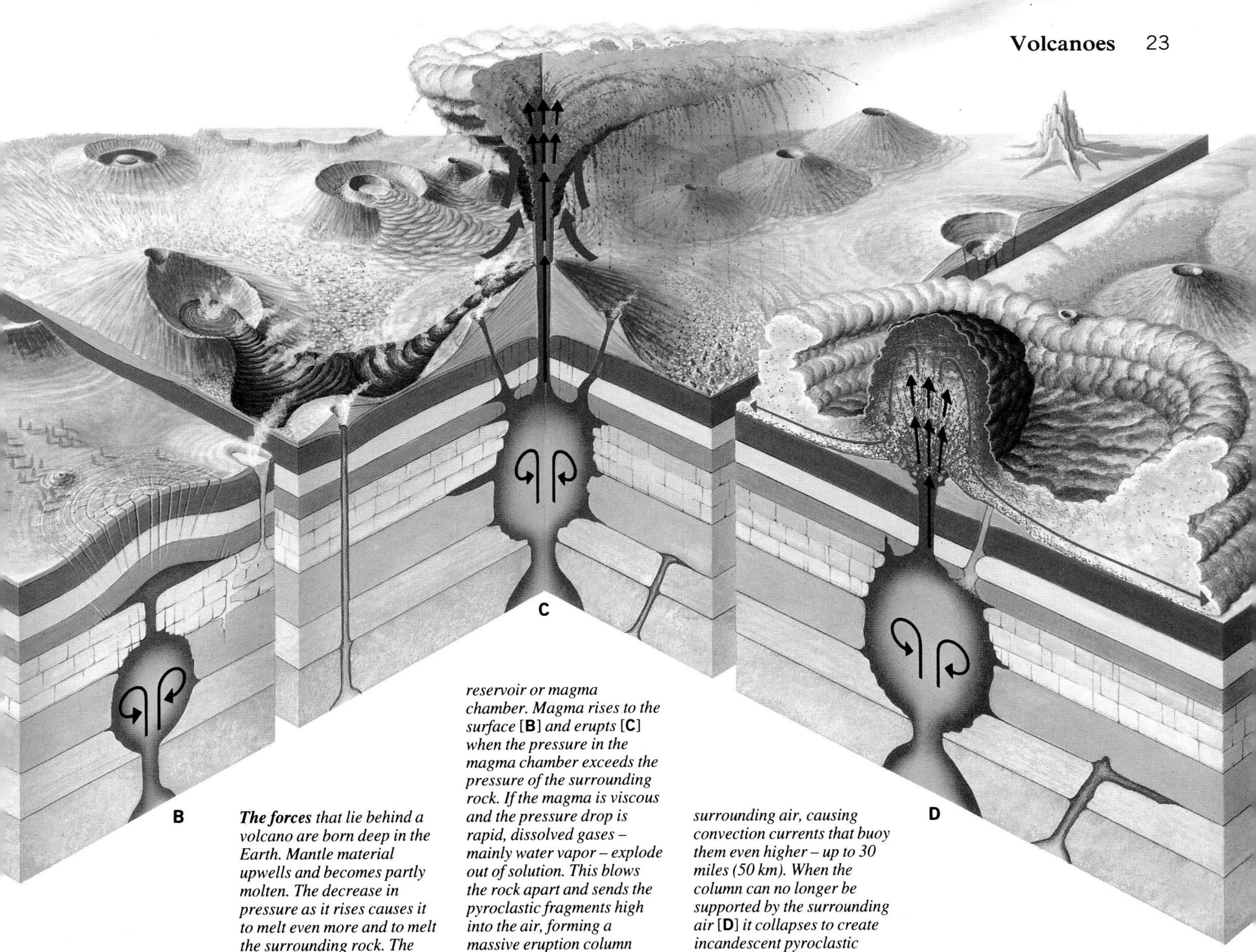

*The forces that lie behind a volcano are born deep in the Earth. Mantle material upwells and becomes partly molten. The decrease in pressure as it rises causes it to melt even more and to melt the surrounding rock. The magma then ponds [**A**] about two-thirds of a mile below the Earth's surface and forms a reservoir or magma chamber. Magma rises to the surface [**B**] and erupts [**C**] when the pressure in the magma chamber exceeds the pressure of the surrounding rock. If the magma is viscous and the pressure drop is rapid, dissolved gases – mainly water vapor – explode out of solution. This blows the rock apart and sends the pyroclastic fragments high into the air, forming a massive eruption column composed of hot gases and incandescent pumice and ash. The particles heat up the surrounding air, causing convection currents that buoy them even higher – up to 30 miles (50 km). When the column can no longer be supported by the surrounding air [**D**] it collapses to create incandescent pyroclastic flows that race outward at velocities of more than 200 mph (360 km/h).*

Paricutin

Imagine the surprise of the Mexican farmers who, on 20 February 1943, stood in their fields and watched as first a crack in the earth appeared and then the ground heaved up, belching out smoke and sparks that set some of the nearby trees afire. Their curiosity would quickly have turned to despair, because a volcano was coming to life on their farms. An hour later red-hot molten rock was flowing out of the crack, and by midnight glowing clumps of lava were flying into the air. The next day the cone was 160 ft (50 m) high. By the following day the first farm had disappeared. Paricutin continued to climb, spewing out lava and volcanic ash to form a steep-sided mound that was 500 ft (150 m) high in a week. During the next year the volcano rose to form a 1,500 ft (450 m) mountain.

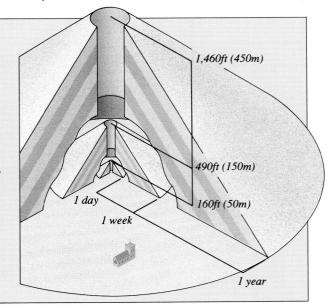

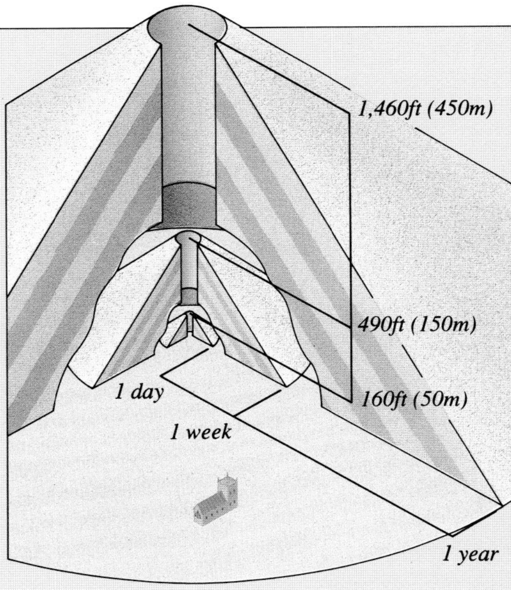

Turning up the Heat

How igneous and metamorphic rocks are formed

The total continental crust of the Earth may be increasing by up to 0.25 cubic miles (between 0.1–1 cubic kilometer) every year. This growth is the difference between the rock being added to the continents by upwelling magma and the rock being lost through attritional forces. There is a constant progression between igneous rocks formed from magma, metamorphic rocks caused when existing rocks are subject to heat and pressure, and sedimentary rocks formed from the erosion of both, which may in turn be melted or metamorphosed to begin the cycle anew.

The great variety of igneous rocks results from the two main features of these rocks – the size and the chemistry of their constituent crystals. In most cases they are composed of an interlocking mosaic of crystals, as magma cools. Formed inside a volcano, lava cools relatively quickly on reaching the surface, especially if the surface is the sea bed, to make *extrusive* rocks. As a result the size to which the crystals can grow is small. Crystals rarely grow larger in diameter than 0.02 inches (0.55 millimeters), and such rock is described as fine grained. Basalt is derived from the Earth's mantle by a quite rapid upwelling of magma. It is at very high temperature when erupted and contains minerals that crystallize at high temperature. These are dense, heavy minerals like olivine and pyroxene. Granite, on the other hand, forms at a much lower temperature and contains minerals such as mica, amphibole, quartz and potassium-rich feldspar, which crystallize at these lower temperatures. Granite is usually *intrusive* (it solidified beneath the surface).

Changing nature

Regional *metamorphism* involves changing the nature of already existing rock in specific areas of the Earth's crust, generally during the processes contributing to mountain formation (*orogeny*). These processes correspond to stress, high pressure, high temperatures and migrating, chemically active fluids. The rocks formed in this way range from those that are altered only slightly to those so altered that their original makeup has become difficult to determine.

Around the margins of an orogenic belt and in parts where rocks have not been buried too deeply in the crust, low-grade metamorphic rocks occur. Slate is typical of this grade. It is formed in conditions of relatively low pressure and temperature and acquires *cleavage* – an ability to split easily into thin plates. This property derives from the alignment of flaky minerals like mica and chlorite at right angles to the directions of pressure. Under such low-grade conditions, only weak rocks like shale, clay and volcanic ash can be affected.

Nearer the center of the orogenic belt, medium-grade rocks are produced. Schist forms, with characteristic wavy banding. This rock contains larger crystals than the fine-

***Granite** is among the most common igneous rocks. It is formed when old, deeply buried rocks partially melt and migrate up through the remaining (usually close to melting) masses of rock* [1]. *The liquid is at higher pressure than the rocks above, and so rises and intrudes them. When igneous rock intrudes into preformed strata in the Earth's crust it may cool and solidify to form thin sheets. If these cut through the existing rocks they are called* dikes [2]; *if they are flat and follow geological structures such as sedimentary bedding planes they are called* sills [3]. *A* laccolith *is a dome-shaped mass that has arched the rock above* [4]. Lopoliths *are saucer shaped* [5]. *Such bodies of once-molten and now highly crystalline rock may be relatively small – only 13–16 ft (4–5 m) thick – and contain rocks of medium grain size – between 0.02 and 0.2 in (0.5 and 5 mm) in diameter, because although they have cooled quite quickly they have solidified more slowly than the volcanic rocks. Dolerite and microgranite are typical rocks of sills and dikes. Other bodies of magma may stretch for 500 miles (800 km) or more. These are* batholiths [6] *or plutons, a contain rocks of coarse gr size. Here the molten magn may have taken many year to cool and crystallize, so t grain size of these rocks is more than 0.2 in (5 mm) in diameter: granite is typica Existing rocks that have*

Connections: Earth's Structure 12 Minerals 32 Mountains 18 Ocean Floor 50 Plate Tectonics 16 Sedimentary Rocks 26 Volcanoes 22

grained slate, and may have garnets set into its mica-rich surface. Under these higher temperatures and pressures a larger variety of rocks is altered, including limestones and sandstones and some igneous rocks. The high-grade metamorphic rock gneiss is formed at the greatest depths in the central region of the orogenic belt.

Much of the Earth's original continental crust is made of this rock, characterized by a coarse, often dark-and-light banding. Dark bands represent heavy minerals like hornblende and biotite; pale bands are rich in quartz and feldspar.

Movement of the Earth's crust may take existing rocks as far as 450 miles (700 km) below the Earth's surface. Metamorphic rocks – or even sedimentary rocks – that sink deeply enough into the Earth may be melted again into igneous rocks, renewing the continuous cycle of rock change and movement.

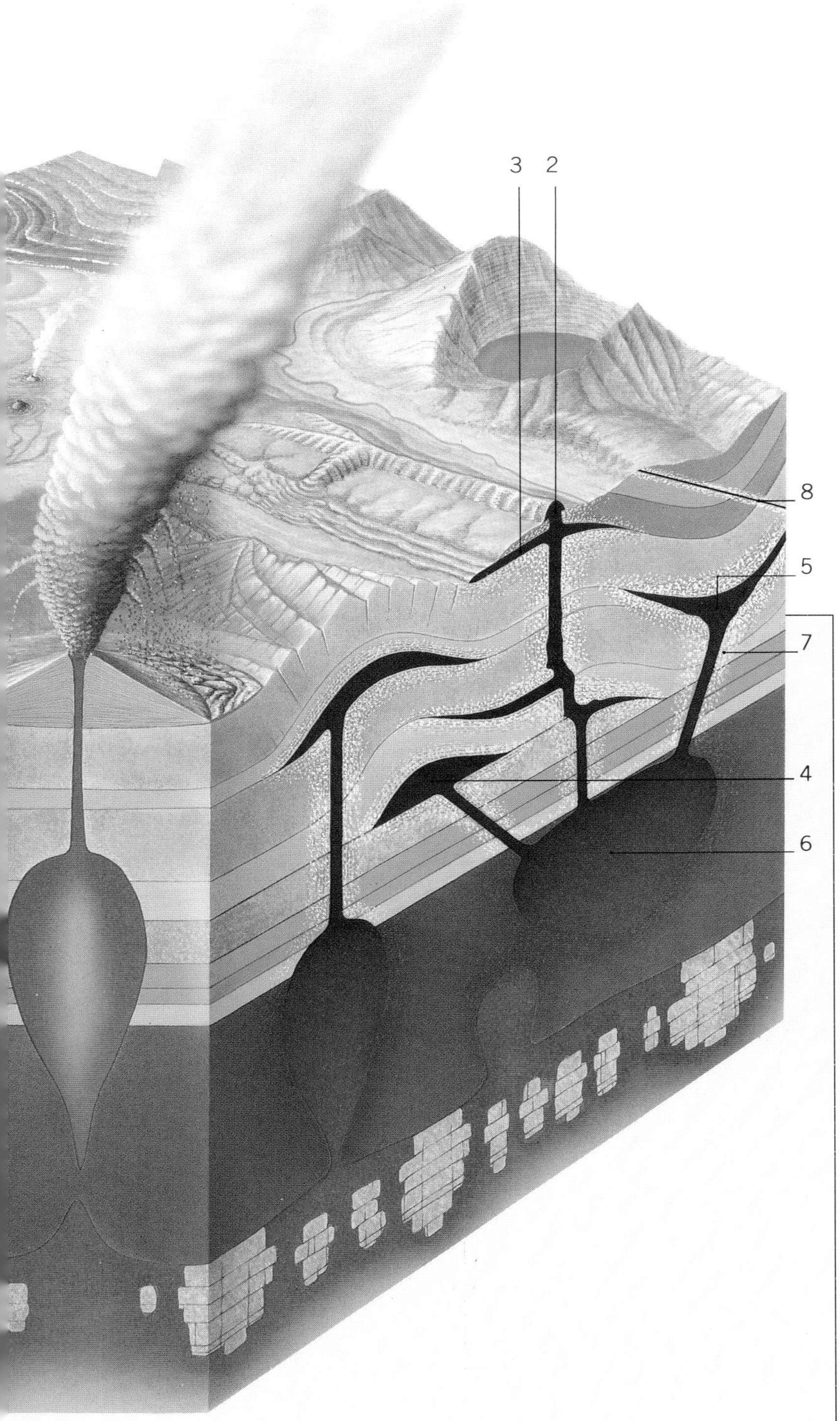

een altered by the heat of he intruding magma are aid to have undergone ontact or thermal netamorphism [7]. This akes place when magma nvades strata underground r a lava flow spreads across the surface. When granite magma at around 1,830°F intrudes into a mass of shale or clay, the minerals in the rock being invaded by the magma (the "country rock") are recrystallized, and bedding planes and fossils disappear. New minerals such as chiastolite and cordierite form and may give the rock a spotted appearance. A tough, flinty rock called hornfels forms immediately next to the intrusion, with the spotted rocks farther away, in zones of minerals, each formed at a different temperature.

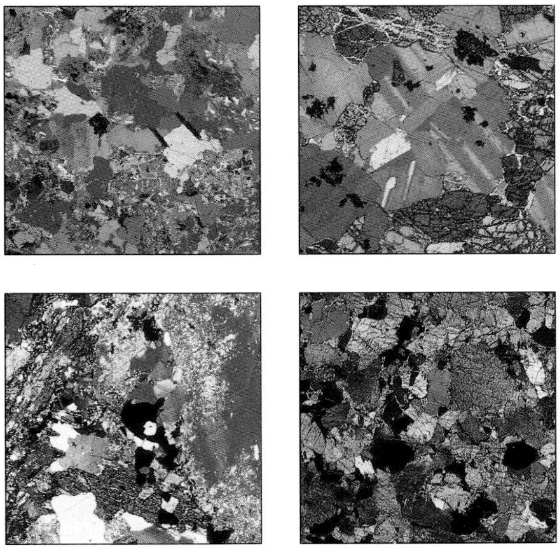

When fault movement *occurs on a large scale, and the rocks near the fault plane (the surface where movement occurs) are pulverized and changed [8], they are said to undergo dynamic or dislocation metamorphosis. Along the fault plane pressure and temperatures are greatly increased. Rocks that are shattered by the pressure are partially melted by the heat. The grains of rock are streaked out in the direction of fault movement. The rocks resulting from this process are called mylonite. Mylonite is a comparatively rare type of metamorphic rock and only occurs in narrow strips, such as the Lizard in Cornwall, England.*

The silica content *of igneous rocks is indicated by the terms acid (over 65 percent silica), intermediate, basic and ultrabasic (under 45 percent silica). Thin rock sections examined by polarized light reveal different minerals in distinctive colors, which helps classify the rock. The acid rocks contain over 10 percent quartz (silicon dioxide). They include granite (top left), potassium-rich feldspar and mica. Diorite (bottom left) is an intermediate rock. The basic rocks like basalt and gabbro (top right) and ultrabasic rocks like dunite (bottom right) are rich in calcium-feldspar and pyroxene, with little quartz.*

Layer Upon Layer

How sedimentary rocks are formed

The vast white chalk cliffs of Britain and Europe are made from countless billions of microscopic plates that were formed from plankton. Yet each individual plate is only a few hundred thousandths of an inch long. This is only one of the many ways in which sedimentary rocks have been made. Some were formed, grain by grain, of preformed rock that was weathered, eroded and deposited elsewhere either by wind, rivers or glaciers. Others settled as vegetable matter or dead organisms on river or sea beds. And yet others arise through chemical reaction.

Sedimentary rocks comprise separable particles, often quartz or rock fragments, cemented together by minerals. They are the products of weathering and erosion of preformed rocks of all types. Weathering is the decomposition of rock by chemical reaction (as in rain action) or by mechanical means (as in ice expansion in rock joints). Local climates determine weathering rates and types: erosion is the movement of weathered particles.

Most *fluvial* (river) erosion will deposit at the coast, but inland deposits, particularly at gradient discontinuities and in the river's senile stages, are also significant. Fine desert-weathered particles (vulnerable because of a lack of binding vegetation), frost-pulverized glacial outwash and soil from mismanaged farm land are often blown away and deposited as loess. Huge glacial deposits, or *till*, are common in glaciated zones of the last 2,000,000 years; icebergs calving off glaciers have also deposited thousands of square miles of sediment far out to sea.

Sediments with large rounded particles – some boulder sized – are *conglomerates*; angular particled sediments are *breccias*. Particle shapes often betray origins, rounded ones being water formed and angular ones the result of frost action or mountainside scree deposition. Smaller-grained rocks like sandstones are usually formed in deserts, rivers or seas.

Three main processes form sedimentary rocks. All three are shown in a coastal context in the stylized diagram opposite. Organic sedimentary rocks **[A]** *are formed from once-living matter. For example, corals extract lime from seawater to make their skeletons. A reef's living coral is found only underwater near the reef top* [1], *and beneath an atoll may be hundreds of yards of dead corals* [2]. *Bits of coral, broken off and eroded away by the waves, and shells of dead crustaceans left on the sea floor will form reef limestone* [3], *a typical organic rock.*

Some sediments are formed by chemical rather than physical means **[B]**. *Economically important deposits such as rocksalt, gypsum and potash are known as* evaporites, *and are deposited by precipitation from saline water, often in warm, arid climates. This is common in the tropics, where the seawater in a partially enclosed basin* [1] *evaporates* [2], *depositing its salts* [3] *to form these chemical rocks* [4]. *Another chemical process that gives rise to sedimentary rocks is that in which calcium carbonate is precipitated around minute nuclei to form spherical particles called oolites, which are the constituents of oolitic limestone. Oolites tend to form where the wave action is strong, by a "snowballing" buildup of the calcium carbonate. Oolites tend to grow to about one twenty-fifth of an inch in diameter.*

The rocks arising from the more familiar mechanical process – which are produced as a result of erosion from older rocks – are called clastic *types. Whether flowing from mountains to adjacent lower land or when reaching its ultimate destination at the sea* **[C]**, *a river tends to first deposit stones and large particles, dropping finer material later. As a result of this process a gentle slope of increasingly fine sediment is created downstream or out to sea* [1]. *Inland, this feature is called an alluvial fan; if conditions are right, river mouth sediments form deltas* [2]. *The fine clays farther out will form mudstone, which has no internal structure, or, if buried by new sediments, will be squeezed into shale, which has its flat crystals well aligned, and so breaks easily into thin slices.*

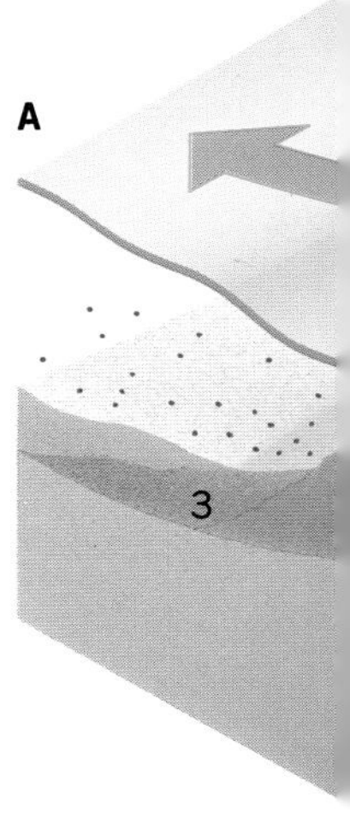

Deep-water dust-bowls

Generally, the finer the grain size the weaker the current that carried the sediment and, often, the deeper the water in which it was deposited. Clays and mudstones are often found in deep oceans, where river-borne sediment is rarely encountered. Much material deposited so far from land is volcanic dust and other material carried long distances by the wind. It is now possible to study sedimentation processes at great depths, thanks to advances in submersible technology.

Sediments often exhibit features that indicate how they formed. For example, a layer of sandstone that has ripple marks and desiccation cracks was evidently once located in shallow water and later appeared above the water level for some time. Sands deposited as a stratum by the wind often show large-scale dune-bedding – where the layers are deposited in sweeping curves and ripples.

Loose sediment is turned into rock by three basic processes. Sandstone results from cementation, where water percolates between sand grains and deposits thin layers of binding iron oxide, calcium carbonate or silica. Clay is turned to mudstone by compaction, when water between the grains is squeezed out by the weight of more sediment accumulating above. Marbles often show the marks of colossal mountain-building forces, which cause the rock minerals to recrystallize in a solid mass with no spaces between them.

Rock once alive

Fossil coral reefs containing numerous shellfish and other fossils cemented by lime mud occur in many parts of the world. These limestones are of great economic significance, used in building and the making of cement. Chalk is made of innumerable microscopic shells, and coal is fossilized vegetable matter.

Connections: Caves 36 Coal, Oil and Gas 34 Coastlines 56 Coral Reefs 336 Fossilization 28 Glaciers 40 Rivers 58 Rock Formation 24 Soils 30

Erosion and Economics

Many important minerals extracted from the sea are taken from sedimentary layers around coastlines or the shallow waters of continental shelves. The limestones, sands and gravels so important to the building industry are taken from beaches and coastal waters. There are titanium ores in the black sands that run along the United States coast from Florida to New Jersey, and the black sand of New Zealand's west coast beaches is rich in iron. However, not all valuable sediments are found in the sea. Building sands and gravels are commonly extracted from glacial and river deposits. So-called "placer" deposits occur where heavier particles in sediment settle at low-energy points, like the bends of streams. It was the discovery of gold placers that caused the great Klondike gold rushes in the nineteenth century.

Three distinct layers *of rock, characteristic of sedimentary processes, are clearly visible in these cliffs (above). Lowest is carstone, a coarse, lower Cretaceous sandstone. The reddish rock above is a pink limestone containing iron and studded with pebbles. The comparatively hard calcite-bearing chalk on top belongs to the upper Cretaceous period.*

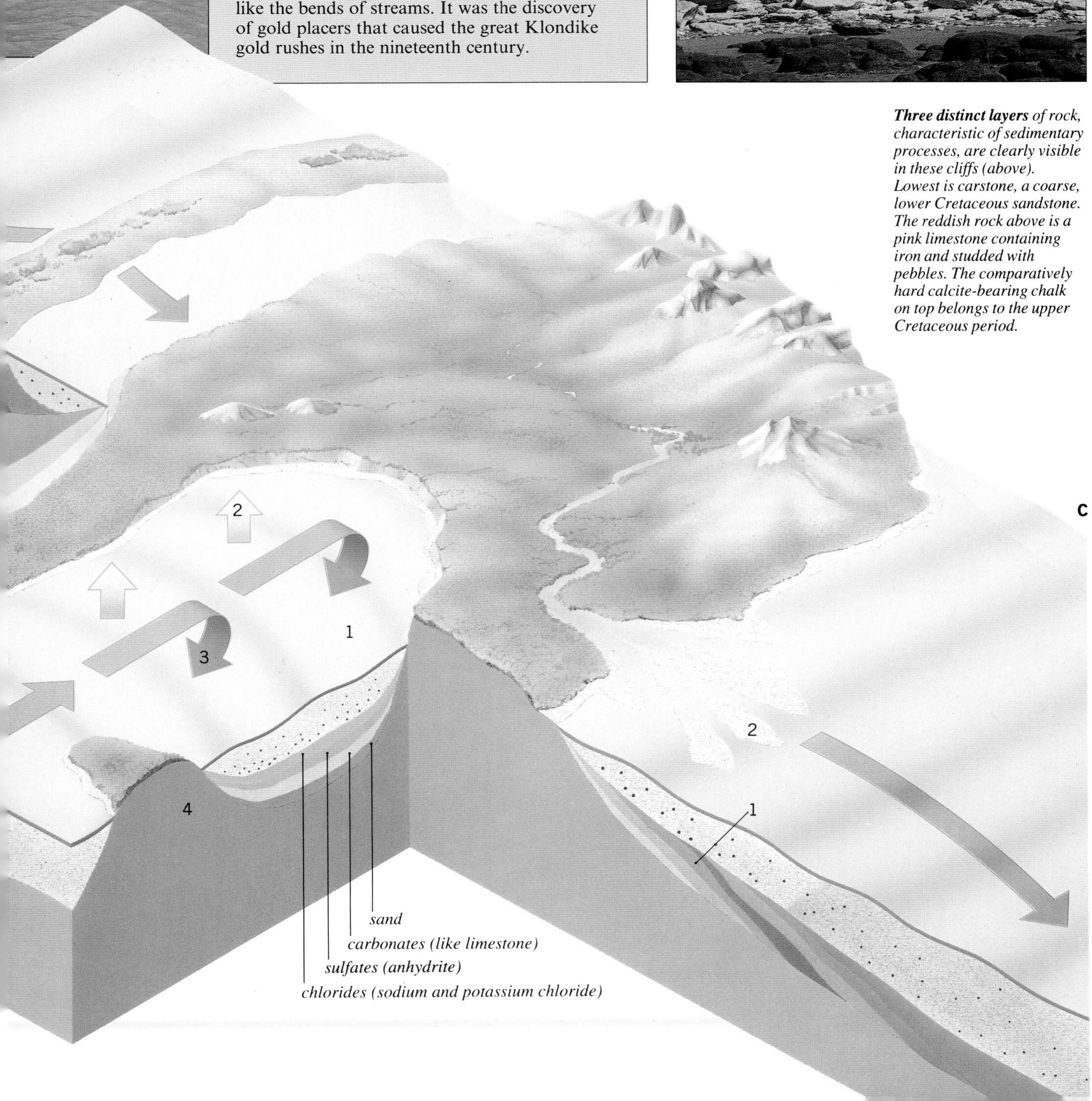

Embedded in Time

How fossils are formed

Only a tiny fraction of all the billions of plants and animals that have ever lived on the Earth are preserved as fossils. These ancient remains hint at the diversity of past life and give us a unique insight into the process of evolution. Fossils range in size from huge dinosaur skeletons to the casts of bacteria visible only under a powerful microscope. The composition of a fossil depends on how it was formed: it may be an entire organism encased in rock, amber or ice, or the faintest impression of a dragonfly wing in soft mud, later compressed into mudstone.

The process of fossilization begins when the remains of an organism or traces of its passing, such as footprints, are buried in mud or sand. Then, given the right circumstances, the buried animal or plant fragment may undergo a number of changes as the mud and sand become compressed into rock. These changes make the fragment physically and chemically similar to the surrounding rock, allowing it to be preserved indefinitely as a fossil.

Secrets in the strata

Although fossilization cannot be predicted accurately in space or time, it is favored by certain conditions. Various rock types contain fossils, but none more so than the sedimentary rocks, consisting of strata of sand, mud or clay, formed on or near the Earth's surface. Of these, limestones are possibly the most productive and commonly contain a wide variety of well-preserved bivalve mollusks, other hard-shelled creatures and corals. Fossils are also occasionally found in shales and clays, although they are often crushed by the great pressures that build up within these rocks unless they are protected within rounded nodules or concretions.

Fossils in rocks formed from shales and clays are of particular value because the sediments are fine grained and able to preserve delicate, soft-bodied organisms in great detail. The Burgess shale, high in the Rocky Mountains of British Columbia, is a classic site of this type.

Probably the most significant single factor promoting fossilization is rapid burial. Speed reduces decay, scavenging by other organisms and physical destruction, and it is for this reason that most fossils are of sea-dwelling creatures, for sediment is most rapidly deposited in marine environments. The remains of terrestrial organisms are rapidly eroded and are less likely to be buried by sediment. But this does not mean that land plants and animals are totally absent from the fossil record. Given the right conditions – for example, a flooded river that traps many creatures beneath great masses of mud – excellent fossils can be formed.

Candidates for fossilization

A creature is more likely to become fossilized if it has some hard parts, such as a shell or bone structure, than if it is made up only of

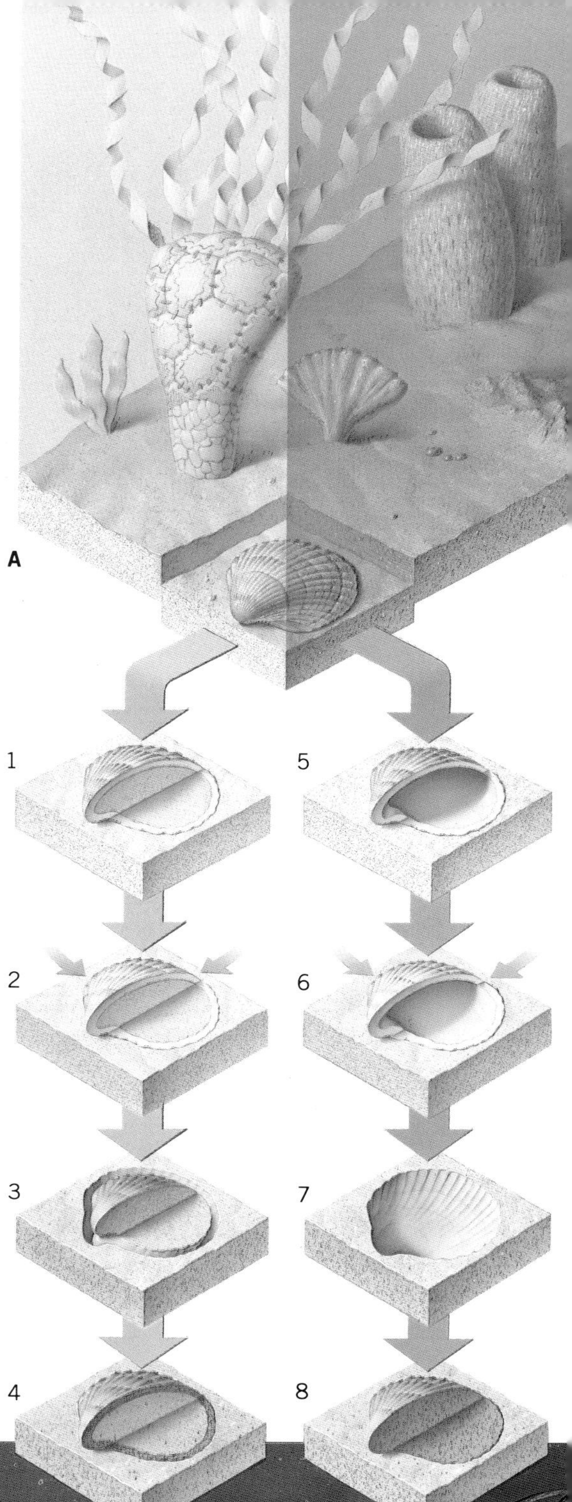

***The type of fossil** formed in a rock is dictated by the chemistry of both the organism and the rock itself* **[A]**. *The harder parts of animals and plants contain a number of materials that do not decay, such as phosphates in bone and calcium carbonate in shells. During fossil formation such minerals are often replaced by others that are better able to survive the rigors of the subterranean environment. For example, iron pyrites, hematite or quartz commonly replace minerals in a shell or a bone molecule by molecule, so a calcium-based coral may be preserved as a hematite fossil. The organic part of the shelled creature is always the first to decay. This leaves a gap in the shell which is often filled with sediment* [1]. *The shell is more soluble than the rock, and slowly dissolves* [2], *leaving a sedimentary-rock cast of the inside of the shell* [3]. *Under the right circumstances the shell material is replaced by some other substance – often a type of silica – to give a cast of the shell's exterior* [4]. *The*

Connections: Coal, Oil and Gas 34 Decomposers 204 Early Evolution 92 Evolution to Date 94 Ice Ages 38 Minerals 32 Origins of Life 90 Rock Formation 24

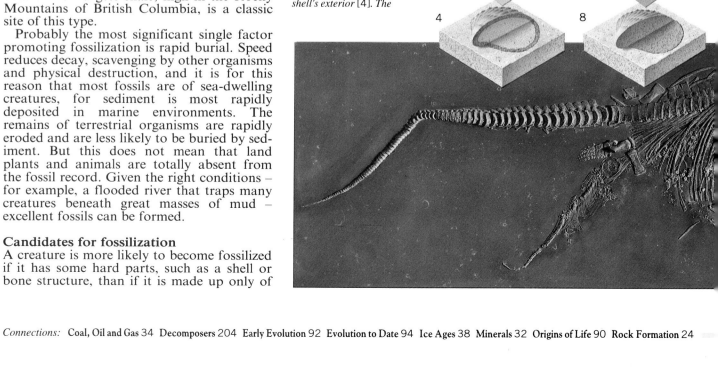

"replacement" may even be the shell material, recrystallized. Sometimes, no sediment fills the empty shell [5]. The solutions that permeate the strata [6] may totally dissolve a buried shell, leaving a "mold" behind in the rock [7]. A replica of the original shell may later be cast, as another mineral or sediment is deposited in this mold [8].

The tracks *of living creatures are often fossilized [**B**], as well as their bones or shells. Typically, a footprint [1] – or the burrow of a prehistoric worm, for example – is left behind in soft mud, which partially hardens to form a cast. If the mud is flooded by a river, or the sea [2], sediment is laid over the mud especially quickly [3], helping to preserve the shape of the footprint. Over the course of time, the mud and the overlaid sediment become compressed and turn to rock [4]. The original mud-based rock forms a mold of the footprint [5]. The sediment-based rock forms a cast [6].*

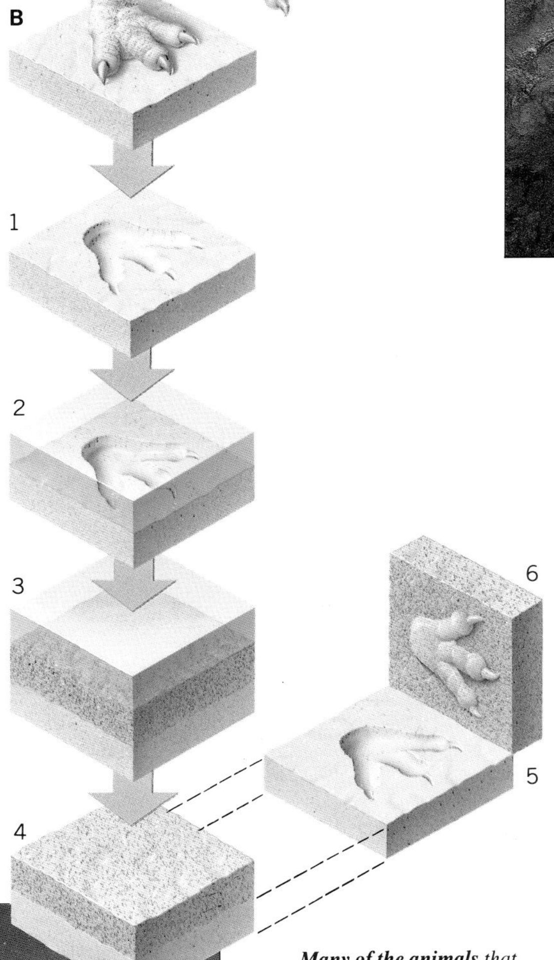

Many of the animals *that become fossils clearly died – and were preserved – suddenly and unexpectedly. They were eating, being eaten, or, like this* ichthyosaur, *in the act of giving birth. Fossil evidence like this provides valuable clues about dinosaur physiology. Scientists were puzzled about how the dinosaurs could return to the water when they needed to lay their hard-shelled eggs on land. However, evidence such as this shows that at least some dinosaurs produced live young.*

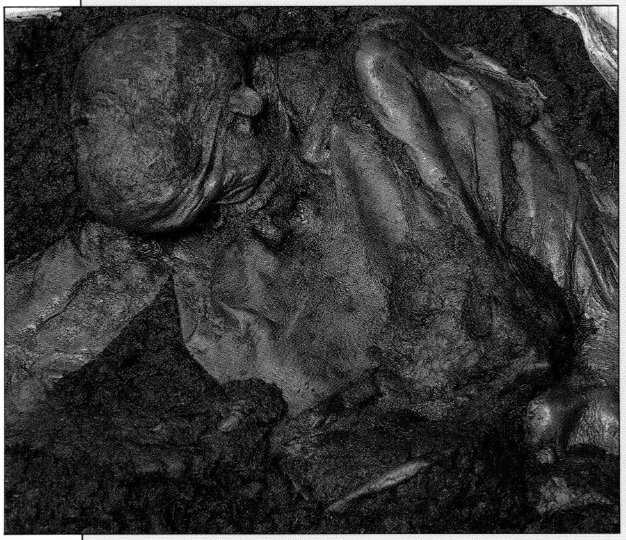

The Bog People

Peat cutting and swamp drainage in several marshy regions of western Europe has unearthed hundreds of human "fossils." The corpses of the bog people have been exceptionally well preserved for around 2,000 years because the oxygen-deficient environment of the peat bog has prevented bacterial decomposition. In addition, burial in peat actually tans the skin of the preserved bog-dweller. These remarkable fossils allow detailed study, and in some cases scientists have even been able to determine the contents of the buried person's last meal. Probably the best known of the bog people – nicknamed Pete Marsh – was found in Cheshire, England, and is now in the British Museum. Many archeologists believe he may have been the victim of a ritualistic ceremonial sacrifice.

soft tissue. However, not just large, bony organisms are preserved. Soft-bodied creatures can leave a characteristic black carbon film on a stratum, and tiny microfossils (visible only through a powerful microscope) are very common in many strata. These microscopic fossils have recently become the focus of considerable attention because they enable geologists to accurately determine the age and relative positions of rock layers that are useful – or even critical – indicators in the search for oil and gas.

Occasionally, an entire creature, or part of its original tissue, may be preserved unaltered: an insect may crawl into the fragrant resin from a pine tree which later hardens into amber, or an animal may become trapped in a naturally occurring pool of tar. Such rare fossils are preserved because they are locked away in an oxygen-free environment where decomposition cannot take place.

Sedimentary Rocks 26 Soils 30

Earth's Outer Skin

What causes the different types of soil

The Egyptian pharaohs built monuments to last, among them graceful granite obelisks that, left in the arid climate of Egypt, are hardly changed after 4,000 years. Granite is generally considered a pretty tough rock, but obelisks transplanted from the Nile Valley to London or New York start to crumble and decompose in less than a hundred years. This is because the damp and wet climate and the polluted atmosphere speed up the breakdown of the rock. This insidious process of weathering is the first step in creating most of the Earth's soils.

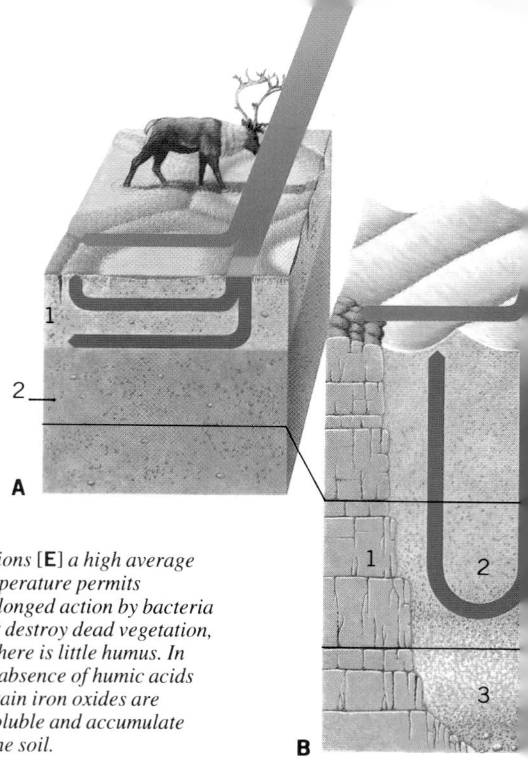

Weekend gardeners tend to think of soil as the first few inches below the Earth's surface – the thin layer that needs to be weeded and that provides a firm foundation for plants. But the soil actually extends from the surface down to the Earth's hard rocky crust. It is a zone of transition and, as in many of nature's transition zones, the soil is the site of important chemical and physical processes. In addition, because plants need soil to grow, it is arguably the most valuable of all the mineral resources on Earth.

Youth means fertility

Soils can be young or old, depending on the time it takes for them to form, and it is the young ones that are often the most fertile because they retain almost all the minerals that were in the parent rock. Such fertile soils form when a powerful natural force rapidly breaks up rocks. In young mountain belts the mountains' high relief causes rapid erosion, and thick soils that keep virtually all their mineral content form at the bases and in the foothills of these ranges. Such soils are commonly found on the west coasts of North, Central and South America.

Young fertile soils also occur near recently glaciated areas, where the rock pulverized by the glacier accumulates. Fertile soils also occur on *loess*, a wind-blown, fine-grained deposit derived from glacial and river-valley sediments. Loess deposits in central Europe and the central United States derive from periglacial areas. The loess deposits of China were blown in from desert valleys or plains. The flood plains of big rivers like the Nile, the Mississippi and the Ganges are also covered by young, mineral-rich sediments on which fertile soils can develop. But a soil's fertility depends on many factors besides its mineral content. The climate and the local flora and fauna all have an effect. Fertility will increase with the ease with which roots can penetrate the soil and with its aeration.

Breaking rocks

Weathering is usually a long, slow process, although rates of weathering vary depending on climate. In the western United States it takes about 100,000 years for granite to break down enough to be crumbly to the touch. Temperature variation in the presence of water is an efficient agent of mechanical

Soil type *owes as much to climate (especially rainfall) and vegetation cover as to the nature of the parent rock. Some rainfall may flow into the groundwater. Some will be transpired back into the air by plants, or will evaporate directly from the soil. Some of the rainfall will not penetrate the soil at all, but will evaporate in the air, be intercepted, or will run off along the surface. In this illustration, the fate of the water is roughly indicated by the blue arrows.*

Tundra soils [**A**] *typically have a peaty upper layer overlying a thick layer of sticky structureless clay of a bluish gray color* [1]. *This* glei horizon *is generally saturated by groundwater, because the flow of water downward through the soil is barred by a layer of permafrost* [2]. *In hot deserts* [**B**], *weathering does not go deep and soils usually have very flat foundations, with bare rock* [1] *at the surface. Where there is a soil covering* [2], *the soils are typically pale, coarse and poor in humus. Rainfall is insufficient to leach out calcium and magnesium ions* [3], *so they remain in the soil. The brown soil beneath a deciduous forest* [**C**] *consists of a moderately acid humus layer underlain by a grayish brown zone leached by the humic acids. The B horizon is often thick, and the minerals required by the trees are concentrated here. Trees bring the minerals up from the B horizon and recycle them. Savanna soils* [**D**] *are subject to heavy but seasonal rainfall. As in deserts, calcium and magnesium ions remain in the B horizon, to be recycled by grasses. Microbe activity is restricted and humus may be abundant in the A and B horizons. In wet equatorial regions* [**E**] *a high average temperature permits prolonged action by bacteria that destroy dead vegetation, so there is little humus. In the absence of humic acids certain iron oxides are insoluble and accumulate in the soil.*

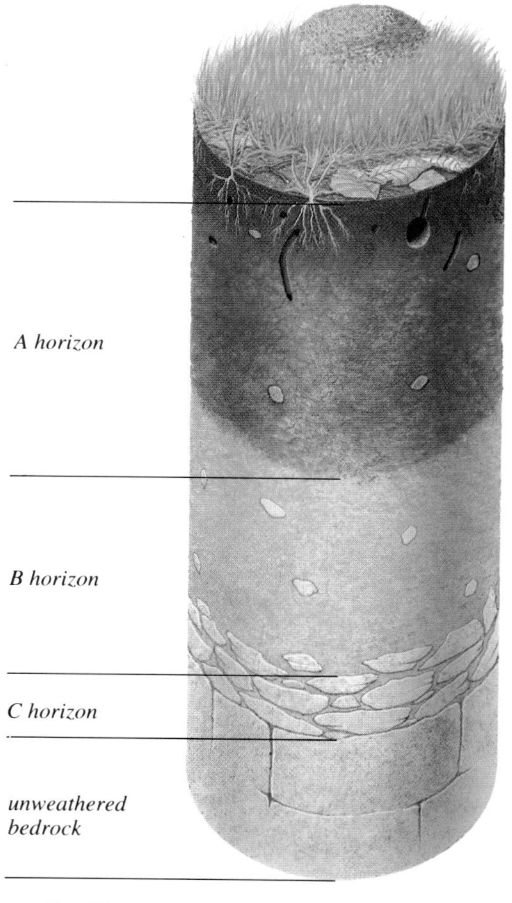

A soil profile *(above) consists of layers, or* horizons, *known as A, B and C horizons. The A horizon is nearest the surface. It is usually dark and rich in humus. The B horizon is where any minerals, in an ionic form useful to plants, usually accumulate. Where a soil develops on bedrock (sometimes called the D horizon), the C horizon is derived from this bedrock, but in a weathered-down state. It is infertile. Where a soil has developed on unconsolidated material deposited in flood plains or alluvial cones, the C horizon is the unaltered parent material.*

Connections: Cycles of Life 206 Decomposers 204 Deserts 54 Glaciers 40 Land Pollution 338 Rivers 58 Sedimentary Rocks 26

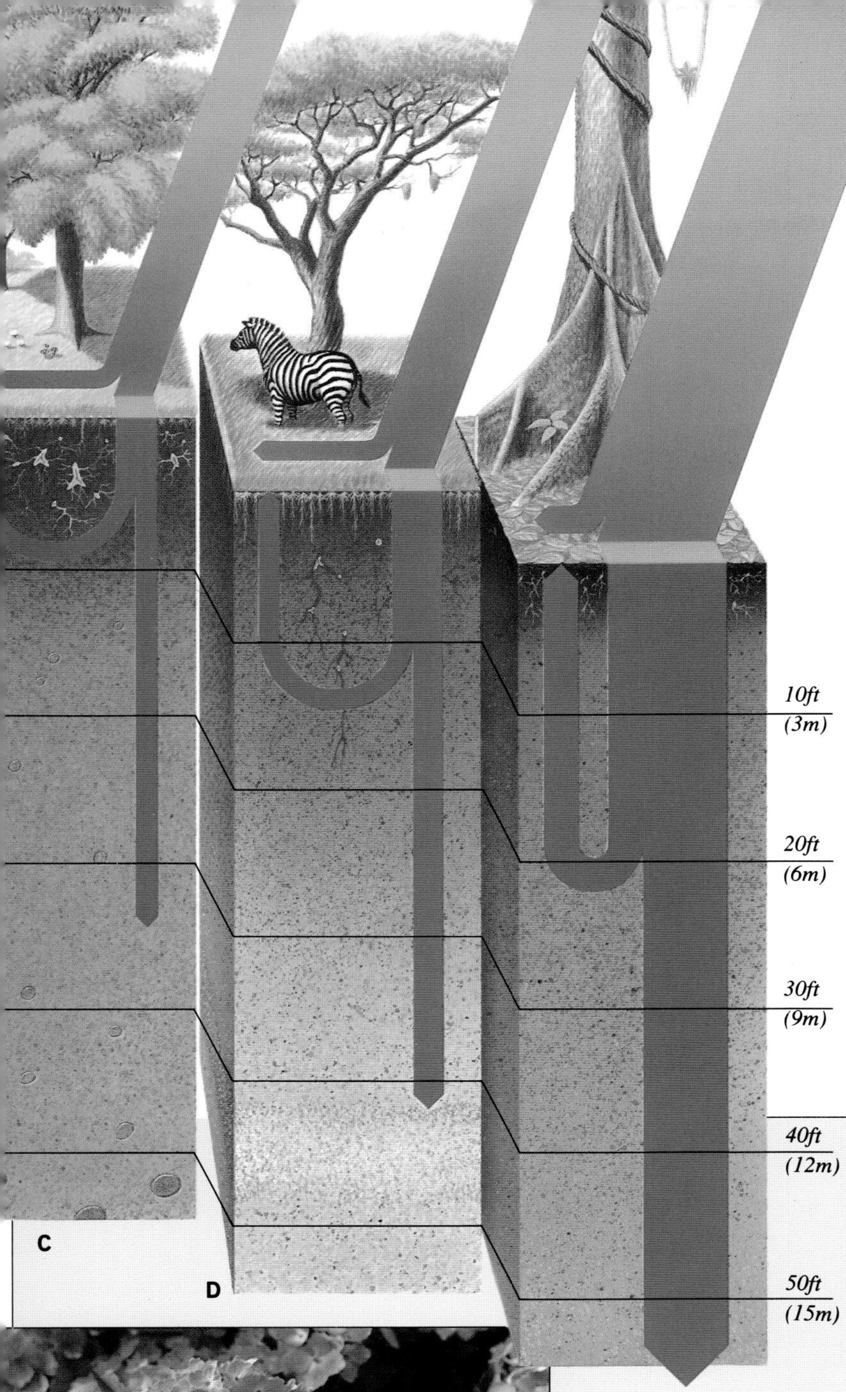

weathering. When water freezes it expands, and alternate freezing and thawing of water in the cracks and pores of a rock can gradually disintegrate it. The roots of plants can also grow into and widen the cracks in rocks, and even burrowing animals can cause weathering on this scale. Chemical weathering, in which rocks decompose from exposure to chemicals in the air and in rainwater, is most common in tropical, humid climates. Chemical and mechanical weathering are often combined, as in the case of salt crystallization in arid regions. As salt crystals grow in cracks in a rock they eventually exert enough pressure to break pieces off.

Diminishing minerals

While the rock is disintegrating, some of its chemical constituents are being dissolved by water and leached from the rock, leaving a less fertile soil behind. Highly weathered old soils, such as the thick red lateritic soils that underlie the tropical rain forests of South America, are the least fertile of all. In the million years or so that they take to form, all the calcium, sodium and potassium have been leached away by water percolating through the decomposing rocks. Even the silica in the minerals that make up the rock will have been removed, leaving mostly iron oxides derived from any iron-bearing minerals that were in the rock.

Living Litmus

Plants can be very choosy about soil chemistry. Rhododendrons, for example, will not grow in limy or chalky soils – the so-called alkaline soils. In the Appalachian Mountains of the eastern USA it is easy to spot the location of limestone bedrock even if there are no outcrops of rock to be seen, because where there is limestone at depth there will be no rhododendron bushes growing at the surface.

Hydrangeas, the big bushy plants that are especially common in and around seaside resorts, are a kind of vegetative litmus paper because the color of the plant's flowers changes according to the acidity or alkalinity of the soil in which the plant grows.

The flowers vary in color from crimson and pink to white, but in acid soils they will become a shade of blue. The acidity or alkalinity of a soil is measured on the pH scale. The pH value measures the concentration of hydrogen ions in a solution.

In acidic solutions the hydrogen ion activity is high. The pH scale increases with increasing alkalinity. The more acidic the solution, the lower the pH. The pH of a perfectly neutral solution is 7 and highly acid soils, for example soil that has come from a peat bog, can have a pH as low as 2.5, whereas soils derived from limestones can have a pH as high as 8.2.

Treasure Trove

How minerals are formed

More than 90 elements occur naturally on the Earth, but just eight of them – oxygen, silicon, aluminium, iron, calcium, sodium, potassium and magnesium – account for 98 percent of the weight of the Earth's crust. And two of them, oxygen and silicon, make up close to 75 percent of the crust. The silicates, which contain these elements, are the most common group out of all the 2,000 minerals that are formed when elements combine. Minerals vary hugely depending on their atomic structures, from slippery talc to diamond to the fan-shaped crystals of molybdenite.

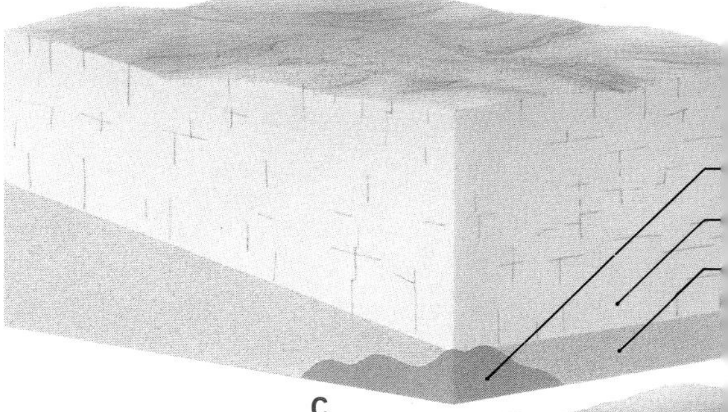

C

Silicates and most other minerals form by crystallizing out of some sort of liquid. Salt forming from pools of sea water on a hot day, quartz slowly crystallizing in a cavity a few hundred yards into the Earth's crust, and crystals that form when slowly cooling magma freezes many miles below the surface all grow in a similar fashion. But the form the crystal takes depends on the environment in which it grows. The milky white lumps of quartz occurring in veins that cut across the natural grain of many rocks at the surface are the same as the beautiful hexagonal prisms of quartz that are usually only seen in museums and books. The difference is that the prisms grew where the chemistry of the solution, cooling rate, temperature, and space were all just right.

The crystal maze

Well-formed crystals occur in a bewildering variety of shapes, from the neat cubes of pyrites, or fool's gold, to the curved fans of silver-purple molybdenite. But all these shapes can be placed into categories which reflect the different regular arrangements of the constituent atoms.

Slippery talc is the softest mineral whereas diamond is the hardest, all because of their atomic structures. Another crystal characteristic related to atomic structure is *cleavage*, or the ability of a mineral to split more easily along certain planes. Natural glasses, like the volcanic glass obsidian, break along curved irregular surfaces because they are amorphous solids that have no underlying regular structure. They are said to have a *conchoidal* fracture because the broken surfaces look a bit like a curved sea shell. But in some minerals the planes of weakness are so pronounced that the crystal breaks much more easily in one direction than another. Crystals of calcite, the main constituent of limestone, always break along three characteristic surfaces to form parallel-sided rhombohedra.

Because minerals' physical properties depend on their structure, many of them, like cleavage or hardness, are different in different directions. This characteristic, called *anisotropy,* means that when energy, such as light or sound, passes through most crystals it travels at surprisingly different speeds when moving in different directions. When white light falls on a mineral some of the wavelengths are absorbed and some are reflected. A mineral's color corresponds to the wavelengths that are reflected. Sometimes this depends on the mineral's chemical composition, which is why copper compounds are usually blue or green. In other instances, for example diamond, the color is related to the crystal structure. But in quartz, color is related to the presence of chemical impurities, which give the lovely purple amethyst, pink rose quartz and brown smoky quartz, although pure quartz is colorless.

So crystal color can relate to a number of different factors. But if the mineral is ground up, the intrinsic color is revealed. Haematite and magnetite, two minerals composed of iron and oxygen, can be difficult to tell apart, but if they are rubbed on a piece of unglazed porcelain (a process called a streak test), the haematite gives a reddish streak whereas the magnetite leaves a black mark.

A

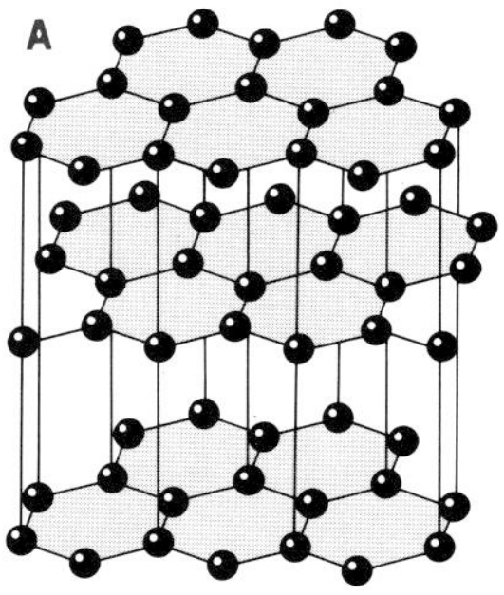

B

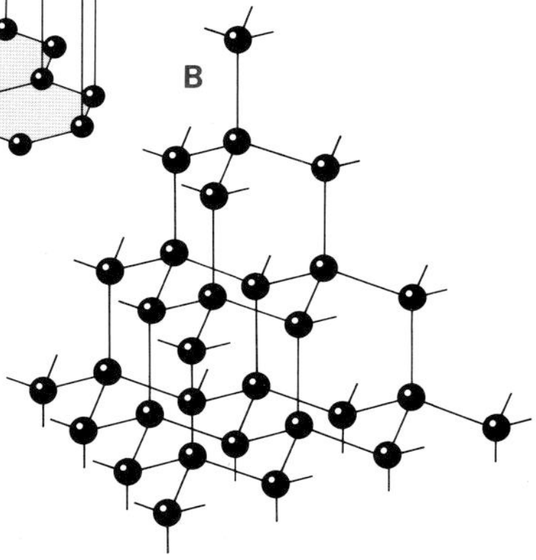

A mineral derives *its physical properties not only from its chemical composition but also from the way its chemical subunits are arranged. For example, the hardest naturally occurring mineral known to man – diamond – and one of the softest – graphite – are both made of the same element – carbon. In graphite* **[A]** *the carbon atoms are arranged in parallel layers. Within the layers, strong chemical bonds join the carbon atoms into interlocking hexagonal rings; but the layers are joined to one another by weak bonds, so they tend to "slide" across each other, a property that makes graphite a good industrial lubricant. In diamond* **[B]** *all the atoms are linked together by strong chemical bonds into a single, resilient molecule. Each atom in the molecule is joined to four others, all of which are an equal distance away from it. This is known as* tetrahedral linkage, *and it creates a very dense and closely knit crystal.*

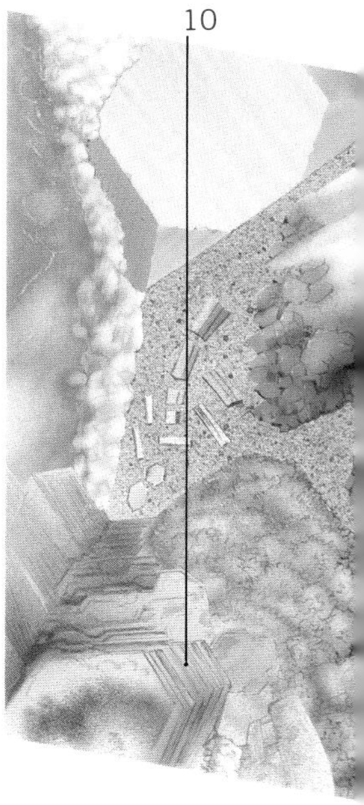

Connections: **Caves** 36 **Coal, Oil and Gas** 34 **Earth's Structure** 12 **Fossilization** 28 **Rock Formation** 24 **Soils** 30

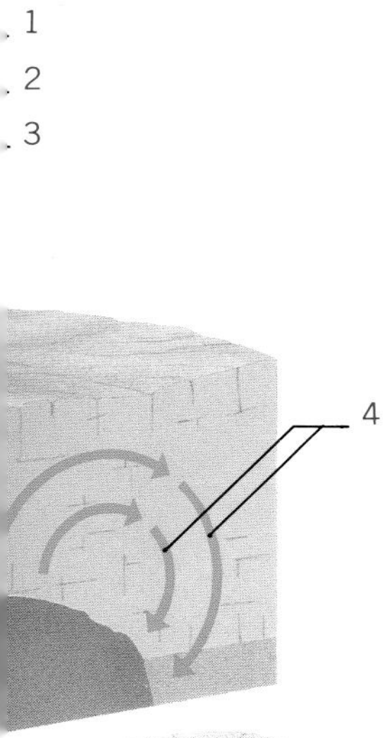

***Ores – rocks rich** in valuable minerals – often form near large bodies of hot, molten rock or magma. In the sequence shown here* [**C**] *a mass of molten granite* [1] *abuts against limestone* [2] *and sandstone* [3]. *The hot magma causes the water present in the limestone to circulate; the hot, circulating solution* [4] *"picks up" elements from the surrounding rock. These elements are subsequently deposited and concentrated into pods of metal ore. At the same time, the heat of the magma body alters, or metamorphoses, the neighboring limestone, forming a zone around the cooling granite known as the* skarn zone [5]. *Ore bodies* [6] *are therefore often associated with such areas of altered limestone. As the overlying limestone is eroded away, the skarn and ore bodies become exposed at the surface* [7] *and are further altered by the action of groundwater. The movement of this water through the ore body causes some minerals to be dissolved and redeposited – occasionally as large crystals* [8].

As the magma cools and begins to crystallize, its water content increases and it becomes less dense. This lighter solution migrates away from the main body of cooling magma and cools to form veins [9] *(known as* pegmatite *veins) that contain rare minerals. When the veins begin to crystallize, the presence of water promotes the growth of large and often precious crystals. A pocket within a pegmatite vein may contain crystals of muscovite* [10], *quartz* [11] *and tourmaline* [12], *amongst other minerals.*

Diamond from Coal

Most attempts to create diamonds have concentrated on reproducing the high temperatures and pressures found beneath the Earth. An early scheme involved melting carbon at the center of a ball of iron, then abruptly cooling the iron so it would contract and squeeze the melted carbon. Explosives have been used to generate high pressure and temperature, but because the pressure falls immediately any diamonds formed quickly revert to graphite. "Grit" diamonds for industrial use can be made by heating specially prepared carbon compounds up to 4,900°F at a pressure of 1,500,000 lb/in² (105,000 kg/cm²).

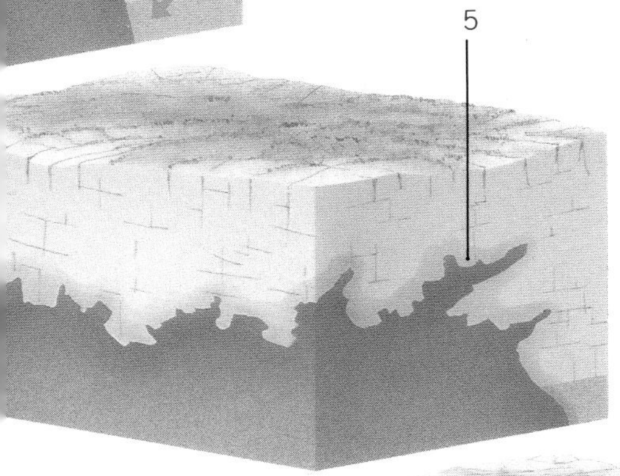

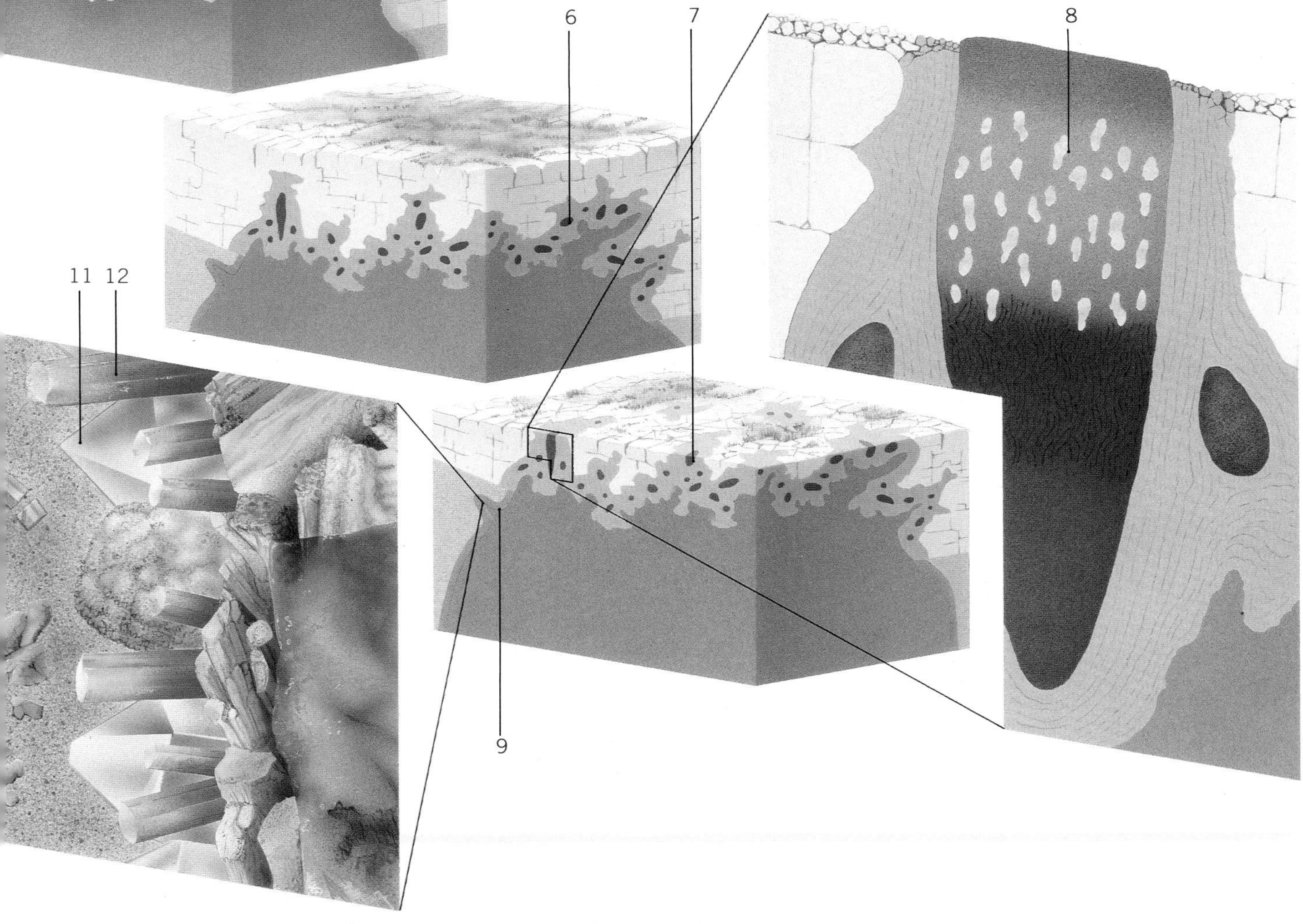

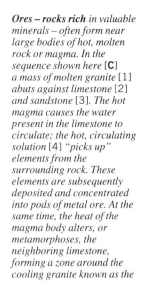

Fossil Fuels

How coal, oil, and gas are formed

The modern world depends on plants and animals that died millions of years ago. Our industries, powered by coal, oil, and gas, are burning the fossil remains of prehistoric life. Over 40 percent of the energy consumed in Western Europe comes from coal, and what took millions of years to create will be consumed in centuries. Predictions about how long fossil fuel reserves will last are colored by political expediency, as are government responses to the charge that fuel combustion builds up sulfur-rich gases and carbon dioxide which threaten the environment.

Oil, gas and coal are all formed from the decay of once-living organisms under heat and pressure. Over 80 percent of the oil and gas currently exploited formed in Mesozoic or Tertiary strata between 180 and 30 million years ago, from marine microorganisms deposited as sediment on the seabed. The basic components of oil and gas are created when the organic remains are not completely oxidized, leaving a residual mass of carbohydrates, hydrocarbons, and similar compounds. As layers of sediment bury this residue, temperatures and pressures increase and the liquid hydrocarbons, are segregated into pore spaces in the rock. Coal deposits come from many epochs, but the best and most abundant are from forests in the warm swampy river deltas of the Carboniferous period (360–280 million years ago). The process by which forest peat is turned to coal is complex. Initially the carbohydrates and waxy materials of the plant are attacked by swamp bacteria and fungi to give volatile gases like methane and carbon dioxide. Gradually the mixture gets richer in sulfides and hydrocarbons. Layers of strata build up, gradually squeezing out water from those lower down.

Cooking up the perfect fuel

The temperature of the Earth's crust increases with depth by about 2°F (1°C) for every 100 feet (30 meters). The increasing temperature causes chemical reactions that turn the peaty matter into coal. A temperature of 400°F (200°C) – meaning burial at a depth of over 3 miles (5 kilometers) – is needed to make top-quality coal. Variations in temperature give a range of coals of different "rank" or carbon percentage, from the peaty brown coals or *lignites* at 60 percent carbon, through the *bituminous* coals at 92 percent to the *anthracites* at over 92 percent.

Optimum oil production

The quantity of oil and gas formed depends on temperature and on the speed of subsidence of the strata. Ideally, up to 2 miles (3 kilometers) of overlying strata are needed. If subsidence is too rapid the temperature may rise too high and oil and gas are lost. Neither is abundant below 4 miles (6 kilometers). In fact they are rarely found in the rocks where they were formed, either. They migrate up –

Oil and gas *form as layers of plankton are buried under thick piles of sediment* [**A**]. *Increasing heat and pressure at depth first cause fats and oils from the bodies of the marine organisms to "link up" into a thick compound called kerogen. As temperature increases with burial, long chains made of hydrogen and carbon atoms break away from the kerogen, giving a viscous heavy oil. With even more heat, valuable light oils and natural gas are formed. The oils accumulate in "reservoir" rocks – permeable rocks such as sandstone which hold the oil like a sponge. To form an oilfield, the oil must be trapped between layers of an impermeable rock, like shale* [**B**]. *Faults, where such rock has sheared to form a seal* [1], *can trap oil and gas, as can convex domes* [2]. *The making of most coal* [**C–G**] *was begun in the middle of the Carboniferous period. The Earth's equatorial regions were hot and wet and lush tropical forests grew in extensive swamps* [**C**]. *In these types of environment, thick layers of peat were laid down* [1]: *typically they were sandwiched between layers of sediment, such as shale* [2], *deposited when the waters temporarily retreated. The seas receded during the Permian period and many of the tropical coastal plains turned to desert* [**D**]. *Other sedimentary rocks, such as sandstones* [3], *were laid down over the shale and peat. With increasing temperature and pressure the buried peat began its metamorphosis into coal. Around 150 million years ago* [**E**] *the deserts were covered over by shallow tropical seas in which limestone* [4] *was deposited.*

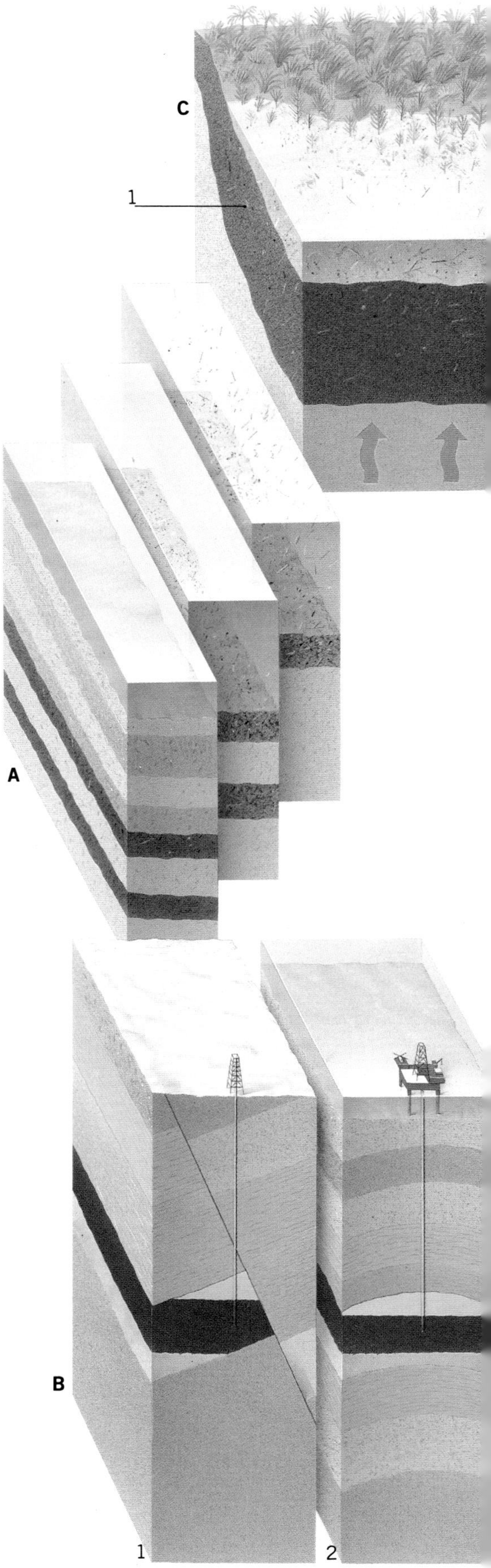

Connections: Air Pollution 342 Cycles of Life 206 Earth's Structure 12 Fossilization 28 Land Pollution 338 Minerals 32 Ocean Floor 50

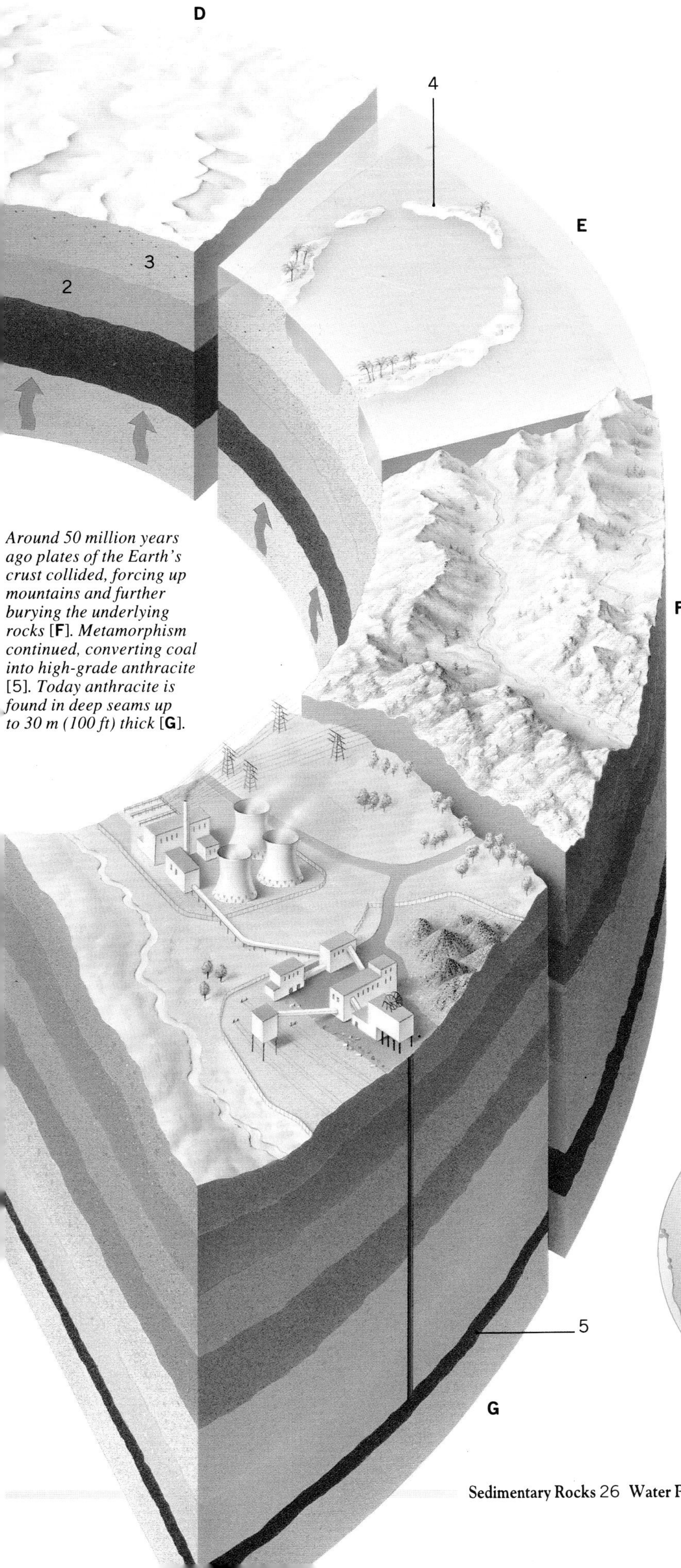

Around 50 million years ago plates of the Earth's crust collided, forcing up mountains and further burying the underlying rocks [**F**]. *Metamorphism continued, converting coal into high-grade anthracite* [5]. *Today anthracite is found in deep seams up to 30 m (100 ft) thick* [**G**].

the gas and lighter oils first, with heavier bitumens staying closer to their source – until they are trapped by overlying impermeable rocks like shale and clay. Considerable quantities of oils and gas, however, may be lost by migration to the surface, where they oxidize and evaporate.

Petrochemistry to petrodollars

Petroleum fluids are a complex mix of organic compounds with a range of specific gravities and boiling points. Natural gas is essentially methane and can be piped directly from the source or may be liquefied for transport. The oils are classified as light distillates, medium distillates and heavy residuals. Light distillates are the motor fuels like gasoline and benzine. Medium distillates make up paraffin, diesel, jet engine and power plant fuels. Heavy residuals are used to fuel power plants and ships.

Fuel Resources

The predicted future life of the world's coal reserves was once put as low as a few decades, but the real figure is actually nearer 300 years. The Soviet Union and the United States have the greatest resources. Western Europe, India, China, Brazil, South Africa, and Australia also have large stocks of top-rate coal from the Carboniferous or Permian periods. Brown tertiary lignates can be found in Central Europe and the Ukraine. Antarctica also has large unused reserves of coal, which could only be exploited at disastrous environmental cost. The best estimates are that, without a major switch to alternative fuels, the world's oil reserves will last a little over 40 more years.

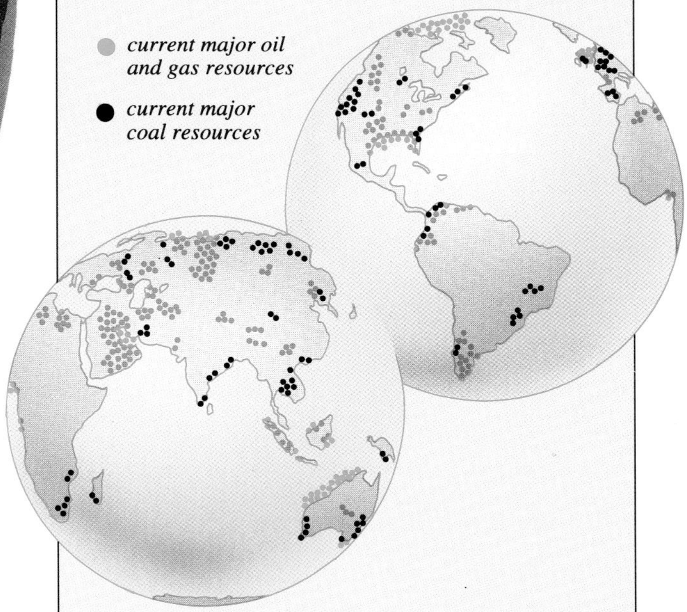

Sedimentary Rocks 26 Water Pollution 340

Hidden Worlds

How caves are formed

A room more than 1,800 feet (550 meters) long and 100 feet (30 meters) high would be a remarkable feature in any building. It is even more remarkable 700 feet (220 meters) below ground. But this is the size of the Big Room – the main chamber of the Carlsbad Caverns in New Mexico. The Caverns were discovered in 1901 when a cowhand on his way home investigated a curious black cloud that appeared to be emanating from a hole in the ground. The cloud was in fact a huge swarm of bats. The hole in turn was the entrance to the vast caves that now form part of a National Park.

Carlsbad's infinitely varied stalactites and stalagmites are almost all composed of calcium carbonate in the form of calcite. Their graceful shapes attest to the crucial role of water in decorating the caves. But water is not only essential to building up such fantastic forms in the caves it is also usually one of the basic factors in the process by which the caves themselves are created.

Water hollows caves out of the calcium carbonate that makes up limestone strata by dissolving it away. But water, by itself, is by no means able to do it all: nearly a gallon (four liters) of pure water can dissolve only 0.001 ounces (0.028 grams) of limestone. What enables water eventually to dissolve proportionately between 10 and 60 times more limestone than this is the fact that as the surface water seeps downward through the soil it passes over decaying organic matter. The plant and animal remains give off carbon dioxide, which dissolves in the water and forms carbonic acid. The acid completes the first step in the process of cave formation.

But if the dissolution of limestone by pure water and the formation of carbonic acid were in turn all that happened, the reactions would grind to a halt once completed, when equilibrium was achieved. There is, however, a third chemical reaction, which uses the products of these first two reactions to continue the process of erosion.

Dissolved in water, calcium carbonate breaks up into charged particles (*ions*) of calcium and carbonate, and carbon dioxide combines with water to form a bicarbonate ion and a hydrogen ion. The third reaction occurs when the carbonate combines with the hydrogen to form more bicarbonate: it is this reaction that allows the groundwater to dissolve up to a tenth of an inch (2.5 millimeters) of limestone in a year.

The opposite happens when water rich in limestone drips off the ceiling of an empty cavern. The carbon dioxide gas comes out of solution, the reactions go into reverse, and calcium carbonate precipitates out.

Agents of erosion

Additional factors may have a significant effect on the creating of caves. Impurities in the limestone may cause inherent weakness in the newly forming stratum. To be strong enough to form a roof over large cavern

Stalactites and stalagmites *form in caves that have been drained of water following a fall in the level of the water table. Drops of water charged with calcium bicarbonate hang from the ceiling of the cave and lose carbon dioxide, so depositing a tiny amount of calcium carbonate. Further drops deposit more layers of calcium carbonate in the same place, and a stalactite slowly develops. Growth rates vary greatly, but stalactites have been known to increase by 3 in (7.6 cm) in 10 years. They are brittle and rarely reach great lengths. Drops of water may fall to the floor and lose carbon dioxide, leaving deposits of calcium carbonate that grow upward into stalagmites. These may meet to form continuous columns.*

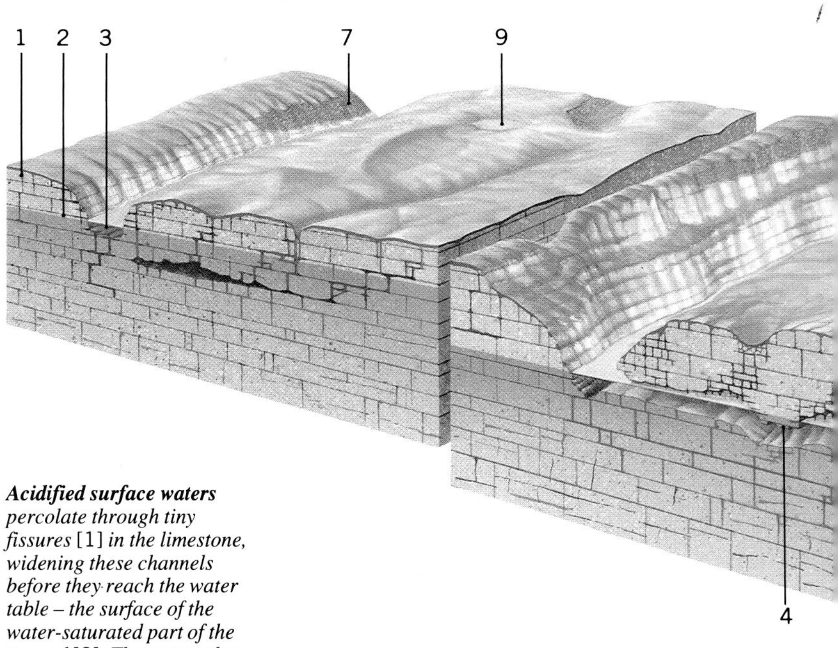

Acidified surface waters *percolate through tiny fissures* [1] *in the limestone, widening these channels before they reach the water table – the surface of the water-saturated part of the ground* [2]. *The waters then flow horizontally toward a natural outlet – in this case a river* [3] *– dissolving away limestone in their path.*

Karst Landscapes

As acidic rainwater works on a limestone area, certain typical surface features become apparent. Regions that display these distinctive formations are known as Karst landscapes, named after the Karst region on the Dalmatian coast of Yugoslavia. Underground, complicated systems of shafts and caves can extend through hundreds of yards of limestone. On the surface, the terrain is peppered with sinkholes, which often coalesce into large depressions known as *poljen*.

The ground resembles a pavement, with blocks of bare limestone separated by enlarged joints. The collapse of caves may leave behind irregular pits surrounded by pillars of rock, or even bridgelike structures. Through deforestation and soil erosion vegetation is often sparse and restricted to dips and hollows.

Connections: Coal, Oil, and Gas 34 Coastlines 56 Cave Life 308 Mountains 18 Rock Formation 24 Sedimentary Rocks 26

chambers, the limestone has to be fairly pure. If the rock contains too much sand and clay the stratum will be flaky or crumbly – caves occur only in limestone that is more than 50 percent pure. Even then the limestone must also have the quality of impermeability so that the flow of water is concentrated along isolated fissures and cracks.

Cave varieties

Other agents help to create caves. The surface of volcanic lava flows commonly solidifies rapidly to form a rocky conduit through which molten lava flows. After all the molten lava has drained away, a curving network of tunnels is often left behind. Elsewhere, hydrogen sulfide brines rising from oil and gas fields may generate sulfuric acid, which eats away limestone, and, on the coast, the action of wind and water may erode caves.

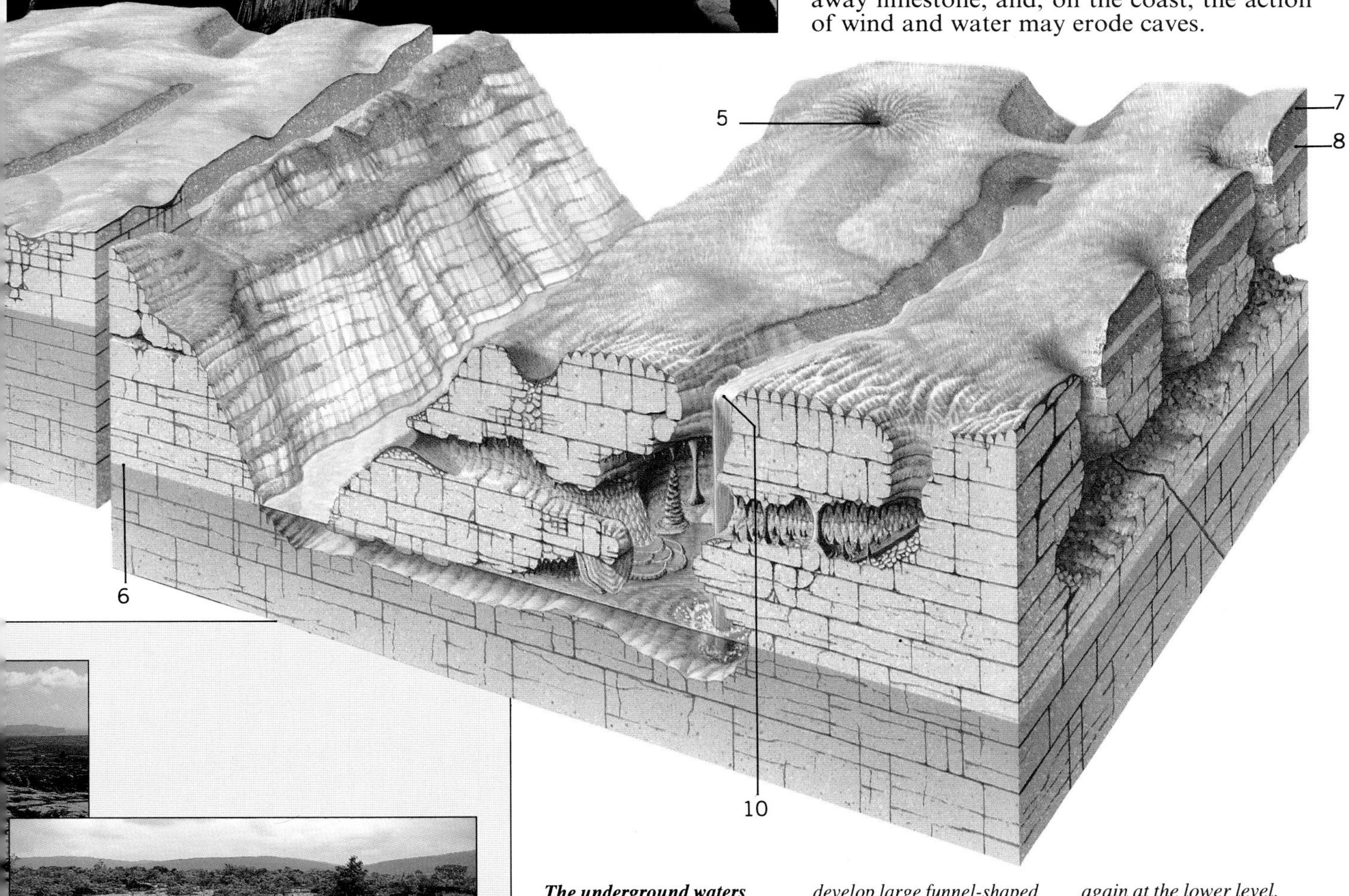

The underground waters *eventually carve out a main horizontal channel* [4] *at the depth of the water table. As it widens, this channel draws an increasing volume of water, thus accelerating its growth. Some of the vertical shafts also begin to attract more than their fair share of surface drainage and may develop large funnel-shaped hollows, or sinkholes, around their mouths* [5]. *On the surface the river cuts through the limestone and the water table drops* [6]. *The water in the main underground channel drains out, seeking new paths to the water table, and the process of channel carving begins again at the lower level. Wherever other rocks cap the limestone, such as sandstone* [7] *and impermeable clays* [8], *water may be trapped on the surface in hollows and depressions* [9], *or may give rise to rivers that run along the surface, high above the river that defines the water table* [10].

The Frozen Past

What causes ice ages

Even in the severest ice ages the global temperature drops by no more than 7–21°F (4–12°C). But this may be enough to plunge 40 percent of the Earth's surface into a 100,000-year winter. There have been as many as 20 ice ages during the past 2.5 million years. At the peak of the last of these glacial periods, around 20,000 years ago, one-third of the world's land resembled the ice-covered landscapes of Antarctica. The breaks between ice ages typically last for only about 10,000 years, so the warm period of the last 10,000 years may be ending, and the next ice age could be on its way.

Ice ages are known as *glacials* and the periods between them as *interglacials.* The periodic fluctuation between glacials and interglacials in the past 2.5 million years is thought to be caused by long-term variations in the Earth's orbit and in the behavior of its axis. In the 1920s the Yugoslavian scientist Miluti Milankovitch computed how orbital changes alter the seasonal amounts of solar energy received by each latitude. He proposed that the strength of the Sun at middle latitudes in the Northern Hemisphere is the key to ice-sheet growth and decay because the quantity of energy received, in the form of heat from the Sun, determines whether a permanent snow cover can persist or not.

Once summers have become cool enough to favor a persistent snow cover, ice accumulates and glaciers and ice sheets expand rapidly because the ice itself reflects solar energy and lowers regional and global temperatures even further.

Alternative theories

Although the Milankovitch theory of ice ages is widely accepted and is, in fact, supported by mathematical analysis, other scientists have put forward their own alternative theories. Some, for example, believe that ice ages are caused by huge volcanic eruptions releasing massive amounts of volcanic dust into the atmosphere, which subsequently blocks out sunlight and lowers the surface temperature of the Earth. Others believe that the Earth's natural radiation output fluctuates. This is thought to be brought about by the Earth's core heating up or cooling down. If this is true, then it is possible that the core cools sufficiently to give rise to ice ages. Other theories include clouds of cosmic dust blocking out the Sun's rays, or even a drop in the Sun's solar radiation output.

Ice power

The land buried beneath an ice sheet is transformed dramatically by the power and weight of the ice. Each summer, raging torrents issue forth from ice sheets carrying tremendous amounts of sands and gravels. In winter, the ice's progress is unstoppable. Strong winds flowing from the tall ice sheets carry away with them some very fine silts and clays, which are then laid down in distant lands as thick layers of fertile soil.

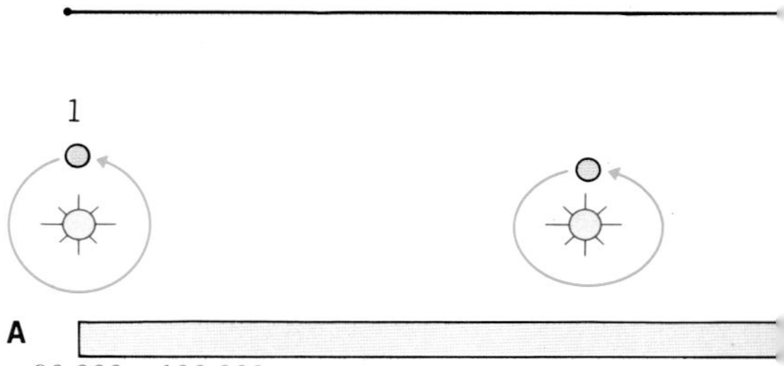

The astronomical theory of ice ages, the "Milankovitch Model," is based on the three changes [A, B and C] in the Earth's movements through space which affect the distribution of the Sun's heat on the Earth's surface. The Earth's orbit around the Sun [A] is constantly changing. It varies from an almost perfect circle [1] (known as an orbit of low eccentricity) to a distinct ellipse [2] (high eccentricity) and back again [3]. This whole cycle takes between 90,000 and 100,000 years. Currently the orbit is becoming increasingly circular. There is also a 400,000-year cycle over which the maximum eccentricity increases and decreases. When in a circular orbit, the Earth receives an even spread of heat from the Sun. However, in the more elliptical shape, at certain times the Earth is farther from the Sun and therefore cooler. The second change [B] involves the tilt of the Earth's axis in relation to the plane of its orbit. At present the axis is tilted at about 23.5° off the perpendicular. This means that the pole tilted toward the Sun receives more hours of sunlight than the pole pointing away from the Sun – this is why we have seasons. However, at the moment the tilt is decreasing, which means the seasonal changes will become less dramatic. The Earth takes about 40,000 years to go from the minimum tilt of 21.8° [4] to the maximum tilt, 24.4° [5], and back again [6]. The third variation [C] is called the "circle of precession." This is the imaginary circle made by the "wobbling" of the Earth's axis. It takes about 22,000 years for the Earth's axis to describe a complete circle. The combination of these three variants [D] – and particularly a high eccentricity – are thought to create the preconditions for severe ice ages.

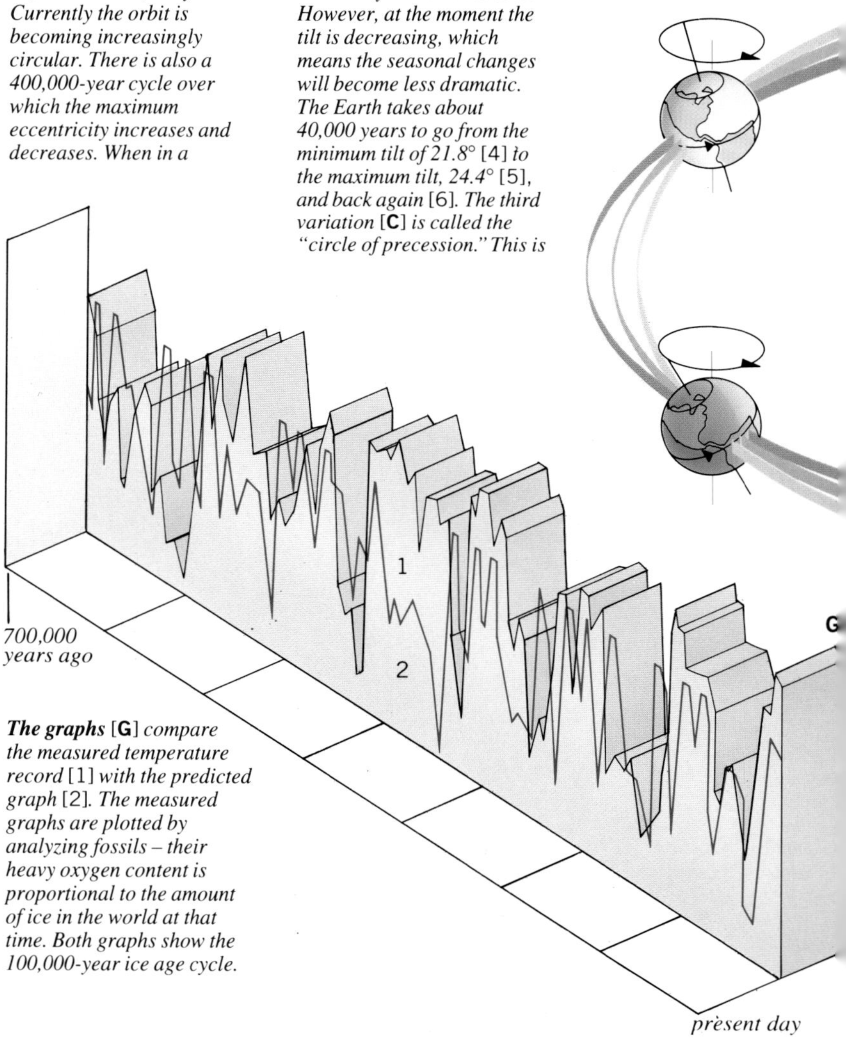

The graphs [G] compare the measured temperature record [1] with the predicted graph [2]. The measured graphs are plotted by analyzing fossils – their heavy oxygen content is proportional to the amount of ice in the world at that time. Both graphs show the 100,000-year ice age cycle.

Connections: Climatic Change 82 Fossilization 28 Glaciers 40 Global Warming 84 Polar Regions 42 Seasons 86 Weather and Climate 70

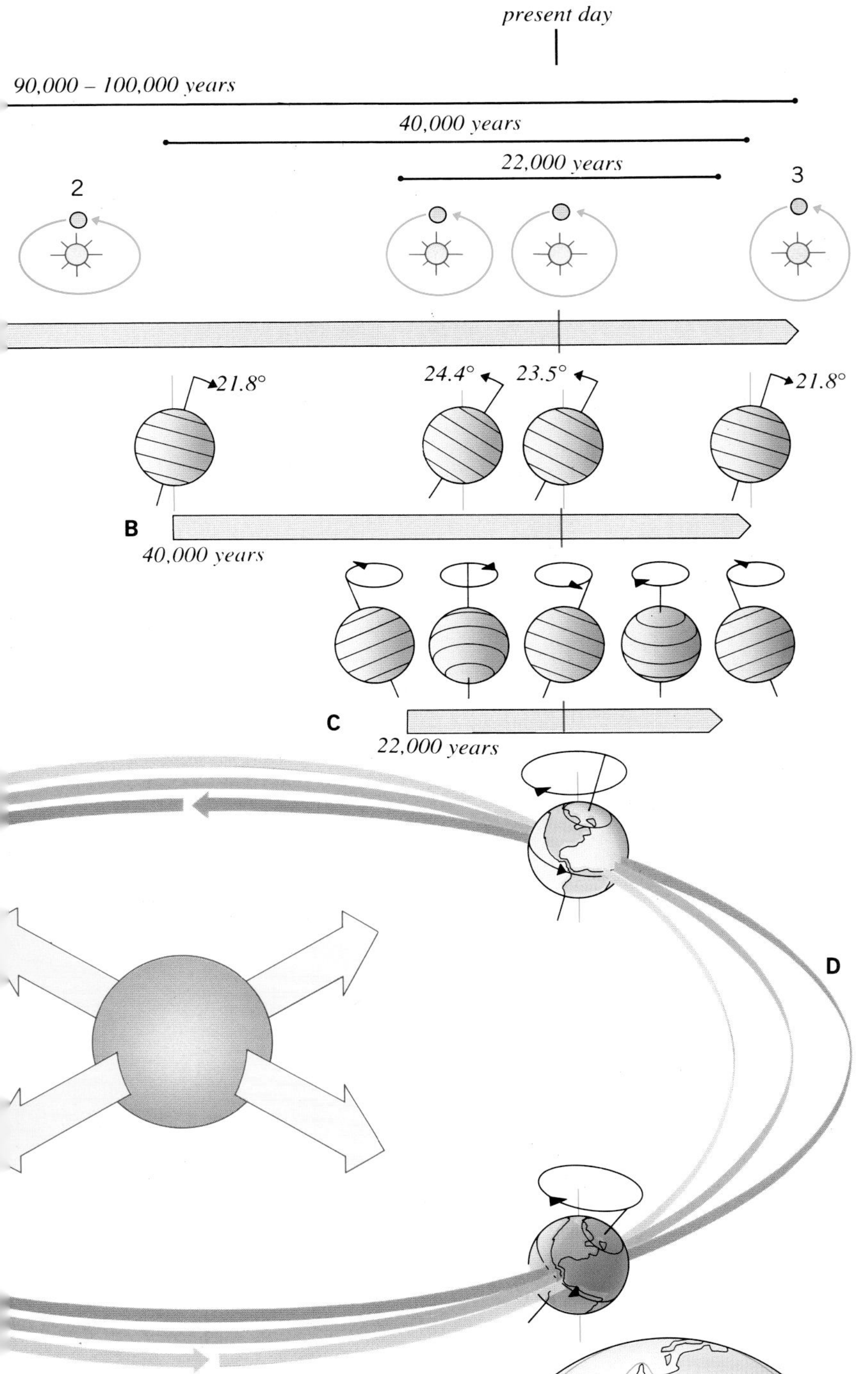

Evidence of the Ice Ages

Advancing ice sheets distort or destroy the signs of any previous advance so it is difficult to determine exactly how many glacials have occurred. Recent research has concentrated on investigating deep ocean floors, which were unlikely to have been disturbed by the ice. Cores drilled from ocean sediments contain fossils of tiny planktonic organisms called *foraminifera.* The preference of each species for warm or cold waters allows a continuous record of past climate to be inferred. Such evidence reveals 13 long-lasting glacials to have occurred in the past 1 million years alone. This has now been confirmed by a range of other indicators of past environmental conditions such as cave stalactites, coral reefs, pollen and insects preserved in sediments in peat bogs, and fossilized fauna in loess deposits.

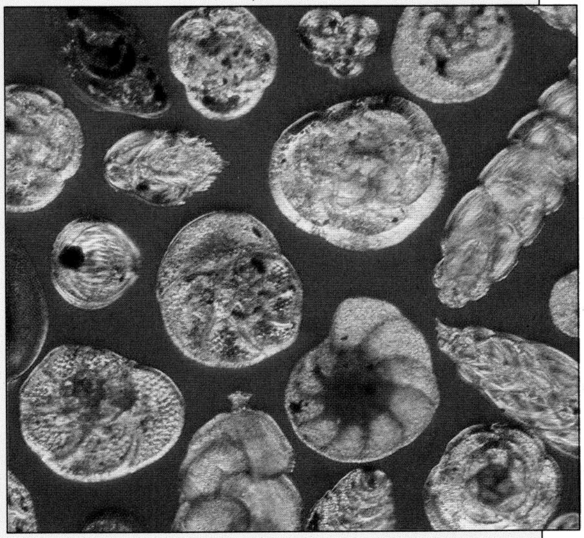

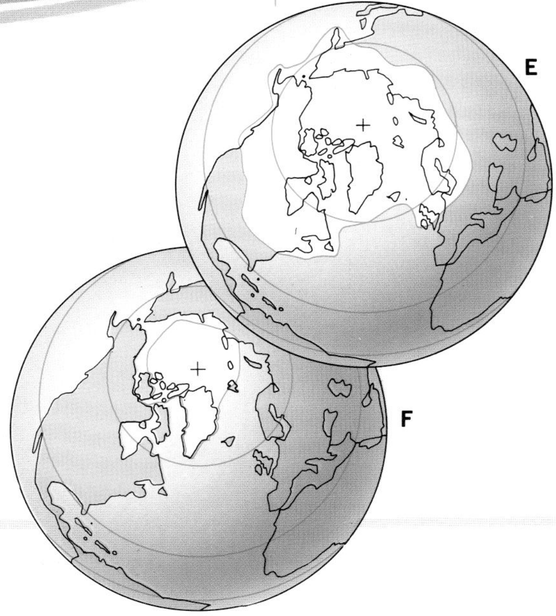

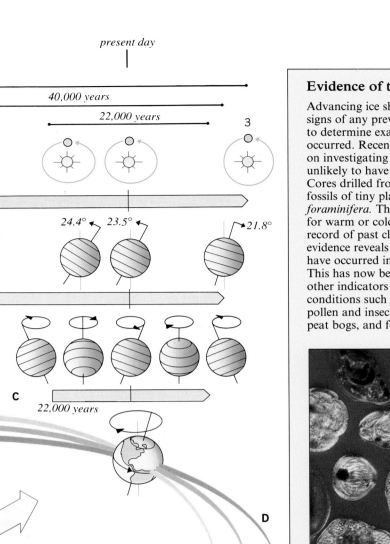

***About 20,000 years ago* [E]** *permanent ice sheets covered 16 million sq miles (42 million sq km) compared with the present* **[F]** *6 million sq miles (15 million sq km). They locked up vast amounts of water, causing world sea levels to fall by 400 ft (120 meters), revealing many new landmasses.*

Retreat of the ice

When the ice sheets eventually retreat, plants and animals begin to colonize the newly exposed lands, returning from their refuges in milder, ice-free climates. During the last glacial, for example, islands off the coast of British Columbia remained ice free (because of dry, cold air which did not bring snowfall) although the mainland was covered by ice up to 6,500 feet (2,000 meters) thick. Not only did the islands provide sanctuary for flora and fauna but they formed a migration corridor during the glacial for the early entry of humans into North America from Asia via the Bering Strait. Subsequently, the land-bridge between the continents was submerged as the ice sheets melted and caused sea levels to rise again. Today, in some countries such as Sweden, new lands are emerging from the sea as the land recovers from the effects of being depressed under the vast weight of ice.

Rivers of Ice

How glaciers form and move

The power of a glacier can rip away the sides of mountains, carve deep U-shaped valleys, and transport huge boulders hundreds of miles. The vast sheet of ice grinds rock to powder and scores deep grooves in the solid rock floor over which it moves ever so gradually. In past ice ages glaciers extended over much of the continents of Europe and North America. The vast heaps and mounds of debris left behind by them as they melted today provide geologists with essential clues to the former extent of the ice and to the direction in which it traveled.

A glacier develops high in a mountainous region, or in polar regions even at sea level, when the accumulation of snow exceeds the rate of its melting. The bottom layers of snow gradually change to ice as compression by the overlying mass compacts and rearranges the crystals. When it first falls snow has a very low density and contains a good deal of air. The delicate snow crystals turn into ice grains as their air content is reduced and the water molecules of which they are made change their position.

Snow in this coarse, granular state is called firn. This changes into glacier ice as the pores containing air separate into bubbles and the resulting ice loses its permeability. The whole process, which may take 50 or more years, usually occurs in a small rock hollow. As ice accumulation continues, the small new glacier flows under gravity.

Sculpting the terrain

A valley glacier like those in the European Alps is characteristically a few tens of miles long, a few hundred yards wide, and several hundred yards thick. It has tremendous erosive power and easily increases the depth of the valley that it occupies, giving it a typical U-shaped cross section and a long profile marked by many basins and steps. After the ice has melted these deep valleys are frequently flooded and occupied by ribbon lakes. Where the glacier has cut a valley down to the sea a fjord forms when the sea takes the place of the melted ice.

The erosional processes of a glacier vary. Rock fragments are frozen into the ice at the base and sides of the glacier and as it moves these are plucked out and carried along. Such fragments of rock are powerful sculptors of the land over which the glacier moves, and in turn are gradually ground down to powder. The rock materials incorporated in the glacier ice thus range in size from house-sized boulders all the way down to fine dust. By studying where the transported boulders (called *erratics*) have come from it is possible to reconstruct the direction of movement of long-vanished glaciers.

Water on the glacier's bed helps it to flow by providing lubrication. This water results from the melting under pressure of some of the ice, and from the presence of subglacial streams. As the volume of water increases (in summer) below the ice, so the speed of the glacier may increase; if the glacier has much rock debris in its basal regions, the friction generated slows the ice flow down. In addition to valley glaciers and piedmont glaciers – formed where valley glaciers spread out at the feet of mountains – ice caps are the most widespread glaciers.

The speed of a glacier

Movement can be measured by planting a series of stakes in the surface of the ice and studying their displacement. By inserting sensitive instruments into holes drilled in the ice, the internal flow of the ice can also be calculated. The central parts of a glacier move fastest, in the case of an alpine glacier on a shallow slope maybe 330 feet (100 meters) each year, while the margins of the glacier are restricted by friction to speeds under 100 feet (30 meters) each year.

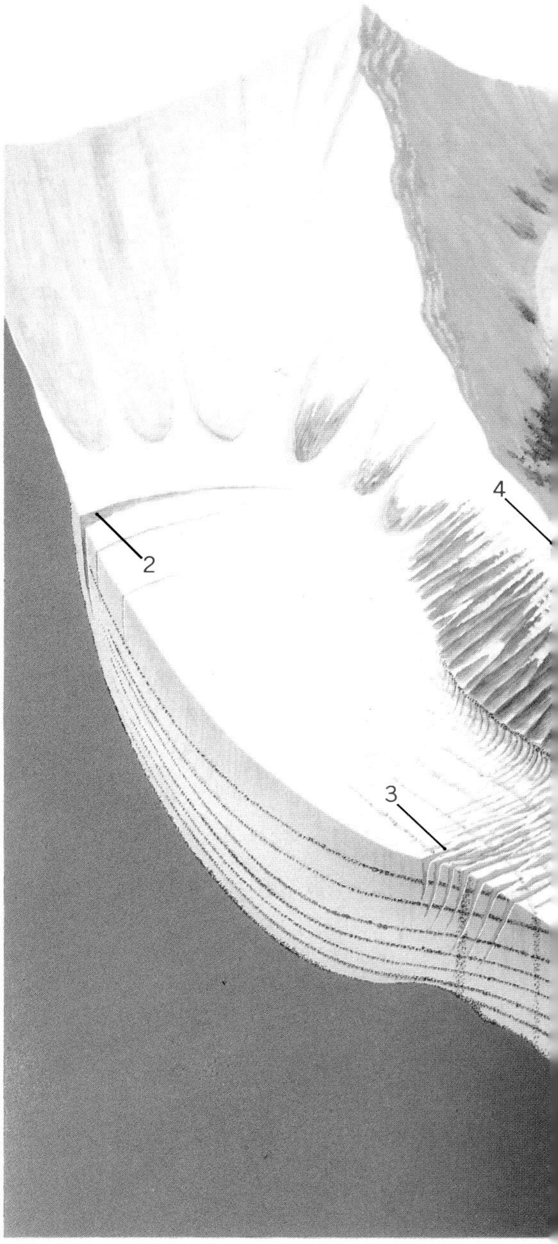

***The snow that gathers** in a cirque [1] is known as névé. As it compacts to form a glacier it pulls from the valley head. The glacier is separated from the permanent snow by a deep crevasse called a bergschrund [2]. Because glaciers move at different speeds across their widths, and because they move over uneven ground, cracks or crevasses also appear in the glacier [3]. Moving glaciers grind mountains into sharp ridges or arretes [4]. The matter they scrape off as they move against the valley sides makes moraines. Moraines concentrated on the sides of the glacier are lateral moraines [5]; when two glaciers merge, the combined lateral moraines become a medial moraine [6] in the new larger glacier. Where a glacier has deepened a main valley, and subsequently retreated, any of the former tributary valleys are left "hanging" [7] and often drain into the main valley beneath by means of a waterfall [8].*

***A pinnacle of ice** may form beneath a rock that shelters it from the melting of the Sun's rays. This is a glacial table [9]. The glacier's snout [10] will move forward if ice is brought from uphill faster than it melts. If the two rates are balanced the snout is stationary, but rubble is continually brought down and deposited as till. When the snout retreats, melting faster than it is replenished, this rubble is left behind as a terminal moraine [11], which dams the ribbon lake [12] formed by the melting ice. Caves and tunnels in the glacier [13] can be caused by crevasses or by meltwaters at the surface or beneath the glacier.*

Connections: **Coastlines** 56 **Rain, Hail, Snow and Fog** 76 **Ice Ages** 38 **Lakes** 60 **Mountains** 18 **Polar Regions** 42 **Soils** 30

A roche moutonnée *[right, and 14] is created underneath a moving glacier. As the glacier passes over a resistant rock hummock the debris that it carries smooths the shape of the hummock's uphill side while scratching striations in the rock. The downhill side of the roche moutonnée is rough, having been plucked at by repeatedly melting and refreezing ice. Drumlins, on the other hand, are mounds of till deposited by the glacier. As such they tend to have their streamlined "tails" pointing downhill, not up as with the roche moutonnée. Drumlins may be several miles long.*

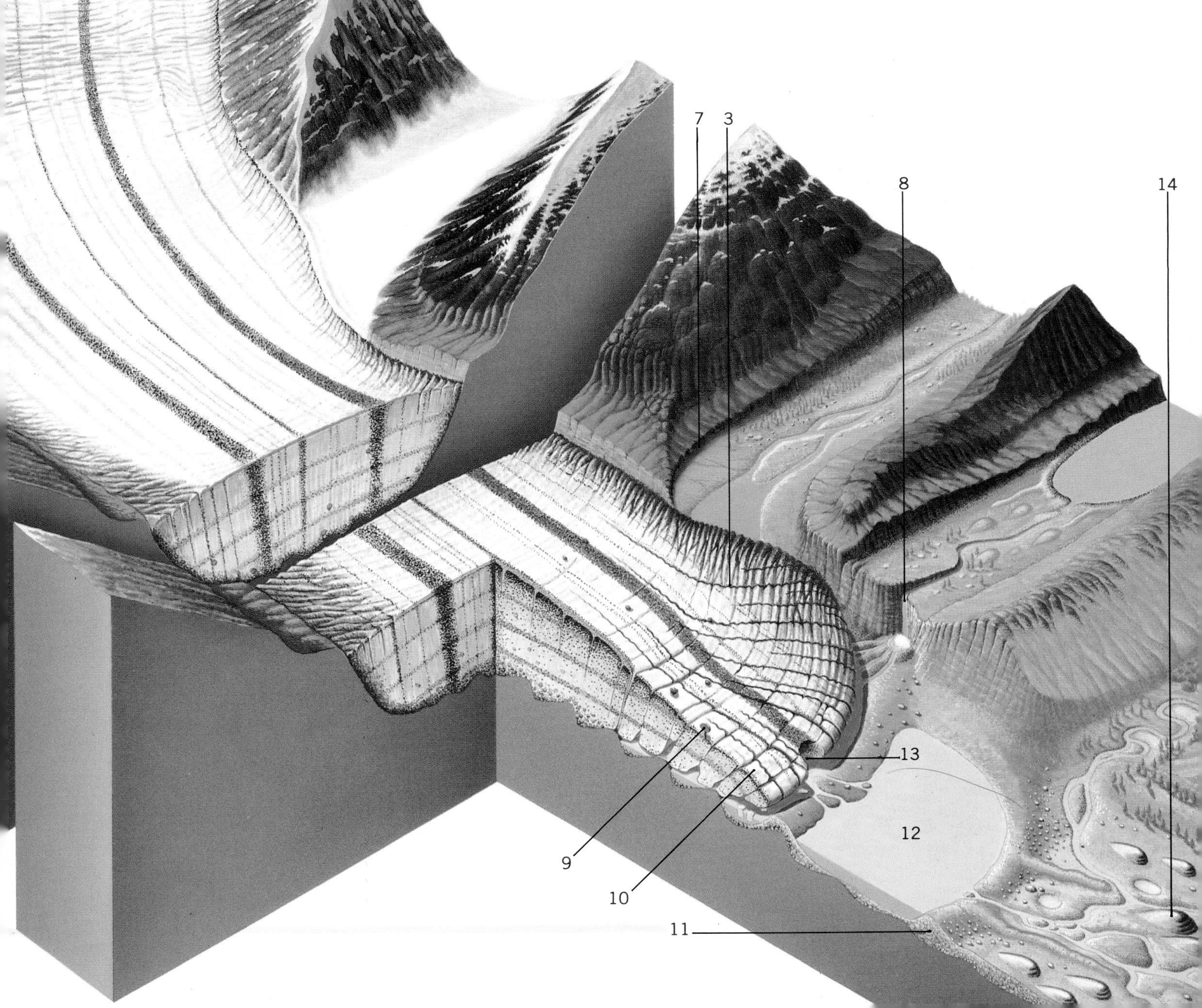

At the Ends of the Earth

What makes the polar regions cold

It may take several thousand years for a snowflake falling on the Greenland ice cap to reappear as ice on the coast of Greenland. Both the Arctic and Antarctic regions are in a constant state of change – a dynamic balance between growth and shrinking. The ice coverage of the Arctic increases by a third in its winter, whereas the ice coverage of the Antarctic grows almost sevenfold. Despite their apparent similarities, the two regions are fundamentally different – the Arctic is largely ocean, while the Antarctic is a high and mountainous continent covered in ice.

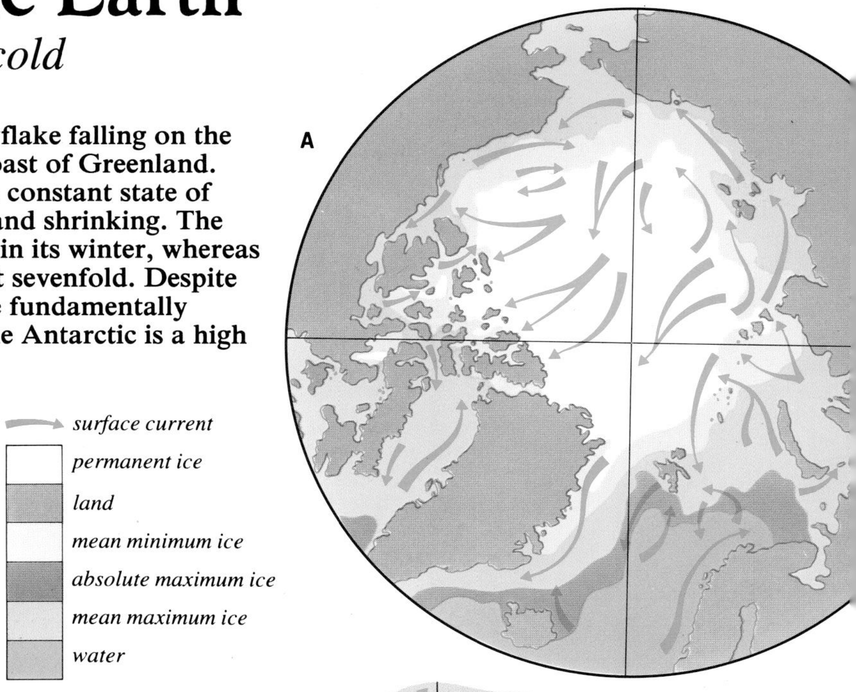

At the north pole the summer temperature is near freezing and the winter can get as cold as –89°F (–67°C). But this is a real scorcher by the standards of the south pole. There an observer could expect winter temperatures of –128°F (–89°C) or even lower. The Antarctic is much colder than the Arctic because it is so much higher. At the north pole you would find yourself at sea level on slowly shifting ice under a cloudy sky. Near the south pole you could be standing on about 15,000 feet (4,500 meters) of ice and the sky would probably be clear. In addition, the Arctic ocean helps to warm up the northern polar region. But not only is the Antarctic colder than the Arctic it is also much bigger. Using the *Antarctic Convergence* – a zone separating the cold waters of the Antarctic from the warmer, less biologically productive mid-latitude waters – as the boundary of the south polar region, the Antarctic has an area greater than 20 million square miles (50 million square kilometers). The boundary of the Arctic is taken to be the line where mean temperatures rise to 50°F (10°C) in July. Its area then is 10 million square miles (25 million square kilometers), of which two-thirds are sea and one-third land.

The glancing Sun

But these comparisons do not explain why the polar regions are so cold in the first place. The most basic reason is that, at the Earth's extremities, the Sun never rises very far above the horizon. It remains low in the sky so that less of its radiant energy reaches the polar regions. Secondly, during their six-month-long sunless winters the polar regions lose much more heat than they gain during the equally long six-month summer seasons.

The year-round ice caps are the third factor that keeps the Arctic and the Antarctic so cold. They reflect away 95 percent of the solar energy that reaches them, so what little warmth the polar regions receive is sent straight back into the atmosphere. In fact the polar regions would be much colder than they are now if it were not for the oceans and the atmosphere, which moderate the polar climate and warm things up by transporting heat, eventually, from the equatorial regions.

The Antarctic ice cap and the Greenland ice sheet have been very stable in recent history. Their growth – through the addition of

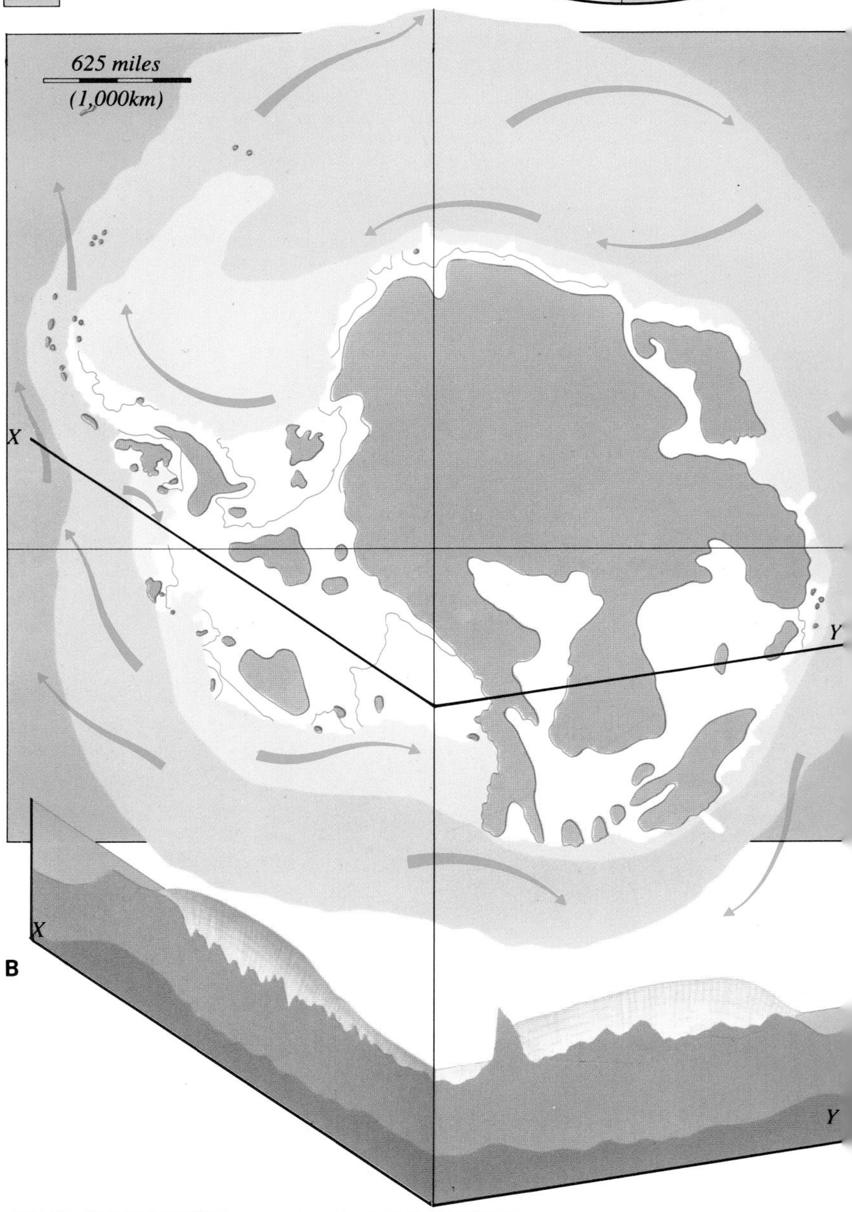

The Greenland ice cap **[A]** *constitutes 90 percent of the ice in the northern hemisphere (but only 9 percent of the Earth's total), and provides 90 percent of the hemisphere's icebergs. The ice belt around the north pole is distorted by its ocean currents. Because of its own insulation, the ice in the north polar sea can get little thicker than 10–13 ft (3–4 m), except where it is layered as a result of movement. Over much of the 2 million square miles (6 million sq km) of permanent pack ice, ice coverage is over 70 percent of the sea's surface. Icebergs also float along with the pack ice, and with the ice floes that surround it.*

new snow and ice – and shrinkage – through *sublimation* (the passage of ice directly into water vapor), evaporation, melting, and the formation of icebergs – are more or less balanced. Whether they remain stable, grow or shrink depends on how the Earth's global climate changes.

Ice giants

Each year about two inches of ice accumulates on the Antarctic ice cap. It falls as either snow or as hoar frost. Glaciers, the "rivers of ice" which move faster than ice sheets as a whole, usually originate in areas of higher than average precipitation. The very low precipitation of the high Arctic means that glaciation may be less around the shores of the polar sea than in mountainous regions farther south. Much of Peary Land, in Greenland, for instance, is ice free even though it is the most northerly land on Earth.

Glacier bergs *(top) break away from socalled tidewater glaciers, or glaciers that end in the sea. Up to nine-tenths of a glacier berg may be under water. Tabular icebergs (above) break off from large ice shelves, and have around five-sixths of their total volume submerged.*

Polynyas **[C]** *are open areas of water among sea ice. Because there is no insulating layer of ice, they act as heat vents from the ocean to the air. The convection currents set up by this heat loss influence the patterns of local ocean circulation. As well as open-ocean polynyas* [1] *there are coastal polynyas* [2]. *As well as causing convection currents these polynyas affect ocean circulation because their surface water has a high concentration of salt – left behind when the salt water freezes. This denser water helps the formation of the bottom-water currents.*

C

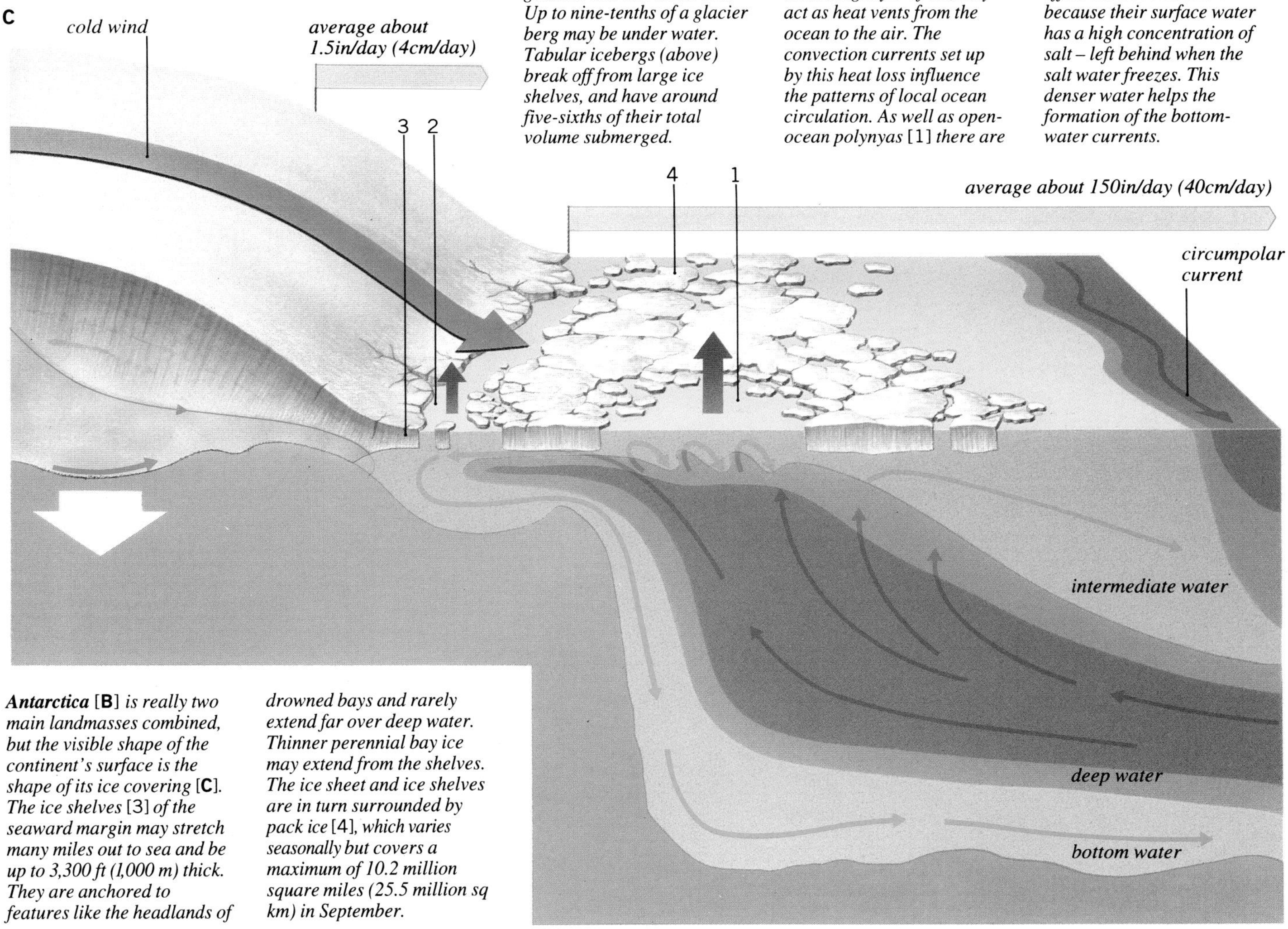

Antarctica **[B]** *is really two main landmasses combined, but the visible shape of the continent's surface is the shape of its ice covering* **[C]**. *The ice shelves* [3] *of the seaward margin may stretch many miles out to sea and be up to 3,300 ft (1,000 m) thick. They are anchored to features like the headlands of drowned bays and rarely extend far over deep water. Thinner perennial bay ice may extend from the shelves. The ice sheet and ice shelves are in turn surrounded by pack ice* [4], *which varies seasonally but covers a maximum of 10.2 million square miles (25.5 million sq km) in September.*

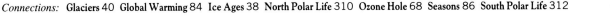

Connections: Glaciers 40 Global Warming 84 Ice Ages 38 North Polar Life 310 Ozone Hole 68 Seasons 86 South Polar Life 312

Mysteries of the Deep

What makes the seas and oceans

If life in the oceans stopped, the carbon dioxide content of the Earth's atmosphere would triple. This is because of the vast numbers of marine plants that use carbon to form their bodies, thus reducing the carbon dioxide in the surface water and hence in the air. In fact, the sea is the major source of atmospheric carbon dioxide – it has dissolved within it 45 times as much carbon as there is in the atmosphere. The sea covers over 70 percent of the surface of the Earth and most of its incredible volume is still little known and unexplored.

The ocean is divided into two distinct zones, or reservoirs. The upper reservoir is warmed by the Sun and stirred by the wind. Water from the sea's surface evaporates, condenses and falls as rain. Rain water that falls on land then percolates through the land, dissolving and carrying into the ocean some of the chemical constituents of sea water. Volcanic activity, on land and beneath the ocean, supplies others. In the upper reservoir microscopic phytoplankton convert the Sun's energy into organic matter which supports a food chain of animals and bacteria throughout the oceans. The systems that control the distribution of carbon between the sea and the air involve the components of sea salt that are used by life in the sea: oxygen, calcium ions, phosphates, and three carbon-based mixtures: carbon dioxide, carbonate ions and bicarbonate ions. Marine animals use oxygen for respiration and carbon and phosphorus for various biochemical functions. Some surround themselves with shells made of calcite, or calcium carbonate. Somehow, through a series of interconnected cycles, the ocean chemically balances the demands of marine organisms with the input of material from erosion or volcanic activity and the loss of material by burial as sediments on the ocean floor.

The thermal barrier

The upper reservoir, which accounts for only about 2 percent of the oceans' volume, is separated from the lower reservoir by the main *thermocline*, a zone in which water temperature decreases rapidly with depth. The temperature starts to decrease at depths ranging from 300 feet (100 meters) to 1,000 feet (300 meters), and continues to drop for the next few hundred yards down to just a few degrees centigrade. Since cold water has a higher density than warm water, the thermocline acts as a very effective barrier to prevent mixing between the two layers. So efficient is the thermocline that a water molecule starting off in the lower reservoir takes about 1,000 years to enter the upper layer.

The elemental store

The lower layer comprises most of the oceans' volume and here the temperature of the water is generally constant and lies within the range of 30–41°F (–1 to 5°C). The waters of the lower layer originate in the polar regions where the surface water is denser than water at the lower latitudes, because it is colder. These waters of the lower layer also have a constant concentration of dissolved material or *salinity*. Most of the naturally occurring elements are included in the 5,000 trillion tons or more of solids that are dissolved in the sea – if only in minute amounts – and sea water is the only source of one element, iodine, which is absolutely essential for human life.

Although the oceans' composition is often thought to have been constant through time, there is now strong evidence that significant changes have occurred in the concentrations of ocean salts. Some changes appear to be related to major cycles of glaciation, which changed the volume of the oceans, while others may relate to the change in the locations of the continents caused by plate tectonic couplings and splits.

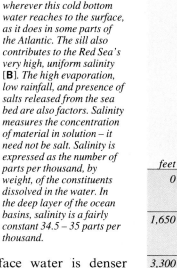

*The **Bab-el-Mandeb** sill divides the Red Sea from the more typical open-ocean waters of the Gulf of Aden. This explains the very uniform temperature profile* [**A**], *which lacks a thermocline. Open-ocean basins are filled with cold water like that found on the bottom of the Gulf, with a thermocline. The thermocline disappears wherever this cold bottom water reaches to the surface, as it does in some parts of the Atlantic. The sill also contributes to the Red Sea's very high, uniform salinity* [**B**]. *The high evaporation, low rainfall, and presence of salts released from the sea bed are also factors. Salinity measures the concentration of material in solution – it need not be salt. Salinity is expressed as the number of parts per thousand, by weight, of the constituents dissolved in the water. In the deep layer of the ocean basins, salinity is a fairly constant 34.5 – 35 parts per thousand.*

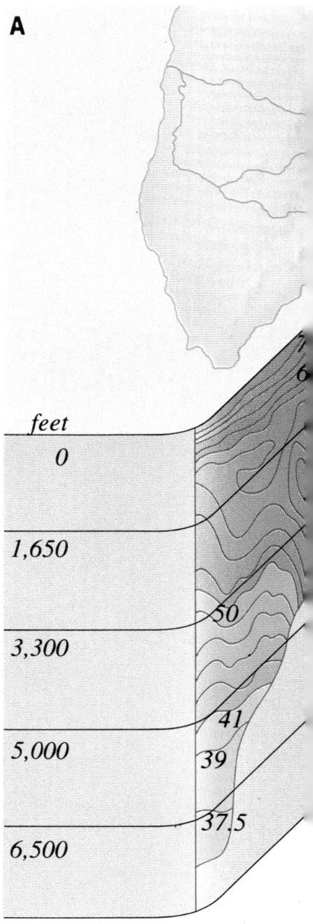

Shades of the Sea

The Earth is often called the blue planet because the oceans give it a blue color as seen from space. Clean, clear ocean water appears blue because the shorter wavelengths of blue light are absorbed less and travel farther through the salt water than the long red wavelengths. Thus the blue color is due to a greater proportion of blue light being scattered and returned back to the surface without being absorbed. But even in the clearest seas only 1 percent of light – of all wavelengths – penetrates a few hundred yards.

Phytoplankton in their teeming millions make ocean water appear green and decrease the depth to which light penetrates. The yellow-brown color of many shore waters can be caused by various pollutants dumped by man, or by sand and mud that have been stirred up by the action of the waves and tides.

Connections: Atmosphere 64 Coastlines 56 Cycles of Life 206 Ocean Circulations 46 Ocean Floor 50 Ocean Life 332

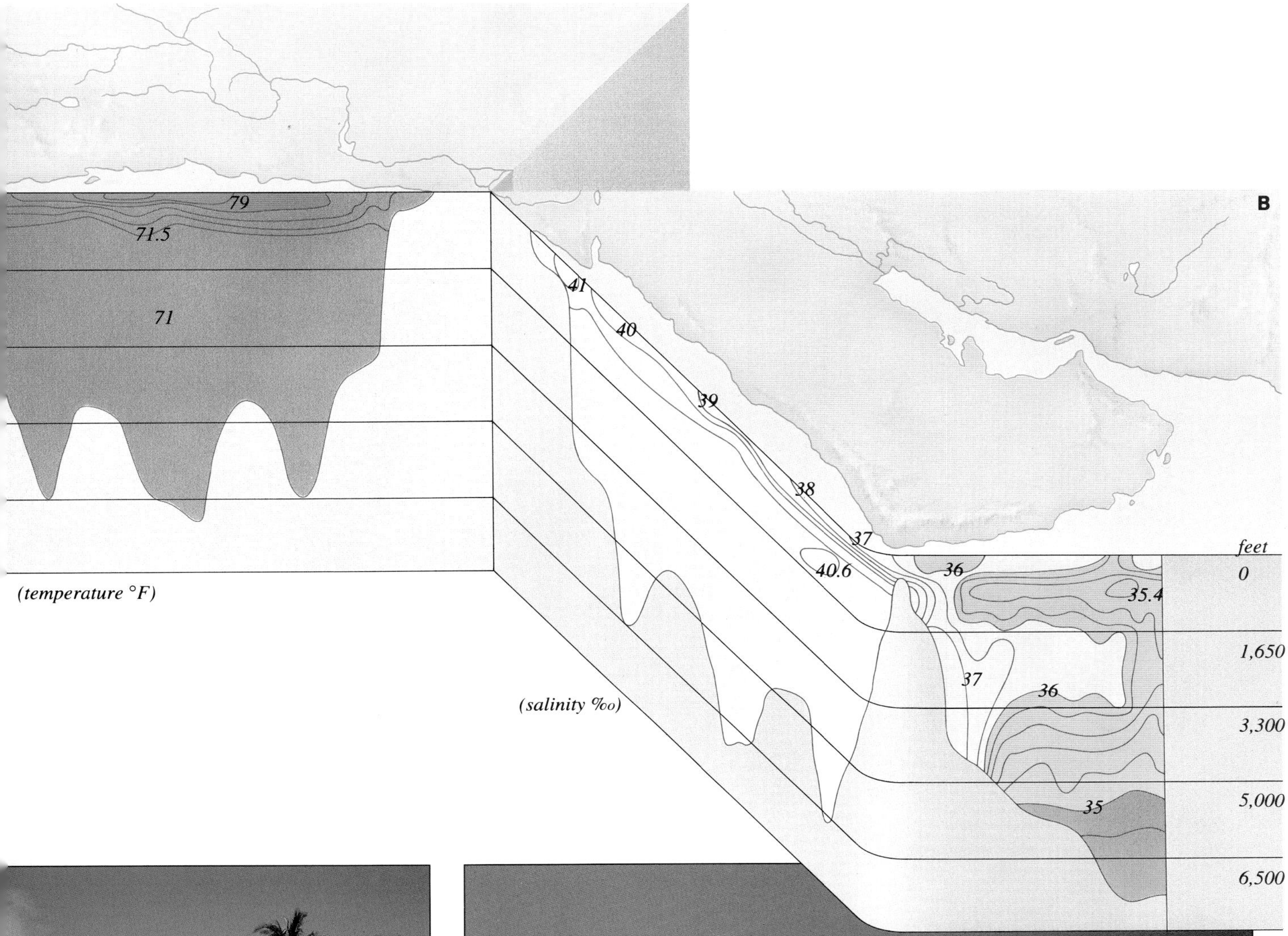
B
79
71.5
71
(temperature °F)
41
40
39
38
37
40.6
36
35.4
37
36
35
(salinity ‰)
feet
0
1,650
3,300
5,000
6,500

Torrents in the Sea

What causes ocean currents?

There are many awesome currents on the Earth's surface. Venezuela's Angel Falls has a 0.6 mile (1 kilometer) vertical drop and the Amazon River pours forth 4 million gallons (200,000 cubic meters) of water every second. But even this is dwarfed by the deep-ocean cataracts that sweep across the sea floor. Perhaps the biggest of these submarine waterfalls lies below the Denmark Strait that separates Greenland and Iceland. Here, 1 billion gallons (5 million cubic meters) of water fall 2 miles (3.5 kilometers) into the North Atlantic Ocean every second.

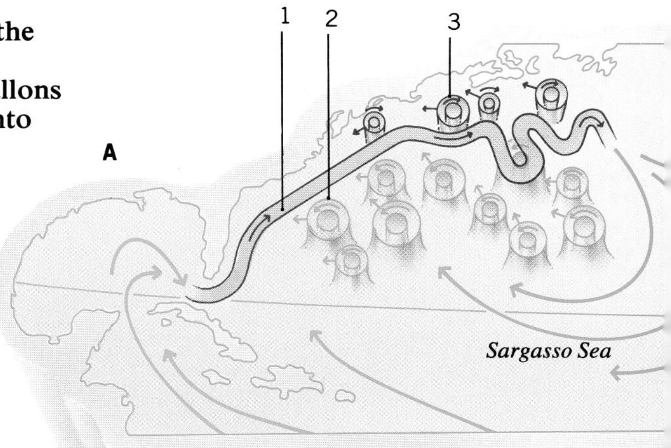

Deep-ocean currents are fed by water from the polar regions, which sinks and flows beneath the warmer, lighter water at lower latitudes. Because these currents are driven by the density differences caused by the temperature and salinity of the waters they are called thermohaline currents. The cold bottom water that eventually cascades over the Denmark Strait starts its journey in the North Atlantic. As it moves northward near the surface it becomes colder.

Eventually it enters the Norwegian Sea, where the formation of sea ice leaves a greater concentration of salts in any water that has not frozen. The increased salinity means that the water can continue to cool to below 32°F (0°C) without freezing. This makes the water dense and so it sinks and begins its return trip across the ocean floor. As the dense water flows toward the equator it hugs the sea floor and plummets over cliffs at the bottom of the ocean, dropping down just like a waterfall and displacing the resident, warmer water upward.

Planetary influences

Oceanic circulation, like atmospheric circulation, is controlled by the rotation of the Earth and the effects of the Sun's radiation. Density differences drive the deep ocean currents, but surface currents – like the Gulf Stream, which carries warm water from the southern tip of Florida northward along the east coast of North America as far as the Grand Banks of Newfoundland – get driven by the major wind circulation systems.

The trade winds of the Northern and Southern Hemisphere both drive westward-flowing equatorial currents. These currents would circle the Earth, but the continents block the flow of water and force the currents into circulation cells called *gyres* which are centered near the subtropical atmospheric high-pressure cells. Water tends to move in toward the centers of the major ocean gyres. It piles up and produces a pressure gradient to oppose the Coriolis force that results from the Earth's spin. The forces balance out – giving socalled *geostrophic* currents – and water flow continues around the gyre. Warm currents flow poleward on the western sides of the oceans in relatively narrow bands of fast-moving water, while the cold currents in the sea are broader and more slow moving.

The rings of water *[**A**] that form around the Gulf Stream [1] can have warm-water cores or cold-water cores. The cold-core rings [2] of the Sargasso Sea may be 190 miles (300 km) across and stretch to the sea floor 16,000 ft (5,000 m) down. The warm-core rings [3] of the slope waters are shallower. The Gulf Stream is the fastest surface current in the North Atlantic gyre, traveling around 125 miles (200 km) in a single day. It carries warm water from the northeast coast of North America, but beneath it flows a large cold current heading south from the Arctic.*

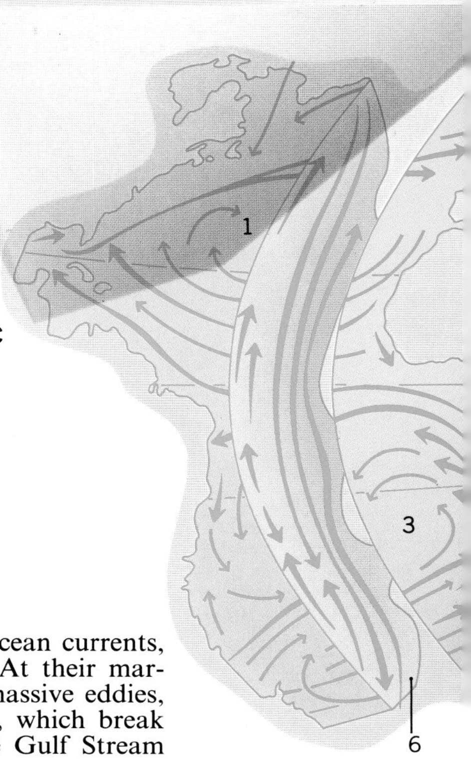

Water rings

The surface currents, like all ocean currents, are not smooth steady flows. At their margins the currents give rise to massive eddies, hundreds of miles in diameter, which break away from the current. In the Gulf Stream the eddies form when the current starts to wind or meander as it flows north. If the meander becomes a loop it pinches itself off from the main flow and forms a ring. These big rings are important because they transport water from different regions across frontal boundaries, like the Gulf Stream. The oceans are continuously in motion: the wind and submarine disturbances whip up waves and eddies and swirls in addition to the powerful currents at the top and bottom of the sea. All this movement – which still awaits thorough investigation – transfers vast amounts of heat, and helps to moderate and control the Earth's climate.

Surface current *directions in January [**C**, and the blue arrows in the detailed map, **D**] are very different to the current directions in July [**D**, red arrows]. They vary with the prevailing winds, but in general currents flowing equatorward are usually slow and broad, while poleward flows are rapid and, like deep currents, heaviest near the western margins of ocean basins.*

Connections: Climatic Change 82 Ocean Floor 50 Ocean Life 332 Polar Regions 42 Seas and Oceans 44 Tides 48 Weather and Climate 70

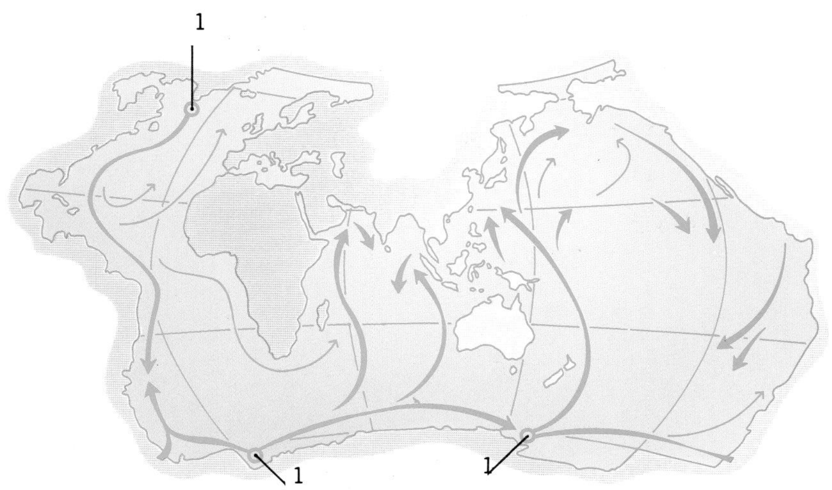

Cold deep currents [**B**] *which originate near the poles are deflected westward as they move toward the Equator by the Coriolis force arising from the Earth's eastward spin. Thus bottom currents are pressed against the western edge of the ocean basins, where they are as a consequence strongest, and may travel at 0.5 mph (0.8 km/h), over 80 times faster than elsewhere. The principal sources of bottom water* [1] *are in the Weddell Sea and the Ross Sea in the Southern Hemisphere, and in the Arctic Ocean and in the seas around Greenland in the Northern Hemisphere.*

In the Northern Hemisphere *there are two large gyres* [**C**], *in the Atlantic Ocean* [1] *and in the Pacific Ocean* [2], *which rotate clockwise. In the Southern Hemisphere there are three major gyres, in the Atlantic Ocean* [3], *Pacific Ocean* [4], *and Indian Ocean* [5], *which all rotate counterclockwise. Because surface currents are largely wind driven they often have completely different directions to deep currents. A cross section of the ocean* [6] *would show currents in many different directions at many different depths. When surface waters are converged together by currents* [7] *they accumulate and are forced downward. Where surface currents diverge* [8] *there is an upwelling. These vertical current movements can extend 3,300 ft (1,000 m) below the surface, and are very important because the cold water that rises up from the depths is rich in important nutrients.*

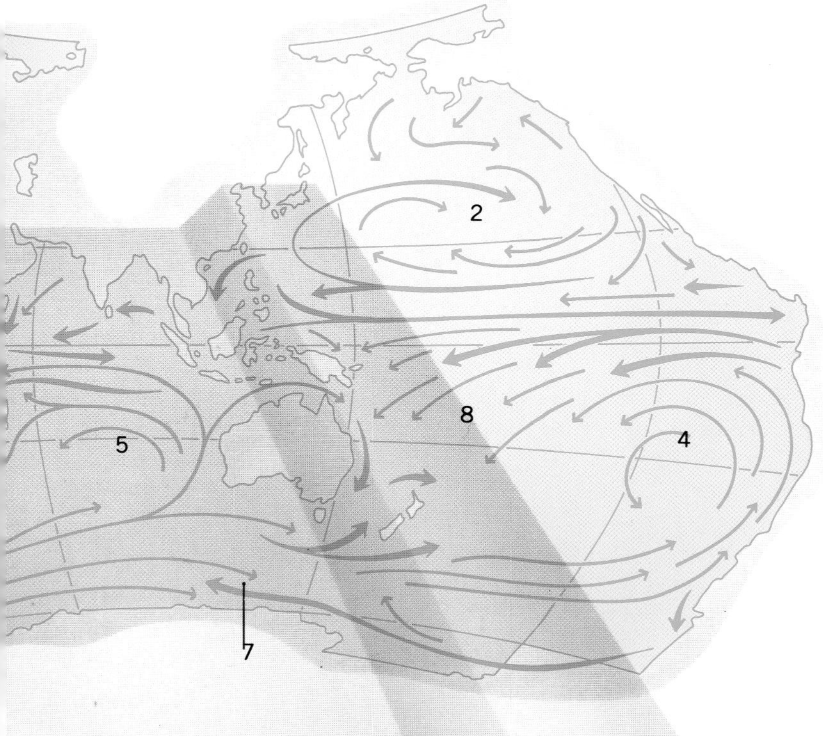

The Corkscrew Current

A wind will tend to drag ocean surface water after it. However, as soon as the water starts to move the Coriolis force throws it off at an angle. Each successive thin sheet of water beneath the surface layer begins to move because it is linked by friction to the layer above. But each layer is thrown even farther off course by the Coriolis force. So the surface current moves at an angle of 45° to the wind at about 2–3 percent of the wind speed. Lower layers move more slowly, at an increasingly eccentric angle.

The resulting vertical profile of the motion is the Ekman spiral. At a depth called the Ekman depth the water flows in the opposite direction to that of the surface current and at 0.043 times the speed of the surface current.

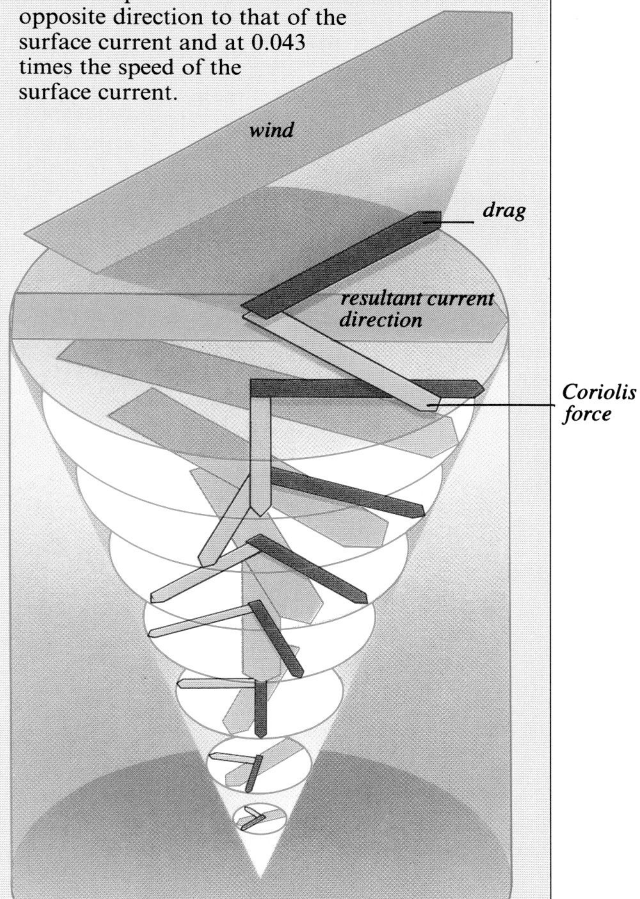

Celestial Forces

What causes tides and waves

Tides are not unique to the sea. The force of gravity exerted by the Sun and the Moon can raise the Earth's crust by over 18 inches (half a meter) and cause widespread cracking in the planet's surface. These colossal forces have variable effects. On some coasts spring tides can cause the sea to rise and fall by a vertical distance of over 50 feet (16 meters), while some shores experience no tides at all. Both tides and waves contain vast amounts of potential energy. If harnessed, the available power could more than satisfy the total demand for electricity in many countries.

Tides rise and fall – flow and ebb – as a result of the gravitational pull of the Moon and the Sun. The extent of the tidal fluctuations varies on both a monthly cycle, corresponding to the orbits of the Moon around the Earth, and on an annual cycle, because of the way in which the Earth orbits the Sun. The height of the whole body of water in the oceans fluctuates, but it is generally only at the margins – the places where the sea meets the land – that such variation is apparent in the form of the tides.

Global variations

Tides are affected by the shapes of ocean floors and landmasses. In Vietnam and the Caribbean there is one tidal high and one tidal low every lunar day (slightly longer than a normal day): each is known as a *diurnal* tide. In places such as the North Sea, however, there are two high tides and two low tides every lunar day: this is called a *semi-diurnal* tide, and the two high- and two low-water marks are normally much the same per day. The angle between the Moon and the equator varies between 28°N and 28°S, tilting the tidal bulge at an angle to the Earth's rotation and causing different heights in the tides. Diurnal tides generally occur where the smaller of the two high tides is insignificant.

Some areas of the world experience the characteristics of both diurnal and semi-diurnal tides: the Pacific and Indian ocean coasts mostly have these mixed tides. At the other extreme there is little tidal fluctuation in the nearly enclosed Mediterranean and Baltic seas. A feature of estuaries is the tidal bore, funneled by the estuaries into a high body of water with a wall-like front.

For effective tidal power generation a constant tidal range in excess of 10 feet (3 meters) is necessary. To this limitation must be added the fact that suitable sites for harnessing tidal energy tend to be environmentally sensitive areas. Barrages inevitably affect tidal movement and pollution may become an increasing menace.

Waves and the weather

Understanding of the mechanism by which the tides are generated is not yet complete, but enough is known to allow relatively accurate tidal predictions. One complicating factor, though, is that atmospheric and climatic

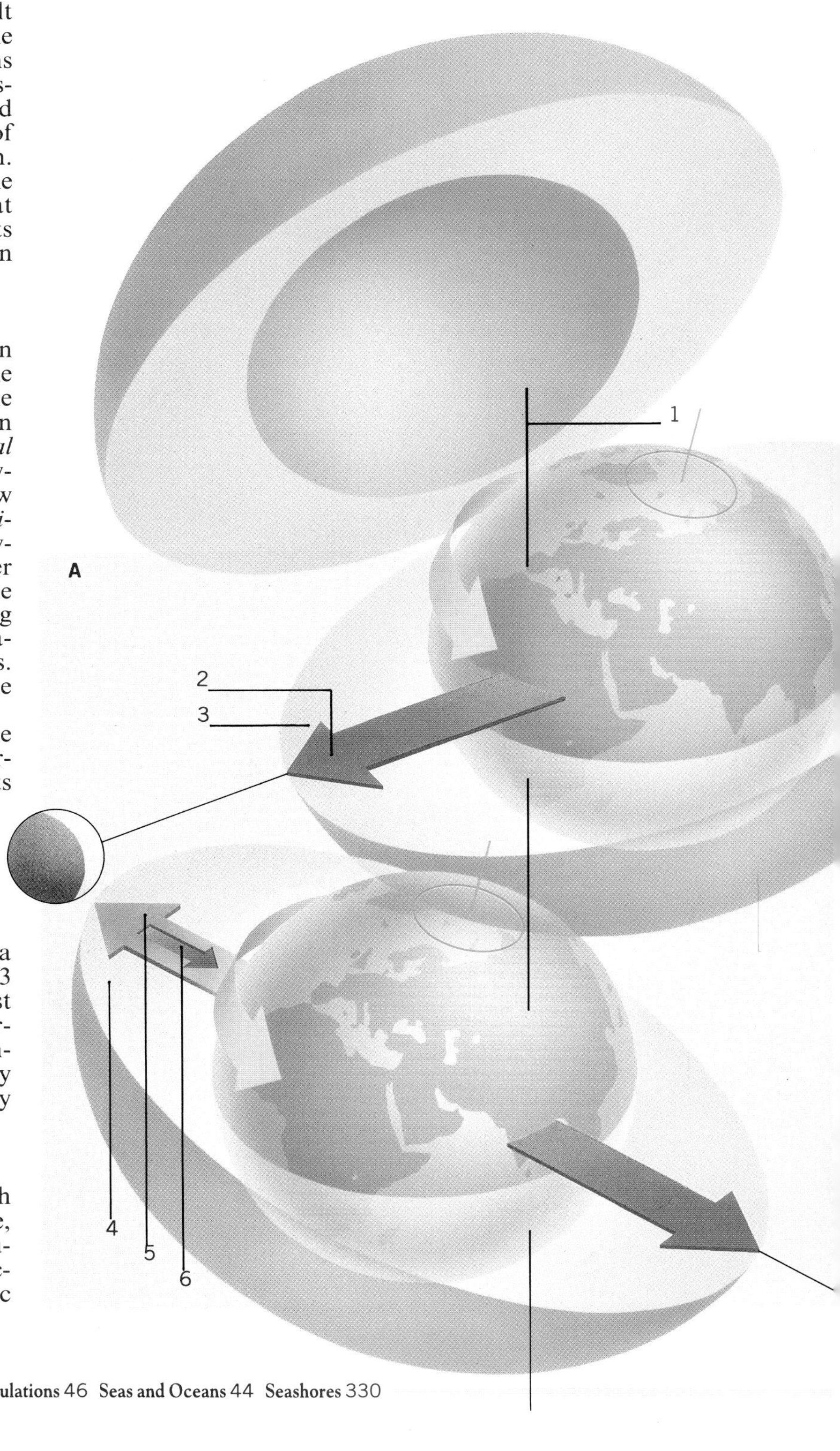

***As the Earth and Moon* [A]** *rotate about their common centre of gravity* [1] *they are in effect continually falling toward – and past – each other through space. The gravitational pull of the Moon* [2] *makes one tidal bulge* [3]. *But there is a puzzling second bulge* [4], *making it look like there is also an "antigravitational" force* [5]. *The other bulge exists because the Moon's pull on the Earth is stronger than its pull on the farthest body of water* [6]. *So the Earth accelerates faster toward the Moon, leaving the water behind as a second bulge that appears to be the result of a pulling force.*

Connections: **Coastlines** 56 **Global Warming** 84 **Ocean Circulations** 46 **Seas and Oceans** 44 **Seashores** 330

In fact no such force exists. The tidal water-bulges rotate at roughly the same speed as the Moon, once in 29 days. But the Earth is also independently rotating on its own axis, within this parcel of water, once every 24 hours. So once a day a single piece of coast will pass through two high and two low points on the parcel: the high and low tides. However, in the day it takes the Earth to revolve once, the Moon has moved on and dragged the water parcel after it through an angle of some 12°. The Earth therefore has to travel a little farther to catch up. This is why the tides occur some 50 minutes later every day.

conditions can have strong positive or negative effects. Fierce winds or extremes of atmospheric pressure may alter a tidal height by up to 10 feet (3 meters). The wind may actually blow water toward or away from the shore, whereas high atmospheric pressure weighs down the water and prevents it from rising, and low pressure allows water to rise up more than it otherwise would.

A difference of 1 millibar (up or down) in atmospheric pressure can make a difference of about 0.4 inches (1 centimeter) in the height of the sea. The maximum tidal surge caused by atmospheric fluctuation is of the order of 20 inches (50 centimeters).

The swelling seas

Waves are mainly caused by the wind blowing over a *fetch*, or open stretch of water. They may persist – in the form of a swell – long after the forces that created them have subsided. All waves, including those that reach the shore, are the product of interactions of the swell from distant weather systems, oceanic currents, and prevailing winds.

The swell that reaches the Californian coast could have come from the stormy region south of New Zealand, and waves that began off Cape Horn may crash against the coasts of western Europe 6,000 miles (10,000 km) away. Thus ocean waves are major transporters of energy around the world.

Watts from Water

La Rance power plant in northwest France delivers over 1,000 tons of water per second to a set of four turbines which generate 65 megawatts of power. Tidal power plants need a large difference between high and low tides in order to work, and at La Rance this difference is 27 ft (8.4 m). The plant uses excess power to pump water up into its dam. The water stored by this process is then used to run the turbines in order to maintain the plant's power production at low tide.

4

5

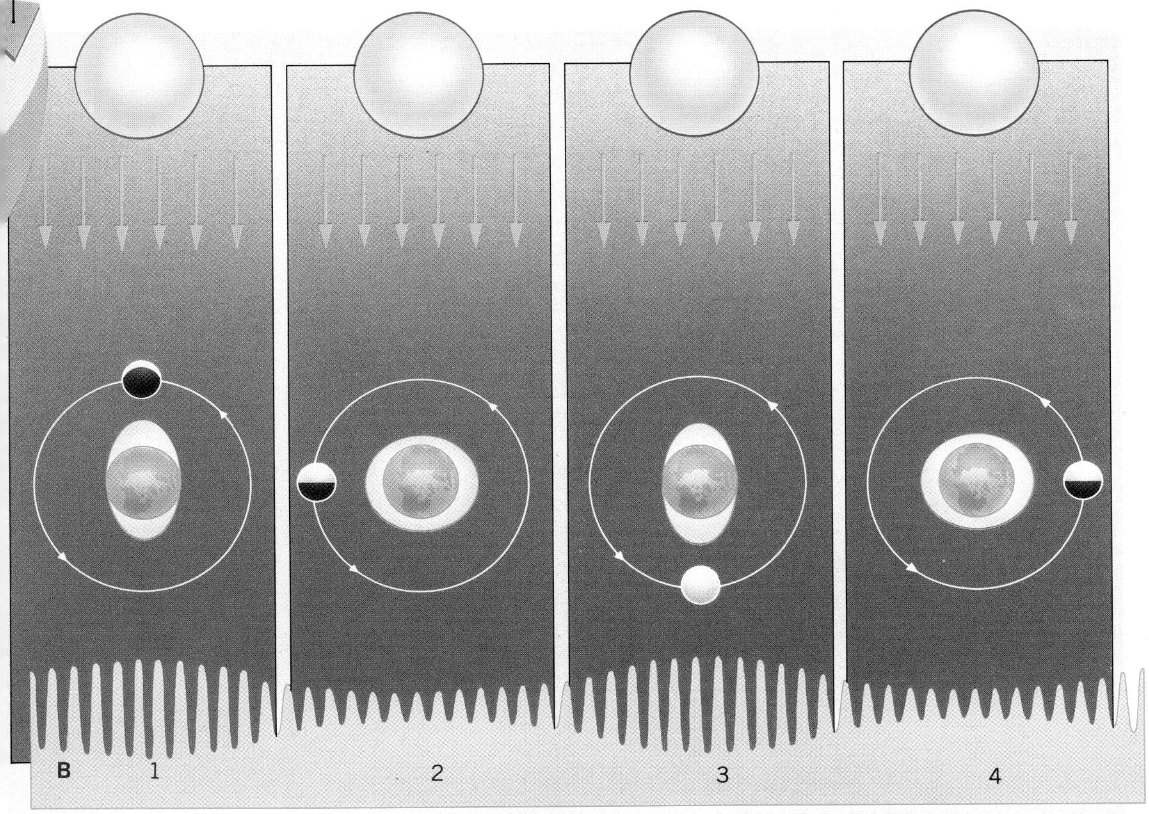

***The Sun's** gravitational force on the Earth* [**B**] *is less than half that of the Moon, but it still has a major influence on the tides. The spring tides (spring meaning "to rise up," and not the season) occur when the Sun, Moon, and Earth are in a straight line* [1 *and* 3], *so the forces exerted by the Sun and Moon add up. The forces are less effective when acting at 90° to each other* [2 *and* 4]. *Then we get the lower, neap tides. The tides do not coincide exactly with the Moon's phases, however. There is a time lag of as much as a day and a half – called the age of the tide – between the Moon's position and the tidal range. The angles between the Sun, the Moon, and the Earth also vary. At the equinoxes (21 September and 21 March) the Sun is over the equator. Then the Sun-Moon-Earth line is very straight and the spring tides are highest. At the solstices (21 June and 21 December) the Sun is at its greatest angle and the spring tides are smallest.*

The Depths of the Sea

What lies on the ocean floor

The greatest mountain chains on Earth are beneath the sea, some stretching almost all the way from pole to pole. So are the mightiest volcanoes and deepest valleys. From the undersea trenches to the continental shelves the ocean floor has features as striking as any landscape on Earth. The basaltic ocean floor is the youngest part of the Earth's crust, at most only 200 million years old; the oldest rocks on land are over 4 billion years old. The sea bed also holds some of the richest mineral and metal deposits – although reclaiming them is a hazardous operation.

__The lithosphere__ "floats" in the Earth's mantle [__A__], and this defines where the surface of the crust – oceanic or continental – will be. So thicker, less dense continents "float" higher than the thinner, denser oceanic lithosphere. The continental shelves are the transition.

Extensive volcanic mountain ranges cross the vast abyssal plains on the ocean floor between the continental shelves. From many, lava still issues, creating fantastic shapes as it cools. A few mountains broach the sea's surface, as in Iceland and the Azores; beneath the water the flanks of the mountain chains may plunge more than 3 miles (5 kilometers) and the ridges may be nearly 600 miles (1,000 kilometers) wide.

In some oceans, especially the Pacific, clusters of volcanoes have been formed where the moving floor of the ocean has crossed a hot plume of lava rising from the Earth's mantle. In Hawaii, for example, the greatest volcano on Earth, Mauna Loa, rises some 30,000 feet (9,100 meters) from its undersea base to 14,000 feet (4,250 meters) above the sea.

Much of the sediment that filters down to the ocean plains is wind-blown dust (often volcanic), rocks and sand carried out to sea on icebergs, space debris, and microscopic skeletal remains of plankton. This all generally settles into a soft, fine-grained clay. Depending on the water depth, pressure, and temperature, such clays vary from being calcium rich above about 15,000 feet (4,500 meters) to silica rich at greater depths. The cutoff line is the *calcite compensation depth* (CCD), below which calcium carbonate dissolves in the carbon dioxide-rich sea water. The CCD varies from 14,000 feet (4,300 meters) to 17,000 feet (5,200 meters) from ocean to ocean. Iron-rich red clays, from volcanic dust borne on the wind, cover enormous areas and occur in the deepest waters.

Below the abyss

Even deeper than the abyssal plains are the ocean trenches. These can be more than 6 miles (10 kilometers) deep. One of the best known is the Peru-Chile trench, which, in common with other trenches, has been formed by the relentless subduction of the ocean floor below the adjacent continent.

Such trenches may be over 600 miles (1,000 kilometers) wide and thousands of miles long.

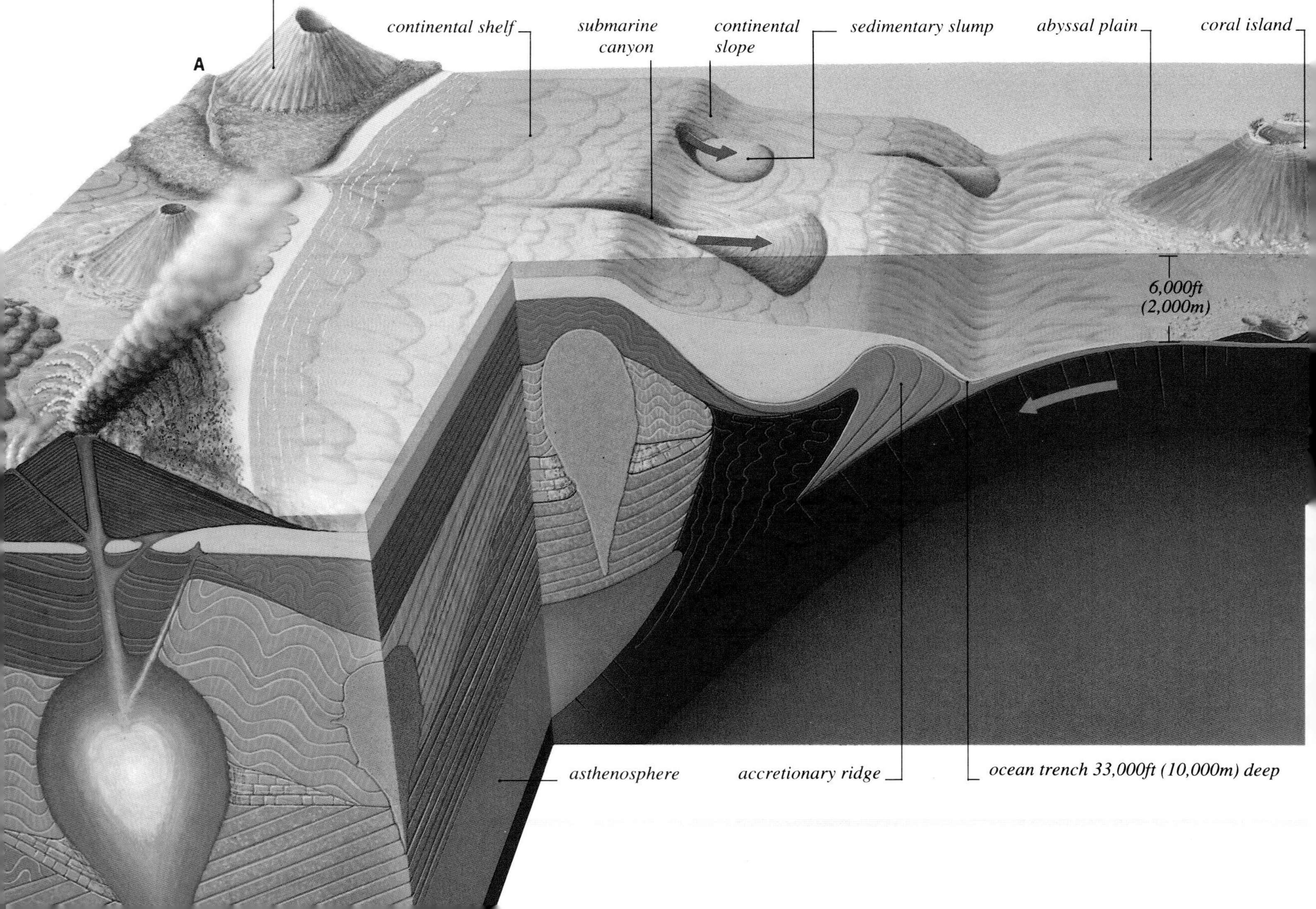

Great thicknesses of sediment accumulate in the trenches, derived from the attrition of rocks on nearby landmasses and from the silt of river estuaries.

On and off the shelf

Between the ocean plains and the coastline are two further important regions: the continental slope and shelf. Climbing up from the deep is the continental slope. This rises from the plains at a depth of about 2 miles (3.5 kilometers) to a depth of only about 500 feet (150 meters) below sea level.

Thorough investigation of the shelves has shown that the solid rocks of each shelf are similar to those of the continent and very different from the basalts of the ocean floors.

Two types of continental shelf have been recognized. The "Atlantic" or passive type tends to be wide – extending over 60 miles (100 kilometers) from the coast – stable and covered with sand and mud; the "Pacific" or active type is much narrower and prone to earthquake activity. Both descriptions derive from the names of the oceans in which they tend to be most common. The Atlantic type occurs around oceans where seafloor spreading is taking place; the Pacific type has subduction zones nearby.

The Grand Canyon [B] *of the Colorado River and the Monterey submarine canyon* [C], *which lies off the coast of California, have remarkably similar profiles, indicating that they were probably formed by similar erosional processes.*

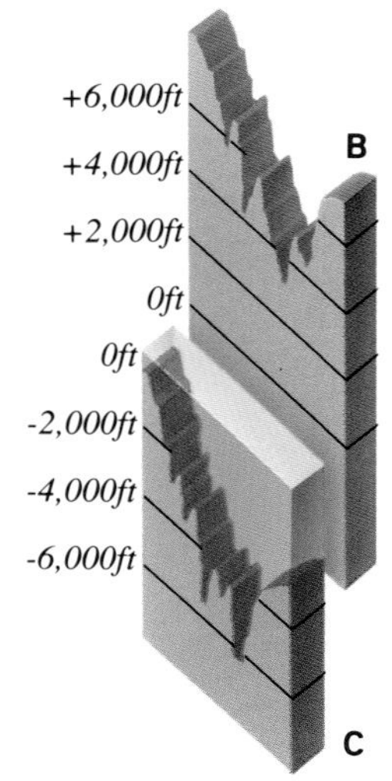

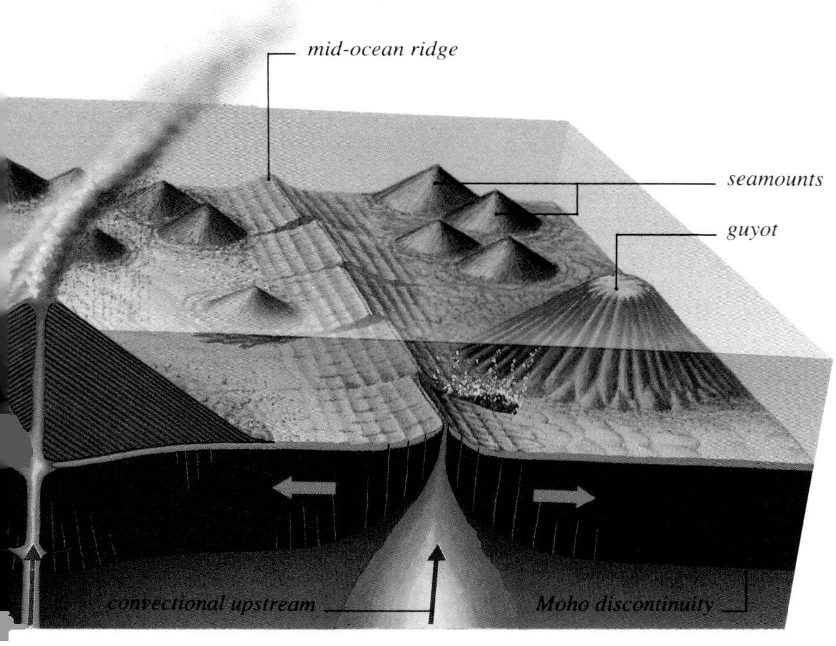

***Chains of volcanoes** or volcanic islands are often found at active margins, as shown in the – vertically exaggerated – diagram. The sediment scraped back when the oceanic plate subducts beneath the continental plate forms the – often highly structured –* accretionary wedge. *Submarine canyons are more common on passive margins, but can be found here, too. They were probably caused by surface erosion during periods of low sea level, and subsequently widened by turbidity currents, or "submarine avalanches" of sediment. Many parts of the ocean floor are highly fractured, especially around the mid-ocean ridge. There are many volcanoes. Those that never emerge are* seamounts. Guyots *are volcanoes with their tops cut off flat, probably by wave erosion when the sea level was lower. Their great weight has sunk them into the ocean crust.*

Mineral Riches

Hydrothermal vents in the seabed create local areas of hot water. The mineral-rich (iron, zinc and copper sulfides, among others) columns of water that shoot from the vents – at temperatures up to 660°F (350°C) – are called "black smokers." Potato-sized manganese nodules (below), especially common in the Pacific, average 18 percent manganese, alongside iron, zinc, copper, cobalt, and nickel. They have an onion-skin structure, precipitated from the sea water at only 0.04–0.16 in (1–4 mm) thickness/million years around a nucleus such as a sand grain, or fish tooth.

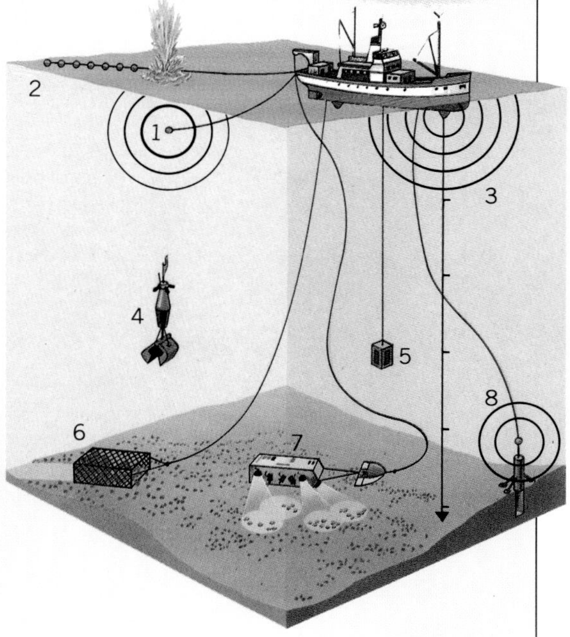

***The commercial mining** of manganese nodules lies in the future, but work is centered on two systems – hydraulic suction dredges and mechanical dredges. There are, by contrast, many ways of exploring for abyssal ore nodules, including seismic surveying, with an airgun* [1] *and a string of hydrophones* [2], *echosounders* [3], *grab samplers with cameras attached to them* [4], *oceanographic probes like the bathysonde* [5], *dredges* [6], *deep-sea cameras* [7], *and wire-bound samplers like the box corer* [8].

Connections: **Animal Adaptation to Pressure** 146 **Coastlines** 56 **Island Formation** 52 **Ocean Circulations** 46 **Ocean Life** 332 **Plate Tectonics** 16 **Volcanoes** 22

Surrounded by the Sea

How islands form

Not a single one of the 1,300 coral atolls that make up the Maldive Islands is higher than 30 feet (9 meters) above sea level. Many are much lower. If the sea were to rise just a few yards they would all vanish. Rising seas or sinking continents can create islands – such as Britain, for example, where the dry land connection with continental Europe was severed by rising water. The long thin islands off the coast of Yugoslavia are the tops of a flooded mountain range, and the Thousand Islands in the St. Lawrence River of Canada are the remains of a drowned river valley.

The Hawaiian Island–Emperor Sea Mount chain [**B**] *extends for over 3,700 miles (6,000 km), with more than 107 volcanoes. The volcanoes become older from the southeast to the northwest and the islands to the northwest are lower and more eroded. Rocks on the island of Hawaii* [1] *are less than 1 million years old, rocks from Oahu* [2] *are 2 to 3 million years old, and the lava flows of Kauai* [3] *are about 5 million years old. The islands in an island arc, such as the Aleutians, are by contrast all of the same age, having been created by the same process – subduction – at the same time.*

The islands age *as they do because they are formed on a sort of plate tectonic conveyor belt* [**C**]. *A plume of hot material from deep in the Earth's mantle* [1] *rises beneath the present location of the island of Hawaii.*

Hawaii consists of two large volcanoes, Mauna Kea and Mauna Loa, one of the most active on Earth. The plume, or hot spot, is the source of the magma that feeds the Hawaiian volcanoes. It is stationary. The Pacific plate [2], *on the other hand, is moving to the northwest, driven by ocean floor spreading* [3], *so that after the islands form they are carried away by the plate motion. As the islands*

Chains of islands such as those in Japan and Hawaii, and isolated landmasses such as Australia – the "island continent" – are all the result of fundamental plate tectonic processes. The production of molten magma at hot spots and subduction zones – where one plate is squeezed beneath another – builds up volcanic islands and chains, which of course can also be destroyed at the subduction zones. The movement of the Earth's plates can also slice fragments from a large continent and carry them away, thus creating an island. This is exactly what is happening in western North America, where the San Andreas Fault is removing a piece of the continent.

Unsinkable islands

If an island is too buoyant to be subducted, it will eventually be accreted – or slapped on – to a continent. The western Pacific Ocean is a site where accretion appears to be taking place today. As the Pacific plate moves west and the Australian-Indian plate moves north, islands like Fiji, Java, and the Philippines will probably eventually be joined to continents, so forming what is called a terrain. A similar process is responsible for the western margin of North America. It is composed of hundreds of different exotic terrains, some of which have traveled distances of over 3,000 miles (5,000 kilometers) before finally being accreted.

On a larger scale, plate movement resulted in the final splitting of Australia from Antarctica during the Eocene epoch, between 55 and 38 million years ago.

A necklace of volcanoes

Volcanic island chains that are formed when one tectonic plate is subducted beneath another are called island arcs, because they form graceful curves a few thousand miles in length. Island arcs occur about 60–120 miles (100–200 kilometers) landward of the trench where the subducting plates meet. Along its length the arc usually consists of separate groups of about five or ten volcanoes, each of which forms a more or less straight line around a few hundred miles in length. Bends occur between the straight segments. Island arc volcanoes usually occur about 30–40 miles (50–70 kilometers) apart, which is also about the thickness of the lithospheric plate on which they form. The volcanoes erupt above a site where the subducting plate is about 60–120 miles below the Earth's surface. Here the water and carbon dioxide in the sediments dragged down with the subducted plate act as a flux that lowers the melting temperature of the rocks, so creating fluid magma. The water and carbon dioxide also cause the volcanoes above the site to erupt explosively. The volcanoes of such an arc – like the Aleutians off southern Alaska – are all of the same age.

Conditions for life

Rain that falls on an island may form a lens-shaped reservoir of fresh groundwater beneath the surface. The freshwater lens floats in the denser sea water. To maintain a stable freshwater zone beneath it, an island must be about 1,300 feet (400 meters) across and receive sufficient rainfall.

Connections: Coastlines 56 Coral Reefs 336 Ocean Floor 50 Plate Tectonics 16 Volcanoes 22

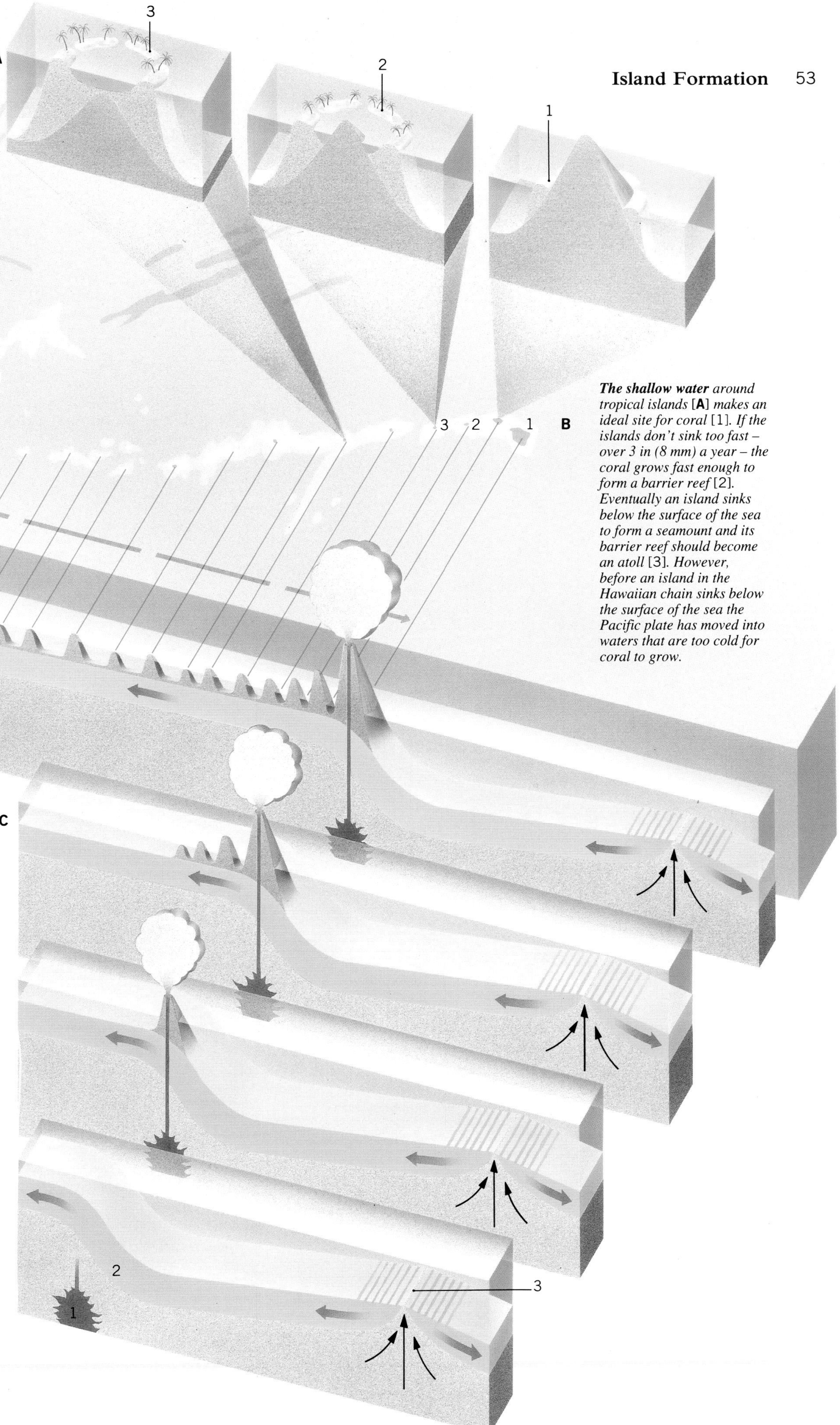

The shallow water *around tropical islands* [**A**] *makes an ideal site for coral* [1]. *If the islands don't sink too fast – over 3 in (8 mm) a year – the coral grows fast enough to form a barrier reef* [2]. *Eventually an island sinks below the surface of the sea to form a seamount and its barrier reef should become an atoll* [3]. *However, before an island in the Hawaiian chain sinks below the surface of the sea the Pacific plate has moved into waters that are too cold for coral to grow.*

move away from the hot spot and the broad region of upward-domed sea floor created by the rising mantle material, they sink. The Pacific plate conveyor belt also controls the lifecycles of coral reefs that form around the islands. The islands are doomed to destruction, however, because the Pacific plate carries them ever nearer to the Aleutian trench [4], *where they are subducted and may contribute to the Aleutian island arc. Twenty-five to thirty million years after the initial outpouring of lava the island has become a seamount. It will take twice this long for it to reach the Aleutian trench.*

The Driest Lands

What makes a desert

The largest and driest desert in the world is the Antarctic. Most of Greenland, parts of Canada and Siberia, and many high-altitude mountain ranges are also technically desert. Deserts may be defined by low precipitation, high evaporation, or by the landforms or vegetation they contain. But most people think of deserts as rocky or sandy wastes, stretching under a burning Sun. There are deserts in all the continents except Europe, and an estimated 20 percent of the Earth's surface is covered by arid land. Human activities are helping to expand these barren environments every day.

Among warm deserts there are huge variations to be found in landscape, vegetation, and other natural factors. Deserts can be completely bare rock (called *hamada*) with, perhaps, a sprinkling of pebbles and stones (*desert pavement*); or they can be covered with sand dunes. Their surfaces can consist of ash, lava, or salt; and they either have very little vegetation of any kind, or plants that are classified as "desert" types. These either have very deep taproots, or are xerophytes (plants such as cacti, specially adapted to life in a dry environment). Lichens, algae, and fungi may also grow under stones.

All of the warm deserts are dry, lying within the arid and hyperarid regions of the world. Average annual precipitation is under 10 inches (25 centimeters) and evaporation rates are high. Dry lands are not necessarily desert lands, however: they may be productive, like the vineyards of southern Spain.

Arid and hyperarid lands cover between 15 and 30 percent of the world's land area (depending on how they are defined). Hyperarid regions may not receive any rainfall at all in a year and are thus indisputably deserts. Semiarid and arid regions may be classified as "desert" according to the amount and type of their vegetation.

The shifting sands

Sandy deserts, in which the surface is composed of wind-blown dunes, are often devoid of vegetation because the sand particles are constantly being moved on by the wind and plants have nothing on which to anchor. Although the Sonoran Desert in Arizona and northern Mexico is teeming with plant life and supports a magnificent desert ecosystem, deserts normally have a lower *biomass* (the weight of living plants and animals) than any other landscape. Typically, a sand-dune desert has a biomass under 220 pounds/acre (250 kilograms/hectare), compared to a savanna's biomass of 9,000 pounds/acre (10,000 kilograms/hectare), or a tropical forest's biomass of over 220,000 pounds/acre (250,000 kilograms/hectare).

The pressure cooker

Climatic factors are the natural creators of arid and hyperarid lands. Prevailing dry conditions are often the result of atmospheric stability associated with the presence of great

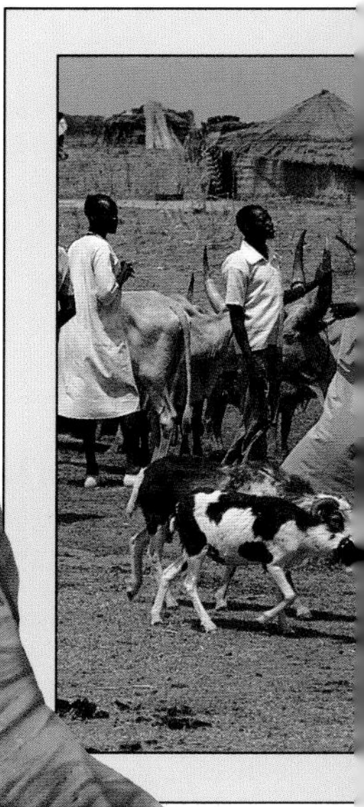

A

***Where rain** is infrequent but heavy* [**A**] *it cuts deep, steep or vertically sided valleys called* wadis [1], *which isolate* buttes [2] *and the larger* mesas [3]. *Wadis develop when flash floods widen and deepen random depressions in mountain areas or arid plateaus. Rocky surfaces are much more common than sand seas in the desert. About 30 percent of Arabian deserts are sand covered, whereas only 1 percent of desert surface in North America is sandy. Any rain that falls on the mountains may drain into porous rock layers* [4] *that can stretch several miles beneath the desert. Wherever these porous layers come to the surface they give rise to an* oasis [5]. *Finding and utilizing water is one of the most urgent problems in the desert, and one ancient solution that is still used today is the* qanat [6]. *First a head well* [7] *is sunk down to the porous rock, which may be 300 ft (100 m) deep. Then a line of ventilation shafts* [8] *is dug and finally a channel* [9] *is begun from the qanat mouth* [10]. *When the qanat is completed, gravity brings water to the settlement and canals* [11] *take it where it is needed.*

Connections: Animals in Heat and Drought 142 Clouds 74 Desert Life 304 North Polar Life 310 Plants in Heat and Drought 138 Polar Regions 42

Desertification

Human activities over recent centuries have caused and are causing extensive tracts of arid lands to be degraded. Changes normally occur slowly, although sand-dunes can rapidly encroach on farmland if they are not anchored by fast-growing, drought-resistant trees, or stabilized by waist-high windbreaks across their crests. Shelter belts of trees and shrubs can also cut down wind-erosion, which, along with water-erosion, is exacerbated by overgrazing, deforestation and bush-burning. If the topsoil is blown away, degradation of the land is permanent. Even irrigation can cause problems. Water evaporates quickly in arid lands, leaving behind any salt that it may carry. This salt is not leached away into the deep soil, so its concentration rises rapidly in the upper layers, until at last the land may have to be abandoned.

high-pressure cells around 30° latitude. Only occasionally do rain-bearing depressions spread into these zones. Deserts also form in the lee of mountains, or where land is subject to very dry continental winds, or even at the coast, when sea breezes do not release their moisture until they get farther inland. When rain does come, it is frequently heavy.

The huge sand seas that cover about 25 percent of the world's desert areas were largely formed by the water erosion of sandstone, shells, or whatever else the sand particles are composed of. Sand need not originate where it is found. Frequently it is created on the borders of desert areas and is carried by wind or water into low-lying regions. In many cases it may have been created during earlier, wetter climate cycles. The pebbles, cobbles, and bedrock of desert pavement are often smooth and shiny from the polishing effect of particles in the wind.

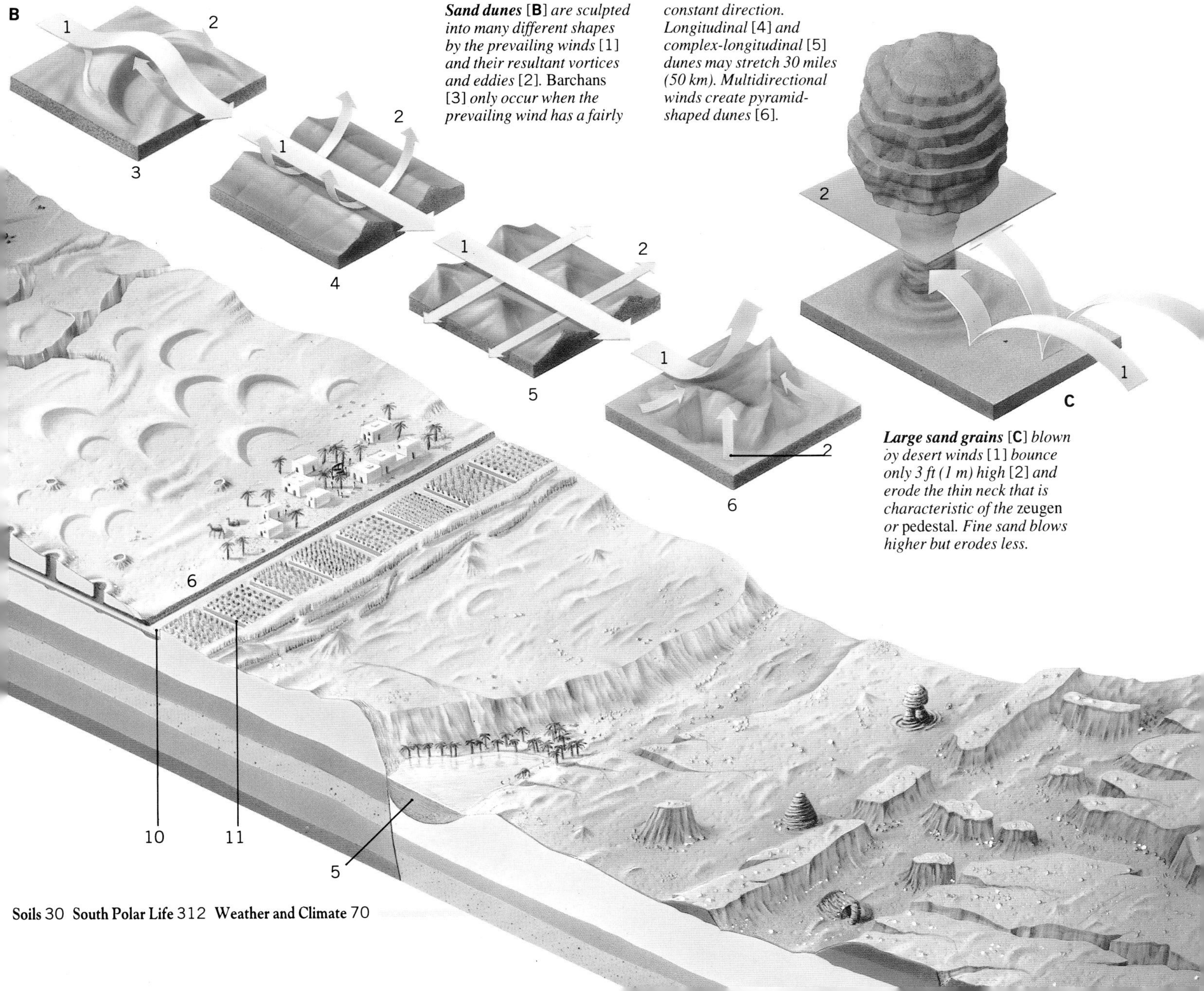

Sand dunes **[B]** *are sculpted into many different shapes by the prevailing winds* [1] *and their resultant vortices and eddies* [2]. Barchans [3] *only occur when the prevailing wind has a fairly constant direction. Longitudinal* [4] *and complex-longitudinal* [5] *dunes may stretch 30 miles (50 km). Multidirectional winds create pyramid-shaped dunes* [6].

Large sand grains **[C]** *blown by desert winds* [1] *bounce only 3 ft (1 m) high* [2] *and erode the thin neck that is characteristic of the* zeugen *or* pedestal. *Fine sand blows higher but erodes less.*

Soils 30 South Polar Life 312 Weather and Climate 70

At the Edge of the Land

How coastlines are formed

Over 100 million years ago the Mississippi Delta started at Cairo, Illinois, but since its birth the mouth has migrated south to the Gulf of Mexico – over 1,000 miles (1,600 kilometers) away – at a rate of more than a third of an inch a year. Much of the beach that runs up the eastern coast of North America, from the Gulf of Mexico to New Jersey, is every bit as mobile. It consists of 295 barrier islands – sedimentary deposits flanking the coast – that have migrated, split up and re-formed under the action of storms for the last 6,000 years.

On some barrier islands as much as 25 feet (8 meters) of land a year can be eroded. But as long as there is a surplus of sediment, the islands will remain. And there is always a surplus of sediment dumped into the Atlantic by rivers like the St. Lawrence and the Hudson. Beaches, however, are not just made of sand. Whatever sedimentary material is available from rivers, eroded cliffs, glacial deposits, or coral and shell fragments can wind up concentrated on a beach. In Hawaii, where the only material is broken lavas, many of the beaches are black, and in western Scotland, where the old eroded roots of the Highlands expose garnet-bearing metamorphic rocks, some of the beaches are pink.

Winds and waves

Cliffs are attacked by waves and suspended sediment in the area between high and low tides. A notch is usually cut in the base of the cliff and as it deepens the cliff face above it collapses. Eventually a wave-cut platform is created between the high and low tide marks which is visible at low tide. Cliff debris may be carried along the coast or deposited offshore, but the removal of debris is rarely continuous and destructive wave action usually alternates with periods when the waves deposit material on the shore.

Although storms do the major work of modifying beaches, each wave that runs up the beach also plays a role. Breaking waves throw sand into suspension and fling it farther up the beach before the wave retreats. In addition, because waves approach the coast at an angle the wash and backwash is asymmetrical and although most of the movement is up and down, some of it is along the beach – a process called *longshore drift*. The angle of approach of the waves determines the longshore drift of sand and this angle is in turn determined by the location of the sea storm from which the waves emanate.

The longshore current produced as a result of the angular approach of the waves and the return of the backwash water can transport large quantities of sand. This is also the current that moves swimmers rapidly down shore. But if waves approach with their crests parallel to the beach the longshore currents are weak. Instead, under the pressure of the incoming waves the backwash piles up and is periodically released as *rip currents*, which flow for tens or hundreds of yards out to sea as dangerous and often lethal undertows.

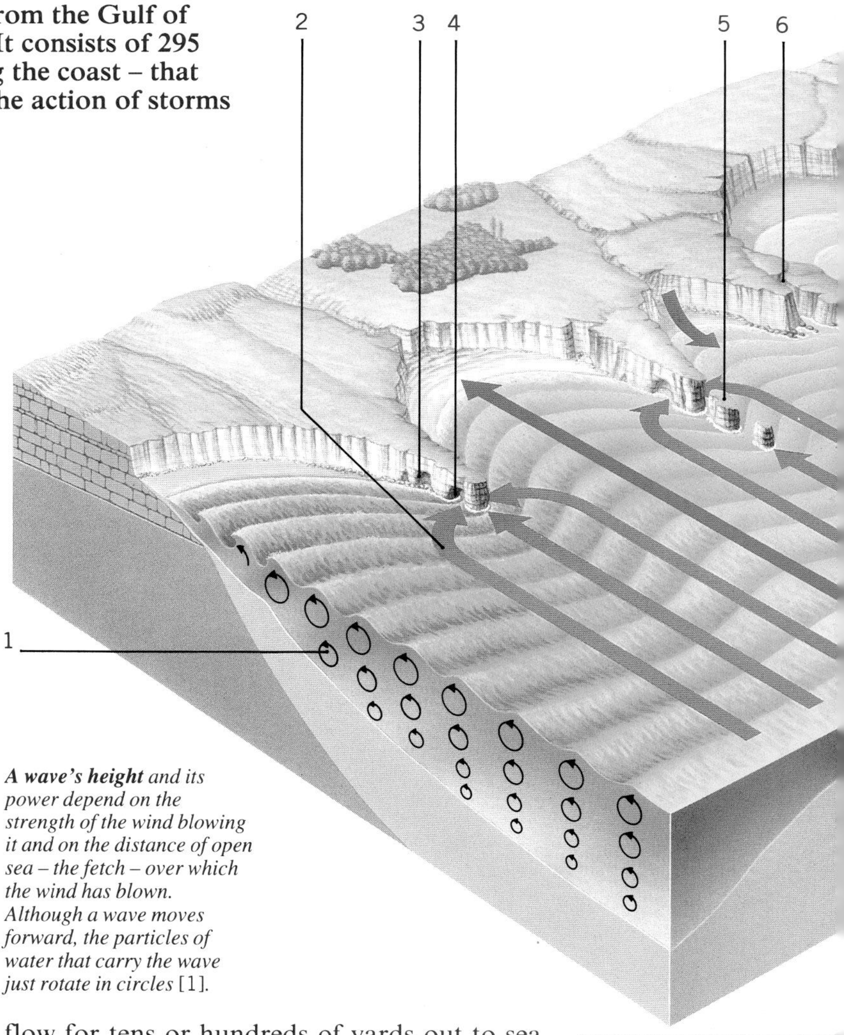

__A wave's height__ and its power depend on the strength of the wind blowing it and on the distance of open sea – the fetch – over which the wind has blown. Although a wave moves forward, the particles of water that carry the wave just rotate in circles [1].

Delta deposits

Rivers are the main transporters of sand and before the sand finds its way to a beach it often forms a delta. When a river meets the ocean the characteristics of the delta that forms are controlled by the speed of the river, the size of its channel, depth of the water and of course the nature of the sediment. As the moving river water enters the sea the main process that occurs is the mixing of still water with fast water at the edges of the current. When fresh water enters the sea it often floats out over the top of the denser salt water. However, if the amount of sediment in the river is high enough, the river water and its suspended sediment is sufficiently dense that no salt wedge forms and the muddy water flows directly into the sea.

Connections: Glaciers 40 Island Formation 52 Seashores 330 Sedimentary Rocks 26 Tides 48

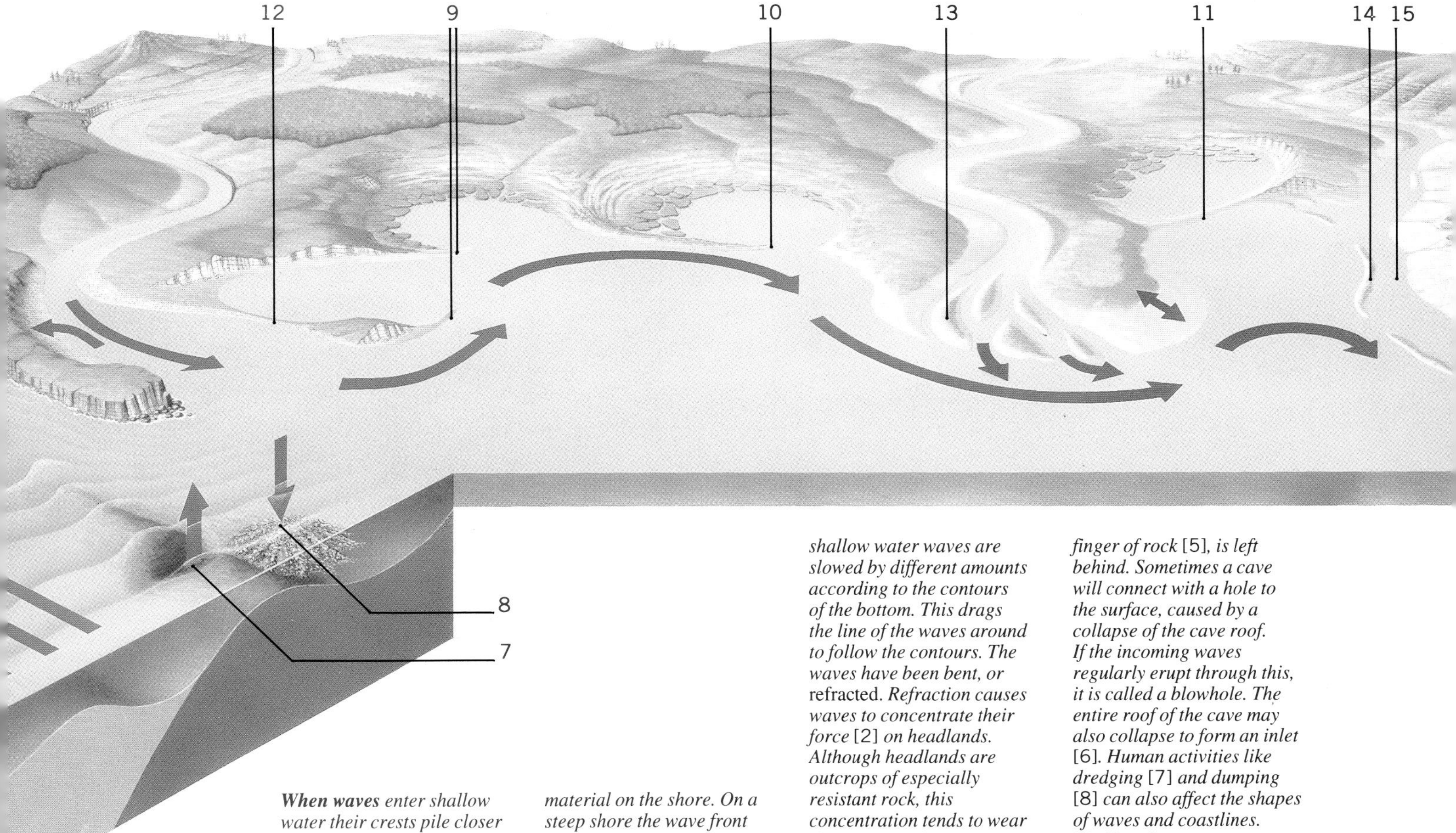

When waves enter shallow water their crests pile closer together and topple over, causing the water to race forward. Coming up a gently sloping shore, the front of a wave gets steeper and its crest starts to spill over the top. This is a spilling breaker and it creates an uprush, which tends to deposit material on the shore. On a steep shore the wave front also steepens violently. An unstable hollow appears in front of the wave and the crest curls over and plunges down into it. A plunging breaker will create a backwash that scours sand and other material away from the shore. As they enter shallow water waves are slowed by different amounts according to the contours of the bottom. This drags the line of the waves around to follow the contours. The waves have been bent, or refracted. *Refraction causes waves to concentrate their force* [2] *on headlands. Although headlands are outcrops of especially resistant rock, this concentration tends to wear them away and, over time, "flatten out" the coastline. Cracks in a rock are easily eroded, forming caves* [3]. *Caves on opposite sides of a headland may eventually meet to form an arch* [4], *and when the roof of the arch collapses a stack, or isolated finger of rock* [5], *is left behind. Sometimes a cave will connect with a hole to the surface, caused by a collapse of the cave roof. If the incoming waves regularly erupt through this, it is called a blowhole. The entire roof of the cave may also collapse to form an inlet* [6]. *Human activities like dredging* [7] *and dumping* [8] *can also affect the shapes of waves and coastlines. Mudflats form at estuary mouths. Longshore drift deposits sediment as spits or ridges* [9], *which may be curved by wave action. When the drift is strong and any flow into the bay is weak, the spit may extend right across the bay* [10] *or even cut off the bay, as a bay-mouth bar* [11]. *A* tombolo *is a ridge of sediment linking the mainland to a rocky island* [12]. *The shape of a delta* [13] *depends on whether the sea water is quiet or active. It may be modified or curtailed, as shown, by tides or currents. The Mississippi delta is unaffected by strong tides or currents and consequently consists of many lobes. On the other hand, rivers like the Congo and the Amazon have no surface deltas because strong ocean currents and waves remove the sediment before such a delta can form. Sediments may also generate barrier islands* [14], *which will eventually grow to isolate lagoons* [15].

Fjords

Fjords are deeply glaciated valleys partly filled by an arm of the sea. These sinuous and elongated basins reach depths of over 1,000 feet (300 meters) and give evidence of deep glacial erosion. Because the effect of the glacier is greater farther inland, fjords are often more shallow nearer the sea. The Sognefjord in Norway is over 3,000 feet (900 meters) deep, but near its seaward end it shallows to only 450 feet (140 meters). Fjords are so deep because the glaciers that made them were descending to a lower sea level than today. Glacial erosion determines the U shape of the fjord but the former ice load also helped to cause the high, steep-sided cliffs in another way. During glaciation the weight of the ice causes the land to sink. When the glacier is gone, the land rises at a rate too fast to be rounded off by erosion.

The Race to the Sea

How a river shapes the land

Both the River Nile and the Amazon are more than 4,000 miles (6,400 kilometers) long, but whereas the Nile has few tributaries, the Amazon has around 15,000, four of which are themselves more than 1,000 miles (1,600 kilometers) long. In its course. the Amazon descends no fewer than 17,700 feet (5,400 meters), but the Nile actually descends for less than a quarter of that height, despite its comparable length. The great and evident differences between these waterways and others around the world belie their underlying similarities.

The principal feature of all rivers is unidirectional, downhill water movement. The rate of flow of the water determines the physical, and hence the biological, characteristics of each particular region of the river as it wends its way to a greater waterway or to the sea.

Rivers most often begin at a mountain spring or lake, from which one or more headstreams flow. Such streams typically are swift running, have a steep gradient, follow a relatively straight course, and are subject to stretches underground or to waterfalls. Waterfalls result when a river's course passes over strata of markedly differing hardness: where the water passes over hard and then soft material, the latter is eroded more quickly, causing a sharp change in gradient at that point. These fast-flowing headstreams tend to have a stony bed on which few plants can survive; they tend also to have a typical but limited fauna adapted to withstand such arduous conditions.

The merging waters

Rivers gather strength as their headstreams merge, combining their flows to form wider, deeper, swifter courses. The large amounts of sediment being carried by these waters cannot settle onto the bed in the fast flow, and the water that tumbles over the rocky bed is rich in oxygen and maintains a comparatively regular temperature throughout the year. Again, few plants can survive these conditions, and only trout and migrating salmon make their nests at this level.

As the gradient of the river valley declines the river broadens and the speed of flow diminishes. At this level the relatively horizontal course results in a meandering path and the selective deposition of suspended solids. The river bed becomes less stony and more muddy, particularly on the insides of meanders where the river flow is least and accumulation takes place. Because of the slow-moving water, and because the muddy deposition usually has a high nutrient content, such areas are ideal for plants.

Finally, the river continues at a much more stately pace to the coast, where it flows into the sea. Here it may form an estuary, or a delta. Estuaries are protected arms of the sea into which rivers flow. The most common are drowned river valleys. Deltas mostly arise where the offshore waters are shallow and sheltered, so that the transported sediments may be deposited far out to sea. But there are exceptions, such as the Mississippi Delta.

Tides of change

Few, if any, major rivers remain that have not suffered the ravages of humankind. Their waters have been diverted or dammed for power generation, their lower channels have been dredged for navigation, their waters polluted by industrial chemical cocktails. Activities on land have also changed the nature of many rivers: improved drainage and the increasing use of fertilizers have meant river flows are less controlled – floods occur more frequently and the waters are generally more acidic. Although the impact of these influences is finally being recognized, it is important that the river system – from source to end – is seen as a whole that is indivisible, and remedial actions devised accordingly.

__Most river systems__ [__A__] are fed by rain falling on high ground. The amount of water that reaches a river depends on the nature of the underlying rock. Steep mountains of impervious rock deliver about 75 percent of the total rainfall to a surface river system, whereas gently undulating porous rock may deliver as little as 15 percent. The pattern of precipitation may also influence the size and shape of the river. Many rivers are seasonal and remain dry for part of the year. Even on permanent rivers, seasonal variations far upstream can affect the downstream course. This can be seen in the criss-cross pattern of old and new channels in a braided river (below). Braiding occurs when periodic increases in water level, perhaps due to annual snowmelt, cause the river to overflow its existing banks onto the flood plain. While the plain is inundated the current cuts new channels into which the river retreats when the water level falls.

__Rivers flow very slowly__ across flat plains, and small variations in the speed of flow can alter the river's course. Water flowing at the outside of the bend moves a little faster and cuts away the bank as it picks up sediment [1]. On the inside of the bend the water moves more slowly and tends to deposit sediment. Over time these variations form a migrating meander [2], an exaggerated looping curve. Subsequently, the river cuts an alternative course across the neck of the meander [3]. Sediment builds up at the ends of the loop, which eventually becomes separated, forming an oxbow lake.

Connections: Caves 36 Coastlines 56 Fish 124 Lakes 60 River Life 328 Sedimentary Rocks 26 Swimming 210

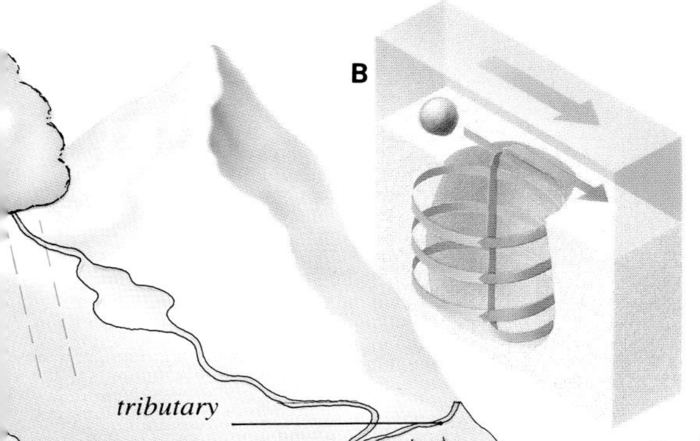

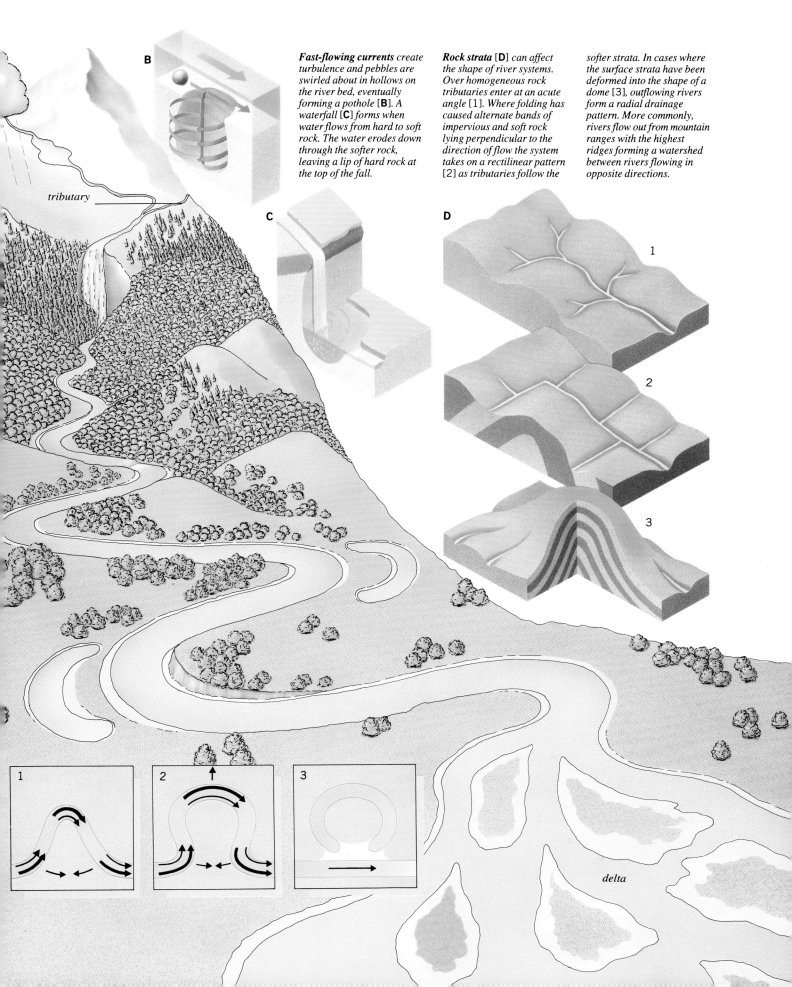

Fast-flowing currents *create turbulence and pebbles are swirled about in hollows on the river bed, eventually forming a pothole* [**B**]. *A waterfall* [**C**] *forms when water flows from hard to soft rock. The water erodes down through the softer rock, leaving a lip of hard rock at the top of the fall.*

Rock strata [**D**] *can affect the shape of river systems. Over homogeneous rock tributaries enter at an acute angle* [1]. *Where folding has caused alternate bands of impervious and soft rock lying perpendicular to the direction of flow the system takes on a rectilinear pattern* [2] *as tributaries follow the softer strata. In cases where the surface strata have been deformed into the shape of a dome* [3], *outflowing rivers form a radial drainage pattern. More commonly, rivers flow out from mountain ranges with the highest ridges forming a watershed between rivers flowing in opposite directions.*

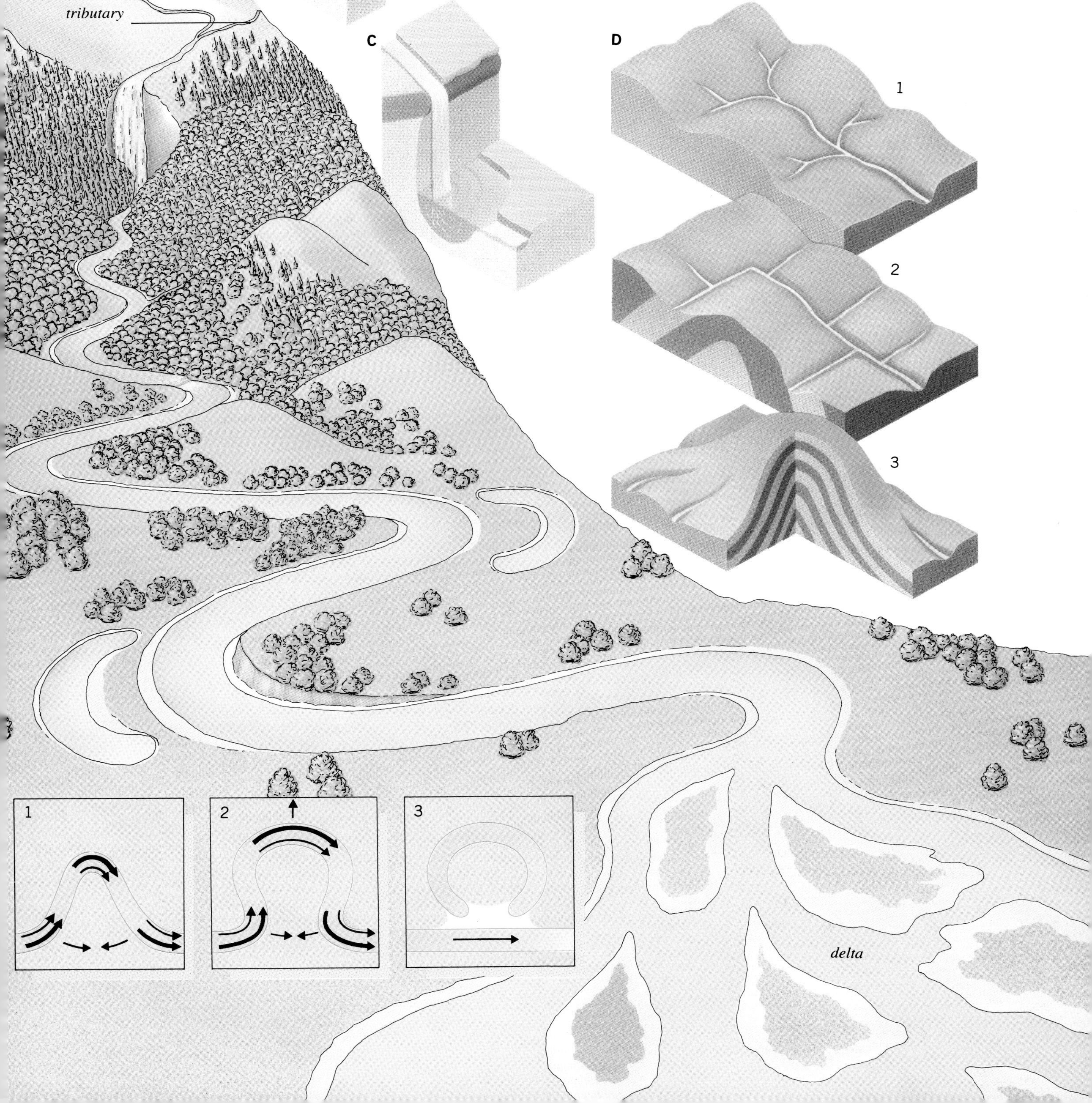

Still Waters

How lakes form and die

Almost a fifth of the world's fresh water is contained in just one lake: Lake Baikal in eastern Siberia, which is in parts just over a mile deep. Not only is it the deepest, it is also the oldest lake on Earth. The vast rift valley that the lake fills began to form 80 million years ago, during the age of the dinosaurs, and the valley has been full of water for about 25 million years. Most lakes are geologically temporary phenomena, lasting a maximum of hundreds of thousands of years and sometimes, in the instance of a small beaver pond, for only a few days.

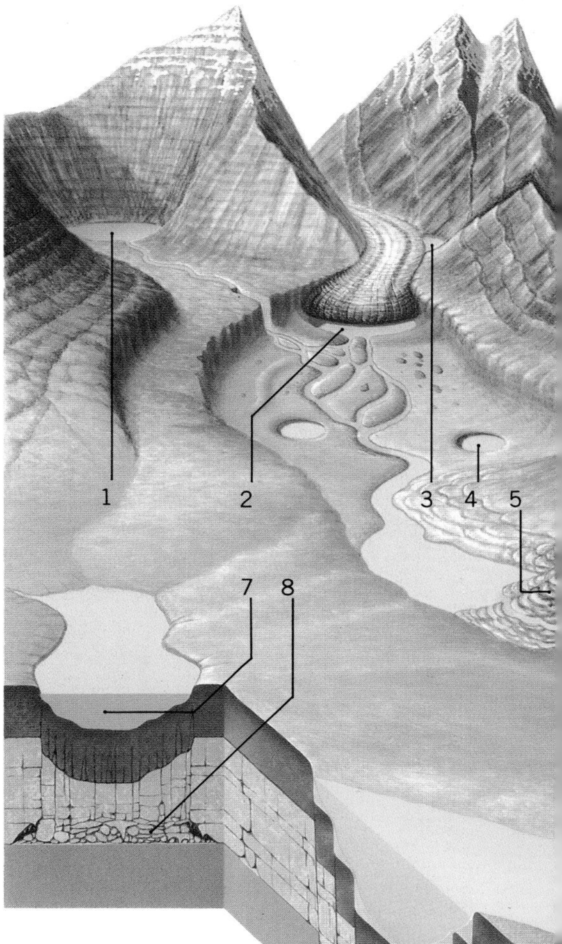

***Apart from lakes** that form in meteor craters, lakes* [**A**] *are created by various geophysical processes that occur above and below the Earth's surface. Glaciation during the last ice age accounts for most lakes.* A cirque *lake* [1] *forms as water collects in the hollow left at a glacier head. At the snout of a glacier a lake may form as meltwater is trapped by the* terminal moraine [2], *a bank of rock and debris left behind as the glacier retreats. Lakes may also form behind* lateral moraines [3] *at the glacier's sides. Blocks of ice buried in glacial debris eventually melt, leaving hollows that become* kettle *lakes* [4].

Lakes may also be created by volcanic activity, as when larva flows into a river and solidifies into a dam [5]. *Subsequent cooling of the volcanic cone may permit a crater lake to form in the caldera. An* oxbow *lake* [6] *forms when accumulated sediment cuts off an exaggerated loop of a river. In regions where the underlying rock is limestone, a surface lake* [7] *may form in the sinkhole created by the collapse of an underground cave system* [8]. *Large-scale earth movements may form a lake by creating surface depressions in which water can accumulate. Faulting of the rock strata* [9] *tends to produce deeper lakes than folded rock strata* [10].

Lakes grow old and die because they eventually fill up with sediments and wastes. The lifetime of a lake is controlled by its origin, size, and shape, together with the climate and, most importantly, the area drained by the lake (its watershed). The natural aging process that affects lakes is called *eutrophication.* Although the process is gradual and continuous, it can be divided into stages according to the amount of nourishment the lake offers any potential life. At first, when the lake is "young," there is *oligotrophy* – nourishment is sparse; next, *mesotrophy*, when there is moderate nourishment; and finally *eutrophy,* or abundant nourishment.

Barren waters

Young lakes in temperate climates have, at all depths, clear oxygenated water that is poor in the nutrients essential for plants to grow, such as nitrogen and phosphorus. This lack of fertility limits the growth of *phytoplankton,* the tiny plant organisms that form the basis of the food chain. As a result, relatively few plants and animals grow and live here. Rapid replacement of the water in a lake slows eutrophication, because the discharged water carries the phytoplankton away with it. In some lakes in northern Canada the inflow and discharge rates are so great that the entire volume of water is replaced in less than a week. At the other extreme, Lake Tahoe in California takes 700 years for complete water exchange.

Bedrock variations

Lake Tahoe's water changes only slowly, but it is still pure and nutrient free because it lies on crystalline rocks. Algae and phytoplankton require 21 different elements in order to flourish. In general, small amounts of these elements are carried into the lake by rain and snow, but the main source is the lake's watershed. Rivers and streams that flow into the lake leach these elements from the rocks and soil. Where lakes occur on hard crystalline rocks, the supply of nutrients is low and the lake stays "younger" for longer. In lakes formed on a sedimentary terrain or on glacial deposits there are more nutrients and aging takes place more quickly.

As a lake becomes old its fertility increases at an accelerating rate. In old lakes plants grow luxuriantly near the shore: mats of

Saltwater Lakes

Most salt lakes, such as the Dead Sea or the Great Salt Lake in Utah, start as fresh water that becomes saltier with time. But in the Caspian Sea, the lake with the largest surface area in the world at 139,000 sq miles (360,700 sq km), the water is getting fresher. About 290 million years ago the Caspian Sea was cut off from the Mediterranean, since when the Volga River has been diluting it. Lakes usually become salty when a geological upheaval cuts off the lake's outflow and slows the streams feeding it. When lake water evaporates, only fresh water is lost – any salts remain. The waters of Lake Natron, in the African Rift Valley, are salty not just through evaporation but also because the lake is fed by volcanic springs which produce hot solutions rich in sodium carbonate, which in places is thick enough to walk on.

Connections: Algae 112 Caves 36 Glaciers 40 Lake Life 326 Rivers 58 Volcanoes 22 Water Pollution 340

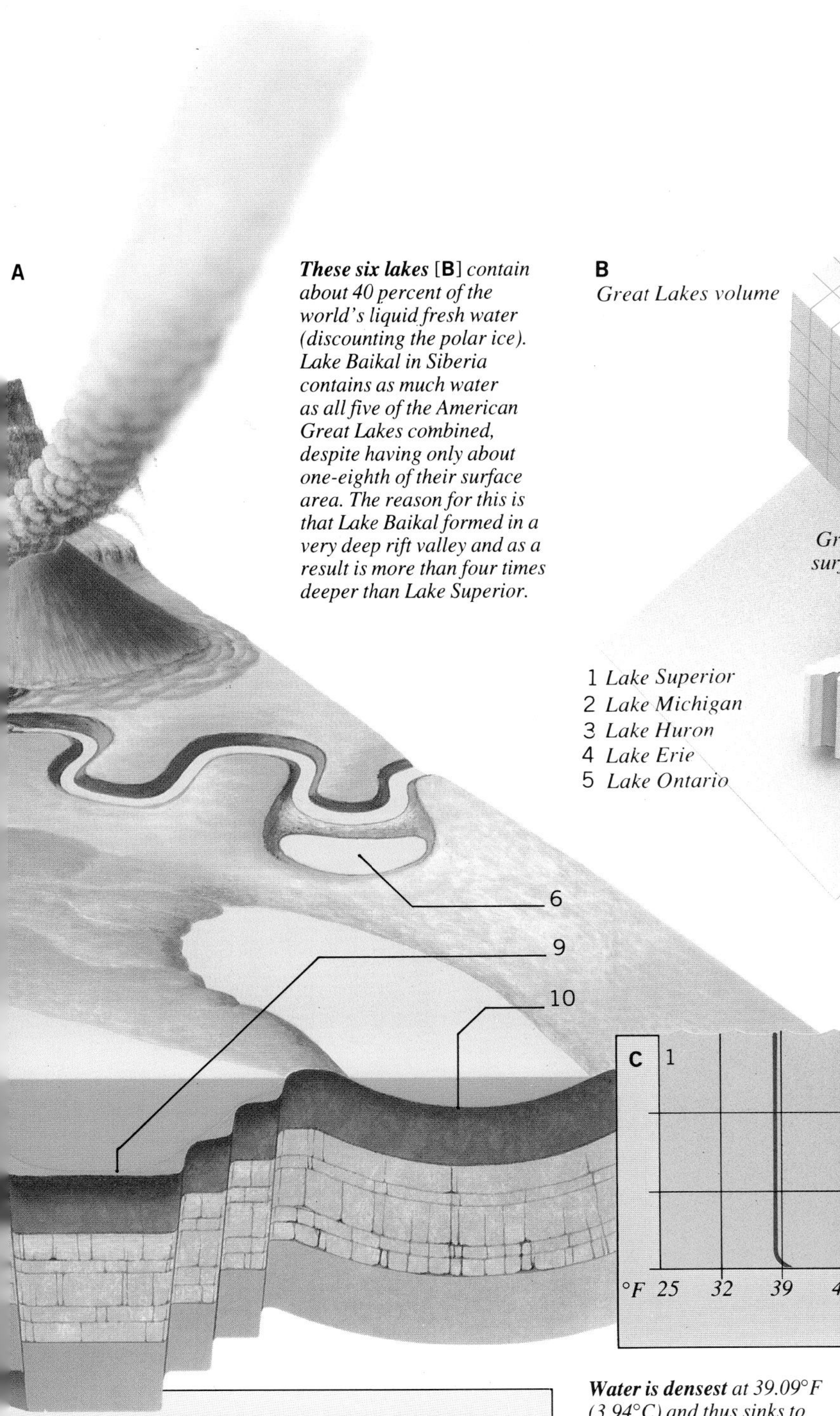

These six lakes [**B**] *contain about 40 percent of the world's liquid fresh water (discounting the polar ice). Lake Baikal in Siberia contains as much water as all five of the American Great Lakes combined, despite having only about one-eighth of their surface area. The reason for this is that Lake Baikal formed in a very deep rift valley and as a result is more than four times deeper than Lake Superior.*

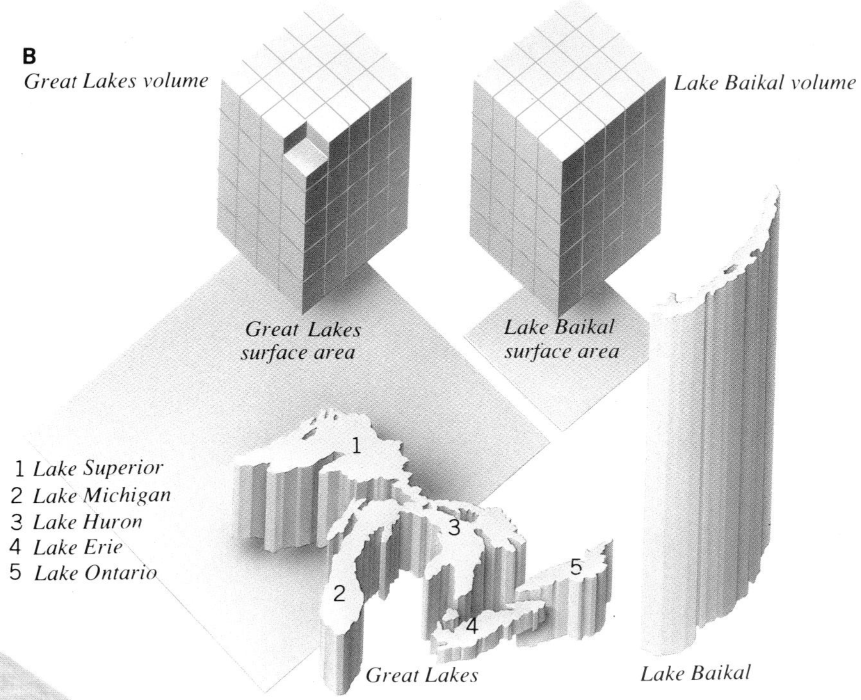

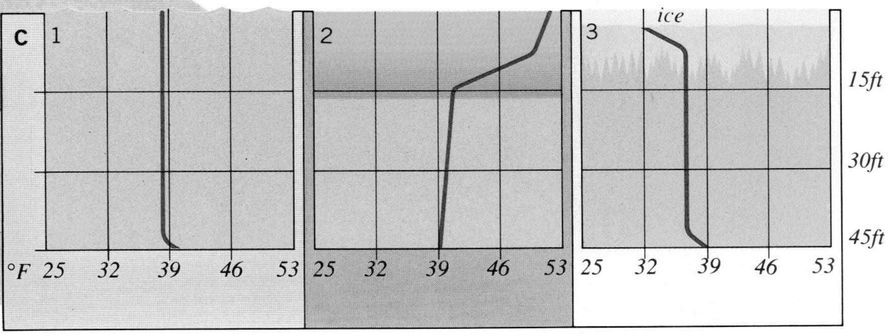

Water is densest *at 39.09°F (3.94°C) and thus sinks to a lake bottom, where it cannot cool further without becoming less dense and trying to rise again. Thus lakes freeze from the surface downward, allowing living organisms to survive in warmer water below the ice. Shallow lakes may freeze all the way down. In temperate regions lakes over 33 ft (10 m) deep show seasonal cycles of thermal layering* [**C**]. *In early spring* [1] *water at all depths (apart from the bottom) is at a similar temperature. In summer* [2] *surface heating creates three distinct thermal zones. In winter* [3] *surface ice leaves water below unfrozen.*

algae cover the surface in green slime. At depth the waters of a eutrophic lake are cloudy with decaying organic matter – and the decomposition of the organic matter uses up the oxygen dissolved in the lake water. At the surface, interaction with the air replaces the oxygen. However, at the lower levels of the lake, especially in deep lakes where different well-defined thermal layers prevent the mixing of the surface and bottom waters during the summer, the water becomes permanently oxygen depleted. In the absence of oxygen, *anaerobic* bacteria attack any organic matter, releasing hydrogen sulfide gas. Eventually the lake turns into a marsh, which in turn will finally disappear.

Lakes remain numerous in regions that were once glaciated, such as northern Canada and Scandinavia, but far more numerous are the ones that formed, existed for a period, and have since filled in and died.

The Atmosphere
Atmospheric Phenomena
The Ozone Hole
Weather and Climate

2 The Power of the Atmosphere

Beaufort Wind Scale

Beaufort Wind Number		mph	km/h
0	Calm	less than 1	less than 1
1	Light air	1-3	1-5
2	Light breeze	4-7	6-11
3	Gentle breeze	8-12	12-19
4	Moderate breeze	13-18	20-28
5	Fresh breeze	19-24	29-38
6	Strong breeze	25-31	39-49
7	Moderate gale	32-38	50-61
8	Fresh gale	39-46	62-74
9	Strong gale	47-54	75-88
10	Whole gale	55-63	89-102
11	Storm	64-73	103-117
12-17	Hurricane	74 and over	118 and over

Winds and Weather
Clouds
Rain, Hail, Snow and Fog
Lightning and Thunder
Cyclonic Storms
Climatic Change
Global Warming
The Seasons

Blanket Around the Earth

How the atmosphere was formed

Life on Earth is supported by an extraordinarily thin shell of atmospheric oxygen – at an altitude of only 13 miles (20 kilometers) even the world's highest-flying bird, the Ruppell's vulture, would not only be unable to breathe in the low-pressure air, but its blood would boil in its veins. However, the Earth is unique in the solar system in having an atmosphere capable of supporting any life at all. In fact, the evolution of life was intimately linked with the evolution of the atmosphere, most crucially 2 billion years ago, when plants began producing free oxygen.

The cloud of dust and gases that formed the Earth 4.6 billion years ago created a dense molten core enveloped in cosmic gases. This proto-atmosphere, composed mainly of hydrogen, carbon dioxide, and carbon monoxide, did not last long before it was stripped away by a tremendous outburst of charged particles from the Sun. As the outer crust of the planet solidified a new atmosphere began to form from the gases outpouring from gigantic volcanoes and hot springs. This created an atmosphere of carbon dioxide, nitrogen oxides, hydrogen, sulfur dioxide, and water vapor. As the Earth cooled, water vapor condensed into highly acidic rainfall, which collected to form lakes and oceans.

Birth of the ozone layer

For the first half of the Earth's existence, only trace amounts of free oxygen were present. But then green plants evolved in the oceans and they began to add oxygen to the atmosphere as a waste gas. The addition of large amounts of oxygen was critical for the further evolution of life because of the role that ozone (tri-molecular oxygen) plays in protecting plants and animals from lethal ultraviolet radiation. Because the early ozone layer was very thin and close to the surface, living organisms had to rely on alternative protection and could only develop under about 33 feet (10 meters) of water.

The evolving atmosphere

As oxygen increased to 1 percent of the atmosphere, the required depth of protective water became only 12 inches (30 centimeters) and complex multicellular marine life-forms could develop. The atmosphere itself continues to evolve, but human activities – with their highly polluting effects – have now overtaken nature in determining the changes.

Oxygen levels *began to increase 2 billion years ago, as shown in the formation of extensive "red bed" sediments – sands colored with oxidized (ferric) iron. Previously, ferrous formations had been laid down showing no oxidation. Already 4.5 billion years ago the carbon dioxide in the atmosphere was beginning to be lost in sediments. The vast amounts of carbon deposited in limestone, coal, and oil indicate that carbon dioxide concentration must once have been many times greater than today, when it stands at only 0.04 percent. The first carbonate deposits appeared about 1.7 billion years ago, the first sulfate deposits about 1 billion years ago. The decreasing carbon dioxide was balanced by an increase in the nitrogen content of the air. The forms of "respiration" practised advanced from fermentation 4 billion years ago to anaerobic photosynthesis 3 billion years ago to aerobic photosynthesis 1.5 billion years ago. The aerobic respiration that is so familiar today only began to appear about 500 million years ago.*

Light and Air

The concentrations of neon, krypton, and xenon in the atmosphere are much lower than in the Sun and in cosmic dust clouds. Because these elements do not combine with anything else and are too large to diffuse away into space, they could not have been lost gradually and must have been removed. We know about the chemistry of the Sun and other stars because of spectroscopy. The unique electron structures of the elements mean that each element absorbs light at a different energy – and thus wavelength. So light shone through various elements and then split into a spectrum will show black lines or gaps where the light has been absorbed. Each line will indicate the presence of a particular element.

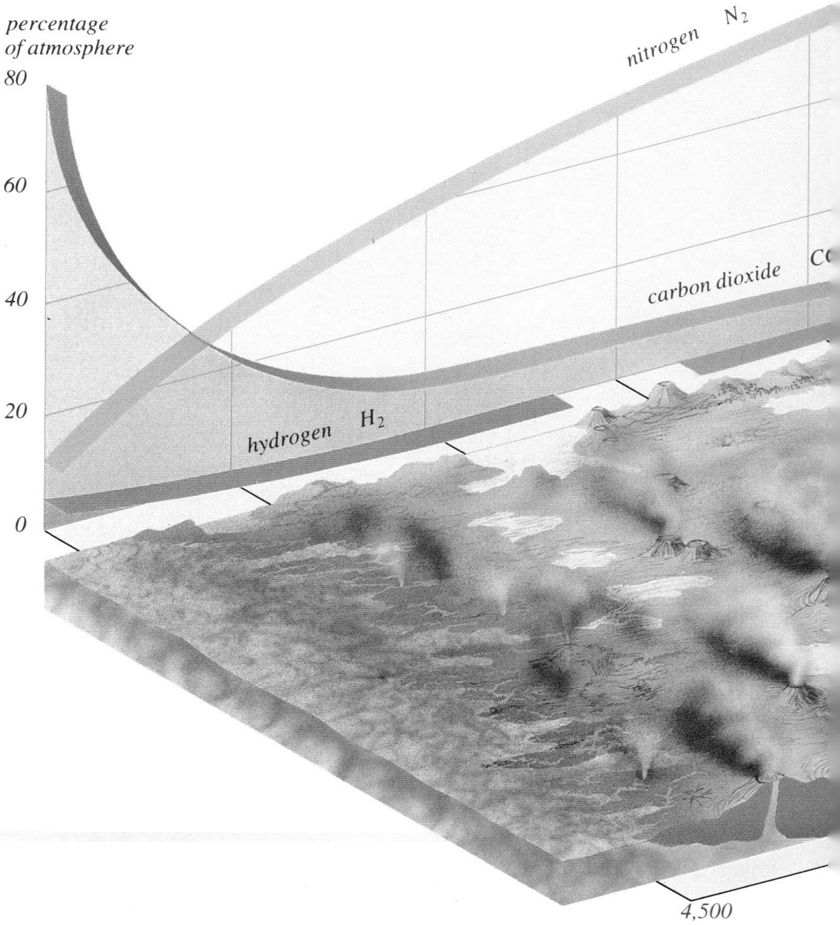

Because the stratospheric ozone layer absorbs ultraviolet radiation from the Sun, it creates a warm lid to the troposphere beneath. Consequently, virtually all moisture is trapped below about 5–10 miles (8–16 kilometers) and this, in turn, sets the altitude limit for clouds and precipitation. However, special clouds do occur. Around sunrise and sunset mother-of-pearl nacreous clouds may sometimes be seen at around 13–19 miles (20–30 kilometers). They are composed of spherical droplets either too small to freeze at the exceptionally low temperatures or prevented from freezing by the presence of sulfuric acid released from volcanic eruptions.

The atmospheric thermometer

From an average surface value of 60°F (15°C) the temperature of the air falls to around –75°F (–60°C) at the top of the troposphere and warms to 50°F (10°C) at the top of the stratosphere. It then plunges to –185°F (–120°C) in the mesosphere, after which it warms through the thermosphere, where a layer of oxygen around 125 miles (200 kilometers) high absorbs some solar ultraviolet radiation and is heated in the process.

Aurorae *[1] can be seen in the thermosphere, which extends from 50 miles (80 km) to 250 miles (400 km) up. Noctilucent clouds [2] only occur around the mesopause – the line between the thermosphere and the mesosphere. Some meteors [3] reach the surface of the Earth, but most burn up in the mesosphere. Cosmic rays [4] penetrate to the stratosphere. Most of man's activities, and the weather that directly affects him [5], occur in the troposphere.*

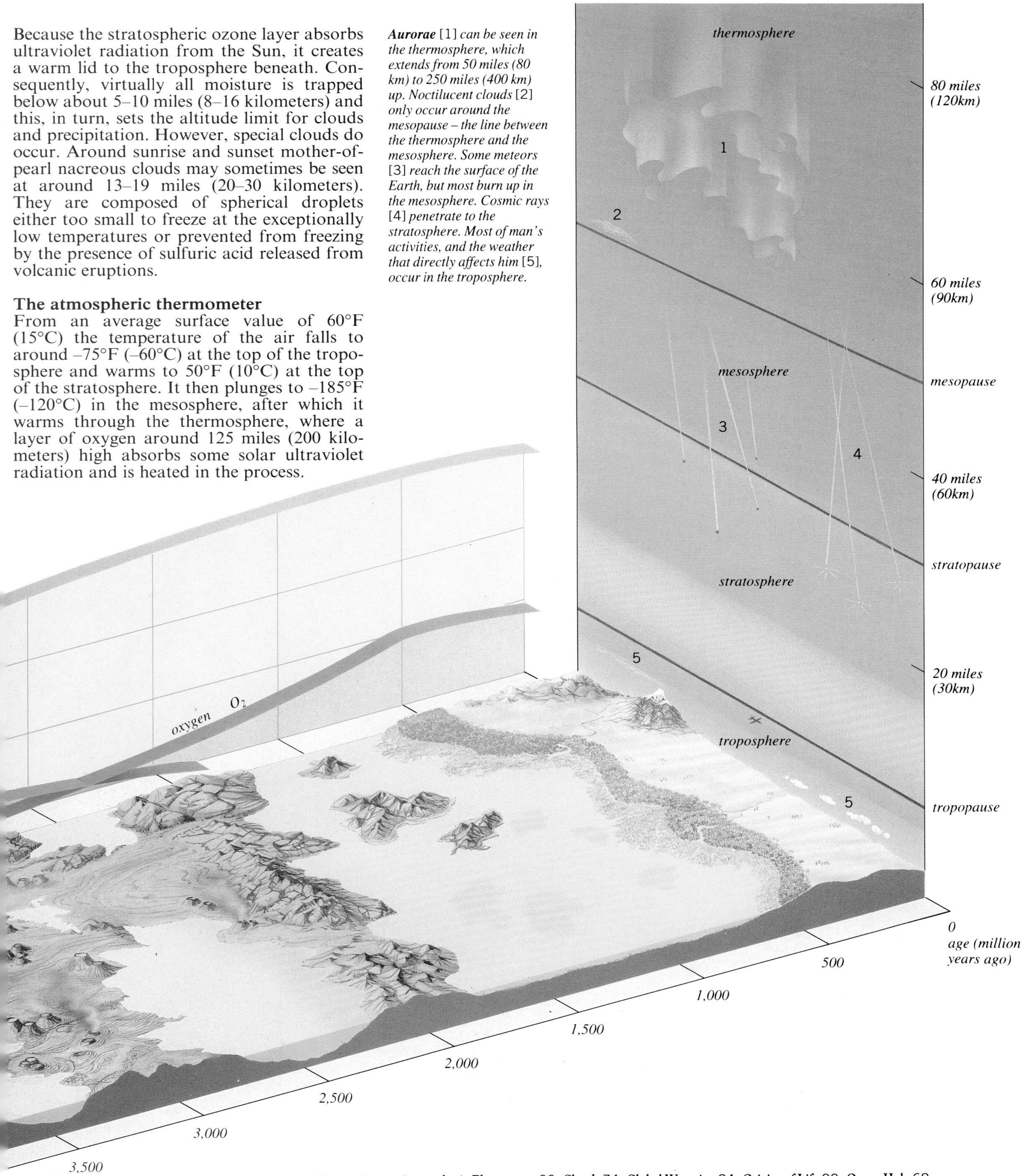

Connections: **Atmospheric Phenomena** 66 **Clouds** 74 **Global Warming** 84 **Origins of Life** 90 **Ozone Hole** 68

Lights in the Sky

What causes atmospheric phenomena

An Arctic explorer once reported seeing nine Suns in the sky. He had fallen victim to one of the many optical illusions caused by light rays being bent or reflected as they pass through the atmosphere. The towering, spectacular Fata Morgana is another such illusion, which fools sailors into seeing mountain ranges floating over the surface of the sea. The atmosphere provides an ever-changing backcloth of color, with dynamic vistas of blue sky, white clouds, gray storms, and red and yellow sunsets. But one of its eeriest phenomena is the albino rainbow sometimes seen in fog.

Molecules of air scatter the shorter wavelengths of light (which are at the blue end of the spectrum of colors) more than the longer wavelengths (at the red end) – which is the reason that the sky appears blue in color. The cleaner the air, such as after a rain shower, the darker the shade of blue the sky appears. When the Sun is near the horizon its light passes through large dust particles, which scatter longer wavelengths, giving a red sky at night and in the morning. Fog and cloud droplets, with diameters larger than the wavelengths of light, scatter all colors equally and make the sky look white.

A green flash from just above the Sun's disk is sometimes seen at the moment the disk disappears below the horizon at sunset. It occurs because different wavelengths of light are bent (*refracted*) in the atmosphere by differing amounts. Because green light is refracted more than red light by the atmosphere, green is the last to disappear (blue light having been scattered by air molecules and lost from view).

An airborne prism

When sunlight enters a raindrop, refraction and reflection take place, splitting white light into the spectrum of colors from red to blue and making a rainbow. Rainbows come in pairs. Red is on the outside bend of the stronger, primary rainbow, but the secondary rainbow has its red band on the inside. Between the two rainbows little light is scattered and this creates a dark area known as Alexander's dark band. As many as four faint pink, green or violet bands are sometimes visible on the inside of the primary rainbow. They are called supernumerary arcs and are caused by interactions of light waves arriving at the observer's eye. When light waves are out of phase (so that the crest of one wave coincides with the trough of an adjacent one) they cancel each other out, but when in phase (with crests and troughs matching) color intensity is enhanced.

Living halos

Interactions of light waves can produce a "glory." People standing on a mountain, with the sun on their backs, may cast shadows on the fog in the valley which appear to be surrounded by colored halos. Similarly, the shadow on the cloud-tops of an airplane

Halos are rings *of white or colored light* **[A]** *around the Sun or Moon. The commonest halos are seen at radii of 22° and 46° out from the Sun or Moon. They are formed when sunlight gets refracted through small ice crystals. Most halos are produced by hexagonally shaped plate or column crystals* **[B]***, but some rare halos may be formed by ice crystals with triangular faces. Halos are usually white, but if the refraction is clear enough, spectral colors may also be seen. Halos appear between the spectator and the Sun, whereas rainbows* **[C]** *always appear on the side of the spectator opposite to the Sun. They are arcs of a circle whose (imaginary) center is as far below the horizon as the Sun is above it. The violet of a primary rainbow* [1] *is always at an angle of 40.4° to the spectator. Its red is at 108°F. The red of a secondary rainbow* [2] *is always at an angle of 50.2°, with its violet at 53.2°. The size of the reflecting rain droplets affects the intensity of rainbow colors. Small droplets cause colors to overlap so that orange and pink tones occur at the edges instead of reds and violets.*

Connections: **Atmosphere** 64 **Clouds** 74 **Color** 262 **Earth's Magnetic Field** 14 **Rain, Hail, Snow, and Fog** 76 **Lightning and Thunder** 78

__Primary rainbows__ are caused [__D__] by light that, after it is split by refraction into the spectrum of colors, is reflected only once from the back of a water droplet [1]. The light for a secondary rainbow is reflected twice within the droplet, giving it a greater angle and reversing the spread of colors [2].

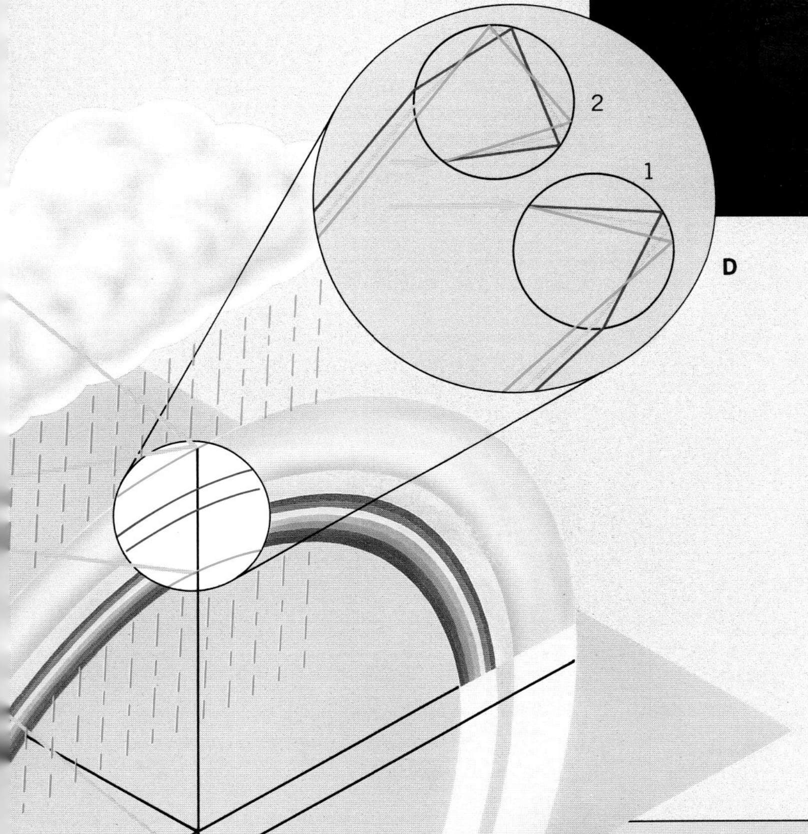

__Coronas are caused__ by the bending of light around small cloud droplets. These bent or diffracted waves of light interfere with the undiffracted waves to give alternating rings of light and dark. As the different wavelengths corresponding to the various colors of light bend slightly differently a red ring may appear in the dark region between two blue rings. The size of the rings of light depends on the size of the droplets. The smaller the droplets, the larger the rings. The more uniform the droplets, the more regular the rings. Typical coronas have a 4° – 10° radius, corresponding to cloud droplets of 4–10 µm.

flying above the clouds may be surrounded by rings of colored light. The glory is caused by light entering the edges of tiny droplets and being returned in the same direction from which it arrived. These light waves interfere with each other, sometimes cancelling out and sometimes adding to each other. A glory can seem even more impressive when the shadow of the observer is enlarged (a Brockenspectre).

Multiplying Suns

Reflection and refraction of light by ice crystals can create bright halos in the form of rings, arcs, spots, and pillars. Sun dogs (mock Suns) may appear as bright spots 22° to the left and right of the Sun. Sun pillars occur when ice crystals act as mirrors, creating a bright column of light extending above the Sun. Such a pillar of bright light may be visible even when the Sun has set.

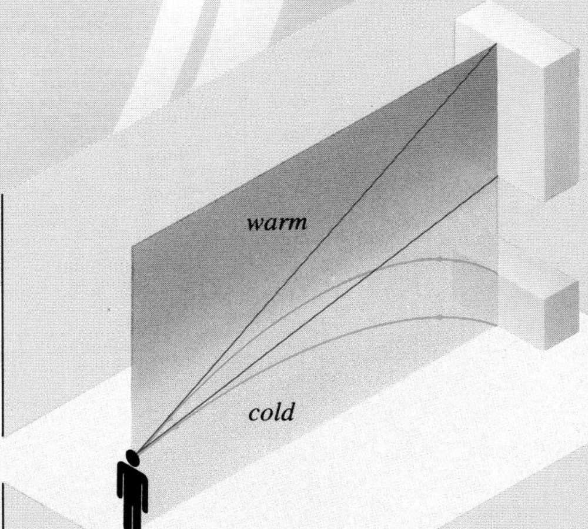

Mirages

The mirage is an image created when rays of light from a distant object are bent (refracted) by a difference in air density which is caused by varying temperature. The observer imagines the light has traveled in a straight line and sees the object displaced. When the air temperature is greatest near the ground, the image of a distant object is displaced downward creating an "inferior" mirage. An image of the blue sky on a hot day can seem to cover a road with a shimmering pool of water. When temperature increases with height, as is common over a cool lake on a summer afternoon, the image is displaced upward, creating a "superior" mirage. Sometimes the light from the top and bottom of a distant object are refracted by different amounts, giving magnified or compressed images.

The Global Sunscreen

What is causing the ozone hole

If it were not for a diffuse gaseous layer 10–30 miles (14–45 kilometers) up, life on Earth could not exist. The ozone layer is all that protects us from the Sun's harmful ultraviolet radiation. Without the ozone layer terrestrial life could not exist. But recent observations have shown that pollution of the atmosphere could be destroying this vital layer. Scientists have detected an "ozone hole" over Antarctica, which appears seasonally, extending over an area the size of the United States. Were this hole to spread, the consequences would be disastrous for both animal and plant life.

Without sunlight ours would be a barren planet. Solar radiation warms the Earth's surface and is harnessed by plants in photosynthesis. But some wavelengths of light emitted by the Sun are less beneficial to life: indeed, ultraviolet light is an effective sterilizing agent and is used in hospitals to kill microorganisms. Much of this radiation is filtered out in the stratosphere before it can wreak its destructive effects. Under normal circumstances, the concentration of the ozone that absorbs the ultraviolet light is constant.

Punching a hole in the heavens

In the 1970s and 1980s, it became clear that human activity was disturbing this delicate equilibrium: gases – most notably chlorine – emitted into the atmosphere had depleted the levels of stratospheric ozone. Chlorine atoms are released in the form of *chlorofluorocarbons* (CFCs), which are widely used as coolant fluids in refrigerators, to make bubbles in foamed plastic, and in aerosols.

After their release CFCs take many years to rise into the stratosphere. At these altitudes, ultraviolet radiation is sufficiently intense to split CFC molecules and liberate chlorine atoms, which can then attack ozone. Chlorine can attack many thousands of ozone molecules before it is finally removed from the stratosphere. Small increases in the amounts of chlorine released can therefore cause enormous changes in the chemistry of the upper atmosphere.

Emissions of CFCs have been rising rapidly in recent decades. Because CFCs remain in the stratosphere for anywhere between 65 and 130 years, immediate reductions in their release are essential. International agreements may succeed in phasing out CFC production, and alternative ozone-harmless chemicals are being developed. Even so, a vast reservoir of ozone-depleting chlorine will remain in the stratosphere for many decades.

In 1985 scientists discovered that the ozone layer was being depleted, not gradually all around the globe, but massively and seasonally in one particular area: the Antarctic. The amplified impact of CFCs in Antarctica is a result of the unique meteorology of that region. During the southern winter the cold, heavy air high over Antarctica is prevented from mixing with the warmer, strong winds that whirl around the continent. Within this thin, cold air, at temperatures around –130°F (–90°C), icy particles form clouds known as polar stratospheric clouds (PSCs). In the dark polar winter, atmospheric chemistry is at a virtual standstill. But some reactions still take place on the surface of the cloud particles, slowly preparing the chlorine held in "reservoirs" to make a rapid escape when the Sun begins to shine in the southern spring. This sudden springtime "flush" of chlorine rapidly depletes the Antarctic ozone layer.

Life in a destructive light

Increased levels of ultraviolet radiation, especially ultraviolet-B (290–320 nanometer wavelengths), are expected to cause marked increases in sunburn, in skin cancer, and in eye problems such as cataracts. For every 1 percent decrease in ozone there may be a 5 percent increase in the number of nonmalignant skin cancers in humans each year.

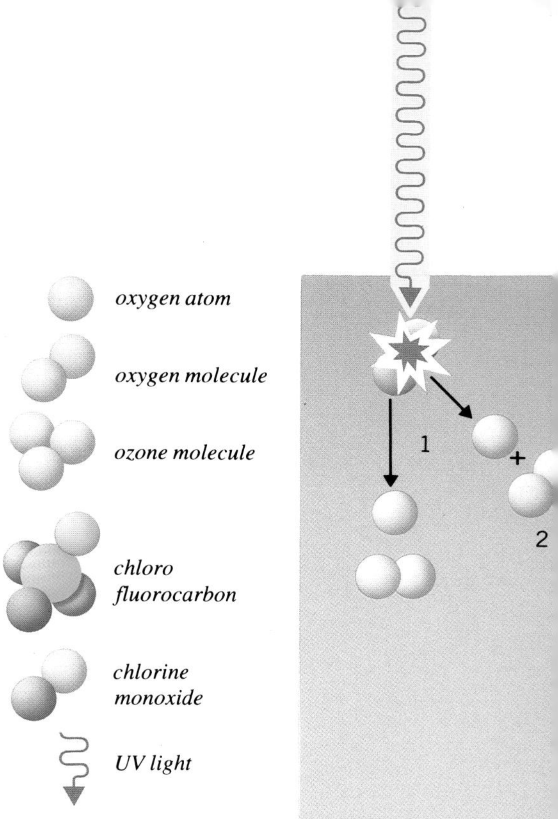

***Ozone** is a form of oxygen with three atoms instead of two in its molecules. Ozone molecules can become oxygen molecules by losing one oxygen atom. The process is reversible. Chlorofluorocarbons are made of one carbon atom, one fluorine atom, and three atoms of chlorine.*

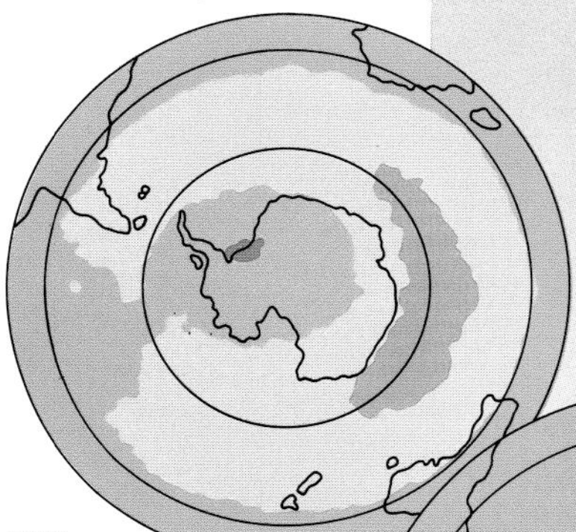
1979

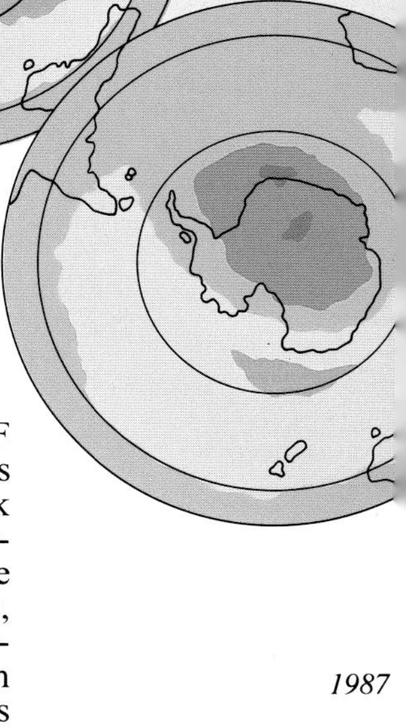
1982

1987

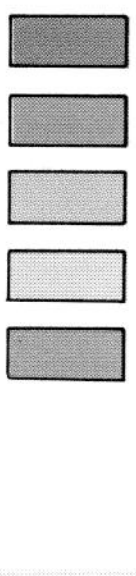

***Each October** since 1979 the thickness of the ozone layer over the Antarctic has decreased by up to 50 percent or more. The resulting ozone "hole" is constantly expanding. The thickness of the ozone layer is measured in Dobson units. A hundred Dobson units correspond to 0.04 in (1 mm) of compressed ozone. In the southern summer, the thickness of the layer is about 350 Dobson units. The ozone hole is repaired in this way because the polar stratospheric clouds evaporate in the warmth of the summer, and fresh air returns to the Antarctic from other latitudes, allowing ozone levels to recover.*

Connections: **Air Pollution** 342 **Atmosphere** 64 **Climatic Change** 82 **Global Warming** 84 **Polar Regions** 42

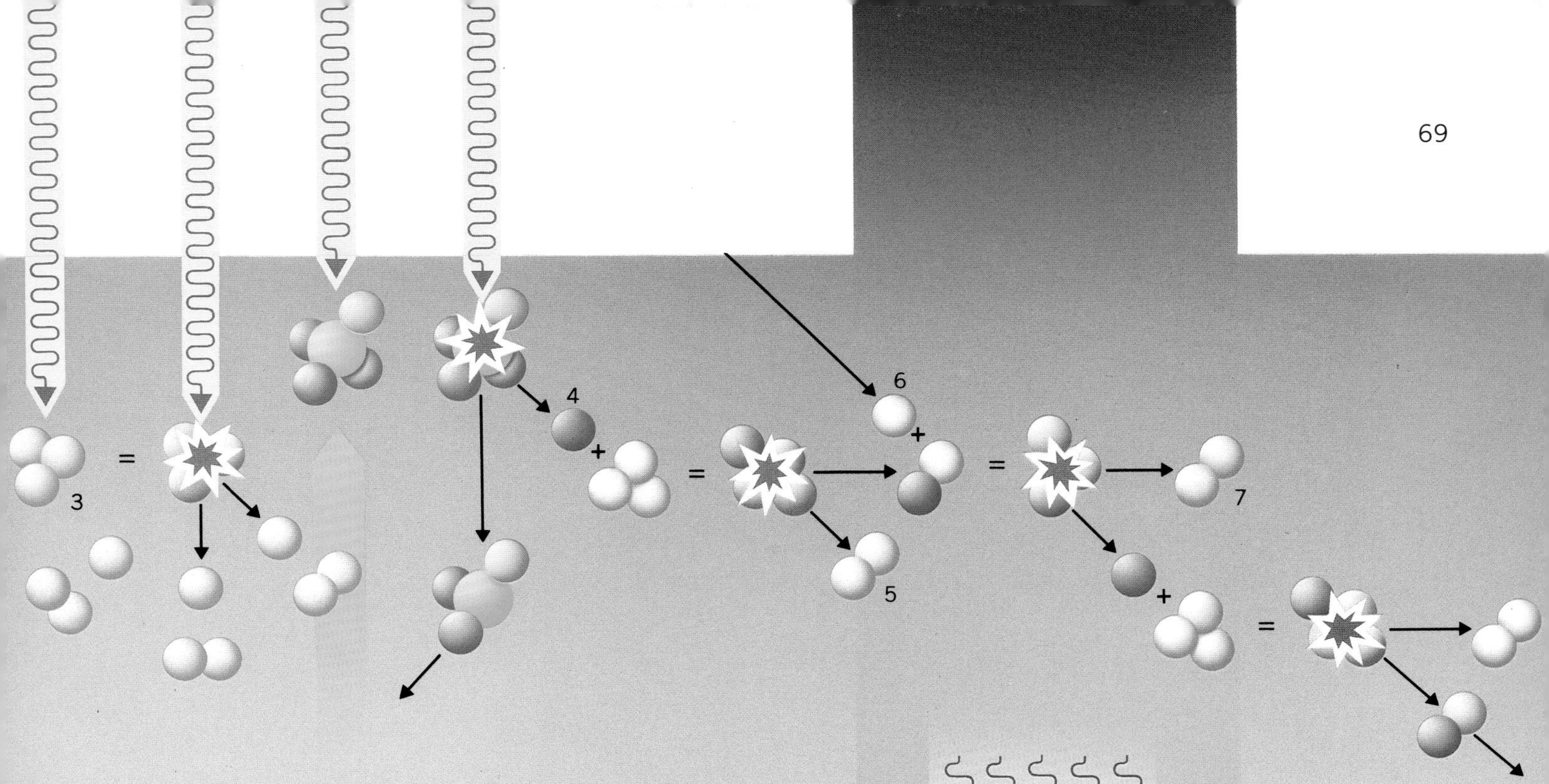

Ozone *– and oxygen – absorbs ultraviolet light by using up the energy of the light to fuel chemical reactions. When ultraviolet light strikes an oxygen molecule (O_2), it splits it into two free oxygen atoms* [1]. *These are highly reactive and in turn combine with other oxygen molecules* [2] *to form ozone (O_3). The ozone can then absorb more ultraviolet light* [3] *and be broken down into an oxygen molecule and a reactive oxygen atom. The oxygen atom will combine with another free oxygen atom to form an oxygen molecule, or it may collide with an ozone to give two oxygen molecules, and the whole process can begin again. In this way a chain reaction – and an equilibrium – is established in which ozone formation keeps up with ozone loss. In the absence of any interference from man, the result is a constant concentration of ozone in the stratosphere.*

When a ***$CFCl_3$*** *molecule is struck by ultraviolet light, it loses a chlorine atom* [4]. *The free chlorine can steal an oxygen atom from an ozone to form a chlorine monoxide (ClO) molecule, leaving behind an ordinary molecule of oxygen* [5]. *When the ClO collides with a free oxygen atom* [6], *the oxygens combine into a molecule* [7] *and leave the chlorine free to go and destroy another ozone. This is known as a "catalytic" process, because the chlorine can destroy many ozones without itself being consumed in the process.*

As stratospheric ozone *decreases, more of the Sun's energy reaches the Earth. Ultraviolet-B may kill marine phytoplankton, which absorb large quantities of carbon dioxide from the air. As atmospheric CO_2 increases, less of the heat radiated from the Earth's surface escapes. Thus the ozone hole may also contribute to global warming.*

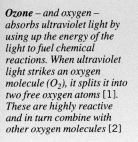

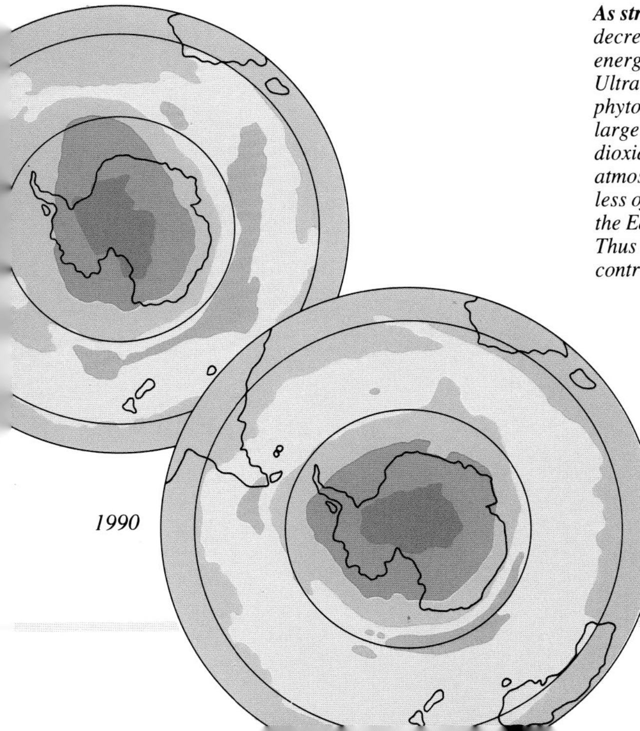

1990

The Weather Machine

What generates the world's weather

"A butterfly flapping its wings in Brazil can cause a tornado in Texas," according to one eminent meteorologist. He was pointing out how the avalanching influence of tiny effects makes accurate long-range weather prediction almost impossible. Nevertheless, the real engine of the Earth's weather is the Sun. Its heat generates powerful movements of air within the Earth's atmosphere that circulate for tens of thousands of miles back to their starting point, providing the world's winds with a total kinetic energy greater than a trillion jumbo jets flying at top speed.

***Regional climates* [A]** *change enormously with seemingly small changes in global mean temperature. This happens because polar latitudes are highly sensitive to alterations in global temperature compared with the tropics. As the global temperature falls, the area of snow and ice in higher latitudes expands rapidly, causing greater amounts of the Sun's energy to be reflected back into space, so chilling these areas even more. In contrast, the tropics change very little. As the temperature difference increases, it forces a rearrangement of atmospheric circulation.*

A

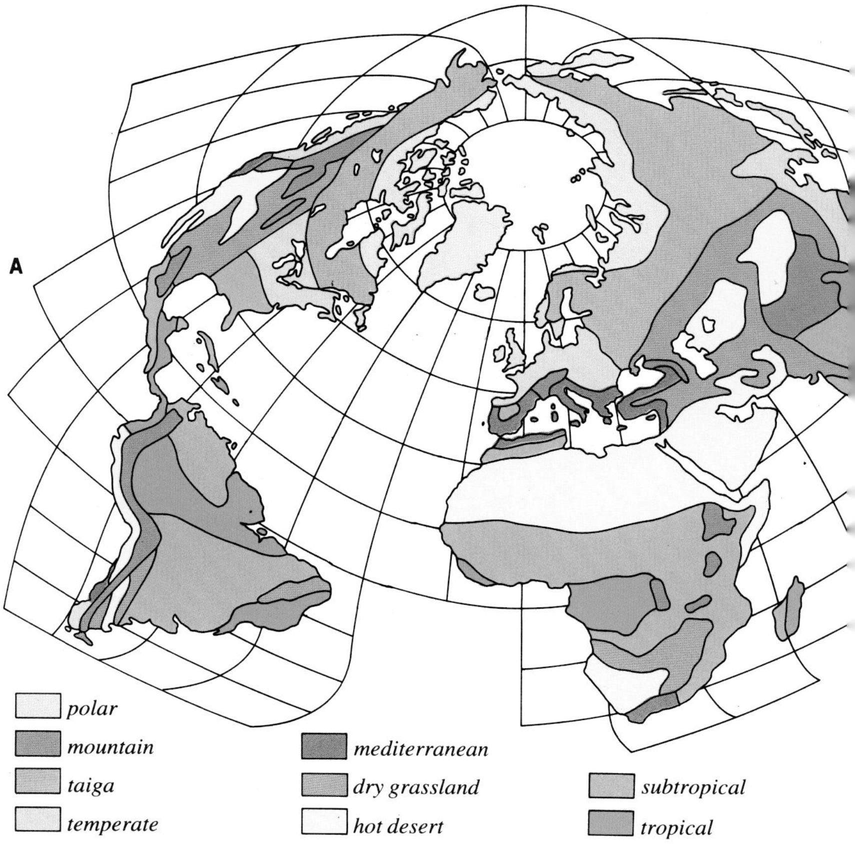

All the world's weather is driven by the Sun's intense heating of equatorial regions, which sends air currents surging high into the atmosphere. The rising air cools and the moisture it carries condenses to form clouds and heavy rainfall, creating the characteristic wet climate zone of Africa, Asia, and South America. But the air currents themselves continue moving poleward. Eventually, this warm tropical air sinks at around 30° latitude and encounters increased pressure at the lower altitudes. It is compressed and heated and moisture is squeezed away, so creating the subtropical climate zone of permanent deserts in places like Australia and the Sahara region of Africa.

On reaching the surface, the sinking air is pushed outward, with some returning to the equator as the northeast trade winds (southeast trades in the Southern Hemisphere). This circulation of air between the equator and 30° latitude is called the *Hadley Cell.*

Mixing up a storm

The rest of the descending air is sent poleward to form the warm winds of the mid-latitudes. At 50°–70° latitude, these winds, now moist from their journey over the oceans, encounter cold dry winds spilling from the bitterly cold polar and taiga climatic regions. A battle takes place where these winds meet, resulting in the creation of the large swirling storms called frontal depressions that are typical of mid-latitude temperate climates.

Classifying climates

When they occur again and again over a long period, these geographical patterns in the weather – the balmy doldrums near the equator; the steady trade winds of tropical latitudes; the calm, cloud-free regions of subtropical latitudes; the stormy winds of the mid-latitudes, and the light, bitterly cold easterly winds blowing from the poles – can all be used to define the climate of a region. In addition to their precipitation and temperature, climates have been classified into zones by moisture index, vegetation and even estimates of human discomfort. But all classifications consider only the broad similarities between climates at various places around the world. Local differences are ignored and boundaries are approximate, and shift over time with the Earth's changing temperature.

Reflected glory

Photographed from space, the clouds and oceans and the various climatic zones with their characteristic landscapes make a spectacular global patchwork of color. This is because not all the energy the Earth receives from the Sun is absorbed. A lot is reflected, and the proportion of solar energy the Earth bounces back into space is its *albedo.* Although the Earth's average albedo is about 30 percent, a wide range of values exists throughout the world. The albedo of forests is 9–18 percent, varying according to tree type and foliage density. The albedo of grasslands is 25 percent, of cities 14–18 percent, and of deserts 25–30 percent. Fresh snow may reflect as much as 85 percent of solar energy. The albedo of oceans rises from 2 to 3 percent, when the Sun is high in the sky, to 50 percent, when the Sun drops below 15° as it descends toward the horizon.

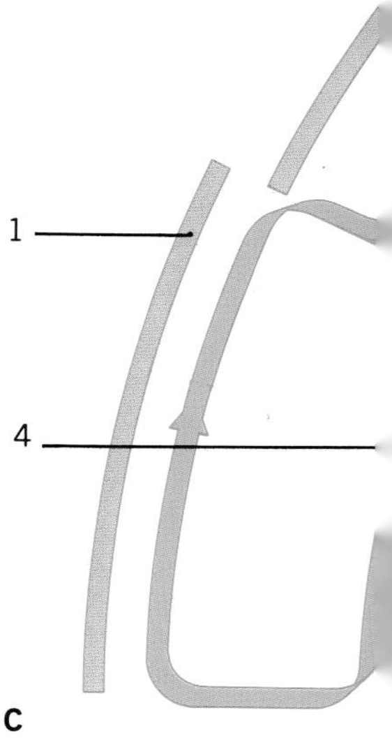

C

Connections: Climatic Change 82 Clouds 74 Cyclonic Storms 80 Hail, Snow and Fog 76 Seasons 86 Winds and Weather 72

B

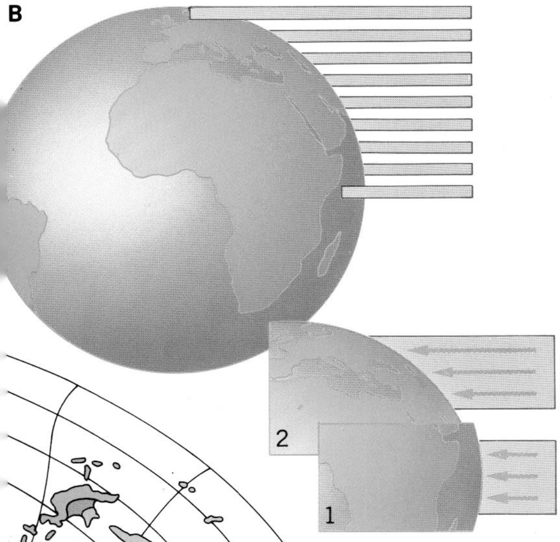

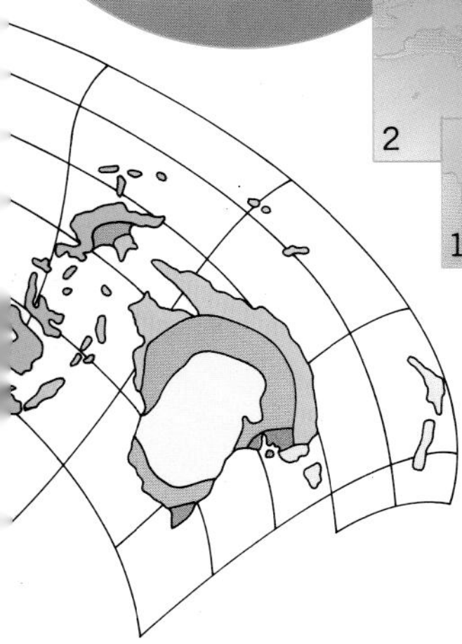

The Earth [**B**] *is spheroidal in shape. As a consequence, the equatorial latitudes* [1] *receive far more heating from the Sun's rays per unit area than do the poles* [2]. *Heat is also lost at the poles because the rays must travel through a greater thickness of air before reaching the ground. Since lower latitudes are clearly not overheating and higher latitudes not cooling, an annual balance between the energy received from the Sun and the heat lost by the Earth is achieved. It is the atmosphere and ocean currents that accomplish the task of transferring any excess energy between the equator and the poles.*

The hemispheric pattern *of air currents* [**C**] *forming the Hadley Cell* [1], *Ferrel Cell* [2] *and Polar Cell* [3] *change only slowly with the seasons, but surface winds and weather fluctuate markedly from day to day in many parts of the world. The subtropics suffer devastating hurricanes* [4], *embedded in the trade winds, while mid-latitudes experience squally frontal depressions. Each depression (cyclone), driven along by a jet stream* [5] *racing aloft, contains bands of rainfall along the cold front* [6] *and warm front* [7], *followed by a short period of clear and dry conditions associated with a ridge of high pressure* [8].

Forecasting the weather tomorrow, next week or next month is enormously complicated – which is why some forecasts are inaccurate. Weather forecasting involves the most powerful computers in the world. Forecasters express what is happening in the atmosphere around the world in the form of equations for the computers to solve using up-to-the-minute meteorological information from weather stations, balloons, radar, aircraft, ships, and satellites.

Jet Streams

Pilots flying high over the Pacific Ocean in World War II were often astounded to encounter high-speed winds that held their aircraft stationary. They were among the first people to discover jet streams – narrow bands of air traveling at speeds up to 300 mph (500 km/h). There are many jet streams in the atmosphere, and their presence is often revealed by ribbons of cirrus (high-level clouds of ice crystals) many hundreds of miles long.

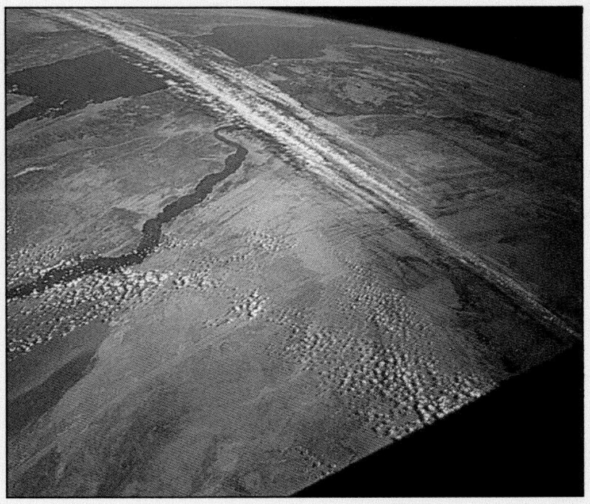

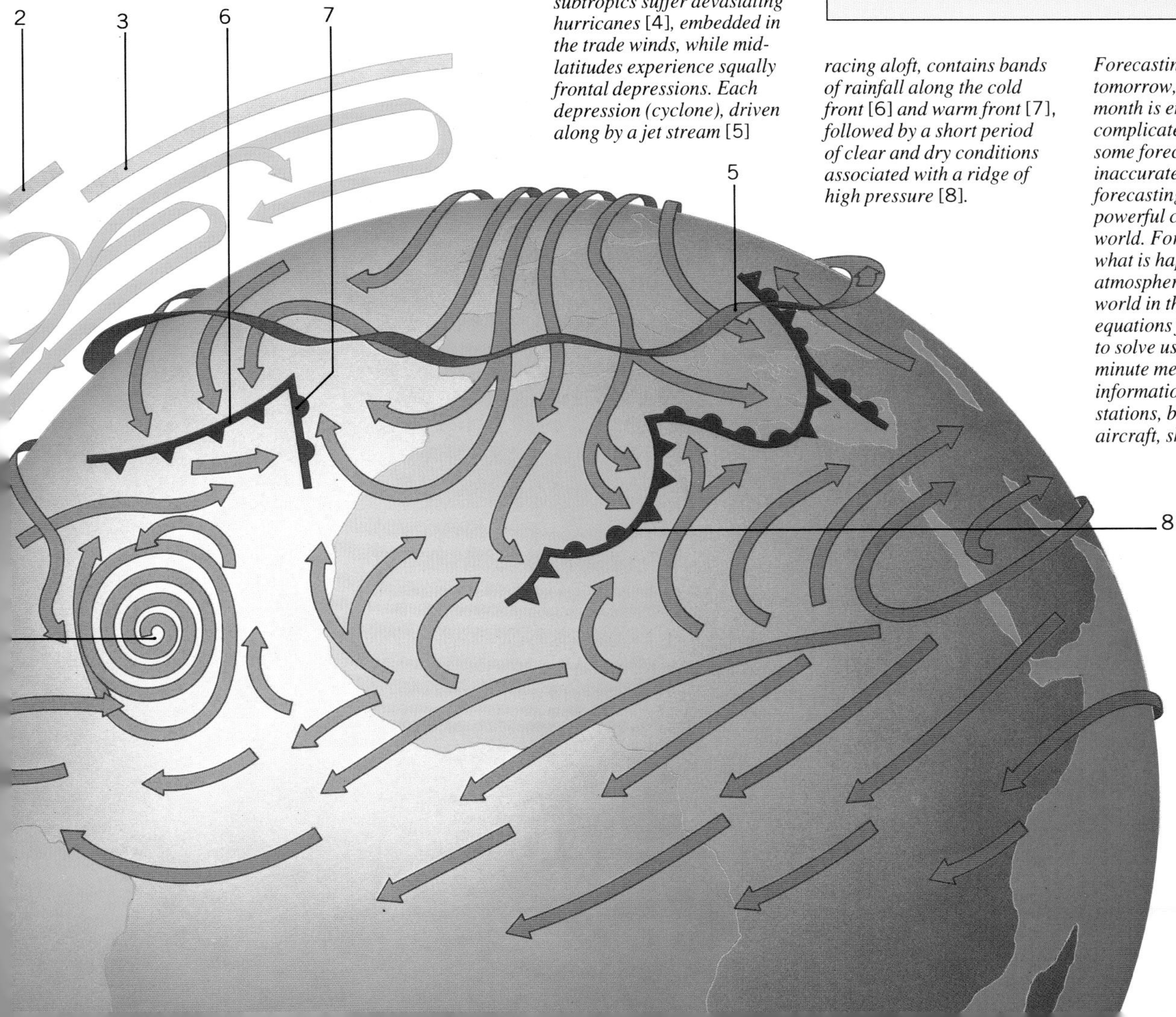

Borne on the Wind

How winds affect the weather

Many of the world's most important and predictable winds have been given their own names. They may bring life-giving rain like the Monsoon winds of India and Southeast Asia. Or they may sour tempers, like the irritating dry sirocco roaring north from Africa. Among the most important are the polar jet streams, because of the role they play in the creation of the huge areas of high and low pressure that dominate the weather of the mid-latitudes as they sweep eastward, swirling counterclockwise in the Northern Hemisphere and clockwise in the Southern.

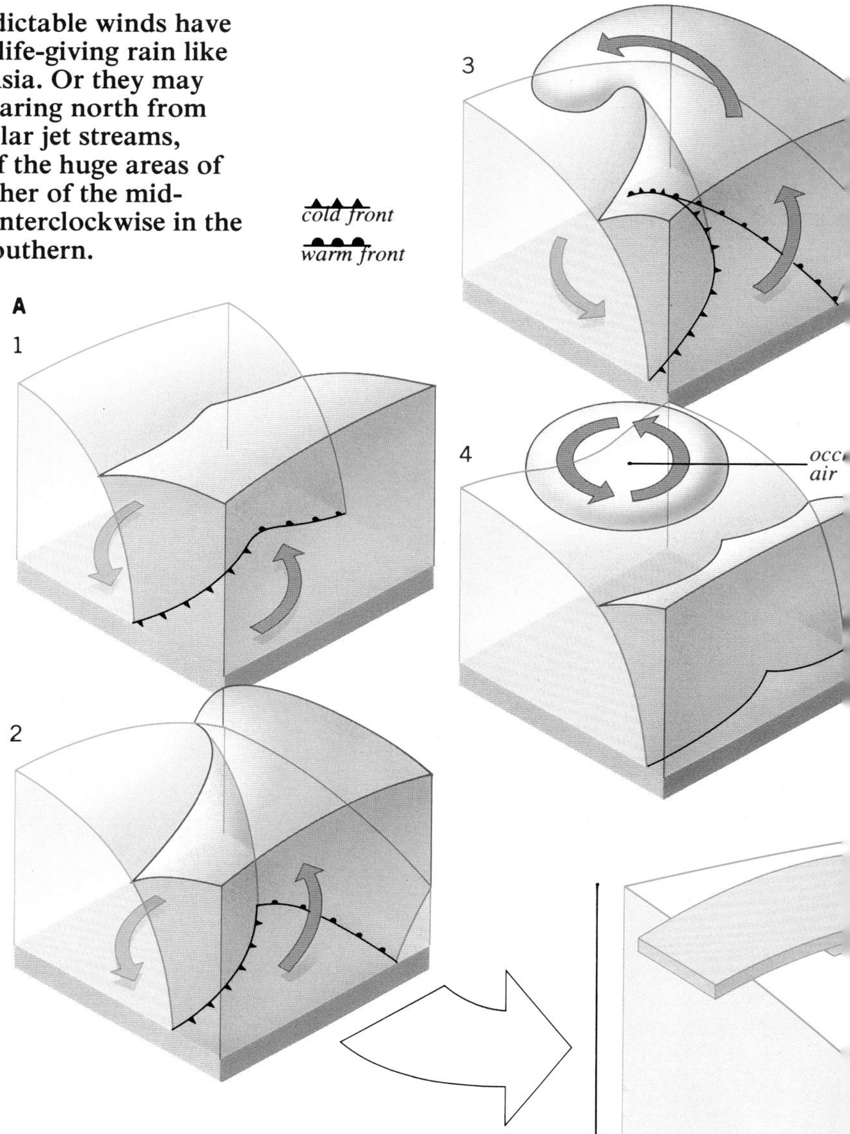

Winds occur when air masses of different pressures meet and air flows from the higher to the lower pressure. The air flow continues from the higher pressure to the lower until a balance is reached. The greater the pressure difference, the faster the wind. Many day-to-day weather changes are caused by the passage of air-mass boundaries – called fronts – within a *depression* (an area of low air pressure). They create a typical sequence of shifts in the weather as they pass overhead during the 4–7-day life cycle of an average frontal depression. The fronts do not stretch straight up: the warm front slopes forward at an angle of less than 1° from the horizontal, so the clouds in its highest part may herald the front's arrival by 12–24 hours. The cold front leans backward at an angle of 2°, which results in rainfall being briefer but heavier than with the warm front. Rainfall ceases abruptly with the passage of the front; temperature falls, air pressure rises, and the wind veers sharply – generally toward the northwest in the Northern Hemisphere.

Speeding air streams

Accurate weather forecasting depends on precise knowledge of where and when depressions form. This in turn comes from an understanding of jet streams. Although there are several in the atmosphere, the most important is the eastward-flowing jet stream found at a height of 6–9 miles (9–14 kilometers) along the polar front. It is formed because the strong temperature gradient that exists across the front causes in its turn a steep pressure gradient that can generate very fast winds. As a consequence, a frontal depression often forms beneath the jet stream as it heads poleward.

Regional winds

One of the most common local winds occurs in coastal regions. Here surface temperature differences between land and sea often generate daytime onshore sea breezes and nighttime offshore land breezes.

The French *mistral* is a cold, violent wind that funnels down the lower Rhône valley, killing vegetables and fruit when it brings frost during the early spring, and threatening people's lives in late summer as its cool, dry air spreads forest fires. It is triggered when air gets pulled southward from central Europe by a storm that intensifies in the Gulf of Genoa. As wind whips down the Rhône-Saône corridor it may reach speeds of up to 125 miles an hour (200 km/h).

The *bora* of the Balkans is another much-hated mistrallike cold wind that sweeps down out of coastal mountains to inflict damage on the Adriatic coast between Trieste and the Albanian border.

Dry and deadly heat

The scorching *sirocco* (or *scirocco*), called "the breath of the Sahara," occurs when the leading edge of an eastward-moving storm draws air northward from the hot interior of the Sahara. The strong, gusty winds are heavily laden with fine iron-stained sands, which are sometimes carried across the Mediterranean to produce "red rains." The sirocco is reviled as the *chili* in Tunisia, the *ghibli* in Libya, and the *khamsin* in Egypt.

Connections: Atmosphere 64 Climatic Change 82 Clouds 74 Cyclonic Storms 80 Weather and Climate 70

The mid-latitude depression *(or extratropical cyclone), although far less severe than its tropical counterpart, is more common, lasts longer, and affects the weather over a larger area. In the development of a frontal depression* [**A**]*, the early stage* [1] *occurs when a mass of cold polar air meets a mass of warm subtropical air, forming a front. As the depression develops to the open stage* [2]*, a wave forms, around which two, now distinct, fronts begin to spiral. As the depression matures, the warm air rises above the advancing cold front and the cold air slides beneath the warm, forcing it off the ground, to form the occluding stage* [3]*. Eventually, in the dissolving stage* [4]*, the cold areas have merged and the warm air has been completely cut off, or occluded.*

Vital Deluge

Cherrapunji, in India, once experienced 35 inches (897 millimetres) of rain in 24 hours. A record 350 inches (9,000 millimetres) fell in just the month of July, and a single monsoon season brought 642 inches (16,305 millimetres) in 1899. The term "Monsoon," now used for heavy rain-bearing winds worldwide, first referred to the seasonal reversals of wind direction that occur in and around the Indian Ocean and Arabian Sea. These winds blow from the southwest for one half of the year (summer) and from the northeast for the other (winter). The cycle is caused by the differential heating between ocean and land owing to the seasonal movement of the Sun. Monsoon rains are not always reliably regular, and regional variations can produce catastrophic floods or, alternatively, disastrous droughts in areas where the rains have failed.

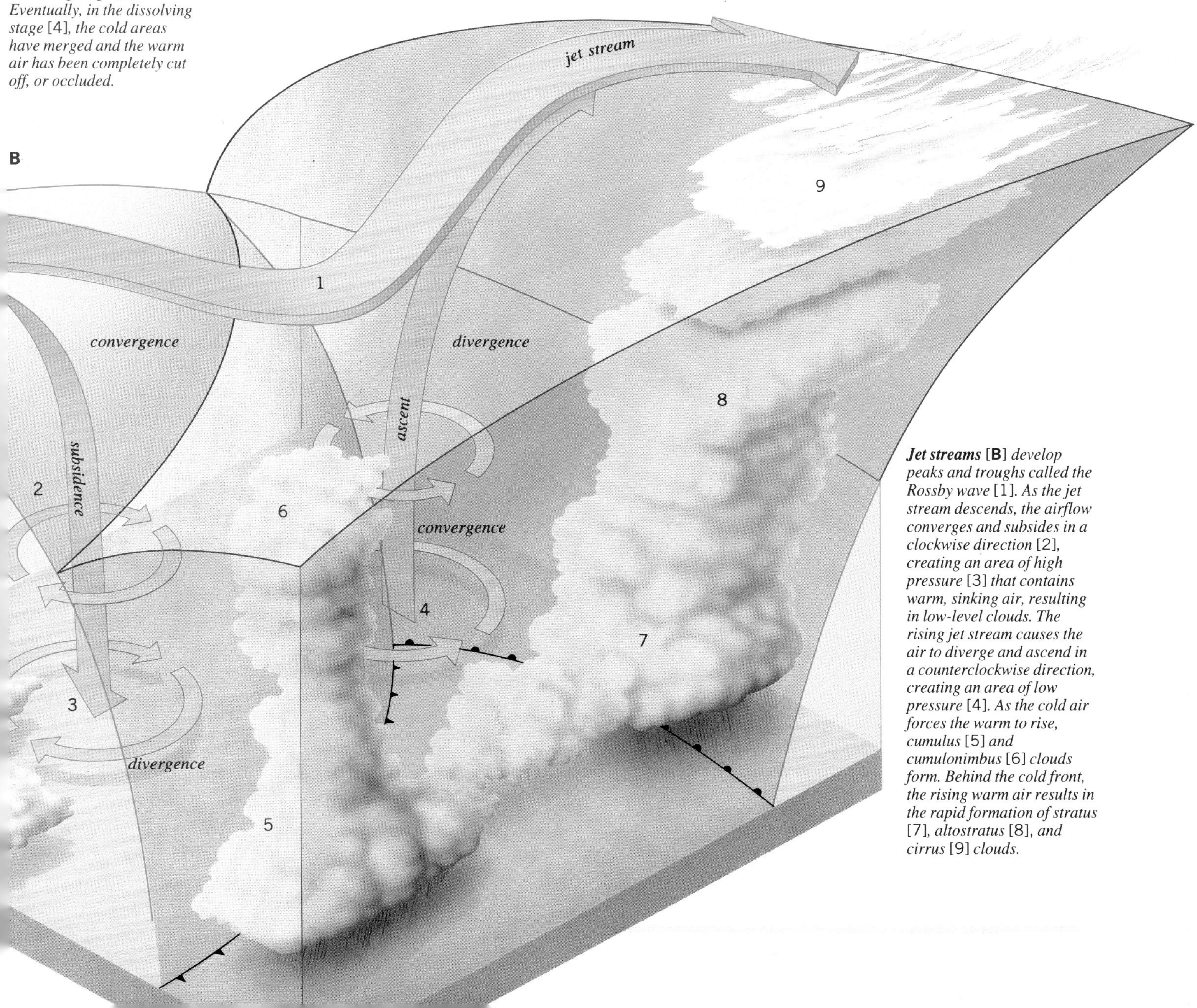

Jet streams [**B**] *develop peaks and troughs called the Rossby wave* [1]*. As the jet stream descends, the airflow converges and subsides in a clockwise direction* [2]*, creating an area of high pressure* [3] *that contains warm, sinking air, resulting in low-level clouds. The rising jet stream causes the air to diverge and ascend in a counterclockwise direction, creating an area of low pressure* [4]*. As the cold air forces the warm to rise, cumulus* [5] *and cumulonimbus* [6] *clouds form. Behind the cold front, the rising warm air results in the rapid formation of stratus* [7]*, altostratus* [8]*, and cirrus* [9] *clouds.*

Saturation Coverage

How clouds are formed

There are countless millions of tons of moisture in the Earth's atmosphere. The air of the driest desert is still 0.1 percent water vapor, which, in the right conditions, becomes visible as clouds. Although there seems to be an infinite variety in cloud shape, color, and texture, there are actually only ten main types. They can be created by the Sun heating up a hillside, by factory chimneys, or jet aircraft exhausts. And as the visible signs of motion and change in the atmosphere they reveal what weather may be expected up to 48 hours ahead.

Clouds come into existence wherever moist air is cooled to its dew point – the temperature at which it becomes saturated with water vapor. This cooling usually occurs as the moist air is driven higher up into the atmosphere by convection currents, turbulence, or physical obstacles like mountains. Atmospheric pressure decreases with height, causing the water-bearing air to expand and cool. At the dew point, water condenses around minute airborne particles, called condensation nuclei, to make tiny cloud droplets. Some nuclei, such as sea salt particles, have a strong affinity for water vapor and encourage droplets to form even before the air has reached moisture saturation.

Carving out clouds

The form and size of clouds is determined by the nature of the forces that propel the moist air upward, and by the temperature structure of the atmosphere. For example, when converging bodies of air in the lower atmosphere force a large, stable mass of air to rise, broad, featureless cloud tends to form. When the atmosphere is unstable, small "bubbles" of warm, moist air are forced up by local convection currents, giving rise to clouds with a small base.

The vertical growth of a cloud is influenced by the origin of the surrounding air. The moist cloud-forming air will rise through the atmosphere as long as it is warmer than the air around it – the same principle as a hot-air balloon. So if the surrounding air is from a cold, polar source, the cloud will rise higher than in warm air from the tropics.

Celestial classification

The basic process that forms clouds belies the huge variety of shapes, sizes and colors. Clouds can be grouped into several main types: a classification that takes into account their height and shape. Low-level clouds such as cumulus are small and dome shaped; if joined together, they become stratocumulus. Stratus is an amorphous, layer cloud often covering hills and coasts, while cumulonimbus (thunderhead) is a billowing white cloud tower with a striking cauliflower- or anvil-shaped top. Middle-level clouds include altostratus, through which the Sun appears "watery." Altocumulus forms rafts of cloud (called a "mackerel sky"), while nimbostratus usually appears as a dark formless mass. Cirrus are the high-level, feathery clouds composed of ice crystals and often resembling mares' tails. Cirrocumulus are thin, high-level clouds of dappled or rippled appearance. Cirrostratus is a thin cirrus veil often revealed only by the optical effects that it produces.

Mountains making waves

Mountains create clouds because passing air is forced upward, so cooling it to its dew-point temperature. Mountaintops and windward slopes are frequently shrouded in ragged stratus. In their lee, mountains often generate rising and sinking waves in the airflow. The waves appear stationary relative to the ground, but the air is moving through them at 6–12 miles an hour (10–20 km/hour). Frequently, air rising into the crest of a wave cools sufficiently to create striking thin lens-shaped clouds stacked one on another.

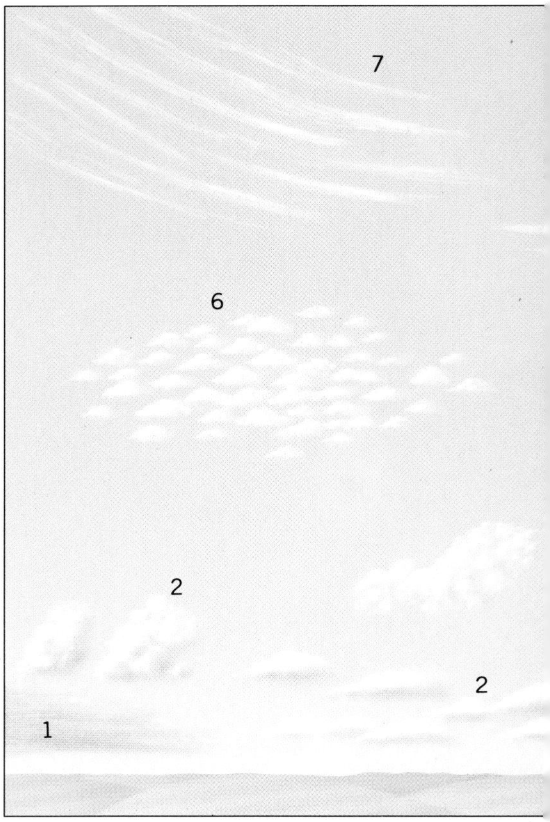

***The major cloud types** have characteristic shapes and occur within broad altitudinal boundaries* [**A**]. *Stratus* [1] *is a low-level cloud, usually featureless and gray. Its base may obscure hilltops or occasionally extend right down to the ground, and because of its low altitude, it appears to move very rapidly on breezy days. Stratus can produce drizzle or snow, particularly over hills, and may occur in huge sheets covering thousands of square miles.*

Cumulus [2] *clouds also seem to scuttle across the sky, reflecting their low altitude. These small "fleecy" clouds are typically short lived, lasting no more than 15 minutes before dispersing. They are typically formed on sunny days, when localized convection currents are set up: these currents can form over power plants or even stubble fires, which may produce their own clouds. Cumulus may expand into stratocumulus* [3], *or into giant cumulonimbus* [4], *which are up to 7 miles (10 km) in diameter. These clouds typically form on summer afternoons: their high, flattened tops contain ice, which may fall to the ground in the form of heavy showers of rain or hail. Rising to middle altitudes, stratus and cumulus cloud (then termed altostratus* [5] *and altocumulus* [6]) *appear to move more slowly because of their greater distance from the observer.*

Cirrus clouds [7] *are named after the Latin for "tuft of hair," which they resemble. Relatively common over northern Europe, they sometimes become associated with a jet stream and then appear to move rapidly despite their great height. Cirrocumulus* [8] *is often present with cirrus cloud in small amounts, while the presence of cirrostratus* [9] *is given away by halos surrounding the Sun or Moon.*

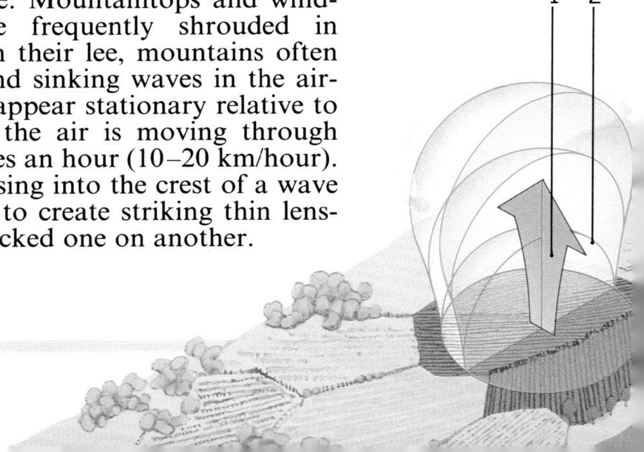

A

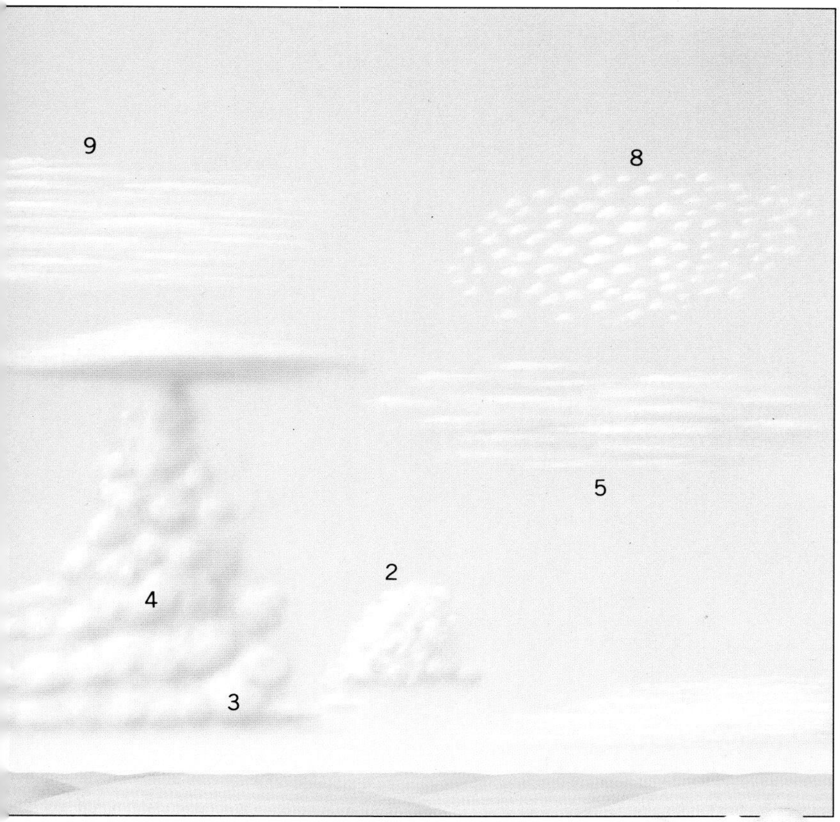

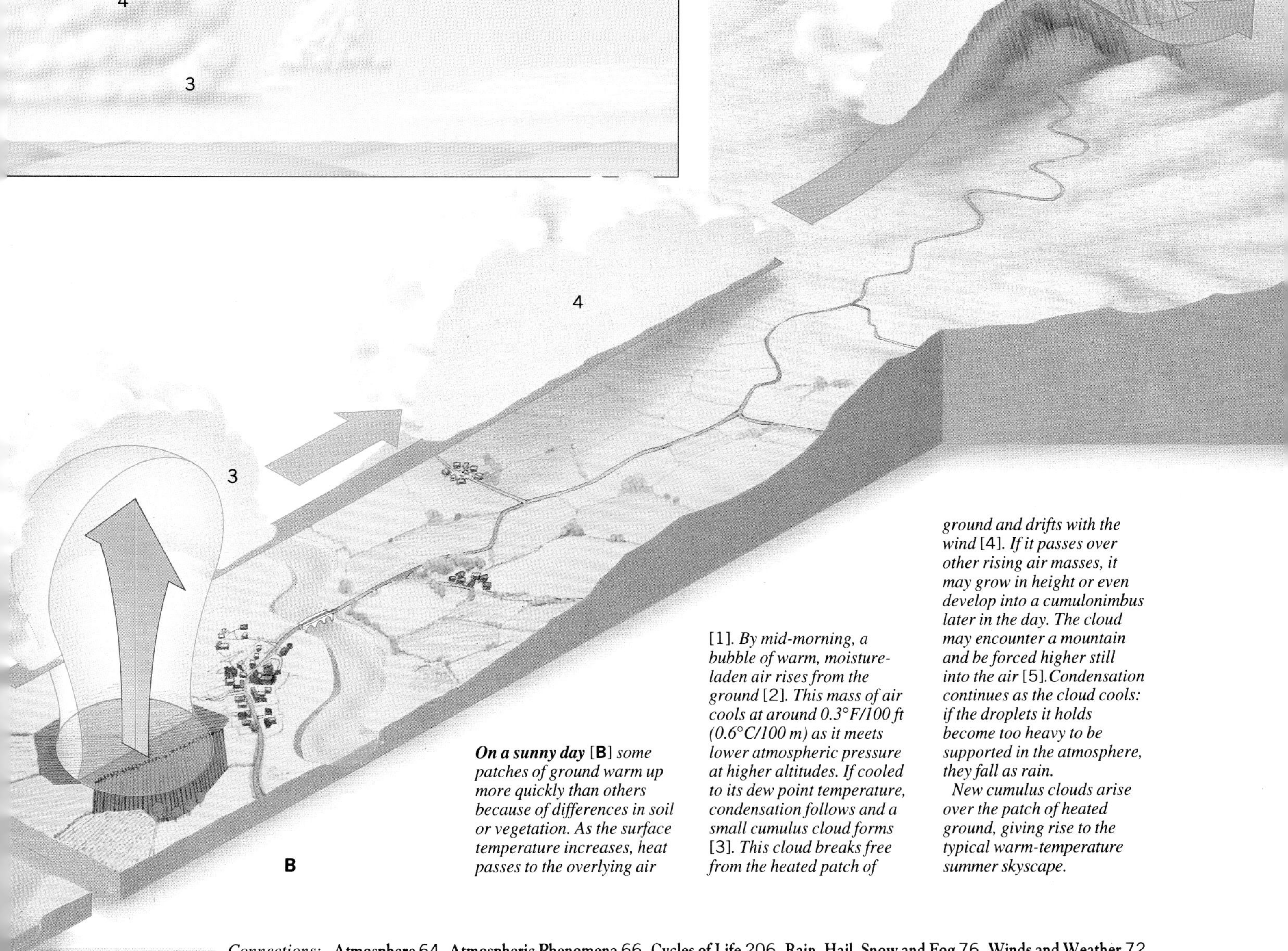

B

On a sunny day [**B**] *some patches of ground warm up more quickly than others because of differences in soil or vegetation. As the surface temperature increases, heat passes to the overlying air* [1]. *By mid-morning, a bubble of warm, moisture-laden air rises from the ground* [2]. *This mass of air cools at around 0.3°F/100 ft (0.6°C/100 m) as it meets lower atmospheric pressure at higher altitudes. If cooled to its dew point temperature, condensation follows and a small cumulus cloud forms* [3]. *This cloud breaks free from the heated patch of ground and drifts with the wind* [4]. *If it passes over other rising air masses, it may grow in height or even develop into a cumulonimbus later in the day. The cloud may encounter a mountain and be forced higher still into the air* [5].*Condensation continues as the cloud cools: if the droplets it holds become too heavy to be supported in the atmosphere, they fall as rain.*

New cumulus clouds arise over the patch of heated ground, giving rise to the typical warm-temperature summer skyscape.

Connections: **Atmosphere** 64 **Atmospheric Phenomena** 66 **Cycles of Life** 206 **Rain, Hail, Snow and Fog** 76 **Winds and Weather** 72

Out of the Skies

What causes rain, hail, snow, and fog

Mount Wai-'ale-'ale in Hawaii has 350 days of rain a year. It is the wettest area of "dry land" on Earth. But not even the driest desert is free from some sort of precipitation. The Earth's surface can be chilled by snow or sleet; moistened by drizzle; obscured by fog or mist; and pelted by hailstones – or by giant ice blocks. These blocks – which can weigh many pounds – can form from the cistern overflows of aircraft, or from electrically charged cloud droplets that are fused together by a bolt of lightning. Or they may be meteors of ice, falling to Earth from the depths of space.

***Snowflakes** have a characteristic six-sided symmetry which reflects the internal hexagonal bonding of the water molecules of ice. Variations in the basic shape arise because some crystal faces grow faster in some directions than they do in others. As a result, no two snowflakes are identical. At high altitudes, where temperature is very low and the amount of water vapor small, the simpler kinds of ice crystals grow, such as columns, needles and plane hexagonals. Where the temperature is warmer and more moisture exists, the extremely beautiful and complex feathery six-sided stars are formed.*

The base of a cloud is the air's condensation level, where water droplets form from moisture. Many clouds rise much higher than this, above a "freezing level" where the temperature drops below 32°F. Such clouds contain a mixture of ice crystals, *supercooled* water droplets (in a liquid state even though the temperature is below freezing point) and water vapor. When enough ice crystals have formed, not only does water vapor in the cloud freeze directly onto the ice but droplets evaporate and add their water vapor to the ice crystals. The tiny ice crystals grow to snowflake size and fall from the cloud, often melting on the way.

Colored snow

Dust particles in the air may stick to snowflakes as they fall. If the dust has been stained by the presence of iron, pink or red snow can result. The French alpine resort of Isola experienced bright pink snow in December 1975. Yellow snow fell in Vienna in February 1979. Snow pellets (also called soft hail or graupel) are white, brittle grains of ice, spherical or sometimes conical, 0.08–0.2 inches (2–5 millimeters) across. They are formed when tiny supercooled water droplets freeze onto ice crystals (a process called riming). Ice pellets are hard transparent or translucent lumps of ice about 0.04–0.2 inches (1–5 millimeters) across. They are frozen raindrops, formed when rain falls through a freezing layer of air.

Danger on the roads

When drizzle or rain falls through freezing air, it does not always freeze but may reach the ground as a supercooled liquid which, if the ground is also below freezing, then freezes instantaneously, coating highways, sidewalks, trees, plants, buildings and overhead cables with thick "ice rain" or glaze. This clear smooth ice is called "black ice" by motorists because it is not readily visible and produces treacherous road conditions. Mist and fog are formed by suspended cloud water droplets at ground level which reduce visibility to 3,250–16,500 feet (1,000–5,000 meters) and less than 3,250 feet respectively.

Perhaps the most generally destructive form of precipitation is hail. Most hailstorms last only 10 minutes and leave a damage pattern or "hailstreak" about 5–7 miles (7–10

***Shallow clouds** and those in the tropics do not reach the freezing level, so ice crystals do not form* [**A**]. *Instead, a larger-than-average cloud droplet may coalesce with several million other cloud droplets to reach raindrop size. Electrical charges may encourage coalescence if droplets have opposite charges. Some raindrops then break apart to produce other droplets in a chain reaction which produces an avalanche of raindrops.*

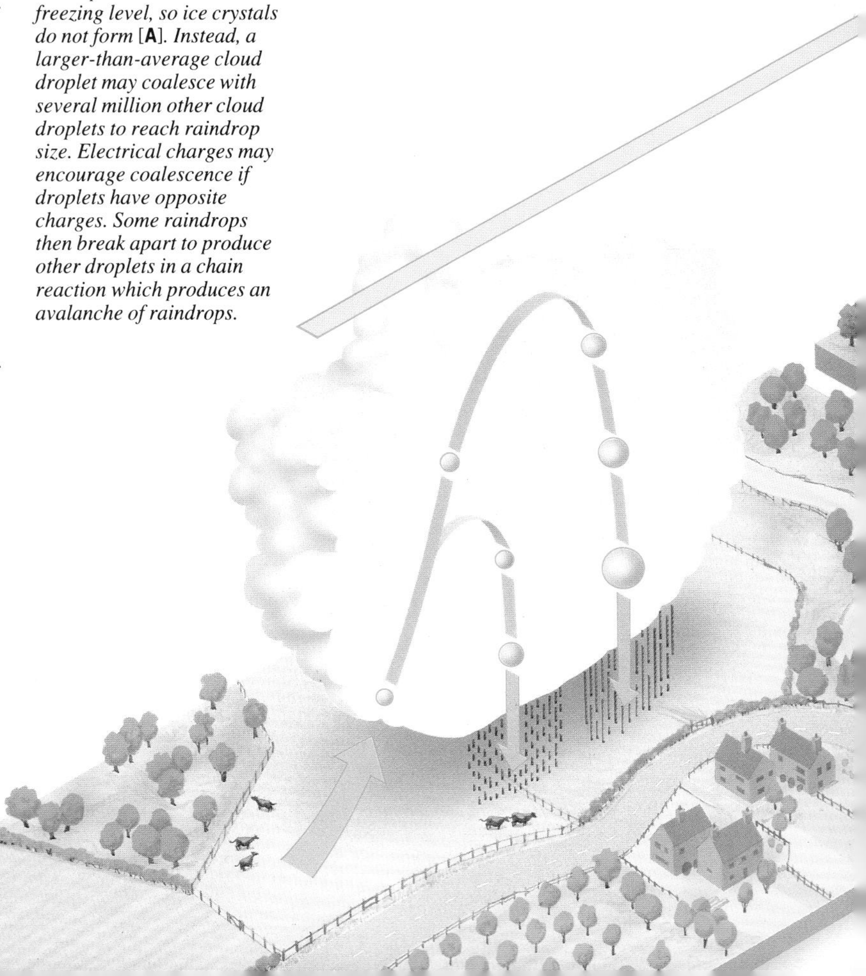

kilometers) long and about a mile (1–2 kilometers) wide. Even small hailstones can pulverize crops and smash windows.

Artificial precipitation
Modern attempts to increase rainfall rely on seeding sub-freezing clouds with substances similar to ice crystals, such as silver iodide or dry ice, which encourage ice crystals to develop. Seeding only increases precipitation by 10–20 percent, but can cause international problems because a country may be accused of "stealing" its neighbor's precipitation.

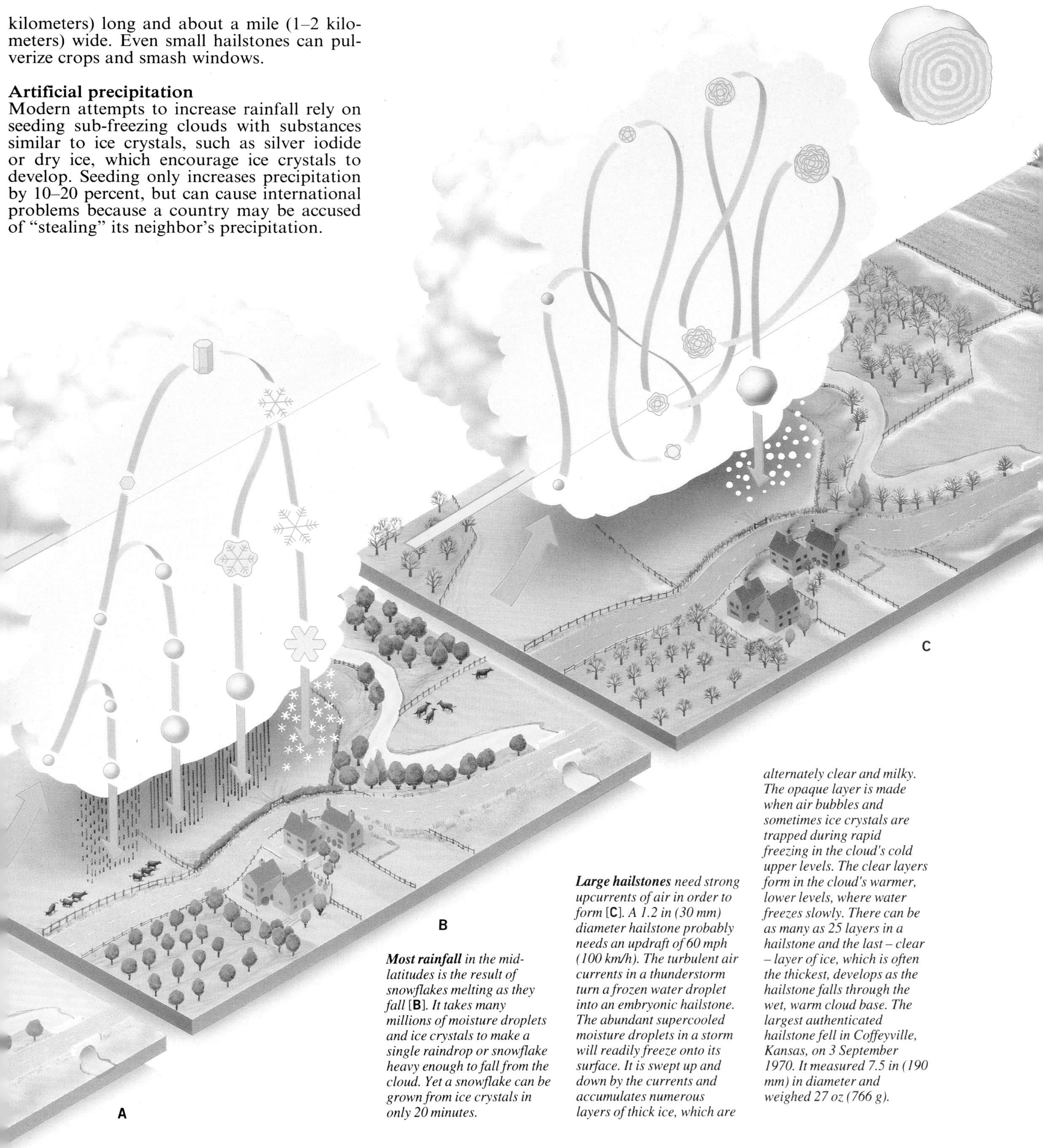

Most rainfall *in the mid-latitudes is the result of snowflakes melting as they fall [***B***]. It takes many millions of moisture droplets and ice crystals to make a single raindrop or snowflake heavy enough to fall from the cloud. Yet a snowflake can be grown from ice crystals in only 20 minutes.*

Large hailstones *need strong upcurrents of air in order to form [***C***]. A 1.2 in (30 mm) diameter hailstone probably needs an updraft of 60 mph (100 km/h). The turbulent air currents in a thunderstorm turn a frozen water droplet into an embryonic hailstone. The abundant supercooled moisture droplets in a storm will readily freeze onto its surface. It is swept up and down by the currents and accumulates numerous layers of thick ice, which are alternately clear and milky. The opaque layer is made when air bubbles and sometimes ice crystals are trapped during rapid freezing in the cloud's cold upper levels. The clear layers form in the cloud's warmer, lower levels, where water freezes slowly. There can be as many as 25 layers in a hailstone and the last – clear – layer of ice, which is often the thickest, develops as the hailstone falls through the wet, warm cloud base. The largest authenticated hailstone fell in Coffeyville, Kansas, on 3 September 1970. It measured 7.5 in (190 mm) in diameter and weighed 27 oz (766 g).*

Connections: **Atmospheric Phenomena** 66 **Clouds** 74 **Cycles of Life** 206 **Lightning and Thunder** 78 **Winds and Weather** 72

Sound and Fury

What causes lightning and thunder

The power of a lightning strike is awe inspiring: 10,000–40,000 amps of current surge to Earth down a narrow air channel which reaches temperatures of 54,000°F (30,000°C) – over five times the temperature at the surface of the Sun. But lightning doesn't just strike downward. A lightning flash consists of alternating upward and downwards strokes. Each stroke may be over 20 miles (30 kilometers) long under certain conditions. Although 70–80 percent of the people who are struck survive, around 100 people are killed directly by forked lightning every year in the United States alone.

Lightning is a massive electrical discharge which neutralizes the charges that build up in a thunder cloud. The discharge either occurs within the cloud as sheet lightning, or between the cloud and the ground as forked lightning. The flash is the same in both cases, and sheet lightning only appears to be different because of the intervening cloud. At any time there are 1,500–2,000 thunderstorms active on our planet, triggering some 6,000 lightning flashes a minute, the majority of which are sheet lightning. Nearly 250 years after Benjamin Franklin, the American scientist and statesman, proved that lightning was a form of electricity, scientists still lack a complete understanding of how it works.

To initiate the avalanche of electrons that marks the start of a lightning stroke a build-up of electrical charge (potential difference) of about a million volts/yard is needed in the thunder cloud. Incredibly, the thunderstorm generates this voltage potential in only half an hour, as a result of strong adjacent rising and falling air currents.

Large falling raindrops, hailstones and ice pellets frequently touch and collide with smaller water droplets and ice crystals being swept aloft, and electrical charge is transferred. The falling stream gains a negative charge, while the rising stream gains a positive charge, resulting in an accumulation of negative charges at the bottom of the cloud and positive charges at the top.

When the lightning strikes it travels along an air channel 0.5–2 inches (1–5 centimeters) in diameter. The evidence for this comes from lightning strikes on sand, which fuses into a hollow tube of silica glass known as *fulgurite*. When lightning strikes a moist surface such as a tree, highway, or wall, the heating effect is so great that the moisture boils explosively and may blow the object to pieces.

The dangers of lightning

The downward-flowing pathway of electrons, or stepped leader, of lightning always seeks out the best conducting route to the ground. Tall buildings, television aerials and trees are struck most frequently, but people outdoors also face a serious risk, and the danger is increased if they are holding a metal object such as a golf club or umbrella. One of the safest places to be if lightning strikes is inside a vehicle, since the current flows safely

Once an electrical potential *of 300,000 volts/ft (1 million volts/m) has been created in a thunder cloud, the lightning process begins* **[A]**. *A stream of electrons flows down, colliding with air molecules and freeing more electrons and in the process giving the air molecules a positive charge (ionizing the air). This intermittent low-current discharge forms a highly branched pathway and is called the stepped leader* [1]. *As the leading branches of the stepped leader, carrying large negative charges, near the ground, they induce short upward streamers of positive electrical charges from good conducting points on the ground* [2]. *When a branch of the stepped leader contacts an upward positive streamer, a complete channel of ionized air has been created. This allows a huge positive current called the return stroke to flow upward into the cloud in the form of a bright lightning stroke* [3]. *The different strokes have been given different colors to distinguish between them. In nature, all lightning is colorless. The return stroke causes the first of the shock waves we hear as thunder* **[B]**. *The flash effectively reaches the eye instantaneously, whereas the sound of thunder travels at approximately 360 yd/s (330 m/s). So multiply the number of seconds between the flash and the thunder by 360 to find the distance in yards.*

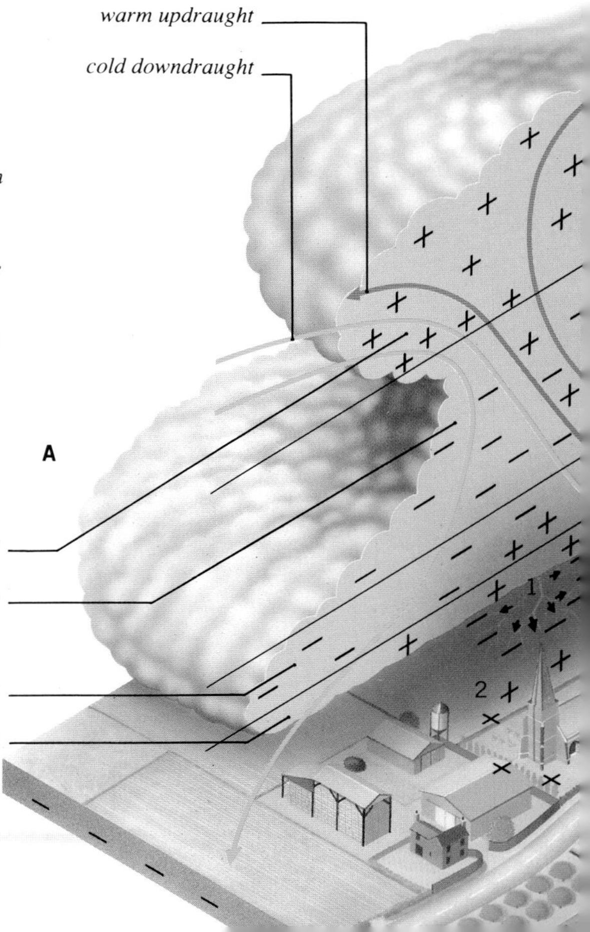

Connections: **Amosphere** 64 **Atmospheric Phenomena** 66 **Clouds** 74

through the metal of the bodywork, around the occupants, before earthing to the ground across the wet, often steel-wired tires. Taking shelter beneath a tall, isolated tree is making a mistake: the lightning is initially attracted to the tree because of its height, but may side-flash through or over the body of anyone underneath, because the human body offers a better path. About a quarter of all lightning victims are struck while misguidedly standing under a tree.

There are sometimes signs that lightning is about to strike. Positive electrical charges streaming upward from pointed objects such as trees and church spires may be visible as faint, luminous glows and may produce buzzing noises. People's hair may stand on end because of the same effect, and if this happens, it is probably a good idea to leave the area immediately.

***A fraction** of a millisecond* [**A**] *after the return stroke* [3], *a negatively charged dart leader passes down the ionized channel* [4] *and triggers another upward return stroke. The process is repeated several times within fractions of a second until the charge in the cloud is completely neutralized.*

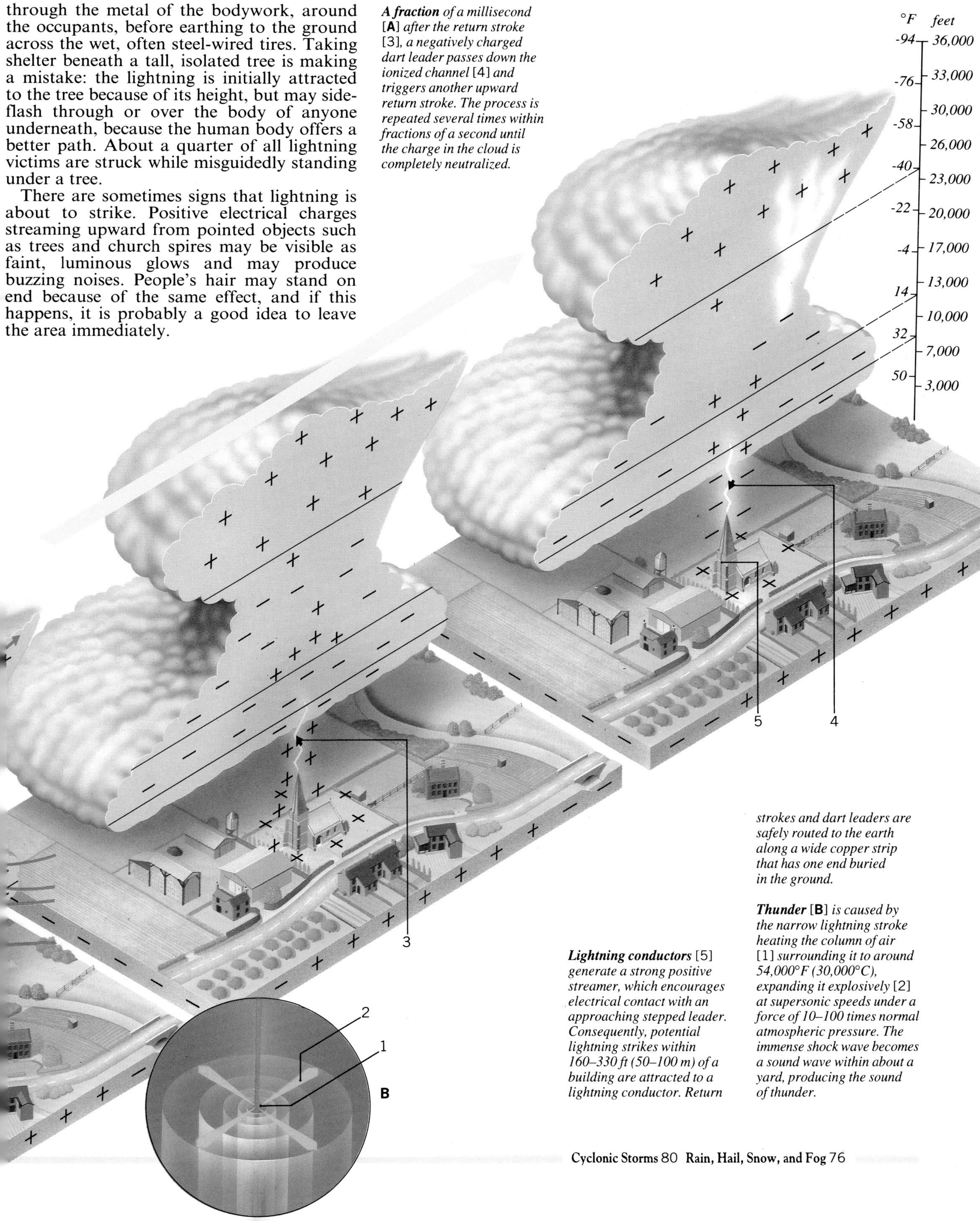

Lightning conductors [5] *generate a strong positive streamer, which encourages electrical contact with an approaching stepped leader. Consequently, potential lightning strikes within 160–330 ft (50–100 m) of a building are attracted to a lightning conductor. Return strokes and dart leaders are safely routed to the earth along a wide copper strip that has one end buried in the ground.*

Thunder [**B**] *is caused by the narrow lightning stroke heating the column of air* [1] *surrounding it to around 54,000°F (30,000°C), expanding it explosively* [2] *at supersonic speeds under a force of 10–100 times normal atmospheric pressure. The immense shock wave becomes a sound wave within about a yard, producing the sound of thunder.*

Cyclonic Storms 80 Rain, Hail, Snow, and Fog 76

Hurricanes and Tornadoes

How cyclonic storms form

A cyclone that struck Bangladesh in 1970 caused half a million deaths. "Cyclones" and "hurricanes" are, in fact, exactly the same phenomena – the most violent and destructive storms on Earth. Although tornadoes are less than one-hundredth the size of hurricanes their winds can reach similarly devastating speeds. Two-thirds of the world's tornadoes last less than 3 minutes, however, and under 2 percent are true killer tornadoes, creating swathes of destruction hundreds of yards wide and over 100 miles (160 kilometers) long.

***Most tornadoes** have wind speeds of about 110 mph (180 km/h), which can damage roofs, uproot trees and fill the air with lethal debris. Some tornadoes exhibit wind speeds up to 290 mph (470 km/h) and can crush sturdy buildings and toss vehicles, people and animals through the air.*

The storms that are collectively known to science as tropical cyclones are given many different popular names in different areas of the world. In the Atlantic and eastern Pacific Oceans, for example, they are called hurricanes. In the western Pacific they are known as typhoons. Around the Philippines they are called baguios. And in the Indian Ocean they are cyclones. Only when wind speeds in a tropical storm exceed 36 yards/second (75 mph, 120 km/h) should it be called any of these names.

Hurricane force

Hurricanes develop over ocean water that is at or above a temperature of 80°F (27°C). A hurricane is composed of bands of thunderstorms and cumulus that spiral around the storm center, the calm, cloud-free eye, which is typically 20–25 miles (30–40 kilometers) across. In the cloud wall around the eye exceptional wind speeds occur: the smaller the eye, the higher the wind speeds. Hurricanes vary greatly in their intensity, and in meteorological terms are classified according to their "damage potential" on a five-point scale, ranging from minimal (1) to catastrophic (5). Such measurement is based on air pressure in a storm's eye, because the lower the pressure, the greater the wind speeds in the hurricane. A hurricane that measures 5 on the scale is characterized by air pressure below 920 millibars, producing wind speeds of over 155 miles an hour (250 km/h) and a coastal storm surge in the sea of over 18 feet (5.5 meters) above the normal level. Fortunately, fewer than 1 percent of all hurricanes fall within this category.

One hurricane that did, however, was Hurricane Gilbert, which swept through the Gulf of Mexico in September 1988. As this hurricane, 950 miles (1,500 kilometers) in diameter, passed over Jamaica, it generated as much energy as the country would need for the next 1,000 years, at current rates of consumption. Although Hurricane Gilbert traveled at a mere 10–15 miles an hour (18–25 km/h), the wind around the eye at the center, where air pressure was a record low of 885 millibars, reached speeds of over 200 miles an hour (320 km/h).

A tidal surge of 20 feet, flash floods from torrential rainfall that totalled 10–15 inches (250–380 millimeters) in a few hours, and the creation of 24 tornadoes as the storm reached land, all together contributed to the devastation caused: 318 people lost their lives, 100,000 people were evacuated from the Mexican coast, and 500,000 people were rendered homeless in Jamaica.

Forecasting the power and path of a hurricane remains problematical, although satellite monitoring since the 1970s has allowed far more of a warning to be given to those in the path than was once the case.

Taming a hurricane

When a hurricane travels across the ocean and finally reaches the coast, it begins to die as its supply of energy from the warmer ocean is cut off, and as the increased surface friction upsets the storm's circulation and reduces the wind speeds. Even the terrible Hurricane Gilbert was eventually laid low.

An American attempt to weaken hurricane winds by seeding a storm with silver iodide or dry ice some distance outside the eye-wall clouds – an enterprise called Project Stormfury – met with limited success. The intention was to produce rainfall, so releasing latent heat that would otherwise have sustained the high wind speeds in the eye wall. The decrease in wind speeds should in theory have caused the hurricane simply to fizzle out. Instead, a new eye wall was created farther out, with lower wind speeds.

***Tornadoes* [A]** *are mainly found in the mid-latitudes. They are produced by thunderstorms, such as the cyclones of Asia and Japan. The USA suffers many tornadoes, especially at its center, where humid air streaming north from the Gulf of Mexico meets the cold, dry air moving westward from the Rocky Mountains. The basic reason for a tornado's spin is the Coriolis force caused by the Earth's rotation. The thermal* [1] *does not begin to spin at ground level but within the cloud. The rotation gradually extends to the surface as a funnel* [2] *stretching from the base of the parent cloud.*

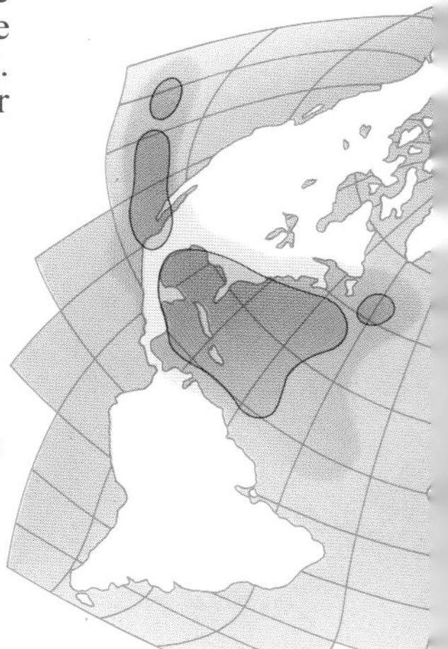

***Tropical cyclones* [B]** *are powered by the latent heat released when vast quantities of water – evaporated from the oceans – condense to form towering clouds. A seasonal variation in cyclone formation occurs because high sea-surface temperatures are needed to encourage strong evaporation. Tropical cyclones do not form close to the equator because the Coriolis force (due to the planet's rotation) is not powerful enough to cause the air to spiral wildly. The light shaded areas suffer on average 0.1–1 storm of force 8 or stronger/year. The darker, ringed, areas have over 1 such storm/year.*

Connections: **Atmosphere** 64 **Earthquakes** 20 **Ocean Circulations** 46 **Weather and Climate** 70 **Tides** 48 **Winds and Weather** 72

A

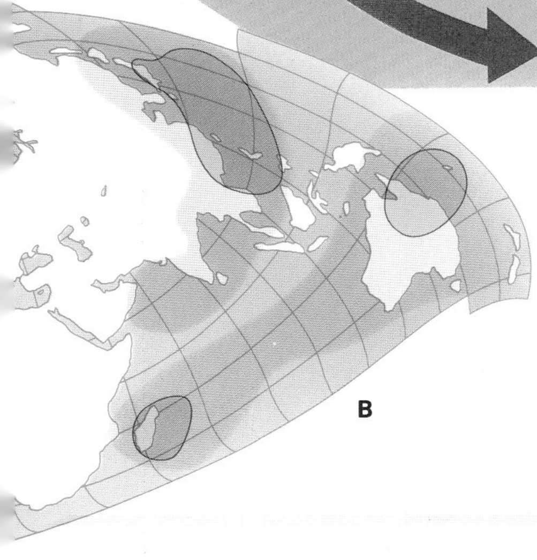

B

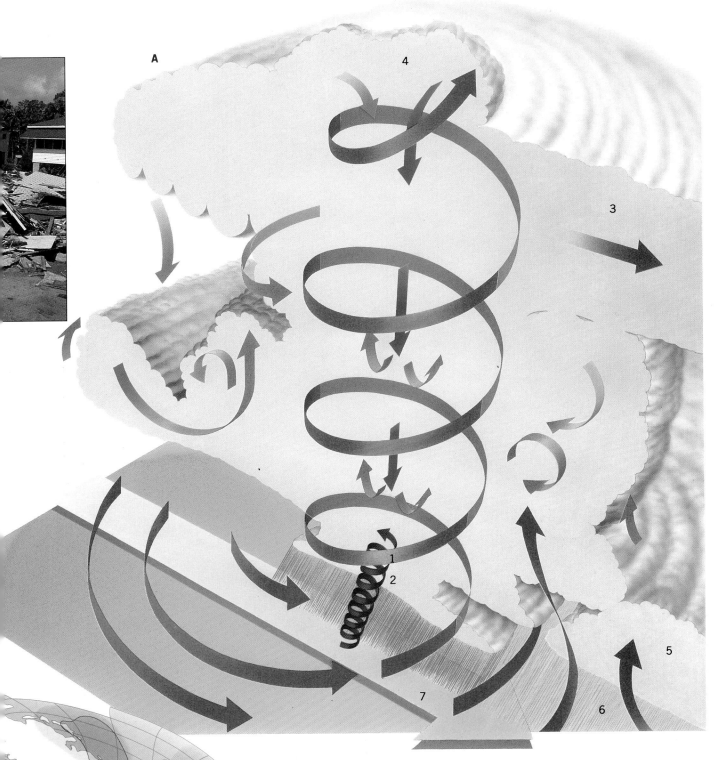

The flat top *of a tornado's parent cloud* [**A**] *is at the equilibrium level* [3]*, where the temperature within the cloud is 32°F. But the updraft of the tornado is so strong that rising air overshoots this level* [4] *before falling back into the cloud. The shelf cloud* [5] *forms where a cool downdraft caused by rain* [6] *undercuts warm air ahead of the path* [7] *of the storm. The highest wind speeds in a tornado are in a tight spiral at its center. However, powerful tornadoes can generate miniature air spirals around the edge of the main twister, with even greater wind speeds. These mini-vortices account for some freak damage, such as where one half of a house is totally wrecked while the other half is left with only minor damage. Water spouts are weaker tornadoes that form over seas or lakes from cumulonimbus or towering cumulus. The strongest have 100 mph (160 km/h) winds.*

The Changing Climate

What is causing Earth's climate to change

During the last ice age, parts of what is now the Amazon jungle were verging on desert. From year to year the Earth's climate varies, gradually bringing about many dramatic changes. The variations are governed by how much energy reaches the planet from the Sun, how much is absorbed by the atmosphere and the Earth's surface, and how much is radiated or reflected back into space. The climatic changes that follow these variations occur naturally, but human factors are playing an increasing and unpredictable role in influencing the world's weather patterns.

Many factors influence the global climate. Most important among these are variations in the Sun's energy output, the distance between the Sun and Earth, and what happens to this solar energy when it reaches our planet. A great deal of it is simply radiated back into space, and we rely on this to keep the planet cool. But even the energy that reaches the Earth's surface varies in intensity at different times, and there are additional terrestrial factors that affect the climate. These include the movement of the continents, the direction and speed of ocean currents, variations in the composition of the atmosphere, cloud coverage and volcanic activity, among others. During the 20th century, human activities have played an increasingly significant role in changing the climate, and are a major cause for concern.

Changes in the past

During the past two million years, the global climate has undergone dramatic changes. A sequence of alternating cold glacial and warm interglacial periods, triggered by slight orbital changes of the planet, have influenced how much of the Sun's energy reaches the Earth's surface. Each glacial period lasted tens of thousands of years, during which ice sheets – some as thick as 6,500 feet (2,000 meters) – submerged many regions of the high and middle latitudes of the Northern Hemisphere. The last glacial period ended about 10,000 years ago, since when global temperature has increased by 11–18°F (6–10°C) in the relative warmth of our present interglacial. Following the introduction of agriculture about 8,000 years ago, the reflective brightness (*albedo*) of the planet's surface has increased by about 10 percent – which has offset the recovery from the last glacial period by about 2°F (1°C).

The major climatic change in recent centuries was the occurrence of the relatively cold period of the "Little Ice Age" (AD 1430-1850), when European temperatures were lowered by about 2–4°F (1–2°C). This cooling was probably caused by a slight reduction in the Sun's activity. Despite its name, the Little Ice Age was not a true glacial period but rather a time of severe winters and violent storms.

Since the 19th century, global temperature has risen by about 1°F (0.5°C) in response to increasing concentrations of "greenhouse"

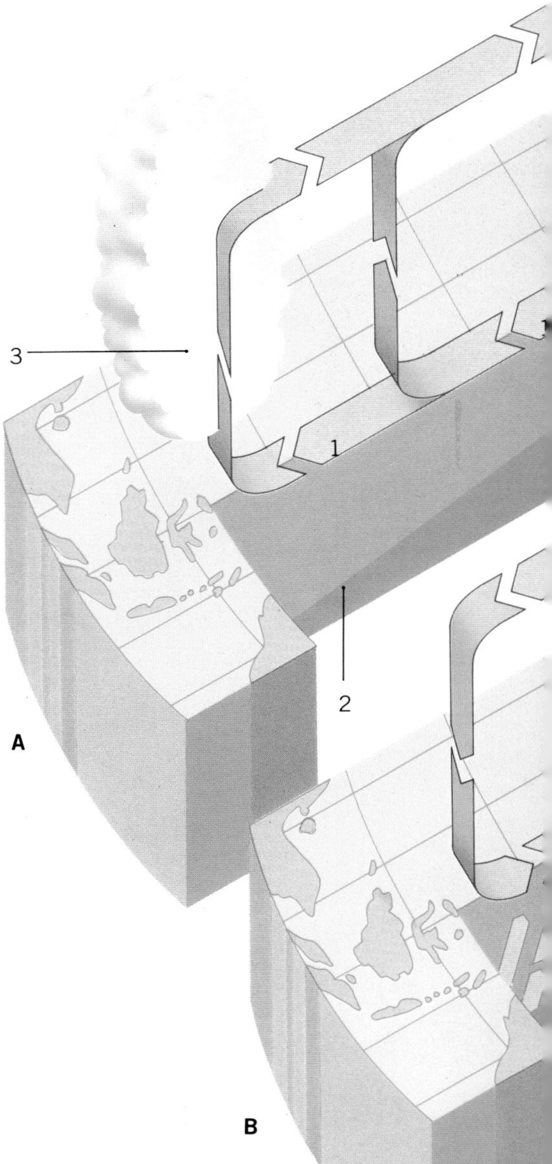

***The combined influence** of land, sea and air on weather conditions can create a global climate rhythm. In the Pacific Ocean, for example* **[A]**, *trade winds normally blow from east to west* [1] *along the equator, "dragging" sun-warmed surface waters into a pool off northern Australia and thereby depressing the* thermocline – *the boundary between warm surface waters and the cooler layers beneath* [2]. *High cumulus clouds form above these warm waters, bringing rain in the summer wet season* [3]. *Cooler, nutrient-rich waters rise to the surface off South America* [4], *supporting extensive shoals of anchovies, on which a vast fishing industry has developed. The weather over this cold-water region is dry.*

Every 3–5 years a change occurs in the ocean–atmosphere interaction. The climatic pattern is reversed **[B]** *– an event known as El Niño (the Spanish for "boy," as it begins around Christmas).*

Patchwork Climate

By modifying prevailing regional climates, surface features such as fields, lakes, orchards and urban areas create their own distinctive climates within the lower layers of the atmosphere. In an urban area, for example, there are many artificial heat sources such as factories, cars, furnaces and air conditioners, and the city's canyonlike terrain of tall buildings traps solar energy more efficiently than open country. As a result, the city center [1] may become several degrees warmer than the outskirts [2]. Warm air in the center is forced to rise by cooler, denser air flowing in from the country [3]. Under certain conditions the rising columns of air [4] may concentrate airborne particles into a dust dome over the city. If the air is driven up high enough [5], this effect may also cause local rainfall.

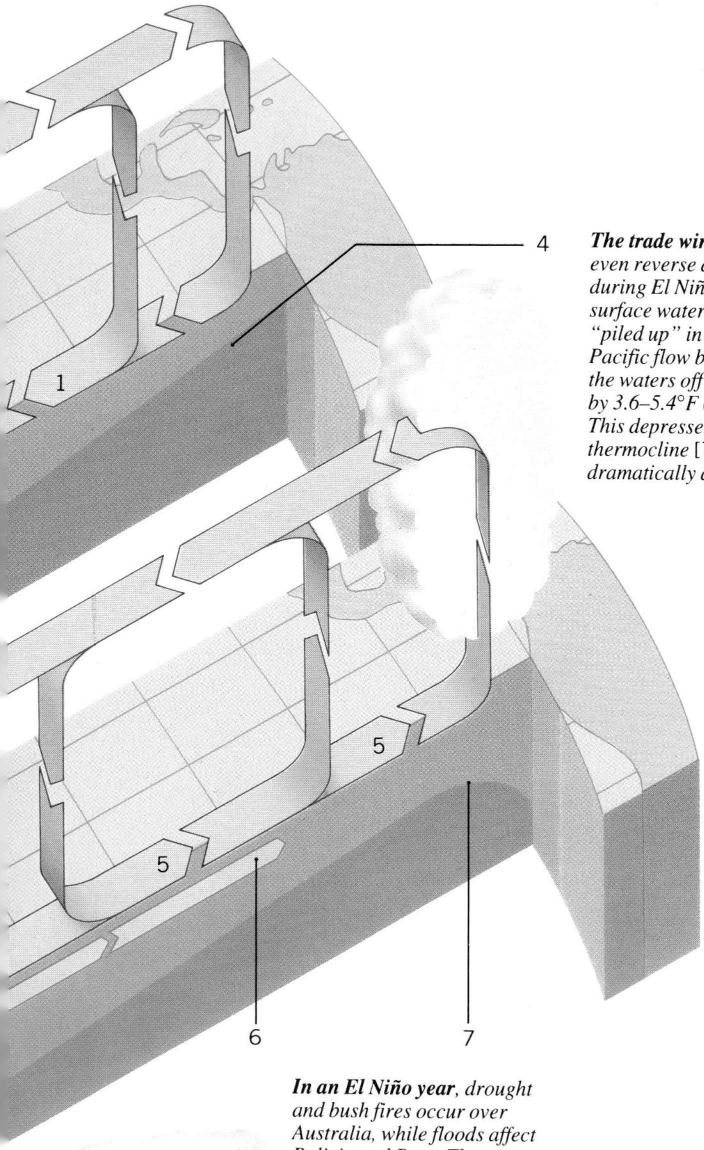

The trade winds *ease, or even reverse direction* [5], *during El Niño and the warm surface waters that have "piled up" in the west Pacific flow back to warm the waters off South America by 3.6–5.4°F (2–3°C)* [6]. *This depresses the eastern thermocline* [7] *and dramatically affects climate.*

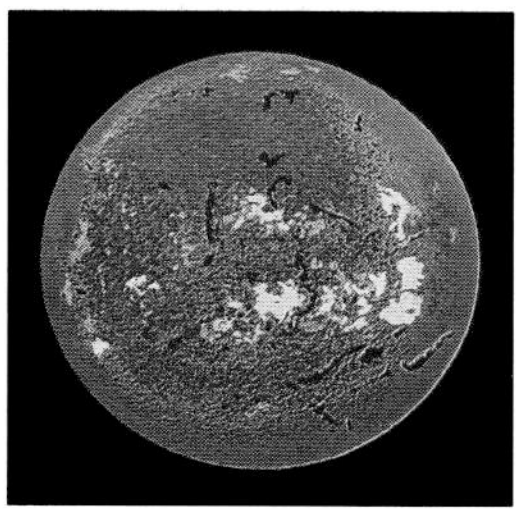

The Sun's activity *varies over an 11-year cycle, with high activity marked by an increase in the occurrence of sunspots. At such times, a greater number of charged particles are emitted by the Sun. These particles hit the Earth's upper atmosphere, influencing the strength and course of storms.*

gases in the atmosphere. These gases – notably carbon dioxide and methane – absorb some of the Earth's outgoing radiation that would otherwise be lost to space. This leads to a warming effect on the world's climate.

Changes that we may be causing

Burning fossil fuels such as coal delivers more greenhouse gases into the atmosphere. At the same time, cutting down vast areas of tropical forest reduces the amount of oxygen being recycled into the atmosphere. So, unintentionally, human activities are combining to increase an atmospheric accumulation of the heat-trapping gases. Furthermore, the destruction of the forests – leading in many cases to an expansion of urban areas or of arid, eroded deserts – also modifies the land surface and thus affects the climate by altering the amounts of the Sun's energy that are absorbed and reflected.

In an El Niño year*, drought and bush fires occur over Australia, while floods affect Bolivia and Peru. The warm waters off South America suppress upwelling of the cold, nutrient-rich waters, bringing disaster to the fishing industry. The most severe El Niño this century was in 1982–83.*

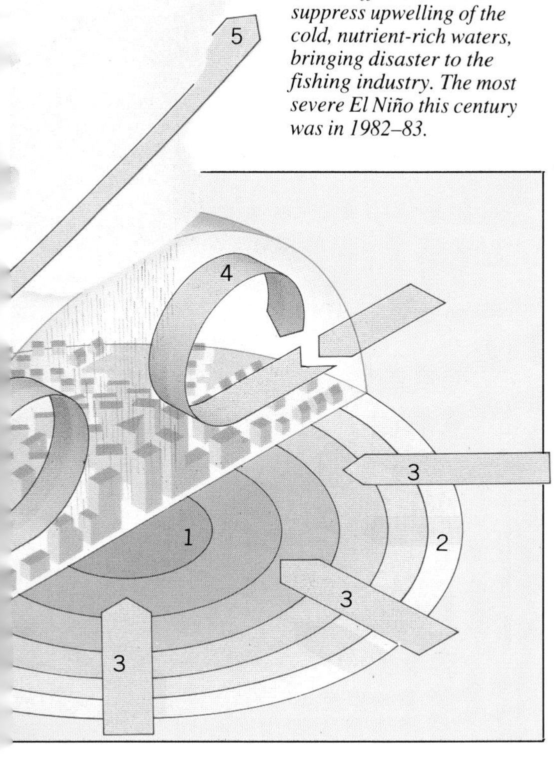

Many physical*, chemical and biological processes influence the climate system. Heat from the Sun is radiated or reflected back into the atmosphere by land, sea and ice (pink arrows). The oceans store heat energy: currents in the water can transfer this energy over great distances (purple arrows).*

Carbon dioxide is absorbed by green plants and trapped as limestone, and is emitted in respiration and when fossil fuels are burned (green arrows). Water falls as rain and evaporates into the atmosphere (blue arrows). The complexity of these interactions makes climate change hard to predict.

C

Weather and Climate 70 Winds and Weather 72

The Insulated Earth

How the greenhouse effect warms the globe

If there was no moisture or carbon dioxide in the atmosphere, the Earth's average temperature would only be 0°F (–18°C). But the presence in the atmosphere of small quantities of the above, and several other gases, helps to trap the Sun's heat. This natural insulation is vital to our survival. But the gases involved seem to have increased during the last century, trapping more heat and warming up the atmosphere. The increase may be due to human activities, and the gases may well double in concentration in the coming decades, with dire consequences.

According to the worst predictions, the global temperature could rise by 4–6°F (2–3°C) over the next couple of decades. If this happens, the world will be warmer than at any time during the past two million years. But no one has been able to say definitively whether the global warming that many scientists claim to be threatening the world really does represent an alteration in the world's climate. There is a possibility that the unusual weather patterns of the 1980s and 1990s ought to be regarded as no more than normal, cyclical, climatic variation.

Insulation materials

The dominant greenhouse gases are water vapor and carbon dioxide, although the Earth's atmosphere has a carbon dioxide content of a mere 0.03 percent. What is currently causing concern is that human activities seem to be increasing concentrations not only of carbon dioxide but also of other greenhouse gases too, including methane, nitrous oxide, chlorofluorocarbons (CFCs) and low-level ozone. There are many different estimates of how much these gases will increase by the mid-21st century, but it seems likely that, without legislation, atmospheric nitrous oxide (from fertilizers, fossil-fuel burning and land use) will increase by 50 percent and atmospheric methane will more than double. Carbon dioxide, the dominant greenhouse

Greenhouse gases *[**A**] are fairly transparent to the near-infrared, visible and shorter wavelength light that brings most of the Sun's energy [1], though about 25 percent is reflected by the atmosphere [2], and 25 percent is absorbed by it [3]. About 5 percent is reflected from the Earth [4], which absorbs the rest [5]. Some of this absorbed energy rises again in thermals [6] or in the heat of evaporated moisture [7]. The rest is reradiated [8] as long-wavelength infrared rays.*

Natural vegetation zones *are closely related to climate and research in Europe suggests that the major vegetation zones may shift northward by as much as 700 miles (1,100 km). Deciduous forests will be replaced by temperate evergreen forests. The temperature changes would differ greatly in different parts of the world. The diagram [**C**] shows computer predictions for mean regional temperature increases by the year 2050.*

A

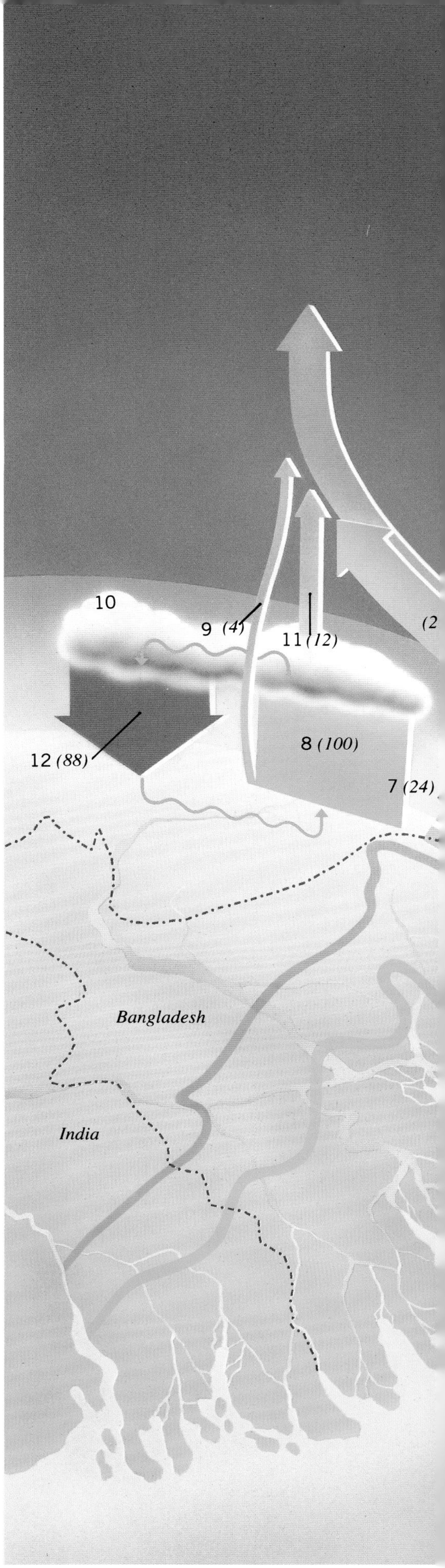

C

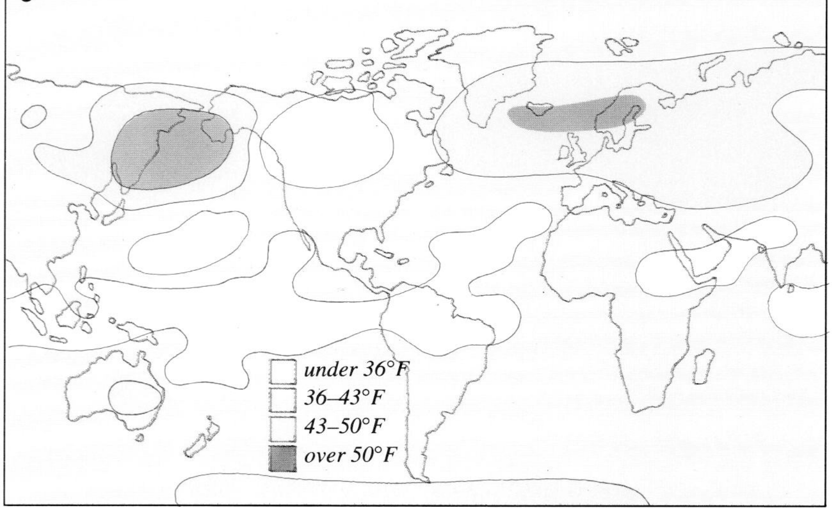

Connections: Air Pollution 342 Atmosphere 64 Climatic Change 82 Cycles of Life 206 Deserts 54 Ice Ages 38 Ozone Hole 68 Polar Regions 42

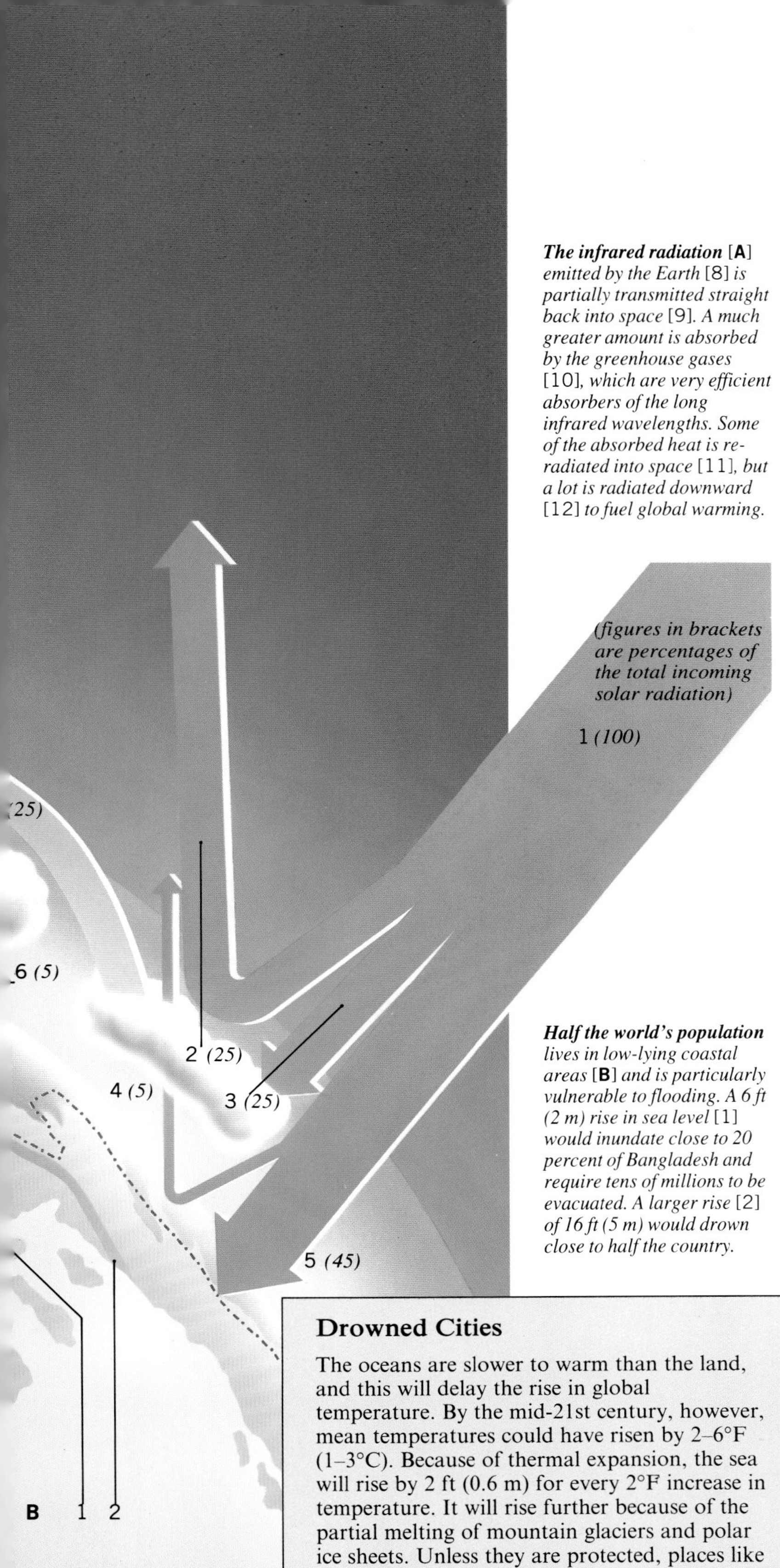

The infrared radiation **[A]** *emitted by the Earth* [8] *is partially transmitted straight back into space* [9]. *A much greater amount is absorbed by the greenhouse gases* [10], *which are very efficient absorbers of the long infrared wavelengths. Some of the absorbed heat is re-radiated into space* [11], *but a lot is radiated downward* [12] *to fuel global warming.*

Half the world's population *lives in low-lying coastal areas* **[B]** *and is particularly vulnerable to flooding. A 6 ft (2 m) rise in sea level* [1] *would inundate close to 20 percent of Bangladesh and require tens of millions to be evacuated. A larger rise* [2] *of 16 ft (5 m) would drown close to half the country.*

gas, has been increasing steadily during the past century as the burning of fossil fuels releases over 40,000 tonnes of carbon dioxide every minute. In addition, tropical deforestation releases the carbon stored in the trees. Each hectare of tropical forest burned releases up to 700 tons of carbon dioxide, and as much as 14 percent of the Amazon basin has been destroyed in Brazil and Colombia.

By drilling into polar ice, which has remained in place for thousands of years, and extracting sample cores, it is possible to measure the concentration of carbon dioxide in the atmosphere many years into the past. For the last 100,000 years that concentration was 180–280 parts/million. Today it is 350 parts/million and by 2030 it may have reached 560 parts/million. This would eventually warm the planet by 4–9°F (2–5°C).

Different effects in different areas

If world temperature is genuinely rising, higher latitudes are expected to warm up more than lower latitudes, and the effects are likely to be greater in winter than summer. The warming of the tropical oceans would increase the geographical spread, frequency and intensity of both hurricanes and droughts. The tracks followed by rain-bearing storms will alter and crop yields in the United States, central Europe and the Ukraine are expected to fall dramatically, causing worldwide shortages of food. Warmer conditions may expand crop production in Canada, Siberia and Scandinavia, but they have poorer soils and yields will be low.

Warmer conditions in the Arctic Ocean may mean the pack ice becomes a seasonal feature, melting in the summer and re-forming during the winter. Some Arctic ice sheets have melted by 15 percent since 1980, and even the vast west Antarctic ice sheet may suffer substantial melting.

Drowned Cities

The oceans are slower to warm than the land, and this will delay the rise in global temperature. By the mid-21st century, however, mean temperatures could have risen by 2–6°F (1–3°C). Because of thermal expansion, the sea will rise by 2 ft (0.6 m) for every 2°F increase in temperature. It will rise further because of the partial melting of mountain glaciers and polar ice sheets. Unless they are protected, places like Amsterdam, Bombay, Hong Kong, Los Angeles, New York, Tokyo and Sydney could disappear under a 10 ft (3 m) rise in sea level by the year 2100. Existing coastal defenses will have to be raised and reinforced at a cost – for example – of $3–9 billion in the Netherlands and $6–7 billion in Britain. Poorer nations cannot afford such sums and will be dependent on international assistance.

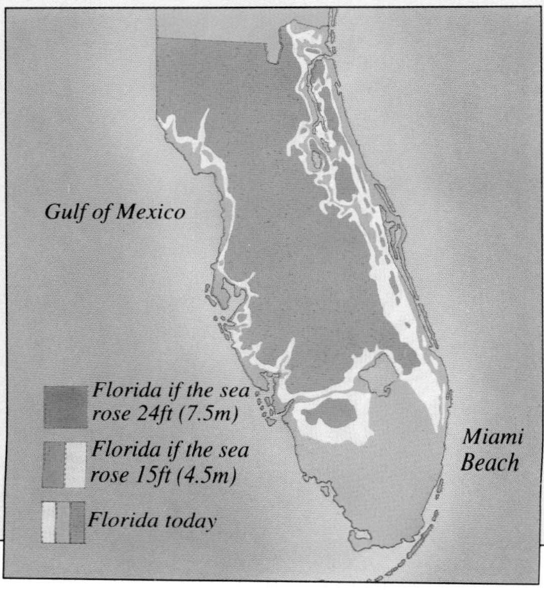

The Rhythm of Life

How seasons change

Parts of Siberia can vary in temperature from –108°F (–78°C) in winter to over 97°F (36°C) in summer. This is the world's most extreme temperature variation, and the farther away from the equator and the closer to the poles, the more life is influenced by seasonal change. Plants reproduce on a seasonal cycle. Some animals change fur color in winter, others migrate to escape winter's ravages. Many Arctic and temperate animals, however, reduce all bodily functions to the absolute minimum to sustain life, when, to the casual observer, a hibernating animal appears dead.

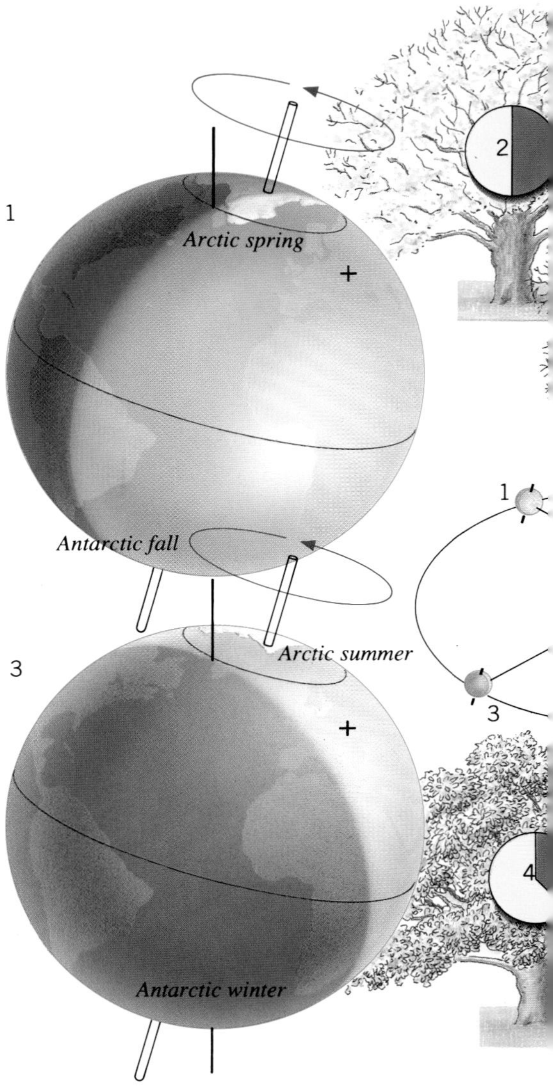

We have seasons mainly because the Earth's rotational axis tilts 23.5° from the perpendicular to its solar orbital plane. Since the tilt is constant as the Earth orbits the Sun, the angle at which the Sun's rays strike various latitudes changes through the year. Summer daylight hours are longer and there is a stronger warming effect, for the Sun's rays strike the atmosphere less obliquely than in winter, when much of their heat dissipates.

Polar zones experience the greatest annual variation in attitude relative to the Sun, and thus the greatest seasonal change, from long, dark and cold winters to short, lukewarm summers. The contrast between seasons gradually becomes less marked closer to the equator, where tropical zones, which experience minimal change in attitude to the Sun, have little annual temperature change, although they display great seasonal rainfall variation. Between the equatorial and polar extremes, mid-latitude temperate zones exhibit four well-defined seasons.

Sleeping the winter away

Extreme seasonal variation demands greater animal adaptability to cope with the changes. Many animals simply migrate to avoid cold and lack of food. Small mammals use insulating snow-trapped air to survive; snow-covered soil may be 18°F (10°C) warmer than air temperatures; having eaten as much as they can in times of plenty, mice or voles can find enough food beneath the snow to sustain them through the long winter.

Each of the four seasons *begins and ends at one of the two* solstices *(the shortest and longest days of the year), which are separated by six months, or one of the* equinoxes *(when day and night are equal), also separated by six months. The solstices mark the days when one pole is at its closest position to the Sun and the other is farthest away from it. Equinoxes mark the days when each pole is the same distance from the Sun. For example, spring in the Northern Hemisphere at the spot marked by a cross on the globe commences on the equinox of 20 March* [1], *when day and night are of equal length* [2], *and is followed by the summer solstice on 20 or 21 June* [3] *(the dates vary because of leap years), when the North Pole is at its closest to the Sun and the day is the longest of the year* [4]. *The autumn equinox commences on 22 or 23 September* [5], *when day and night are once again equal in length* [6]. *Winter in the Northern Hemisphere begins on 21 or 22 December* [7], *when the day is the shortest in the year* [8]. *This seasonal sequence is reversed in the Southern Hemisphere, where, for example, winter starts on 20 or 21 June, when the South Pole is farthest away from the Sun but the North Pole is closest and the Northern Hemisphere summer begins.*

The start and finish of each season can thus be defined in very precise astronomical terms by the Earth's position relative to the Sun.

Obviously, the climate does not suddenly change on a specific astronomically determined date – we usually mark seasonal change by factors such as changes in animal behavior and leaf and flower growth or decay. In

Connections: Atmosphere 64 Biological Clocks 274 Climatic Change 82 Global Warming 84 Ice Ages 38 Migration Patterns 228 North Polar Life 310

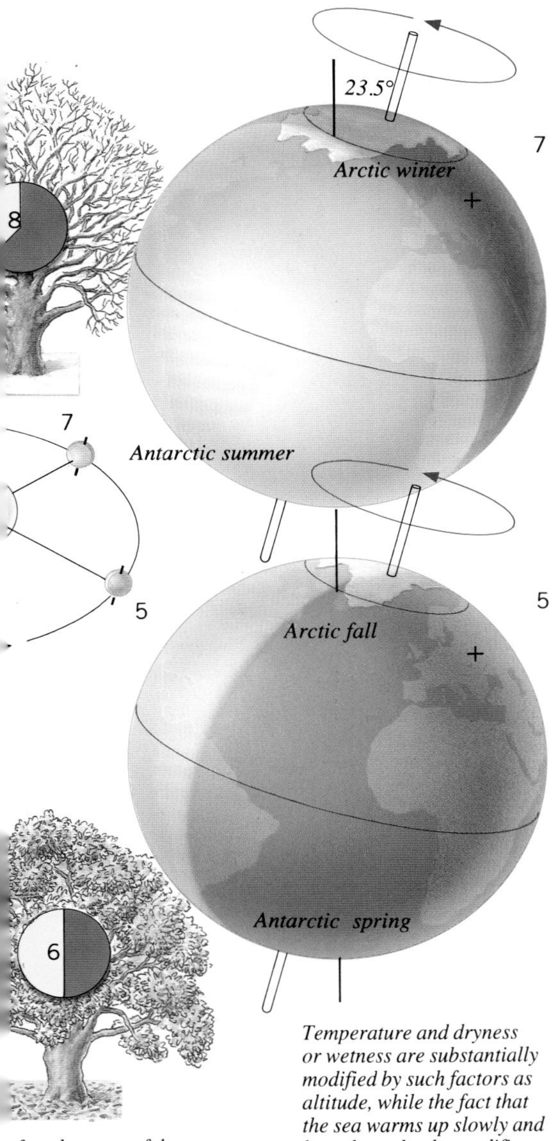

fact, because of the greenhouse effect, seasonal temperature extremes occur two months later than purely astronomical factors would suggest. Many other terrestrial factors modify astronomical effects.

Temperature and dryness or wetness are substantially modified by such factors as altitude, while the fact that the sea warms up slowly and loses heat slowly, modifies purely astronomical factors and delays the onset of warm or cold periods. Warm currents, such as the Atlantic Gulf Stream, have a general modifying effect on adjacent landmasses, resulting in far warmer climates in, for example, parts of northern Scotland, where purely latitudinal factors would suggest much colder winters than is actually the case. In general, the farther away from the sea, the more a climate is determined by purely astronomical factors.

The Earth's solar orbit is nearly circular, but its slight ellipticity means that the Sun is not exactly at the center of the orbit; thus, at present, the Earth is closest to the Sun in January, reinforcing the Antarctic summer and weakening the effect of winter in the Arctic. The effect of solar radiation on air circulation also affects the weather, producing, for example, phases of high pressure and long spells of hot weather in summer but unpredictable bouts of wind and rain in spring.

***The Earth's** tilted rotational axis causes the phenomenon of "The Land of the Midnight Sun" (below). Each polar extreme has six months of continuous day and then six months of continuous night. Inside the polar circles the effect is less extreme, the regions alternating between six-month periods of very long nights and short days and very long days and short nights. Regions inside the polar circles experience one day when the Sun never sets. This time-lapse photograph, taken inside the Arctic Circle at the Lofoten Islands, shows the summer Sun's progress in the sky at hourly intervals over a 24-hour period.*

One of the most fascinating ways of coping with extreme cold, however, is by sleeping to conserve energy. Large animals, like badgers, sleep much of the time, readily awakening if disturbed. But this winter "doze" uses considerable energy, which must be replenished by hunting. True hibernation is a state of suspended animation when a mammal slows its metabolism, lowering its temperature from perhaps 90°F (32°C) to 39°F (4°C) and its pulse and breathing rates until barely discernible; a hibernating hedgehog, for example, breathes once every six minutes. As well as minimizing energy needs, a mammal stores energy by gorging itself before winter – dormice resemble furry balls in autumn, and some sheep store food in fatty tail deposits. In fact, a herbivore's fat deposits, not the cold, may trigger hibernation; an artificially warmed animal still hibernates at the same time as wild animals.

A hibernator can be picked up without awakening, but it wakes up if it is so cold it risks freezing solid, shivering violently and even running around to raise its body temperature. Usually, however, it simply stretches occasionally in hibernation (which may help prevent cramp in its muscles), slowly beginning to stir or twitch its tail as spring nears. Fully awake, the mammal is ravenous after losing perhaps half its body weight during its sleep.

The long, dry siesta

Animals such as kangaroo rats reduce water loss by becoming nocturnal in droughts. Others avoid fatal dehydration by *estivating*, using similar techniques to hibernators to "sleep" through droughts. Male Californian ground squirrels estivate but the females stay awake to feed and tend the young. The Jersey tiger moth estivates in Greece and, having hibernated as a caterpillar, spends over half its life asleep. Some lungfish aestivate in dried-up ponds, using a mucus-lined mud cocoon to reduce skin evaporation.

South Polar Life 312 **Temperate Forests** 318 **Tundra** 314 **Weather and Climate** 70

3 Evolution and Adaptation

How Plants and Animals are Classified

	Example: Sweet Pea		Example: Tiger
Class:	Anthophyta (angiosperms)	Phylum:	Chordata (chordates)
Subclass:	Dicotyledonae (dicotyledons)	Subphylum:	Vertebrata (vertebrates)
Order:	Rosales	Class:	Mammalia (mammals)
Family	Leguminosae (leguminous plants)	Order:	Carnivora (carnivores)
Genus:	*Lathyrus* (pea)	Family:	Felidae (cats)
Species:	*Lathyrus odoratus* (sweet pea)	Genus:	*Panthera* (big cats)
		Species:	*Panthera tigris* (tiger)

Primordial Soup

How life began

Nobody knows what first caused the spark of life to ignite in the primeval oceans over four billion years ago. Even today the leap from inorganic chemicals to primitive "cells" remains a mystery. But however early life evolved, it did so under incredibly inhospitable conditions, on an Earth scoured by intense ultraviolet light, where volcanic eruptions and violent thunderstorms stirred up a choking atmosphere of hydrogen, methane, ammonia and water vapor. For this reason it took two billion years for more complex cells to develop.

The first cells *probably arose about 3.5 billion years ago, and may well have been the result of spontaneous molecular aggregations. The development of a delimiting outer membrane was a crucial step in the evolution of complex cells, because it allowed the first self-replicating molecules selective control of the environment. Proteinoid microspheres* [**A**] *are small spherical aggregates of protein that can be artificially made by heating amino acids. These spheres have certain properties of cells suggesting that aggregates like these may have been involved in that first step.*

All living things are composed of carbon-based organic molecules and, crucially, are capable of reproducing themselves. These characteristics, which typify life, first developed in simple molecular systems some 600 million years after the Earth's formation.

Some clues as to how these molecules were first formed on the early Earth come from laboratory experiments. But only a few of the simpler building blocks of life – such as the *amino acids* that make up long protein molecules – have to date been produced in such experiments. These simple organic compounds accumulated in the ancient seas and, warmed by the Sun, this "prebiotic soup" formed the larger and more complicated molecules – the nucleic acids, proteins, lipids and polysaccharides, for example – that make up living cells. Larger organic compounds formed molecular systems capable of storing information about their structure in such a

Experiments [**C**] *by Stanley Miller and Harold Urey in the 1950s imitated conditions on the primitive Earth 4,000 or more million years ago, to show they could have favored the formation of simple organic molecules which were the precursors of life. A flask of boiling water provides the apparatus with heat and water vapor* [1]. *This is mixed with hydrogen, methane, and ammonia* [2] *and then subjected to an electrical discharge* [3] *(simulating lightning). The resultant liquid is cooled to condense it* [4] *and drips back into the U-bend. The chemicals generated in this liquid are the building blocks of life, the four major small organic molecules found in all cells: amino acids, nucleotides, sugars, and fatty acids. Although these chemicals are in a very simple form, they could have come together to form larger molecules –* polymers. *The bonding together of amino acids to form proteins, for instance, involves the removal of one molecule of water, something which could have happened in the intense, dry heat. Polymers, once formed, can influence the formation of other polymers. After thousands or millions more years this may have enabled some polymers to become self-replicating molecules – the origins of life.*

C

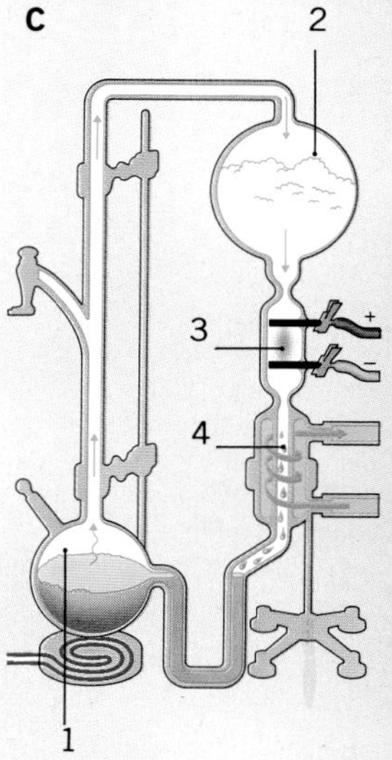

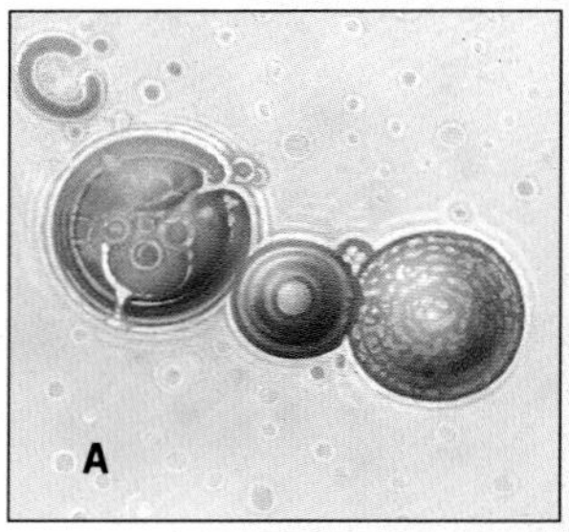

Planet Earth [**B**] *coalesced from a cosmic cloud about 4.6 billion years ago, but it took hundreds of millions of years for conditions to stabilize enough for organic molecules to accumulate. The molecular precursors for life arose in conditions that were wholly unsuitable for life itself. The atmosphere was devoid of oxygen, made up mostly of hydrogen, ammonia, methane and water vapor. This thin layer was no shield against the Sun's powerful radiation. Lightning storms, volcanic eruptions and meteorites were commonplace, but all provided energy vital to the evolution of life.*

Connections: Algae 112 Animal Cells 102 Bacteria 106 DNA 96 Early Evolution 92 Fossilization 28 Heredity 98 Plant Cells 100 Protozoa 108

way that identical systems could be reproduced. Just what such early self-propagating systems were like is unclear, because this primary stage of life has left no fossils behind in the rocks.

Simple cells

The first "cells" might have been formed when hollow spheres of self-sealing fatty membrane coalesced around groups of self-replicating molecules.

For almost two billion years simple unicellular microorganisms were the only form of life on the planet. Their remains are sometimes found in fossil stromatolites, structures laid down from successive layers of cells and trapped debris. Some of these early cells developed the ability to photosynthesize, giving out óxygen as a waste product, and in time producing an oxygen-rich atmosphere.

The next milestone in evolutionary history was the appearance of much more highly developed cells – *eukaryotic* cells – around 1.5 billion years ago. From these cells, which have a nucleus and complex internal structure, evolved the single-celled protozoa and algae, and all multicellular life.

The Cambrian explosion

The earliest traces of multicellular animals are rare imprints of soft-bodied invertebrate animals in rocks from around 600 million years ago, toward the end of the Precambrian period. They resemble jellyfish, segmented worms and sea pens. Some scientists think they may represent an entirely different form of body organization corresponding to a failed evolutionary experiment.

In general, only the hard parts of an organism – shells, scales, spicules and, later, bones – become fossilized. So the fossil record is incomplete and highly selective, incorporating only very rare traces of the many entirely soft-bodied animals and plants that must have existed. Invertebrate animals with hard parts started to appear at the beginning of the Cambrian, the period that saw an explosion of animal life in the oceans.

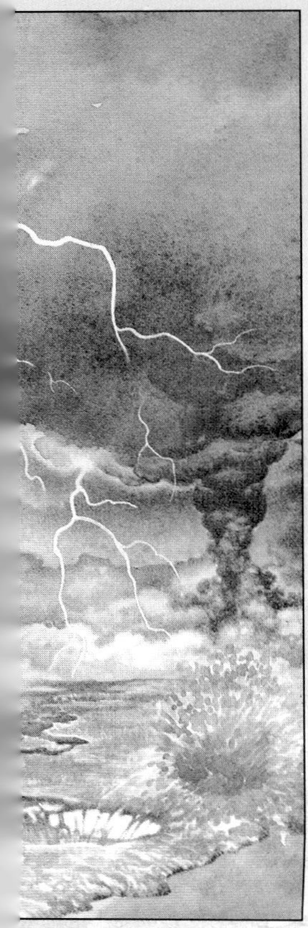

D

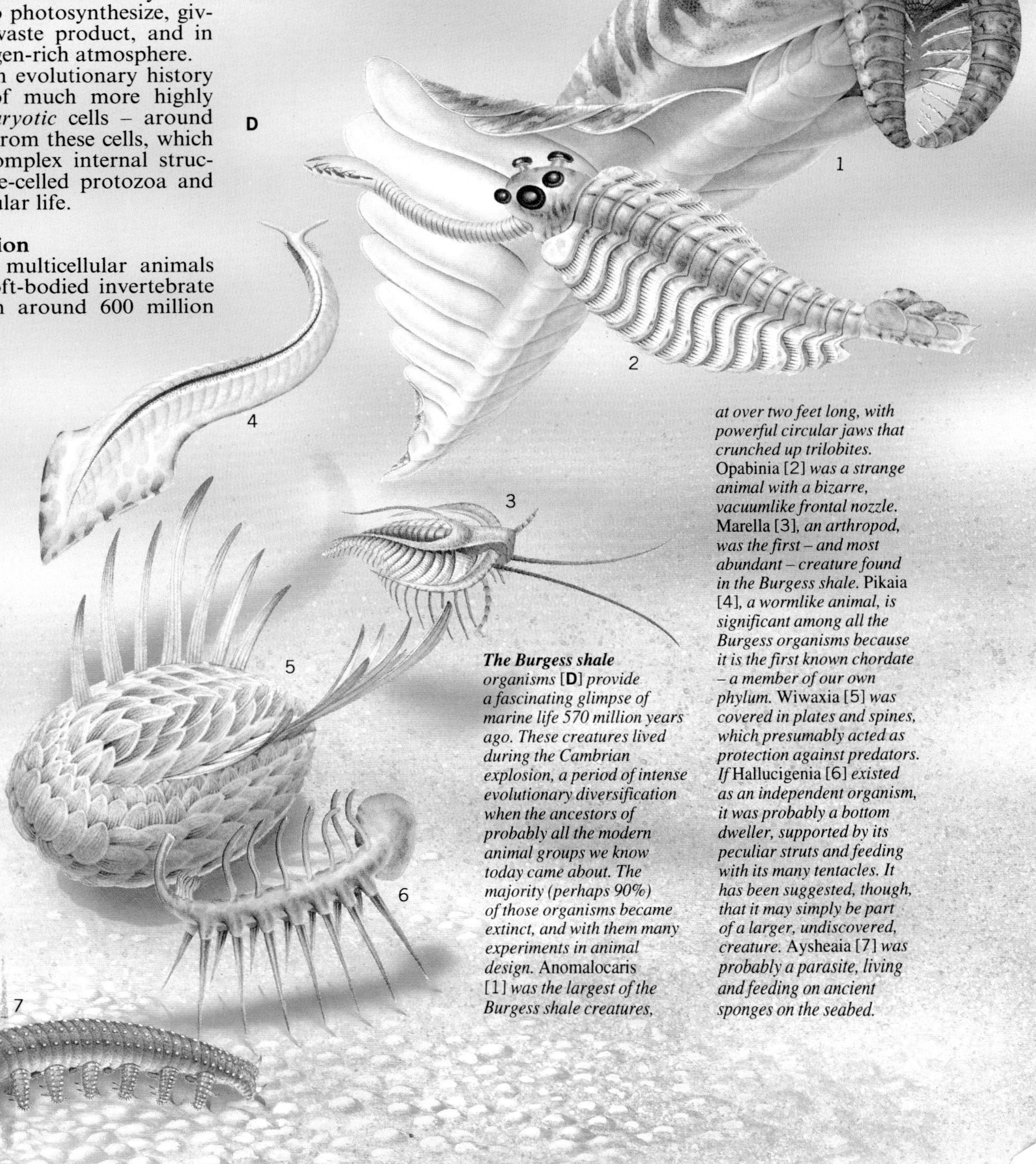

***The Burgess shale** organisms* [**D**] *provide a fascinating glimpse of marine life 570 million years ago. These creatures lived during the Cambrian explosion, a period of intense evolutionary diversification when the ancestors of probably all the modern animal groups we know today came about. The majority (perhaps 90%) of those organisms became extinct, and with them many experiments in animal design.* Anomalocaris [1] *was the largest of the Burgess shale creatures, at over two feet long, with powerful circular jaws that crunched up trilobites.* Opabinia [2] *was a strange animal with a bizarre, vacuumlike frontal nozzle.* Marella [3], *an arthropod, was the first – and most abundant – creature found in the Burgess shale.* Pikaia [4], *a wormlike animal, is significant among all the Burgess organisms because it is the first known chordate – a member of our own phylum.* Wiwaxia [5] *was covered in plates and spines, which presumably acted as protection against predators. If* Hallucigenia [6] *existed as an independent organism, it was probably a bottom dweller, supported by its peculiar struts and feeding with its many tentacles. It has been suggested, though, that it may simply be part of a larger, undiscovered, creature.* Aysheaia [7] *was probably a parasite, living and feeding on ancient sponges on the seabed.*

Rulers of the Earth

How animals evolved to live on the land

The first tiny arthropods crawled out of the water onto the land 440 million years ago, heralding the second great burst of animal diversification after the Cambrian explosion. Some 200 million years later, the mighty dinosaurs evolved, rulers of the land for 140 million years before being mysteriously wiped out. They ranged from enormous lumbering herbivores like *brachiosaurus*, up to 100 feet (30 meters) long and weighing 80 tons, to the scurrying ratlike reptiles that spawned the mammals and included familiar creatures like flying *Pterodactyls* and the fearsome *Tyrannosaurus*.

About 520 million years ago, within a mere 50 million years of the beginning of the Cambrian period, representatives of most of the main groups (phyla) of animals had appeared. This explosion of animal life, however, remained confined to water until past the end of the Cambrian, some 500 million years ago. The first forms of life on the land were mats of algae, lichens and bacteria. They managed only to colonize the edges of shallow pools. But around 400 to 500 million years ago they were followed by the first true land plants. This simple vegetation was in turn colonized by the first known wave of air-breathing land animals: tiny millipedelike arthropods, protected from drying out by their hardened outer skeletons. The first animals with backbones – the fish – evolved in the oceans of the late Cambrian. One line of fish with a bony skeleton developed air-breathing lungs and "limbs" strong enough to support them on land. They gave rise to the first four-legged vertebrates – the amphibians – from which all future vertebrates evolved.

The first amphibians to emerge from their freshwater habitat found low-lying, swampy, open forests of treelike horsetails and club mosses, liverworts and other small plants. Reptiles evolved from one of the amphibian groups and were able to make much better use of the land, filling every available habitat and ecological niche. They became adapted to many different ways of life, taking to the air as pterodactyls, and even returning to rule the water for a time, as did the plesiosaurs, ichthyosaurs and other forms.

The Mesozoic era, stretching from the end of the Permian period (250 million years ago) to the end of the Cretaceous (65 million years ago), is often called the Age of Reptiles. The earliest mammals also appeared even as the reptilian dinosaurs were rising to prominence, but they remained small and inconspicuous for millions of years.

Evolving to extinction

Throughout evolutionary history many new species have appeared, and many animals and plants have also disappeared, becoming extinct, so that the many millions of present-day species represent only a small part of all the living things that have ever existed. Extinction, however, is not always gradual. The history of life has been punctuated by several periods of mass extinctions, when large numbers of species became extinct over a very short time – in geological terms. One such mass extinction occurred at the end of the Permian period, when up to 96 percent of marine species are estimated to have become extinct. But the best known example of extinction happened at the end of the Cretaceous period, when the dinosaurs and many other species vanish from the fossil record.

The most likely explanation for the demise of the dinosaurs is that the Earth's climate went through a major change and got much cooler. It has been suggested that this was due to the impact of a gigantic meteorite that threw up clouds of dust into the atmosphere. Dinosaurs lacked the sophisticated system that mammals possess to maintain a warm body temperature, and probably could not stand the change.

__Darwin's theory__ of "natural selection" states that members of the same species can differ from each other genetically. In a particular situation some inherited feature may give its possessors an advantage in the struggle for survival, so that they have a better chance of living to reproduce and pass on the feature. Those of their descendants carrying that trait will similarly benefit and become more numerous over subsequent generations. Thus new characteristics may accumulate in a population over several generations. The acquisition of different characteristics may take place to such an extent that animals are no longer able to reproduce with each other then a new species is formed. Separate lines of descent develop from a common ancestor as the descendants grow distinct from each other and from the common ancestor. The genetic variation that makes diversification possible comes from several sources, including mutation. DNA can mutate, either naturally, by copying itself wrongly, or because of the action of radiation or chemicals. Most mutations, though, are not beneficial, and result in disease (such as cancer or hemophilia) or in freakish, frequently harmful, variations.

1 *Drepanaspis (jawless fish)*
2 *Platysomus (lobe-finned fish)*
3 *Eusthenopteron (transition to amphibian)*
4 *Icthyostega (early amphibian)*
5 *Diadectes (early amphibian)*
6 *Meganeura (prehistoric insect)*
7 *Pareiasaurus (early reptile)*
8 *Icarosaurus (gliding reptile)*
9 *Thrinaxodon (early mammal)*
10 *Archaeopteryx (gliding reptile)*
11 *Tyrannosaurus (largest carnivorous dinosaur)*

Connections: Algae 112 Amphibians 128 Atmosphere 64 Bacteria 106

By Devonian times *vertebrate fishes had evolved a number of separate groups. Extinct placoderms (plated fish), such as* Dunkleosteus, *swam alongside ray-finned fish, such as* Platysomus, *which have survived to the present day. Lobe-finned fish, for example* Eusthenopteron, *made the transition to air-breathing amphibians like* Ichthyostega.

The lush vegetation of the Carboniferous swamps supported a profusion of animal life. Amphibians exploited the new environment and some became increasingly terrestrial. One such group evolved the trick of reproducing away from water by means of tough-shelled eggs, and these became ancestral reptiles (denoted by the suffix "saurus"). Animals increased in size, with large carnivores such as Diadectes *feeding on herbivores, like* Pareiasaurus. *Insects took to the air and gave rise to air-borne predators like* Meganeura *and the gliding* Icarosaurus.

The Triassic saw the beginning of the Age of Dinosaurs, and for more than 150 million years no other animal larger than a hen walked the Earth. Other reptile groups emerged at this time, one of which evolved into warm-blooded mammals.

As dinosaurs grew larger, some, such as Kentrurosaurus, *evolved spiked protection against predators; others, including* Compsognathus, *relied on speed and developed a bipedal gait.* Archaeopteryx *was an early feathered, birdlike creature from the Jurassic capable of primitive flight. It shared the sky with the* Pterodactylus. *Around the beginning of the Cretaceous, flowering plants evolved, perhaps in response to overgrazing by dinosaurs like* Iguanodon. *By the late Cretaceous, dinosaurs had developed a wide range of efficient forms.*

Breathing 128 **DNA** 96 **Evolution to Date** 94 **Fish** 124 **Fossilization** 28 **Heredity** 98 **Mosses and Ferns** 116 **Origins of Life** 90 **Reptiles** 130

The Warm-blooded World

How birds and mammals evolved

If the dinosaurs had not disappeared so suddenly at the end of the Cretaceous period, man would not exist today. For as the dinosaurs became extinct, mammals flourished, doubtless as a result of their major evolutionary achievement – homoiothermy – the ability to maintain a constant body temperature. For the next 30 million years mammals diversified greatly and spread all over the world. In time they pursued several different "experimental" evolutionary channels, related chiefly to the manner in which the animals' young were brought to birth.

From a present-day point of view, the world and its flora and fauna would have begun to look increasingly familiar after the end of the Cretaceous period (65 million years ago). Flowering plants and trees flourished, insects had diversified into their modern forms, birds flew in the air and small mammals walked, climbed, scurried, ran and hopped over the land. During the succeeding Tertiary period (up to 2 million years ago), rain forests, temperate broad-leaved forests and, later, great grasslands provided new habitats for these novel creatures to spread into.

But mammals did not have the land entirely to themselves. Predators believed to have preyed on small mammals included giant flightless birds, such as *Diatryma* (during the early Eocene period) and the South American *Phorusrhacos* (in the Miocene).

During the Tertiary, mammals spread all over the world and evolved into many different types. Large, fleet-footed, hoofed mammals roamed grassy plains, preyed on by swift carnivores. Bats took to the air. The ancestors of dolphins and whales returned to the oceans from whence their remote ancestors had emerged several hundred million years before. The early primates took to the trees, where their precarious life-style led to the evolution of sharp, stereoscopic vision, delicate control of hands and feet, and an enlarged brain. From their descendants evolved the line leading to the great apes and to human beings.

Monotremes and marsupials

The very first mammals probably laid eggs, like their reptile ancestors and like the primitive egg-laying monotremes, which survive today – the duck-billed platypus and spiny anteater. Different species of spiny anteater are also found in New Guinea. These early mammals gave rise to the marsupials, which once lived mostly in South America, where 70 species still exist. Australia, however, supports the greatest number of marsupials. On this island continent, which became isolated by continental drift before the later placental carnivores could reach it, marsupials were able to evolve further without competition.

As mammalian orders diversified, many progressively grew larger in size. Good fossil records for horses, for example, show how they developed from relatively small animals

Mammals first appeared *during the Triassic period, but until the end of the Cretaceous they remained small and insignificant.* Zalambdalestes *is a typical mammal of the late Cretaceous. Since then mammals have diversified greatly in size and body form, with each group evolving at a separate rate. Several of the mammalian orders are shown here. Elephants, dugongs and hyraxes are related, and may share a common ancestor in some form of primitive paenungulate. The hyrax line has retained a fairly consistent body form and the modern hyrax is a recognizable descendant. Ancestral elephants and dugongs came from a different line, which perhaps first led to a kind of early tethythere at the beginning of the Oligocene. Thereafter their lines diverged as they became adapted to a different environment. The dugong branch became increasingly aquatic and during the Miocene moved entirely into the water –* Rytiodus *had flippers instead of legs. On land, the elephant line made further branches: one leading to the extinct mammoths, another to modern elephants. Cats emerged as a separate group much earlier, from an ancestor much like* Zalambdalestes, *and their body form has changed little over the last 25 million years. From the beginning, the cat was a beautifully adapted hunting machine.* Nimravus *looks remarkably similar to a modern leopard. Birds' body form has generally varied much less than that of mammals, because the basic design is less flexible. The Oligocene* Osteodontornis *bears a close resemblance to modern forms.*

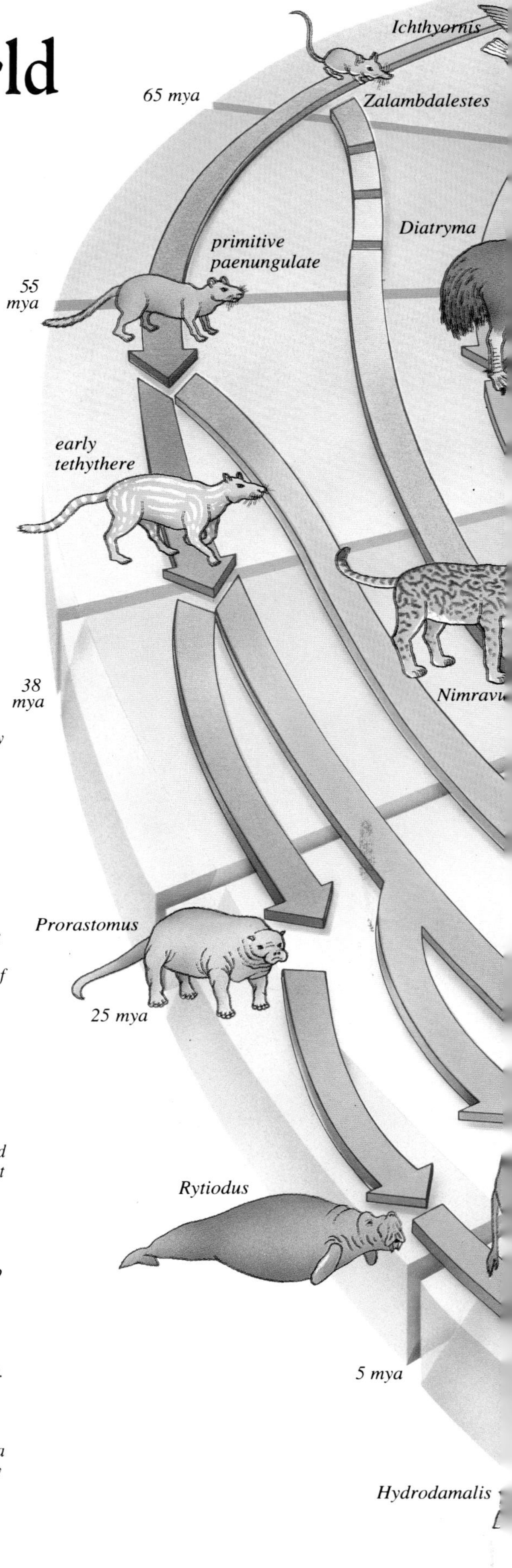

The Ascent of Man

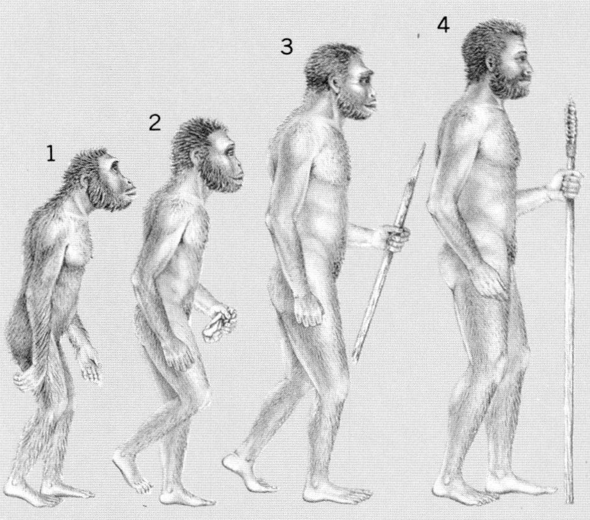

Humans evolved in Africa from hominid ancestors – the australopithecines – who walked upright but retained many features of their apelike ancestors. The hominid line probably split off from the line of the great apes – chimpanzees, gorillas and orangutans – between 15 and 5 million years ago. *Australopithecus afarensis* [1], from West Africa, lived between around 3.5 and 2.5 million years ago and is believed to be on the direct line of human descent. The genus *Homo* – to which humans belong – appeared first in Africa in the early Pleistocene period, with hominids like *Homo habilis* [2] and then *Homo erectus* [3]. Modern humans – *Homo sapiens sapiens* [4] – probably emigrated from Africa 35,000 years ago and spread across the world, replacing the Neanderthal people and surviving the end of the Pleistocene Ice Age.

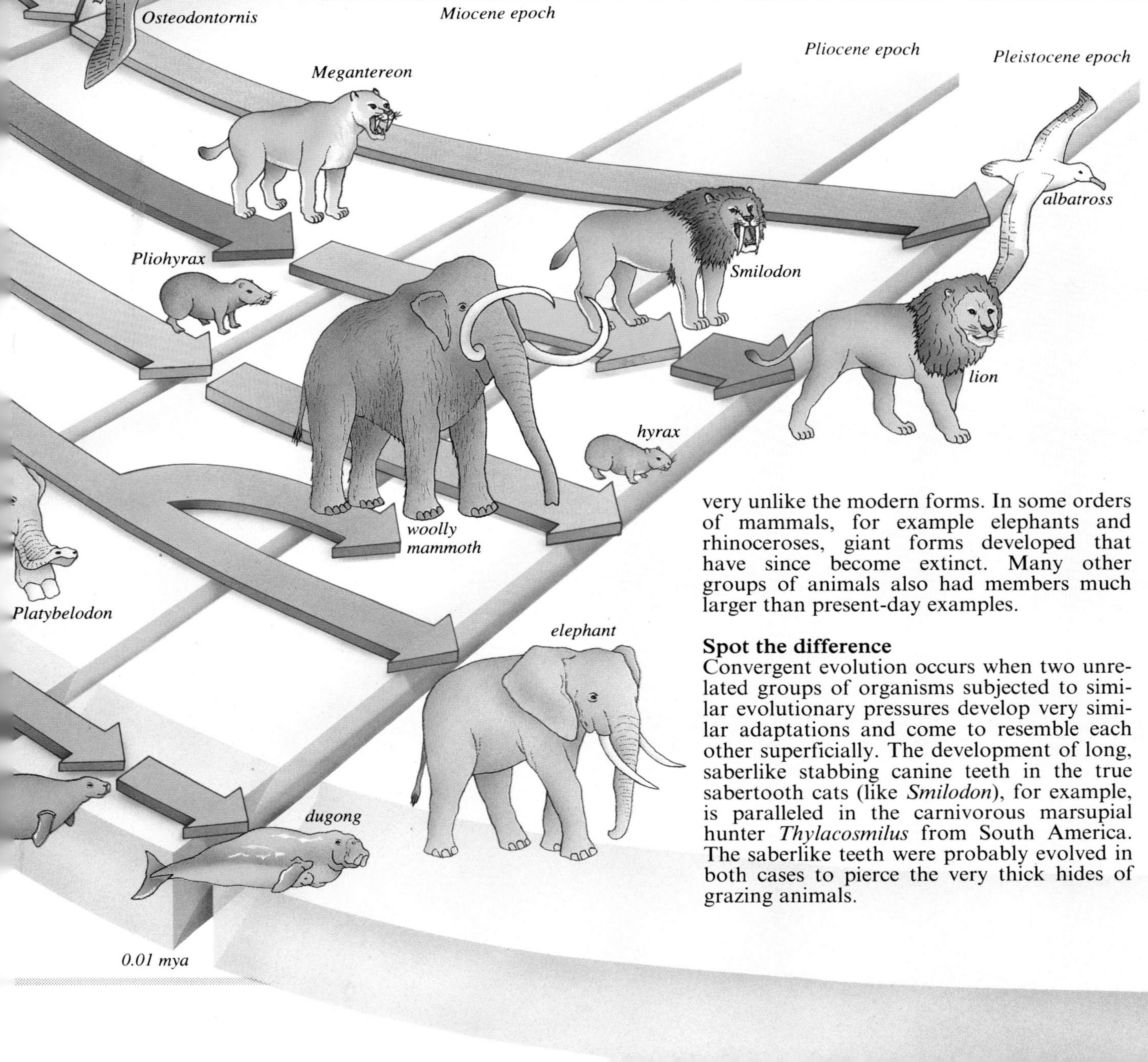

very unlike the modern forms. In some orders of mammals, for example elephants and rhinoceroses, giant forms developed that have since become extinct. Many other groups of animals also had members much larger than present-day examples.

Spot the difference

Convergent evolution occurs when two unrelated groups of organisms subjected to similar evolutionary pressures develop very similar adaptations and come to resemble each other superficially. The development of long, saberlike stabbing canine teeth in the true sabertooth cats (like *Smilodon*), for example, is paralleled in the carnivorous marsupial hunter *Thylacosmilus* from South America. The saberlike teeth were probably evolved in both cases to pierce the very thick hides of grazing animals.

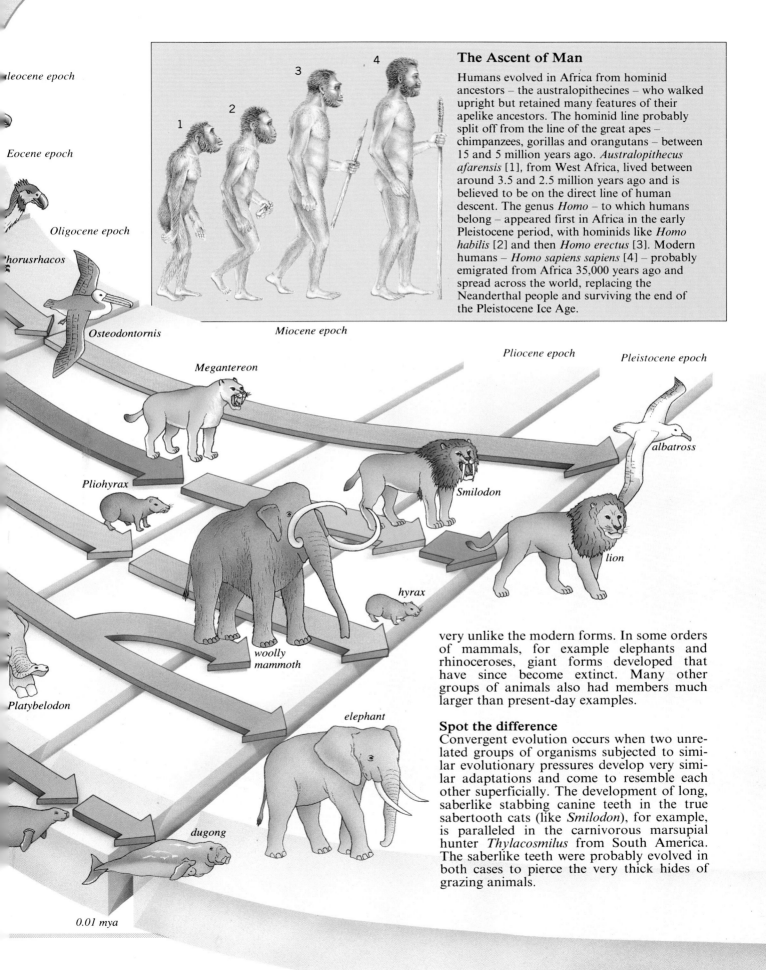

Blueprint for Life

How DNA works

A single human cell contains 13 feet (4 meters) of DNA (deoxyribonucleic acid), packed into a nucleus less than a thousandth of an inch across. In this mass of tangled threads is contained all the information needed to make a human being. DNA directs development and maintains the life of an organism by instructing cells to make proteins – the versatile molecules on which all life depends. The cell's DNA is a vast library of coded commands: the long molecules are packaged into chromosomes on which genes are arranged like beads on a string.

***Viewed under the light** microscope, a chromosome of a dividing cell has a simple crosslike shape* [**B**] *that belies the complex but elegant way in which it "packages" DNA. Magnifying a small section* [**C**] *reveals a tightly coiled strand of* chromatin *– DNA closely associated with protein. Further enlarging a segment of chromatin* [**D**] *shows it to be a tight coil of* nucleosomes *– beadlike subunits composed of a protein core wrapped by the DNA molecule* [**E**]. *The protein core is positively charged, allowing it to bind to the negatively charged DNA molecule* [**F**], *with its double helix structure* [**G**]. *It is essential to the organization of the cell that the DNA is condensed. If it was not, the DNA double helix would occupy thousands of times more space. By keeping DNA in compact bundles, the cell is much better able to manage it, uncoiling certain lengths as the genes contained in them are required.*

Each *chromosome* in the cell of an animal or plant is composed of a single long DNA molecule, coiled many times on itself and wrapped round a protein superstructure. Each chromosome is thought to involve some one hundred thousand different genes – shorter, functional units of DNA – each representing one of the instructions needed to make and maintain a human being. The complete set of genes of a living organism is its *genome*, and every cell of that organism carries at least one copy of the basic set. When cells are not dividing, the chromosomes are extended into long threads running throughout the nucleus. But when a cell is about to divide, they condense and become visible under the optical microscope as short rods.

The cell's library

All information, whether written, spoken or held in a computer, is coded in some form of language: the language in which DNA encodes instructions for making proteins – the genetic code – is remarkably simple. Each strand of the DNA double helix is a chain of chemical subunits linked together, and the order of these subunits along the DNA strand is the genetic code. There are only four different subunits (or bases), and therefore only four letters in the genetic alphabet – adenine (A), thymine (T), guanine (G) and cytosine (C). Three consecutive letters along a DNA molecule make up a "word" that codes for one amino acid – the building blocks of protein molecules: for example, the word GUC specifies the amino acid valine. Individual words may repeat letters. A sequence of words along a DNA molecule can thus be translated into a sequence of amino acids, which is then assembled into a protein. Each of the 20 types of amino acids found in proteins is specified by one of the 64 possible three-letter words. Other words act as punctuation. Each gene consists of a stretch of DNA thousands of letters long, specifying a protein chain that is typically several hundred amino acids long. The DNA "story" that is stored in even the smallest bacterium is millions of letters long, and the number of combinations of bases, and thus proteins it can code for, is nearly infinite.

DNA is the permanent information store of a cell and never leaves the nucleus. Its job is to store the genetic blueprint safely and pass it on unchanged from cell to cell and generation to generation.

Cracking the code

Biochemists can now read the sequence of bases in DNA almost as easily as you are reading this page. The complete DNA sequences of many viruses, which are of the order of thousands of bases long, have already been determined, as have the sequences of many human genes. There is now an ambitious project to read out the complete sequence of human DNA – all three billion bases of it. But even with the great advances in DNA technology over recent decades, it is estimated that such a project would take up to 20 years to complete and cost billions of dollars. The benefits, however, would be very great: the project could lead to the discovery of the genes that are thought to predispose us to diseases.

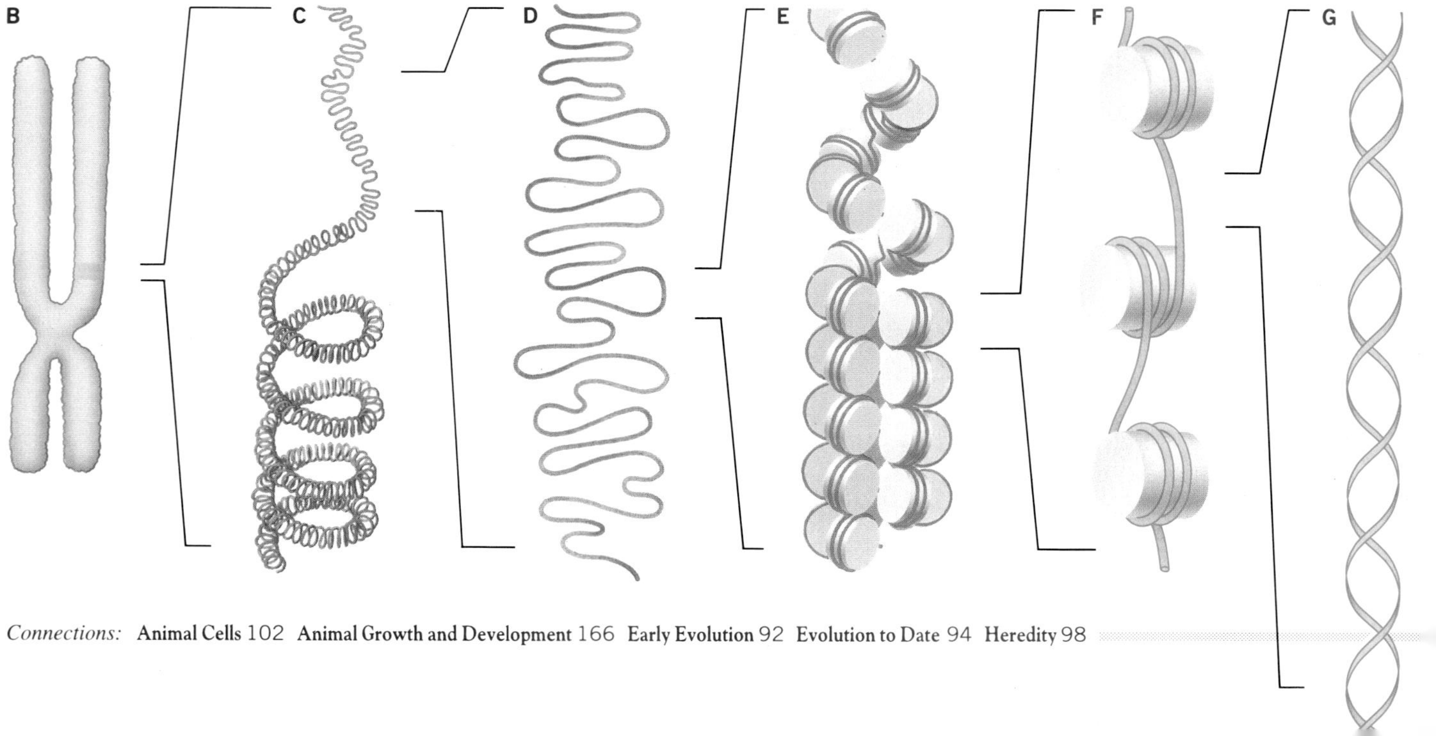

Connections: Animal Cells 102 Animal Growth and Development 166 Early Evolution 92 Evolution to Date 94 Heredity 98

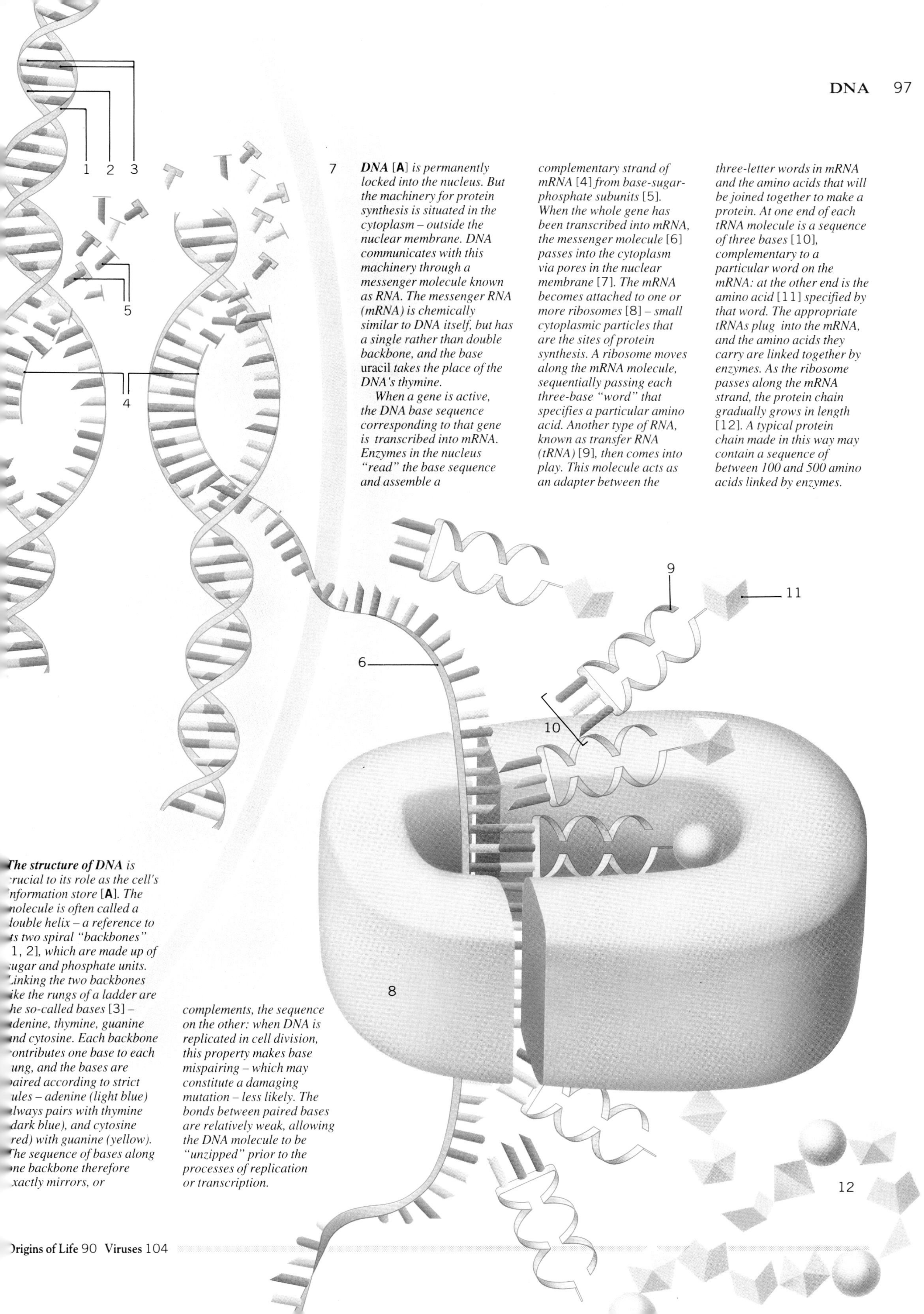

DNA **[A]** *is permanently locked into the nucleus. But the machinery for protein synthesis is situated in the cytoplasm – outside the nuclear membrane. DNA communicates with this machinery through a messenger molecule known as RNA. The messenger RNA (mRNA) is chemically similar to DNA itself, but has a single rather than double backbone, and the base* uracil *takes the place of the DNA's thymine.*

When a gene is active, the DNA base sequence corresponding to that gene is transcribed into mRNA. Enzymes in the nucleus "read" the base sequence and assemble a complementary strand of mRNA [4] from base-sugar-phosphate subunits [5]. When the whole gene has been transcribed into mRNA, the messenger molecule [6] passes into the cytoplasm via pores in the nuclear membrane [7]. The mRNA becomes attached to one or more ribosomes [8] – small cytoplasmic particles that are the sites of protein synthesis. A ribosome moves along the mRNA molecule, sequentially passing each three-base "word" that specifies a particular amino acid. Another type of RNA, known as transfer RNA (tRNA) [9], then comes into play. This molecule acts as an adapter between the three-letter words in mRNA and the amino acids that will be joined together to make a protein. At one end of each tRNA molecule is a sequence of three bases [10], complementary to a particular word on the mRNA: at the other end is the amino acid [11] specified by that word. The appropriate tRNAs plug into the mRNA, and the amino acids they carry are linked together by enzymes. As the ribosome passes along the mRNA strand, the protein chain gradually grows in length [12]. A typical protein chain made in this way may contain a sequence of between 100 and 500 amino acids linked by enzymes.

The structure of DNA *is crucial to its role as the cell's information store* **[A]**. *The molecule is often called a double helix – a reference to its two spiral "backbones" [1, 2], which are made up of sugar and phosphate units. Linking the two backbones like the rungs of a ladder are the so-called bases [3] – adenine, thymine, guanine and cytosine. Each backbone contributes one base to each rung, and the bases are paired according to strict rules – adenine (light blue) always pairs with thymine (dark blue), and cytosine (red) with guanine (yellow). The sequence of bases along one backbone therefore exactly mirrors, or complements, the sequence on the other: when DNA is replicated in cell division, this property makes base mispairing – which may constitute a damaging mutation – less likely. The bonds between paired bases are relatively weak, allowing the DNA molecule to be "unzipped" prior to the processes of replication or transcription.*

Origins of Life 90 **Viruses** 104

From Generation to Generation

How offspring inherit their characteristics

Only one in 20,000 human babies is born albino – though one in 70 people carry the albinism gene. This is because human characteristics are due to the complex interaction of many different genes. Yet, despite its individuality, a child possesses all the complex organs, tissues and biochemical pathways necessary to support life. This delicate balance between constancy and variability is achieved by the controlled "reshuffling" of parental characteristics when the gametes – the sperm and ova – come together and fertilization occurs.

An organism's growth, overall appearance and day-to-day functioning are ultimately controlled by its genes – the coded biological instructions in every cell of its body. The genes achieve this by controlling the types and quantities of proteins made in every cell of the body. It is the protein molecules themselves that form the structures and mechanics of the body.

When an organism grows or replaces old or damaged tissues, new cells may be formed by cell division: the chromosomes are duplicated and another complete set of genes is passed on unchanged to the "daughter" cell. This process is known as *mitosis*. Some single-celled microorganisms, plants and simple animals reproduce asexually by simple cell division. All the offspring of asexual reproduction are genetically identical to each other and to their single parent. But most familiar multicellular plants and animals reproduce sexually. Specialized sex cells (gametes) in the form of sperm from the male parent and ova (eggs) from the female parent in animals, or pollen and ovules in plants, are brought together to form a zygote, the initial single cell from which an embryo develops.

Sexual segregation

The sex cells are formed by a special type of cell division known as *meiosis*. A normal cell formed by mitosis is *diploid* – it contains two copies of each chromosome (and therefore of each gene), one inherited from its mother and one from its father. But a sex cell formed by meiosis is *haploid* – meaning it contains only one copy of each chromosome. Meiosis therefore keeps the number of chromosomes constant in each generation.

Meiosis also serves another important function, for this form of cell division allows the two sets of parent genes to be "reshuffled," albeit in a controlled way, resulting in genetically different offspring. Sexual reproduction therefore gives rise to genetic variation, which together with random mutation is the raw material for evolution.

Hiding in the genes

Within the population as a whole there are many alternative forms of genes (or *alleles*): for example "blue," "brown," and "green" alleles can all code for the same characteristic: eye color. But an individual only carries two

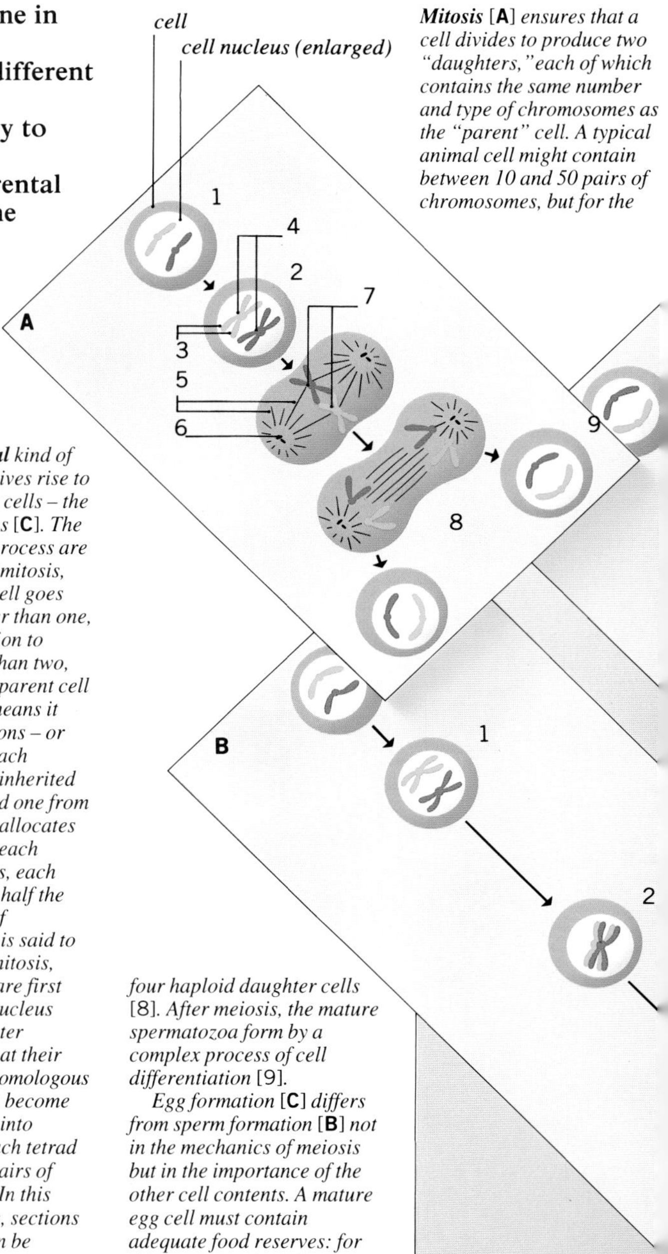

***Meiosis is a special** kind of cell division that gives rise to the gametes or sex cells – the sperm* [**B**] *and eggs* [**C**]. *The mechanics of the process are similar to those of mitosis, but the "parent" cell goes through two, rather than one, round of cell division to form four, rather than two, "daughters." The parent cell is diploid, which means it contains two versions – or* homologues *– of each chromosome; one inherited from its mother and one from its father. Meiosis allocates one homologue to each daughter cell. Thus, each daughter contains half the parental number of chromosomes and is said to be haploid. As in mitosis, the chromosomes are first duplicated in the nucleus to form pairs of sister chromatids joined at their centromeres* [1]. *Homologous chromosomes then become closely associated into* tetrads [2], *with each tetrad composed of two pairs of sister chromatids. In this tightly paired state, sections of chromosome can be exchanged between homologous pairs – a process known as* recombination [3]. *This brings together different versions of genes in new combinations. This genetic reassortment is followed by spindle formation* [4] *and movement of the two, paired sister chromatids to opposite ends of the cell* [5]. *The parent cell then splits, producing two diploid daughters* [6]. *After a recovery period, each daughter goes through a second round of cell division* [7]. *This creates a total of four haploid daughter cells* [8]. *After meiosis, the mature spermatozoa form by a complex process of cell differentiation* [9].

Egg formation [**C**] *differs from sperm formation* [**B**] *not in the mechanics of meiosis but in the importance of the other cell contents. A mature egg cell must contain adequate food reserves: for this reason meiosis is temporarily arrested at the tetrad stage to allow time for growth* [10]. *At this stage the* cortical granules *develop* [11]. *These will ensure that fertilization by only one sperm goes ahead normally. In addition, cell divisions in egg formation are unequal, with one of the daughters getting more than its fair share of cytoplasm* [12, 13]. *This one daughter develops into the egg itself* [14]; *the other three smaller cells, known as* polar bodies [15], *eventually degenerate.*

***Mitosis** [**A**] ensures that a cell divides to produce two "daughters," each of which contains the same number and type of chromosomes as the "parent" cell. A typical animal cell might contain between 10 and 50 pairs of chromosomes, but for the*

Connections: Animal Cells 102 Animal Growth and Development 166 Bacteria 106 DNA 96 Early Evolution 92

sake of simplicity only two paired chromosomes are shown here [1]. *Before mitosis begins, the chromosomes are duplicated in the cell nucleus* [2] *by special replicating enzymes. They then coil up and condense, a process that prevents them becoming tangled. Each chromosome now consists of two identical units – or sister* chromatids [3] *– joined together by special regions known as* centromeres [4]. *The nuclear membrane then breaks down and a network of* microtubules *begins to develop in the cell* [5]: *the microtubules radiate from structures called* centrioles [6], *one of which is located at each end of the cell. Each chromatid becomes attached to one or several microtubules at its centromere* [7]. *The microtubules are involved in the movement of the chromatids to opposite ends of the cell. The sister chromatids first become aligned along the cell's central axis* [7]. *They then separate at their centromeres and one chromatid from each pair is drawn toward one end of the cell, while the other moves to the other end. Each end of the parent cell now possesses a full set of chromosomes. The parent cell constricts along its central axis* [8], *"pinching off" two daughter cells* [9]. *The cell membrane regrows around each daughter and a nuclear membrane re-forms, enclosing the chromosomes.*

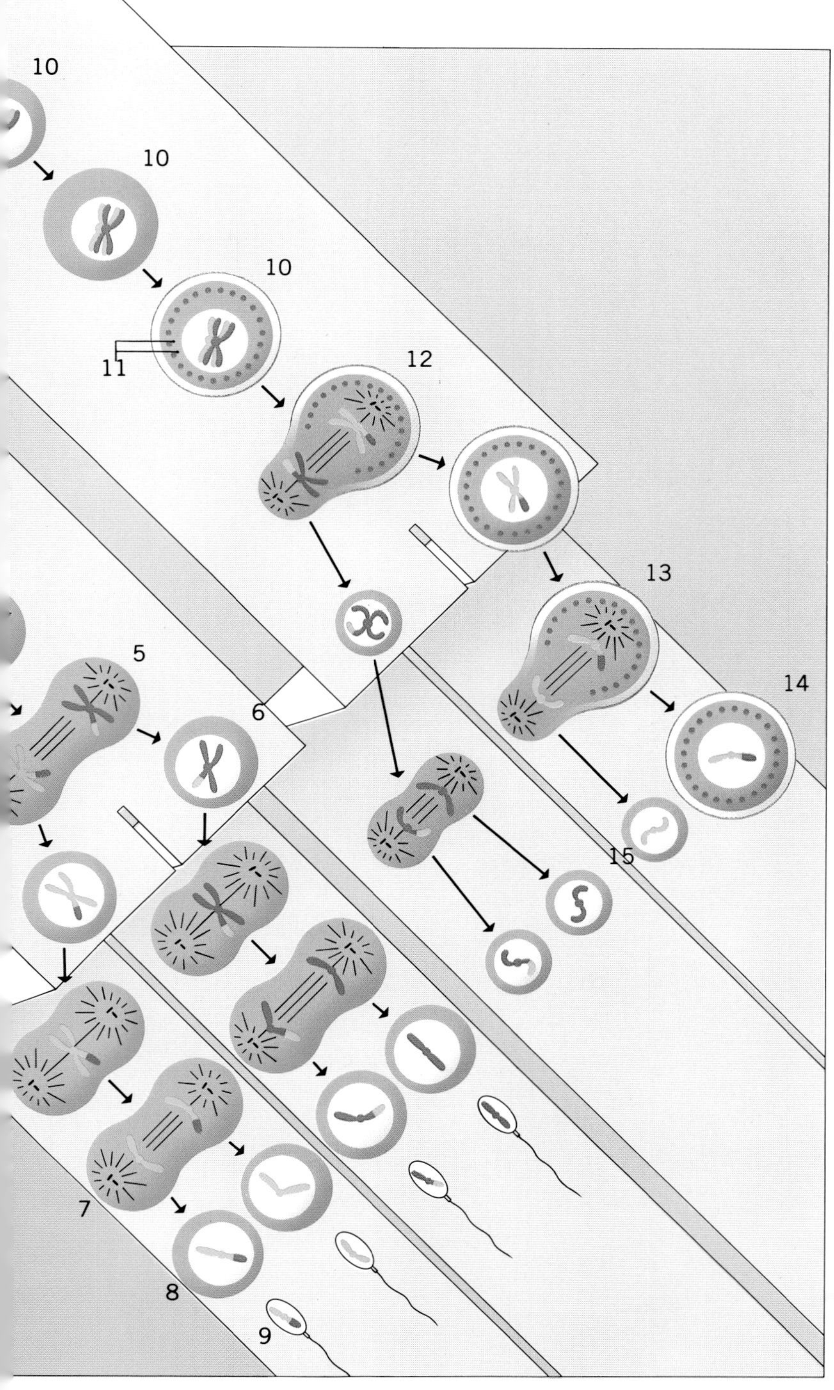

Good Breeding

Simple breeding experiments reveal the dominance of some alleles (gene forms) over others. In some mice the allele coding for black face color [B] is dominant over the white-face allele [b]. If black-faced mice possessing only the B allele are bred with white-faced mice (which must possess two b alleles), all the offspring have black faces since they all possess one [B] allele. If the Bb mice then interbreed, an average of 25 percent of their offsping will have two b alleles and therefore have white faces. It was by carrying out experiments such as these that the Austrian abbot Gregor Mendel discovered the basic laws of inheritance in the 19th century.

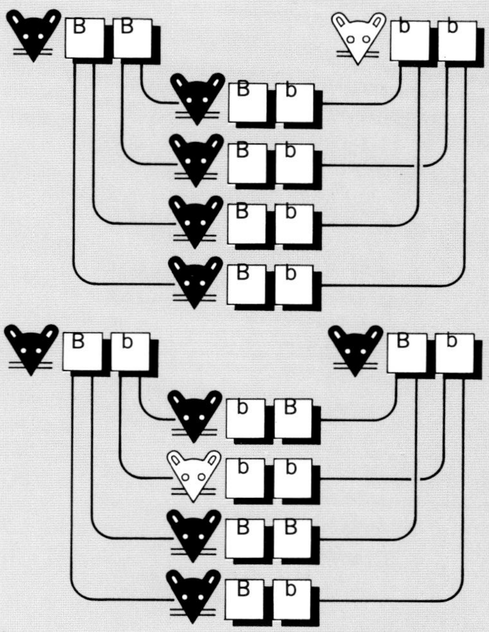

copies of each gene, so its eye color depends on which pair of alleles it has inherited. If a boy inherits a "blue" eye-color allele from his mother and a corresponding "brown" allele from his father, he will have brown eyes, because the brown allele "masks" the blue. Many alleles are like this, one form being "dominant" over another, "recessive," form. This is how a trait such as blue eyes can skip a generation: the "blue" gene is hidden inside brown-eyed parents and only becomes apparent when they both pass their recessive "blue" genes on to a child.

The final appearance and behavior of an organism is determined both by its genes and by an unknowable variety of outside influences, including how much food it eats, what climate it lives in, and whether it has suffered disease or injury during its development. But only those characteristics directly determined by the genes can be inherited.

Mammal Reproduction 176 **Plant Reproduction** 150

Plant Building Blocks

How plant cells function

Thre hundred feet (100 meters) above the forest floor in the crown of a redwood tree, a plant cell captures the energy contained in sunlight and converts it into sugar. Supported by a mass of living and dead cells forming a trunk weighing up to 2,000 tons and supplied by elongated vessels adapted to carry water and minerals to these dizzy heights, this cell is truly at the summit of the plant kingdom. Yet the characteristics of plant cells that allow them to form such spectacular structures and occupy the most diverse habitats have also imposed limits on plant evolution.

The earliest plant cells are thought to have formed more than a billion years ago, when cells that fed on the nutrient "soup" of the primeval seas were colonized by small bacteria capable of photosynthesis – the process by which sunlight is harnessed by the green pigment chlorophyll and converted into sugars. Over time the bacteria lost their independence and developed into chloroplasts, the *organelles* – the specialized "organs" of a cell – that carry out photosynthesis in plant cells. The sugars made in photosynthesis can be broken down in the *mitochondria*, releasing energy to fuel the cell's activity, or used as a source of carbon for larger molecules from which new plant material is made. The presence of structures that produce and store food is just one feature that distinguishes plant from animal cells.

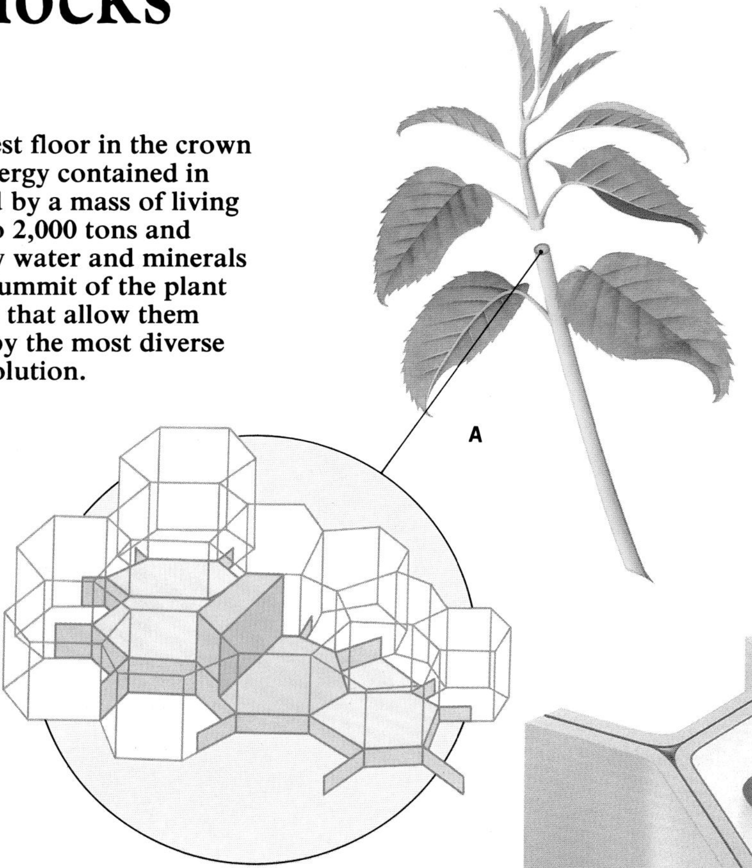

***A parenchyma cell** in a young shoot* [**A**] *contains all the structures that can be observed in an unspecialized plant cell. The inflexibility of the cell walls, together with the hydrostatic pressure within the cell, dictates the way that cells are packed together. A fairly regular array made up of more-or-less hexagonally shaped cells is typical.*

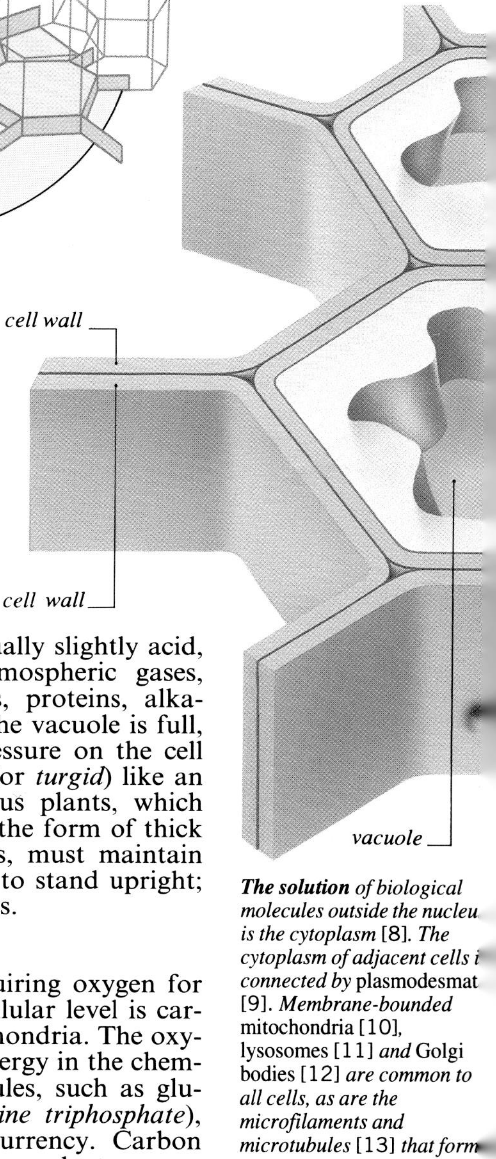

***The solution** of biological molecules outside the nucleu is the cytoplasm* [8]. *The cytoplasm of adjacent cells i connected by* plasmodesmat [9]. *Membrane-bounded* mitochondria [10], lysosomes [11] *and* Golgi bodies [12] *are common to all cells, as are the microfilaments and microtubules* [13] *that form a cell's internal "skeleton."*

Skeleton or straitjacket?

Plant cells come in a great variety of shapes and sizes, ranging from independent single-celled algae to the specialized cells that combine to form a multicellular land plant, and not all cells contain all the features in the "typical" cell opposite. One key feature common to every plant cell is the inflexible cellulose cell wall laid down on the outside of the cell membrane. The walls of adjacent cells are "cemented" together by the *middle lamella*, which helps give shape and support to the plant's tissues. However, the rigid walls restrict movement and communication between cells, and for this reason plants have never developed the mobility and responsiveness that typifies animal life.

Strength under pressure

The organization of a plant cell is shared to a large extent by other higher cells, such as those of animals and fungi. All possess a nucleus and various elaborate organelles, each of which carries out a particular task, just as the organs of the human body perform different functions. These organelles are surrounded by a single or double membrane, similar to the cell membrane that delimits the cell's *cytoplasm*. The membranes control what enters and leaves the organelle, thus keeping it distinct from the rest of the cell. An important component of the plant cell that is not found in animal cells is the vacuole – a membrane-bound sac filled with a watery solution. This *cell sap* is usually slightly acid, and contains dissolved atmospheric gases, salts, organic acids, sugars, proteins, alkaloids and pigments. When the vacuole is full, the cell's contents exert pressure on the cell wall, making the cell rigid (or *turgid*) like an inflated football. Herbaceous plants, which lack mechanical support in the form of thick cell walls and woody stems, must maintain this internal water pressure to stand upright; if they cannot, the plant wilts.

Cellular respiration

Plant cells are aerobic, requiring oxygen for respiration, which at the cellular level is carried out chiefly in the mitochondria. The oxygen is used to liberate the energy in the chemical bonds of "fuel" molecules, such as glucose to make ATP (*adenosine triphosphate*), the cell's internal energy currency. Carbon dioxide is released as a waste product.

Connections: Algae 112 Animal Cells 102 Germination 158 Photosynthesis 182 Plant Architecture 114 Protozoa 108

***The most prominent feature** of a plant cell* [**B**] *is the nucleus, which contains the cell's genetic material or DNA, normally arranged in thin strands called* chromatin [1]. *Messenger molecules copied from the DNA pass through pores in the nuclear membrane* [2]*: they then attach themselves to* ribosomes [3], *where they direct the synthesis of new cell proteins. Ribosomes are anchored to parallel membranes – the* endoplasmic reticulum [4] *– that form a mazelike network in the cell.*

Specialized Cells

Not all plant cells carry out photosynthesis. Wide, elongated xylem cells [1] have thickened, rigid cell walls. Aligned end to end they form hollow vessels that carry water and minerals from the roots to the growing shoots. Unlike xylem, phloem [2] cells are alive when mature. These highly specialized cells form structures known as sieve tubes, which together with small companion cells actively transport dissolved sugars to wherever they are needed. Much plant tissue is composed of polygonal thin-walled parenchyma cells [3], which can take on a variety of functions, including photosynthesis, sugar storage and support, depending on their location in the plant body. Collenchyma [4] and sclerenchyma [5] are simple tissues whose cells have thick walls and thus help to support the plant.

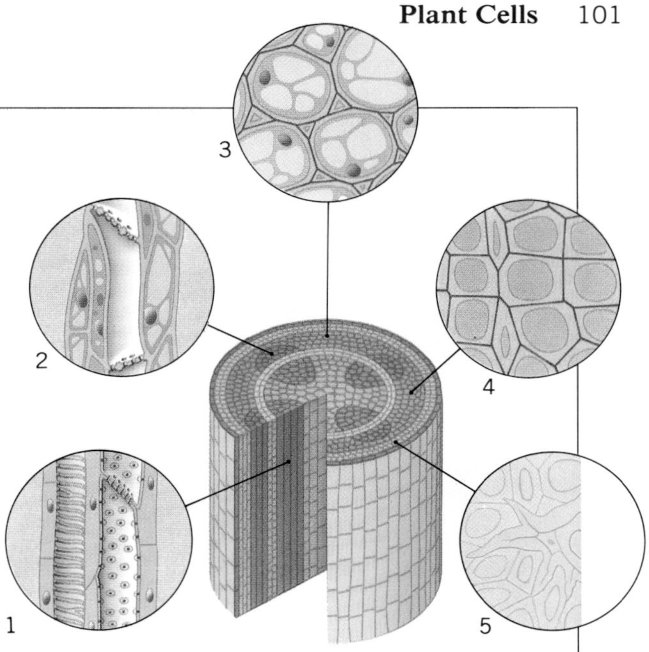

B

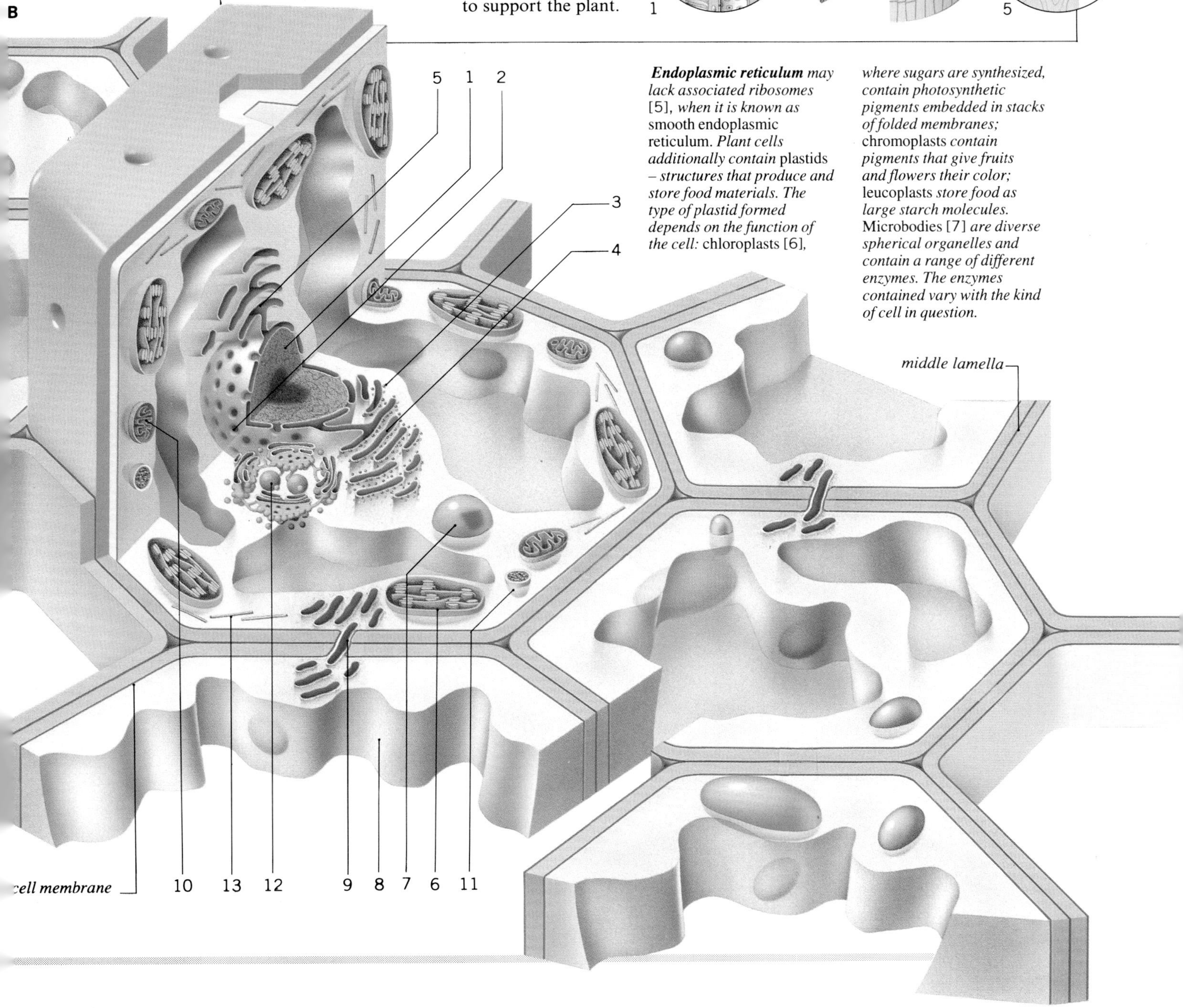

***Endoplasmic reticulum** may lack associated ribosomes* [5], *when it is known as* smooth endoplasmic reticulum. *Plant cells additionally contain* plastids *– structures that produce and store food materials. The type of plastid formed depends on the function of the cell:* chloroplasts [6], *where sugars are synthesized, contain photosynthetic pigments embedded in stacks of folded membranes;* chromoplasts *contain pigments that give fruits and flowers their color;* leucoplasts *store food as large starch molecules.* Microbodies [7] *are diverse spherical organelles and contain a range of different enzymes. The enzymes contained vary with the kind of cell in question.*

Animal Building Blocks

How animal cells function

The human body is made up of about 10 trillion cells. Although each cell has its own function to perform, they also cooperate with one another to make our bodies work. When we think of the millions of different types of animal, it is surprising to learn that the cells of which they are made are remarkably similar in structure. There are more than 250 different types of cell in the body of a rabbit, for example, but all of these cells – so varied in form and function – share the same basic internal organization as those in the simplest of animals, such as sponges.

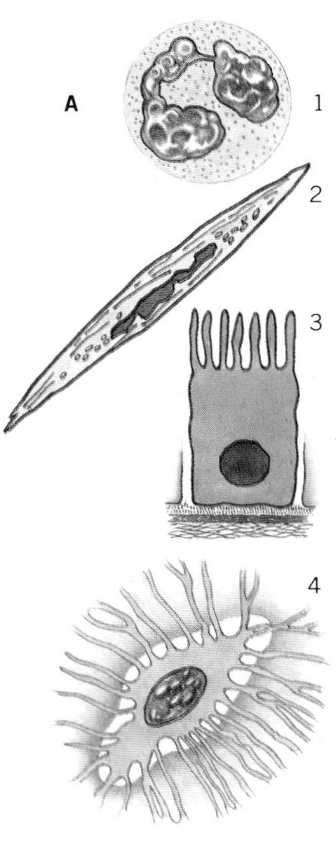

The body's cells [**A**] *come in many shapes and sizes. White blood cells* [1] *are much less numerous than red blood cells, but they play an essential role in the immune system. In vertebrates, several types of muscle cell contract to produce mechanical force. Thin elongated cells* [2] *are present in smooth muscle, which is found in the digestive tract and in blood vessels.* Epithelial *cells form the inner and outer surfaces of the body. These absorptive cells* [3] *have hairlike projections called* microvilli *to increase the surface area for absorption of molecules in the gut.* Osteocytes [4] *are found in bones.*

As recently as 200 years ago, scientists imagined the cell to be no more than a shapeless jelly: we now know it to have a highly organized internal architecture. Modern microscopy has revealed internal structures responsible for maintaining the cell's shape, assembling and transporting complex molecules and controlling cell division. Within each cell there is a distinct division of labor: different cell processes occur within different types of compartment (*organelles*), which together may occupy over half the volume of the cell. Many of these organelles are common to all plant and animal cells, but, significantly, animal cells lack chloroplasts and therefore cannot photosynthesize. Unable to make their own food from inorganic chemicals, they rely on a constant supply of ready-made organic compounds – sugars for energy, amino acids to build proteins, and fatty acids to make lipids (which are an important component of cell membranes).

The flexible frontier

To stay alive, all cells have to create a stable set of internal conditions different from those outside. To do this a cell has to isolate itself physically and chemically from its surroundings, which it achieves by means of the cell membrane, a thin layer of *lipid* (fatty molecules) and protein at the surface of the cytoplasm. All cells – whether bacteria, plant or animal – possess a cell membrane; similar membranes also enclose the organelles. Traffic through these membranes, including the flow of essential nutrients and of ions (such as sodium, potassium and calcium), is regulated by special proteins in the membrane, although the membrane does in fact allow water and small molecules to leak in or out. It is also peppered with receptor proteins that allow the cell to detect and respond to chemical "cues" from its environment and from other cells. As a result, animal cells are great communicators, continually signaling back and forth to each other.

Cell mobility

Compared to the thick plant cell wall, the cell membrane is a much less rigid barrier to the outside world. But the lack of a cell wall permits animal cells to be both more mobile and to display a greater range of shapes than plant cells. They can change shape rapidly, as

Animal cells [**B**] *are compartmentalized into various* organelles. *The most prominent of these is the* nucleus [1], *the information center of the cell, which contains the genetic material in the form of long, threadlike* chromosomes. *It is bounded by the nuclear membrane* [2], *which has many pores* [3] *to allow communication between the nucleus and other parts of the cell. In the center of the nucleus is the* nucleolus [4], *which is responsible for the production of* ribosomes. *The ribosomes* [5] *are the cell's protein factories and are found studded on the outer surface of the* rough endoplasmic reticulum [6]. *This is a system of flattened sacs and tubes of membrane connected to the nuclear membrane. It brings the messenger RNA molecules – which direct protein synthesis – to the ribosomes. Lipids are also produced here which form part of the cell membrane. The* smooth endoplasmic reticulum [7] *– connected to the rough – produces small membranous spheres called* vesicles [8]. *These transport proteins to the* Golgi apparatus [9], *which modifies, sorts and packs many large molecules into other vesicles, which bud off the apparatus* [10]. *They are then sent to other organelles or secreted from the cell. The fusion of such vesicles with the cell membrane allows particles to be transported out of the cell* (exocytosis*)* [11–13]. *Similarly, particles can be brought into the cell* [14–17] *in vesicles* (endocytosis*). Molecules entering the cell may be broken down by enzymes found in special vesicles, called* lysosomes [18].

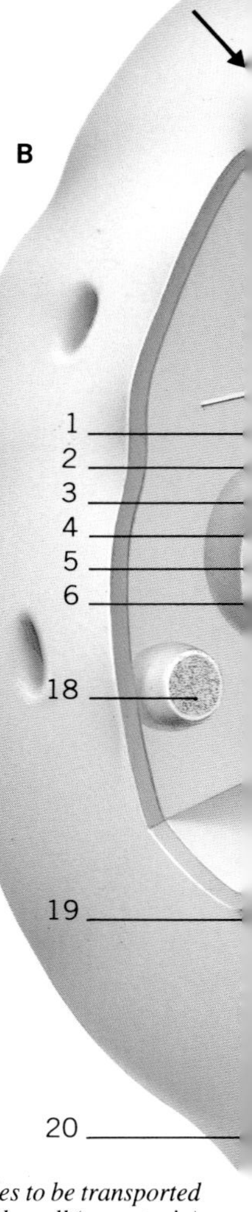

The Proton Powerhouse

Cells are powered primarily by energy released from *ATP* (adenosine triphosphate) as it becomes *ADP* (adenosine diphosphate). ATP is made in the mitochondria [1] by recycling ADP. The first step is to split *pyruvate* [2] – a fuel molecule derived from glucose in the cytoplasm – into carbon dioxide, hydrogen and high-energy electrons. These electrons pass along a line of special proteins in the inner membrane [3], giving them energy to pump out protons [4] into the intermembrane space [5]. As more protons are pumped out, pressure builds up in the space – like air in a balloon – which forces protons back across the membrane. But the protons can only flow back into the matrix via the ATP generator [6] – the enzyme *ATP synthetase* – and as they do so they drive round the blades of this turbine, producing ATP [7].

Connections: Animal Growth and Development 166 Bacteria 106 DNA 96 Hearing 282 Heredity 98 Nervous Systems 288 Origins of Life 90 Plant Cells 100

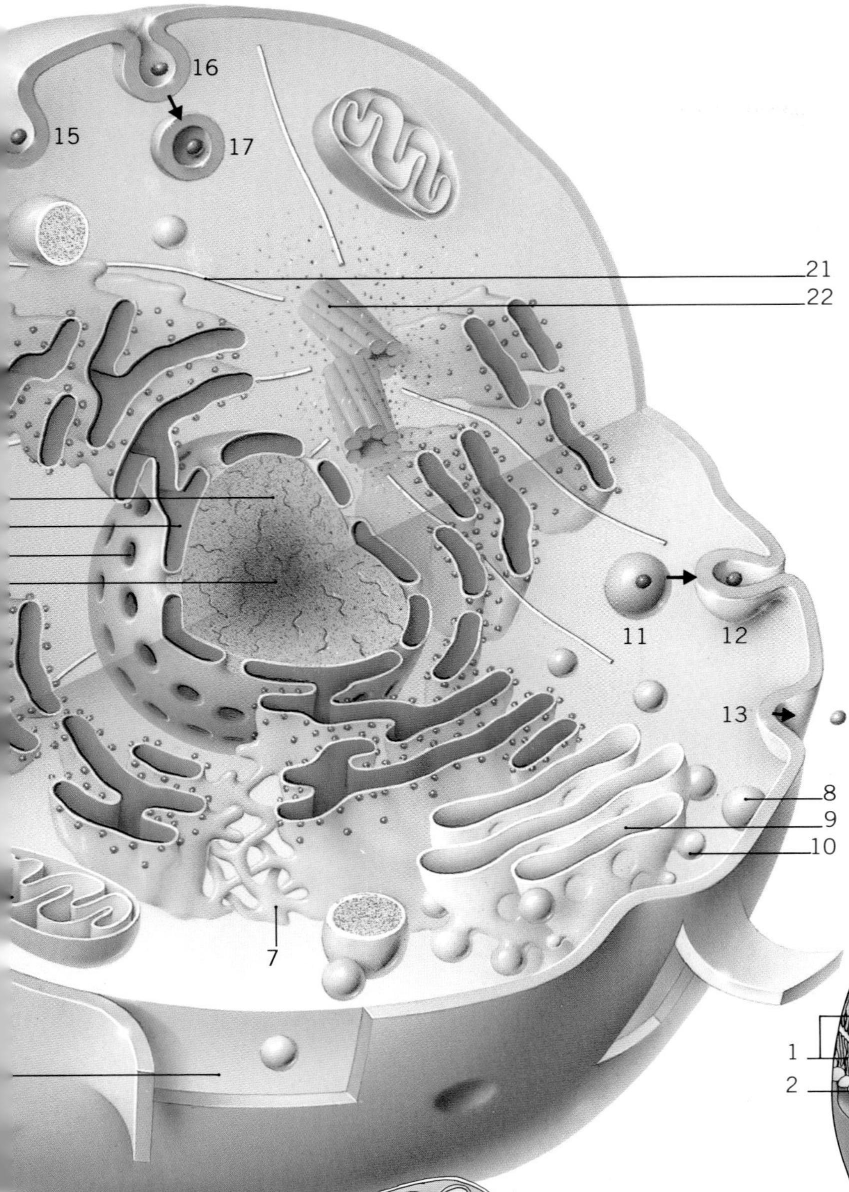

***The* mitochondria** [19] *are the powerhouses of the cell, using oxygen and food to generate energy (as ATP), which is then used in many metabolic processes. The majority of these are chemical reactions, which take place in the aqueous medium of the* cytoplasm [20]. *Running through the cytoplasm there is a matrix of protein filaments* (microtubules) [21] *known as the* cytoskeleton, *which acts like scaffolding, giving the cell shape and also providing a system for transport and movement. The cytoskeleton originates at the* centrioles [22], *which also help the chromosomes line up during cell division.*

The cell membrane [**C**] *is a thin, twofold layer of lipid molecules* [1] *that surrounds the cytoplasm of all cells. Very few molecules can pass through the cell membrane unaided. Special transport proteins and protein-lined channels* [2] *in the membrane allow through sugars, amino acids and essential ions like sodium and calcium. Other proteins* [3] *act as receptors for chemical signals, and provide a chemical signature that allows recognition by other cells, particularly of the immune system. Cholesterol molecules* [4] *are important for the membrane's stability, though too many can cause the membrane to seize up.*

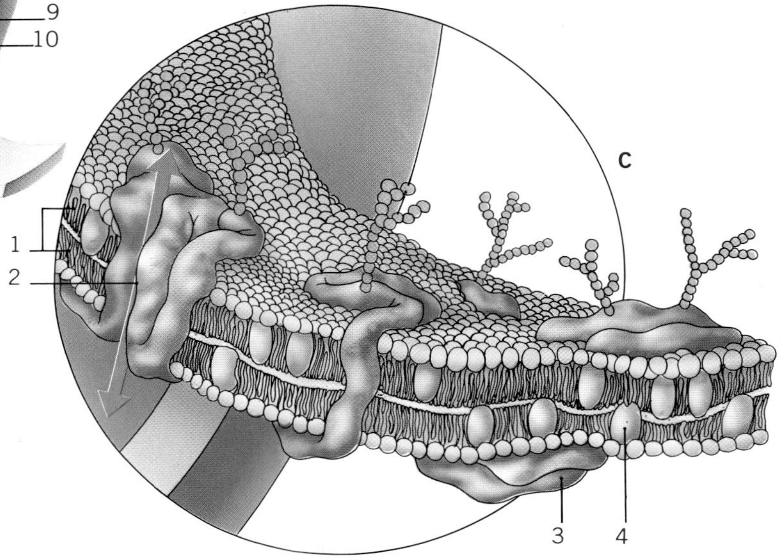

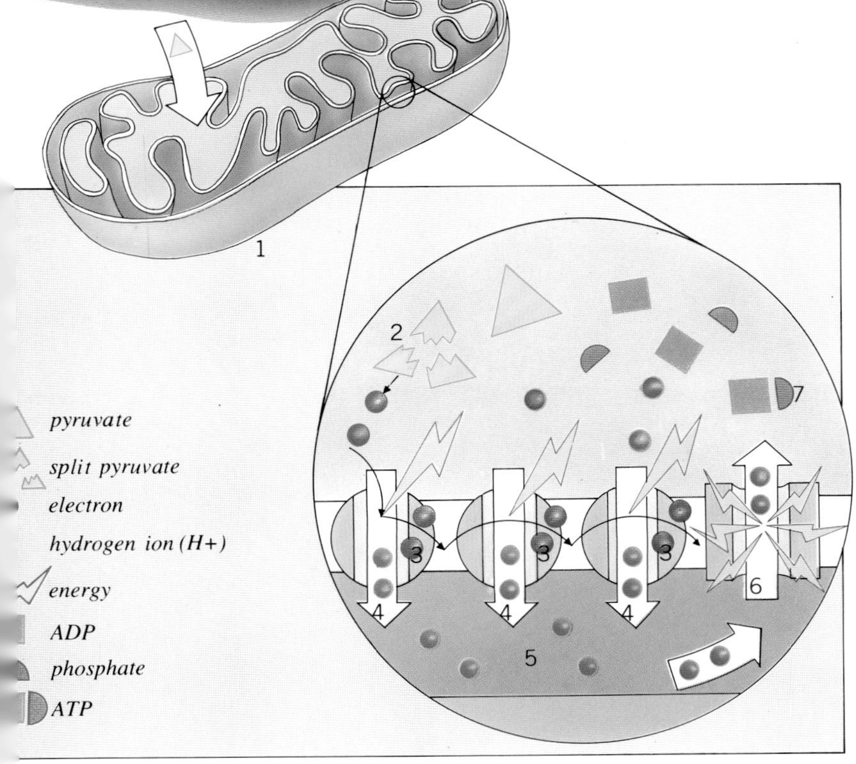

muscle cells do when they contract and relax, and can grow into elaborate and delicately shaped forms, such as nerve cells.

Cellular semaphore

All cells receive signals from the outside world and from other cells, at their cell membrane. Animal cells are specialized at this. The ability of a nerve cell, for example, to pass an electrical signal down its length is due to the properties of the cell membrane – in particular its ability to control the movement of sodium and potassium ions across it.

The cells of multicellular animals act together in tissues and organs to form incredibly complex systems. For example, nerve cells from cockroaches and humans are much the same, but the billions of nerve cells connected in orderly arrays in the human brain give us a range of abilities denied to the cockroach, with its tiny brain.

Protozoa 108 Smell and Taste 284 Symbiosis 192 Viruses 104 Vision 278

Hidden Invaders

How viruses cause disease

On the mysterious border between the living and the nonliving, viruses are the ultimate molecular hijackers. Outside a living cell they seem to be nothing but inert chemicals. But once inside they come to life, taking over the cell's biochemical machinery for their own ends – to make more viruses. Too small to be seen under the ordinary light microscope, they are revealed only under the much greater magnification of the electron microscope. This shows them to be packets of genetic material – DNA, or the closely related RNA – surrounded by a protective coating of proteins.

Viruses are most familiar as the cause of human, animal and plant diseases. Smallpox (now eradicated), rabies and polio are ancient scourges. Other viruses, such as the human immunodeficiency virus (HIV), the cause of AIDS (Acquired Immune Deficiency Syndrome), seem to have arisen in recent times. Viruses are also responsible for many diseases of childhood, such as measles, mumps and chickenpox, and for irritating but minor complaints like the common cold. There are other viruses that cause cancer.

Viruses are unlike any other form of life in that they are not made up of cells – not even of a single cell – but represent simply a core of genetic material wrapped in protein. They have often been likened to "rogue genes" that have somehow escaped from cells and taken on an independent existence.

Many viruses have a regular geometric shape because identical protein units cluster and stack around them, forming a coating. Rods, spheres and polyhedrons of various sorts are common. Because of such regular shapes, many viruses can be crystallized and stored like any chemical.

Hijacking a cell

Viruses often gain entry to cells by latching onto proteins on the cell surface – proteins that are normally used by the cells for quite different purposes. The AIDS virus, for example, attacks only a particular set of white blood cells, T-lymphocytes. These carry a special protein on their surface to which the virus sticks. Once attached to its "receptor" protein, the virus is then taken into the cell.

Upon entering a cell, a virus discards its protein coating. The cell, which normally produces proteins and replicates itself by copying and translating its own genetic material, then begins copying or translating the virus's nucleic acid instead. The genes encoded in viral nucleic acid direct the manufacture of more virus-coating protein, and the nucleic acid itself is also copied many times. So the cell is tricked into making new virus particles that are eventually shed from the cell, sometimes destroying the cell in the process. Most viruses have relatively few genes, all directed to replicating the virus.

The nucleic acid in viruses is either DNA or RNA. Most viruses that infect plants carry RNA – like the tobacco mosaic virus, the

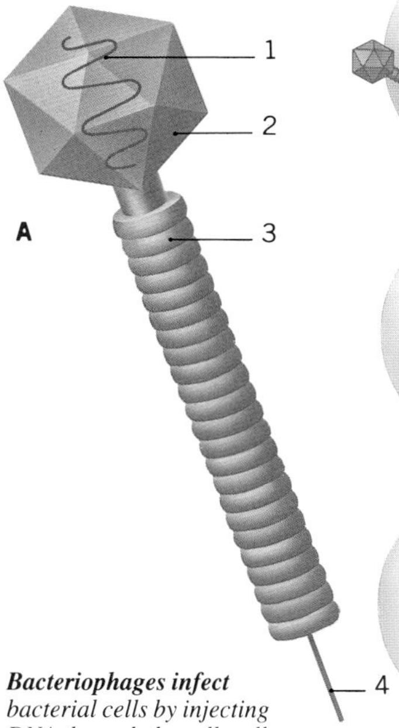

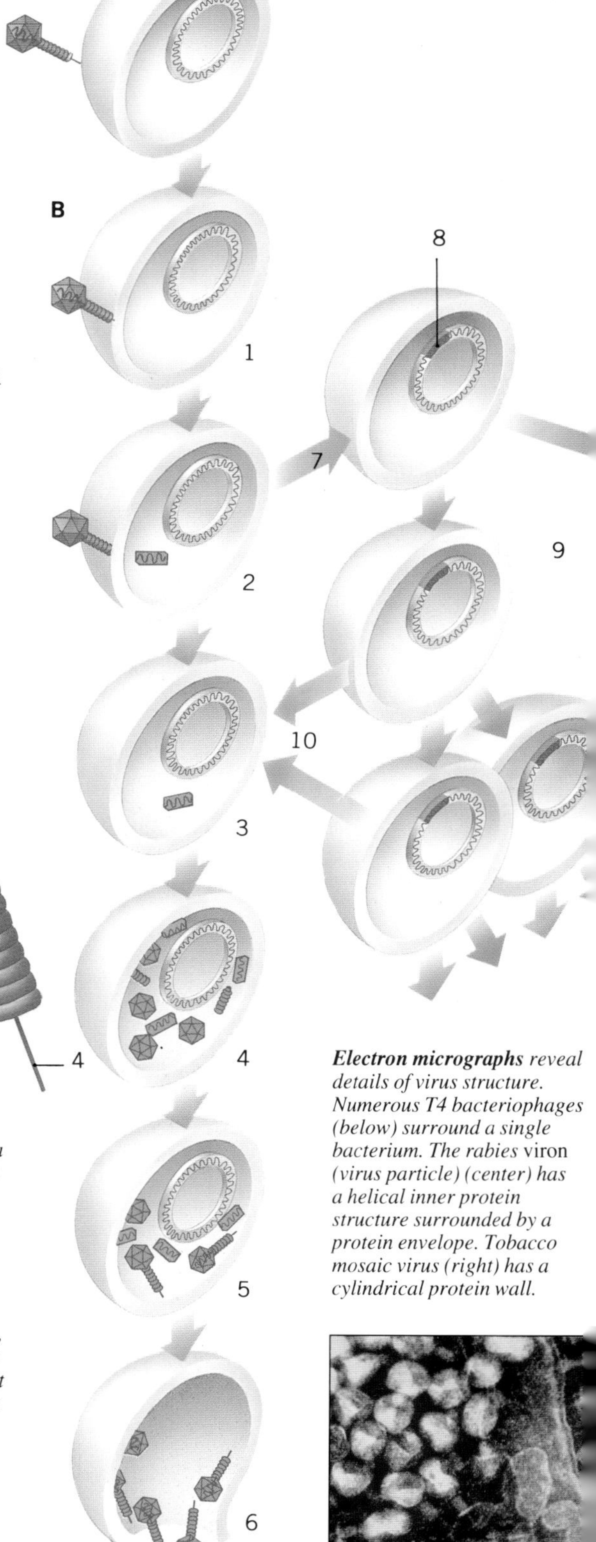

***Bacteriophages infect** bacterial cells by injecting DNA through the cell wall. Bacteriophage* lambda [**A**], *which infects* E. coli *bacteria in the human intestine, has a simple structure. The DNA* [1] *is stored in a polyhedral protein head* [2] *attached to a hollow tail* [3] *with a single tail fiber* [4]. *The phage attaches itself to a bacterium* [**B**], *at the tail* [1], *and injects its DNA* [2]. *This can result in an infection that causes the cell to* lyse *(break open), releasing replica phages. During such a* lytic *infection, the phage DNA remains separate from the bacterial DNA* [3] *and manipulates the cell's enzymes to synthesize the proteins that form the components of new phages* [4]. *The phage DNA replicates and large numbers of phages are assembled* [5]. *In the process the bacterial DNA is used, and by the time the cell lyses* [6] *may be completely destroyed. In some cases the injection of phage DNA results in a* lysogenic *infection* [7], *when the phage DNA becomes part of the bacterial chromosome* [8]. *Cell division* [9] *then produces numerous replicas of the phage DNA. During the course of lysogenic growth, damage to the cell may result in a lytic infection* [10] *by causing the phage DNA to be ejected from the bacterial chromosome.*

***Electron micrographs** reveal details of virus structure. Numerous T4 bacteriophages (below) surround a single bacterium. The rabies* viron *(virus particle) (center) has a helical inner protein structure surrounded by a protein envelope. Tobacco mosaic virus (right) has a cylindrical protein wall.*

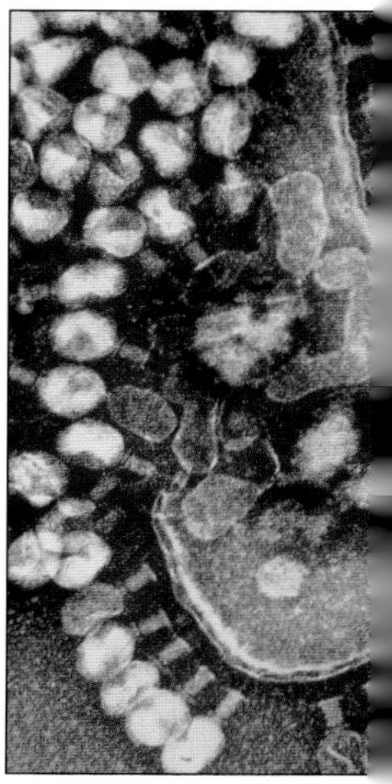

The human adenovirus [**C**] *is of the type responsible for colds and sore throats. The virus is color-coded for ease of identification. The casing consists of 252 protein molecules* (capsomeres) *arranged into a regular icosahedron (20 faces). This structure occurs in many viruses, representing the most economical packing arrangement around the DNA inside. Twelve of the capsomeres, located at the points of the icosahedron, are five-sided penton bases [yellow]. The remaining 240 are six-sided hexons [green]. Five of these [green-yellow] adjoin each penton base, from which extends a single fiber [red] tipped with a terminal structure [blue] that begins cell entry.*

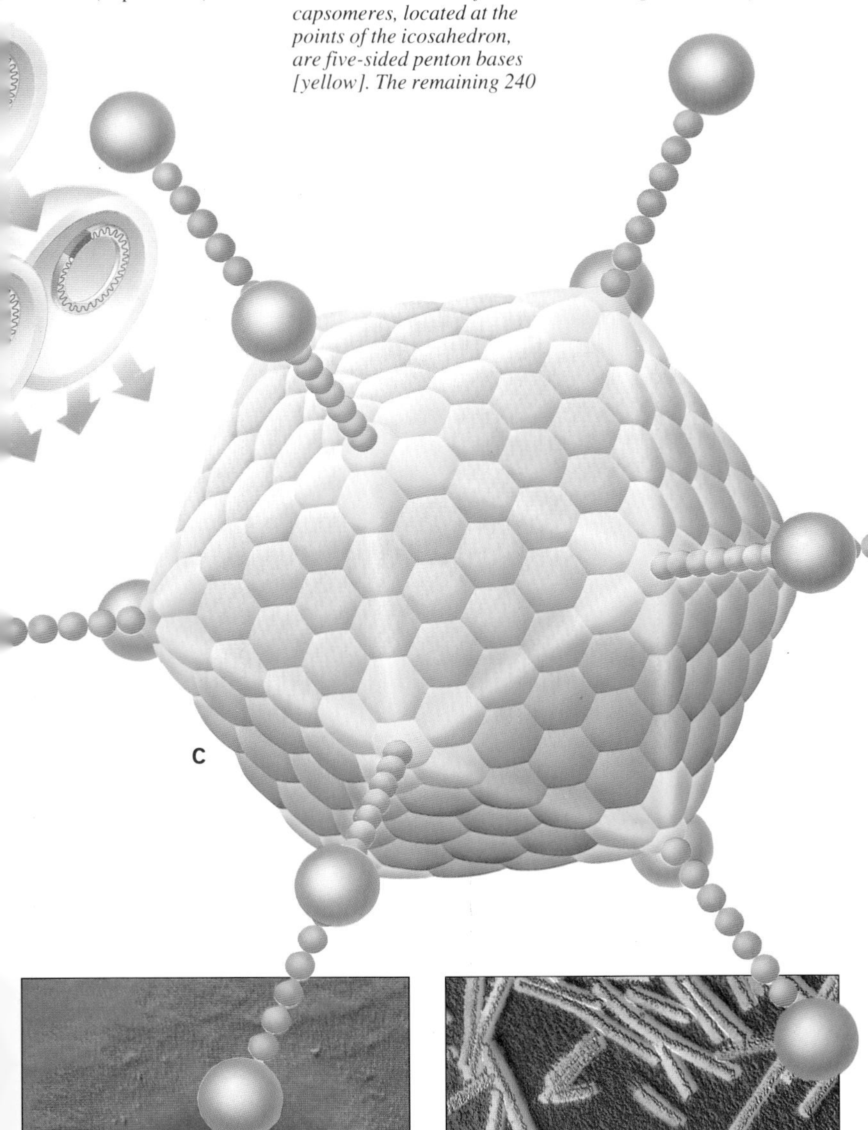

Fighting Viruses

Viral diseases, unlike bacterial diseases, are very difficult to treat with drugs. Because viruses take over the cells' own biochemistry, it is difficult to interfere with their multiplication without destroying the patient's healthy cells as well. The main line of attack against viral diseases in humans and animals is vaccination – which primes the body's immune defenses to fight the virus as soon as it enters the body and before it becomes established. But when, like HIV, the virus infects the cells of the immune system itself (infected cell shown below), developing an effective vaccine is even more difficult. Also some viruses, like flu viruses, mutate rapidly, changing the face they present to the immune system and making previous vaccines ineffective.

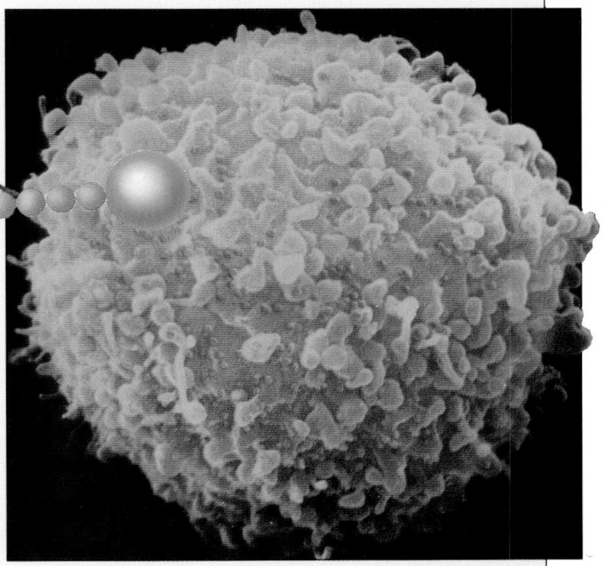

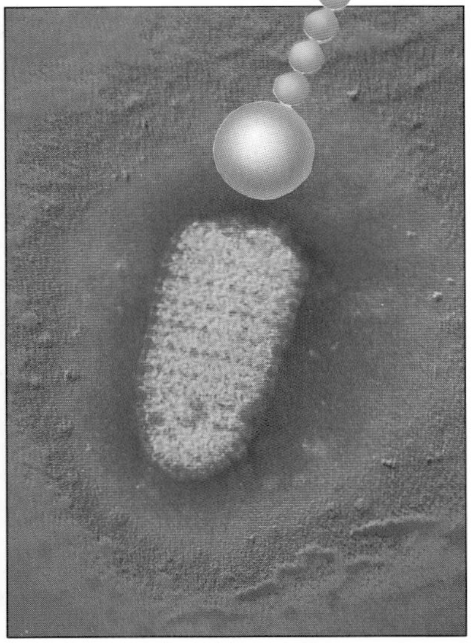

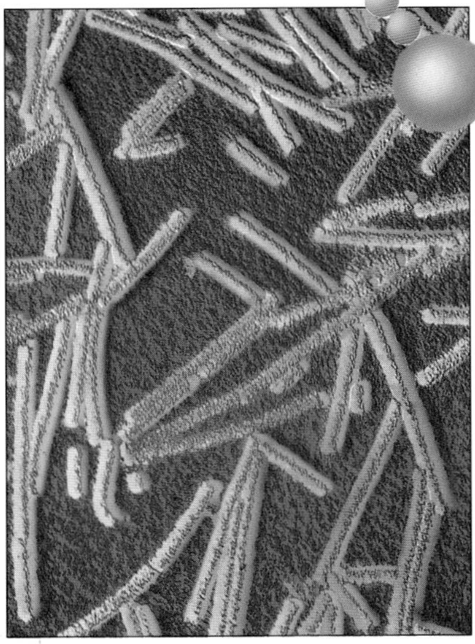

first virus discovered. Viruses that infect animals carry RNA or DNA. Bacteria have their own viruses, called *bacteriophages* or phages, which can carry DNA or RNA.

Genetic hitchhikers

The so-called retroviruses, to which HIV belongs, are RNA viruses. Instead of simply making more RNA copies of their own RNA, they carry an enzyme that enables them to make DNA copies of their genes – reversing the normal cell process of DNA to RNA. This DNA is then inserted into the cell's chromosomes, where it can be passed on, just like the cell's own genes, when the cell divides. Some retroviruses become inactive in this state and appear to have become permanent and harmless passengers in animal and human chromosomes. But others are considerably less benign and may carry *oncogenes*, which can make cells become tumor-

Hidden Life

How bacteria live

The most abundant life-forms on Earth are invisible to the human eye. Bacteria are all around us – they grow on and inside all other living things, and also in the soil, in ponds, rivers, lakes and oceans, in boiling springs and in the cold dark depths of the sea. They were probably the first forms of life on this planet, and will probably outlast everything else. Some are lethal to humans; but without others life could not continue, because they break down dead organic material and recycle essential elements that all plants and animals need.

All living things are made of cells. They can either be single-celled like bacteria, protozoa and some algae, or multicellular like plants and animals. Bacteria and their close relatives are termed *prokaryotes* to describe their simple internal structure and to distinguish them from the more complex plants, animals and fungi (*eukaryotes*).

Individual bacteria range in size from less than several hundred thousandths of an inch to several hundredths of an inch long. They occur in variations on three basic shapes: spherical, rodlike and spiral. Although bacteria are exclusively single-celled, some do aggregate into complex temporary structures. Bacteria multiply by simple cell division.

What bacteria live on

Every living thing needs not only a source of energy but also a source of carbon atoms (on which all organic molecules are based) and other essential atoms such as nitrogen, phosphorus, hydrogen and oxygen to live. Bacteria are much more versatile than plants and animals in the materials that they can use for these purposes. For almost any natural organic compound, and even some man-made pesticides and industrial chemicals, there exists somewhere a bacterium that can break it down and use it as food. In the *cyanobacteria* (blue-green algae) there is a photosynthetic process very similar to that of green plants. In contrast, *heterotrophic* bacteria, like *Escherichia coli,* a common inhabitant of the human gut, need ready-made organic materials, which they break down, extracting energy and reassembling the chemical subunits into new building material for their cells. Some heterotrophs need oxygen to respire in a very similar way to plant and animal cells. Others use fermentation to break down their nutrients without oxygen.

The cause of decay

Bacteria absorb small nutrient molecules such as sugars, amino acids and fatty acids directly through their cell membrane. Larger molecules such as proteins, starches, the cellulose of plant cell walls, and so on, are first broken down into simpler compounds outside the cell. This job is done by digestive enzymes that are released by the bacterium. The everyday rotting of meat, fruit and vegetables is often due to the action of bacteria.

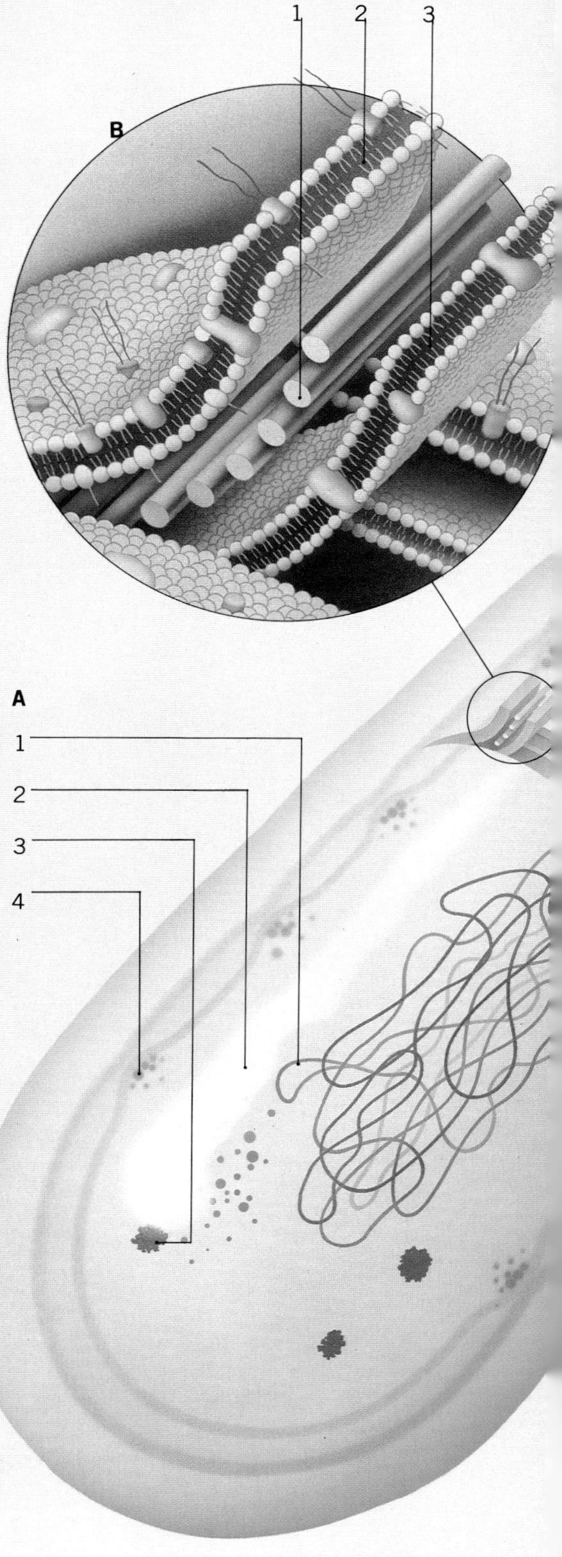

***Bacteria* [A] *have no nucleus. Instead, they have a nucleoid* [1]*, a single loop of DNA. This carries the genes, chemically coded instructions that define the bacterium. The average bacterium has about 3,000 genes, compared to a human's 100,000. The cytoplasm* [2] *also contains glycogen (food) granules* [3]*, and ribosomes* [4]*, which give the cytoplasm a grainy appearance and are the site of protein production. In many bacteria it also contains minute genetic elements called plasmids. Most, but not all, bacteria have rigid, protective cell walls* [B]*. There are two main types. One has a single thick (10–50 nm) layer. Bacteria with this type of cell wall are called Gram-positive, because they stain bright purple with the Gram stain. Gram-negative bacteria, as shown, have a thinner wall* [1] *with an extra layer of proteins and lipids on the outside* [2]*. This type of cell does not stain purple, a basic distinction useful in medicine. The defensive cells of the body recognize bacteria by their cell walls. A cell membrane* [3] *surrounds the cytoplasm. It is a few molecules thick, made of proteins and lipids, and is the barrier at which a living cell controls what enters and leaves it. Some bacteria move* [C] *using whiplike flagella* [1] *whirled about by a hook* [2]*. The motion is powered by a steady flow of protons across the cell membrane* [3]*, which rotates a disk of protein molecules* [4] *in the membrane. A rod* [5] *connects this protein "rotor" to the hook via another disk* [6] *that seals the cell wall.*

Connections: **Animal Cells** 102 **Cycles of Life** 206 **Herbivores** 202 **Internal Animal Parasites** 190 **Origins of Life** 90 **Protozoa** 108 **Symbiosis** 192 **Viruses** 104

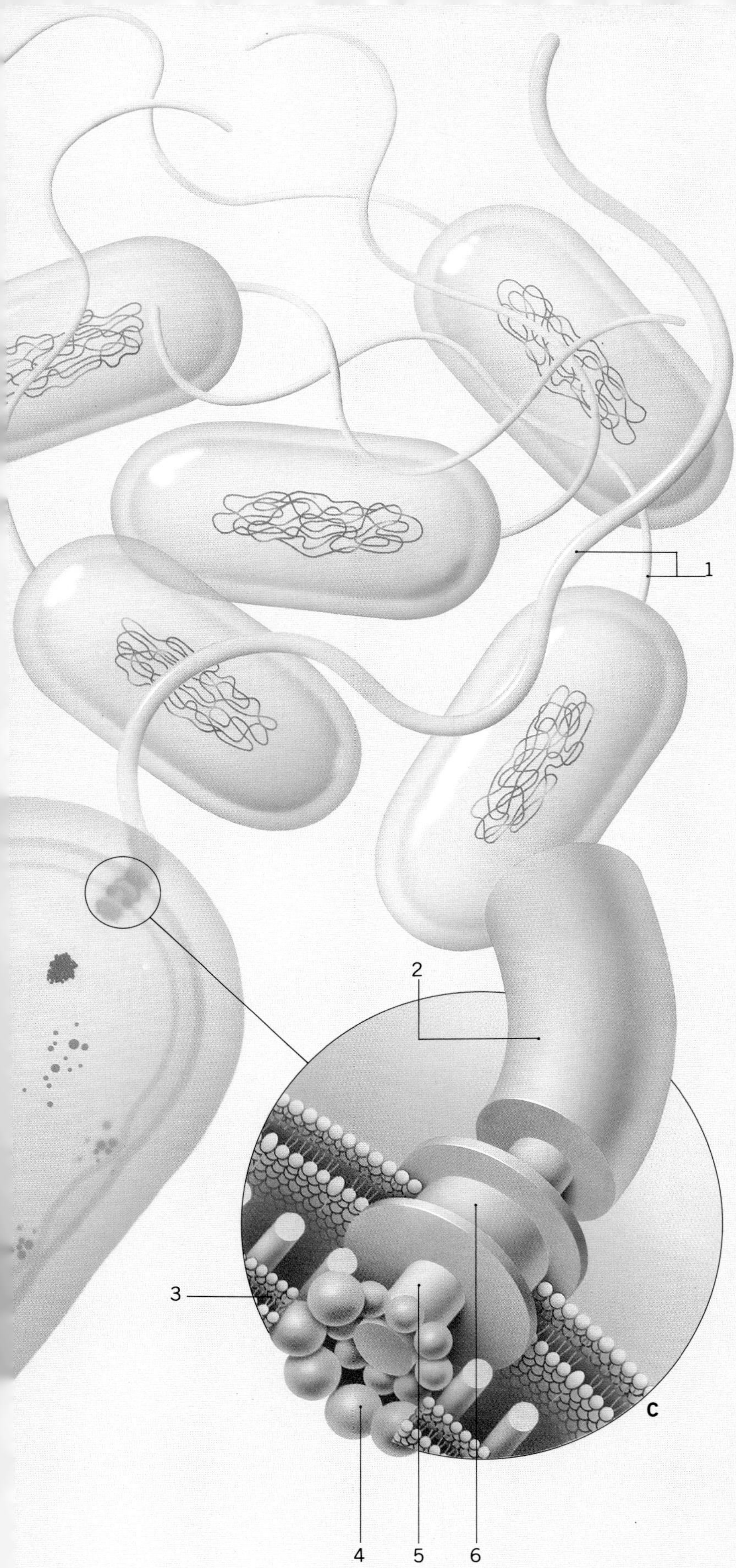

Another feature of bacteria is their ability to "fix" atmospheric nitrogen by extracting nitrogen from the air and converting it into ammonium compounds that plants can use. But what bacteria can do naturally chemists have only been able to reproduce at extremely high temperatures and pressures.

Speed in reproduction

One key to the success of bacteria is their ability to reproduce rapidly, a property that is used to full effect in commercial genetic engineering. In favorable conditions, *E. coli* can duplicate itself every 30 minutes. In a day a single cell can theoretically give rise to more than eight million new cells. But in the real world bacteria have not taken over because of various natural predators – such as protozoa – that reduce their numbers by billions per day, and because their growth is limited by the available food and space.

Bacterial Killers

Before the advent of effective sanitation and the discovery of antibiotics, recurrent epidemics of serious bacterial diseases swept Europe. The symptoms of many bacterial diseases are caused by toxic proteins (*toxins*) produced by the bacteria. Botulinum toxin, produced by the food-poisoning bacterium *Clostridium botulinum,* is one of the most powerful poisons known. Tetanus toxin, produced by the related *Clostridium tetani* [1], infects deep, dirty wounds. When a nerve impulse [2] tenses a muscle cell, the toxin blocks the relaxing part of the signal so the muscle stays tensed.

Most of the real killers among bacteria are now under control in the developed world: there, tuberculosis is rare and diphtheria seldom a problem, but in the Third World bacterial diseases still take a dreadful toll.

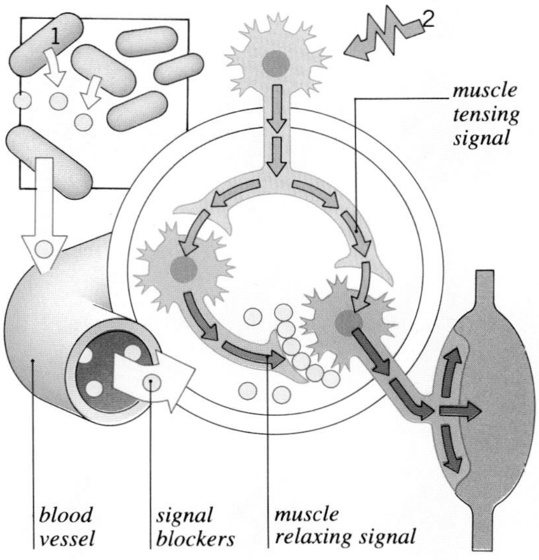

Neither Plants nor Animals

What protozoa are and how they function

There are nearly 30,000 different species of protozoa, single-celled microorganisms that live mostly in water or watery liquids. Abundant throughout the world, they may drift in their liquid environments, or actively swim or crawl along; a few remain relatively static and some live as parasites in animals. Many are microscopic, although some of the larger ones are visible to the naked eye. In form the protozoa are also amazingly diverse, from the simple bloblike amoeba to those that are equipped with elaborate structures for catching prey, feeding and moving.

***An amoeba* [A]** *moves by pushing out projections called* pseudopods *from its body. Cytoplasm – the fluid content of the cell – flows into the pseudopod, constantly enlarging it until all the cytoplasm has entered and the amoeba as a whole has moved. Pseudopods are also used in feeding: they move out to engulf a food particle* [1]*, which then becomes enclosed in a membrane-bound food vacuole* [2]*. Digestive enzymes enter the food vacuole, which gradually shrinks as the food is broken down* [3]*. Undigested material is discharged by the vacuole and left behind as the amoeba moves on* [4]*.*

A

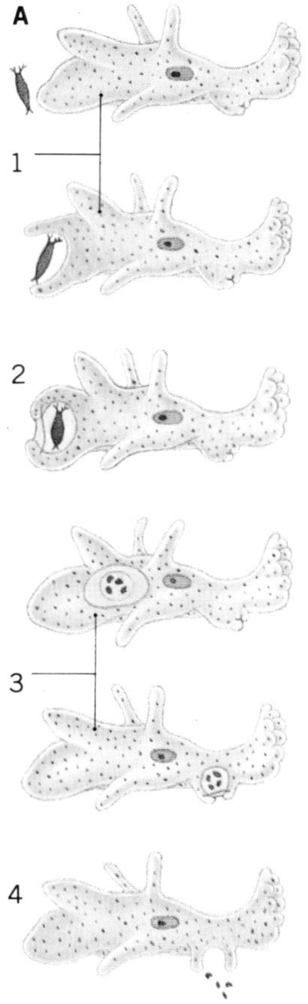

Among biologists, there is no real consensus as to what defines a protozoan. These organisms are classified in a kingdom of their own – the *Protista* – because they differ in some respect from bacteria, fungi, animals and plants. They have a more advanced organization than the bacteria in that they possess distinct compartments such as nuclei and mitochondria. But they are distinguished from plants, animals and fungi in that they are unicellular, not multicellular. Some of them are plantlike, having the ability to photosynthesize, but most are nonphotosynthetic, gaining nourishment by absorbing organic detritus or other microorganisms.

The Kingdom *Protista* is not a "natural" grouping – some protozoa may be related more closely to animals or plants than they are to other protozoa. It has served as a convenient pigeonhole for unicellular organisms that are otherwise difficult to classify.

The versatile protozoan

The actual sizes and shapes of protozoa are extraordinarily diverse, proving that protozoa represent a peak of unicellular evolution. The familiar amoeba, continually changing shape, is one type of protozoan. Others have contractile stalklike elements, and yet others include the foraminiferans, which are encased in coiled shells (*tests*) often impregnated with calcium carbonate. These chalky shells sink to the bottom of the ocean when the cells within them die and eventually become part of sedimentary rocks.

Some ciliate protozoa (those with tiny "hairs") have a distinct "mouth" and "gullet" down which bacteria, protozoa and algae are swallowed whole, whereas *suctorians* have long "tentacles" through which they suck out the contents of the cells they prey on. Protozoa lack rigid cellulose walls like those of plant cells, although *Euglena* and its relatives have a thin layer of flexible protein plates embedded just under the surface of the cytoplasm. Many protozoa have a protective outer coat. Some radiolarians and amoebas, for example, make casings for themselves out of sand grains or other debris. Although they cannot normally live out of water, many protozoa can survive the seasonal hazard of a pond or watercourse drying up by forming a tough coat or cyst around themselves and entering a state of dormancy.

***Stentor* [B]** *is one of the largest protozoa known, reaching lengths of up to 0.08 in (2 mm). It can attach itself to a rock or large alga by means of a holdfast at its base* [1]*, but may also swim free, propelled by its many cilia* [2]*. At the broad end of its body, the cilia are arranged in a spiral* [3]*: when these cilia beat, they create a vortex that draws particles in the water toward the oral pouch* [4]*. Here, suitable food – bacteria and small algae – is selected and passes into the buccal funnel or "gullet"* [5]*. Once packaged into vacuoles* [6]*, the food particles pass through the protozoan's body and are broken down. Undigested material is expelled through the cell wall. Like other ciliated protozoa,* Stentor *has a large string-of-beads-shaped nucleus* [7]*.*

B

D

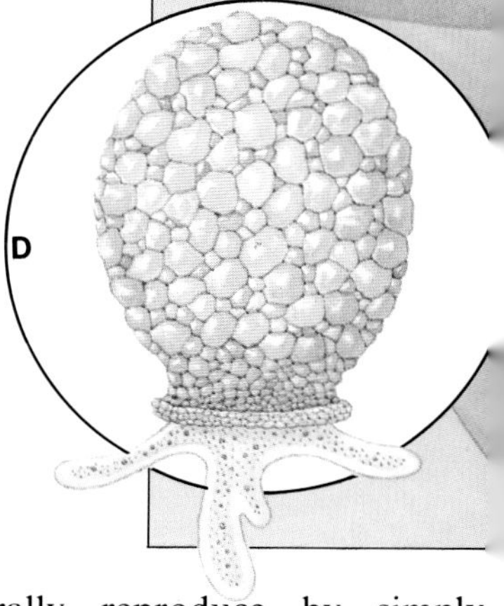

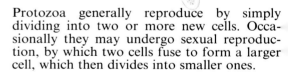

Protozoa generally reproduce by simply dividing into two or more new cells. Occasionally they may undergo sexual reproduction, by which two cells fuse to form a larger cell, which then divides into smaller ones.

Agents of good and ill

Protozoa are responsible for various human diseases, including malaria and sleeping sickness (*trypanosomiasis*), and also for many diseases in other animals, notably in cattle, fish, game and poultry. However, protozoa can be beneficial, and even essential, to some animals. Ciliates are part of the microbe life in the *rumen* (stomach) of cud-chewing animals such as cattle, helping to digest the enormous amount of cellulose present in the animal's diet, which it cannot digest itself. Protozoa are useful to humans in sewage treatment works, where they help to remove bacteria during processing.

Connections: **Animal Cells** 102 **Bacteria** 106 **Fossilization** 28 **Herbivores** 202 **Internal Animal Parasites** 190 **Origins of Life** 90 **Symbiosis** 192

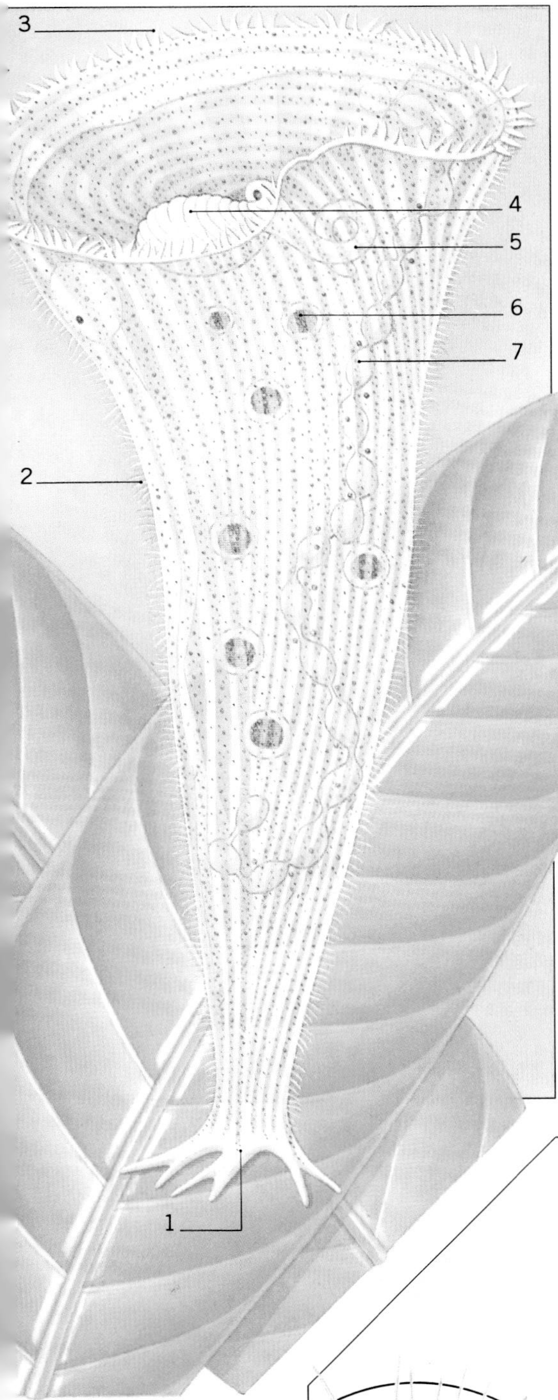

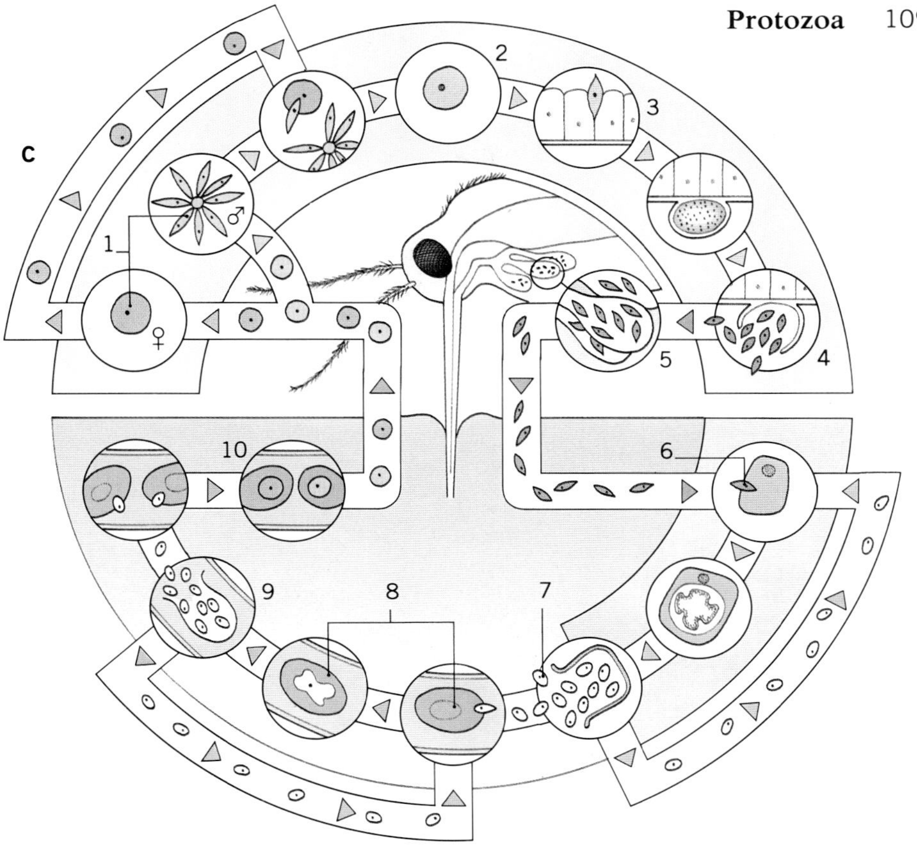

__The protozoan__ Plasmodium [**C**] *causes malaria, one of the most common and widespread human diseases. The organism is transmitted by female* Anopheles *mosquitoes, which begin to feed on human blood soon after mating. When a mosquito bites an infected person, it may take up sexually reproducing forms of* Plasmodium [1]. *These mate inside the mosquito to produce a zygote* [2], *which embeds itself in the stomach lining* [3] *and starts to divide. After 1–3 weeks, the swelled zygote bursts open, releasing many new* Plasmodium *cells known as* sporozoites [4]. *These migrate to the mosquito's salivary gland* [5], *from where they may be passed on to the insect's next human victim. Once in the human bloodstream, sporozoites infect liver cells* [6], *where they divide to form* merozoites [7]. *These may infect more liver cells or red blood cells* [8]. *Within the blood cells the merozoites multiply, and upon release* [9] *may attack yet more blood cells. Some of the merozoites eventually mature, forming sexually reproducing cells* [10]. *If these are picked up by another feeding mosquito, the cycle begins again.*

The symptoms of malaria may include liver failure and anemia, a reflection of the growth of Plasmodium *in the liver and red blood cells.*

__In appearance__ the protozoa are highly variable. Difflugia [**D**], *for example, has a similar biology to* Amoeba, *but makes a "shell" out of sand grains cemented together with organic matter.* Actinophrys [**E**], *like* Amoeba, *has pseudopodia, but they are long, slender and spinelike.* Ammodiscus [**F**] *lives in marine environments: it produces a hard, inorganic shell with many chambers.* Cementella [**G**] *builds a composite "shell" from the skeletons of other protozoa. Much of the deep ocean floor is covered by the hard shells of organisms like* Cementella: *over time they may become compacted into limestone.*

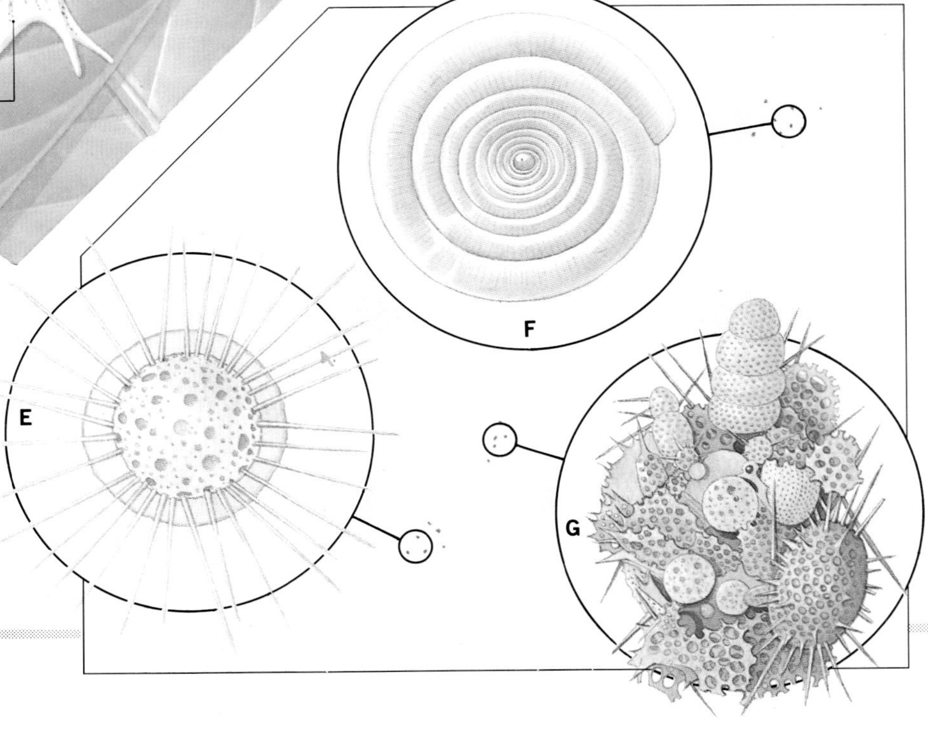

Mushrooms and Molds

How fungi live and grow

Blue cheese and Dutch elm disease at first appear to have very little in common. Yet both are, in fact, the result of fungal activity. Fungi have had an immense influence on human history – their fermentations give us bread, wine and beer, and many antibiotics. On the other hand, fungal infections cause crop failure across the globe. Fungi play a vital role in the natural cycle of life. By decomposing dead animal and plant material they release lockedup nutrients which can then be recycled in a new generation of plant and animal life.

A

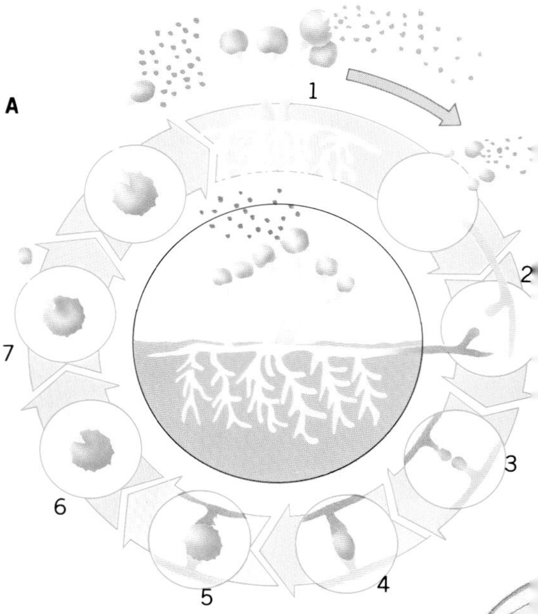

Traditionally regarded as plants, fungi are now usually classified by biologists in a kingdom of their own because they are so different in their structure, in how they grow, and in the way they feed. Unlike green plants, fungi cannot harness the Sun's energy to make their own food. Some fungi grow only on simple sugars, using them as sources of carbon, and get their nitrogen in the form of inorganic nitrates or ammonium compounds. Other species release enzymes that digest the complicated molecules present in dead plant and animal matter, turning them into a solution of simple nutrients that can then be absorbed. Still others are parasitic or symbiotic, gaining their nourishment from living plants or animals.

Threads of life

Some fungi are unicellular or consist of only a few cells, but most grow as fine threads (*hyphae*) that extend and branch at their tips, forming a network or *mycelium*. Although individual hyphae can only be seen under a microscope, the fluffy mycelia of common household molds are a familiar sight. Hyphae of simple fungi – or *zygomycetes* – are no more than continuous tubes of cellular material (cytoplasm) containing many nuclei, all enclosed in a cell wall. In contrast, the hyphae of higher fungi – the *basidiomycetes* and *ascomycetes* – are divided by crosswalls into compartments, and make up the fleshy tissue of a mushroom, toadstool, or truffle.

Mobile molds

All fungi reproduce by spores. A spore is a single cell, often surrounded by a protective coat, from which a new organism can develop. Some simple fungi produce *zoospores*, which look like tiny sperms and are propelled by one or two whiplike flagella. Fungi that produce zoospores are either aquatic, like the water molds, or parasitic, living inside the cells of plants and releasing their zoospores into the film of water that covers the surfaces of leaves, stems and roots. The common household molds produce simple spore-bearing containers (*sporangia*), which burst when mature, releasing a cloud of minute dustlike spores. When they alight on a suitable growing medium, they germinate. From each spore a hypha emerges and rapidly develops into a new mycelium.

Like most zygomycetes, *black bread mold* [**A**] *feeds by breaking down dead matter. The bulk of the mycelium grows within the food substance, but some hyphae grow upright and swell at their tips, forming sporangia, the organs of asexual reproduction. Numerous spores develop within these swellings: the sporangium wall disintegrates and the spores are dispersed by air currents* [1]. *They germinate and give rise to a new mycelium. Zygomycetes also undergo a sexual process (conjugation). Two hyphae of different "sex" (represented by pink and purple colors) grow toward one another* [2], *attracted by chemicals released by the mycelia. Once in contact, the tips of the hyphae swell and become cut off by a cross wall* [3]. *The two adjacent tips, which contain many nuclei, fuse* [4] *and develop a tough wall* [5]. *The resistant spore then germinates* [6], *undergoes genetic reshuffling* (meiosis) *and produces spores* [7].

Most common higher fungi are basidiomycetes – so called because they produce spores externally in structures called basidia. The club-shaped basidia are borne on short-lived fruiting bodies – the toadstools and mushrooms – that raise the spores above the soil and leaf litter, thus aiding their dispersal by air currents. The permanent part of the fungus, which can persist for years, is a mycelium that often spreads for many yards through the soil or wood. Less familiar basidiomycetes such as the rusts and smuts cause disease in many cereal crops.

The morels, truffles and yeasts belong to the third major group of fungi – the ascomycetes. They produce their spores internally in capsules (or asci), which are formed when two hyphae of different "mating type" join. The nuclei of the parent hyphae fuse and divide, giving rise to four or eight spores arranged within the ascus like peas in a pod.

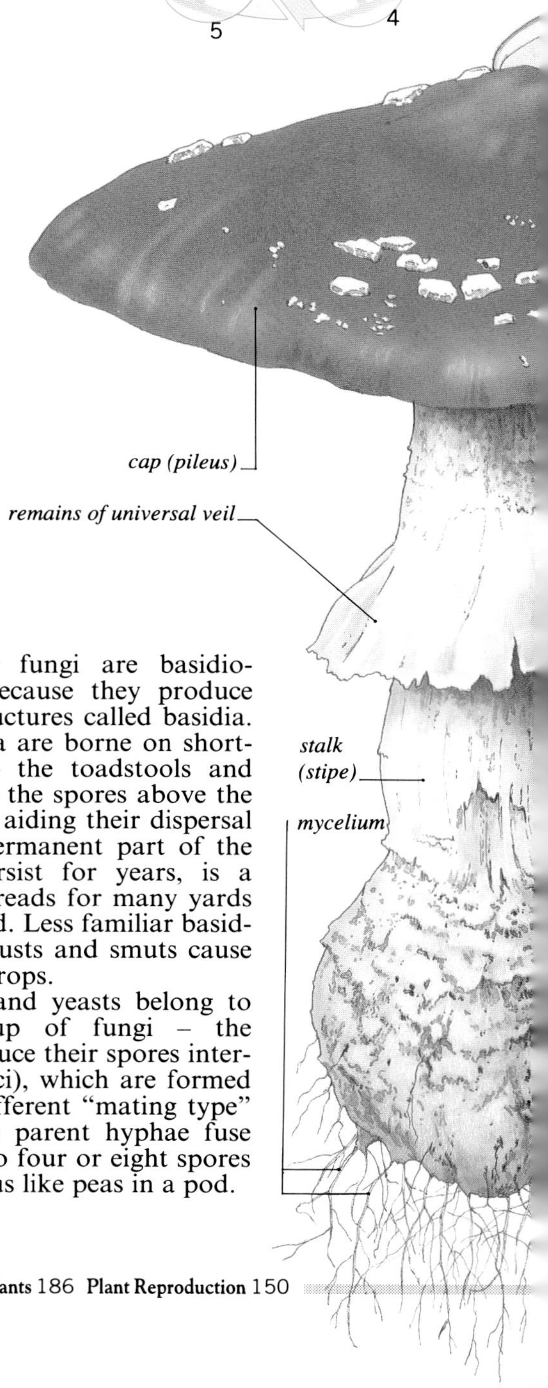

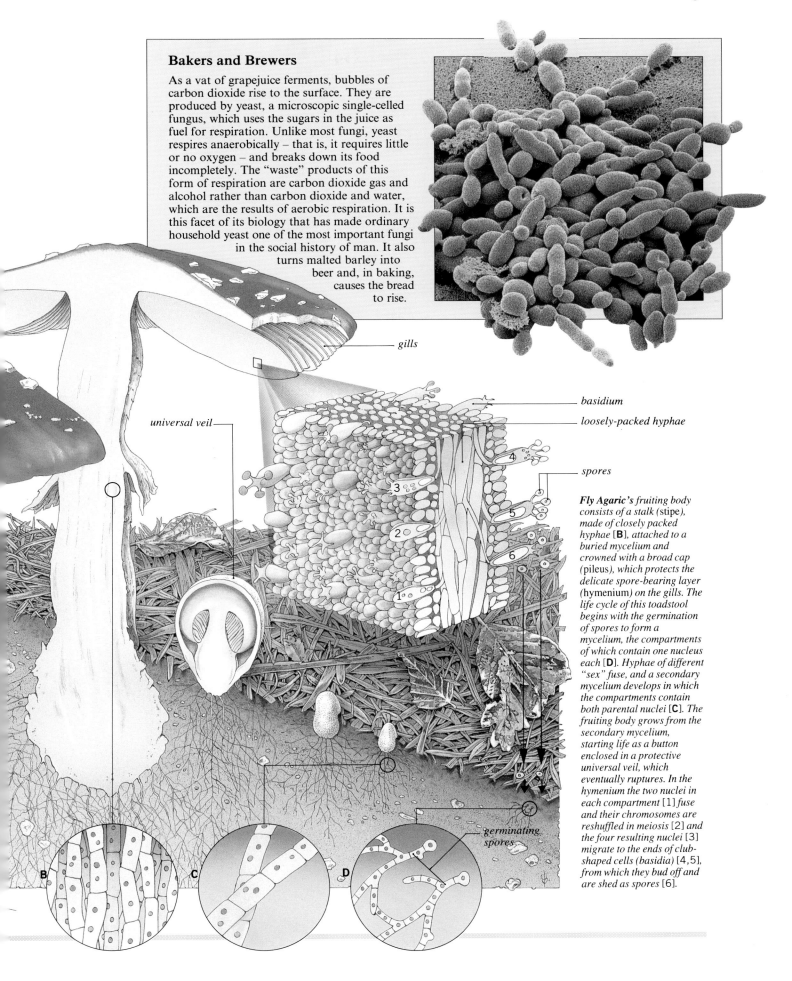

Bakers and Brewers

As a vat of grapejuice ferments, bubbles of carbon dioxide rise to the surface. They are produced by yeast, a microscopic single-celled fungus, which uses the sugars in the juice as fuel for respiration. Unlike most fungi, yeast respires anaerobically – that is, it requires little or no oxygen – and breaks down its food incompletely. The "waste" products of this form of respiration are carbon dioxide gas and alcohol rather than carbon dioxide and water, which are the results of aerobic respiration. It is this facet of its biology that has made ordinary household yeast one of the most important fungi in the social history of man. It also turns malted barley into beer and, in baking, causes the bread to rise.

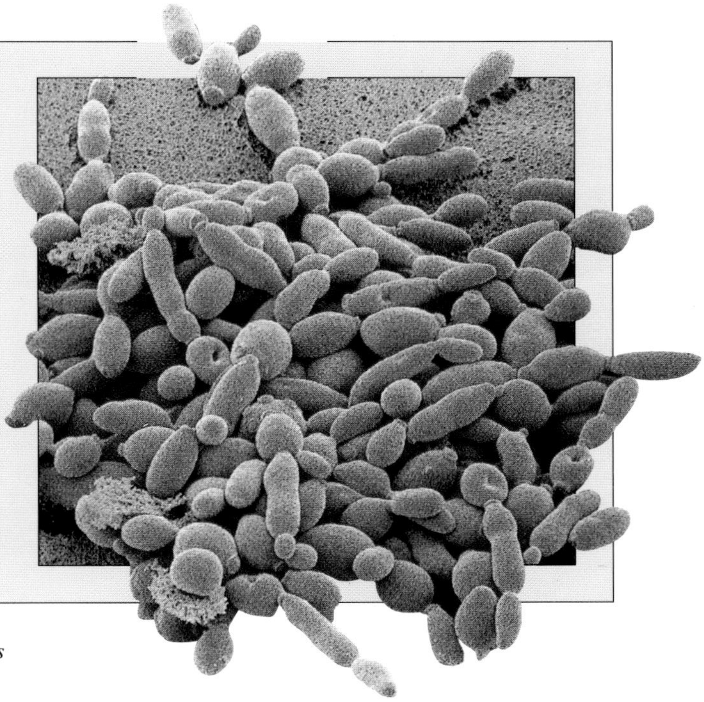

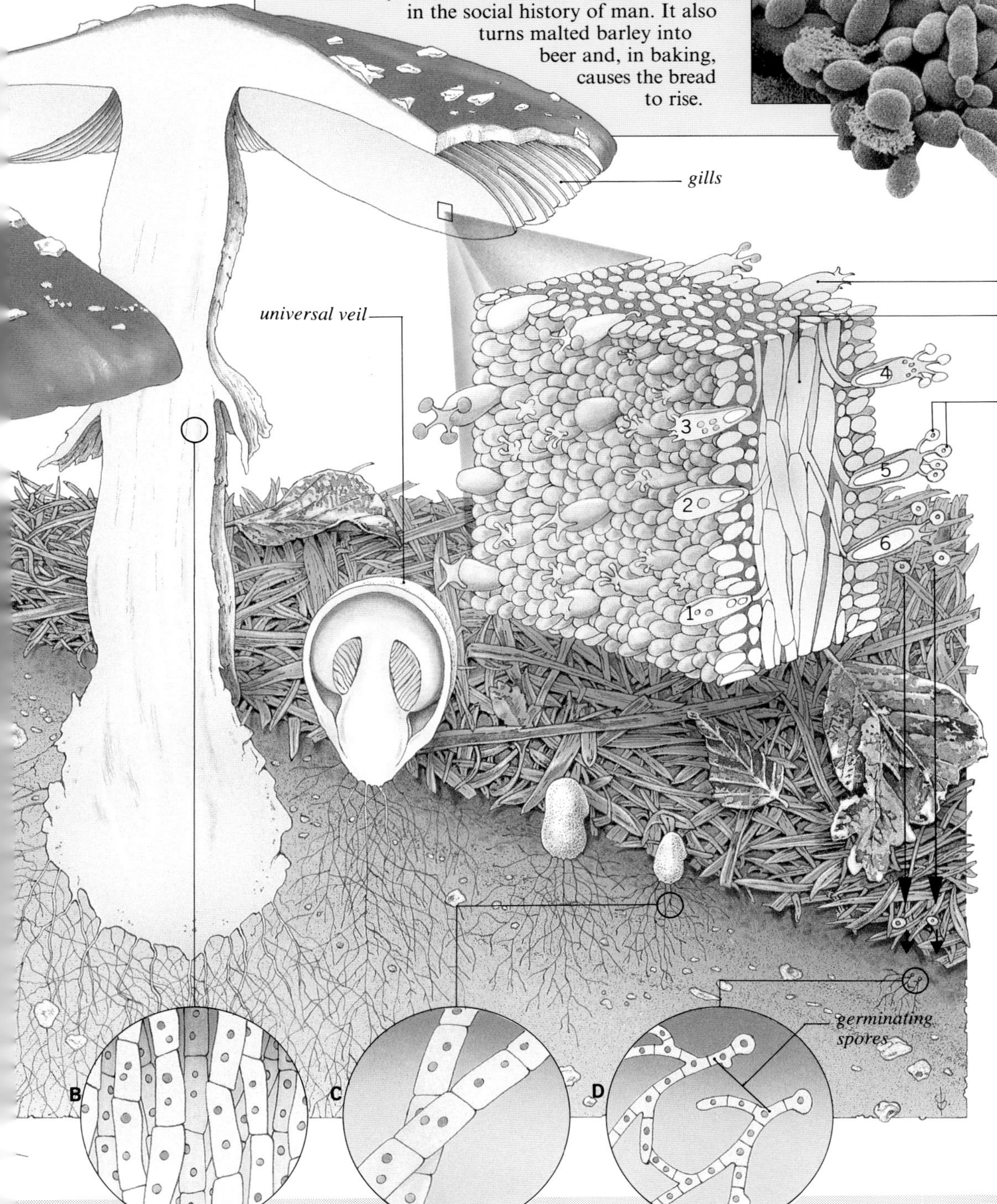

Fly Agaric's fruiting body consists of a stalk (stipe)*, made of closely packed hyphae* [**B**]*, attached to a buried mycelium and crowned with a broad cap* (pileus)*, which protects the delicate spore-bearing layer* (hymenium) *on the gills. The life cycle of this toadstool begins with the germination of spores to form a mycelium, the compartments of which contain one nucleus each* [**D**]*. Hyphae of different "sex" fuse, and a secondary mycelium develops in which the compartments contain both parental nuclei* [**C**]*. The fruiting body grows from the secondary mycelium, starting life as a button enclosed in a protective universal veil, which eventually ruptures. In the hymenium the two nuclei in each compartment* [1] *fuse and their chromosomes are reshuffled in meiosis* [2] *and the four resulting nuclei* [3] *migrate to the ends of club-shaped cells (basidia)* [4,5]*, from which they bud off and are shed as spores* [6]*.*

Water Weeds

How different types of algae live

Leave a jar of clear pondwater on a sunny windowsill and it will soon go cloudy and green. You may even see this green cloud stir, following the light as the Sun moves round. The cloud is made up of thousands of single-celled algae, which are using the sunlight to carry out photosynthesis and in this way manufacture their own food. Not all algae are so tiny, however, and not all are green either. The green microorganisms in pondwater seem to have little in common with the large red and brown seaweeds of the seashore – yet they are in fact all algae.

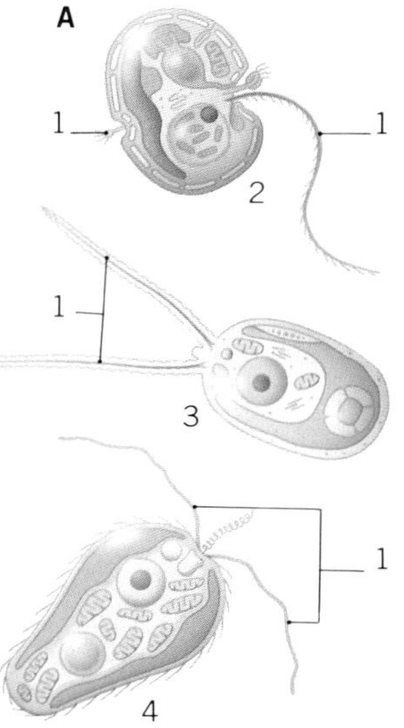

***Many unicellular algae* [A]** *are said to be motile, which means that they move in response to changes in their environment, in particular to light. This is achieved by taillike flagella* [1], *which propel them through the water. The dinoflagellate* Gonyaulax tamarensis [2] *has two such flagella at right angles to each other, one almost hidden in a groove running around the cell. When both of these work together, the cell moves forward with a whirling motion that is characteristic of the dinoflagellates.* Chlamydomonas [3], *a green algae, and the golden-colored* Prymnesium parvum [4] *also have two flagella, though others may have more or less – such as* Euglena, *which has one, or* Platymonas, *which has four.*

Because they can carry out photosynthesis, algae are self-supporting and can live wherever there is light, oxygen and carbon dioxide, sufficient water, and a few essential elements. Like land plants, algae contain the green pigment chlorophyll, by which they capture the energy of sunlight to power photosynthesis. But although they all contain chlorophyll, many also contain other pigments that color them red, brown, or gold.

Algae many yards long

The brown seaweeds, such as bladderwrack, oarweeds and tangleweeds, are the largest and most complicated algae, growing many yards long in some cases. They are multicellular like land plants. Many have a stalk ending in a rootlike holdfast, which anchors them to rock or the sea bottom, and leaflike fronds, which carry out photosynthesis. But inside they are much simpler than almost all land plants. They do not have specialized roots to take in water, nor internal tissues like xylem to transport water and minerals around the plant.

Algae absorb the materials they need over most of their surface. Truly aquatic algae take in the carbon dioxide they need for photosynthesis and the oxygen they need for respiration from the water. So multicellular algae like the seaweeds have to grow in the form of thin sheets only a few cells thick, or flattened blades, or fine strands, so that water, carbon dioxide and other nutrients reach all their cells.

Only the large brown seaweeds, such as the giant kelp, have specialized tissues, like the phloem of land plants, which carry the end products of photosynthesis from the frond near the sea surface to the poorly illuminated stalk and holdfast many yards below.

The most complicated freshwater green algae are the stoneworts, which grow up to a yard (1 meter) high on the bottoms of ponds, with whorls of fine branchlets on a delicate stem. Although they look very similar to "ordinary" plants, their "stems" and "branches" are in fact made up of individual large cells placed end to end. Green algae also form the photosynthetic partner in some lichens, a symbiotic association of algae or photosynthetic bacteria and fungi.

Fossilized single-celled algae have been found in rocks nearly a billion years old.

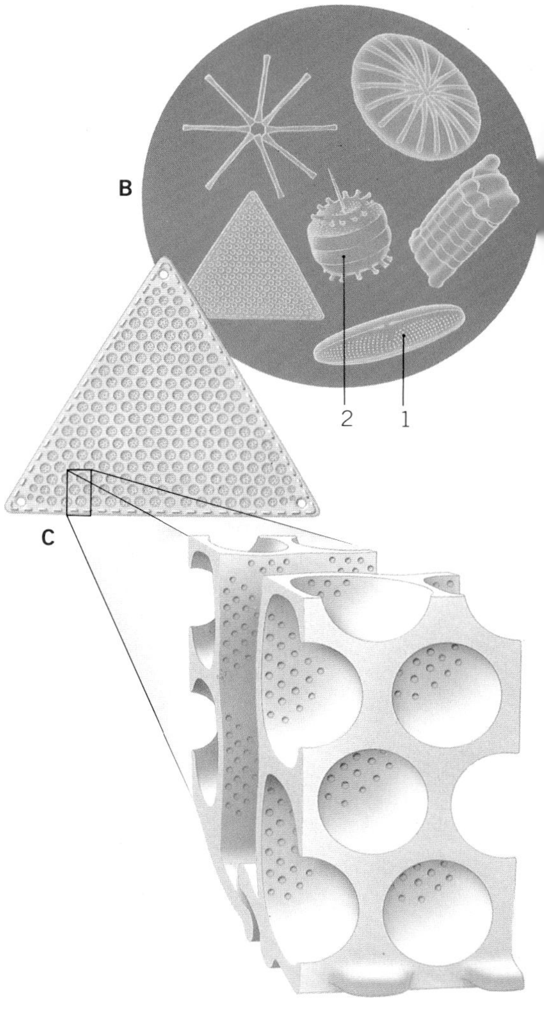

***Diatoms* [B]** *are mostly unicellular, eg* Navicula digitoradiata [1], *though some form simple colonies, eg* Thalassiosira [2]. *Diatoms found in fresh and saltwater include (counterclockwise):* Biddulphia biddulphia, Asteromphalus elegans, Asterionella formosa, Triceratium favus.

The shells of diatoms *are in two symmetrical halves that fit together like a box and lid. These shells have an amazingly elaborate structure: that of* Triceratium favus [C] *has many tiny pores, which provide passageways between the inside of the cell and the environment outside.*

From their origins in the sea the green algae in particular have successfully moved farther afield and colonized fresh waters and moist habitats on land.

The meadows of the sea

But the most important algae in the food chain are the microscopic algae of the plankton. Single-celled algae such as diatoms and dinoflagellates make up a large part of the phytoplankton, the microscopic photosynthetic bacteria and algae that drift in the oceans. The notorious "red tides" off the coast of North America, which poison fish and shellfish, are caused by dinoflagellates. In certain conditions they undergo a population explosion, producing an algal "bloom" that turns the sea red.

From seaweeds come many useful products, including agar, a jellylike medium on which bacteria can be cultured.

Connections: Bacteria 106 Fungi 110 Lake Life 326 Ocean Life 332 Photosynthesis 182 Plants in Cold and Wind 140 Plant Reproduction 150

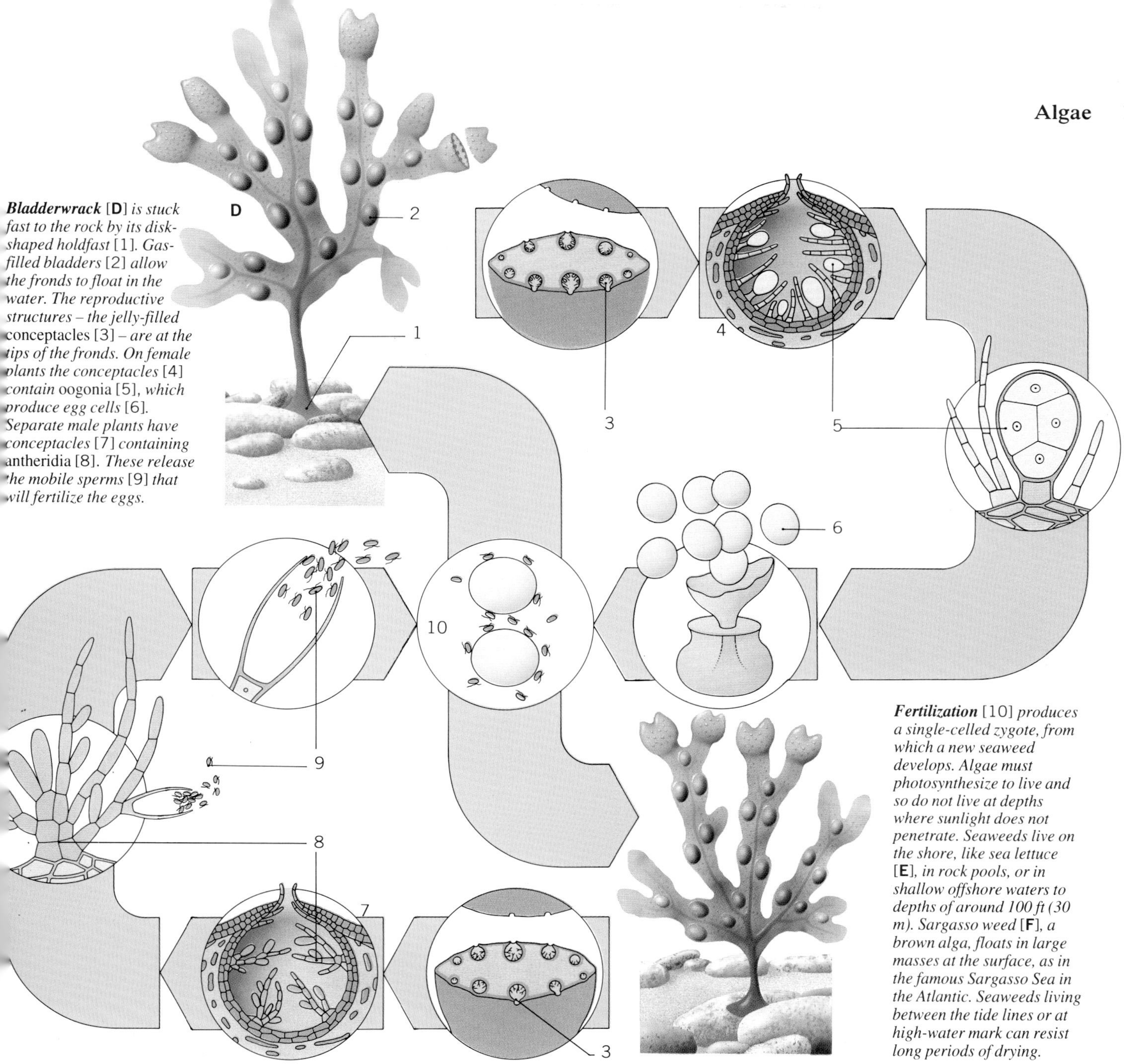

Bladderwrack [**D**] *is stuck fast to the rock by its disk-shaped holdfast* [1]. *Gas-filled bladders* [2] *allow the fronds to float in the water. The reproductive structures – the jelly-filled* conceptacles [3] *– are at the tips of the fronds. On female plants the conceptacles* [4] *contain* oogonia [5], *which produce egg cells* [6]. *Separate male plants have conceptacles* [7] *containing* antheridia [8]. *These release the mobile sperms* [9] *that will fertilize the eggs.*

Fertilization [10] *produces a single-celled zygote, from which a new seaweed develops. Algae must photosynthesize to live and so do not live at depths where sunlight does not penetrate. Seaweeds live on the shore, like sea lettuce* [**E**], *in rock pools, or in shallow offshore waters to depths of around 100 ft (30 m). Sargasso weed* [**F**], *a brown alga, floats in large masses at the surface, as in the famous Sargasso Sea in the Atlantic. Seaweeds living between the tide lines or at high-water mark can resist long periods of drying.*

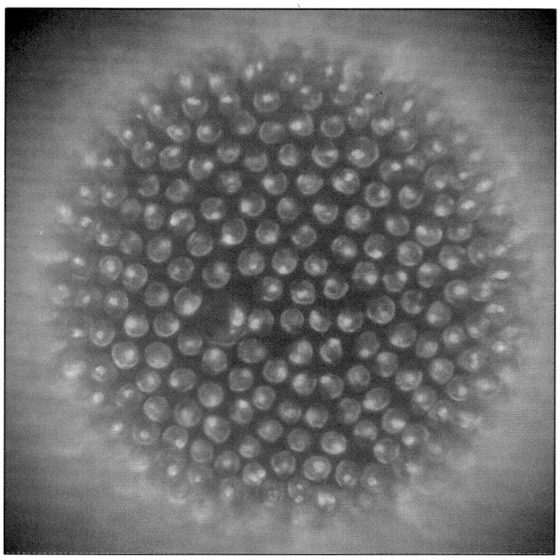

olvox *is a colonial green eshwater alga that forms a ollow sphere just visible to e naked eye, made up of ousands of microscopic dividuals, all joined at eir cell walls by thin rands of protoplasm coated ith a jellylike substance. ach cell has two flagella hose beating spins the olony along. Within an exual colony only some dividuals in the interior of e sphere can reproduce emselves. They lose their agella and divide to form ny new colonies, which are entually released from the arent. There are also sexual olonies whose zygotes are ly released on the death the parent.*

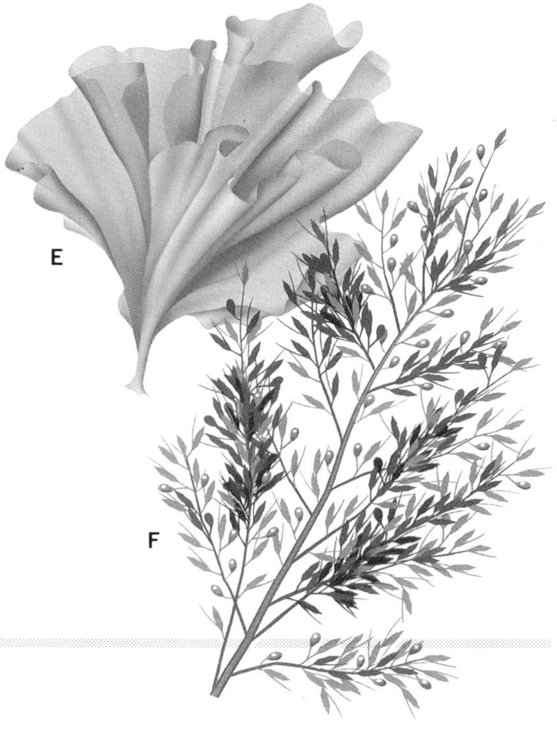

eashores 330 Water Pollution 340

From Root to Shoot

How plants are structured

From the tiny flower of the aquatic duckweed, only a hundredth of an inch across, to the vast redwood trees, of which just the bark itself may be a foot thick, the structures of plants exhibit great diversity. The strongest – trees – spread widely in many directions, the roots of each searching out water and essential minerals. Overhead, supported on a trunk and branches, a canopy of leaves spreads out to collect as much light as possible. Unable to move around, plants have nonetheless evolved to extract all they need from their surroundings.

Each part of a plant is designed to fulfill a particular function: to make food, to take up water and minerals, to transport and store food within the plant, or to reproduce.

Above ground are the green stems and leaves, which capture light and produce food. The leaves of green plants grow in many shapes and sizes but are generally arranged regularly along the stem. Such an arrangement minimizes overlap and ensures that each leaf catches a maximum amount of light. The leaf surface is sealed with a watertight, airtight layer of cells, the *epidermis*, which secretes a waxy *cuticle*. This is pierced by pores in the leaf, *stomata*, through which air enters and waste gases leave. Water vapor also evaporates from the inside of the leaf through the stomata, creating the transpiration pull required to draw water upward from the roots. To avoid excessive water loss on a hot sunny day, stomata are usually much more numerous on the lower surface of the leaf, the side not turned to the Sun. Guard cells on leaf surfaces open and close the stomata to control the rate of water loss.

The stem of a plant holds the leaves up and out toward the light, and holds the flowers out to catch the wind or to where the insects or birds that pollinate the plant can find them. The stems of more *arboreal* plants are supported by a rigid column of wood. *Herbaceous* plants with slender stems and no woody tissue maintain a rigid stem by keeping the stem cells full of watery sap.

Plant feeding

Roots absorb water and mineral salts, and anchor the plant in place. The surface epidermis cells of roots just behind the growing tip have long, tubular extensions – the root hairs – which increase the surface area of the root available for absorption. Plants continually take up water from the soil, for 90 percent or more of the water they take up is always lost through inevitable transpiration: the evaporation of water and water vapor into the atmosphere from the leaves.

The way a plant grows is determined partly by its own heredity but also by its surroundings. Plants growing on the woodland floor are often tall and slender, competing for light, whereas the same plant growing in the open may be shorter and bushier. Plants growing in short grass will often develop

The diagram of a plant **[A]** *shows the tubelike cells used for transporting water and nutrients. The cells form the* vascular bundles, *which begin near the root tip and run up the length of the stem and into the leaves. They consist of two main elements – the* xylem *and the* phloem. *The xylem carries water and dissolved mineral salts up through the plant to the leaf tissue, while the phloem carries sugars up and down the plant. The leaf of a dicotyledonous plant consists of two distinct parts – the* petiole, *or leaf stalk, which prevents the leaves bunching on the stem, and the* blade. *The cross section of the leaf* **[B]** *shows how the veins, themselves vascular bundles, apart from carrying water and nutrients, also support the rest of the blade, including mesophyll tissue, which contains chloroplasts, vital for photosynthesis. The stem supports the plant and contains many vascular bundles* **[C]**. *The roots* **[D]** *have thousands of tiny root hairs growing from them. As well as anchoring the plant in the soil, the root hairs also increase the surface area of the root system, thereby enabling the plant to absorb as much water and as many minerals as possible. The root tip* **[E]** *is divided into three sections. The root cap protects and lubricates the tip of the root, giving the growing roots a smooth passage through the soil. The root cap is also the region in the root that perceives gravity and thereby controls the direction the root grows. The* meristematic *zone is where cells divide, adding new cells to the root cap below it, which is being permanently worn down, and to the zone of elongation above it, the zone that aids rapid growth in root length.*

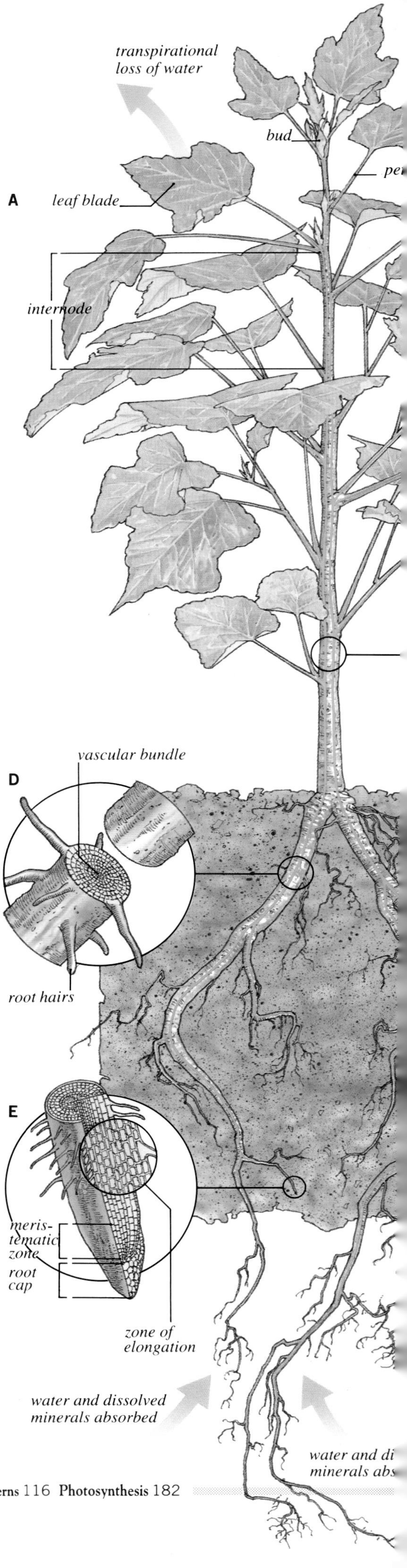

Connections: Conifers and Cone Bearers 118 Flowering Plants 120 Germination 158 Mosses and Ferns 116 Photosynthesis 182

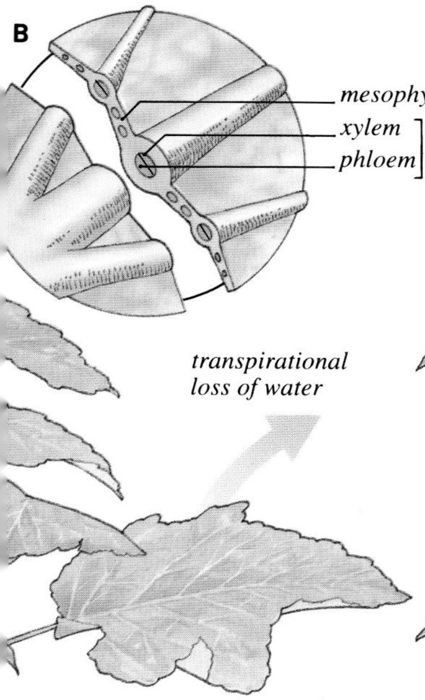

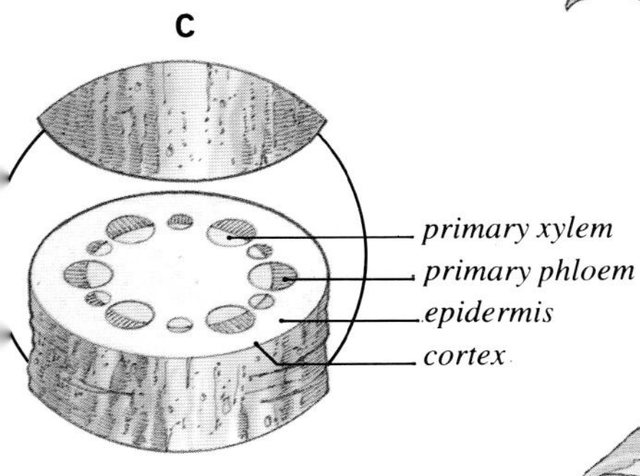

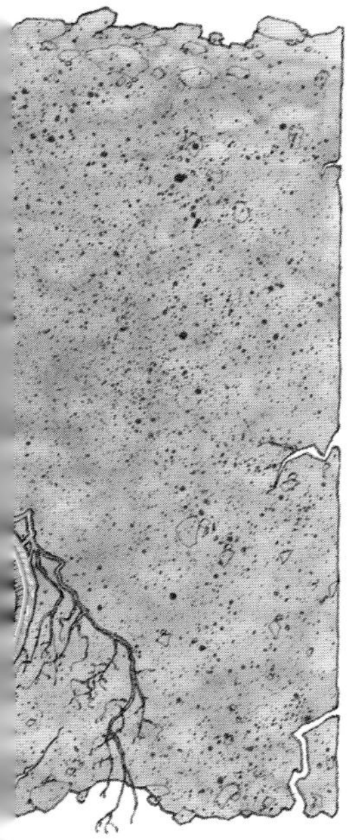

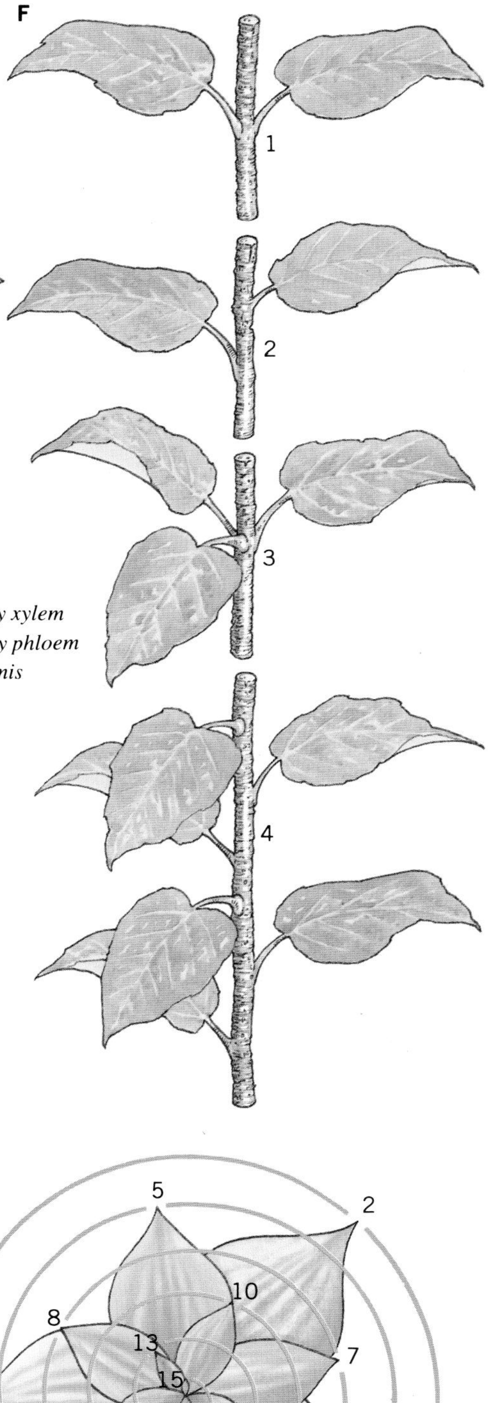

The position of leaves [F] *growing from the stem is not always the same. The growing patterns (base patterns) vary from family to family. There are, however, four distinct patterns – opposite* [1], *of which the butternut of North America is an example, alternate* [2], *such as English holly, whorled* [3], *as seen in hedge bedstraw, and finally spiral* [4], *of which bittersweet is an example. In the case of alternate and spiraled configurations, new stalks grow a set distance apart. However, if a stalk is broken off, its replacement will grow closer to the nearest existing stalk.*

ground-hugging rosettes of leaves to suppress the growth of grass around them and also to prevent the permanent parts of the plant from being eaten by grazing animals.

Shape and form

A fir tree, an oak and a date-palm illustrate three quite different types of plant architecture. The fir tree (a conifer) has a symmetrical conical shape. Its branches come out at regular angles around the trunk and bear small, stiff twigs along their length. The oak (a broad-leaved tree) has branches dividing many times into successively smaller branches and finally into twigs that bear the leaves and flowers, creating its distinctive shape. The date-palm is quite different from either. Its trunk does not branch at all and carries much-divided leaves in a bunch at the top.

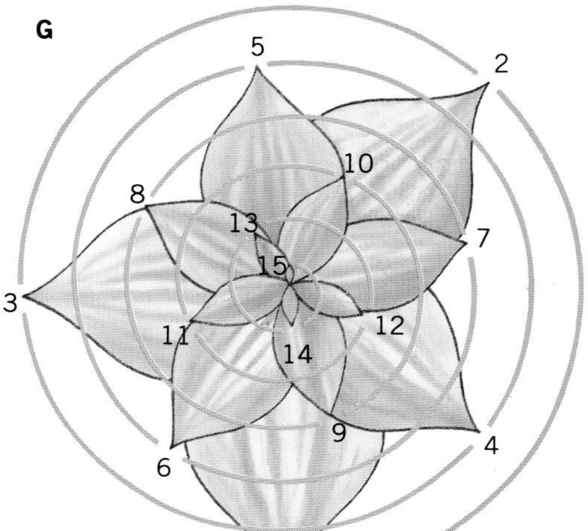

'he leaves of a plant [G] *'hen seen from above grow 'n a spiral pattern. The 'ngles between one leaf and 'he next follow a strict 'nathematical series known 's the Fibonacci series. This 'nsures that each individual 'eaf on the plant stem 'eceives the maximum 'mount of sunlight available.*

Woody Plants

Wood is made of the old xylem of previous years' growth, the walls of which have become heavily impregnated with *lignin*. The cells die and are embalmed in tannins and resins, forming the strong, dark heartwood. This is surrounded by sapwood, which is weaker and makes up most of the thickness of the trunk in young trees. A layer of new sapwood is added to the outer edge of the column of wood with each year's growth, so forming a ring visible among others in the trunk. These are most apparent in temperate regions, where patterns of growth vary from season to season.

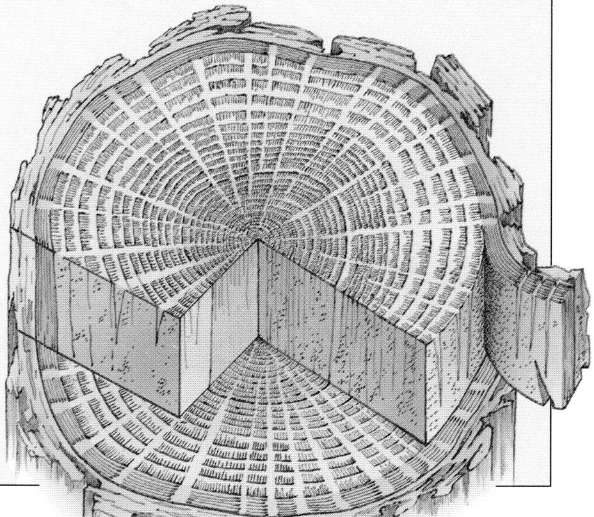

'lant Cells 100 Plant Feeding 184 Plant Reproduction 150 Seeds and Fruit 156

Primitive Plants

How mosses and ferns live and reproduce

The modern liverworts resemble the earliest and simplest land plants that emerged from the water 400 million years ago. Then they were little more than a layer of green cells on the wet mud. The most advanced of the modern spore plants – the ferns – can grow more than 80 feet (25 meters) tall. But all spore plants are survivors from an earlier age, snapshots that recall the evolutionary story of the land plants, and even these tree ferns, found mainly in dense tropical rain forests, are a reminder of the giant ferns and club mosses that once dominated the Earth.

***The dominant** sporophytic form of the lady fern* [**A**] *has upright leaves (or fronds)* [1] *and roots* [2] *that arise from nodes on the underground stem* [3]. *Ferns can reproduce asexually, usually by fragmentation of the underground stem, or sexually, through the formation of the independent, short-lived plant – the gametophyte. The sexual process* [**B**] *begins with meiosis, the results of which are spores. On the underside of fronds structures called* sori [4] *enclose the* sporangia [5] *in which the spores are formed. After its release* [6] *a spore germinates into a heart-shaped* prothallus *(the gametophyte)* [7], *which bears egg and sperm-forming structures –* archegonia [8] *and* antheridia [9] *respectively. In the wet the sperm, propelled by their many flagella* [10], *swim down the neck of the archegonium to fertilize the egg* [11]. *The resulting zygote develops into the fern plant* [12].

The most familiar spore plants are the mosses and ferns, but there are several other types of plants in this group as well, including liverworts, club mosses and horsetails. All produce spores – microscopic particles that can grow into new plants – although they do so indirectly, via a complex life cycle with two distinct stages. One of these stages relies on a wet environment, and this restricts most spore plants to damp habitats. Another limiting factor for spore plants is their system for transporting water from the roots to the leaves, which is less well developed than in higher plants. Indeed, in the mosses and liverworts it is rudimentary, which is why they are limited to a small size.

The two phases of the life cycle are called the *gametophyte* ("gamete producing": gametes correspond to eggs or sperm) and the *sporophyte* ("spore producing"). The gametophyte stage represents the sexual generation of the plant, and the sporophyte stage is the one in which the plant matures, asexually reproducing itself by cloning and by dispersing spores. The cells of the gametophyte have only one copy of each chromosome (they are *haploid*). These haploid cells are produced by *meiotic* cell division. The cells of the sporophyte each have two copies (*diploid*).

Fertilization occurs when a gametophyte's sperm penetrates an egg. In combination they pool their chromosomes, so the fertilized egg has two copies of each chromosome. This grows into the sporophyte. To reach an egg cell, a spore plant's sperm has to swim – just as the sperm cells of animals do – so the gametophyte needs a damp habitat.

The struggle for dominance

The story of plant evolution on land is the story of how the sporophyte gradually took over from the damp-loving gametophyte. This allowed plants to grow in drier and drier places without losing the ability to reproduce sexually. At the beginning of the story, however – with the mosses and liverworts – the gametophyte is the dominant partner. In a moss, for example, the gametophyte is the green plant that we are familiar with. The sporophyte is the slender brown stalk and cap that grows up from the moss plant at certain times of the year. It has no independent existence, relying on the gametophyte to supply all its nourishment.

In the next stage of the story, represented by the ferns and horsetails, the sporophyte and gametophyte are physically separate, but the gametophyte is small and short-lived whereas the sporophyte is the dominant partner. Some club mosses take things one stage further: their gametophytes do not even emerge from the spore case. They stay lodged on the sporophyte plant and fertilization takes place there. But they still need water, because the sperm must swim to the eggs.

Another important change occurred during this evolutionary saga. Whereas the mosses have a single type of spore and a single type of gametophyte, producing both eggs and sperm, the more advanced spore plants – such as the club mosses and horsetails – have separate male and female spores. This was an important step toward the development of the higher plants, in which the male spore evolved into the pollen grain.

***Bracken is an aggressive** and successful species of fern. Its cocktail of toxins and carcinogens deters mos grazers, but crucial to its success are its underground stems, or rhizomes, which spread through the soil. Thi asexual reproduction is so effective that bracken has given up sex.*

Connections: Algae 112 Conifers and Cone Bearers 118 Early Evolution 92 Flowering Plants 120 Germination 158 Plant Cells 100

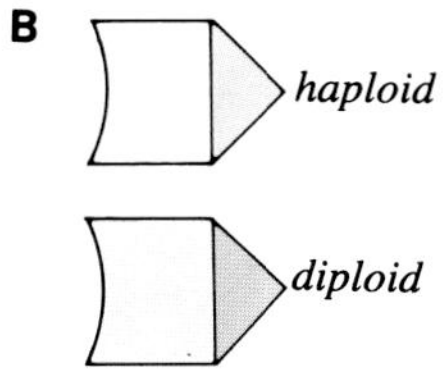

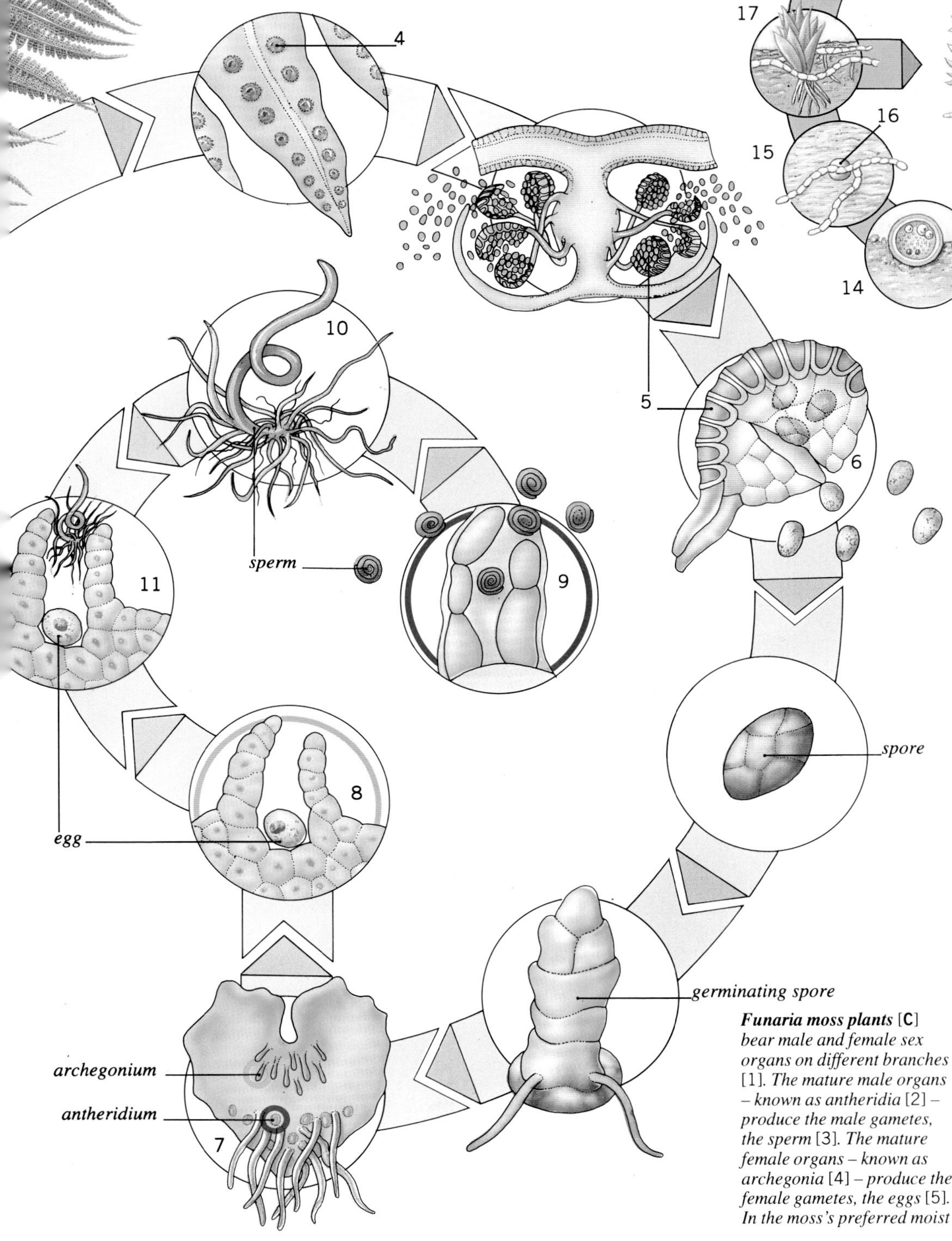

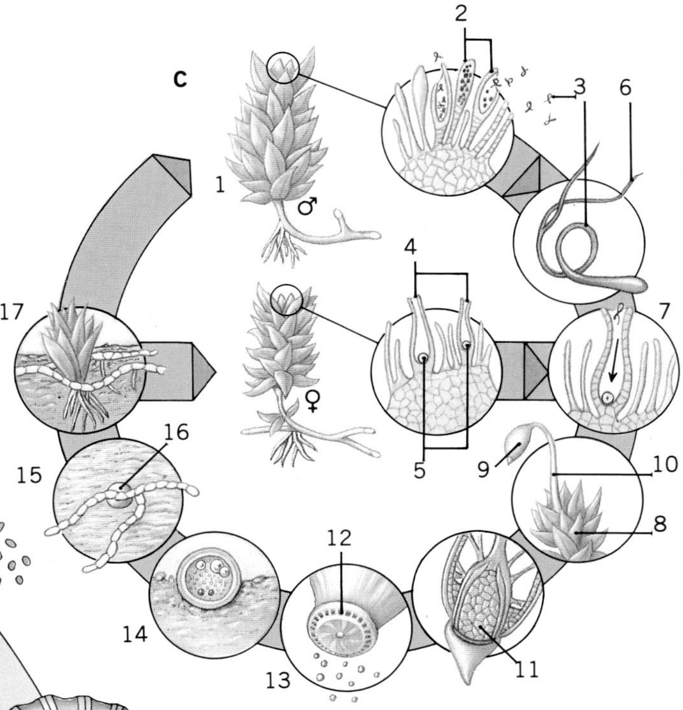

***Funaria moss plants* [C]** *bear male and female sex organs on different branches [1]. The mature male organs – known as antheridia [2] – produce the male gametes, the sperm [3]. The mature female organs – known as archegonia [4] – produce the female gametes, the eggs [5]. In the moss's preferred moist climate there is a constant thin film of rain or dew over the leaves. Using whiplike tails [6] the sperm swim through this to the females. They make their way down the necks of the archegonia to fertilize the eggs [7], producing a diploid zygote. This develops into an embryo from which the sporophyte grows, borne on top of the female gametophyte [8]. The sporophyte consists of a* sporogonium, *or capsule [9], on top of a stalk, or* seta *[10]. Within the capsule repeated meiotic divisions produce many spores [11]. Upon maturation these are released through the openings, or* opercula, *of the capsule [12] and dispersed by the wind [13]. Those spores that land on moist ground germinate [14] to form a branched filament called the* protonema *[15]. The buds of the protonema [16] eventually develop into gametophytic moss plants bearing male and female sex organs [17].*

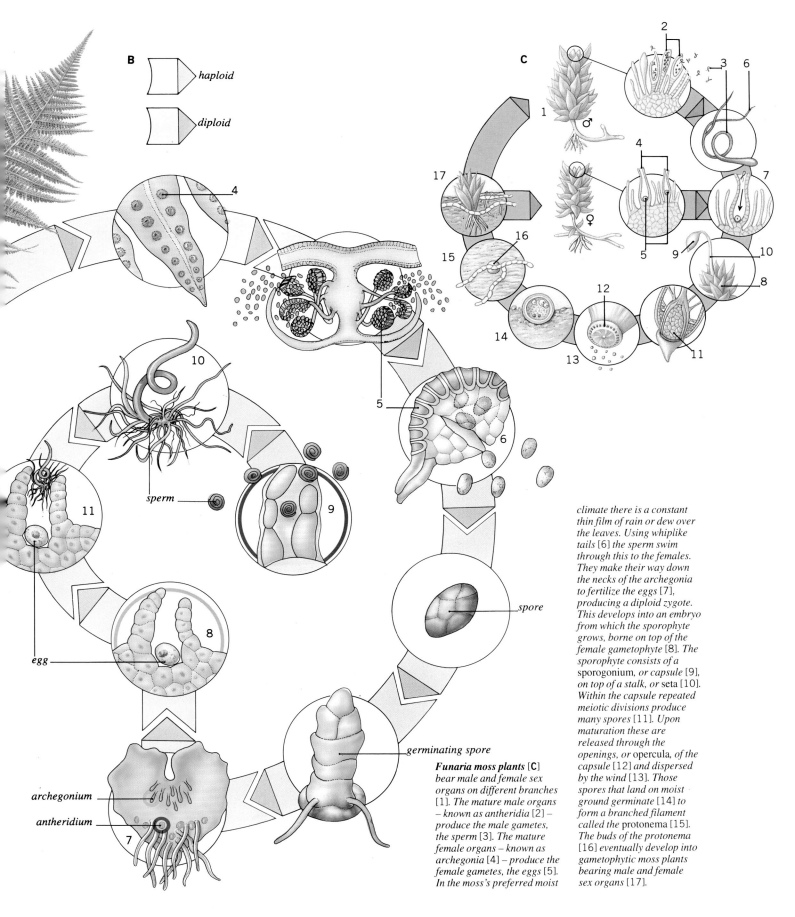

The Ancient Evergreens

How conifers and other cone bearers thrive

The great redwood tree (giant sequoia), found in the mountains of the western American Pacific coast, grows so big that a tunnel can be cut – and cars driven – right through its trunk. The biggest specimens can be 260 feet (80 meters) tall, and weigh an estimated 2,000 tons. Coast redwoods are even taller. The conifers, the group of trees to which the redwoods belong, boast not only the biggest trees but also the oldest. Nearly 4,000 years ago, high in the White Mountains of California, a pine seedling started to grow. That bristlecone pine is still alive, and is one of the longest-lived trees in the world.

The great coniferous forests of the Northern Hemisphere, stretching across the North American continent and much of Siberia, make up a third of the world's forests. They are the home of most of the world's species of conifer trees, although a few relatives such as the monkey puzzle from South America and the kauri pine, found in regions of Australia and New Zealand, are still to be found in the Southern Hemisphere.

The cone bearers were once even more numerous and widespread than they are now, having appeared millions of years before the flowering plants. All present-day conifers are trees or shrubs. They and other cone bearers such as the primitive cycads are known as *gymnosperms* (literally "naked seeds") as distinguished from the flowering plants, or *angiosperms*, in which the seeds develop in a folded *sporophyll*, or ovary.

A new leaf

The leaves of conifers are generally either needlelike, as in pines, firs, spruces and larches, or scalelike, pressed closely to the twigs as in cypresses and false cypresses. The needles are often hard and tough, covered with a thick waterproof cuticle, an adaptation for growing in dry conditions. The needle shape itself is also an adaptation to drought. Compared with a thin, flat leaf, a needle presents a much smaller surface area for the same amount of internal photosynthetic tissue, and so cuts down the amount of water lost through transpiration. The *stomata*, the pores in the leaf through which water vapor evaporates, are concentrated on the underside of the needle, away from the Sun, and are also sunken below the leaf surface to slow down transpiration.

Conifers can therefore flourish in the thin, rapidly drained, stony soils on mountainsides, in the harsh conditions of the northern taiga, and in hot, dry climates such as those found around the Mediterranean.

All year long

With a few exceptions, such as larch and swamp cypress, conifers are evergreen, their somber foliage standing out in mixed forests in winter. Instead of shedding all their leaves at once like deciduous trees, they lose and replace needles throughout the growing season. Since they are always clothed in green leaves, they can start to photosynthesize and grow early in the spring and make the most of the brief northern spring and summer.

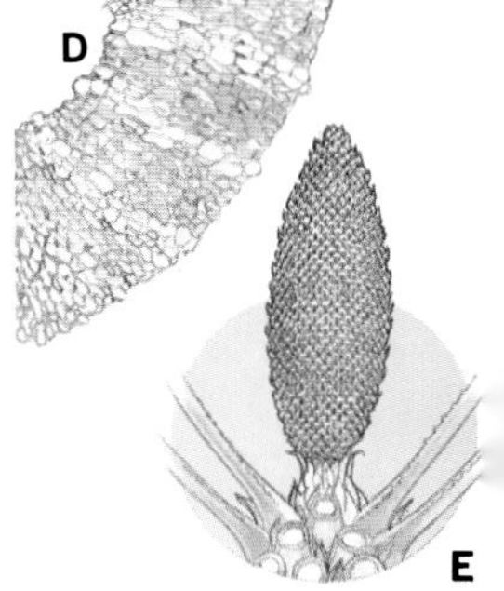

***As plants adapted** to conditions on dry land, reproductive strategies improved. The more primitive land plants have two distinct physical forms,* sporophyte *and* gametophyte, *that alternate in generations. An asexual sporophyte produces spores that grow into a gametophyte, which has one or both sex organs.*

Among mosses [**A**] *the gametophyte form predominates. Club mosses* [**B**] *and ferns* [**C**] *produce spores that grow into tiny hermaphrodite plants.*

Cycads bear either a female [**D**] *or male* [**E**] *cone, similar to conifers. However, in cyads the mobility of the sperm and thus fertilization requires a wet environment. The spores develop into gametophytes. After fertilization the female cones produce seeds that will grow into sporophytes.*

Although the woody cone is the hallmark of conifers, coming in many shapes and sizes, not all conifers bear cones. In junipers, for example, the scales of the cone are fleshy and are fused into a blue-black aromatic berry that encloses the seeds.

One of a kind

The unique ginkgo or maidenhair tree, with its fan-shaped leaves, is related to the conifers. Once widespread, only one species – *Ginkgo biloba* – now survives. Like the true conifer, the dawn redwood, it is a living fossil, having survived unchanged for two hundred million years. The ginkgo bears male and female reproductive structures on separate trees. The male structures resemble small pollen-producing cones. The female trees do not carry cones but large, stalked, fleshy coated seeds.

***The redwood tree,** giant sequoia, despite its height and bulk, bears very small cones. Each female cone produces thousands of seeds, but the germination rate is surprisingly low.*

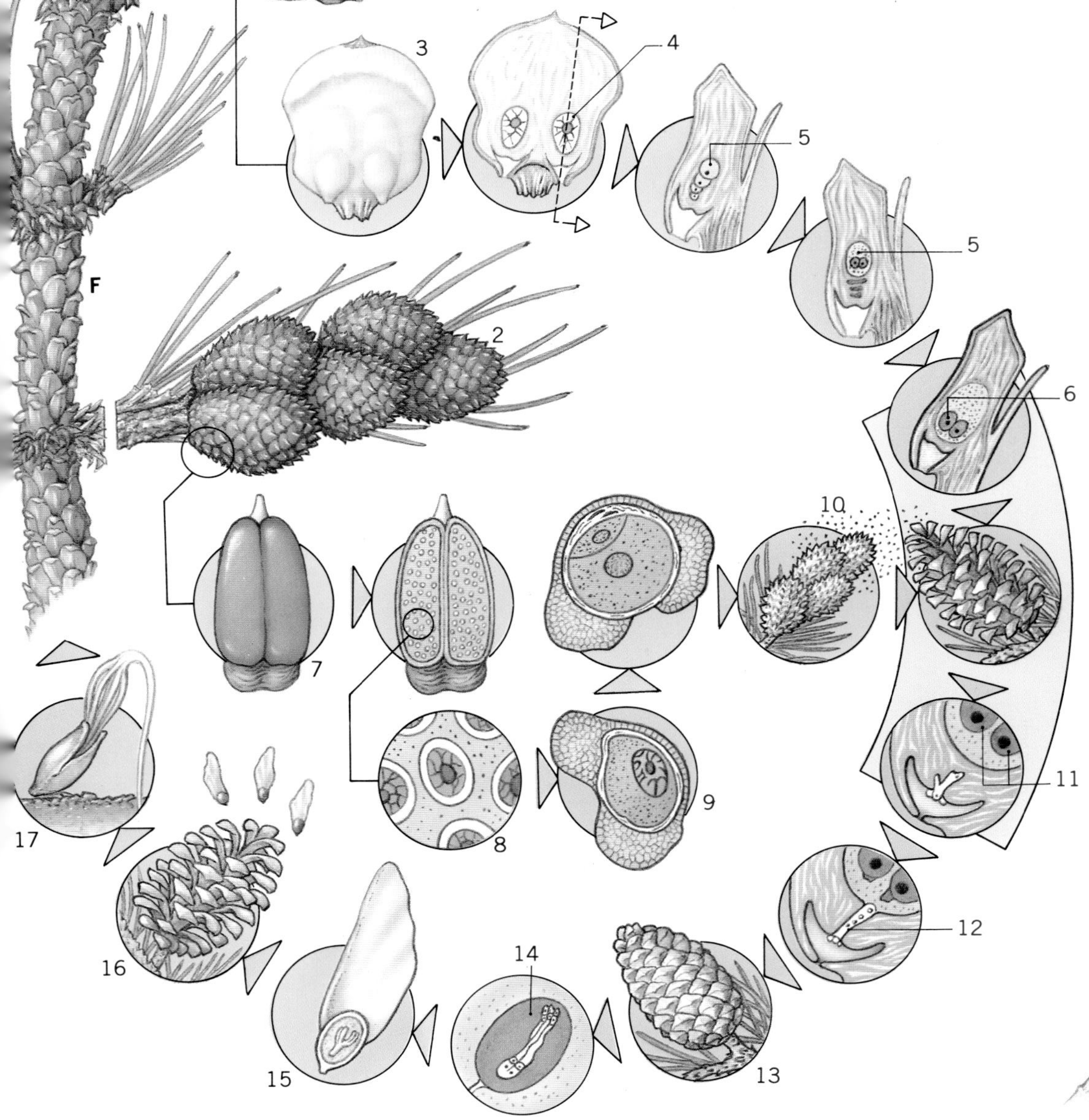

The reproductive cycle *of the ponderosa pine (western yellow pine)* [**F**] *is typical of many conifers. In summer the mature sporophyte tree bears both female cones* [1] *and male cones* [2]. *A scale from the female cone* [3] *contains two* ovules [4]. *Within each ovule a spore cell* [5] *divides first by mitosis and later by meiosis to develop into a female gametophyte* [6]. *A scale from the male cone* [7] *contains many spores* [8]. *Each of these develops into a male gametophyte within a winged pollen grain* [9]. *This process lasts one year.*

Pollination [10] *occurs early the next summer, when female cones open so that airborne pollen grains can enter an ovule. Inside the ovule the female gametophyte develops two ova* [11]. *Fertilization occurs during the spring of the following year, after the male gametophyte has matured and grown a pollen tube* [12] *to one of the* ova. *Sperm nuclei pass down the pollen tube and the cone closes* [13]. *Within the female gametophyte the fertilized ova* (zygote) *develops into an embryo* [14] *and around it, a tough, winged seed case is formed* [15]. *In the autumn of the second year the female cone opens* [16] *and seeds are dispersed by the wind, ready to germinate* [17].

The magnolia [**G**] *is one of the most primitive of the flowering plants and may reflect possible evolutionary linkages between gymnosperms and angiosperms. The carpels of the magnolia flower form a tight spiral "cone," and after pollination this "cone" matures into a collective fruit* [**H**] *in which the seeds are contained.*

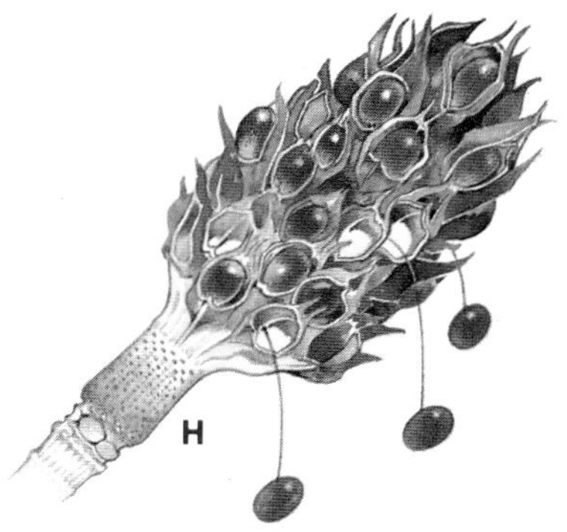

A Budding Future

How flowering plants have flourished

In the rain forests of Indonesia live the parasitic rafflesia, whose red and yellow flowers, at more than 3 feet (one meter) across, are the largest in the world. A foul stench attracts the carrion flies that spread rafflesia pollen from one flower to another, thus allowing them to reproduce. At the other end of the scale are the tiny, stemless, free-floating duckweeds, which are the world's smallest flowering plants, at only a few hundredths of an inch across. Their "flowers" consist of a single anther, which bears the pollen, and a single ovule from which a seed develops.

Wherever you go – through meadow or wood, on bleak mountaintop or in the tropical rain forest – most of the plants you see have one thing in common: they all bear flowers, though some are very small and inconspicuous. In other respects the 235,000 or so species of flowering plants (the *angiosperms*) are a very varied group, and include the familiar wild and garden flowers and flowering shrubs, most trees other than conifers, and the grasses, reeds and rushes.

Difficult as it might be to imagine a world without flowers, the flowering plants evolved only relatively recently. Becoming established some 120 million years ago, during the age of the dinosaurs, they soon came to dominate the world's vegetation, as they still do today. They developed in parallel with the evolution of insects, on which a great many flowers depend for pollination. In many cases insect and flower have co-evolved so closely that the two are completely dependent upon each other – as is the case with yucca moths.

Flowering groups

Botanists have traditionally divided the angiosperms into two groups: the *monocotyledons* and the *dicotyledons*. Monocotyledons generally have long, narrow leaves, with veins that run lengthwise. The leaves often grow from the base of the stem rather than from the tip of a shoot, and the embryo plant curled up in the seed has only one "seed leaf" (hence its description – monocotyledon – meaning "single leaf"). Grasses and palms, bananas and plantains, orchids and bulbs – such as lilies, daffodils and tulips – are all monocotyledons.

The dicotyledons are a much larger group. Their leaves come in many shapes, the veins forming a network through the leaf. The plants themselves grow in a branching fashion, with growth points at the tips of the shoots. The embryo plant has two cotyledons, or seed leaves. Most common flowers and vegetables are dicotyledons, as are all broad-leaved trees (like oak, ash, or chestnut).

Trees are the largest of the flowering plants. Their sheer size often means they provide habitats for a whole community of animals. Insects live in the bark; mosses and lichens grow on the trunk; birds and animals make homes in the roots and branches; plants may even grow in their hollows.

Flower power

The purpose of the flower is to carry the reproductive structures and, in most cases, also to attract pollinators. To achieve this, most species of flower are characterized by particular colors and odors, which are recognized by their pollinators. The bright colors are produced by pigments in the flower's cells: anthocyanins, anthoxanthins, betacyanins and carotenes. The fragrant scents of many flowers are often equally important to pollinators: 30 species of Central American *Coryanthes* orchids are all pollinated by different bees, which discriminate on the basis of smell alone. In many orchids scent is commonly made in osmophores, which are modified inner petal parts. The sugary nectar, made in the flower's nectaries, and the pollen itself are both attractive rewards to insects, which then inadvertently bring about pollination.

***Flower structure** varies enormously between species. But underlying the bewildering diversity of form is a common function – that of producing the male and female gametes (*pollen *and* ovules *respectively) and aiding in their union. A typical complete flower comprises four parts –* sepals, petals, stamens *and* carpels *– attached to the enlarged end of the flower stalk, or* receptacle. *The sepals* [1] *lie outside the showy petals* [2] *and form a protective covering over the developing flower bud. Each male stamen is made up of an* anther [3], *which contains the pollen grains, borne on a* filament [4]. *The female carpels are found at the center of the flower: each contains an* ovary [5], *which bears ovules, and a* style [6], *which supports the* stigma [7] *– the structure on which pollen is deposited. Flowers like the buttercup* **[A]** *show relatively primitive organization in which all the flower parts are separate from one another.*

In many angiosperm groups, certain flower parts have changed or disappeared altogether. In the iris **[B]** *the sepals as well as the petals are brightly colored. The male and female parts of the orchid* **[C]** *have fused to form the* column [8]; *pollen is borne not on anthers but in a mass called a* pollinium [9]. *Dandelion "flowers"* **[D]** *are in fact composed of many individual* florets: *those on the periphery are sterile but attract insects to the fertile central florets. Grasses* **[E]**, *which are wind pollinated, have no need for showy petals and sepals: their reproductive parts are enclosed in green, leaflike bracts. Hazel trees* **[F]** *have separate male and female flowers: male flowers lack carpels and are borne on the long catkins; female flowers lack stamens and form small clusters on the stem. In certain species such as holly* **[G]** *the male and female flowers are borne on separate individuals. (See key for* **D – G** *insets.)*

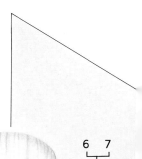

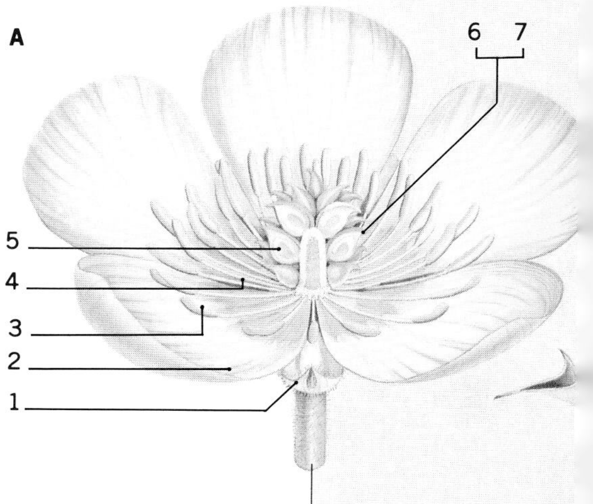

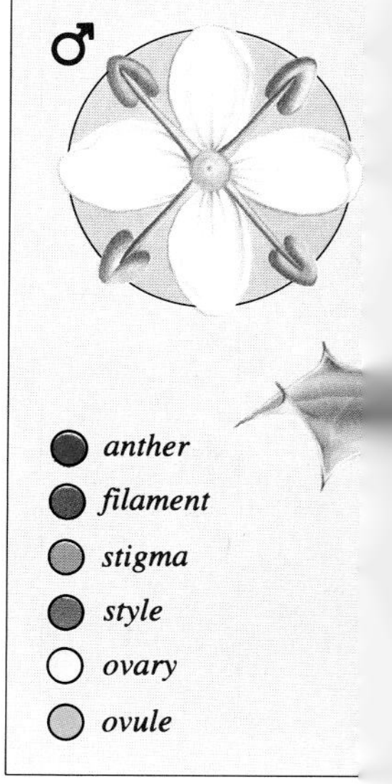

Connections: Plant Architecture 114 Plant Reproduction 150 Pollination by Animals 152 Pollination by Wind and Water 154 Seeds and Fruits 156

***While some plants** bear just a single flower, the majority of species produce a number of flowers in a cluster or* inflorescence. *These inflorescences may be organized in a number of ways. Most grass flowers, like rye (below), grow in* spikes, *with individual groups of small flowers separated into* spikelets. *In an* umbel, *like wild garlic (far left), the flower stalks grow to an equal length, producing a flat-headed inflorescence, while in a* raceme, *like foxglove (left), the main stalk has short branches, each of which terminates in a flower. The oldest flowers are found at the base of the inflorescence and the youngest at the apex. A branched raceme is known as a* panicle. *In a* cyme, *like common chickweed (bottom), the main flower stalk produces a single flower that "blocks" further vertical growth of the stalk. Other flowers are produced farther down the main stalk, giving a branched structure.*

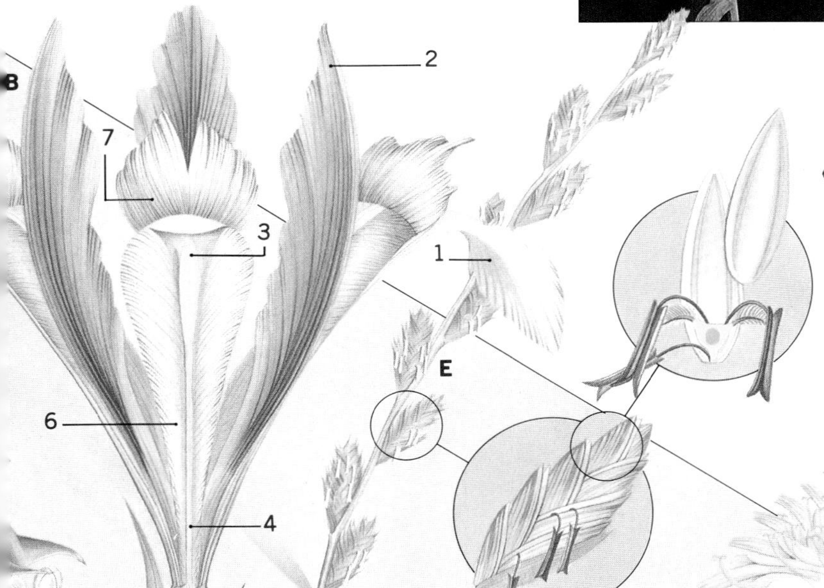

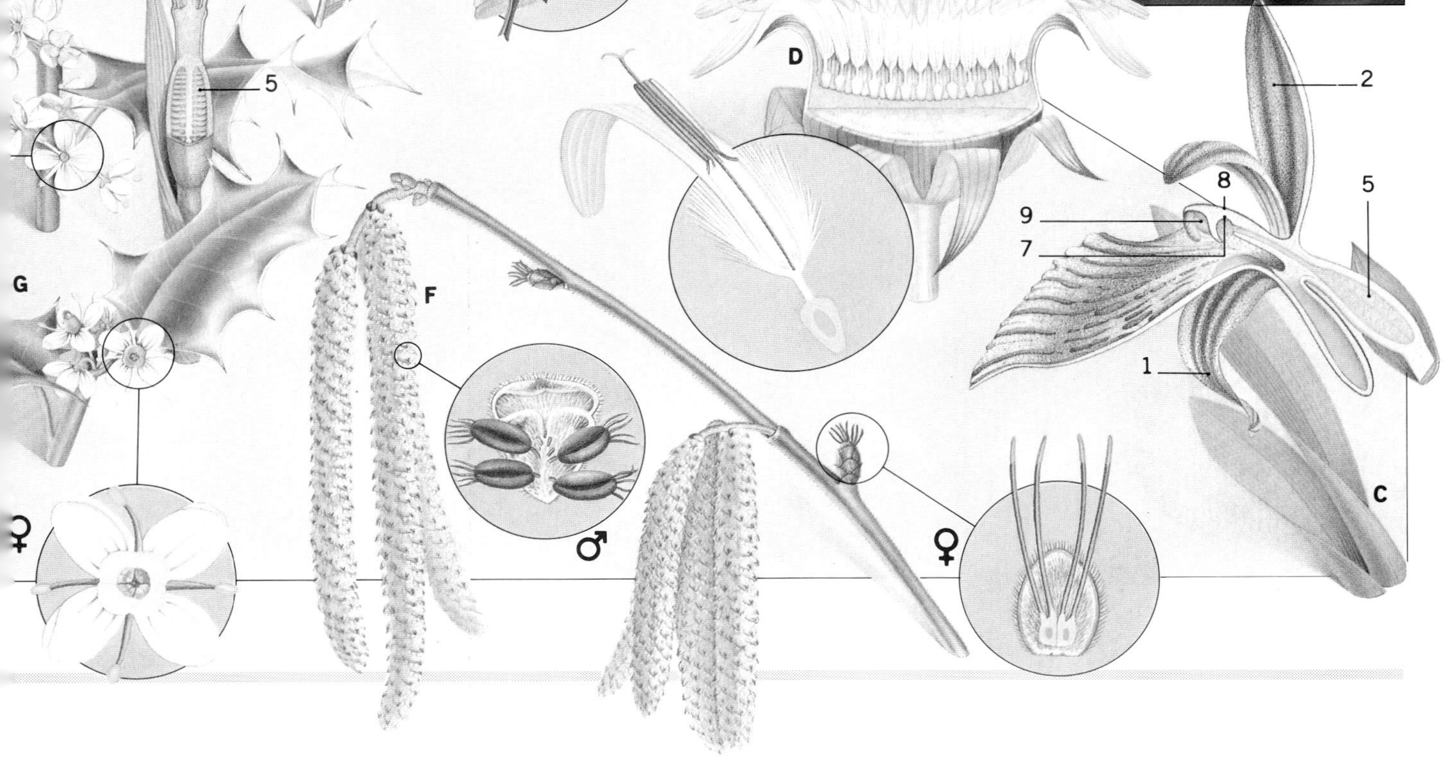

Spineless Life

How invertebrates adapt and survive

The total weight of ants and termites in a tropical forest may well exceed that of the forest mammals, while a single desert locust swarm may cover over 2,000 square miles (5,000 square kilometers) and contain some 250,000 million insects, consuming over 120,000 tons of vegetation a day. There are between 1 and 2 million known species of insect – beetles alone account for some 500,000. With their diversity of form and reproductive methods, insects are found the world over, from the arctic snowfields to desert rocks, from the ocean depths to hot, mineral-laden springs.

Well over 90 percent of all living animals are invertebrates. In fact only one of the 25 or so animal phyla, the *Chordata*, contains vertebrates, and even then several invertebrates are included in this group. Some invertebrates cause considerable problems for the human race – they bore into trees and furniture, undermine houses, damage crops, spoil food and carry diseases; others are invaluable – they pollinate crops, prey on harmful invertebrates, and play a vital role in the decomposition of dead plants and animals, and thus speed the recycling of nutrients.

Invertebrate variations

The diversity of invertebrate shape, size and form is astonishing. Some species, like horseshoe crabs and dragonflies, have remained virtually unchanged for hundreds of millions of years; others, like fruitflies, are still evolving rapidly. Some invertebrates have modified mouthparts, supporting a huge range of feeding habits – from the biting mouthparts of caterpillars to the piercing, sucking mouthparts of mosquitoes. Invertebrates can even tackle such unlikely food as wood and rock.

The evolution of resistant, encapsulated dormant forms has enabled some species to tolerate extreme conditions; freshwater fairy shrimps, for example, can remain dormant as eggs in dried mud for years, hatching only when the mud is submerged again. A wide variety of breathing mechanisms has also evolved, ranging from simple absorption of oxygen through the skin to the complex lungs of spiders, gills of aquatic insects and trachea of terrestrial insects.

Monster invertebrates

Invertebrates are traditionally defined as animals without backbones. However, invertebrates are not without support: they rely instead on hard external skeletons or on the pressure of body fluids for support and as a basis for locomotion. The weight of the external skeleton precludes large size in those terrestrial invertebrates that have them, but in the sea, where water provides support, giants exist. The largest living invertebrate is the Atlantic giant squid, which may weigh up to 2 tons. Such giants can have tentacles almost 50 feet (15 meters) long, and the eyes are the largest ever to evolve – some 16 inches (40 centimeters) in diameter.

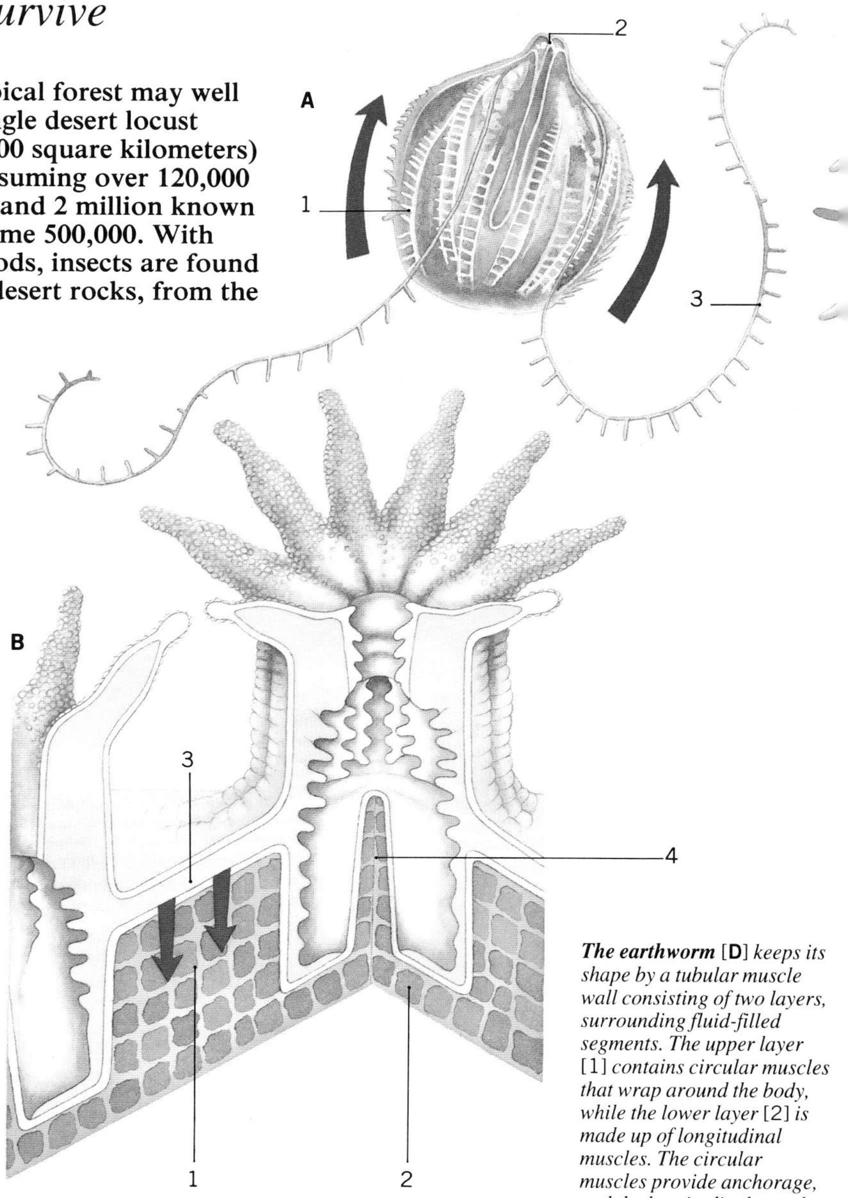

***The free-swimming** comb jelly* [**A**] *maintains its spherical shape through the presence of water in internal canals that lead off the stomach. These canals provide support for the* ctenes *(comb plates)* [1] *with which the comb jelly swims. Each ctene is covered with rows of cilia that beat in rhythmic waves and propel the animal forward. The direction of the effective stroke is away from the mouth* [2], *so that the animal swims mouth first. Simple muscles provide some control over the tentacles* [3].

Stony coral polyps [**B**] *obtain support from a mineralized* theca *(cup),* [1] *which is secreted by the animal. The theca, which is formed of minerals absorbed from seawater, anchors the polyp to the basal structure* [2] *and to its neighbors in the colony via connecting plates* [3]. *All of the mineralization occurs outside the body, although mineral spikes* [4] *may provide additional support. The closely related sea anenome* [**C**] *does not use rigid structures, but supports itself by circulating water around its central cavity* [1]. *Water is drawn in down* siphonoglyphs *(grooves)* [2] *at the side of the cavity and is expelled up the center. The tentacles* [3] *can be withdrawn by individual retractor muscles* [4].

The earthworm [**D**] *keeps its shape by a tubular muscle wall consisting of two layers, surrounding fluid-filled segments. The upper layer* [1] *contains circular muscles that wrap around the body, while the lower layer* [2] *is made up of longitudinal muscles. The circular muscles provide anchorage, and the longitudinal muscles permit forward movement. The intestine* [3], *blood vessels* [4] *and nerve cord* [5] *run right through the segment walls.*

Arthropods, such as the centipede [**E**], *owe their success to the* exoskeleton *(external skeleton) that covers their bodies and allows the development of jointed limbs. Individual body segments are encased in a rigid protein* cuticle [1] *and body flexibility is permitted by an overlapping articulatory membrane* [2]. *The rigidity afforded by the exoskeleton means that muscles* [3] *can be anchored to the inside of the cuticle. Groups of muscles* [4] *are used to articulate the limbs.*

Connections: Animal Parasites 188 Ants' Nests 236 Chemical Defences 268 Insect Flight 226 Internal Animal Parasites 190 Invertebrate Carnivores 200

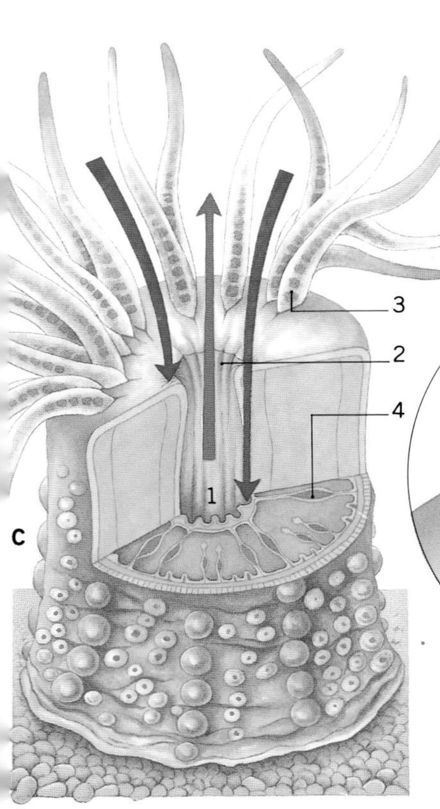

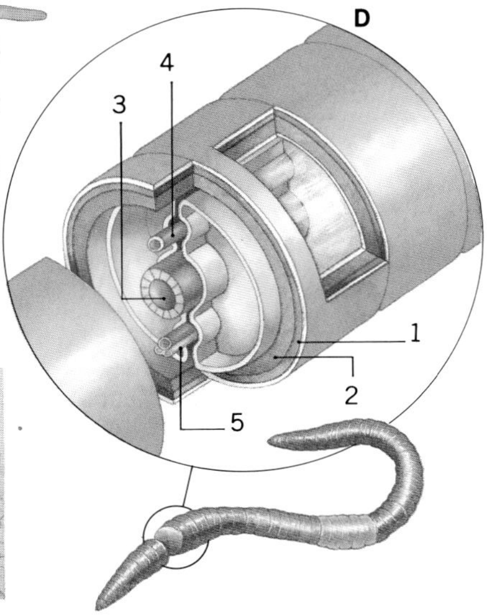

Protective armor

The most successful group of invertebrates – the *arthropods*, or joint-legged animals – have a hard, protective, outer skeleton which can resist desiccation, rain and considerable physical stress. The jointed appendages allow various species to walk, run, jump, swim, fly and seize prey, signal to the opposite sex, impregnate mates and sting enemies.

Mollusks have evolved shells, while tube worms build their own protection, and hermit crabs inhabit the discarded shells of others. Soft bodies, too, have their advantages, especially for internal parasites such as tapeworms. Their soft bodies allow them to absorb the nutrients they obtain from the host through their skin. Invertebrate mobility also varies considerably. Some invertebrates, such as mussels, cement themselves to rocks and the hulls of ships, while others, such as the locusts, are constantly on the move.

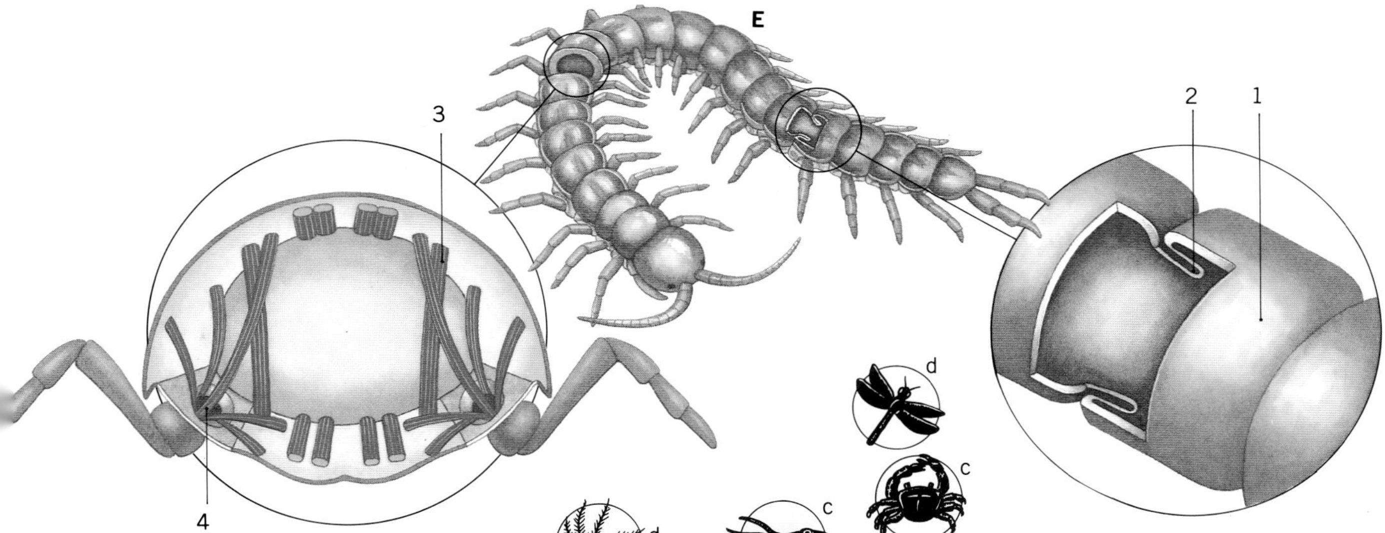

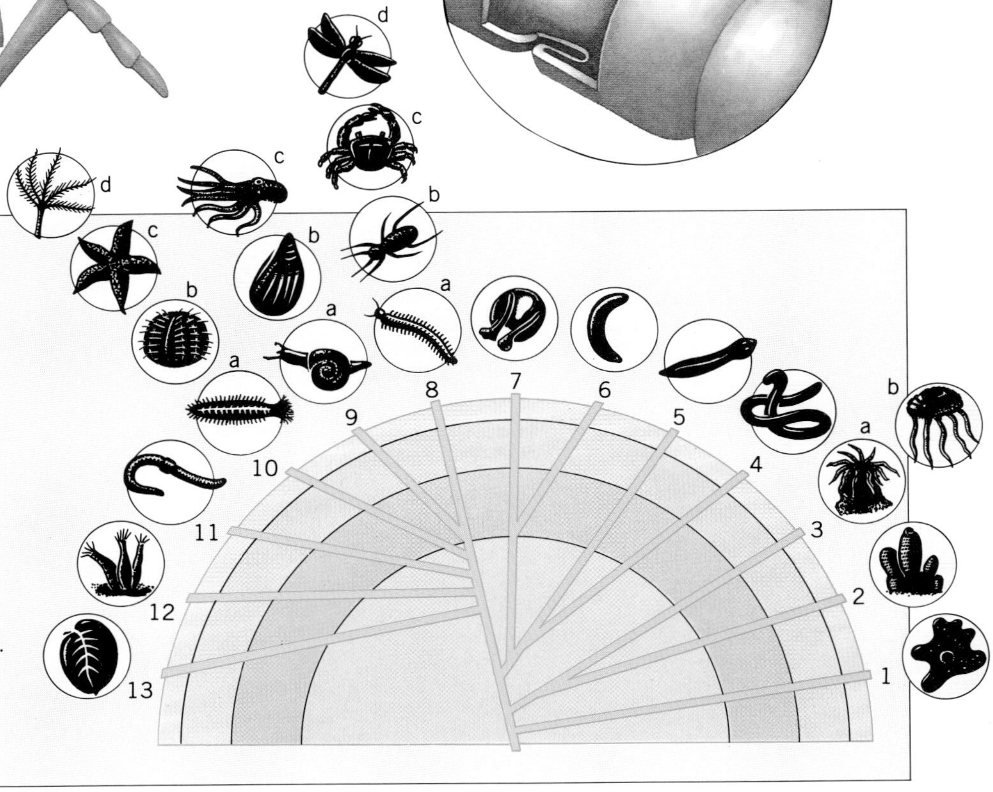

Main invertebrate phyla:
1 ***Protozoa*** *– amoebae, parameciums (motile – able to move independently).*
2 ***Porifera*** *– sponges (sessile – attached to a surface).*
3 ***Coelenterata*** *– 2 groups:* (a) *corals and sea anemones (sessile);* (b) *jellyfish (motile).*
4 ***Nermetina*** *– ribbon worms (motile).*
5 ***Platyhelminthes*** *– flatworms (motile).*
6 ***Nematoda*** *– roundworms (motile)*
7 ***Rotifers*** *(motile)*
8 ***Arthropods*** *– 4 groups (motile):* (a) *centipedes and millipedes;* (b) *arachnids – spiders and scorpions;* (c) *crustaceans – crabs and shrimps;* (d) *insects.*
9 ***Mollusca*** *– 3 groups (motile):* (a) *gastropods – slugs and snails;* (b) *bivalve mollusks – clams and mussels;* (c) *cephalopods – squid and octopods.*
10 ***Echinoderms*** *– 4 groups:* (a) *sea cucumbers (motile);* (b) *sea urchins (motile);* (c) *starfish (motile);* (d) *sea lilies (sessile).*
11 ***Annelids*** *– true segmented worms (motile).*
12 ***Ectoprocta*** *– moss animals (sessile).*
13 ***Brachiopods*** *– lamp shells (motile).*

Invertebrate Herbivores 198 **Invertebrate Reproduction** 160 **Spiders** 252

Dwellers in the Deep

How fish evolved and survive

There are over 21,000 different kinds of fish, ranging from tiny tropical species to sharks 40 feet (12 meters) long weighing over 12 tons. Many live near the ocean surface, but some survive in the abyss; lantern fish migrate up and down hundreds of yards between the surface and the depths every day. The icefish lives under the polar ice, whereas desert pupfish live in hot springs. Both lungfish and walking catfish can survive long periods on land and are capable of breathing air. Salmon and eels travel thousands of miles to spawn.

The evolution of fish is not straightforward. Although in general there is a real progression from the jawless fishes of the oceans some 460-480 million years ago, through the first jawed fish of 450 million years ago and the sharklike fish of 380 million years ago, to the true bony fishes (*teleosts*) that first appeared 175 million years ago, the evolutionary success of the earlier types meant that they did not succumb to the competition and simply die out when newer, improved models came on the scene. Instead, they too went on evolving. The class of fish that contains the most species today – the "ray-finned" fish that have a single dorsal fin, pectoral fins lined with thin radial bones, scales that grow throughout life, a bony skeleton, and a swim bladder for flotation – derives from ancestors that appeared some 390 million years ago: they are a "modern" type of fish. The sharks (which are often described as relatively primitive) evolved later, between 190 and 135 million years ago.

Incomparable adaptability

Although most fish have the same basic body plan, they vary enormously in size and shape. Eels and pipefish are able to glide in and out of crevices in reefs, whereas copperband butterfly fish use a long, narrow snout for probing. Sea horses cling to weeds with their curling tails. Skates, rays, plaice and flounders have evolved flattened shapes for lying in ambush on the seabed. Some cave-dwelling fish may save energy by having no functional eyes or pigments.

Diets vary considerably, from tiny suspended particles of plant and animal material to algae (seaweeds) growing on the rocks, from corals and other invertebrates to other fish and even marine mammals. Some fish are parasitic on or inside other fish.

Sight across the spectrum

For the purposes of finding their food, most fish have color vision and laterally placed eyes that give them a wide field of view. In the deep-water gloom fish often have upward-directed eyes: they detect their prey by spotting its silhouette against the light coming down from above. They also have very large eyes to maximize light-gathering. Fish that live in caves where total darkness reigns have eyes that have degenerated into uselessness. A few surface-dwelling fish have eyes adapted for seeing in both water and air.

Sharks and rays, on the other hand, depend heavily on smell for detecting prey. Eels have the most acute sense of smell, through their very long nasal sacs. They are believed to find their way across the oceans by detecting minute changes in the chemical composition of different stretches of water.

The electrical web

As well as fish that use electrical fields to detect their prey, or navigate, electricity is employed as a means of defense or attack for some species. Using organic batteries that probably evolved from muscles or nerves, skates, rays and electric eels are among the fish that possess the faculty; the latter generating up to 600-volt pulses from batteries that occupy nearly half its body length, which can be as much as 8 feet (2.5 meters).

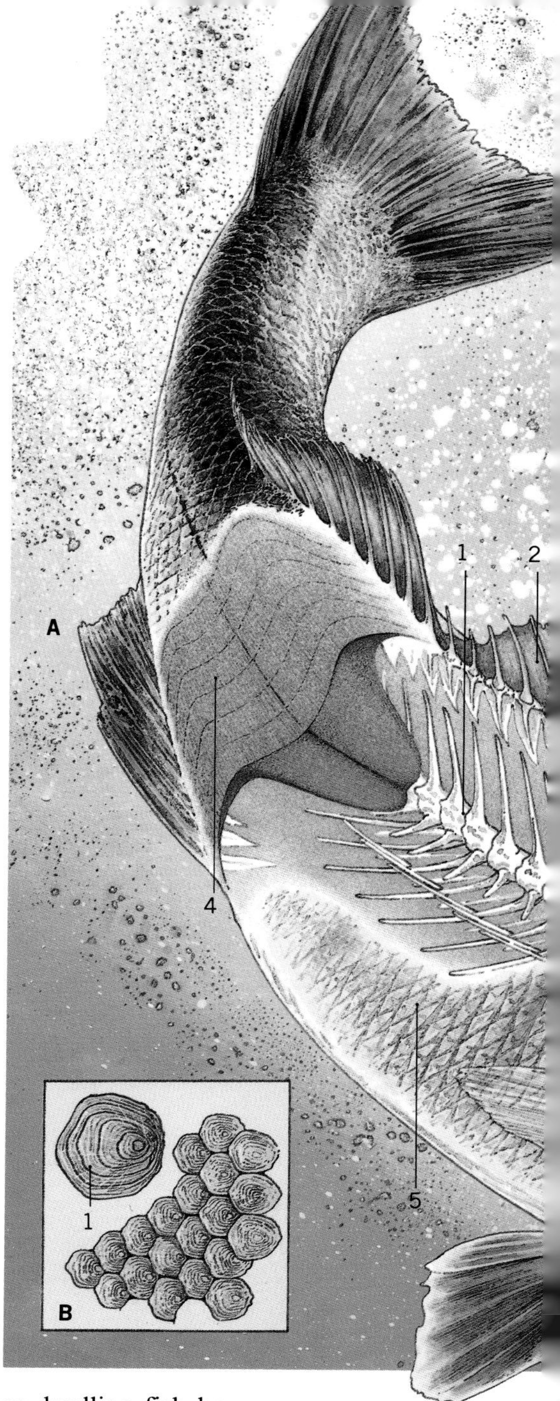

Sea bass **[A]**, *like most modern fish, belong to the class* Osteichthyes, *or bony fish. They have a bony skeleton* [1] *with fins* [2] *supported by bony rays* [3]. *Fins and powerful muscles* [4] *(overlapping in blocks corresponding to a pair of vertebrae) in the flexible body provide propulsion for swimming. The streamlined body, which tapers smoothly at each end, offers minimal water resistance; most fish have scales* [5] *– bony skin outgrowths. Gills* [6], *eyes* [7] *and nostrils* [8] *enable the fish to breathe, see and smell underwater.*

Modern fish *have thin, overlapping* cycloid **[B]** *or* ctenoid *scales. Fish scales are arranged in rows, each one having a series of tiny ring-shaped ridges* [1] *– growth rings that can show the age of a fish. Scales are both a protection and a flexible covering. Because they are translucent, the scales let the pigmentation on the skin of the fish show through from below.*

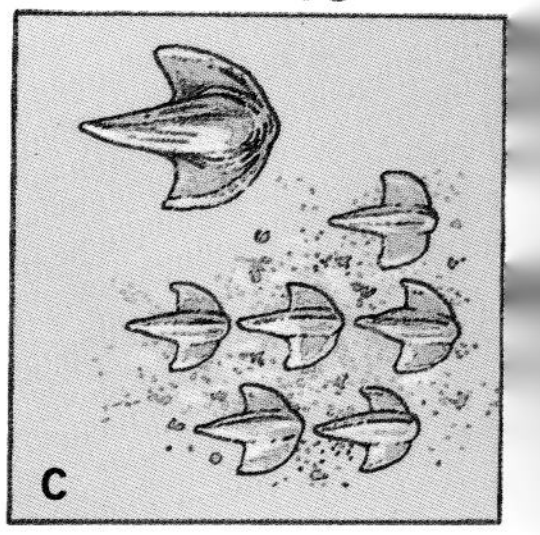

Placoid scales **[C]** *are found on primitive fish with a cartilaginous exoskeleton, such as sharks.* Ganoid *scales are another type of scale found on primitive fish, such as the bony gar. These diamond-shaped scales contain ganoin, which gives a silvery, mirrorlike look.*

Connections: **Animal Adaptation to Pressure** 146 **Breathing** 126 **Fish Reproduction** 162 **Lake Life** 326 **Ocean Life** 332 **River Life** 328 **Vision** 278

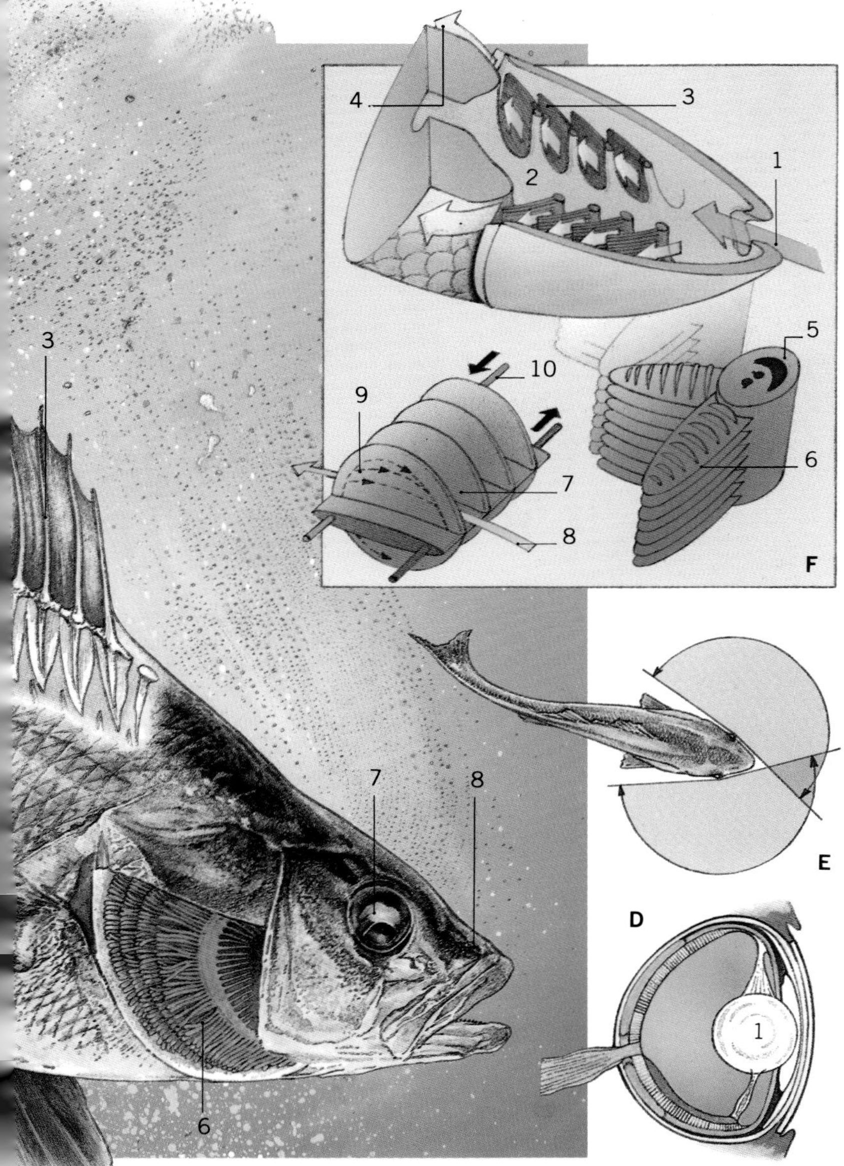

Fish breathe oxygen *dissolved in water and extracted by gills* [**F**]*, which can achieve 80 percent extraction rates, over three times the rate of human lungs from air. Water enters the mouth* [1]*, passes through the gill chamber* [2] *over the gills* [3] *and exits via a gill flap (*operculum*)* [4]*. Flow is maintained by the pumping of the mouth, synchronized to the opening and closing of the operculum. The gills are rows of bony rods* [5]*, to which are attached fleshy filaments* [6] *rich in blood capillaries to absorb oxygen. Each filament has fine secondary flaps (*lamellae*)* [7] *to maximize the gas exchange surface area, which, in active fish like mackerel, can be over ten times the outer body area. Water* [8] *passes over the gills against the capillary blood flow* [9]*; this "counter-current flow" ensures water always passes over deoxygenated blood, maximizing oxygen absorption. Blood vessels* [10] *circulate the blood.*

Cleaner fish *(below) are common on coral reefs and lead a symbiotic life, eating the external parasites on other fish, including large predators, which open their mouths and gill covers to be cleaned inside. Fish recognize a cleaner fish by its very distinctive coloration and its "invitation" dance.*

Fish eyes [**D**] *are adapted to see underwater. Unlike a human lens, a fish lens* [1] *is a perfect sphere, which may reduce image distortion. The eyes protrude somewhat to give reasonable allround vision, but there is very little overlap between each eye's field of vision* [**E**]*, hence 3-D vision is poor. Fish have no eyelids – there is no need to prevent the eye drying up – and they also lack pupils that can vary their size. Most fish have some color vision, but sharks and rays appear to see only in black and white. Unlike cave-dwellers, abyssal fish have functioning eyes, probably used to detect luminous deep-sea creatures.*

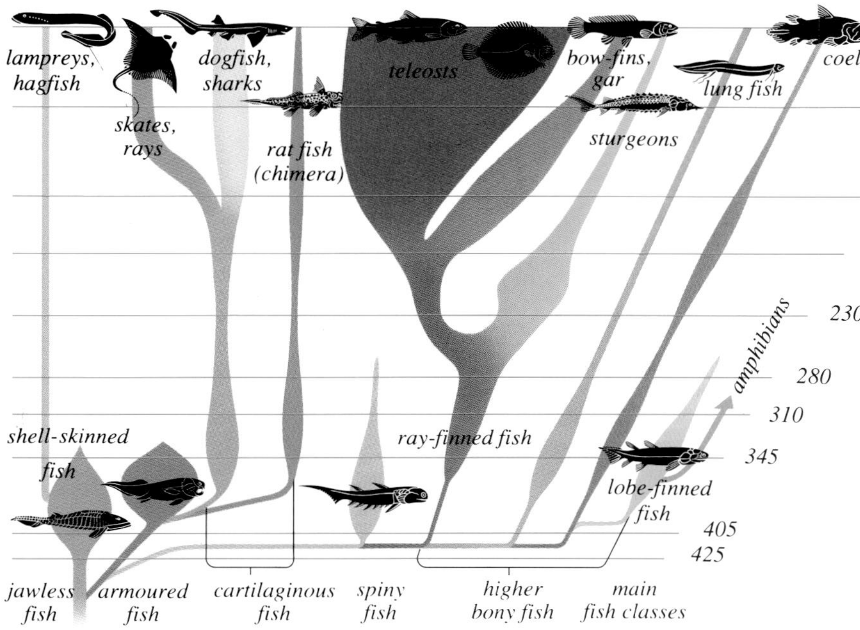

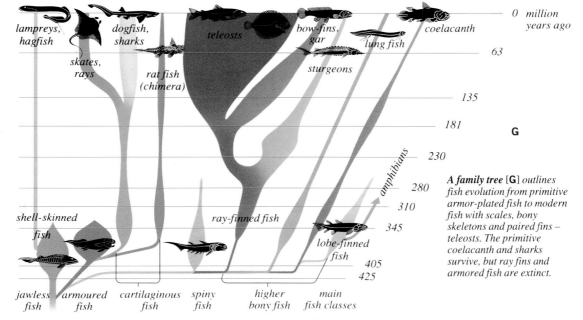

A family tree [**G**] *outlines fish evolution from primitive armor-plated fish to modern fish with scales, bony skeletons and paired fins – teleosts. The primitive coelacanth and sharks survive, but ray fins and armored fish are extinct.*

Taking the Air

How animals breathe

Weddell seals have been known to dive underwater for over an hour, but ultimately they must return to the surface for air. Virtually all organisms require oxygen to turn food into energy, and on land this is in plentiful supply thanks to millions of years of photosynthesis by plants, which turns carbon dioxide into oxygen. A watery environment, on the other hand, is poor in oxygen in comparison to air – but to compensate for that, most animals that live in water have a far more efficient means of extracting the oxygen than their air-breathing equivalents.

For most creatures, respiration means absorbing oxygen and disposing of waste carbon dioxide by breathing. Respiration is also the complex process that breaks down fuel molecules to make energy. It usually uses oxygen, but some bacteria can use sulfur.

In a plant or a very small, primitive animal the mechanism may be fairly simple: air comes into direct contact with the cells of the organism, and gas molecules filter through the cell walls by the process known as diffusion. The oxygen diffuses into the cells, the waste gas diffuses out.

A more complex land animal cannot rely on straightforward diffusion because although it may have a large surface area, a far greater proportion of all its body cells is not in direct contact with the air. Insects have developed the logical solution to the problem: a complex system of branching ducts that carries the air to every part of the body. It works well – but only over short distances; this is one of the reasons why insects have never grown to large sizes.

Mammals, birds, reptiles and many other creatures employ a much more sophisticated arrangement, which exploits the bloodstream as a gas transport system. The oxygen is carried in the red blood cells alongside energy-rich sugars dissolved in the blood plasma. The two are delivered together to each organ, and particularly to the muscles, for storage or for instant use. When the energy in the sugars has been released by oxidation, the blood carries away the waste carbon dioxide.

The blood collects oxygen – and gets rid of carbon dioxide – by flowing through a gas exchanger. In air-breathing animals this is a lung: an air-filled cavity lined with very fine blood vessels (capillaries) that have walls so thin as to allow the flow of gases – but not liquid – through them. As the blood is pumped through the capillaries, carbon dioxide diffuses out into the air and oxygen diffuses in. The oxygenated blood flows away to transport its oxygen to the rest of the body.

Expansion and contraction

Some small animals rely on diffusion through their external surface to refresh the oxygen supply in their lungs, but larger animals actively pump the air in and out. A frog raises the floor of its mouth to squeeze a mouthful of fresh air into its inflatable lungs, relying on

Insects **[A]** *possess a simple, direct respiratory system. Oxygen and carbon dioxide gas are carried through a system of tubes (trachea) [1] that branch repeatedly and extend to all parts of the body. The fine ends of the trachea carry gases to and away from individual cells, and may even extend into the interior of large muscles [2].*

When in flight, birds **[B]** *respire at a very high rate. The oxygen needed to sustain such flight is stored in large air sacs [3]: these are not directly involved in gas exchange but pump fresh air through the lungs when required.*

C

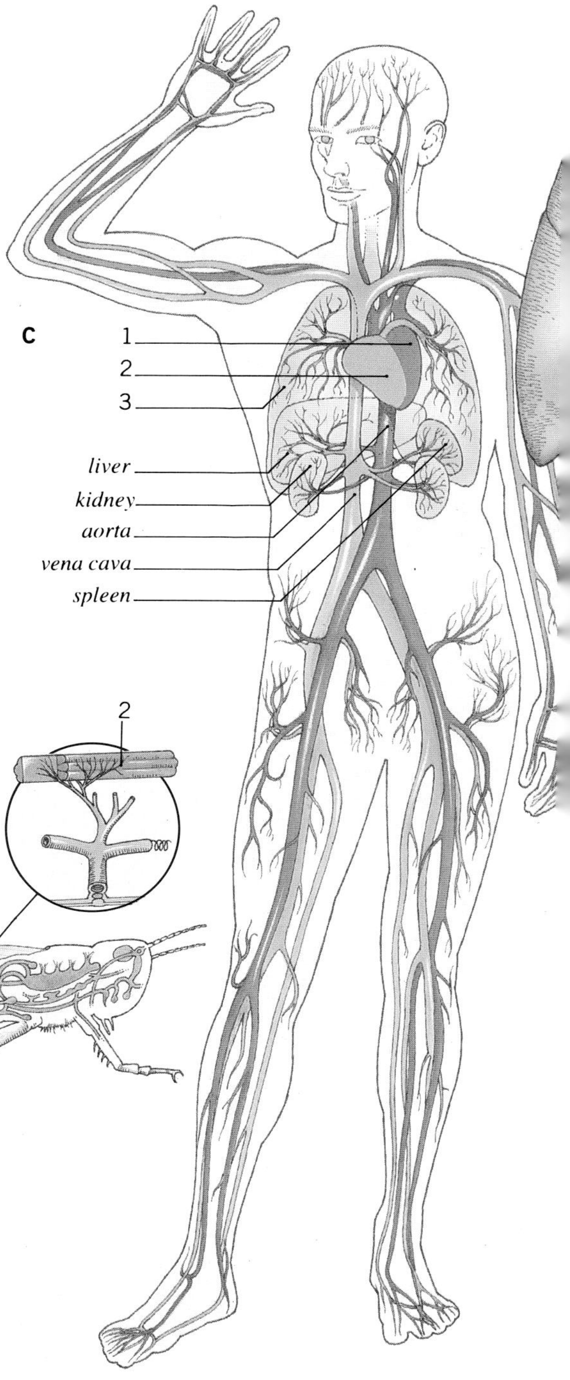

A

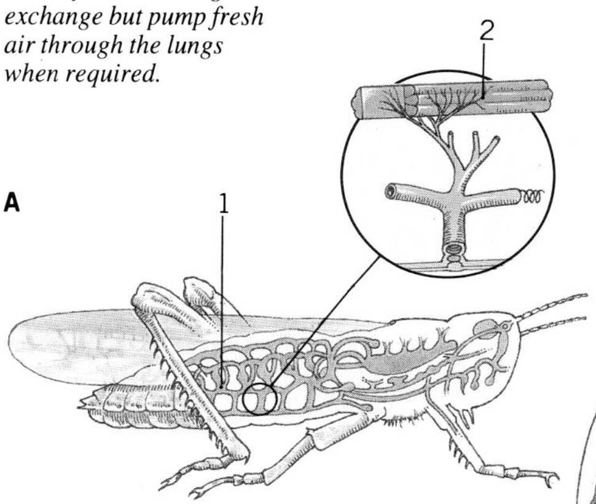

B

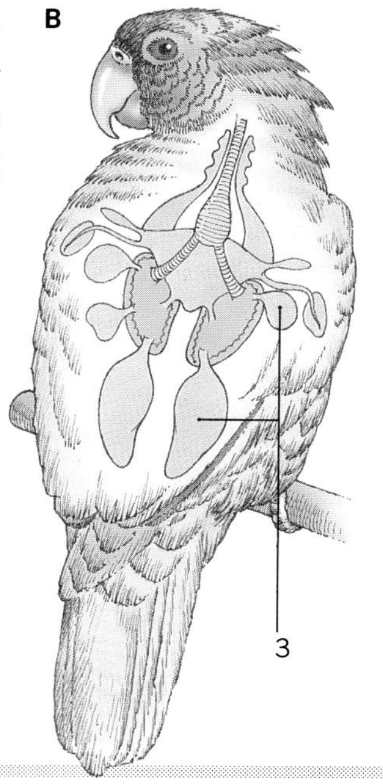

Each contraction *of the human heart drives blood through two distinct circuits in the body* **[C]**. *The left side of the heart [1] pumps oxygen-rich blood (red) via thick-walled, muscular arteries to the capillaries that supply organs and tissues. Deoxygenated blood (blue) passes through a network of capillaries, draining into veins that return it to the right side of the heart [2], from where it is pumped to the lungs [3] to be recharged with oxygen. Arteries may be up to 1 in (2.5 cm) in diameter, while the finest capillaries are just wide enough for a single red blood cell to pass.*

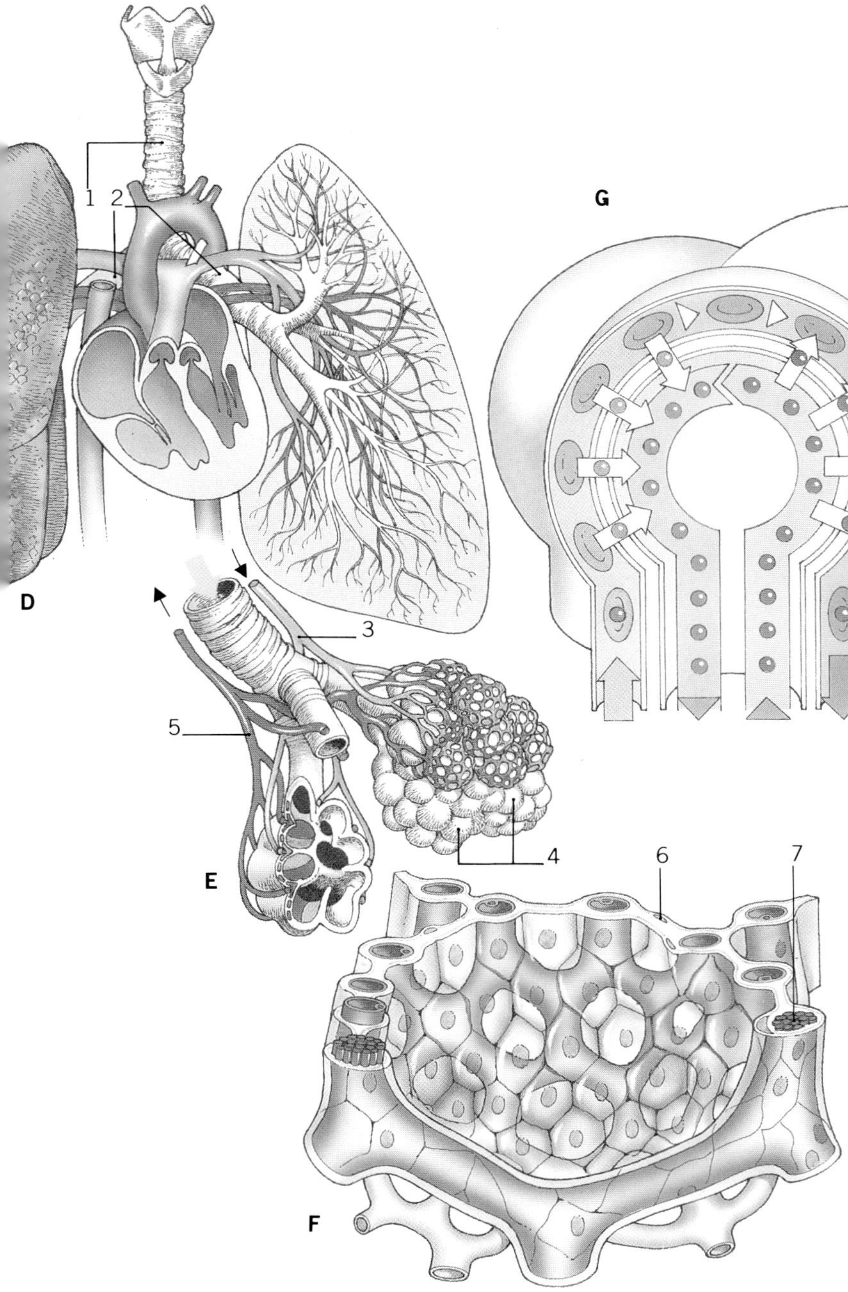

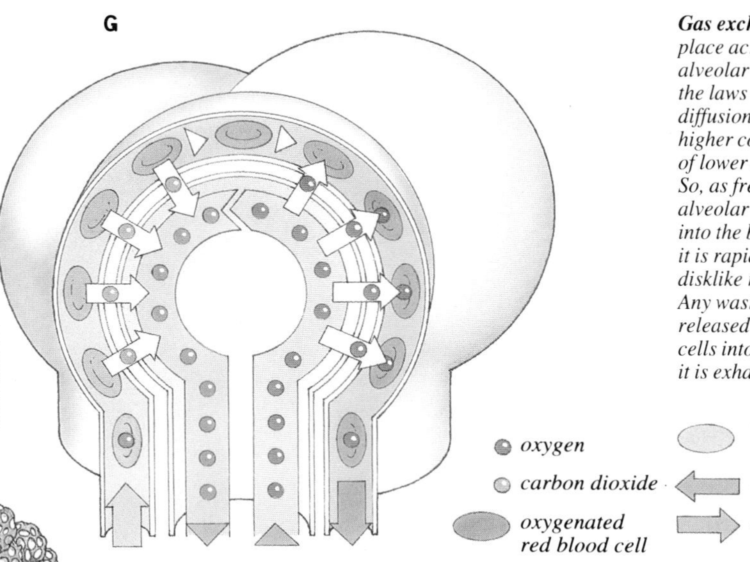

***Gas exchange* [G]** *takes place across the thin, moist alveolar wall. According to the laws of physical diffusion, gases pass from a higher concentration to areas of lower concentration. So, as fresh air enters the alveolar sac, oxygen diffuses into the bloodstream, where it is rapidly bound by the disklike red blood cells. Any waste carbon dioxide is released by the red blood cells into the sac, after which it is exhaled.*

their natural elasticity to force the spent air out again. A tortoise has powerful muscles within its armored body that force its lungs to expand and contract.

A mammal enlarges its chest cavity by expanding its ribcage and lowering the muscular diaphragm beneath its lungs. As the cavity expands, the lungs expand with it, drawing in air. Relaxing the ribs and diaphragm makes the chest cavity smaller, compressing the lungs and pushing the air out.

An animal may consciously hold its breath when it wants to dive underwater, but normally the whole system is automatically controlled by sense organs in the blood and tissues that monitor oxygen levels and transmit the information to the brain. If the animal is working hard, it uses oxygen faster: the levels drop and the breathing and heart rate are accelerated to build them up once more until equilibrium is restored.

***The lungs* [D]** *provide a large surface area over which gas exchange can occur. Inhaled air is drawn through the trachea* [1], *which divides into two bronchi* [2]. *On entering the lungs the bronchi branch repeatedly, eventually leading to blind-ending sacs* **[E]**, *surrounded by a web of vessels that carry deoxygenated, carbon dioxide-rich blood* [3] *to the bulblike alveoli* [4] *and oxygenated blood* [5] *back to the heart. Each alveolus* **[F]** *has walls just one cell thick* [6], *to separate the capillaries from the air within. The alveoli are supported by a framework of connective tissue* [7].

A Personal Air Supply

Most aquatic animals are able to get their oxygen directly from the water, but some have to breathe air. Many of these simply return to the surface at intervals to take a gulp of air before diving again, but the water spider carries a personal air supply, much as a human diver does with an aqualung. In this way it avoids expending energy on trips to the surface and reduces its exposure to aerial and aquatic predators.

The air is trapped by fine hairs to form a silvery bubble around the spider's body. The spider simply breathes the air in the bubble. If the surrounding water contains plenty of dissolved oxygen, this finds its way into the bubble; in this fashion the spider is never in any danger of running short of the air it needs. The creature even builds its nest underwater, rearing its young inside an air bubble lodged among the water plants.

Connections: **Atmosphere** 64 **Cycles of Life** 206 **Early Evolution** 92 **Fish** 124 **Mammals** 136 **Photosynthesis** 182

In and Out of the Water

How amphibians live

Although amphibians are most familiar to us as frogs and newts, they can grow as large as the 4.5-foot (1.5-meter) Japanese giant salamander, weighing 220 pounds (100 kilograms). Some 370 million years ago the first creatures to leave the seas and walk on land were the amphibians. They pioneered the use of lungs for full-time air breathing, and were the first vertebrates to have true legs, tongues, ears and voice boxes. The increase in body area covered by the nerve cells led to those cells invading the brain: a step to intelligence. Today there are some 4,000 species of amphibian.

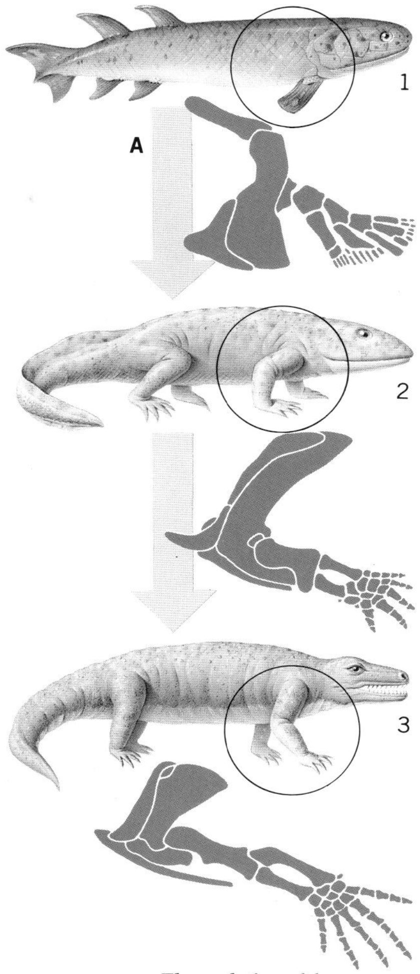

***The evolution** of the* pentadactyl *(five-fingered) limb* [**A**] *can be partly traced through the fossil record. The basic design was already present in lobe-finned fish such as* Eusthenopteron [1]. *Adapting to the terrestrial environment, the limb rotated downward and away from the body. In the first known amphibian,* Ichthyostega [2], *the limb is still compact and the body close to the ground.* Seymouria [3] *developed more extended limbs, which provided greater articulation and ground clearance.*

Despite being cold-blooded, and therefore particularly at the mercy of their environment, amphibians are found on all the continents except Antarctica, from the tropics to north of the Arctic Circle. There are three main groups: the *Urodela* (newts, mudpuppies, sirens and salamanders), the *Anura* (frogs and toads) and the *Gymnophiona* (caecilians or worm lizards). Many features distinguish the amphibians from other vertebrates. They have a three-chambered heart, paired lungs, which are reduced or absent in some salamanders and caecilians, a flattened skull, and teeth that can bend inward.

The breathing body

Unlike reptiles, amphibians have not developed waterproof skins. In fact, many species supplement lung breathing with gas exchange through their skins, which must be kept moist for this purpose. The lining of the mouth is also moist and well supplied with capillaries and can function in gas exchange. A few aquatic salamanders retain their larval external gills in the adult, and a few terrestrial forms have no lungs at all.

The senses also had to adapt to life on land. Eyelids and eye-moistening glands evolved, and true ears (with outer eardrums and two middle-ear bones, neither of which are found in fish ears) for detecting sound vibrations in air. At the same time, the voicebox and vocal sacs evolved – amphibians were the first vertebrates to have a larynx. For the sense of smell, internal nostril openings (*nares*) allow air to be taken into the lungs while the mouth is closed or when only the external nares are above water. This can also help when the amphibian is avoiding land predators.

An aquatic inheritance

Amphibians have retained many of their adaptations to the aquatic life. When moving, many amphibians retain an undulating, fish-like motion, except for the frogs and toads, which evolved long hind limbs for hopping on land, or kicking out to swim. Also, many frogs and toads still discharge unfertilized eggs into the water, where sperm is then shed on to them for fertilization to occur.

Amphibians have kept their streamlined shape, and many species their webbed feet, for locomotion in water. Their inability to generate internal body heat confines them to the warmer parts of continents, but many species survive considerable cold by reducing their metabolic rates

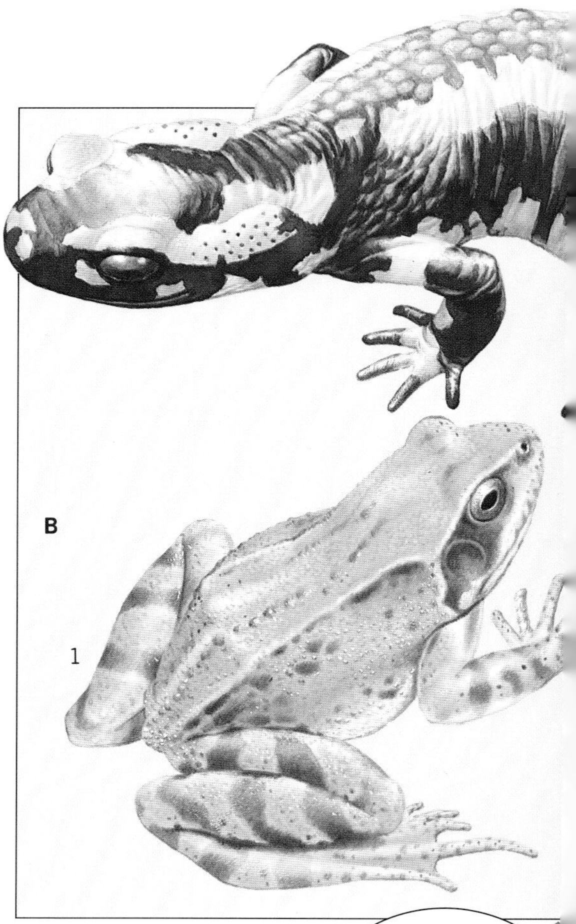

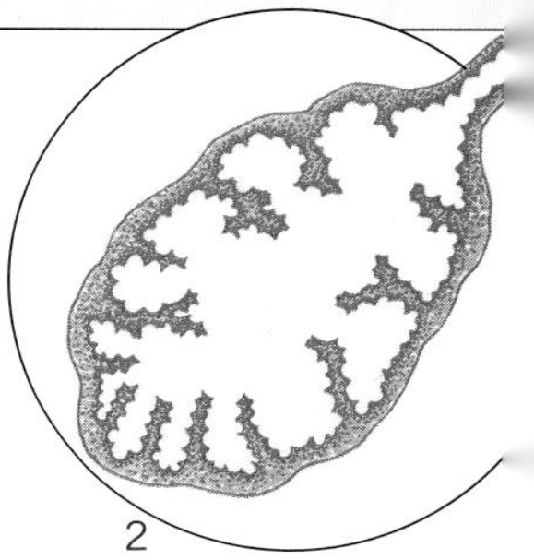

Modern amphibians [**B**] *display considerable variance in the adaptation of their respiratory systems to the terrestrial environment. Only a few aquatic forms retain their gills when adult. The common frog* [1] *has complex lungs* [2], *with intricate infolding of the vascular cavity walls.*

Surviving the desert

Amphibians are relatively common in deserts and other arid regions. They avoid overheating by burrowing in the day and emerging at night. Their main problem is the unpredictable water supply. Certain species can store up to half their body weight of water in the bladder. Some frogs and toads are remarkably tolerant of water loss – the Western spadefoot toad can withstand up to 60 percent reduction in its body water. Another adaptation is to retain urea in the blood, thus permitting the uptake of water by osmosis from even apparently dry soils. Most terrestrial frogs also have a patch of skin rich in blood capillaries in the pelvic region, which takes up water when sitting on damp earth.

Connections: **Animals in Heat and Drought** 142 **Breathing** 126 **Early Evolution** 92 **Reptiles** 130 **Swimming** 210

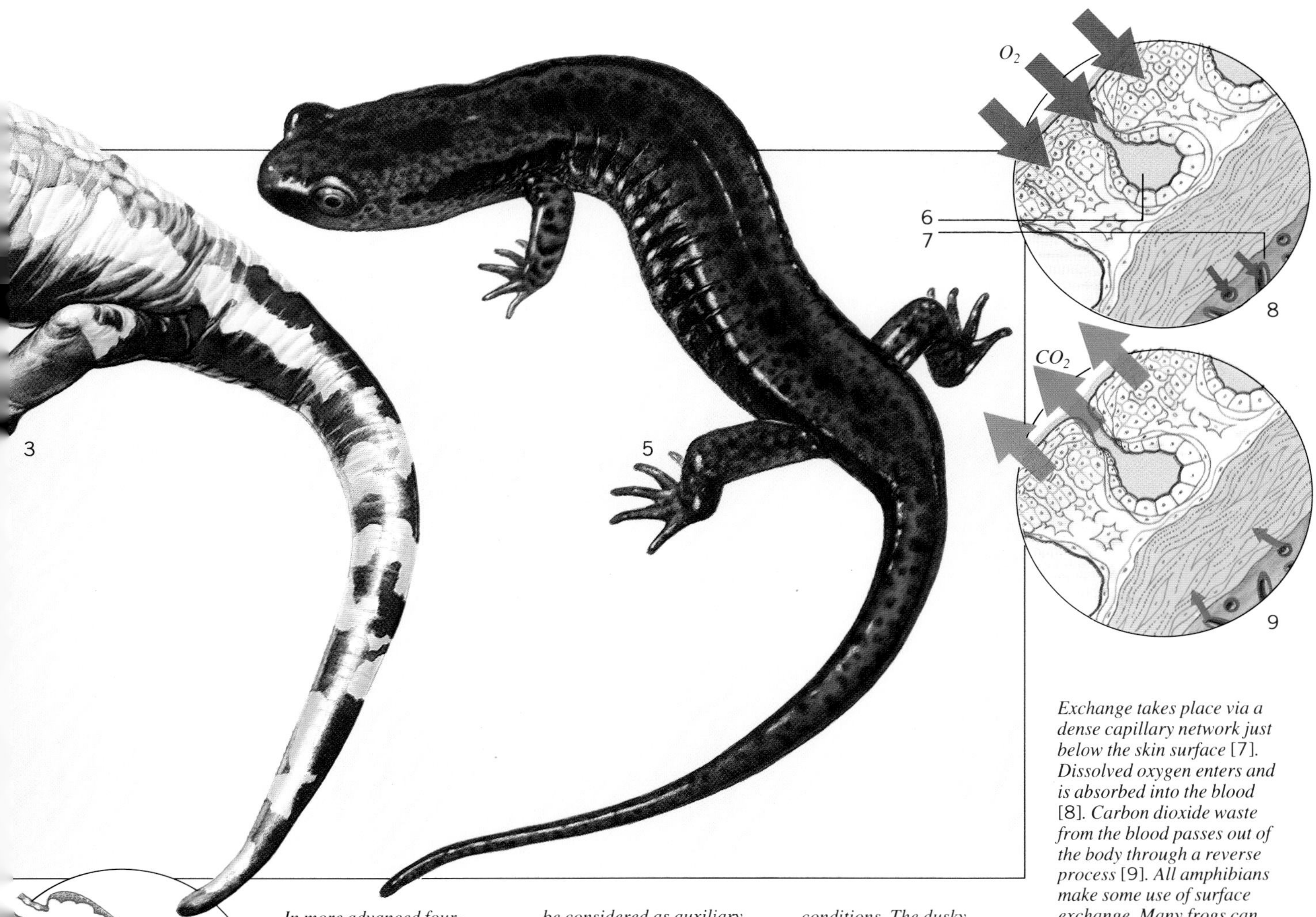

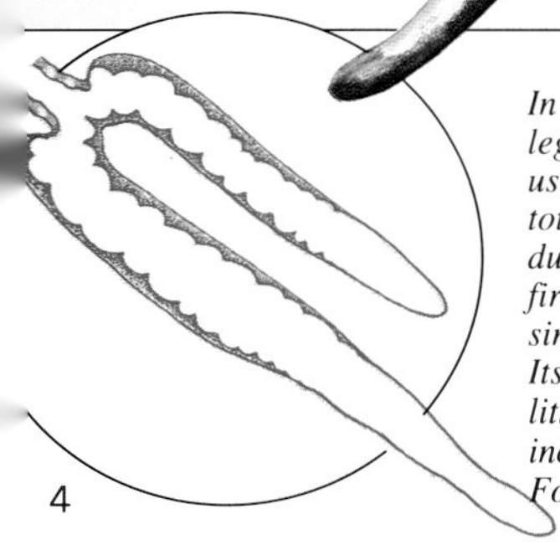

Exchange takes place via a dense capillary network just below the skin surface [7]. *Dissolved oxygen enters and is absorbed into the blood* [8]. *Carbon dioxide waste from the blood passes out of the body through a reverse process* [9]. *All amphibians make some use of surface exchange. Many frogs can meet their total respiratory requirement in this way while inactive in cool and moist conditions. Surface exchange is much less important to other vertebrates, and is usually insignificant. In humans, for example, surface exchange accounts for less than 1 percent of respiration.*

In more advanced four-legged animals the lungs usually provide the animal's total oxygen requirement during active behavior. The fire salamander [3] *has much simpler respiratory organs. Its tubular lungs* [4] *have little infolding with which to increase their surface area. For this reason the lungs can be considered as auxiliary, as oxygen is also obtained through gas exchange over the whole of the body surface. As a result, the fire salamander requires a moist environment and only ventures into the open at night when it is cooler – thereby reducing water loss – and also under very humid conditions. The dusky salamander* [5] *is a* plethodont, *one of a family of lungless salamanders found in North America, for which all respiration takes place through surface gas exchange. This is only possible if the surface is kept moist by secretions from mucous glands* [6].

Missing Links

The first amphibians evolved from lobe-finned fish about 370 million years ago. These fish, like coelacanths, had bony supports in their fins, which might have let them crawl out of the water, and lungs and internal nasal openings for breathing air. The fossil record of amphibians is poor. The earliest amphibians on record, such as *Ichthyostega*, had already evolved hip and shoulder girdles to support the new limbs and ribs to protect the internal organs. One of the largest of them was *Mastodonosaurus*, up to 4 m (13 ft) long. Until the evolution of the dinosaurs, amphibians dominated the land. By about 135 million years ago most were extinct. There are no fossils to link modern amphibians with these ancient forms and no record of the divergence into caecilians [1], newts and salamanders [2], and frogs and toads [3].

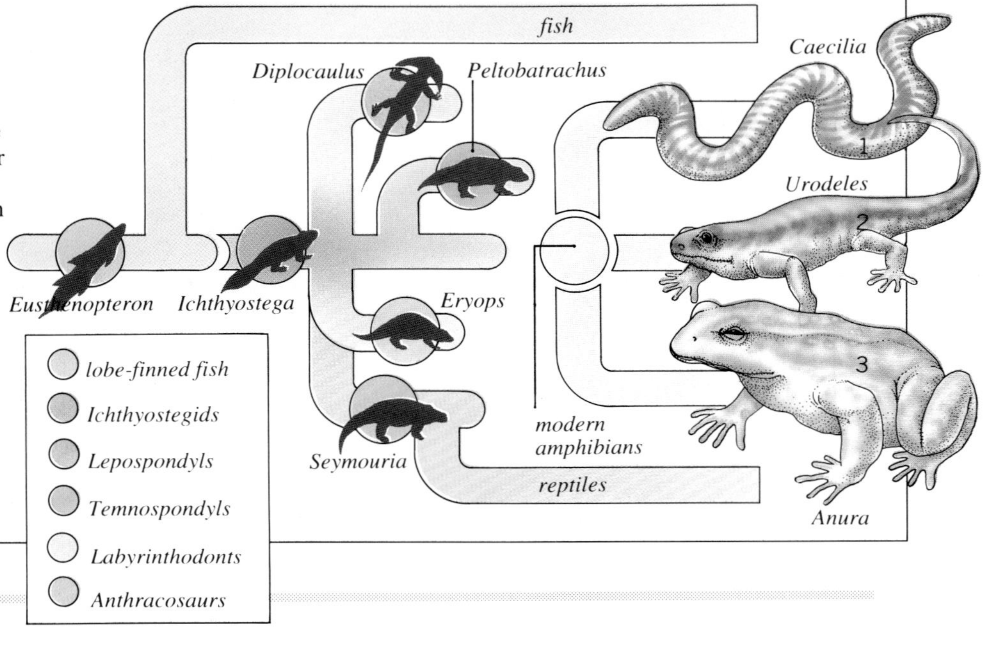

Conquering the Land

How modern reptiles evolved

Today's reptiles are rather like the survivors of a shipwreck – a shipwreck that occurred 65 million years ago, when most reptiles were wiped out by the natural disaster that ended the Cretaceous period. For all that, there remain about 6,000 different reptile species – few in numbers and small in size compared with past ages, but great in diversity. They range from small, legless, eyeless creatures easily mistaken for worms to predatory crocodiles that can grow to lengths of over 20 feet (6 meters), and massive tortoises and turtles that may weigh upward of a ton.

***Heat is essential** for reptiles to survive, but the blisteringly hot, loose sand of desert dunes can get too hot for one of its inhabitants, the Namib dune lizard* [**B**]. *The ideal sand surface temperature for the lizard is 86–104°F (30–40°C), when the reptile can gather food or find a mate. However, since for much of the day the sand surface is so hot that it would quickly kill the lizard, the animal has little time to move around normally and conduct its activities. To increase its "socializing" and feeding time, the lizard speeds up its morning basking – when it warms up – by pressing its underbody*

It was about 300 million years ago that reptiles began to dominate the world of their predecessors the amphibians. Over the millennia many reptiles grew to huge size – notably the dinosaurs. Today, except for the crocodiles and alligators, most are relatively small. There are four major reptilian orders. One, whose skulls most closely resemble those of early reptiles, are the *Chelonia*: tortoises, turtles and terrapins. The *Crocodilia* include crocodiles, alligators, caimans and the gavial; the *Squamata* include snakes and lizards. The final order is represented by the tuatara, the lone member of the *Rhynchocephalia*, now found only in New Zealand.

There were lizardlike amphibians (such as salamanders) and legless amphibians (such as the wormlike caecilians). In evolutionary terms the reptiles carried on where the amphibians left off.

Shapes and sizes

The lizards have five-toed limbs and generally have eyelids, external ears and small scales on top and underneath; most are carnivorous, but a few – such as some iguanas – are vegetarian. A lizard's legs project outward from its body before turning downward, with the unavoidable result that lizards tend to waddle and twist as they walk along.

It may have been as an evolutionary response to such limitations that crocodiles and alligators took to the water, which not only greatly reduced the effect of their own weight and lumbering motion but also allowed them to actually increase in size.

Snakes are legless, have transparent and permanently shut eyelids, and have a row of ridged scales under their bodies. All snakes are carnivores and swallow prey whole, though some suffocate prey first in their coils; others – a minority – are venomous.

The *Chelonia* developed heavy, horny outer shells from a combination of spine, ribs and bony plates deriving from the skin. Land tortoise legs are specially developed to carry the enormous weight of this bony cage; movement is inevitably slow and cumbersome. The front legs of Chelonians assist in tearing food: most land tortoises are vegetarian, but flippered turtles mainly eat fish and other small marine creatures. Terrapins (freshwater turtles) have flatter shells than tortoises, webbed toes, and live in brackish water.

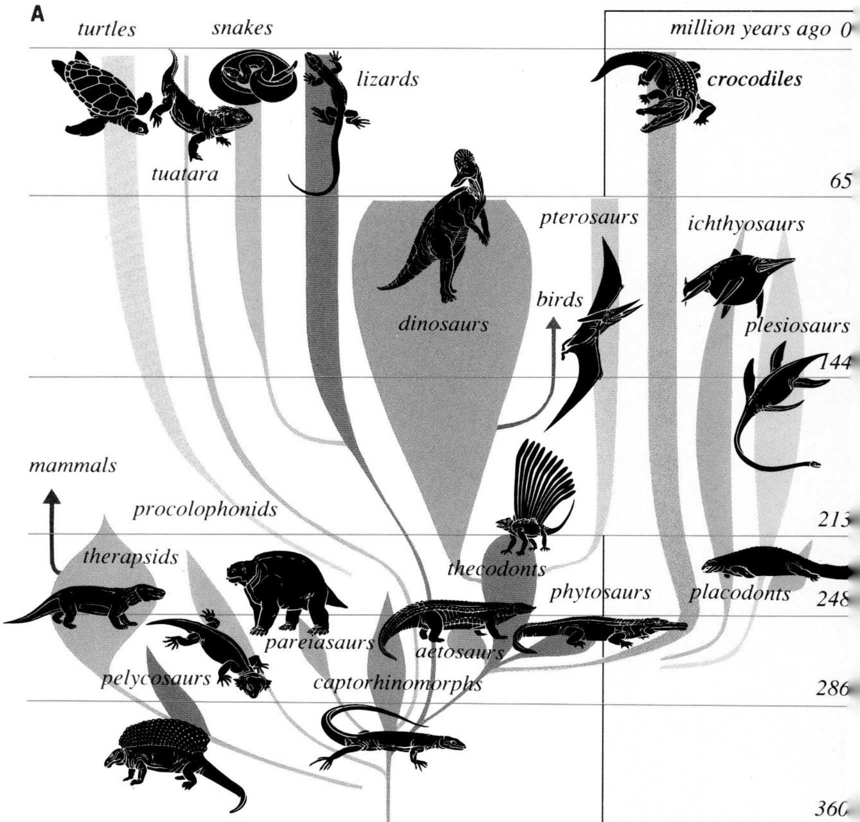

A reptile evolutionary tr [**A**] *indicates how few spe survived from the reptile heyday. Crocodiles and alligators are the closest living relations to dinosa and also have the most advanced brains and he Turtles and tortoises ha changed little in millions years; like many areas o reptilian evolution, their origins are obscure.* Therapsids, *though all n extinct, are a vital branc of evolution, for from the evolved the* synapsids, *w led to the mammals. All, however, evolved from t earliest group of reptile* captorhinomorphs, *or "stem reptiles."*

The cold-blooded lie

A *poikilothermic* animal is often called cold-blooded; this is misleading, for this sort of animal regulates its internal temperature by the temperature of its surroundings, which means that its blood can be "hot" or "cold." Reptiles have optimum temperature ranges for bodily processes; a snake, for example, needs heat to digest food; if too cold it dies because food rots in its stomach. Reptiles have developed various behavioral means to control their temperature, typically basking in the morning sun to warm up, or absorbing heat from the ground below them, using shade to cool down, and modifying posture for "fine-tuning" body temperatures. Few reptiles can survive cold for long, but one reason they have survived strong mammalian competition is that they need less food than mammals to maintain their metabolism and can thus exist on a meager diet.

Connections: **Amphibians** 128 **Crawling** 214 **Desert Life** 304 **Early Evolution** 92 **Poisonous Animals** 250 **Reptile Reproduction** 170

against the sand and holding its legs and tail up in the air [1]. *When sufficiently warm, the lizard is agile enough to move about normally. Later in the day, however, the sand temperature reaches 104°F (40°C) and the lizard avoids overheating by "stilt-walking" on the hot sand with outstretched legs* [2].

Occasionally it lifts two legs at opposite corners of its body [3] *and, using its tail as an extra support, raises its feet off the otherwise intolerably hot sand. By noon surface temperatures reach 113°F (45°C) and the lizard seeks protective, cooler sand by flicking its tail to burrow down into the dune* [4].

B

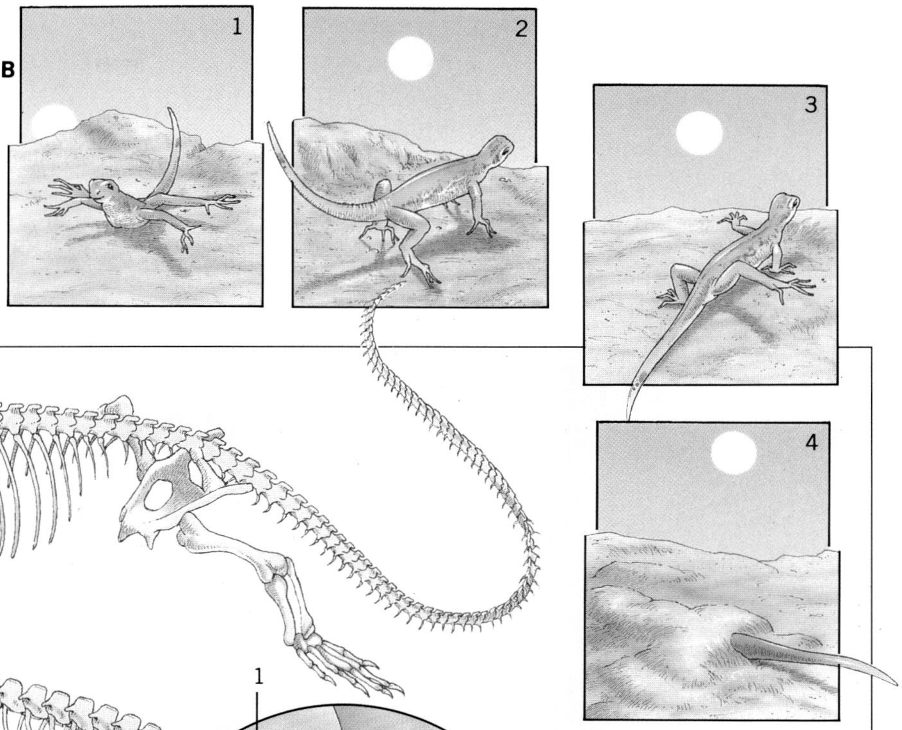

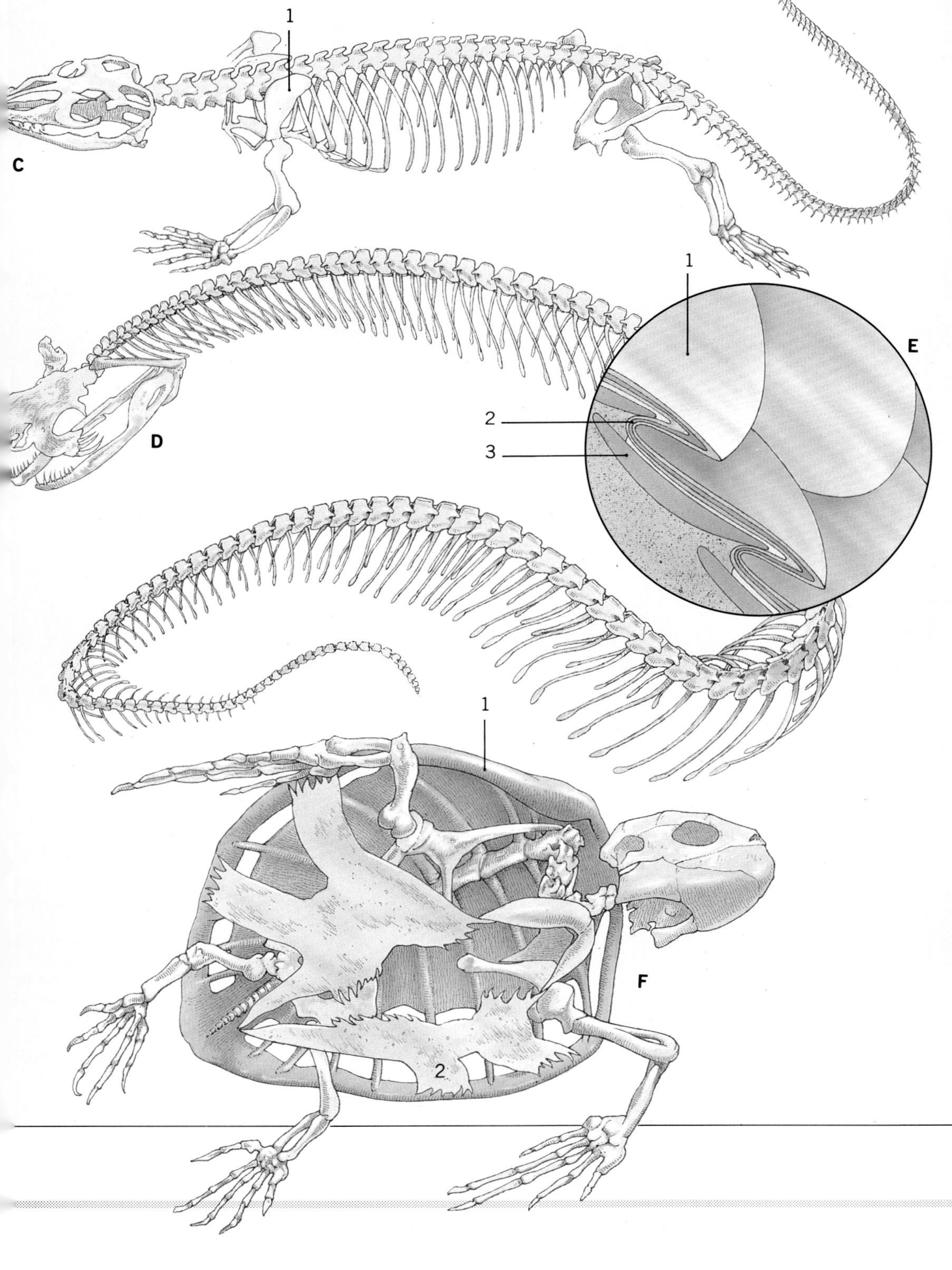

Lizard skeletons [**C**, *right-hand limbs omitted for clarity*] *have a pivoting shoulder girdle* [1] *to permit the typical side-to-side body locomotion. Most reptiles lack bony* epiphyses, *cartilage at bone ends that in mature mammals fuse to the bone to restrict further growth; some old reptiles outstrip normal species size.*

Snake skeletons [**D**] *have no limbs. The spine, however, has an unusually high number of vertebrae, up to 400 in some pythons. A complex ball and socket joint links each vertebra, allowing horizontal movement up to 25°, making a strong but highly flexible support for the ribs, to which join powerful locomotive muscles. A section through reptile scales* [**E**] *shows layers of thick, horny keratin* [1] *hinged at thinner layers* [2] *to allow movement. Pigment cells below* [3] *include black ones that can darken the skin and increase heat absorption when basking.*

Turtle [**F**] *shells consist of a* carapace, *made of an outer keratin layer and an inner bone structure* [1], *fused to the ribs and vertebrae, and a* plastron [2], *evolved from some of the shoulder girdle bones and ribs.*

Flying Machines

How birds are adapted for flight

The smallest known bird is the bee hummingbird, which measures less than 2.5 inches (6 centimeters) in total – of which half is bill and tail. The largest is the flightless ostrich, which can grow up to 9 feet (2.7 meters) high. Many other birds are also flightless, while some can fly for thousands of miles, or reach speeds of over 155 miles an hour (250 km/h) in deadly attacking stoops, or dives. There are an estimated 9,300 species of bird alive today, living in habitats as diverse as the Antarctic ice sheets, the tropical rain forests, arid deserts and the open oceans.

There are 28 orders of birds, but no one is sure how they are related in evolutionary terms. They are all, however, remarkably similar in basic structure – a body plan that evolved primarily as an adaptation to flight. Even the flightless species evolved from ancestors that could fly, and therefore they share many typical features. Birds' bones are light and strong, and the skeleton has the form of a rigid box, with a large breastbone, or *sternum*, usually having a flange as a main attachment point for the powerful flight muscles. The wings themselves are highly modified forelimbs. The unique feature of birds, however, is their covering of feathers. No other vertebrates have these extraordinary outgrowths from the skin, and feathers in all their modifications provide birds with many attributes. As well as forming the wing surfaces needed for flight, feathers also act as insulation to allow birds to maintain a high body temperature, and by their myriad patterns facilitate the exchange of information in display signals, as well as providing concealment and camouflage.

Reptilian forebears

Birds have much in common with reptiles, from which they almost certainly evolved. In particular, the scales on birds' legs and feet are very similar to the scales of reptiles. Both birds and reptiles have a single ball-and-socket articulation between the skull and neck, and a simple middle ear of just one bone (mammals have three bones). Both birds and reptiles have nuclei in their red blood cells (unlike mammals). And, like most of the reptiles, birds reproduce by laying yolky (amniotic) eggs in which the embryo develops on the surface of the yolk.

Ancient ancestors

The birdlike fossil reptile *Archaeopteryx* lived about 150 million years ago and resembled a bird in many respects. It was covered in feathers (just like those of today's flying birds), and it also possessed a distinct wishbone (*clavicles*), both strong indicators of flight. However, unlike modern birds, *Archaeopteryx* had teeth, a reptilelike tail and three claws on each wing. It may have used these claws to clamber in trees. Scientists believe *Archaeopteryx* could fly, but probably only weakly. It was about the size of a crow,

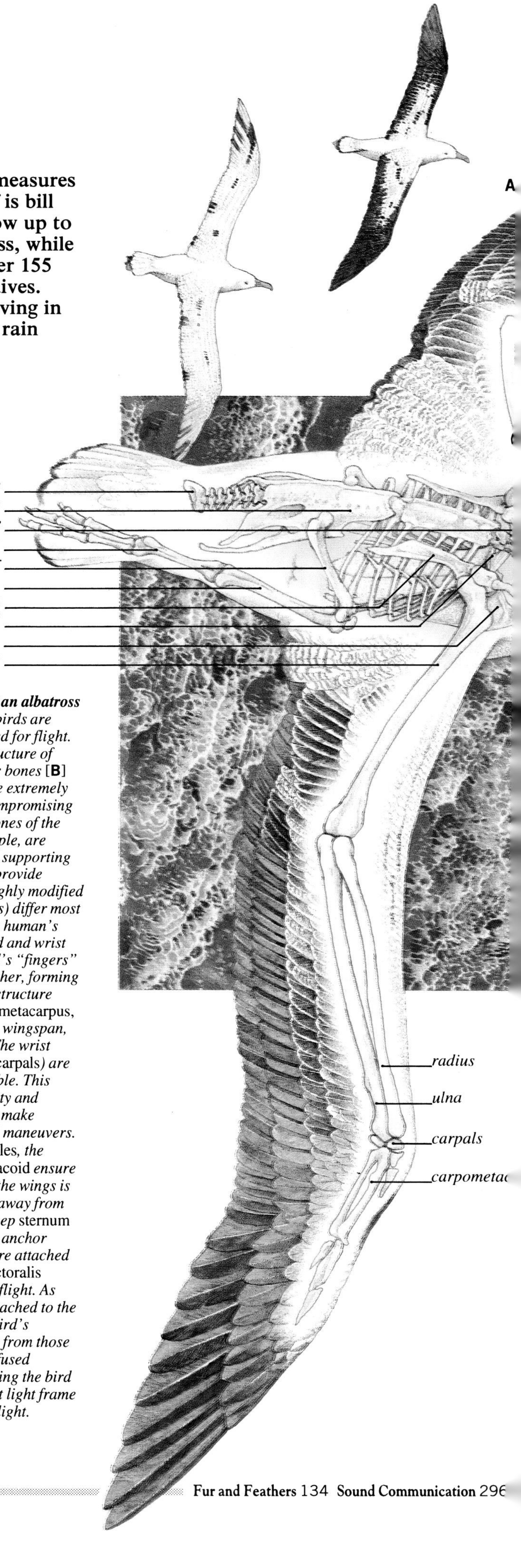

The skeleton of an albatross **[A]** *shows how birds are perfectly adapted for flight. The internal structure of many of a bird's bones* **[B]** *is designed to be extremely light without compromising strength. The bones of the wings, for example, are hollow but have supporting "struts"* [1] *to provide strength. The highly modified forelimbs (wings) differ most radically from a human's arms in the hand and wrist region. The bird's "fingers" have fused together, forming a long, narrow structure called the* carpometacarpus, *which increases wingspan, improving lift. The wrist bones of birds* (carpals) *are strong and flexible. This improves mobility and enables birds to make acrobatic aerial maneuvers. The fused* clavicles, *the* scapula *and* coracoid *ensure that the base of the wings is kept in position away from the body. The deep* sternum *provides a good anchor point to which are attached the powerful* pectoralis *muscles used in flight. As well as being attached to the pelvic girdle a bird's vertebrae, apart from those of the neck, are fused together, providing the bird with a strong but light frame specialized for flight.*

Connections: Bird and Bat Flight 224 Bird Courtship 172 Birds' Nests 238 Fur and Feathers 134 Sound Communication 296

The bones of a bird's skull [**C** *and* **D**] *have a honeycomb structure to reduce weight. Moreover, during their development birds lost the heavy jaws and teeth common to other vertebrates, with the result that they evolved skulls that weigh, in relative terms, much less than the skulls of most mammals. To ensure that the skull remains strong, however, the bones are fused together.*

The various honeycreepers of Hawaii [**E**] *evolved from one species of bird now long extinct (center). Over millions of years the honeycreepers evolved different methods of feeding. This ensured that the islands' large variety of habitat niches could be exploited, resulting in less competition among the birds and allowing more to survive. The main adaptation was the dramatic change in the shape of the beaks. Some species evolved beaks best suited to feed on nectar and insects* [1], *others feed purely on insects* [2], *while some feed on fruit* [3] *or seeds* [4].

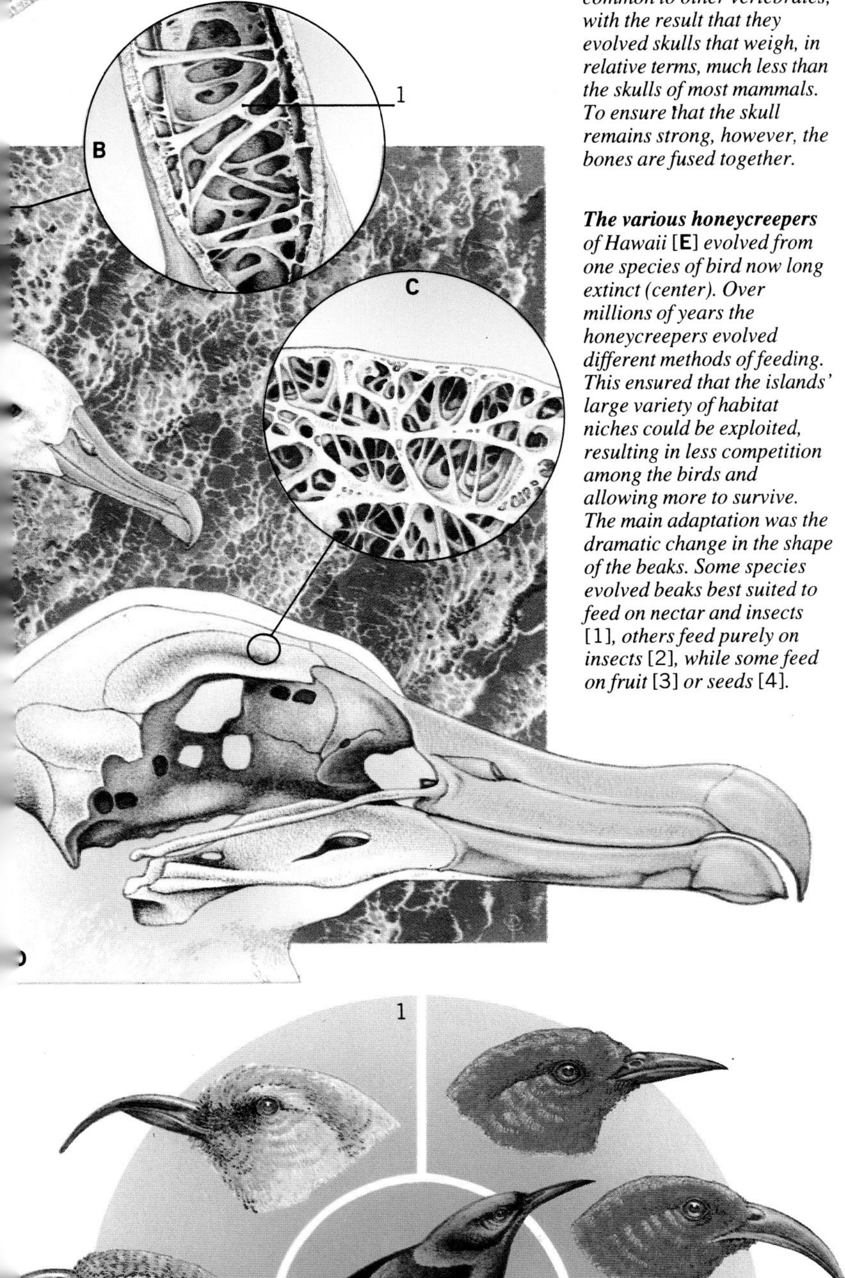

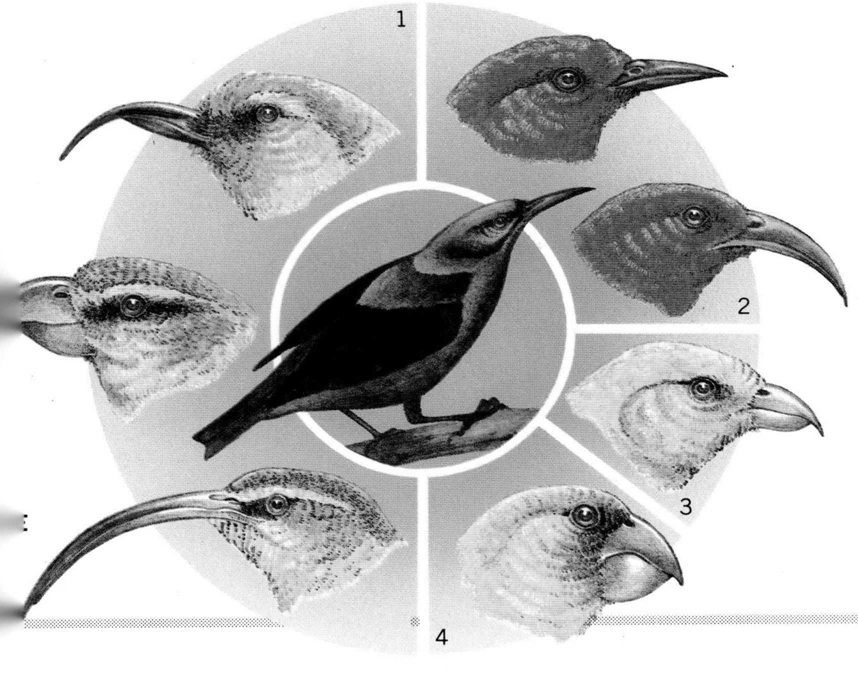

but had a blunt reptilelike snout. In short, it was truly intermediate in structure between birds and reptiles.

In 1990 two birdlike fossils were found in Texas and provisionally named *Protoavis texensis*. If these prove to have been capable of flight, the origin of birds may well have to be pushed back another 75 million years, into the late Triassic period, and *Archaeopteryx* will lose its position as the oldest known bird.

Flight adaptations

Like the reptiles, birds have no external ears. This is an important factor in streamlining. Birds do however have extremely good hearing, as well as the excellent sight necessary for maneuvering, hunting and landing.

When a bird is in flight its legs are held close to the body for aerodynamic advantage, and the large contour feathers also help give the bird's angular body a more streamlined shape. Because of this many birds are capable of very rapid flight. In fact the fastest, such as the peregrine falcon, have baffles in their nostrils to protect their internal organs from the huge air pressures that build up as a result of their speed.

In addition to their lungs birds have air sacs that extend through the body and into the hollow bones. These increase the lightness of the bird, aid its respiration, and help to cool down a metabolism that might easily overheat with the effort of sustained flight. A strong flier may use only a quarter of its air intake for respiration and the rest will be used for cooling.

A fast metabolism

A bird's blood has no higher a concentration of hemoglobin than a mammal's, in general, but its blood pressure is higher and the blood sugar concentration is considerably more, all to power the rapid metabolism. Birds eat high-energy food – like seeds, fruits, fish and insects – and digest it quickly. They are also more efficient than mammals at utilizing food, for immediate energy needs and to put on muscle weight.

A powerfully flapping bird may have half its body weight made up of the breast muscles, which power the wings. A glider such as the albatross has much less muscle than this, but has strong tendons and ligaments that effortlessly hold its wings out and allow it to soar for long periods.

Weight advantages

As well as its light skull and bones, a bird makes many other concessions in the interests of keeping its weight down. There are no sweat glands, which would only moisten their feathers and make them heavy. Female birds have just a single ovary, and outside the breeding season the sex organs of both females and males atrophy.

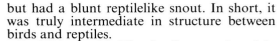

Protective Clothing

How fur and feathers grow

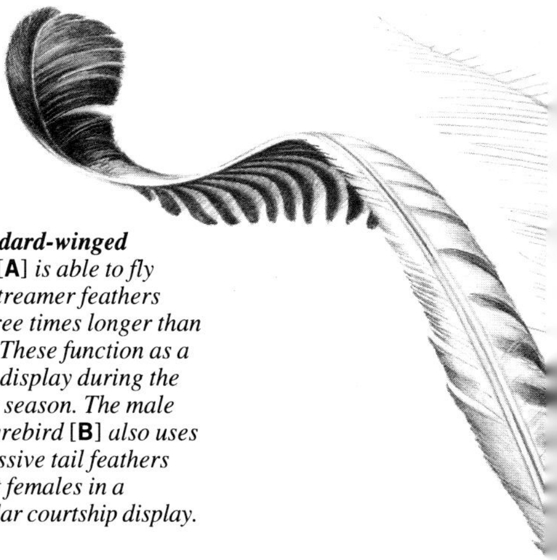

The feathers of the phoenix fowl of Japan can grow up to 33 feet (10 meters) long, but they are purely ornamental. More practically, feathers have equipped birds superbly for the conquest of the air, as well as providing insulation, camouflage and sexual and social displays. Hairs are made of the same protein (*keratin*) as feathers and share many of the same functions, but they also have unique qualities. Hair has contributed to the success of mammals by enabling them to control their temperature easily and efficiently, but some mammals are now endangered by the trade in their fur.

***The standard-winged** nightjar* [A] *is able to fly despite streamer feathers about three times longer than its body. These function as a courting display during the breeding season. The male superb lyrebird* [B] *also uses its impressive tail feathers to attract females in a spectacular courtship display.*

The fur of mammals contains two different sorts of hair. The fine, soft and silky ground – or under – hairs grow close together and form a dense undercoat, the function of which is to insulate and maintain the animal's body temperature. Longer, thicker, tougher, guard hairs extend beyond the undercoat; their function is to protect the underlying ground hairs and skin and to shed rain or snow. Some mammals lack one of the two layers and some, such as humans, have relatively few hairs on their bodies. Others, such as whales, dolphins, elephants and rhinoceroses, have virtually none.

Fur coats

A dense coat of ground hairs, such as the thick wool of sheep, enables very efficient temperature control, because air is easily trapped in thousands of tiny pockets among the hairs. This allows the animal not only to stay warm in winter but also to keep cool in summer. Waterproofing is another function of the hair of many mammals. Reindeer, for example, have long, water-repellent guard hairs above a thick, cold-resistant underfur, and aquatic mammals have particularly efficient waterproofing. The guard hairs of seals are flattened, while their densely packed insulating ground hairs have fine tips, both devices for shedding water. Many mammals living in areas where there are great seasonal extremes in temperature grow extra long, thick winter coats in response to shortening day length and the effects of hormones.

Take to the wing

Feathers, like hairs, are also essentially dead structures, but are linked by a sophisticated system of tiny muscles that control their position with amazing accuracy for insulation, courtship displays and flight. Although a single feather is proverbially lightweight, taken as a whole a bird's plumage generally weighs several times as much as its skeleton. Even so, feathers provide an air-trapping insulating layer for the bird's body at a far smaller cost in weight than would hairs. Feathers would soon become brittle with age and exposure to the elements if the bird did not preen them regularly with the waxy secretions from its preen (or *uropygial*) gland, situated on the rump at the base of its tail. The preen oil (consisting of waxes, fatty acids, fat and water) keeps the feathers flexible, moist and waterproof, as well as protecting them against bacteria, fungi and feather lice. Many birds also produce a substance a little like talcum powder, which is made of minute keratin particles continuously sloughed from specialized feathers called powder downs, which are found throughout the feathers among the ordinary down.

Birds also molt, depending on day length and according to the dictates of the thyroid and gonadic hormones. Birds go through many molts, firstly from the chick's down to its (juvenile) plumage, and as an adult regularly from summer plumage to winter plumage. There is considerable variation between species in the number of molts in a year and their timing. In most birds the pattern of replacement of flight feathers ensures there are only small, brief gaps in the surface of wings, so only a slight loss of flight power.

***Surface feathers** [C] are vaned and all have a strong, light, central shaft, the* rachis *[1], from which hundreds of barbs [2] extend in one plane to form the vane.* Barbules *[3] project from the barbs, and those that point away from the bird's body have tiny hooks,* hamuli *[4], which lock into the hookless barbs. This interlocking construction gives strength and helps the feather to keep its shape. Common types of feathers [D] include the flight feathers [1] and bristle feathers [2], which, in the case of the lyrebird and other ground-dwelling species, function as eyelashes. The loose-webbed barbs running along the length of the short shaft of down feathers [3] provide good insulation. The contour feathers [4] have an interlocking lattice structure. In flying birds they serve as streamlining, but for ground-dwelling species their function is primarily waterproofing. Many contour feathers also have a "downy" lower section that acts as an insulator.*

Filoplumes [5] are hairlike feathers with several soft barbs on the tip. They can be either sensory or decorative. In many birds some body contour feathers [6] (as opposed to those on the wings) have aftershafts, *which resemble a smaller feather growing from the base of the shaft.*

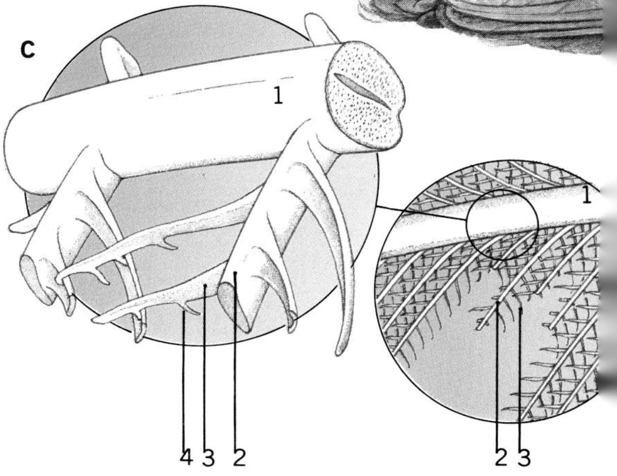

***A cross section of skin** [L*
shows elements that help regulate a mammal's boo temperature. Hair [1] is itself a good insulator, bı if required, hair erector muscles [2] will contract straightening the hairs aı creating an insulating la of trapped air. The sebaceous gland [3] also helps to make the animal waterproof by secreting (the hairs an oily substanc sebum, *that repels water. The* eccrine *and* apocrine *sweat glands [4 and 5] regulate temperature by secreting sweat onto the Finally, a layer of subcutaneous fat [6] als helps to reduce heat loss.*

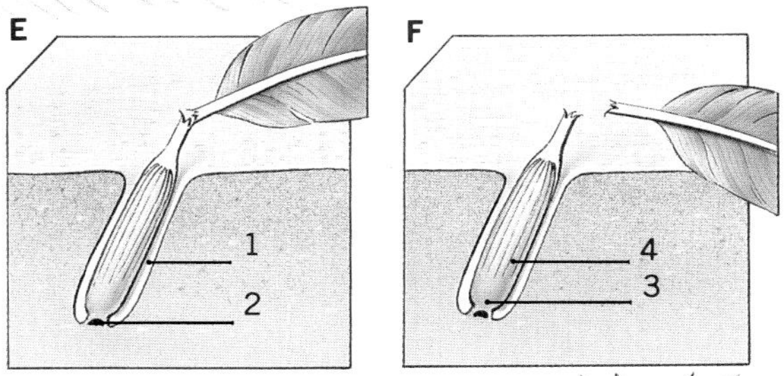

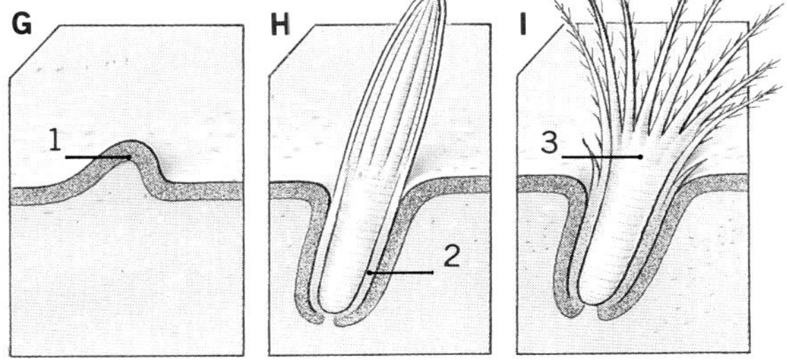

New feathers [**E**] *grow in the follicles of the skin* [1]. *A layer of epidermal cells covering the* papilla [2] *thickens* [**F**] *into a* collar [3], *which in turn develops through cell division into an epidermal cylinder* [4], *the main axis of which becomes the rachis and the secondary barb ridges.*

A down feather [**G**] *begins as a* mesodermal papilla [1]. *As the feather sheaf forms* [**H**], *it sinks into a* follicle [2]. *The internal formation of the feather sheaf takes place as the epidermal cells develop into barbs. When fully developed, the horny sheath splits* [**I**], *releasing the downy feathers inside* [3].

The hair bundle *of a true seal* [**J**], *compared to that of a fur seal* [**K**], *shows the fur seal's dense mat of underfur. The guard hair of a true seal has only a few secondary hairs, while that of the fur seal may have up to 50 secondary hairs, acting as an effective insulation when the animal is in water.*

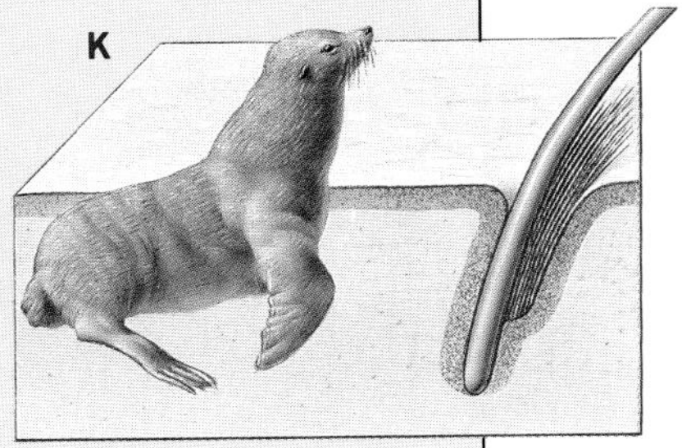

Of Mice and Men

How mammals thrive

Mammals are the largest land animals to rise to prominence since the dinosaurs and are one of the most successful groups in the animal kingdom. From 1.5-inch (4-centimeter) pygmy shrews to 100-ton whales, the 4,500 species of mammals have explored and adapted to most of the planet's different habitats, evolving a wide range of shapes, sizes and life-styles on the way. They owe their success to two essential factors: their warm-bloodedness, which lets them, unlike reptiles, function regardless of temperature, and their development of suckling and looking after the young.

The evolution of *homoiothermy* ("warm-bloodedness") – the internal self-regulation of body temperature – gave early mammals an essential competitive edge over reptiles, because it meant they could be more active at lower temperature and so exploit more niches. The suckling of the young on the female's milk and the investment of parental care were also highly important. Protected and fed, the young could grow rapidly and learn skills from the adults: chimpanzees, for example, learn to use sticks as tools, and many carnivorous mammals learn hunting techniques.

Enough to go around

A wide range of mammalian feeding habits has evolved to make full use of the Earth's food resources. Nocturnal and diurnal species coexist in the same habitat, allowing greater exploitation of the habitat than if all its inhabitants were competing for food and space at the same time. Migrants, such as caribou and the vast herds of African antelope and zebra, move around to exploit seasonal food sources. Other mammals, such as whales and fur seals, migrate to sheltered bays to breed, living on stored fat until they are free to return to more food-rich waters.

Many small rodents, and even a few larger mammals such as the black bear, are able to survive severe winter conditions by hibernating. Others escape environmental fluctuations by retreating underground to burrows or to the depths of caves.

Adaptable mammals

Some mammals with terrestrial ancestors have evolved to exploit the air and the sea. The bats are highly successful and the only vertebrates other than birds capable of sustained flight. Whales and dolphins have become so well adapted to the marine environment that they never come ashore. The baleen whales have evolved huge filters of *baleen* (whalebone), with which they sieve the vast shoals of krill that abound in the oceans, particularly in the Antarctic region.

In areas of scarce prey predatory mammals tend to be solitary, like the polar bear, whereas where food is more plentiful cooperative hunting techniques have evolved, where groups of predators such as hunting dogs or lions cooperate to bring down prey that may be larger than themselves. Many herbivorous

***Feeding offspring* [A]** *on the female's milk is one of the distinctive features of mammals. In most mammals the milk is exuded from nipples situated on the belly. The number of nipples roughly matches the largest number of young commonly produced at one time by that species. Milk production is stimulated by hormones released toward the end of pregnancy. Once the young are weaned, milk production ceases, thus saving the adult energy. Milk provides a complete meal for the young mammal: fats for energy; proteins for tissue development; calcium and iron for bones and the blood.*

A

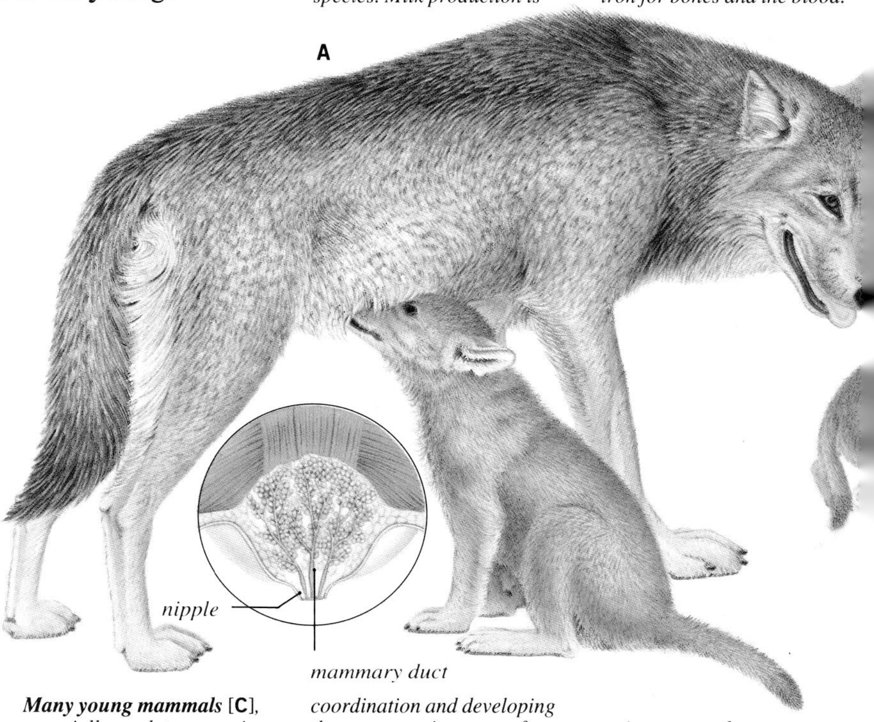

***Many young mammals* [C]**, *especially predatory species, spend a long time in the care of their parents. This allows them to build up adult skills. Play is an important part of growth, improving coordination and developing the responses important for defense (and in some species also for predation). Fur is another special characteristic of mammals, providing insulation, important in the maintenance of temperature. Mutual grooming has an important social function, strengthening the bonds within family groups.*

***In the mammalian heart* [F]** *the right and left sides are completely separate. The left side of the heart receives oxygenated blood from the lungs via the* pulmonary veins [1]. *This is pumped to the body's organs at high pressure via the aorta* [2] *and its branches. Deoxygenated blood returns to the heart via the inferior and superior* venae cavae [3 *and* 4] *and is pumped back to the lungs via the* pulmonary arteries [5]. *This is done by the right side of the heart at a lower pressure, which is important to avoid damaging the lungs. The left side of the heart is frequently larger since it pumps at higher pressure.*

F

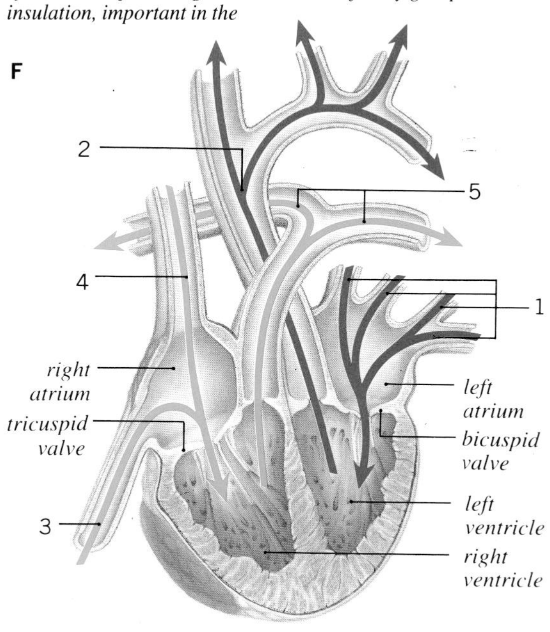

Connections: Animals in Cold and Wind 144 Animals in Heat and Drought 142 Group Carnivores 248 Evolution to Date 94 Fur and Feathers 134

mammals have evolved symbiotic relationships with cellulose-digesting bacteria, so that they can take advantage of relatively indigestible plant food for which there is less competition. Bears and many primates are omnivorous, feeding on fruits, leaves and berries, but hunting other creatures when they get the chance. Other mammals, such as raccoons and hyenas, are mainly scavengers.

Worldwide similarities

Despite geographical isolation, many mammals have adapted in similar ways to similar life-styles in different continents; examples are the anteaters and armadillos of South America, the aardvark of Africa, the pangolin of Asia and the echidnas, of Australia and New Guinea – these creatures are all anteaters with long, tough snouts, long tongues, and powerful claws for breaking into the nests of ants and termites.

B

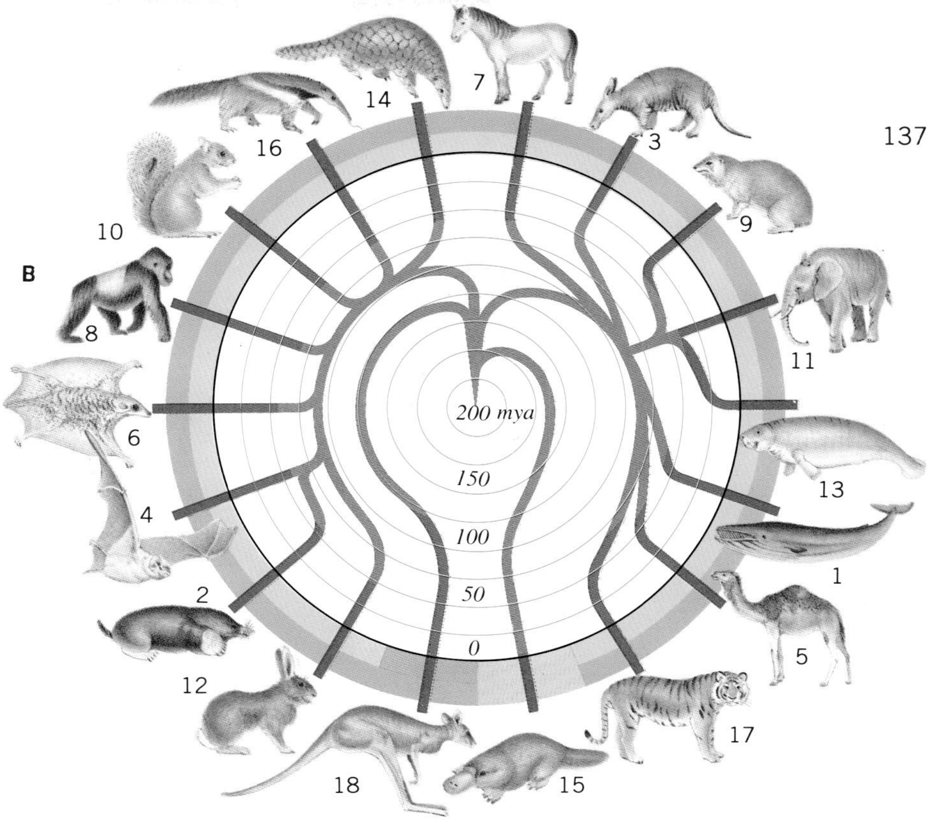

C

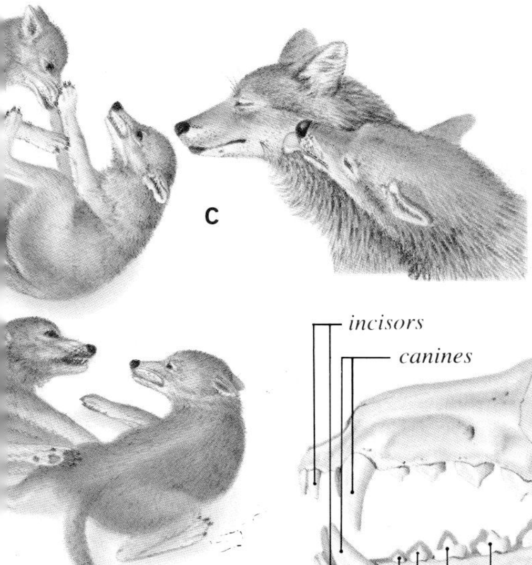

Having specialized teeth [E] *of different kinds is a unique mammalian adaptation. Variety is greatest in the carnivores, which have biting incisors, tearing canine teeth and sharp molars and premolars for crushing bones. Herbivores have more uniform teeth designed for grinding vegetation.*

***Tracing the origins** of mammals* [**B**] *depends on the availability of a good fossil record. For so many of the 18 orders of mammals the record is incomplete. However, possible lines of descent have been suggested. It seems likely that mammals are all descended from lizardlike reptiles known as* synapsida. *The synapsida evolved many millions of years before splitting into two subclasses: the* prototheria *(the monotreme egg layers) and the* theria *(which give birth to live young). The theria themselves divided into two further groups: the* eutheria, *or placental mammals, and the* metatheria, *the marsupials.*

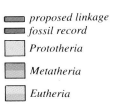

proposed linkage
fossil record
Prototheria
Metatheria
Eutheria

1 *Cetacea (whales)*
2 *Insectivora (insectivores)*
3 *Tubulidentata (aardvarks)*
4 *Chiroptera (bats)*
5 *Artiodactyla (even-toed ungulates)*
6 *Dermoptera (flying lemurs)*
7 *Perissodactyla (odd-toed ungulates)*
8 *Primates (primates)*
9 *Hyracoidea (hyraxes)*
10 *Rodentia (rodents)*
11 *Proboscidea (elephants)*
12 *Lagomorpha (rabbits)*
13 *Sirenia (sea cows)*
14 *Pholidota (pangolins)*
15 *Monotremata (monotremes)*
16 *Edentata (anteaters)*
17 *Carnivora (carnivores)*
18 *Marsupialia (marsupials)*

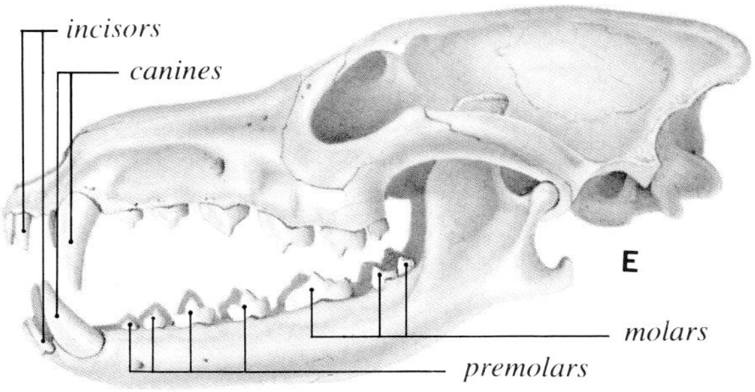

***Mammalian skeletons** [**D**] have become modified in many different ways as animals have adapted to different habitats and life-styles. In primitive mammals limbs tended to be relatively short, as for instance in the opossum. But adaptations, such as running faster, led to changes, like the elongation of limbs. In bats* [1] *the forelimbs have become wings, while in seals* [2] *the limbs form paddle-like flippers. Whales' limbs have a minimal steering function; in baleen whales* [3] *the jawbones are greatly enlarged and curved to accommodate the animal's huge, sievelike plates of baleen (whalebone).*

D

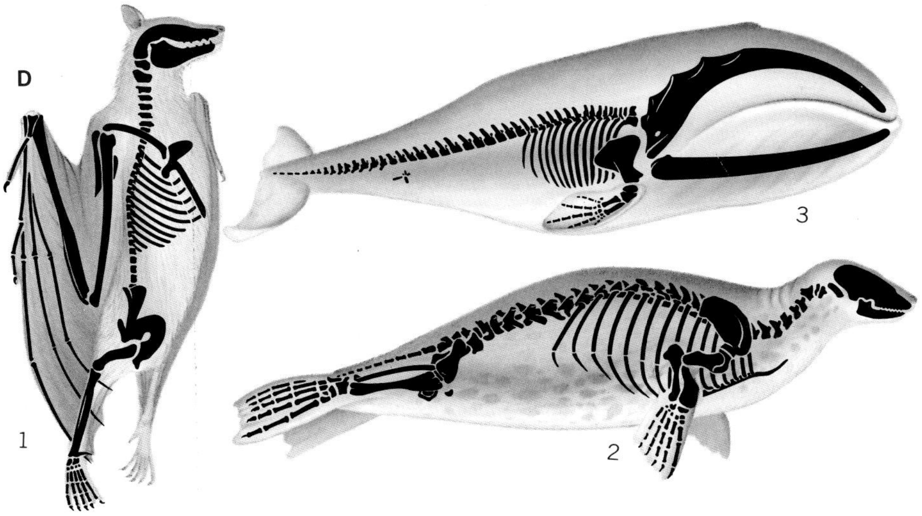

Gliding and Soaring 222 Intelligence 290 Mammal Reproduction 176 Mammal Societies 298 Primitive Mammal Reproduction 178

Hothouse Flowers

How plants survive heat and drought

It may be hard to believe that even in a tropical rain forest there are plants adapted to drought conditions. But drought can strike in the most unexpected places. Conifer trees have evolved to minimize water loss in winter, and many of the plants found near coastlines have features similar to those of desert dwellers. In the neverending battle to conserve their water, plants in arid lands modify their shapes, their leaves and their metabolisms. Up to 90 percent of the mass of some desert plants is below ground, in the shape of huge, water-retaining root systems.

Under the blazing sun of a hot desert many animals hide away, avoiding its desiccating effect by finding shade or burrowing into the ground. But for the plants there is no such escape. They are not mobile and must somehow survive the conditions. Some avoid drought altogether by remaining dormant as seeds. When rains come, they respond by germinating, covering the desert with color for a short time. Many of the plants that live in arid lands, though, remain above ground throughout the year, and survive by means of a whole range of adaptations.

Drinking in the fog

Along the Atlantic coast of Namibia runs the Namib desert, one of the driest places in the world and home to a remarkable drought-adapted plant. Welwitschia's two huge, strap-shaped leaves are ragged at the ends, where they have been moved around against the rough desert ground by the wind.

The leaves are thick, waxy, and efficient at retaining water. In the Namib the only water supplies come from occasional dews and from the fogs that drift in from the Atlantic. It is a popular myth that the fog-shrouded Welwitschia absorbs the moisture in the mist through its leaves. In fact what happens is that water simply condenses onto and then runs along the leaves, drips off the ends, and is finally taken up by the roots.

Water hoarders

The key strategy is to get as much water as possible – especially when there are rains – and then to hold onto it for dear life. Large root systems allow plants to extract every last drop of water from the soil, and to take full advantage when they need to keep it.

Many plants defend their water so that it cannot easily escape. Even so, the plants still have to take in carbon dioxide and pass out oxygen as they carry out photosynthesis, and they do this through pores in their surfaces known as *stomata.* Unfortunately the stomata also provide an exit route for water, so desert dwellers have evolved ways of minimizing this loss. Most plants open their stomata during the day, but some desert plants, including cacti, open them at night instead, when the desert is much less hot and dry. They are able to do this because they have a special photosynthetic pathway that allows carbon dioxide to be taken up and stored in organic form prior to being used in photosynthesis during the day.

Marram grass **[A]** *actively protects its water supply. Its stomata, through which water is lost to the environment, are located on the inside of the leaves. In a drought the marram's leaves curl up, protecting the stomata and conserving water. Marram grass grows in coastal sands.*

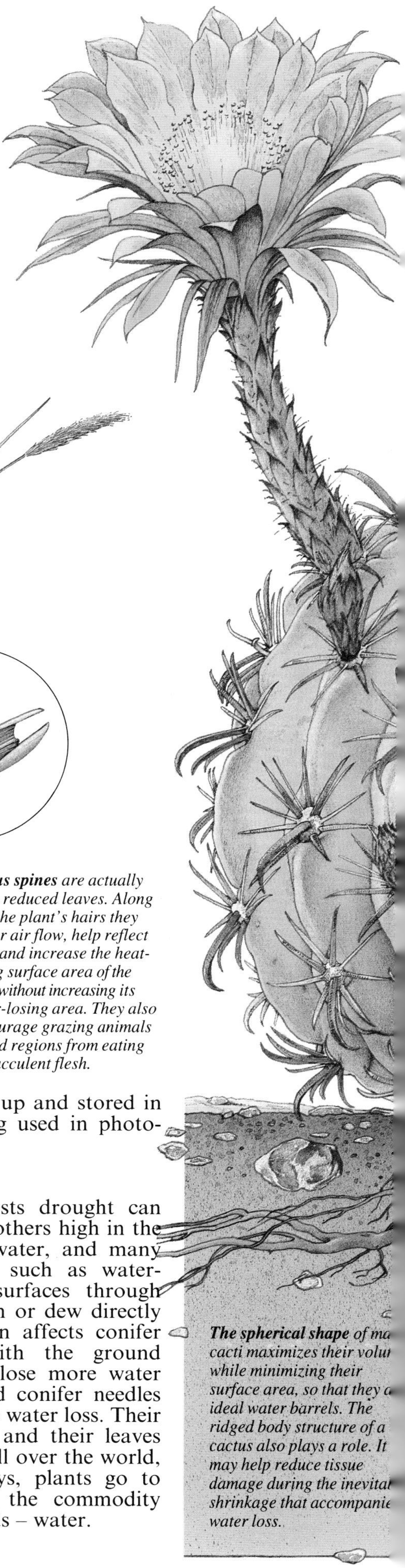

Cactus spines *are actually much reduced leaves. Along with the plant's hairs they hinder air flow, help reflect heat, and increase the heat-losing surface area of the plant without increasing its water-losing area. They also discourage grazing animals in arid regions from eating the succulent flesh.*

The spherical shape *of ma cacti maximizes their volu while minimizing their surface area, so that they a ideal water barrels. The ridged body structure of a cactus also plays a role. It may help reduce tissue damage during the inevita shrinkage that accompanie water loss.*

A universal problem

Even in tropical rain forests drought can strike. Plants that grow on others high in the canopy may receive little water, and many show drought adaptations such as water-absorbing cells on their surfaces through which they can take up rain or dew directly into the leaf. Drought even affects conifer woodlands. In winter, with the ground frozen, the trees stand to lose more water than they can take up, and conifer needles are well adapted to minimize water loss. Their stomata are in buried pits and their leaves have thick outer surfaces. All over the world, and in many different ways, plants go to great lengths to conserve the commodity most vital to their daily needs – water.

Connections: Conifers and Cone Bearers 118 Desert Life 304 Fowering Plants 120 Plant Architecture 114 Plant Cells 100 Plant Feeding 184

Cacti have *their stomata sunk down into pits* [**B**], *which helps cut down transpiration – the evaporation of vital water – because the air in the pit is protected from air movement above the surface of the leaf and becomes humid. Cacti have tough outer skins, which are often waxy, to cut down water loss, but they are fleshy inside and capable of storing a lot of moisture. Cacti flower rarely and briefly, after the often violent desert rains that can turn wastelands into temporary gardens.*

Salt Survival

It is not just in the desert that plants have to deal with drought. Plants that grow in salty places also face the challenge of holding water, because the salty environment around them tends to draw water from them. Such plants are called *halophytes.* Many have adaptations like desert plants, such as succulent leaves and thick outer surfaces. The halophytes can also concentrate salt in their cells, which helps reduce the wilting effects of sea water. Many halophytes have a special root membrane that keeps salt out of the plant. Others have even more ingenious strategies. Some allow the salt in but store it in leaves and stems that are shed at the end of the growing season, thus getting rid of the salt all in one go. Sea lavender actually takes up salt and then excretes it onto its leaves through special glands. The salt dissolves away during fog or rain.

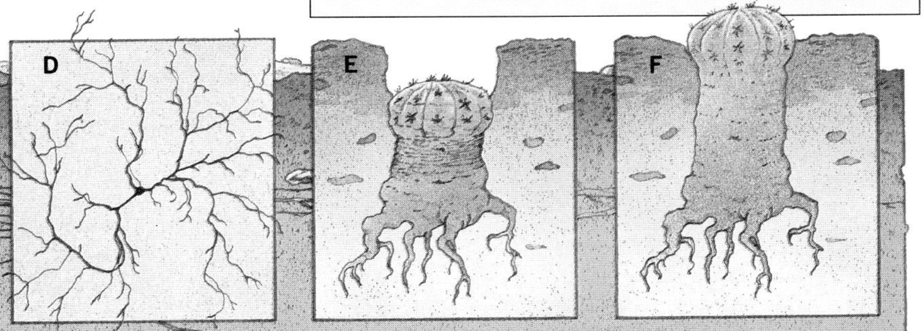

The roots *of xerophytes like cacti can penetrate to a depth of 20 ft (6 m) in their search for water. Alternatively, the root system may become swollen, water-retaining tubers. Most cacti, however, have a wide-ranging system of fine roots, which are equipped with microscopic hairs* [**C**]. *These may only penetrate a short way underground, but often cover a huge area* [**D**] *so that the cactus can quickly replenish its water supplies when water is available. In dry conditions there are many plants that are better at absorbing water than cacti, such as small bushes with long roots.*

Many cacti, *such as the* Echinocereus pulchellus [**E**], *hide themselves away underground in the dry season. Only when conditions are favorable do they extend their green tops above the surface* [**F**]. *In this way they combine drought resistance with effective drought evasion.*

Plants in Cold and Wind 140 Tropical Forest Plants 322

The Chill Factor

How plants survive extreme cold

Dwarf willows often grow branches up to 16 feet (5 meters) long, but the trees are seldom more than 4 inches (10 centimeters) tall. This apparently poor growth is sometimes due to abrasion by wind-blown particles and ice crystals, but it is often genetic – the trees grow stunted even in warm conditions. Slow growth gives the plant a low demand for water and nutrients. Near the ground, wind velocity is much lower, so they are not exposed to such strong forces, and they remain below the snow cover in winter, which protects them from the colder air above.

Red algae *often grows on the surface of ice. The red pigment protects the plant in two ways. Firstly by converting light into heat, and secondly by filtering out light rays that might cause mutations to the plant's DNA. Other cold-climate plants are tinged with similar red pigmentation.*

The vast expanse of tundra that fringes the ice sheets and frozen ocean of the Arctic, the mossy hummocks of Antarctic shores, and the colorful flowers that grow on the precipitous cliffs of alpine regions are all witness to the ability of plants to tolerate cold. High latitudes and high altitudes present some of the most severe climatic conditions faced by plants. Temperatures are extremely low in winter – less than –22°F (–30°C) in many parts of the tundra and in Antarctica – and on cloudless nights in summer, and the strong winds that sweep these regions effectively lower them still further.

The wind increases the evaporation rate and hence the transpiration rate of plants, so imposing a drought situation on many arctic and alpine plants, which are often already growing in thin soils on rocky terrain incapable of holding much moisture. For a large part of the year the low temperatures keep the soil water frozen and unavailable, and what precipitation does fall is in the form of frozen water – snow or hail. The permafrost (permanently frozen soil) also restricts root development and hence the chance to obtain nutrients from the soil.

Keeping warm

Many arctic plants survive the winter underground as roots and stems, bulbs and tubers, swollen with stored food. The moment conditions are suitable for growth they can put out new shoots and start to photosynthesize. Others, with tough, leathery, evergreen leaves, protected by waxes and clustered into low-growing cushions, sit the winter out and do not even have to put out leaves in spring.

Many plants that experience seasonal cold have evolved a kind of frost tolerance. In response to signals from their surroundings, such as the decreasing day length of late summer, changes take place in their cells – known as cold-hardening – which reduce the risk of frost damage. The main damage is done by ice forming between cells, drawing water out of the cells and causing them to dry up. Ice formation inside the cells makes the membranes leaky and so disrupts metabolism. Cold-hardening appears to involve the secretion of certain chemicals, such as alcohols and sugar derivatives, in the cell sap that reduce its freezing point or in some other way protect vital structures.

Connections: Plant Cells 100 Plant Reproduction 150 Plants in Heat and Drought 138 Tundra 314

***Plants adapted** to cold climates often grow close to the ground in order to avoid the wind-chill factor (whereby the cold wind can dramatically lower the air temperature). Although they are trees, the dwarf willow* [1] *and Arctic willow* [2] *are no taller than other plants. The catkins of these trees are extremely hairy. These hairs create a warmer* micro-climate *around the catkin. The catkins are further protected by an internal process called frost-hardening, which allows the plant cells to withstand sub-zero temperatures. The catkins of one dwarf willow tree species resumed growth after being frozen for three weeks at temperatures as low as 14°F (–10°C). The most exposed surfaces are occupied by lichen* [3] *and moss* [4]*. Lichens dehydrate during winter to avoid frost damage. Mosses have rootlike structures called rhizoids that secure the plants in the thin layer of soil. The plant's low cushion shape creates its own micro-climate in which the temperature is a few degrees higher than the surrounding air. Flowering plants, such as mountain avens* [5]*, also exploit the "cushion" effect to keep warm. In addition to its dense, compact form, the plant has hairs on its stem and fruit to increase the insulating effect. The avens' flowers, like those of the Arctic poppy* [6]*, gradually turn to track the Sun. The large petals serve as parabolic reflectors that focus sunlight on the flower to increase metabolic activity during the development of seeds. Whiplash saxifrage* [7] *survives by taking the cushion effect to extremes. Only the sun-trapping flower is exposed to the chill of the wind, the rest of the plant is a compact windproof mass. Moss campion* [8] *spreads over the surface to form a thick bushy mat of vegetation. The roots penetrate deep into the ground to ensure a supply of water during the frozen drought of winter.*

A life of compromise

Life for all plants is a compromise between vegetative growth and the formation of flowers and seeds, but the struggle is especially severe for plants that have only a short growing season. Some arctic and alpine plants have overwintering flower buds, so they have a full growing season in which to flower and set seed. Others spread their reproduction over several years, developing flower buds in the first year, producing pollen the following year, actually flowering in the third year, and releasing seeds in the fourth year or even later. Despite all these adaptations, some species are unable to flower at arctic latitudes. Many arctic plants reproduce vegetatively – producing clones of themselves by the fragmentation of *rhizomes* (rootstock), or the subdivision of bulbs and corms – because this method of reproduction uses around 10,000 times less energy than sexual reproduction.

On the Edge of Survival

How animals are adapted to heat and drought

Devil's Hole pupfish live only in the Devil's Hole, a natural pool in the Nevada desert of the United States which is 50 feet (15 meters) in one direction and 16 feet (5 meters) in the other. This is because they need a constant temperature of precisely 93°F (33.9°C). The fish are the last representatives of a larger community that once lived here. Yet they are exceptions, for in the desert – where rainwater rapidly evaporates from ditches and craters – many animals simply die when the pools dry, leaving nothing but their eggs, which come to life in later years.

Life, for animals that make their homes in the extreme conditions of a baking desert, is a continuous struggle to maintain the correct body temperature and to find enough to eat and drink. Only the most resilient species survive, and those that have adapted successfully are faced with few competitors.

Desert animals survive heat and drought in either of two ways. Some avoid the hottest and driest periods, emerging only briefly after it has rained to feed and reproduce. For example, fairy shrimp can remain dormant for up to 25 years as drought-resistant eggs, then suddenly appear when it rains. Other species manage to remain active throughout the year by adjusting their lifestyles to the arid conditions. Reptiles and rodents are two of the most successful groups of these active desert inhabitants. Both take advantage of microclimates found in the shade of rocks and inside burrows that provide some respite from the heat. Even a few inches below the ground the temperature is lower and the process of evaporation slower.

Most reptiles are able to obtain all the water they need from their food. Geckos and lizards feed on the abundance of insects, scorpions and spiders that are swept off rocks and ledges by the desert winds. Other reptiles feed on leaves, flowers and fruit, but are forced to supplement this diet with grasshoppers and beetles when fresh vegetation becomes scarce.

Rodents that never drink

The kangaroo rat of North America and its counterparts in Asia and Africa – the gerbils and jerboas – live in areas where the temperature can soar above 104°F (40°C). These rodents spend their days in their homes beneath the sand, where the temperature remains more or less constant at 86°F (30°C). They are most active just before sunset, when they emerge to search for food. To survive conditions of almost permanent drought, these tiny creatures have become masters of water conservation. They never drink or sweat. Their kidneys, which have a filtering capacity five times greater than that of a human being, produce very concentrated urine, wasting little water. Kangaroo rats even manage to extract a little liquid from the dry seeds they feed on.

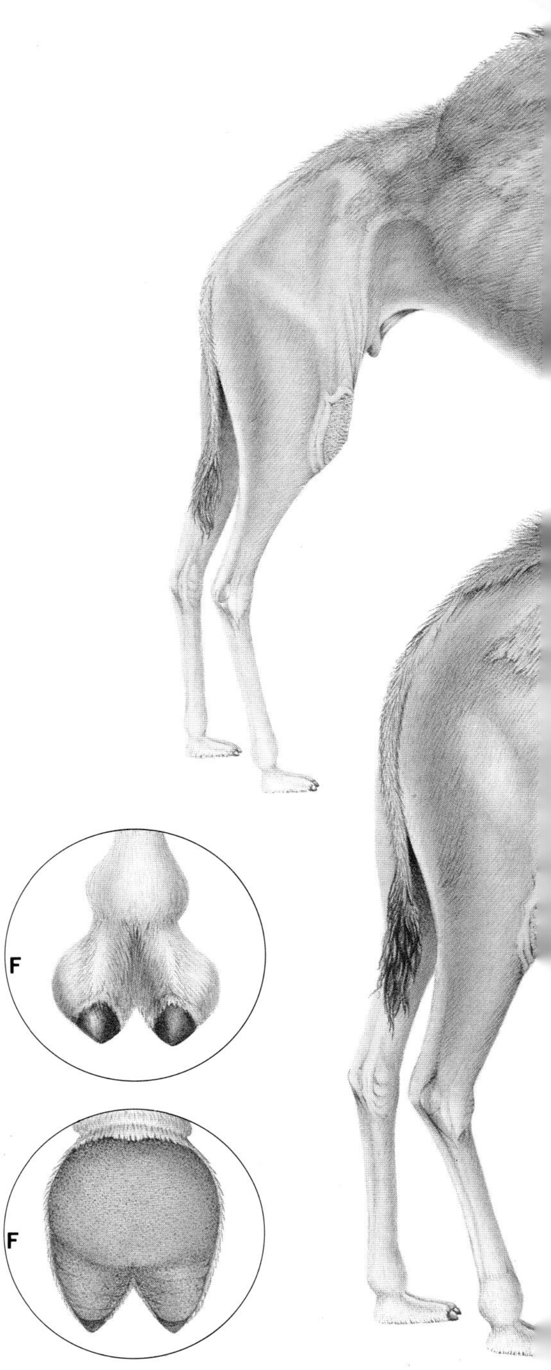

***A laden dromedary** camel can walk across a scorching desert for eight days without drinking or eating. In times of adversity the animal relies on the reserves of fat stored in its prominent hump. It is able to lose one-quarter of its body weight without harmful effects* **[A]**; *but when food and water are available, it can regain its full weight in just two or three days* **[B]**, *and is capable of drinking 33 gallons (150 liters) of water at a time.*

Among its other adaptations to desert life the dromedary has soft fur that traps a layer of air, helping it stay cool during the day and warm at night. Long lashes **[C]** *shield its eyes from wind-blown sand, as do the nostrils, which can be opened and closed at will* **[D]**. *Hard knee pads protect the animal when it kneels on the scorching sand* **[E]**, *and its toes are joined together by a fleshy pad, which acts as a "snow shoe"* **[F]**, *preventing the camel from sinking into the sand.*

An amphibian in the desert

The North American couch spadefoot toad is one amphibian that has habits and behavior perfectly in tune with the desert. It has club-shaped hind feet equipped with horny projections with which it digs a burrow, where it shelters until dusk, when it emerges to hunt for insects and spiders. Throughout the summer months it avoids the heat by remaining in its burrow, without feeding at all. During this time it may lose up to 60 percent of its body weight. When the brief wet season finally arrives, the toads reproduce while there are puddles in which they can lay their eggs. To complete the process before the water has gone, the toads have developed a remarkably rapid reproductive cycle. Mating begins with the rain; tadpoles hatch from fertilized eggs in just two or three days and reach maturity in under six weeks.

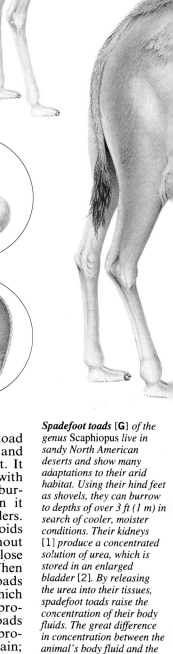

Spadefoot toads **[G]** *of the genus* Scaphiopus *live in sandy North American deserts and show many adaptations to their arid habitat. Using their hind feet as shovels, they can burrow to depths of over 3 ft (1 m) in search of cooler, moister conditions. Their kidneys* [1] *produce a concentrated solution of urea, which is stored in an enlarged bladder* [2]. *By releasing the urea into their tissues, spadefoot toads raise the concentration of their body fluids. The great difference in concentration between the animal's body fluid and the surrounding soil causes water to enter the toad's body across its skin.*

Connections: **Amphibians** 128 **Animals in Cold and Wind** 144 **Desert Life** 304 **Deserts** 54 **Reptiles** 130

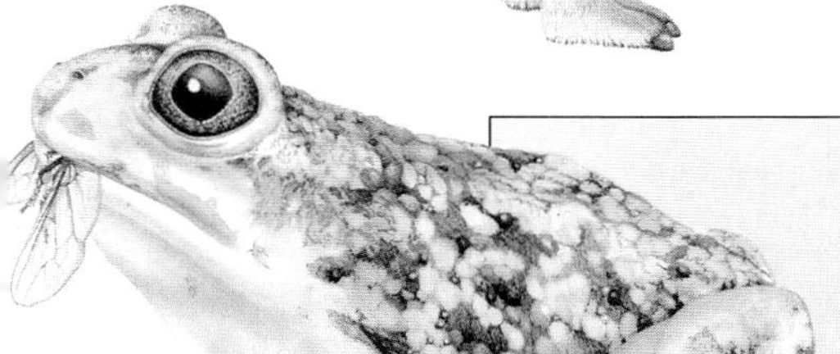

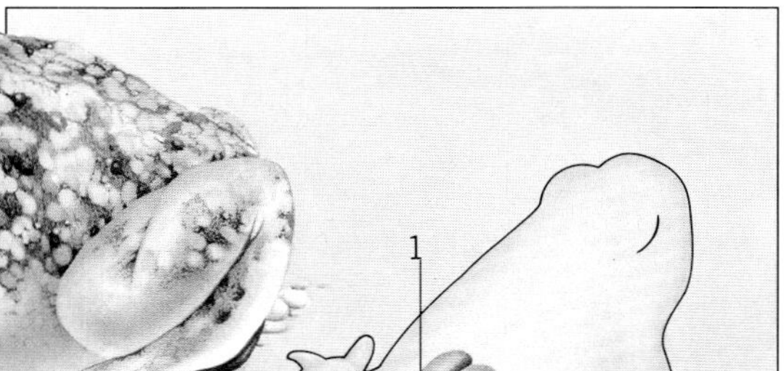

Cooling Ears

The fennec fox (left) of the Sahara desert is the smallest member of the fox family. Although it measures only 2 ft (60 cm) from nose to tail-tip, it has ears that are 6 in (15 cm) long. These enormous ears, with their large surfaces and their network of veins near the surface of the skin, radiate heat and enable the fox to keep cool. They also give the fox a keen sense of hearing, so that it can hear the slightest of sounds made by its prey. The fennec lives on insects, small rodents and fruit. It increases its chances of keeping cool by hunting at night and spending the day underground in its burrow. Although big ears are a clear aid to hunting, their heat-radiating properties make them inappropriate in cold climates. Thus the Arctic fox (right) has small ears, as well as thick fur and a stocky, compact shape that is ideal for heat conservation.

Life on Ice

How animals have adapted to cold and wind

The larva of the chironomid midge can survive after 90 percent of its body water has been turned to ice. Very few animals are so resistant to freezing. Conditions in Greenland are so harsh that only seven species of land mammal can live there. In the Antarctic the amount of snow blowing in the wind and the virtually total lack of vegetation mean that there are almost no terrestrial animals at all. Yet the species that do live in the polar regions generally do so in vast numbers, which indicates how successfully they have adapted to life in such habitats.

Animals of the Arctic and Antarctic have to survive in temperatures that may average –4°F (–20°C) and can plummet to –58°F (–50°C). In order to survive they have had to develop ways of conserving body heat. The most obvious answer – used by most terrestrial animals of such latitudes – is a thick coat to keep out the cold and keep in the heat. Emperor penguins have a double layer of long, dense feathers and large deposits of fat beneath their skin that both provide excellent insulation. The snowy owl also has dense feathers, which not only enclose its body but also cover its legs, toes, claws and beak. Mammals such as polar bears and arctic foxes grow thick coats of warm fur. The insulatory properties of fur depend on its ability to trap air, so it is not as efficient underwater. However, mammals often have two layers of hair: soft underhair that acts as insulation and longer, coarser hair above this that provides a protective outer blanket against the weather and wear and tear.

Thermal architecture

The shape of an animal's body affects the amount of heat lost to the air. Body heat is lost primarily through an animal's external surface, so to minimize heat loss the animal must expose as little surface as possible to the air. Long limbs and extended bodies provide large surface areas, so birds of the icy regions tend to have compact bodies with short legs, flippers and bills. Mammals have shorter legs, ears and muzzles than their relatives in warmer climates. Polar insects, such as the Bolaria butterflies, tend to be dark and hairy to absorb any available heat. They are also small, so that they can quickly warm up enough to be able to move around.

Specialized behavior

Polar bears and arctic foxes dig dens several yards deep to shelter from winds that sweep over the ice in strong gusts, unbroken by trees or vegetation. However, many birds are obliged to remain exposed to the elements. Emperor penguins huddle together in huge gatherings of up to 5,000 in order to keep one another warm and to conserve heat. After a while the birds that are facing the wind on the outside of the group become very cold. They move round and into the center of the crowd, leaving different birds to replace them

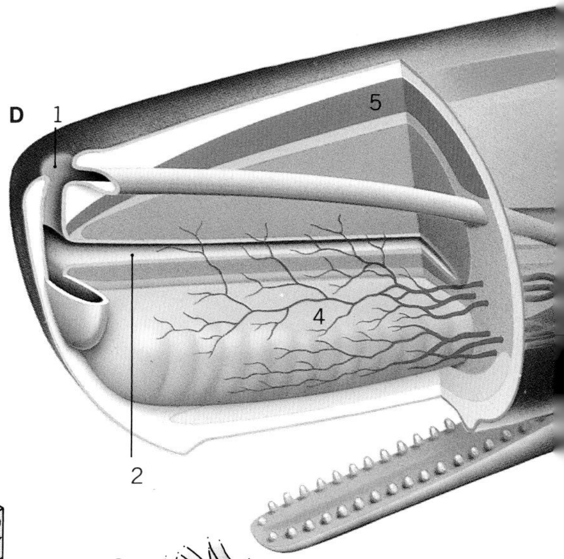

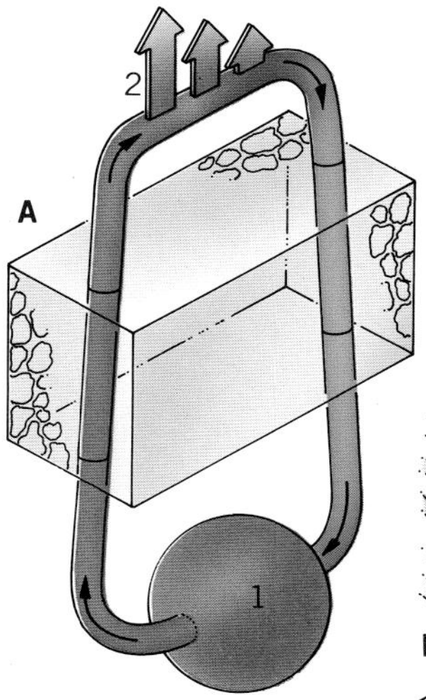

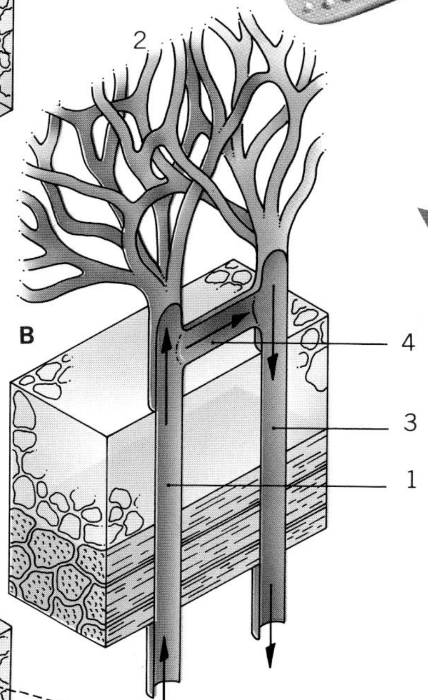

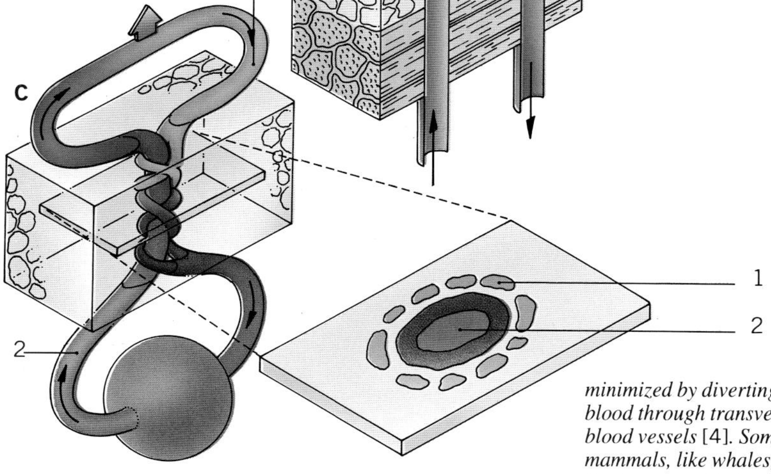

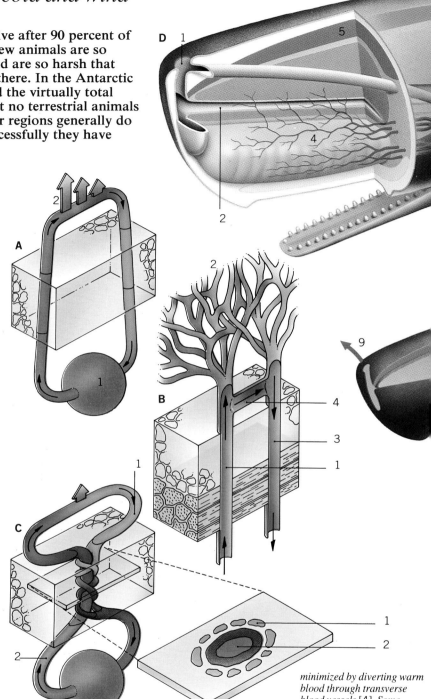

Mammals *are* homoiothermic *– they maintain a constant body temperature, usually at around 98.6°F (37°C), by generating metabolic heat. In the cold, heat loss is reduced by a layer of insulating fur or blubber (fatty tissue), but temperature is further regulated by the circulatory system. A simple heat exchanger* [**A**] *dissipates internal heat* [1] *to the outside* [2], *but most mammals have more complex systems* [**B**]. *The warm blood in an artery* [1] *branches out into capillaries, where heat is lost through the skin* [2]. *Cooled blood returns through the veins* [3]. *Heat loss can be minimized by diverting warm blood through transverse blood vessels* [4]. *Some mammals, like whales, that live in very cold climates have* countercurrent heat exchange *systems* [**C**] *that further reduce heat loss . A vein returning cold blood* [1] *divides and surrounds an artery* [2]. *The arterial blood warms up the venous blood – lessening the shock it might cause to the major organs – and is in its turn cooled down so that less body heat is lost to the surface.*

Connections: **Breathing** 126 **Fish** 124 **Fur and Feathers** 134 **Mammals** 136 **Mountain Life** 306 **North Polar Life** 310 **Ocean Life** 332

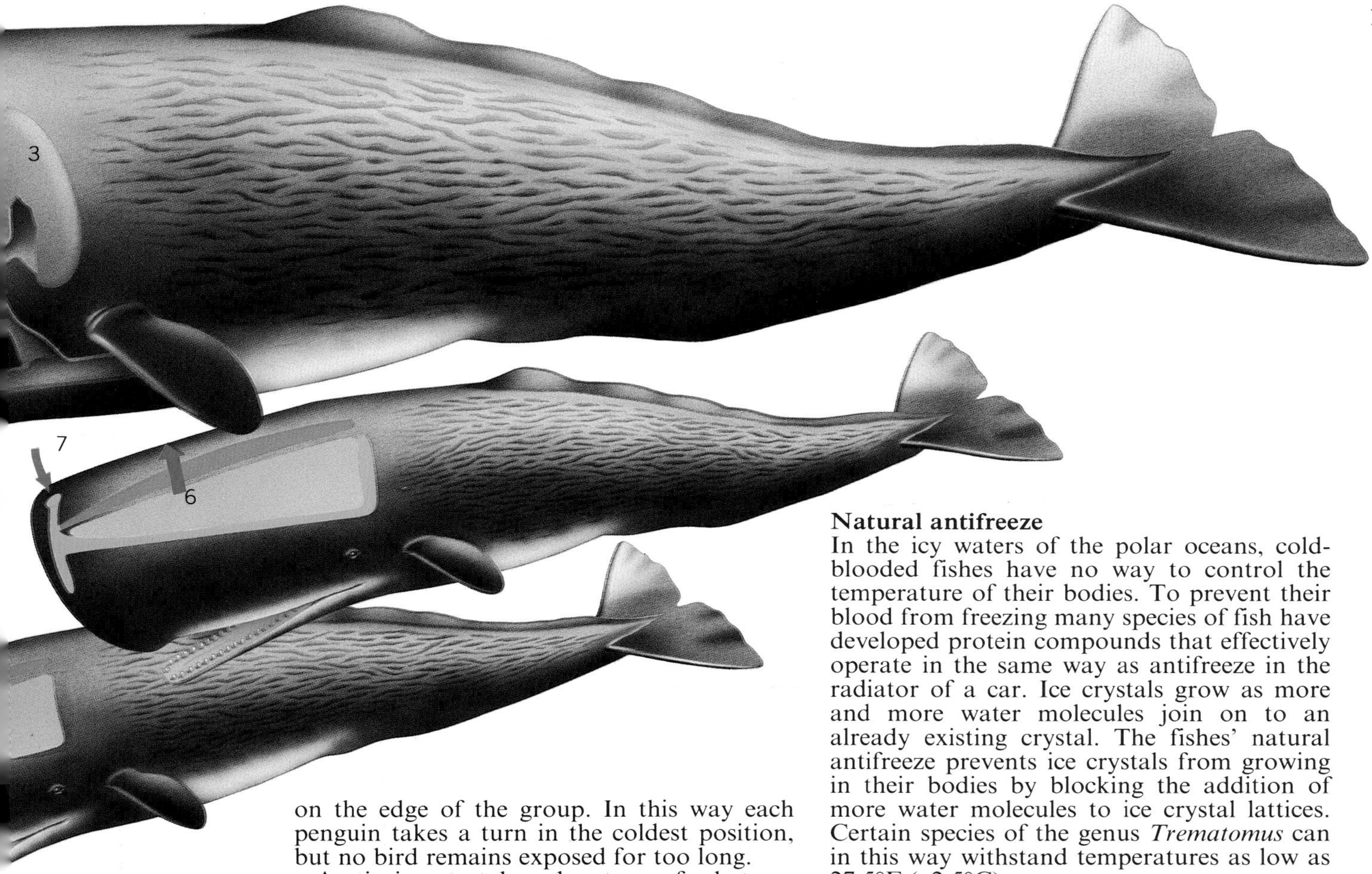

Blubber is such an effective insulator that overheating during vigorous activity can prove to be as great a problem as the cold. The sperm whale [**D**] *can regulate its temperature by dumping heat directly into the sea. When it is swimming it takes in water through the blowhole* [1], *down the right* naris *(nasal passage)* [2], *which terminates in the nasofrontal sac* [3]. *The right naris passes through the* spermaceti [4], *which is richly supplied with blood vessels containing warm blood. The water in the naris and sac thus absorbs heat from the blood. It is then expelled from the blowhole along with the excess heat. The* maxillonasalis *muscle* [5] *controls the flow of water by producing a kind of pumping action. As it contracts, the mass of tissue in the head rises* [6] *and the flattened tube of the nasal passage expands and fills with water* [7]. *When the muscle is relaxed again, the tube flattens* [8] *and the water is expelled* [9].

on the edge of the group. In this way each penguin takes a turn in the coldest position, but no bird remains exposed for too long.

Arctic insects take advantage of whatever sunlight is available by basking on sheltered rocks or inside the arctic flowers, which often track the Sun. This is a mutually beneficial relationship, as the insect warms itself while helping to pollinate the flower.

The mating swarms of winged arctic insects stay close to the ground to avoid the bitter winds. Their life cycles are also longer than their temperate counterparts – often several years compared to one. They lay fewer eggs, which must be especially cold resistant.

Natural antifreeze

In the icy waters of the polar oceans, cold-blooded fishes have no way to control the temperature of their bodies. To prevent their blood from freezing many species of fish have developed protein compounds that effectively operate in the same way as antifreeze in the radiator of a car. Ice crystals grow as more and more water molecules join on to an already existing crystal. The fishes' natural antifreeze prevents ice crystals from growing in their bodies by blocking the addition of more water molecules to ice crystal lattices. Certain species of the genus *Trematomus* can in this way withstand temperatures as low as 27.5°F (–2.5°C).

Many insects that survive in cold climates also produce their own forms of internal antifreeze. In juvenile willow gallflies, for example, the antifreeze can account for as much as half of their entire body weight. Other insects have no such internal protective measures. Midge larvae in Alaska, for instance, have come to rely instead on a metabolism that can be frozen and thawed out several times over without sustaining any permanent ill effects.

Educated Adaptation

Macaque monkeys are among the most adaptable and enterprising of all the primates. Various species are found in many different habitats from North America to India and Malaysia. The most northerly species of macaque lives in the mountains of Japan and is the only monkey found in Japan. The Japanese macaque has a long shaggy coat that serves to protect it from the cold of winter when the snow is thick. One troupe of the monkeys, exploring the forest outside their home territory, came across hot, volcanic springs. One after another the monkeys discovered that the springs were an ideal place to bathe during the cold winter months. Now all the monkeys in the region have learned to take hot baths during cold weather. This learned behavioral adaptation enables them to remain in the mountains all winter.

Plants in Cold and Wind 140 **Polar Regions** 42 **South Polar Life** 312 **Tundra** 314

High Life, Low Life

How animals adapt to extremes of pressure

The Andean hillstar hummingbird survives the extremes of temperature on its mountaintop habitat, albeit in a state of torpor, by allowing its body temperature to fall by as much as 45°F (25°C) from its daytime temperature of 104°F (39.5°C). From mountain heights of 16,400 feet (5,000 meters) to ocean depths of 27,000 feet (8,300 meters), animals struggle to make a home for themselves. Below 3,300 feet (1,000 meters), deep-sea fish live in an environment where there is no light, the temperature is near freezing, and the pressure is crushing.

The world of the deep ocean is a seemingly impossible environment for living things. At 10,000 feet (3,000 meters) the pressure is 300 times greater than at the surface. Despite the conditions, a variety of strange-looking animals do survive and make this place their home. Apart from the very low temperatures and crushing pressure, animals of the deep ocean also live in a world where food is scarce. Without light there is no plant life, so all must either rely on food that drifts slowly down from the sea above, or feed on one another. A few, such as the lantern fish, make nightly migrations to the upper parts of the ocean to feed, but most conserve energy by moving as little as possible and all have adaptations for catching what little food there is. Some fish, such as *Lipogenys gilli,* settle on the ocean floor. This species, with its toothless mouth and very long intestine, sucks in enormous quantities of the ooze that gradually settles on the seabed and extracts what little nourishment it contains.

Life at the top

Life at very high altitudes poses a different array of problems. The air pressure is low and shortage of oxygen can make respiration difficult. Curiously, however, birds do not seem to have any problems flying in thin mountain air. Migrating chaffinches have been seen flying at 5,000 feet (1,500 meters) and some other species have been located by radar as high as 20,000 feet (6,000 meters). No living things, however, can survive permanently on mountain peaks over 23,000 feet (7,000 meters). Although there is plenty of light, mountaintops are subject to extremely fierce winds and temperatures low enough to freeze living cells.

Even at 18,000 feet (5,550 meters), the night temperature falls rapidly to –4°F (–20°C), while during the day the Sun may shine constantly. Trees cannot grow in such conditions and above 17,400 feet (5,300 meters) it is too cold for seed-bearing plants to survive. But some well-adapted, high-altitude plants, such as some mosses and lichens, do survive, growing in low or cushion-shaped forms that conserve heat and moisture.

Coping with cold

Birds and mammals, which produce heat within their bodies, can survive the cold to a certain extent. Birds in particular do not seem to be affected by the shortage of oxygen, but the lack of trees means that they must nest on the ground or in rock crevices. The nests of eagles can be seen set on rocky ledges in many parts of the world, and in Ecuador even the local woodpecker, the Andean flicker, excavates a burrow in which to nest.

Cold-blooded animals find it particularly difficult to survive high altitudes, but one lizard, the iguana, *Liolaemus*, has adapted to do so. This lizard lives in burrows beneath bushes and is able to walk, albeit very slowly, even when its body temperature is as low as 35°F (1.5°C). Minutes after coming out into the Sun, the lizard's body becomes darker to enable it to warm itself to 68°F (20°C). The dark skin rapidly absorbs the maximum amount of radiant heat so that even when the air is at freezing point, the lizard can be as warm as 88°F (31°C).

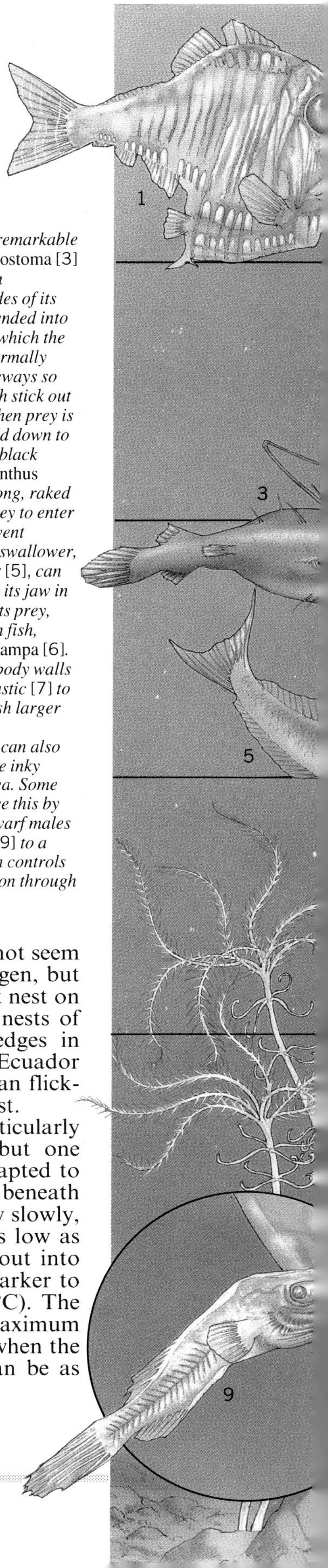

Adaptation *to the high-pressure environment of deep water does not require any special structural development. Internal and external pressures are balanced, and moderate changes in pressure are met through gradual adjustment. Specific adaptations to the ocean depths generally relate to the lack of sunlight and scarcity of food. Bioluminescence is widespread and serves a variety of functions. The hatchetfish,* Argyropelecus olfersi [1], *inhabits the upper portion of the deep-water zone of 650–1,650 ft (200–500 m), but migrates vertically to approach the surface for night feeding.*

*In deeper water, where food is much less abundant, successful predators must take advantage of whatever prey they encounter, including the occasional large fish. The head of the gulper eel ,*Eurypharynx pelecanoides [2], *is massively developed in comparison to its body, and accommodates enormous gaping jaws. The remarkable* Lasiognathus saccostoma [3] *solves the problem differently. The sides of its upper jaw are extended into flexible flaps into which the teeth are fixed. Normally these are held sideways so that the upper teeth stick out at right angles. When prey is taken, the flaps fold down to grip it firmly. The black dragonfish,* Idiacanthus fasciola [4], *has long, raked teeth that allow prey to enter the mouth but prevent escape. The great swallower,* Chiasmodon niger [5], *can partially dislocate its jaw in order to swallow its prey, such as the lantern fish,* Diaphus metopoclampa [6]. *The stomach and body walls are sufficiently elastic [7] to accommodate a fish larger than itself.*

Finding a mate can also be a problem in the inky blackness of the sea. Some anglerfish [8] solve this by mating for life. Dwarf males become attached [9] to a female, which then controls their sexual function through her hormones.

Connections: **Animals in Cold and Wind** 144 **Animals in Heat and Drought** 142 **Mountain Life** 306 **Ocean Life** 332

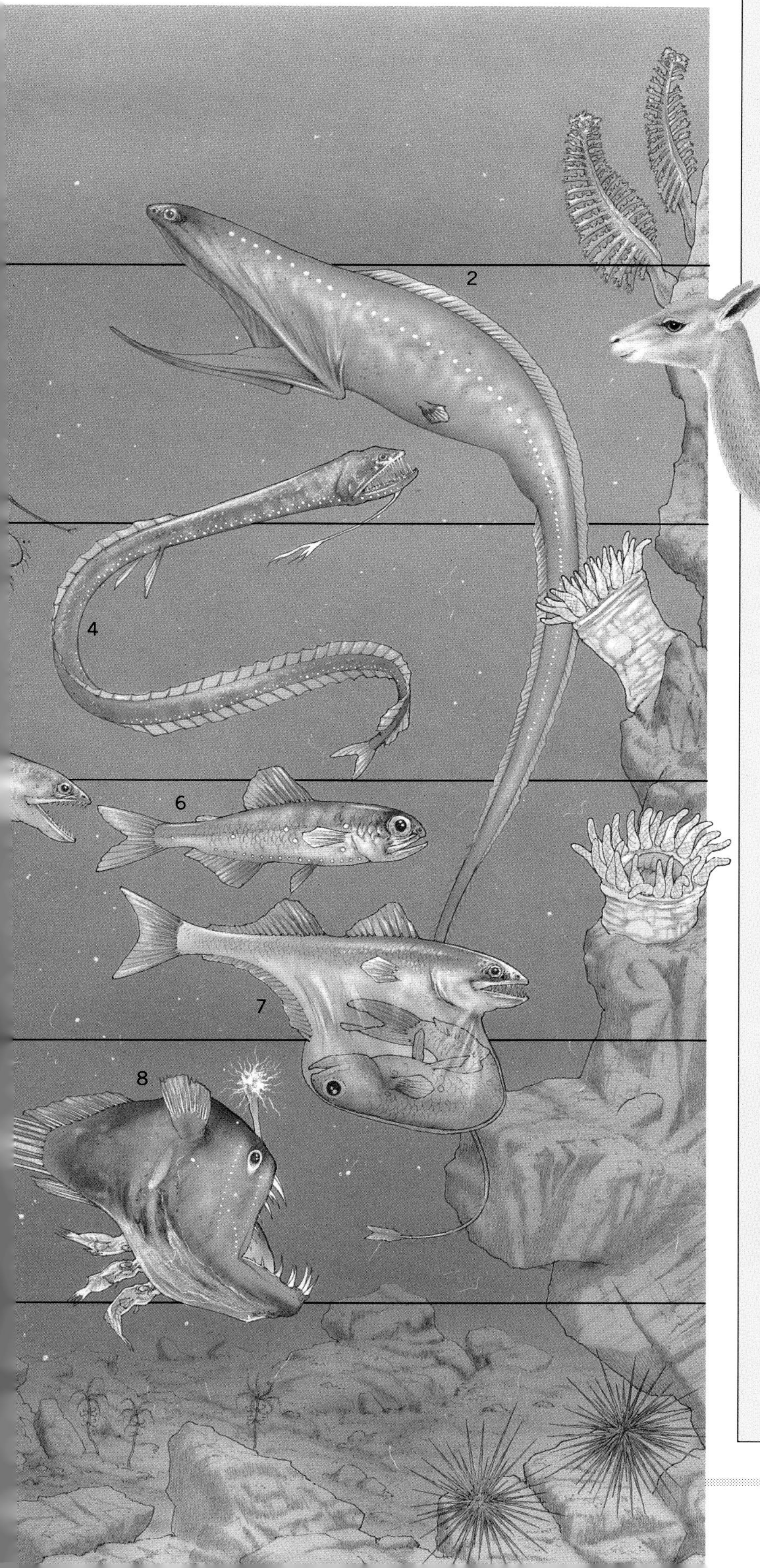

Easy Breathing

There is less oxygen in the rarefied atmosphere at high altitude, making people and animals that visit these areas feel tired and breathless. Oxygen is carried to the brain and muscles by red blood cells. People who live all their lives at high altitude adapt to the shortage of oxygen by producing more red blood cells. Animals such as the llama and the vicuña, which live in the Andes, have very high blood cell counts all the time. The vicuña [1] has about 2.3 billion red blood cells per hundredth of a cubic inch of blood [2] compared only 0.6 to 0.9 billion in humans [3]. Llama red cells live an average of 235 days, more than twice as long as the human cells. They also collect oxygen more efficiently, making the llama and its relatives among the best-adapted animals living at high altitude.

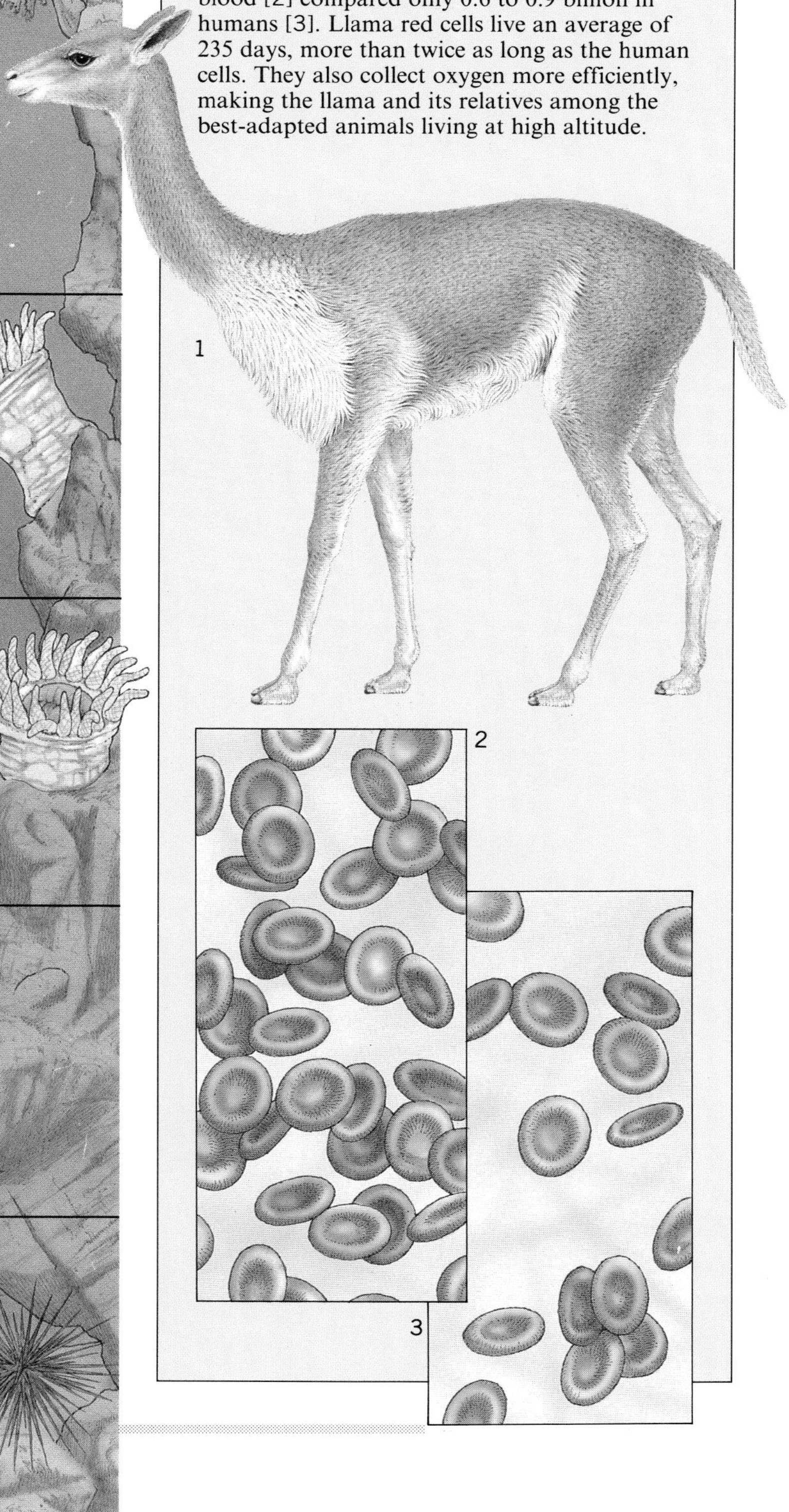

4 Reproducing to Survive

Plant Reproduction
Pollination by Wind and Water
Pollination by Animals
Seeds and Fruit
Germination
Invertebrate Reproduction
Fish Reproduction
Amphibian Reproduction
Animal Growth and Development
Metamorphosis
Reptile Reproduction
Bird Courtship
Bird Reproduction
Mammal Reproduction
Primitive Mammal Reproduction

Green and Fertile

How plants reproduce

One hundred million club moss spores would easily fit into the palm of your hand. Each spore has the potential to produce dozens of club moss plants, but the chances of every spore surviving are tiny. So dependent is each new plant on finding an environment with just the right conditions of moisture, temperature and sunlight, and on escaping the notice of herbivores, that even a hundred million spores might in the end produce only one mature plant. If every spore was successful we would soon find our planet overrun with club mosses.

For most plants the first stage of life is a seed or spore. But despite similar destinies, seeds and spores are quite different. The most important difference is the sexual process that leads to the production of seeds, which does not occur in spore production. Male cells (sperm), packaged within pollen grains, fuse with egg cells (usually in the ovaries of a different plant) to generate seeds. In this process sperm have become passive, non-swimming cells that are safely cocooned within a pollen grain, which is transferred by some outside agent, whether an insect or the wind. This evolutionary step freed most plants from their dependence on water and allowed them to conquer most of the Earth.

Spores, on the other hand, are produced asexually, by a single parent. They are exact genetic copies of the parent and are capable of growing into adult plants without sexual fusion with another cell. However, they have none of the variation that sexual reproduction generates by the combination of two sets of chromosomes in a *zygote*. Variation is essential because it allows adaptability. For instance, if all a plant's offspring were exactly the same, a disease destroying one would probably kill them all. If, however, some of the plants are different, as a result of sexual reproduction, then there is a greater chance that some will survive.

Quick growing

Although asexual reproduction has its drawbacks, it does have some advantages – it is fast and uncomplicated because it does not rely on pollinators. Many plants use asexual reproduction in an opportunistic way, sprouting new plants from the stems and leaves that are broken off by animals or storms. This adaptable form of asexual reproduction – known as vegetative reproduction – is exploited by gardeners when they take cuttings.

Safe storage

Tubers, bulbs, or corms may provide a plant with another method of vegetative reproduction. They consist of the food store together with a dormant bud, which can send up new leaves when conditions are favorable. As the bulb or corm grows larger, it enters its reproductive phase, generating daughter bulbs or corms, with their own buds.

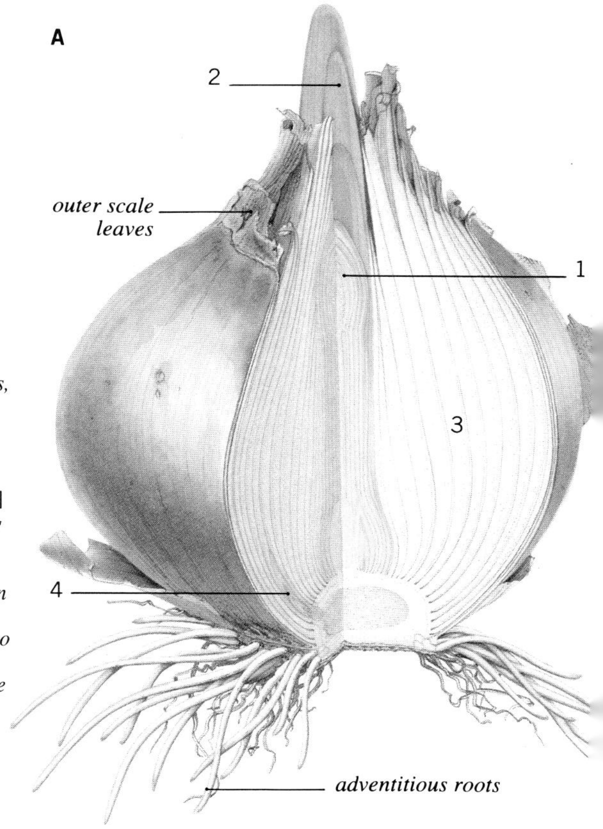

Bulbs *[**A**] – like those of daffodils and many other flowering plants – are underground storage organs, but they may also provide a plant with a means of vegetative reproduction. In spring the flower bud [1] and young foliage leaves [2] will develop into a flowering plant using the food and water stored in the bulb's fleshy scale leaves [3]. When the flower has died, the leaves live on and continue to make food, which is transported downward to the leaf bases. These swell and develop into new bulbs. Axillary buds [4] may develop into daughter bulbs, which break off to form new independent plants.*

Bulbs are one solution to the common problem of getting a new plant started: the parent plant must supply it with enough food and moisture to get over the difficult early stages, before its roots and leaves are properly developed. Some plants with fleshy leaves find this no problem because a single leaf has a good store of water and nutrients. Plants such as stonecrops simply shed their leaves, which can then take root.

Long-distance runners

A different approach to ensure young plants get a good start is taken by strawberries, spiderwort, creeping buttercups and many grasses. These have evolved the equivalent of an umbilical cord to keep their offspring supplied with food. In the strawberry plant this is known as a runner, a long, slender, horizontal stem that spawns new plants 8 inches (20 centimeters) or more from the parent.

Sexual reproduction *in flowering plants [**B**] begins with the production of gametes. Male pollen grains are made in the anther's [1] four pollen sacs [2], which are packed with* microsporocyte *cells [3]. The microsporocytes are surrounded by a layer of nutritive cells that is known as the* tapetum *[4]. Each microsporocyte goes through two* meiotic *(chromosome-reducing) divisions to form first a two-cell* dyad *[5] then a* tetrad *of cells [6], which splits into four* microspores *[7]. Each microspore divides* mitotically *(normally) to form a* generative nucleus *[8] and a* tube cell nucleus *[9], around which a thick, sculptured wall [10] develops. This is now a mature pollen grain. When the mature anther splits open [11], the pollen grains [12] are released to pollinate another plant. The female gametes originate in the* ovaries *[13]. Within each ovary are one or more* ovules, *[14] which will eventually develop into seeds. They are surrounded by two protective layers of tissue, the* integuments *[15], and have a small opening, the* micropyle *[16]. An ovule consists of a single* megasporocyte *[17], which divides meiotically to produce four* megaspores *[18]. Only one of these develops further [19] – by mitotic divisions through two [20], four [21], and finally eight nuclei. Around these eight nuclei are formed the contents of the* mature embryo sac *[22]: the* antipodal cells *[23] and the* synergid cells *[24], which eventually disintegrate; the* endosperm mother cell *[25] with two* polar nuclei *[26]; and the* egg cell *[27]. When the pollen grain germinates on the stigma [28], a*

Connections: Conifers and Cone Bearers 118 Flowering Plants 120 Germination 158 Heredity 98 Mosses and Ferns 116 Pollination by Animals 154

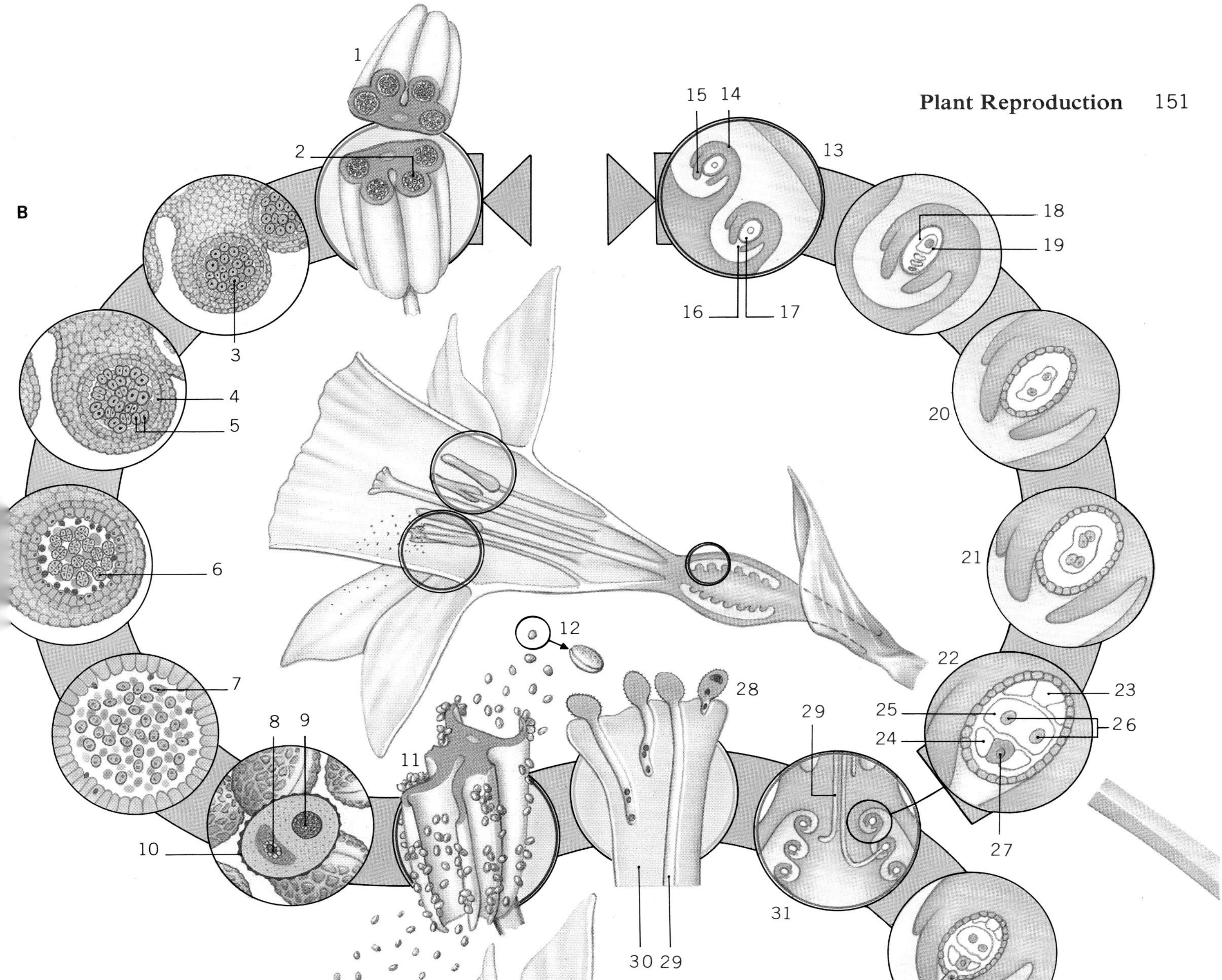

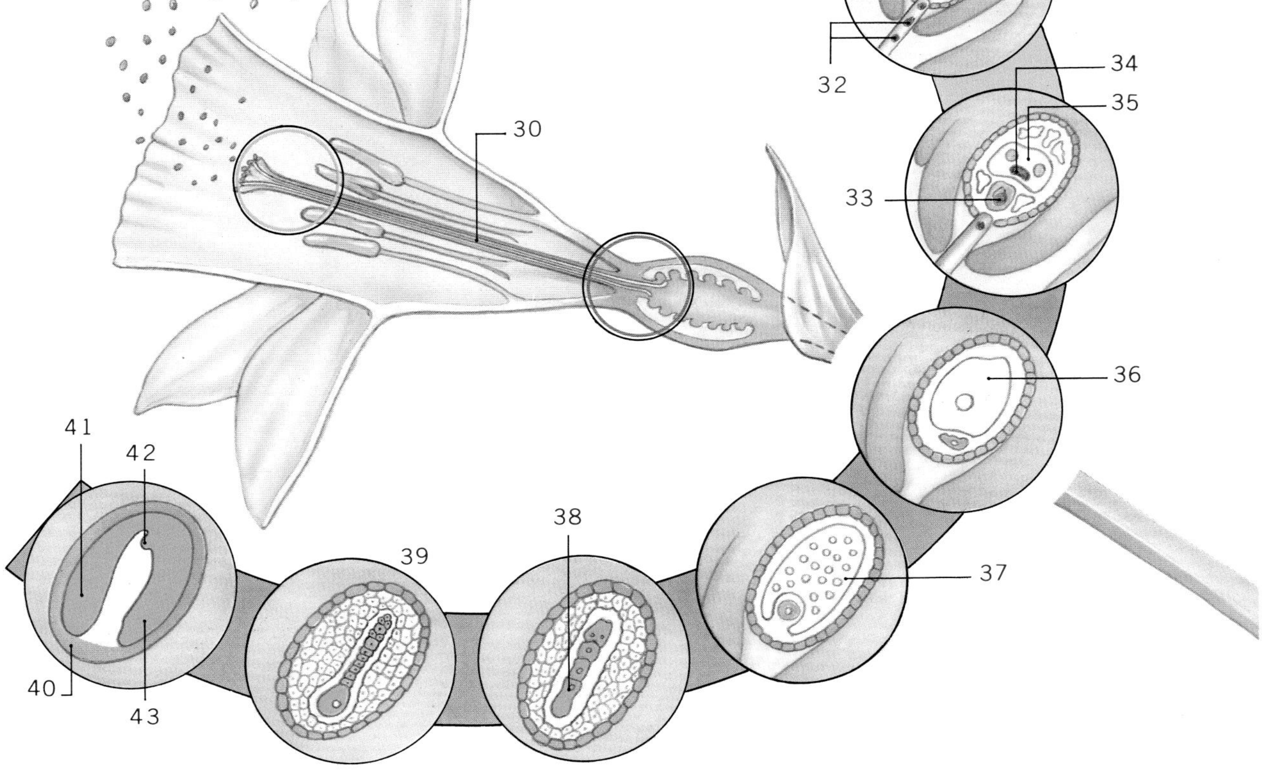

threadlike pollen tube [29] *grows down the* style [30] *into the ovary* [31], *led by the tube cell nucleus. The generative nucleus divides into two* sperm nuclei [32], *which follow down the pollen tube and enter the embryo sac through the micropyle. One fuses with the egg, fertilizing it, making a* zygote [33]. *The other fuses with the two polar nuclei* [34], *producing the* primary endosperm cell [35], *which develops into the* endosperm [36], *the seed's food supply tissue. As the endosperm tissue develops a multicellular structure* [37], *the young embryo* [38] *divides and grows* [39]. *The mature embryo is surrounded by a seed coat (the* testa*)* [40] *– formed from the integuments – and has all the parts of a young seedling:* cotyledon [41], plumule [42], radicle [43].

Pollination by Wind and Water 152 **Seeds and Fruit** 156

The Sexual Lottery

How plants are pollinated by wind and water

The pollen of oak and sorrel from middle North America have been found over the mid-Atlantic after journeying for thousands of miles. Although most pollen grains are tiny enough to be visible only under a microscope, those of wind-pollinated plants are especially small and light so that they may be blown these vast distances to improve the chances of successful reproduction. There are also a lot of them. The North American ragweed can release more than 1.6 billion pollen grains in an hour and its pollen is the most common cause of hayfever in the United States.

A

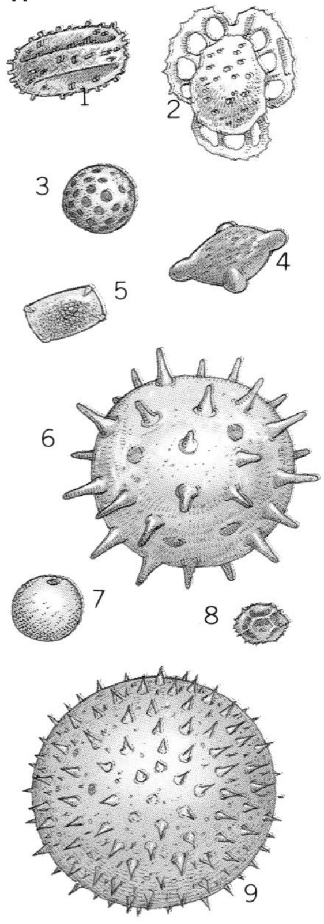

Pollen grains [**A**] *are found in pollen sacs, which are in the* anthers *(part of the stamens). Pollen grains are safe and effective storers of the male gametes. They come in all shapes and sizes depending on the species of plant. The selection shows mistletoe* [1], *venus's-flytrap* [2], *spinach* [3], *honeysuckle* [4], *touch-me-not* [5], *cotton* [6], *rice* [7], *dandelion* [8] *and hollyhock* [9].

Pollination by wind or water is a far more hit-and-miss affair than pollination by insects or other animals. A bumblebee working its way steadily along a flower border is unintentionally transferring pollen from one flower onto the *stigmas* (female organs) of other flowers. By contrast, the wind sends pollen swirling and gusting through the air, and for every grain that reaches its target millions more go to waste. However, there are compensations for the plant: it does not need to produce nectar or showy, colorful flowers to attract insects.

A high-level launch

Wind pollination is often used by trees because they are tall enough to take best advantage of the winds. Some wind-pollinated trees produce their pollen in catkins, which are dangling clusters of male flowers. These are flexible and mobile, so that they readily release their pollen when the wind blows. Other wind-pollinated plants have long, supple *stamens* (male flower parts) for the same purpose, and most have a large, feathery *stigma* to maximize the chance of catching pollen.

Because wind-pollinated plants distribute so much pollen, there is a large quantity of it in the air we breathe. It is these pollens that generally cause hayfever and not the pollen of scented, insect-pollinated plants.

An evolutionary shortcut

The earliest flowering plants were probably insect pollinated, and wind pollination was a later evolutionary development. It may have arisen in certain situations because the insect pollinators were unreliable, or not around at the right time of year. With conifers, however, the evolutionary story is different: they have always been pollinated by the wind. The pollen is produced by small male cones and blown on the breeze to the larger female cones. The vast quantities of pollen produced by the trees of a pine forest in spring months may be so great that a thick yellow scum forms on nearby lakes.

Casting grains upon the water

Flowering water plants evolved from flowering land plants and they are often pollinated by flying insects, pushing their flowers up above the surface of the water to attract their

B

Connections: Conifers and Cone Bearers 118 Flowering Plants 120 Germination 158 Plant Reproduction 150 Pollination by Animals 154 Seeds and Fruit 156

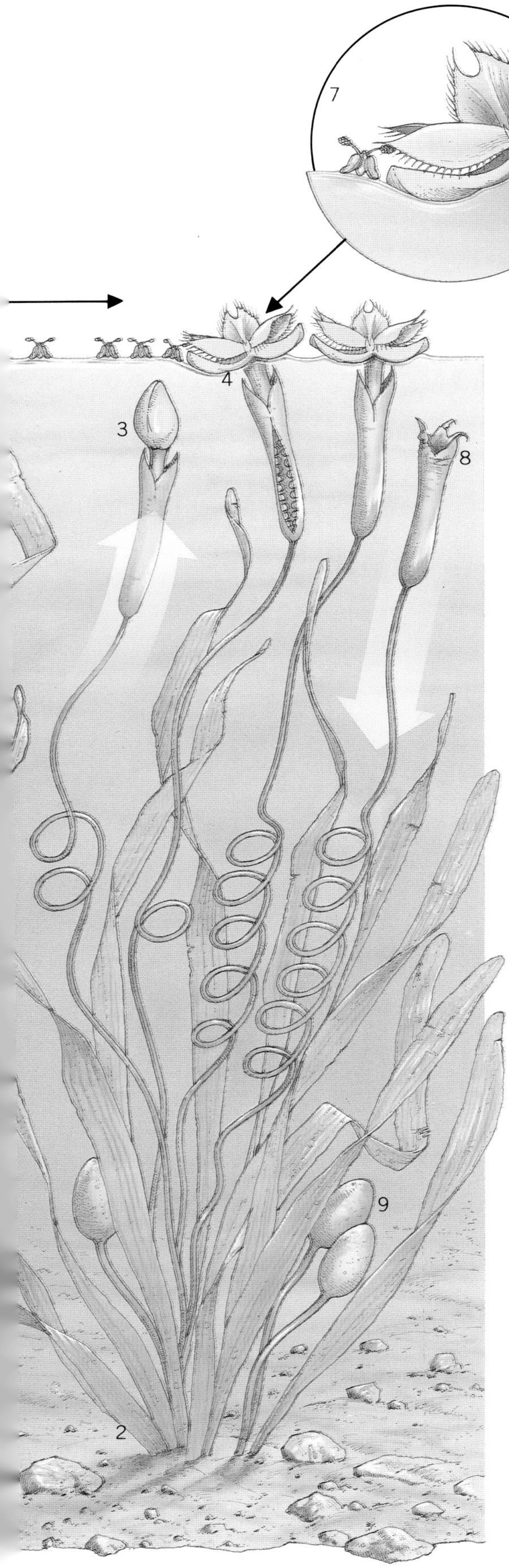

pollinators. But some have opted for water pollination instead, and indeed in the 50 species of sea grass, pollination actually takes place beneath the surface of the water. European eelgrass, an example of sea grass, is found in coastal marine habitats. It has, like most species of sea grass, long, narrow, worm-shaped pollen "grains" which float about on the tides just below or on the surface. The plants have small inflorescences made up of a female flower sitting above two male flowers. These remain just below the surface. When the pollen is released in a cloud of grains, they gently rise to the surface and the threadlike pollen grains become entangled in the two thin *pistils* (central organs) of the female flower. To reduce the possibility of self-fertilization, the pistils of the female flower reach maturity before the stamens of the same inflorescence are ripe and begin to release their pollen.

Pollen grains *(below) have a tough outer shell, which stops the grain from rotting. In certain conditions pollen grains can survive for thousands of years. By studying pollen grains, pollen experts (palynologists) can tell which plants they came from and piece together how vegetation changed with time.*

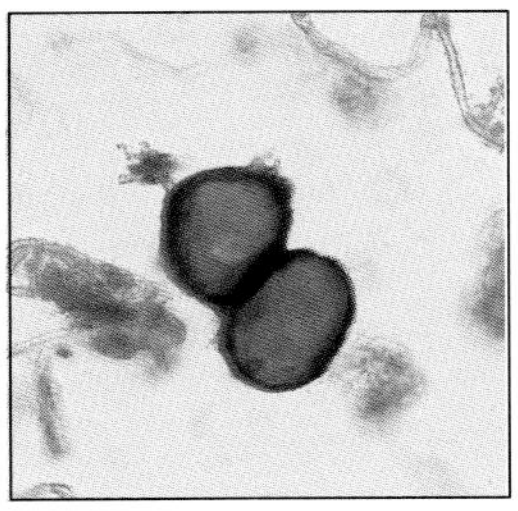

Catkins, *a familiar sight in spring, grow only on wind-pollinated plants. The catkins hold the pollen in their anthers until it is shaken free by wind. Wind pollination is common among plants that do not have brightly colored, scented flowers. Such flowers serve to attract insects, which then carry out pollination. The primary requirements for wind-pollinated plants are that the pollen grains are not only light enough to be carried by the wind but also, perhaps more importantly, they do not stick together. A dry, powderlike pollen ensures effective dispersal.*

Ribbon weed [**B**] *is found in slow-moving fresh water. It has separate male plants* [1] *and female plants* [2]. *For pollination to occur, the female flower* [3] *spirals to the surface on a stem, where it unfolds* [4]. *Meanwhile, the flowers of the male plant are released from the tip of the* spathe [5] *and those that are not eaten by fish float to the surface, where they open. The male flowers consist of two stamens and three petals, which keep the flower afloat* [6]. *They are carried across the water by wind and currents. If they come close to a female flower, the dip in the surface tension of the water created by the female flower* [7] *forces the male flower to accelerate into the female. When they collide clumps of pollen from the stamens catapult over to the stigma of the female flower and pollination takes place. The spiral stem of the female flower then pulls the pollinated flower beneath the surface* [8], *where the fruit* [9] *ripens.*

The Pollen Carriers

How animals pollinate plants

Bumblebees are energetic pollinators: a single bee can visit 240 flowers in a single expedition to collect nectar, each time picking up and leaving behind a little pollen. Only two of those 240 flowers need be of the same species to allow the possibility of fertilization. Many plants dole out their nectar in small amounts throughout the day, encouraging bees and other nectar feeders to keep moving and not stay too long at one flower. With this sort of rationing, the amount of energy invested in nectar can be kept low: a single apple tree produces just 1 ounce (28 grams) a day.

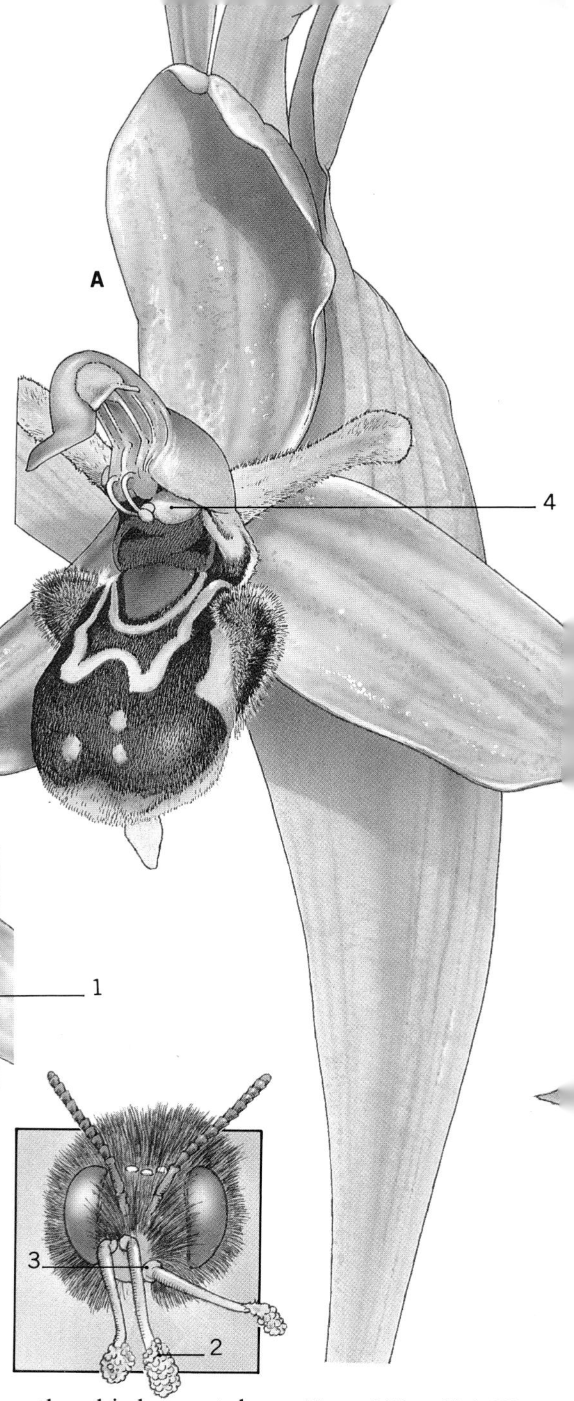

The first flowers probably appeared on Earth about 140 million years ago and were similar to the magnolias of today. They produced pollen that was eaten by beetles, which carried some from flower to flower in the process. This wasteful and inefficient method of pollination was soon improved by the development of nectar, a food reward for insects that reduced the amount of pollen lost. There was already a precedent for exuding sweet fluids: many plants need to offload excess sugar from their sap, because too high a concentration interferes with the transport of proteins and hormones. It was a simple evolutionary step to lay this sugary bait close to the reproductive organs.

Even at this early stage of development it would have benefited the plants to favor some pollinators over others. It is not clear how most plants keep beetles and ants at bay, but there may be trace ingredients in the nectar that are distasteful to the insects. Acquiring features that discriminate between pollinators has been a recurring theme in the evolution of flowering plants. At the same time, a wider and wider range of pollinators has been recruited. Butterflies and moths have evolved closely with the flowering plants, as have the bees. In combination with the hoverflies, wasps and dung flies they are the major insect pollinators.

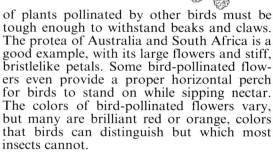

The bee orchid [**A**] *tempts its pollinators with sex rather than nectar. Its elaborate flowers imitate the color, shape, texture and scent of female bees of the genus* Eucera. *Male bees, which emerge before the females, alight on the flower's broad platform or labellum* [1] *and attempt to mate with it.*

Kidnapping a carrier

One Sardinian arum puts out inflorescences that stink like carrion and contain separate male and female flowers. Blowflies are lured into the female flowers and cannot escape. They are fed with nectar and simultaneously divested of the pollen they carry. When the lily's male parts start to produce pollen, about a day after the female flowers stop being receptive, the blowflies are released and acquire a fresh load of pollen. If they are taken hostage by another arum, the cycle begins again.

The birds and the bats

In the tropics birds and bats have become nectar-feeders and pollinators. Such a development was impossible in temperate regions because these animals need a copious supply of nectar all year round.

Although hummingbirds hover while feeding, and so do not cause damage, the flowers of plants pollinated by other birds must be tough enough to withstand beaks and claws. The protea of Australia and South Africa is a good example, with its large flowers and stiff, bristlelike petals. Some bird-pollinated flowers even provide a proper horizontal perch for birds to stand on while sipping nectar. The colors of bird-pollinated flowers vary, but many are brilliant red or orange, colors that birds can distinguish but which most insects cannot.

Bat-pollinated flowers are distinctive for their scents, which are musky and fetid, rather than sweet as in most insect-pollinated flowers. (Because few bird species have any sense of smell, bird-pollinated flowers are generally unscented.) Not surprisingly, bat-pollinated flowers open at dusk or in the night, and are often positioned well away from the body of the plant to avoid damaging the bats' delicate wings.

The orchid's pollinia [2] *– structures containing thousands of pollen grains – become detatched and adhere to the bee's body, anchored by the sticky ends of their stalks* [3]. *The bee transfers the intact pollinia to other flowers, and cross fertilization is followed by the formation of thousands of tiny seeds. In the absence of a suitable pollinator, the bee orchid pollinates itself: the pollinia droop forwards and make contact with the stigma* [4], *one of the female parts of the flower. This is particularly common in northern populations of the bee orchid, where the pollinating bee has become locally extinct.*

Connections: Birds 132 Early Evolution 92 Flowering Plants 120 Insect Flight 226 Plant Reproduction 150 Seeds and Fruit 156 Symbiosis 192

The symbiotic relationship **[B]** *between the fig tree and its pollinator – the fig wasp – is one of the most remarkable known to science. The developing fig "fruit" is in fact a collection of tiny male and female flowers enclosed in a fleshy covering. Female wasps lay their eggs within modified female flowers, which then grow into gall-flowers containing the developing wasps* [1]. *Male wasps emerge first* [2] *and immediately search out young females still within their gall flowers. Inserting their abdomens into the gall flowers, they fertilize the females* [3] *and then complete their second task – gnawing an exit hole through the tough fig wall* [4] *– after which they soon die, never having left the fig. Hours later* **[C]** *the "prefertilized" female wasps emerge* [5] *and the male flowers (with their two anthers)* [6] *simultaneously shed their pollen. After being coated with pollen, the females escape* [7] *by crawling through the tunnels made by their doomed mates. The female wasps then fly to a young fig* **[D]**, *which contains two types of mature female flowers as well as closed male flowers* [8], *and force their way through the* ostiolium [9] *– a small opening in the fig protected by overlapping scales. They transfer the pollen they are carrying to the viable long-styled female flowers* [10], *thus fertilizing them. They also lay eggs* [11] *in the modified short-styled flowers* [12] *with their long ovipositors* [13]. *This completes the reproductive cycle of both fig and wasp.*

Hummingbirds *are capable of highly controlled hovering flight that enables them to draw nectar from delicate flowers and inadvertently pollinate them without landing. Their bills are typically long and slender, and their tongues have brushlike tips to help them extract the nutritious nectar.*

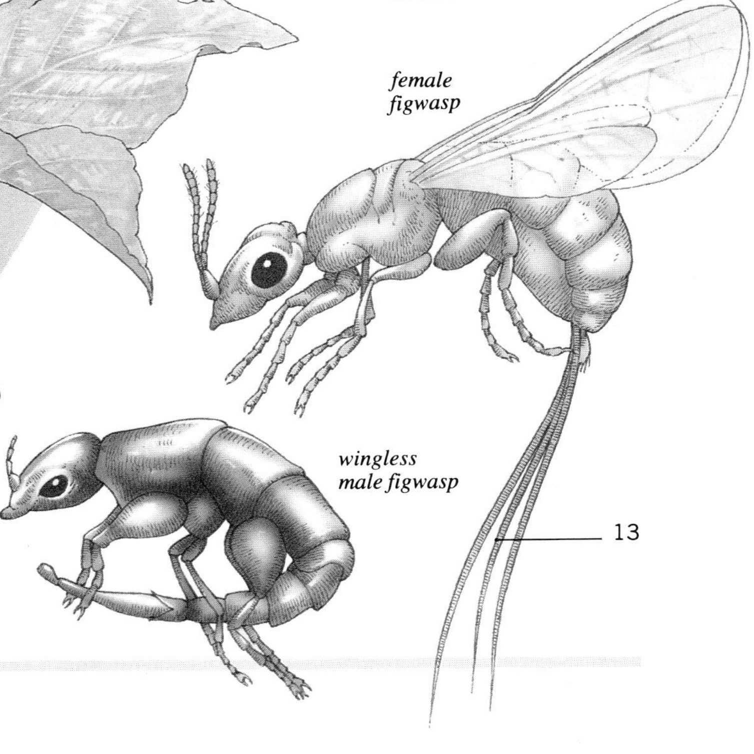

Fruitful Strategies

How plants disperse their seeds

The single-seeded fruit of the double coconut, which grows only in the Seychelles, may weigh as much as 40 pounds (18 kilograms). Fruits and seeds are a plant's means of propagating itself and spreading over a wider area. When blackbirds, for example, strip cherry trees in summer, they are looking for food. But by making its fruit good to eat, the cherry tree also ensures that its seeds are dispersed. For the seed is safely packed away inside a hard coat and passes unharmed through the bird's digestive system to fall to the ground and germinate.

Flowering plants reproduce by seeds formed in the female reproductive parts of a flower after fertilization. The developing seeds are enclosed in a protective case to form a fruit; conifers, on the other hand, bear their seeds on the surface of the scales of the cone, from which they are shed directly.

To ensure the best chance of germinating safely and of surviving to produce their own flowers, plants have evolved ingenious ways of protecting and dispersing their seeds.

Wings and parachutes

Many seeds are borne away on the wind. The small, single-seeded fruits of dandelions and thistles each have their own silky parachute that carries them some way from the parent plant. Conifers, such as pines and spruces, have light, papery-winged seeds. The wings of maple and sycamore fruits keep the nutlike seeds aloft for some time as they gently spiral to earth, allowing the wind to blow them beyond the shadow of the canopy of the parent tree. Orchids have minute seeds that can be blown away like a cloud of dust, and the wind shaking the ripe capsules of poppies scatters their tiny seeds.

Ocean crossings

The seeds of dandelions and thistles usually travel just a few yards, but some fruits are dispersed over hundreds of miles by water. Ocean currents carry the buoyant coconut across the Pacific to colonize newly formed coral atolls. Protected by a fibrous outer coat and a hard inner shell, the coconut embryo can survive these long journeys. The seeds of alder are equipped with a corky knob which ensures they float if they are released over water. They are carried by the stream until they lodge against a muddy bank, where they can germinate.

Animal carriers

Animals can disperse plant seeds and fruits in two main ways. The hooked spines of the burrs of certain plants can become caught in the animal's fur and dispersed as the animal grazes, or, more commonly, the seed may be enclosed inside a tempting juicy fruit. Such fruits often change color dramatically from green to red or purple when they are ripe, indicating to animals that they are now sweet and ready to eat. Red is the most common

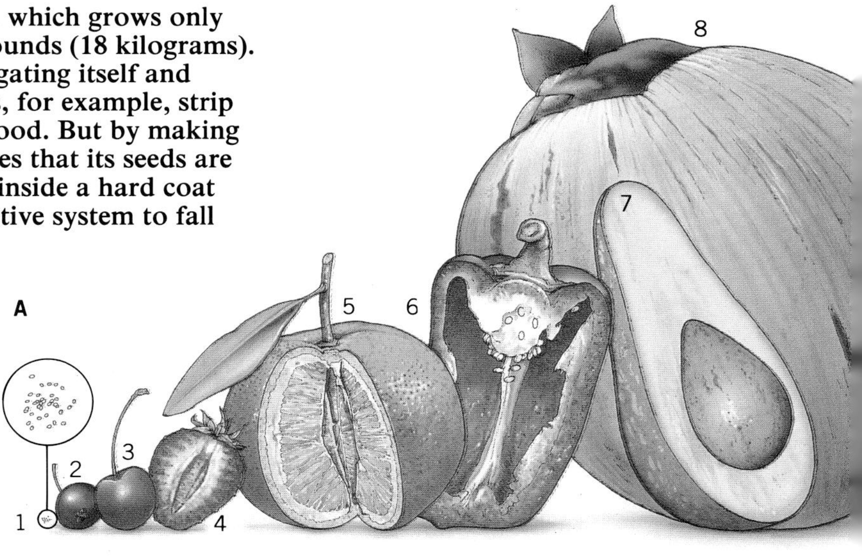

Seeds and fruits **[A]** *come in many varieties of shapes and sizes. The smallest known seeds belong to the epiphytic orchid* [1] *– one million of these seeds weigh only 0.035 oz (less than a gram). Fruits are the ripened ovary of a flowering plant which is enclosed by the* pericarp *(fruit wall) and most, such as the succulent blackcurrants* [2], *cherries* [3], *strawberries* [4], *oranges* [5] *and pepper* [6], *contain more than one seed. However, the avocado* [7] *contains one single, large seed, approximately the size of a golf ball. The largest known fruit is also single seeded – it is the double coconut* [8].

Wind dispersal *is one of the most common ways that flowering plants scatter their seeds. The sycamore, for example, has evolved a "winged" fruit* **[C]** *(the number of wings varies from species to species). When the fruit drops from the tree, the lift created by the air* [1] *causes the fruit to rotate* [2]. *Most of the species of winged fruit, including the sycamore, are designed not to rotate about their own center of gravity but to make a spiral descent. This descent dramatically reduces the terminal velocity of the seed's fall, so that even the slightest crosswind will carry the fruit away from the overshadowing parent tree.*

The dandelion fruit **[D]**, *when mature, becomes easily detached from the plant, and the slightest breeze will catch the "parachute" of hairs* [1], *scattering the fruit. The seeds of the poppy* **[E]** *are stored in ripe capsules* [1]. *When the poppy has matured the ovary dries up and the remains of the stigmas fold up* [2] *revealing the pores* [3], *through which the seeds are shaken out by wind action.*

As soon as the flowers of oats (bottom right) become detached from the plant and fall to the ground, tiny hair-like filaments, known as awns, seek out cracks in the soil in order to bury the seed. The awns are thought to be activated by sunlight.

Animal dispersal *of seeds* **[B]** *does not necessarily involve the ingestion of the seeds through consumption of the fruit, as with, for example, blackberries and strawberries. Some plants, such as* Xanthium occidentale, *known in Australia as noogoora, have evolved burrs* [1] *that are covered with tiny hooks* [2]. *The hooks become attached to the coats of grazing animals such as sheep or horses. As the animals graze, they pick up the burrs, usually near their forelocks, which are then either rubbed off or simply drop off. In Europe other plants, such as burdock, use the same dispersal technique.*

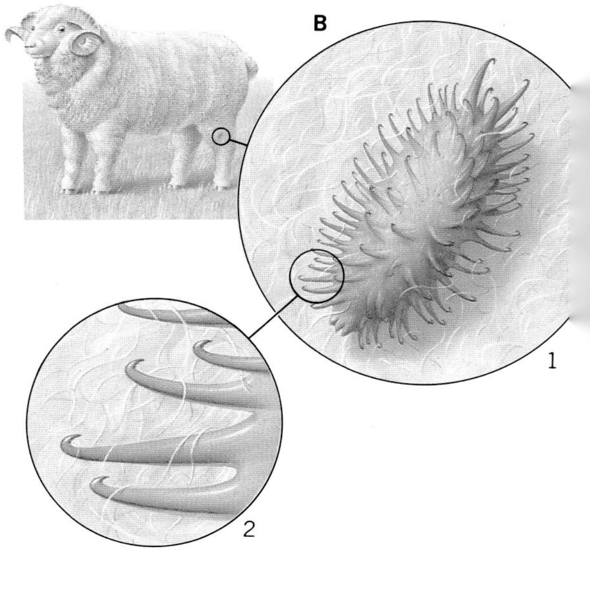

Connections: Conifers and Cone Bearers 118 Flowering Plants 120 Germination 158 Plant Reproduction 150 Pollination by Wind and Water 152

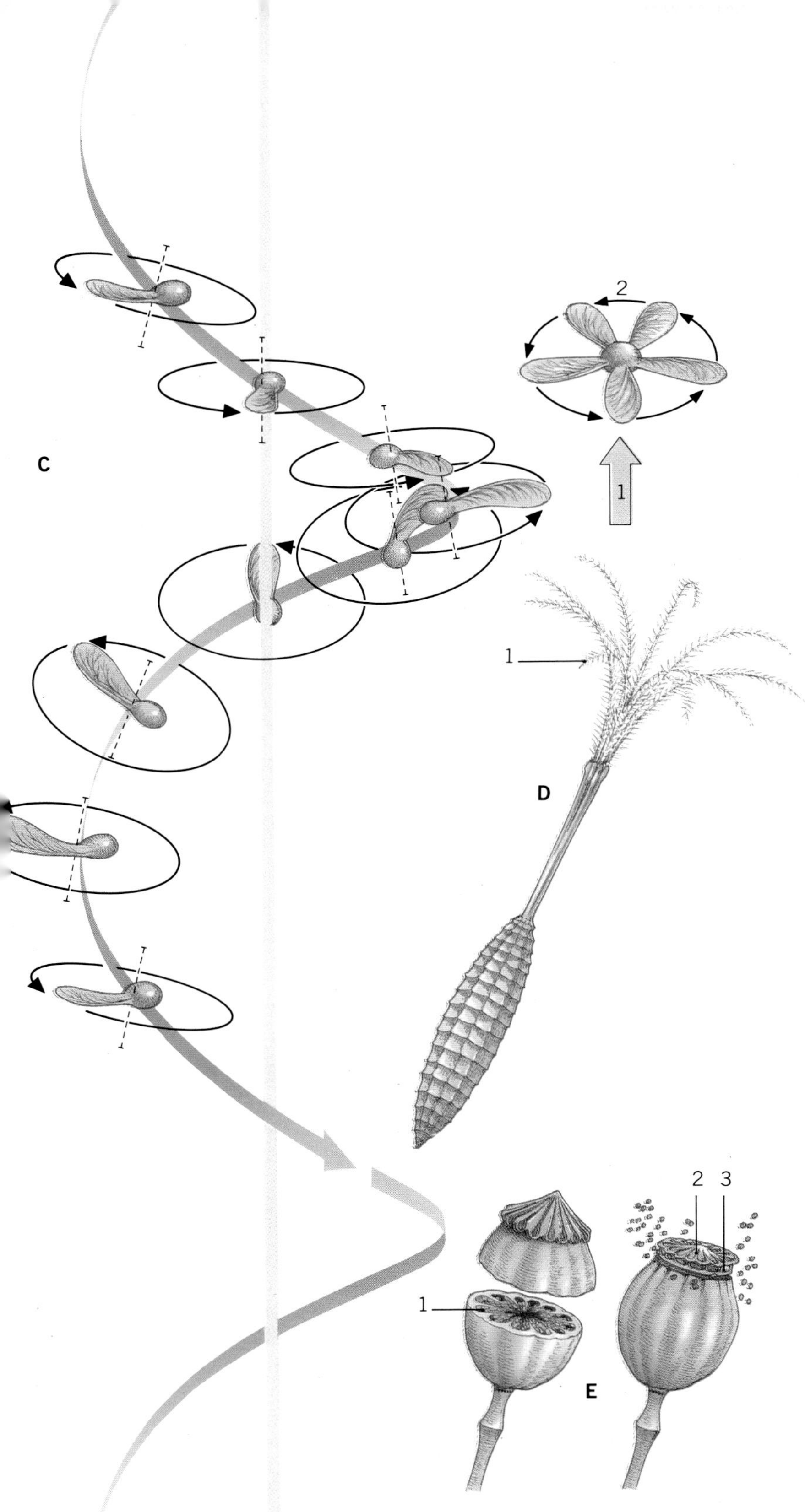

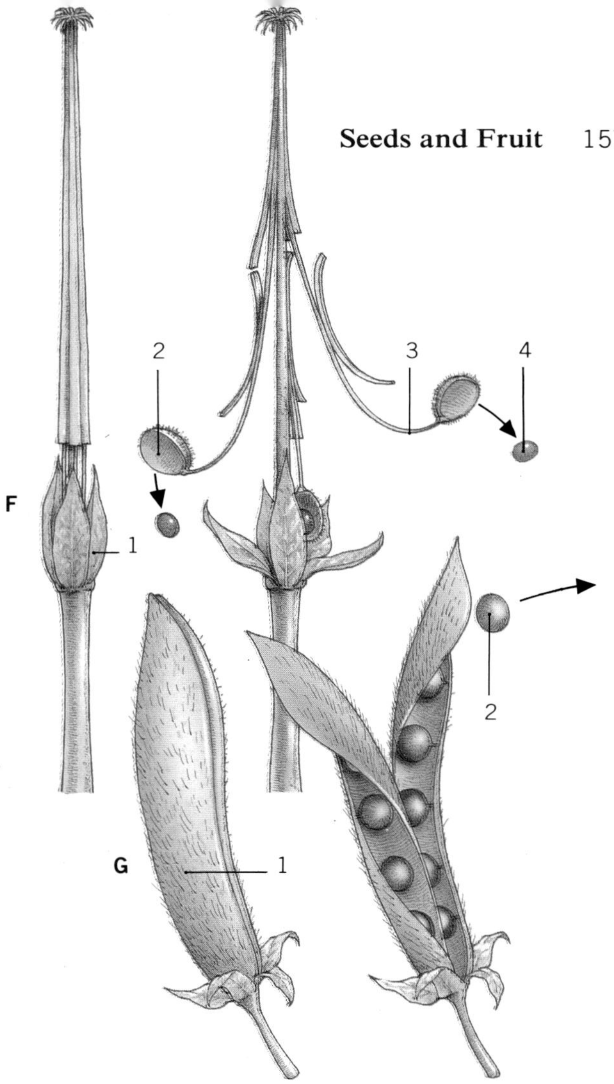

__Some plants__, rather than relying on animals or the wind to disperse their seeds, have self-dispersal mechanisms. These mostly depend on specific parts of the plant drying out and, under tension, releasing the seeds. With the geranium [**F**], *for example, as the sepals* [1] *dry they cannot contain the ovaries* [2], *which are attached to the ends of springy styles* [3]. *In this way the seeds* [4], *contained in the ovaries, are released. In the case of the sweet pea* [**G**] *it is the ripe pod (the ovary)* [1] *that splits open and at the same time twists outward, releasing the seeds inside* [2].

color in fruits dispersed by animals; pure red is invisible to insects, who would nibble the fruit without dispersing the seeds, but makes the fruit more visible to other animals.

One animal dispersal technique that differs slightly from both of these involves the rodentlike agouti of Brazil. The agouti often collects and hoards so many seeds that it often forgets where they are buried. By producing abundant seeds, plants can often rely on the forgetful agouti to disperse them.

The shotgun effect

Not all plants rely on wind, water or animals to disperse their seeds. Within the fruit of the dwarf mistletoe of the United States, water pressure gradually increases until the seeds are blasted out of the fruit over distances of several yards. In other plants, including the Scotch broom, dispersal occurs when particular parts of the plant dry out.

Sprung from the Seed

How plants germinate and grow

A species of lily was once reported as having grown 12 feet (3.6 meters) in 14 days. The roots of a wild fig tree in South Africa grew to a recorded depth of 400 feet (120 meters). Plant growth is much simpler than that of most animals. As the soil warms up in spring, seeds that have lain dormant over the winter begin to germinate. The tiny embryo inside the seed sends chemical signals to the storage tissues to mobilize food reserves. Nourished by these food stores and prompted by its own growth hormones, the embryo plant begins to grow.

All plants grow by the repetition of a simple structural module. The shoot grows by the addition of modules consisting of a length of stem, a leaf and an axillary bud. The point on a stem where a leaf grows is called a node and the length of stem from one leaf to another is the internode. Modules develop successively from the growing point inside the bud at the tip of the shoot.

In plants all growth occurs from localized regions – *meristems* – of actively dividing cells. These are found at the tips of roots and shoots. The root meristem gives rise to new lengths of root, the shoot meristem gives rise to stem, leaves and flowers. The apical meristems give rise to all the internal structures, such as xylem and phloem. In some plants there are also active meristems inside the stem that locally produce new vascular tissue. The increase in diameter of the stem in woody plants each year is due to new wood produced by the *cambium*, a meristem that encircles the trunk just under the bark.

Elongation and expansion

As new cells are formed, they begin to take up water by *osmosis* and to expand under this water pressure. Pressures as great as 10–20 times atmospheric pressure can be developed inside plant cells. The hydraulic pressure developed in the elongation zone behind developing root and shoot tips forces them out of the tough seed coat and pushes the new root and shoot through the soil.

Plants grow both by the addition of new cells by cell division and by considerable elongation and enlargement of cells after they are made. In order to elongate, the cell wall is weakened in places by the action of enzymes, allowing it to stretch to accommodate the expansion of the cell vacuole.

Plant hormones

The course of plant development is controlled by a variety of plant hormones acting in response to the signals a plant receives from its surroundings. Light, temperature and moisture are the main environmental influences on plant growth.

Three important groups of plant growth hormones are the *auxins*, *gibberellins* and *cytokinins*. Each group has several different effects. Auxins, such as IAA (indole acetic acid), promote cell elongation in growing roots and shoots. Auxin produced by the apical bud is also involved in inhibiting the growth of lateral buds. In addition, IAA encourages the development of roots, and it may be used in rooting powders for cuttings. Cytokinins induce cell division, and the balance between auxin and cytokinin in tissues also seems to determine whether roots or shoots are formed. Gibberellins produced by the embryo inside the seed are involved in setting up the events leading to germination. They are also involved in the elongation of cells and the formation of the different plant organs like leaves and flowers.

Unlike most animal cells, plant cells retain the ability to give rise to many different sorts of cells, or even a complete new plant, throughout their life. Roots will grow from the base of a stem cutting, and some plants can be propagated by leaf cuttings, which develop roots and shoots around the edge.

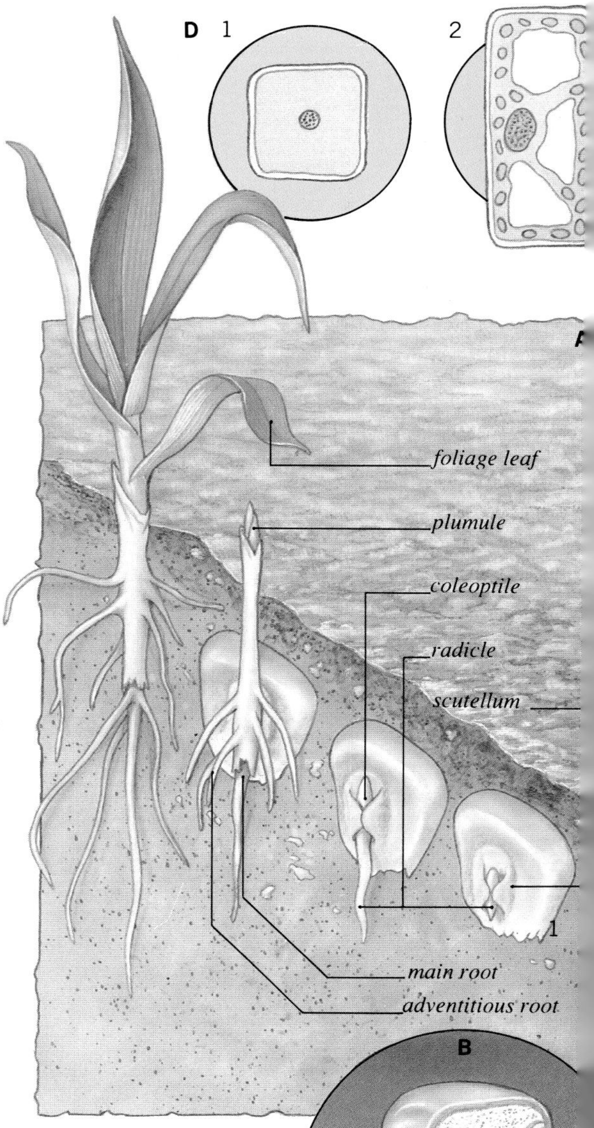

***Germination of the seed* [A]** *is the beginning of all new plants. The* radicle *(young root) is the first part of the young plant to break through the seed case as it absorbs water and rapidly expands. From this root the other roots will develop. The tip of the delicate radicle is protected by a tough root cap. The* plumules *(young shoots), however, do not have the same protection, and different strategies are employed to prevent them being damaged as they push through the soil.*

Many monocotyledon (one seed leaf) plants, like maize (corn) [1] *and other grasses, first send up a protective sheath (*coleoptile*), within which the plumule grows. In maize the single cotyledon stays underground within the seed, absorbing food from the* endosperm *(food store), and is known as the* scutellum. *This provides enough food for the plumules to enlarge and develop into the plant's first green leaves, at which point it can start to photosynthesize for itself.*

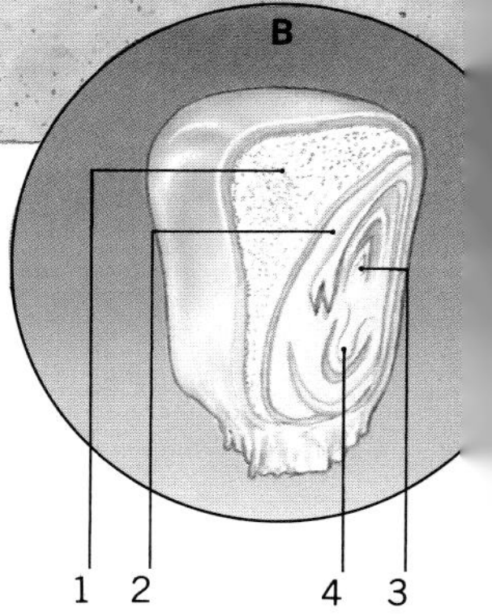

***Within the maize seed* [B]** *food is stored outside the plant embryo as dry, powdery* endosperm [1]. *When environmental conditions trigger the process of germination, the single cotyledon* [2] *absorbs food from the endosperm and delivers it to the developing plumule* [3] *and radicle* [4].

Connections: **Biological Clocks** 274 **Flowering Plants** 120 **Photosynthesis** 182 **Plant Architecture** 114 **Plant Cells** 100 **Plant Reproduction** 150

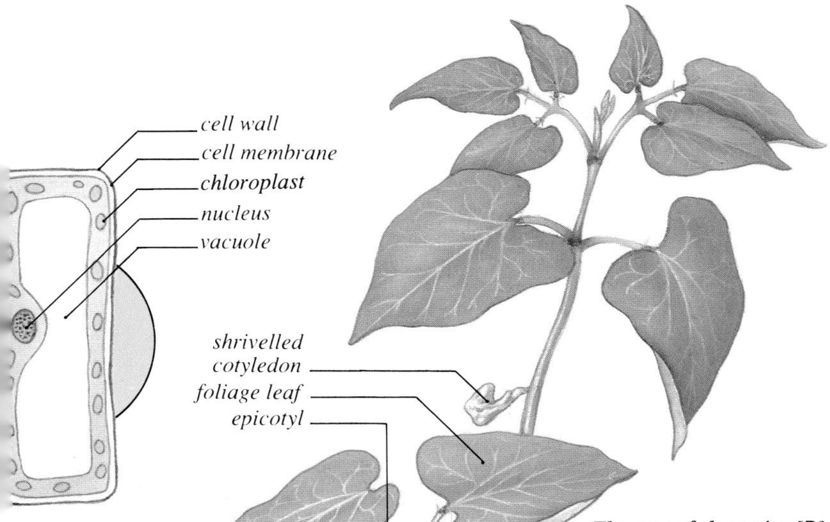

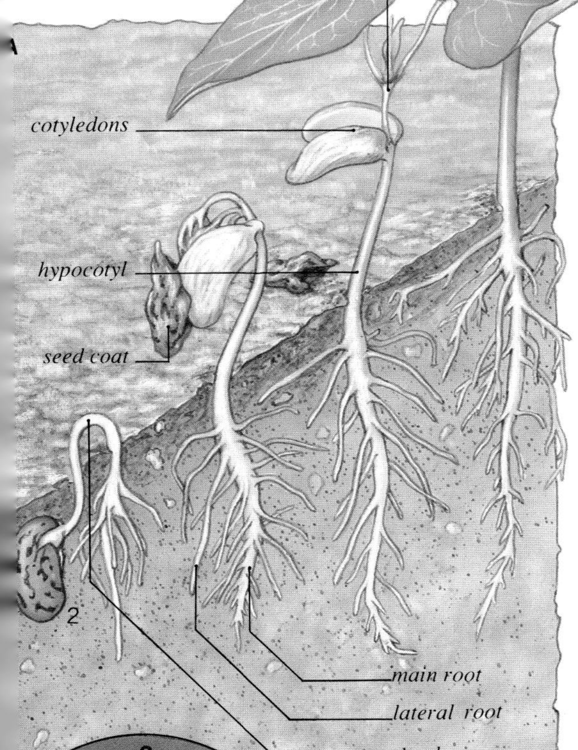

Plant growth occurs in distinct zones [**E**]. New cells are created by cell division in localized meristems. The form and structure of a shoot are governed by the apical meristem [1] within the bud. Growth occurs at three points: the outer horns, which will develop into leaves [2]; the central shoot apex [3]; and the zone of elongation [4] behind the meristem, where cells develop and increase in size. Behind this is a zone where plant cells mature and take on their final form and function. On the stem behind the bud are the remains of lateral shoots [5] that have been inhibited by hormones such as auxin.

The zone of elongation [**D**] *provides the force of growing plants. Newly divided cells* [1] *quickly elongate and form vacuoles as they absorb fluid* [2]. *As the central vacuole forms* [3], *it expands, increasing cell size and pumping up the cell, creating a rigid structure.*

When many dicotyledons *(two seed leaf plants) germinate* [**A**], *like the kidney bean* [2], *a hook structure that protects the plumules, is the first to push through the soil after the radicle. As the hook straightens, it lifts the young leaves free of the surface. In kidney beans this hook is formed by the stem below the cotyledons (the* hypocotyl*), so that when it straightens, the cotyledons are pulled above the ground (this is known as* epigeal *germination). In some other dicots the stem above the cotyledons (the* epicotyl*) elongates into the hook, so that the cotyledons are left below the ground when it straightens (this is called* hypogeal *germination).*

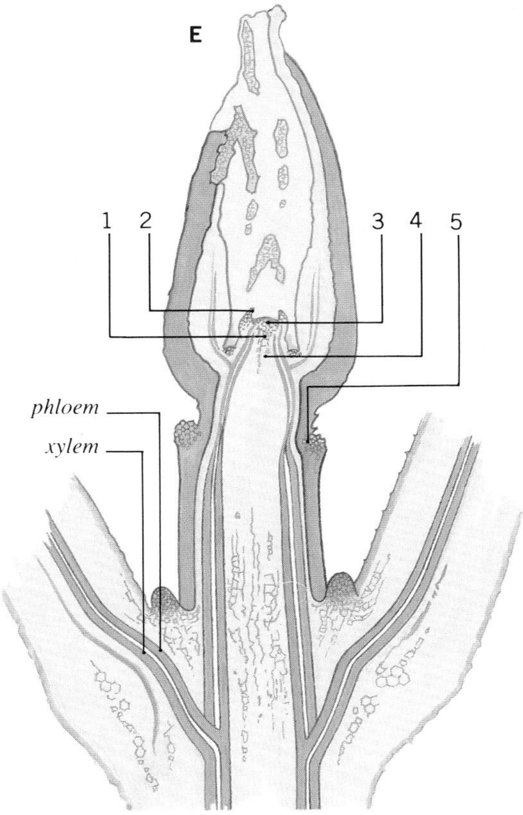

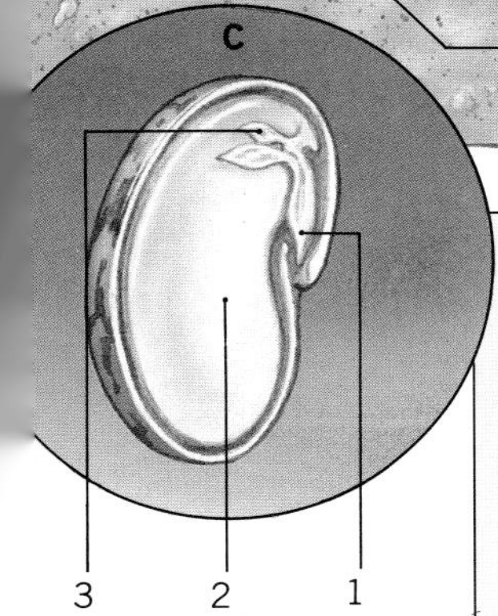

Inside a bean [**C**] *the radicle* [1] *is positioned to burst through the seed case (*testa*) at the start of germination. Food stored within the swollen cotyledons* [2] *will nourish the plant as the plumules* [3] *grow and develop into young leaves. As the nutrients are used up, they will shrivel and wither.*

Short Days and Long Days

Henbane needs a minimum of 12 continuous daylight hours before it will flower. Flowering in this and many other species of plants is in fact determined by the number of hours of darkness they are exposed to. The plant can measure the length of time between the last light signal it receives in the evening and first light the following morning. Some plants, like tobacco, need more than a certain critical number of hours of darkness (which varies from species to species) and are called short-day plants, whereas those like henbane, which require less than a critical dark period, are known as long-day plants. Changes in a photosensitive blue-green pigment called phytochrome in the leaves and stems are responsible for the physiological changes that produce this photoperiodism.

Plants in Heat and Drought 138 Seeds and Fruit 156

Infinite Variety

How invertebrates reproduce

During sexual reproduction a female mantid has occasionally been known to bite off its mate's head. The males are smaller than their mates and must make a hasty getaway after sex, when the female can become violent. The male garden spider also has to approach his larger female mate with caution or he too may be consumed. Attaching a silk thread to her web, he vibrates it with his legs, signaling her to approach. He ejects sperm onto a special miniweb he has woven and then draws it off again into dropperlike appendages that can deposit the sperm inside the female.

Invertebrates have a range of reproductive habits as diverse as their body shapes and sizes. Some can simply regenerate whole animals from fragments of their own bodies. Others require complex courtship, internal fertilization, several larval stages and even parental care for successful reproduction. Many marine species have mass spawnings, even migrating to spawn, while others are thinly dispersed, finding mates with difficulty.

A few invertebrates have adapted to scarcity of mates by becoming *hermaphroditic,* or by changing sex to suit the occasion. Many mollusks, particularly gastropods, and many annelid worms are hermaphrodites – each individual possesses both male and female reproductive organs. Some, such as snails and earthworms, can act as both male and female simultaneously. This has the advantage that fertilized eggs are produced by both partners from each mating. The disadvantage is that each individual expends energy producing both kinds of sex organs and gametes. Slipper limpet larvae only become male when attracted to females.

Some simple invertebrates, such as hydra, avoid the problem of finding mates by reproducing asexually by budding off new individuals. However, asexual reproduction cannot produce the variation needed for evolution and adaptation to new environments, so most also remain capable of sexual reproduction.

A

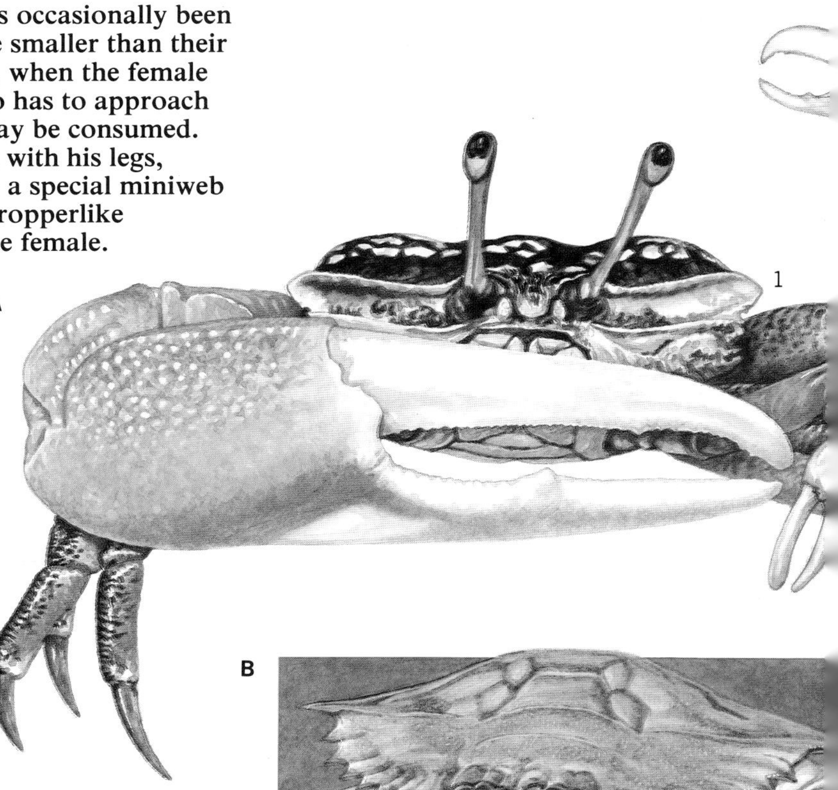

B

Blue crabs *[**B**], like other hard-shelled crabs, mate only after the female has shed her hard shell and her new shell is still soft. In this condition she is extremely vulnerable, so the male hovers in close attendance to guard her. Mating takes place face to face with the female on her back.*

Getting together

Invertebrates that do reproduce sexually use a wide range of signals and senses when courting. Ghost crabs drum in their burrows, fireflies flash special patterns of light, many spiders use a kind of semaphore, while the females of many species of moths and butterflies simply wait patiently as they release irresistibly attractive chemicals. Courtship involves not only attracting another animal but also determining that it is of the same species and of the opposite sex. The timing of breeding is often related to environmental events and signals. Chemical signals are used, sometimes combined with biological clocks, to trigger mass spawnings in many marine species.

The pattern of courtship is also influenced by the mating procedure. Animals that are fertilized internally may have sperm injected into them or else pick up *spermatophores* (packets of sperm). Variation again abounds, including the behavior of some springtails, which simply drop their spermatophores on the ground and leave the rest to chance.

Breeding strategies

As in most groups of animals, there are two main breeding strategies – the production of few but well-provisioned eggs, sometimes combined with parental care, or the production of huge numbers of vulnerable, small eggs. Colonies of social insects such as bees, wasps and ants provide a high degree of care and even have special nurseries in which to rear the young. Parent bugs stand guard over their eggs and young, and female crabs and lobsters carry their eggs around with them. Octopuses glue their eggs into rock crevices and stand guard over them, while water fleas retain theirs inside their bodies until hatching. Parasitic wasps lay eggs in other animals' bodies for a living food larder for the larvae.

The female blue crab, *once fertilized, lays eggs, which she carries around with her* [**C**] *held in place by stiff bristles under her abdomen, which is curled round to form a sort of pouch. In this condition she is said to be "in berry."*

When the eggs are ready to hatch, they are released as larvae. The transparent zoea *larva* [1], *the first stage, is about 0.08 in (2 mm) long and lives among and feeds on plankton, camouflaged in the sunlit surface waters.* Megalops [2], *the next stage, is more crablike and will settle on the seabed. Finally, at 0.1 in (2.5 mm) wide, the young animal gets to the first crab stage* [3].

Connections: **Ants' Nests** 236 **Bees' and Wasps' Nests** 232 **Communication** 292 **Invertebrates** 122 **Metamorphosis** 168 **Termite Towers** 234

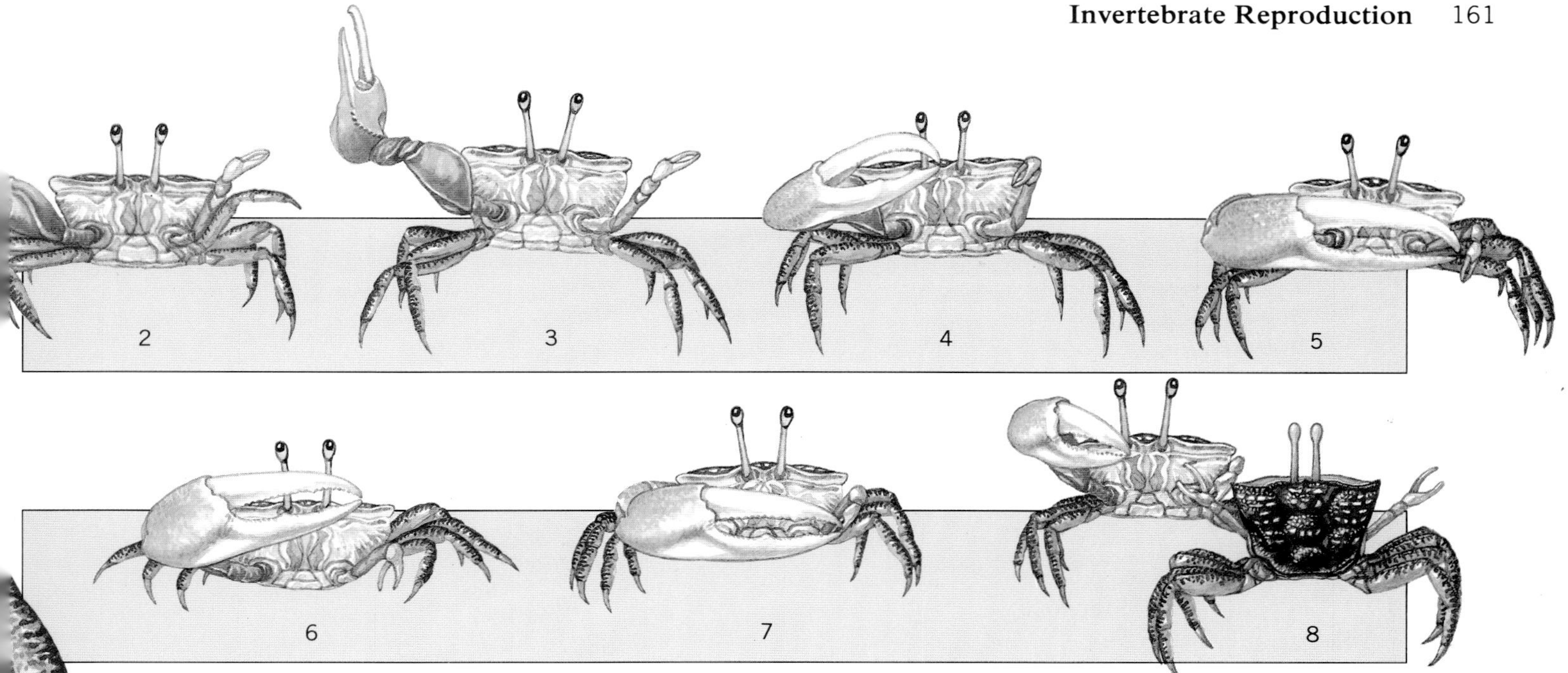

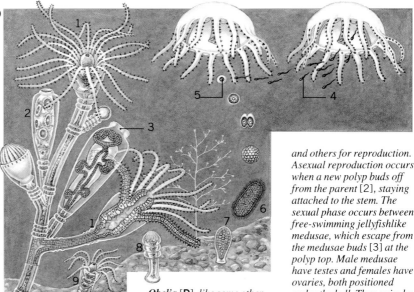

Fiddler crabs [**A**], *like most crustaceans, recognize potential mates by sight. The male crab, however, has one enormous claw (too large, in fact, for feeding with) used solely in courtship or to warn or fight competing males.*

In courtship the male waves his claw in a series of complex semaphore signals. There are several phases to the dance. One [1-5] *is performed before a specific female has been sighted or aroused, when there may be many other males nearby. This is the lateral display, the claw starting in front of the face* [1], *swinging back to an open position* [2] *and then rotating up and back down to the starting position* [3-5]. *While waving its claw the crab is also raising and lowering its body. This stage can last some time and, to conserve energy, is done at medium intensity. The second phase only begins when a female is present* [6-8, *the female shown only in* 8]. *Again, the body bobs up and down, but the claw movement is simpler, consisting of a rapid up and down movement. This vertical display is conducted at maximum intensity.*

These illustrations simplify the crabs' full repertoire, which includes gestures used beside burrows and gestures by unreceptive females.

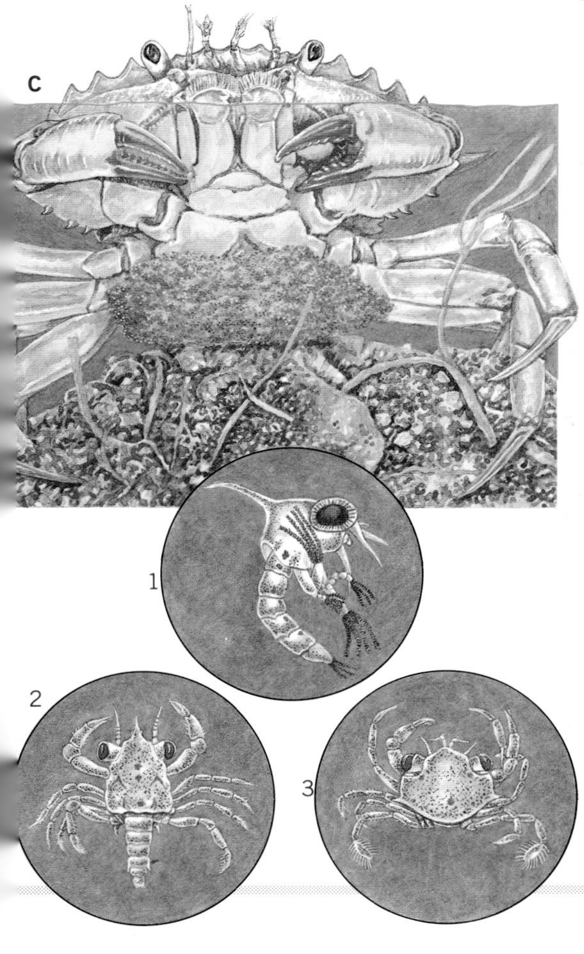

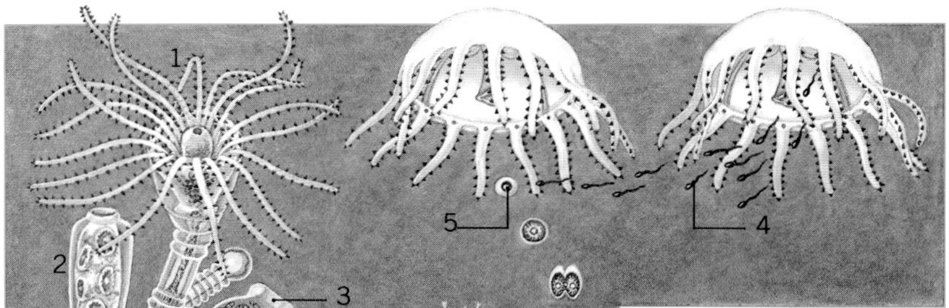

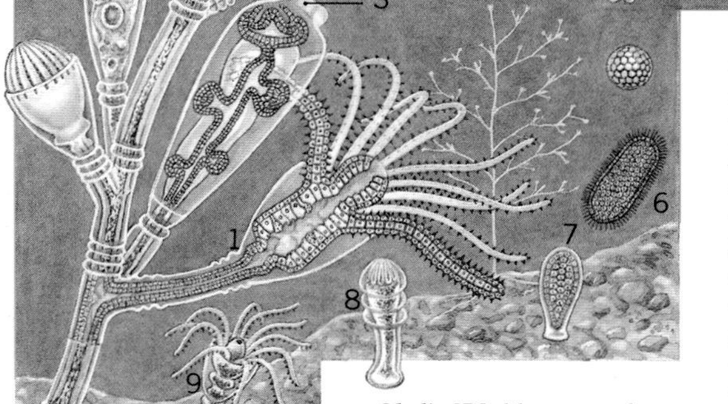

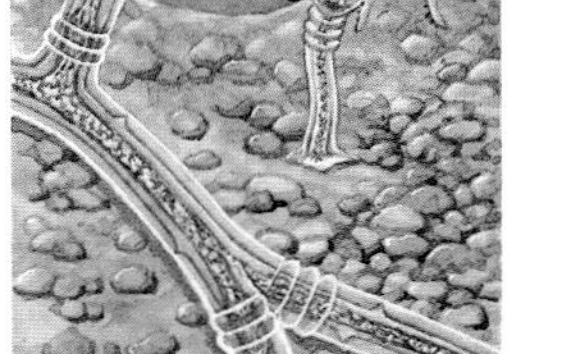

Obelia [**D**], *like some other marine invertebrates, can alternate between asexual and sexual reproduction, often synchronized with particular seasons. Obelia consists of a colony of sub-individuals called* polyps, *some of which are specialized for feeding* [1, *also shown in section*] *and others for reproduction. Asexual reproduction occurs when a new polyp buds off from the parent* [2], *staying attached to the stem. The sexual phase occurs between free-swimming jellyfishlike medusae, which escape from the medusae buds* [3] *at the polyp top. Male medusae have testes and females have ovaries, both positioned under the bell. They swim by pulsating their bells. A male medusa releases sperm* [4] *to fertilize the female's eggs* [5]. *The fertilized egg develops into a larva* [6] *that swims until it finds a surface on which to finally settle* [7], *grow, and eventually commence to reproduce as mature obelia* [8 *and* 9].

Water Courtship

How fish reproduce

During courtship Siamese fighting fish bite and butt each other in a frenzied and spectacular whirl of tattered color. To attract mates, many male fish will change color during the season and twist and turn to display their decoration. Gurnards spread their brilliant pectoral fins like wings. A few species, such as cod and the aptly named grunts and croakers, even use sound as a means of signaling to each other. Male rivals for female attention battle each other in ways individual to their species. Kissing gouramis, for example, engage in mouth-pushing.

***Siamese fighting fish** have two major characteristics: they are able to use oxygen from the air as well as dissolved oxygen in the water; and the males are ferocious. When a female is evidently ready to spawn, a male first builds a large nest of air bubbles at the surface of the water, and then quite literally attacks the female, biting her gills or tearing at her fins. As she lays each clutch of eggs, he wraps himself around her, releasing sperm; he then quickly leaves off to catch the dropping eggs and to place them carefully under the bubbles of the nest. They hatch within 48 hours.*

***The three-spined** stickleback displays complex courtship and nesting behavior. At up to 3 in (8 cm) in length, this freshwater fish is common in Europe. During the breeding season the male develops a distinctive coloration of vividly blue eyes, red underparts and mouth lining and silvery scales on the back. He selects a territory and begins nest building by excavating a hollow in the sand or mud. In this he erects a dome-shaped structure made from weed fragments stuck together with mucus. He waits for a passing female. When a female is ready to mate, her belly appears swollen with eggs.*

To reproduce, most fish simply shed sperm and eggs into the water and rely on water currents to bring them together for fertilization and thereafter to disperse eggs and fry (baby fish). Fish for the most part show little or no parental care, although as a defensive measure they may lay large numbers of small eggs so that it hardly matters if a fair quantity of eggs and fry are lost. However, oceans and lakes are also full of predators, and a relatively small number of fish species through the millennia have evolved further precautions. A few species – such as sharks, swordfish and guppies – depend on internal fertilization, retaining eggs and even larvae inside the mother until they are big enough to fend for themselves. Others lay fewer, larger eggs containing greater amounts of yolk so that the young are better developed on hatching. An even smaller number of species actually take physical steps to look after their offspring. Often they guard the eggs in specially constructed nests.

Among nest-building fish it is usually the males who construct the nest, guarding the nest and its surroundings fiercely against the intrusion of any other males of the same species. Often the same colors that attract females act as a warning to rival males.

A time for breeding

Although some species of tropical fish breed at frequent intervals, most fish have distinct breeding seasons, when the males develop courtship colors and the females become responsive. Seasonal breeding serves to produce young when food is abundant.

Many fish, such as herring, mackerel and plaice, migrate great distances to spawning grounds where conditions are especially favorable for the growth of the young. The timing of some of these migrations appears

The sight of the female stimulates the male into action. He repeatedly zigzags toward her, turning away at the last instant [1]. The female eventually signals her readiness to spawn by assuming a tilted-up position [2] near the surface of the water as she watches the male's display. The male then turns and leads the female down to the nest [3], indicating the

entrance with his snout. In this maneuver he turns on his side to display his red belly [4]. Sometimes a female will not spawn but remains in the area to be courted. Then the male may become frustrated and attack her, nipping her flanks [5]. Eventually the female swims away, or returns to inspect the nest while the male nudges and prods her belly to stimulate egg laying.

Connections: Animal Adaptation to Pressure 146 Communication 292 Fish 124 Lake Life 326 Migration Patterns 228 Ocean Life 332 River Life 328

to be a response to day length and temperature; other migrations follow tidal or lunar rhythms. A few species make the rare transition from fresh to salt water. The European eel breeds in the oceans but returns to the rivers as an elver for instance, and the salmon spends most of its life at sea but migrates up rivers to spawn. How these fish find their way across vast distances of comparatively featureless ocean to their home rivers is one of the great mysteries of the sea. They may use a magnetic sense, or their sense of smell. After their long migration, many salmon fail to reach the spawning grounds to breed, and many more die before they ever make it back to the sea.

Transsexual fish

A few species of fish are able to change sex. Some swordtails start life as females, but after giving birth to several broods of young, they develop a swordlike fin, change color, and become males. A few species of fish, including occasional individuals of herring, mackerel and hake, may even produce hermaphrodite individuals – fish possessing both male and female sexual organs. Astonishingly, sea perch can act as male, female, or both sexes at the same time. Whichever fish is acting male develops broad, dark, vertical bands on its body. The two fish may take it in turns to play male and female.

The female enters the nest and sheds up to 100 eggs, provoked by the male quivering violently while striking his snout against her tail [6]. *The female leaves the nest and the male enters in order to discharge his sperm* [7]. *After the female has served her purpose, the male may chase her away from the nest site in anticipation of another female visitor. Should another male encroach upon his territory, the male stickleback puts on an aggressive display, darting toward the intruder with his dorsal fin erect and mouth agape. After several rounds of fertilization, the male shifts his attention to the care of the eggs. He guards the nest closely, chasing away hungry females, and constantly fans a current of fresh, oxygen-rich water over the eggs with his pectoral fins* [8].

Mouth Rearers

Bizarre tactics are employed by some fish in defense of their eggs. Many cichlids – small fish of tropical America, Africa and Madagascar – brood their eggs in their mouths. After fertilization, the eggs are released into the water and are snapped up in the mouth of one of the parents: they are retained there until they hatch.

Water Babies

How amphibians reproduce

Worm lizards, or caecilians, are able to turn their genital openings (*cloacae*) inside out, and the tailed frog has a taillike cloaca for transferring sperm directly into the female before it is washed away by the fast-flowing streams in which the species lives. Amphibians use the most varied means of reproduction of any vertebrates, ranging from internal fertilization to external fertilization underwater; from egg-laying to live birth. Courting techniques are equally varied, though some species do without courtship altogether and mate in a brief chaotic orgy.

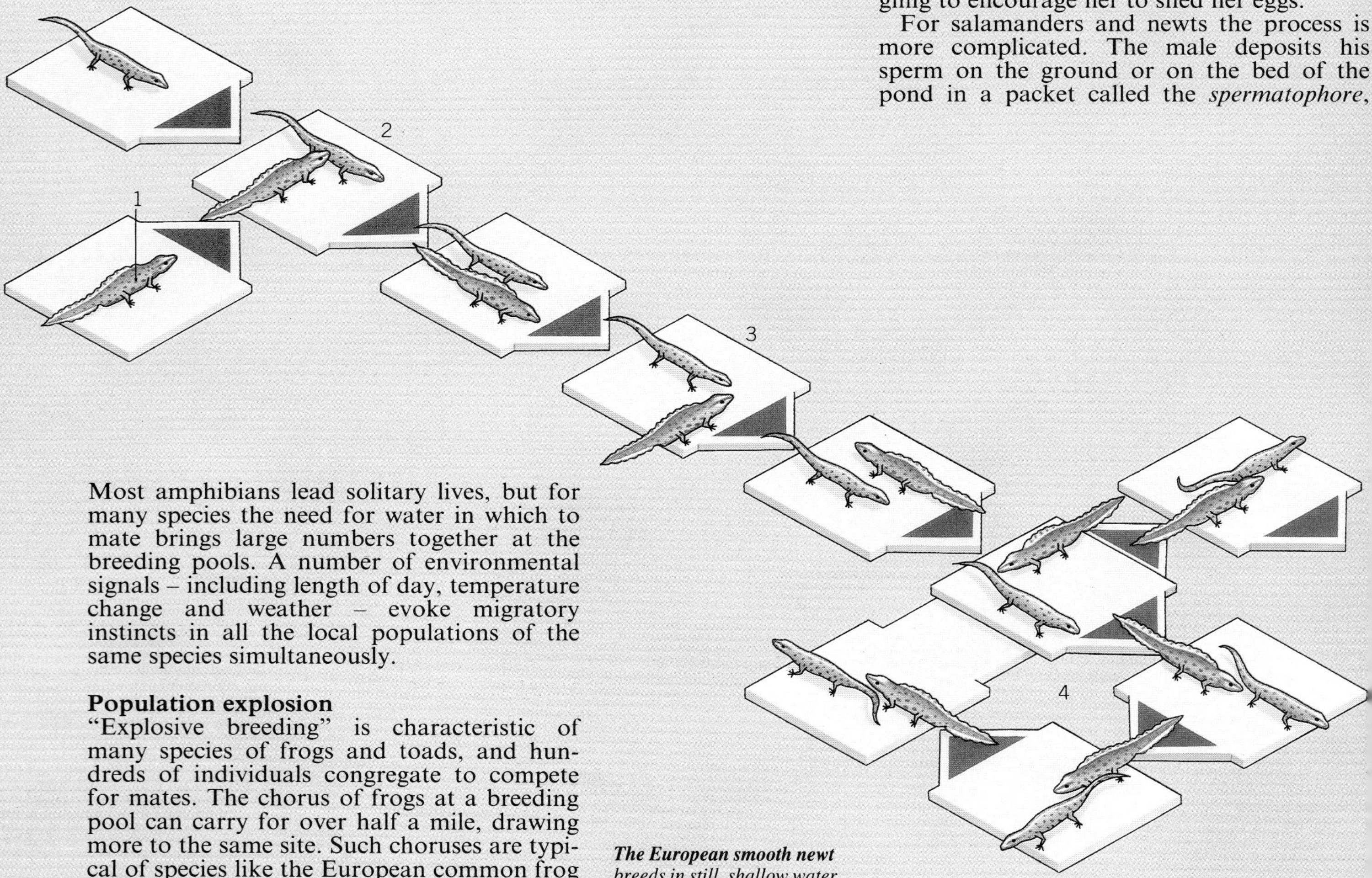

Most amphibians lead solitary lives, but for many species the need for water in which to mate brings large numbers together at the breeding pools. A number of environmental signals – including length of day, temperature change and weather – evoke migratory instincts in all the local populations of the same species simultaneously.

Population explosion

"Explosive breeding" is characteristic of many species of frogs and toads, and hundreds of individuals congregate to compete for mates. The chorus of frogs at a breeding pool can carry for over half a mile, drawing more to the same site. Such choruses are typical of species like the European common frog and common toad, which have a relatively short breeding period. They may need to find a suitable mate within the space of a few days.

In all amphibians it is the female who has to recognize the sex and species of her partner. Males will display to any other frog, toad, or salamander regardless of sex or species. Since the female will probably mate only once, she has a greater vested interest in selecting the correct partner. Female frogs are most strongly attracted to the male with the loudest, deepest croak. Since a frog's croak depends on the size of its voicebox and hence its age, the female is ensuring that she selects a mate who has survived for several years and will, she hopes, pass on his advantageous features to her offspring.

Timing the moment of conception

When mating it is important that sperm and eggs are shed in the same place at the same time. For frogs and toads that mate in water this is achieved by muscular movements of the body that stimulate their partners. The male grasps the female firmly under the arms using special "nuptial pads" on his thumbs. He then presses his body against her, wriggling to encourage her to shed her eggs.

For salamanders and newts the process is more complicated. The male deposits his sperm on the ground or on the bed of the pond in a packet called the *spermatophore*,

***The European smooth newt** breeds in still, shallow water, preferring weed-filled ponds and ditches. In the breeding season the male [1] becomes more conspicuous than the female: a broad crest develops along his back, dark spots appear on his body, and the underside of his tail turns orange. Fertilization is preceded by a complex, nonaggressive courtship dance that incorporates visual, tactile and chemical stimuli. Fertilization is internal, sperm meeting eggs inside the female's body, but the male has no penis: instead the sperm are delivered to the female in a capsule called a spermatophore.*

The male first approaches and sniffs the female [2]. She tends to swim away, so the male follows and overtakes her and positions himself across her path to display his prominent crest [3]. A phase of orientation ensues in which the male follows the female [4]. The two then face

Connections: **Amphibians** 128 **Animal Growth and Development** 166 **Early Evolution** 92 **Metamorphosis** 168 **Swimming** 210

on a tiny stalk. Using scent from glands under his tail to attract a mate, he then performs a dancelike courtship to maneuver the female into position above the spermatophore so that she can take it up into her genital opening.

Coping without water

Although tropical forests appear very moist, there is often very little standing water in which amphibians can breed. Some tree frogs and salamanders lay their eggs in the tiny pools of water that can accumulate at the heart of some plants. Marsupial frogs and Surinam toads retain their fertilized eggs in pockets on the female's back.

The male Darwin's frog from South America broods its young all the way from egg to toadlet in its mouth. A number of frog species, like the South and Central American arrow poison frogs, lay their eggs on leaves or branches, where they are safe from the attention of aquatic predators.

One parent, which is often the male, may guard the eggs and keep them moist with water from its bladder until the tadpoles hatch. Foam nest frogs mate communally on branches overhanging water, using their feet to whip up saliva into a froth of foam, in which the eggs are laid. The foam keeps them moist until they hatch, when the tiny tadpoles drop into the water.

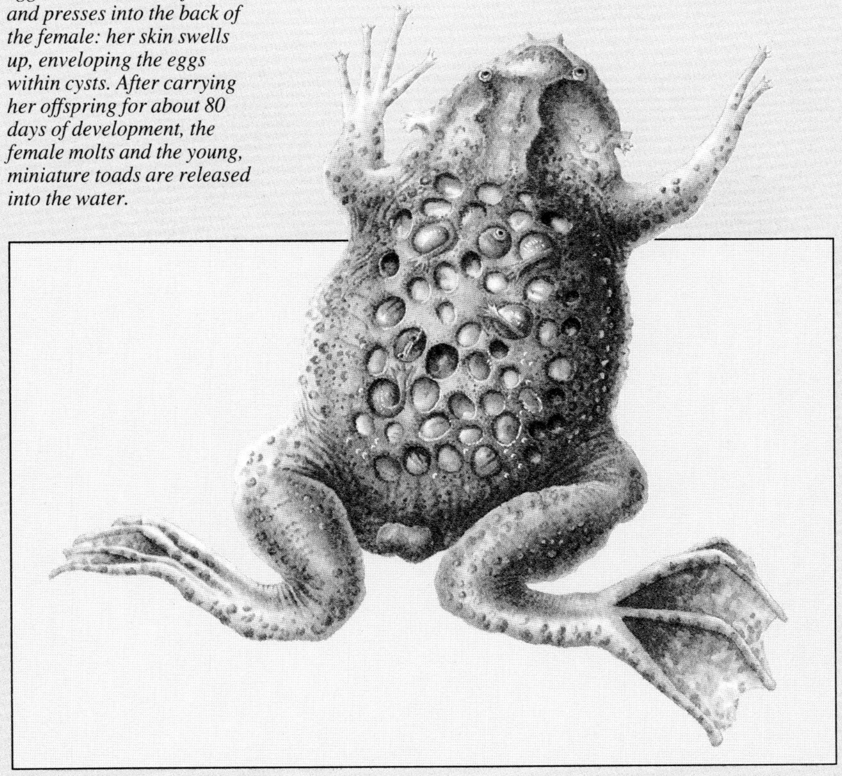

***The Surinam toad** lays 3–10 eggs, which the male fertilizes and presses into the back of the female: her skin swells up, enveloping the eggs within cysts. After carrying her offspring for about 80 days of development, the female molts and the young, miniature toads are released into the water.*

one another and the male doubles back his tail, vibrating it to "waft" secretions toward the female's snout [5]. This stimulus causes a marked change in the female's behavior. She begins to approach him, and now he moves away, continuing to display as he swims backward. Finally he turns and creeps forward, with the female following close behind [6]. The male stops and quivers his tail [7], upon which the female advances until her snout touches his tail [8]. In response to this signal, the male drops his spermatophore [9]. The spermatophore itself consists of a sperm-filled cap sitting on a broad, fluid-filled base, the function of which is to elevate the cap above the bottom of the pond. After depositing the spermatophore, the male advances by exactly one body length and turns sideways: the female moves forward until she is just touching him [10], so her cloaca – the common passageway for waste products and reproductive cells – is positioned over the spermatophore. On contact with the cloaca, the spermatophore adheres to it and is drawn up into the female's body, where fertilization occurs.

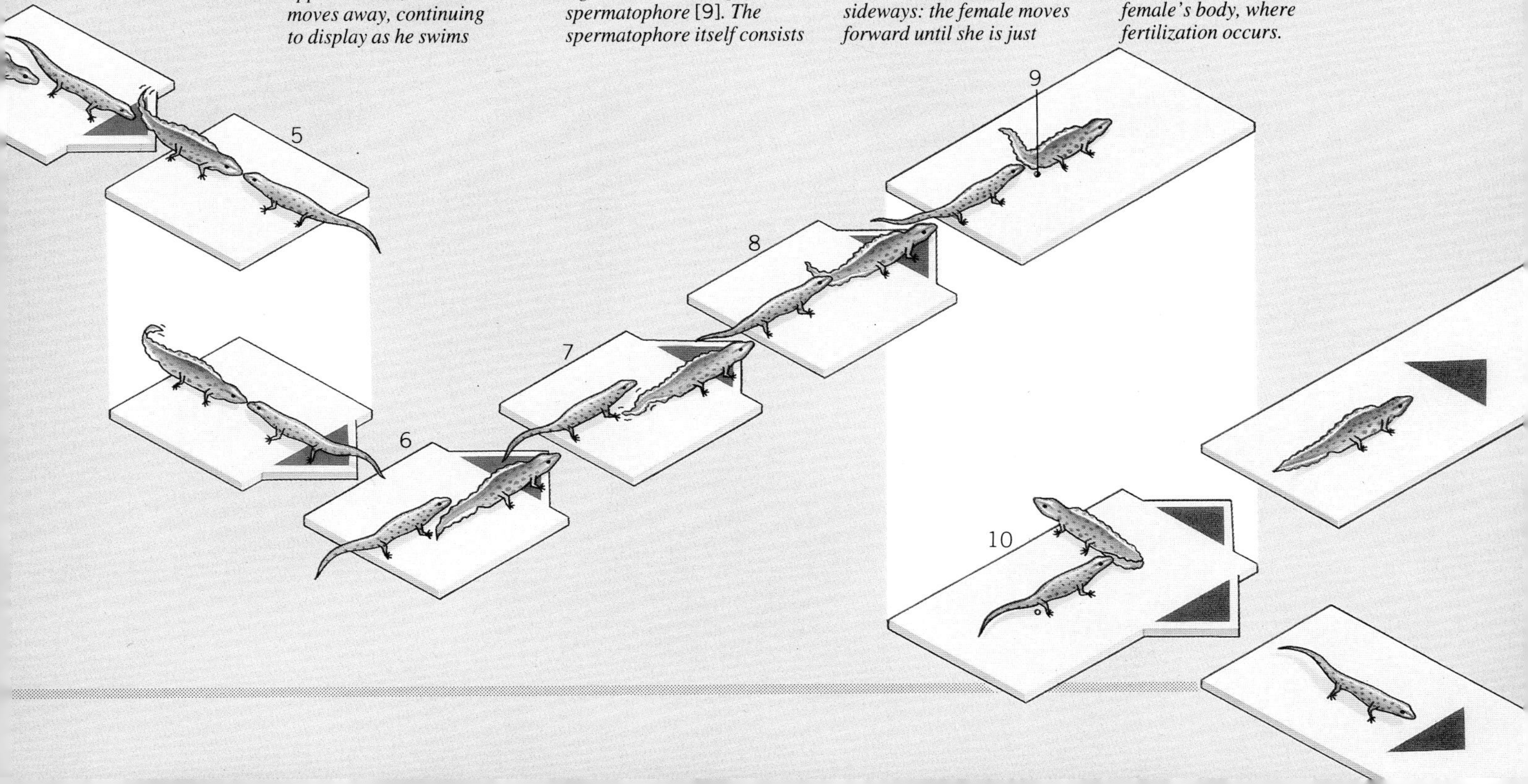

Divide and Multiply

How animals grow and develop

If human beings varied in size as much as breeds of dogs, it would not be uncommon to see people 3 feet (1 meter) tall next to people 9 feet (3 meters) tall. All animals start life as clusters of cells, which multiply to build the tissues of the body. In many animals most of this development takes place before they are born or hatched, but some species, notably insects, undergo startling changes of form at later stages of their lives. These developments are controlled by a genetic program that is built into the animal at the stage of fertilization.

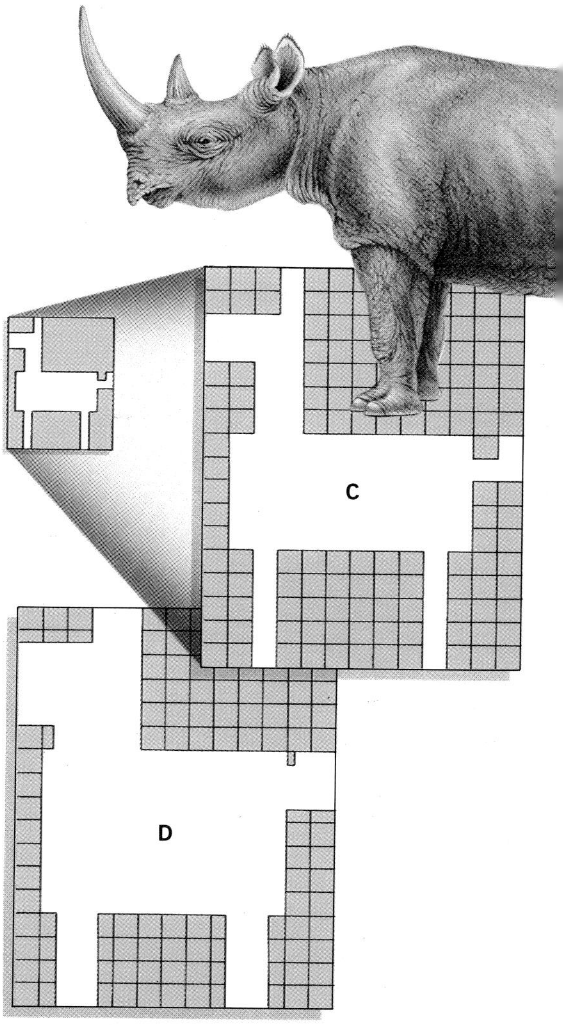

The development of most animals divides into two phases: the embryonic phase, when the complex cells of the egg multiply and differentiate to create all the major tissues of the body, and the juvenile phase, when the young, functional animal grows and acquires the characteristics of sexual maturity.

Privileged youth

We normally associate the embryonic phase of life with the womb or the egg, but some animals must make their own way in life even at this early age. A housefly, for instance, must be self-sufficient while it is still effectively an embryo – as a maggot – since it has no yolk to provide nutrition while in the egg.

By contrast, a bird has a pampered introduction to life. Its mother makes a great investment in each individual embryo, providing it with a sac of nutritious yolk inside the eggshell. By the time the bird hatches it has developed most of its adult features, and it acquires the rest while still in the nest, being fed by its parents. Most birds do not have to find their own food until they are nearly fully grown.

A young mammal is just as carefully nurtured. Its embryonic phase is spent inside its mother's body. When it is born it lives exclusively on its mother's milk, a nutritionally ideal food. Once its digestive system is ready to accept adult food it will stop drinking milk, but even at this stage many carnivores still hunt food for their offspring.

Genetic control

Although the pattern of growth and development varies enormously between different types of animals, the mechanism that controls it is the same: a genetic "program" that is built into the complex molecules of DNA found in the chromosomes of the fertilized egg. The genetic program unfolds like a computer program, organizing the multiplying cells of the embryo into the specialized agglomerations, chains and membranes that make up the body structures.

Among these structures are glands that produce the messenger chemicals known as hormones. In a mammal the activities of these glands are coordinated by a "master" gland called the *pituitary*, which is located at the base of the brain.

***The development* [A] *of a fertilized frog egg is a complex but preset process. The egg* [1] *rapidly cleaves into two smaller cells and then into four until there are 32 cells* [2]. *Cell division maps out an embryonic plan and provides basic building blocks to be organized into organs and tissues. At 32 cells, cleavage becomes less regular* [3–6] *and the cells begin to group into three distinct embryonic layers, each of which is predetermined to eventually form specific body parts, the* endoderm [7], *the* mesoderm [8] *and the* ectoderm [9] *(see key). After the initial three-layer arrangement, a neural plate* [10] *forms out of the ectoderm; this plate becomes a groove* [11] *that rolls in on itself to form the neural tube, the basis for the brain and spinal cord. By this stage the embryo* [12] *has begun to resemble the familiar tadpole that will metamorphose into a frog* [13], *where a color-coded scheme shows the germ-layer origin of the various organs and tissues.*

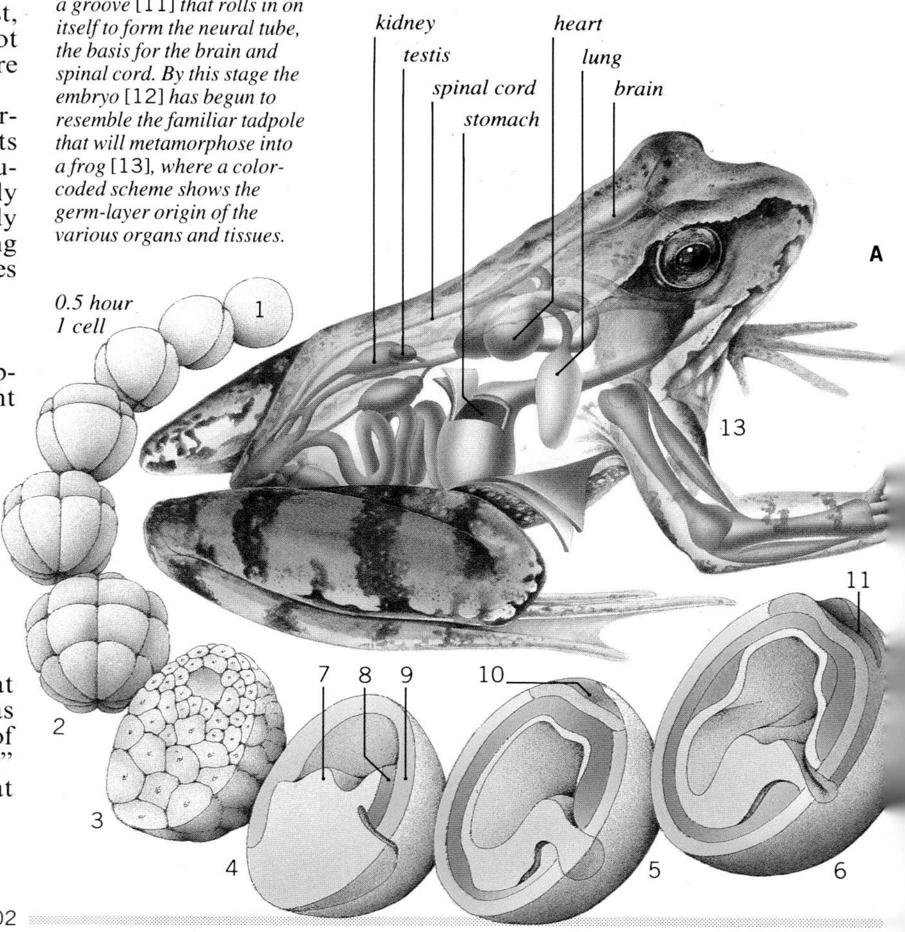

Connections: **Amphibian Reproduction** 164 **Animal Cells** 102

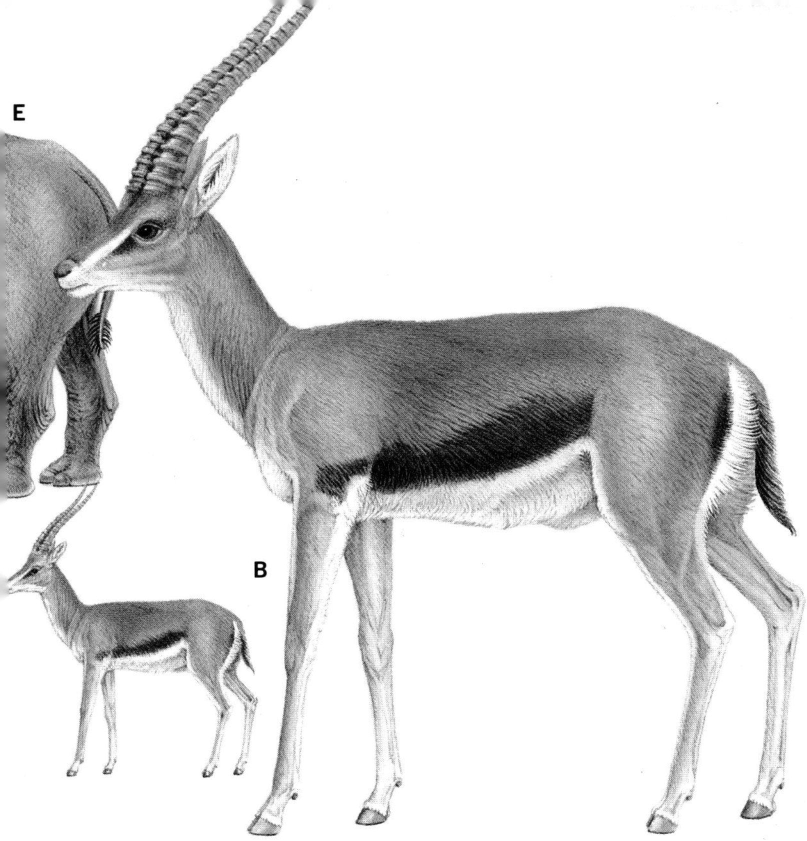

The pituitary produces hormones of its own, including the growth hormone *somatotropin,* which controls the rate at which cells multiply to build the body tissues. But it also secretes substances that affect other glands. As a mammal approaches sexual maturity, for example, the genetic program in the pituitary makes it produce a hormone that stimulates glands in the sexual organs. These then start to generate the sex hormones, which trigger the development of mature male and female characteristics.

The whole process of growth and development is controlled by a complex interaction of genetic and hormonal influences, and malfunctions in the system are not uncommon. The mutations that result are usually fatal, and most never get beyond the embryonic stage. Those that survive birth generally die before they reach sexual maturity – and their problems and defects die with them.

Animal size varies *enormously. The largest land mammal – the African bush elephant at approximately 5.5 tons – is roughly a million times larger than the smallest – the European pygmy shrew – at around 0.18 oz. Sheer size confers many advantages, such as reduced predation, increased ability to travel in search of food, water, or a mate, and the ability to use food far more efficiently than smaller animals. Within limits, a basic animal "design" can be enlarged to keep geometrical similarity, much as one enlarges a photograph* [**B**]. *In* [**C**] *the stylized animal's dimensions are three times larger. But if the animal is imagined as a 1-in cube, it highlights problems of increased scale. The area of one side of a 1-in cube is 1 in*^{2} *and its volume is 1 in*^{3}*; a 3-in cube, however, has a side area of 9 in*^{2} *but a volume of 27 in*^{3}*. As dimensions increase, volume and weight increase far faster than surface area. Since muscle power rises with its cross-sectional area, the "cube" animal's leg muscle power increases nine times but must support 27 times more weight. Many organ "power ratings" relate to surface area, including the kidneys, stomach and lungs, where oxygen absorption depends on the lung lining area. Thus, as dimensions rise, body requirements far outstrip the organs' increased capacity. If a geometrically enlarged "design" outstrips the capacity of the organs to support it, the enlargement must be elastic, so that while height and length increase by the same factor as in* [**C**], *the neck, body and legs enlarge by a greater factor* [**D**] *to accommodate larger organs and muscles, as in, for example, a rhinoceros* [**E**].

32 hours
170,000 cells

12

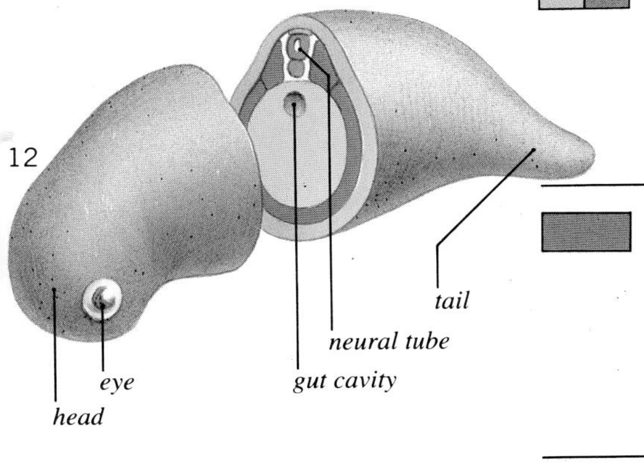

ectoderm: *outer skin layer (epidermis) and associated structures (nails, hair), peripheral nervous system, sensory systems, pituitary gland*

brain, spinal cord

mesoderm: *inner skin layer, skeleton and cartilage, muscles, circulatory, excretory and reproductive systems, outer layers of digestive tract and of systems developed from it (eg respiratory system)*

endoderm: *inner lining of digestive tract and of systems developed from it*

Regeneration Skills

Almost all animals can repair minor tissue damage with relative ease. This is particularly true of superficial damage to skin or muscles, even in mammals, where cuts, grazes, or bruises heal readily. Unlike human beings, many animals have the capacity to regrow lost or damaged adult teeth. Whole organs or limbs, however, are much harder to regrow, though some animals are capable of this. A lizard, for example, may lose its tail to a predator, escape, and grow a new one.

Primitive animals can survive more serious damage. A sponge – one of the simplest of all animals – can be reduced to its component cells, yet these will eventually clump together to form several living sponges. It seems that, as a general rule, the simpler the structure of an animal, the easier it is to rebuild.

Starfish *can shed limbs in defense or to combat injury – the severed limb usually dies but a new limb grows on the stump. Other starfish reproduce by dividing in two, each half growing three extra legs to make it a complete animal.*

Bird Reproduction 174 Heredity 98 Invertebrate Reproduction 160 Mammal Reproduction 176 Metamorphosis 168 Reptile Reproduction 170

All Change

How metamorphosis works

An insect larva's weight can increase 1,000 times before its last, most violent, transformation – of body, diet and life-style – to an adult. On the way from egg to adult, a frog relives its whole evolutionary history, passing through a fishlike stage in a series of dramatic changes on the way to a life on land. Metamorphosis has many uses: a lot of marine invertebrates are sedentary bottom-dwellers, like the giant clam, and before metamorphosis their young are microscopic and may float thousands of miles among clouds of plankton to spread and colonize new areas.

***When common frogs* [A]** *mate, fertilization and egg laying occur in water* [1]. *Within an hour the jelly around the eggs swells to produce frogspawn* [2]. *The eggs develop* [3] *and produce embryos* [4] *that hatch as long-tailed tadpoles with external feathery gills six days after fertilization* [5]. *Mouths and eyes develop later and the tails become powerful means of propulsion. Hind legs are well formed by week eight* [6]*; meanwhile, the tadpole has changed from a herbivore to a carnivore. Via an intermediary gill and lung stage, the tadpole changes from gill to lung breathing, its internal lungs growing as its external gills are absorbed; the process is complete when the gills fully disappear at month three, by which time the forelegs are well developed* [7].

Metamorphosis is not quite complete because the animal still has a long tail [8]*; the final stage is its resorption.*

In all animals the timing of the changes involved in metamorphosis is controlled by hormones, chemical messengers in the body. However, day length, temperature and the animal's physiology may all help to fine-tune the process – timing may be critical to take account of the winter or of dry seasons.

In young tadpoles the main growth-stimulating hormone comes from the pituitary gland. It tends to inhibit metamorphosis, which only starts when the thyroid gland dramatically increases production of its own hormone, *thyroxine*. This causes growth of leg muscles, breakdown of tail tissues, and formation of light-sensitive pigments in the eye. In insects the hormone *ecdysone* initiates growth and development and causes molting. A pair of glands near the brain produce another hormone essential for molting but inhibiting growth of adult characteristics; only if secretion of this second hormone decreases can metamorphosis finish.

As insects undergo metamorphosis, they pass through stages, called *instars*. Most of the reorganization needed to change the larva into the adult is condensed into the final instar stage, the pupa. Until this time the wings, for example, have been developing internally. As the last instar molts to form the pupa, the developing wing buds and limbs become visible for the first time. A big change in musculature is needed to convert from larval locomotion to the flapping flight and powerful legs of the adult; mouths may also radically change for a new diet.

Shelter and protection

All these changes take time, and the pupa needs protection while this happens. Some moth larvae produce a kind of silk and spin protective cocoons around themselves or bind leaves together to form a resting chamber (a *chrysalis*). Others cement soil particles together to form an underground cell, using sticky secretions as glue. In yet other larvae the last larval skin is retained and becomes hardened to form a tough case (*puparium*).

For some the changes to maturity are remarkable. Mosquito larvae hang from the water surface film in ponds and ditches; the pupa lies on the surface film, and the mature adults live in the air.

Not all winged insects undergo such drastic change. Grasshoppers, cockroaches, dragonflies and many other groups have more gradual transitions. After each successive molt the larva, or nymph, more closely resembles the adult insect. There may be as many as 40 stages in some *hemimetabolous* insects (which have no visible change in form between the first instar and the adult).

Fish metamorphosis

Plaice and sole seem flattened from top to bottom, but are really flattened sideways and swim on their sides. The fry are a normal symmetrical fish shape, but after a few weeks' development one eye moves to the other side of the head and the mouth twists round. The fish sinks to the seabed, lies on its side, and develops an undulating swimming style.

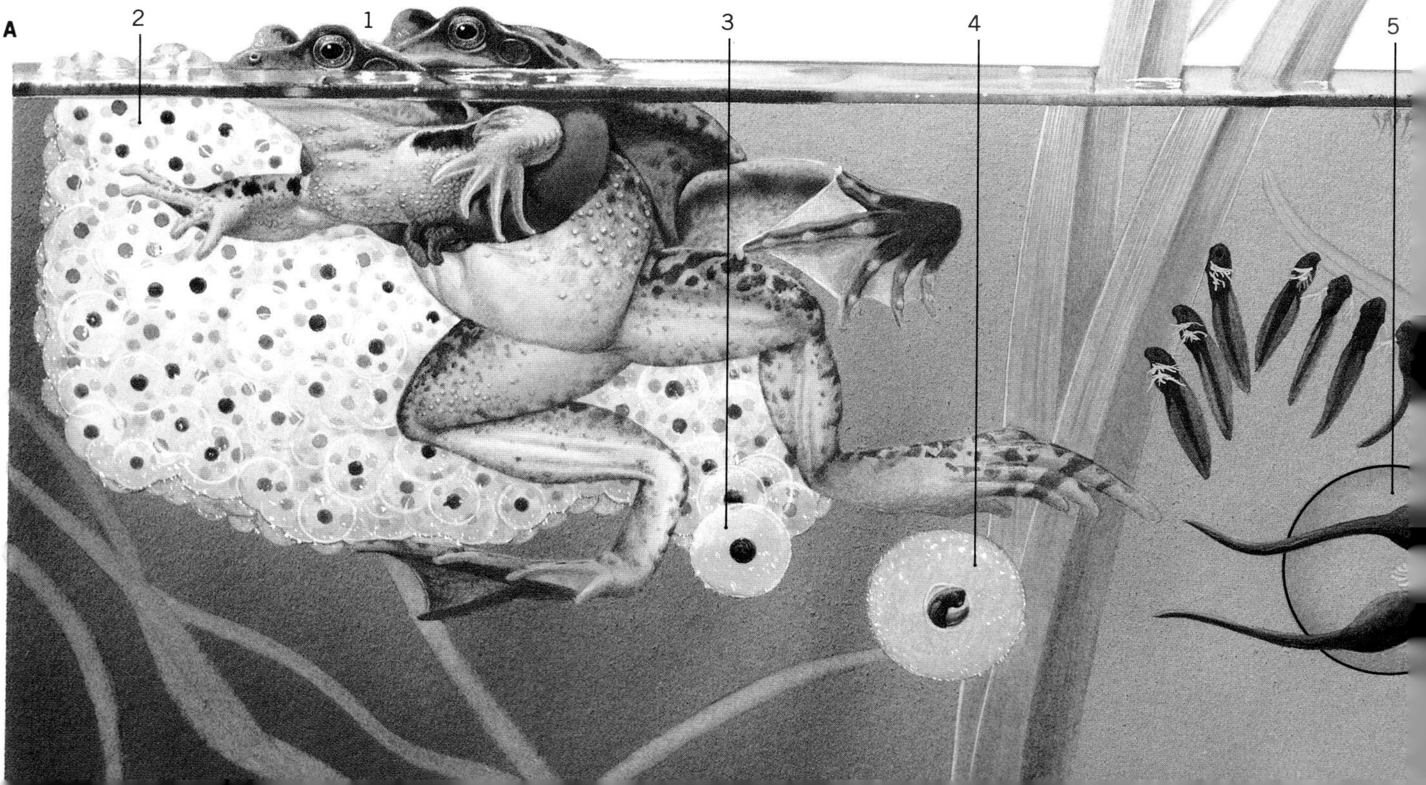

Out of their Skins

Molting does not only occur during metamorphosis. Because an arthropod's cuticle is dead, it is periodically shed, and a new one is secreted as the animal undergoes its next spurt of growth. Land-living vertebrates produce keratin – a hard, water-resistant protein – in the outer skin layer cells. As keratinization causes many cell components to degenerate and eventually die, the layer of keratinized cells is shed from time to time. So snakes and other reptiles literally crawl out of their skins. Birds and mammals slough off small pieces of keratinized skin almost continually. (Most dust in a home is powdered skin.) Even frogs and toads shed the surface of their skin, which they will then often eat.

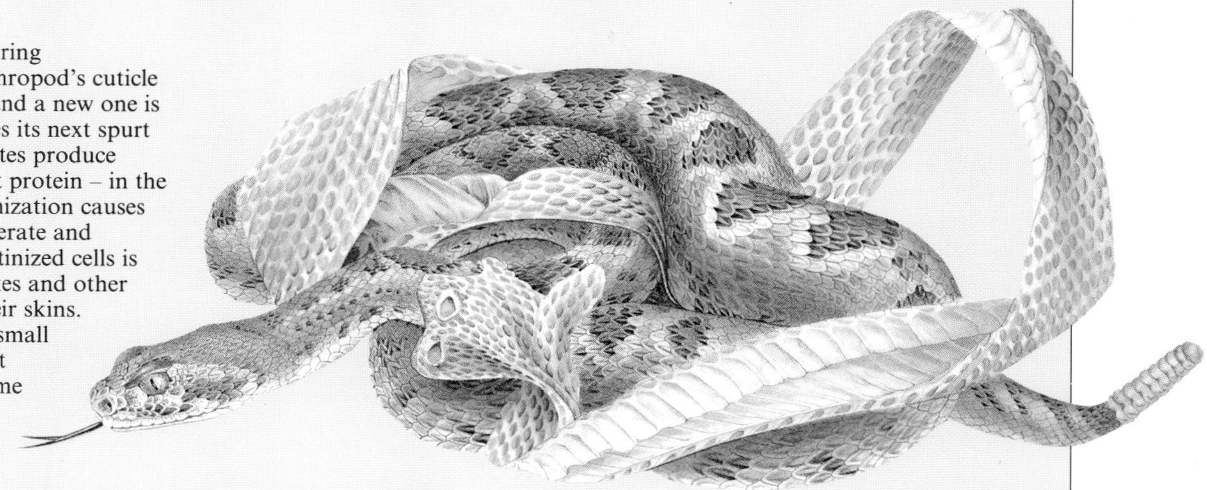

***The Mexican axolotl** is a salamander that should change from a gill to a lung breather. However, it sexually matures without passing the immature gill stage (far left). This condition,* neoteny, *results from change-inhibiting environmental factors, in this case a lack of iodine, which the thyroid gland needs to produce thyroxine, the trigger for metamorphosis. If iodine levels increase, the axolotl matures (left), though very few achieve this.*

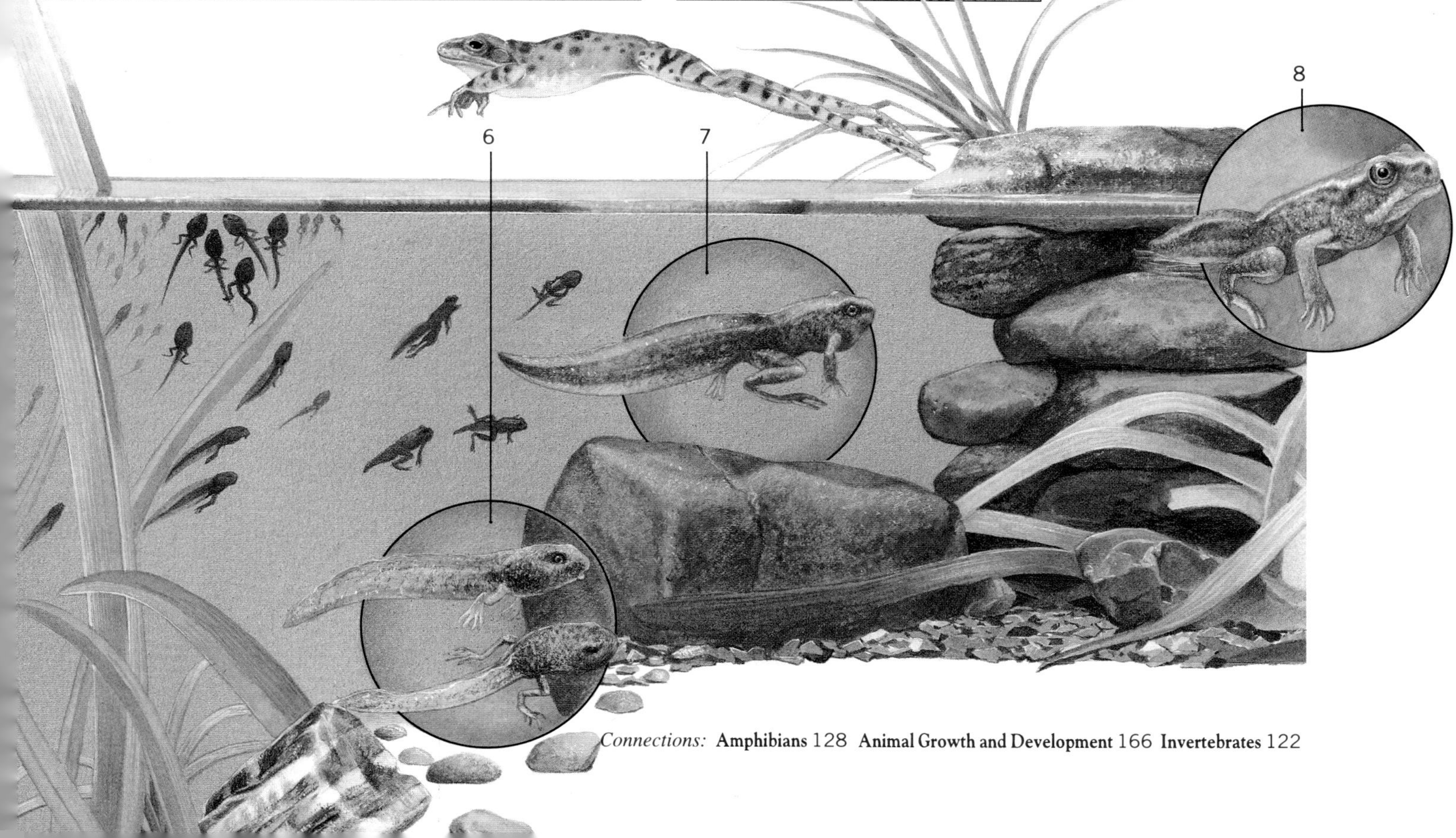

Connections: **Amphibians** 128 **Animal Growth and Development** 166 **Invertebrates** 122

Hatching Out

How reptiles reproduce

Crocodiles, turtles and tortoises can "choose" their offspring's sex by governing the temperature at which their eggs incubate. The waterproof eggs of early reptiles 260 million years ago allowed them to flourish in habitats too dry for most amphibians. As the eggs are so hardy, most modern reptiles hide or bury them and then leave the eggs to hatch and the young to fend for themselves, but there are exceptions. For example, the Great Plains skink not only tends its eggs but also helps the hatchlings to break out of their fetal sacs, and protects and cleans them for ten more days.

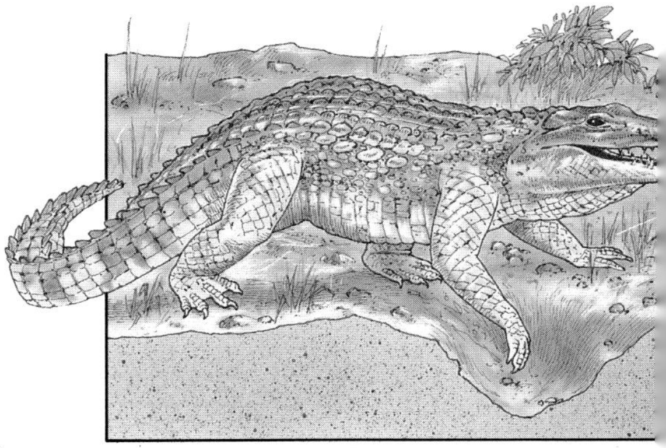

Many reptiles engage in complex behavioral patterns before and during mating. Around this time secondary sexual characteristics distinguishing males and females may also become apparent. In lizards, for example, the males quite often become brightly colored and adopt movements and attitudes that display their temporary splendor. Some merely bob their heads up and down, but others have inflatable pouches, particularly around the throat. A few secrete glandular substances from special openings on the body: these substances are thought to be attractive to members of the opposite sex.

A few snakes also secrete glandular substances to attract mates, but in general snakes live fairly solitary lives and mating often occurs only by coincidence when a male and female meet at the right time; courtship in all species is complex and ritualized. The male snake usually weaves himself against the female, rubbing her chin. In a few species the male and female coil round each other; even more rarely they rear up vertically. Mating is achieved when the male snake inserts one of a pair of organs (*hemipenes*) into the female's *cloaca*, through which sperm flow to fertilize the eggs.

The preliminaries to mating are even more elaborate for tortoises and turtles. The male tortoise may roar and lunge at the female, often snapping at her limbs to force her to withdraw them under her shell; a male turtle may stretch his head out to the female and, while he swims backward in front of her, stroke her cheeks with his forelimbs.

Stroking and rubbing also play an important part in the foreplay of crocodiles and alligators preparatory to mating. The process of mating itself is more violent, as both partners thrash and coil around each other in the water until – sometimes after as long as ten minutes – the male mounts the back of the female so that mating may take place.

Eggs and young

Most lizards lay eggs, although a small number, including a few species of skinks and geckos, give birth to live young. For most of those lizards it is usually a matter of the mother retaining the eggs, which are enclosed only by a thin, membranelike shell, inside her until they have hatched, before "giving birth" (*ovoviviparity*). Most, however, bury

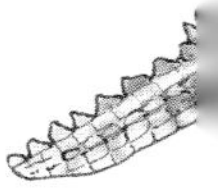

***A female Nile crocodile** digs a hole about 16 in (40 cm) deep in which to lay her eggs* [**A**]. *She picks a site near the water but also near shade, from under which she can watch her young. As her eggs or hatchlings are prey to mongooses, hyenas, large wetland birds and various lizards, she lays up to 50 eggs* [**B**] *to ensure some survive. Having covered the eggs with vegetation and earth, she guards them for up to 90 days* [**C**] *until high-pitched noises from the eggs stimulate her to dig them out. The hatchlings use special egg teeth to break out of their shells* [**D**]. *Gathering the hatchlings into her mouth* [**E**], *the mother releases them in the river to clean themselves. The mother tends her young for about eight weeks, locating lost young by their distress calls; if endangered, she takes them into the safety of her jaws.*

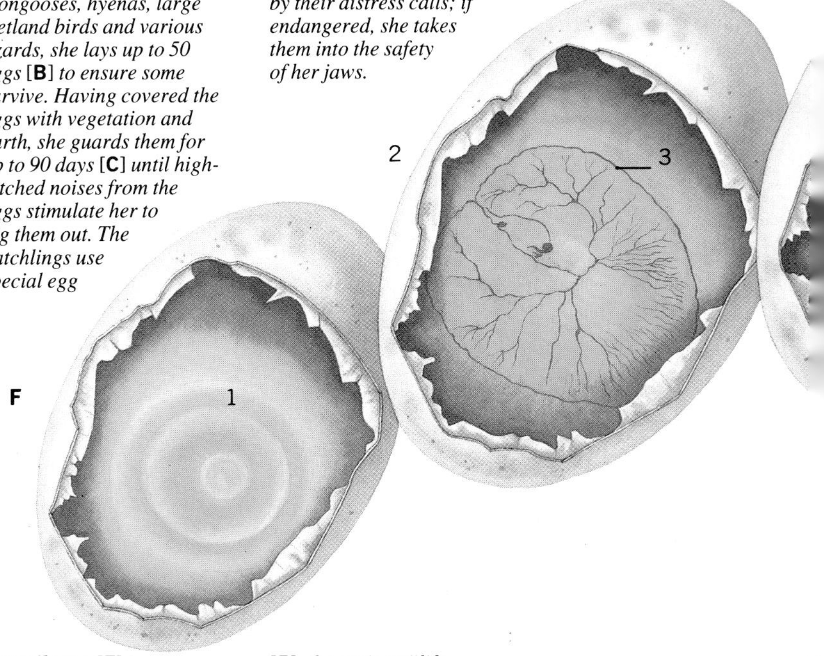

***A reptile egg** [**F**] is a waterproof, self-sufficient entity that lacks nothing apart from an external heat source to keep it at the correct temperature for hatching: 81–95°F (27–35°C). A day or two after laying, a growing disk of cells absorbs nutrient from the rich store of starches, sugars, fats and proteins in the yolk* [1]. *As the early embryo develops* [2], *blood vessels* [3] *begin to cover the yolk surface; these extract nutrient from the yolk to feed the embryo. By the time an extensive blood vessel network develops, the embryo's internal organs are taking form* [4]. *When the embryo is half-developed* [5], *the various "life-support" systems of the egg are clearly differentiated. Between the shell and the embryo are three membranes. The innermost is the* amniotic sac [6], *a fluid-filled shock-absorbing sac surrounding the embryo, which connects to the yolk via the umbilical cord* [7]. *The yolk sac and amnion are enclosed in the* allantois [8], *which collects waste products and separates them from the embryo. It also acts as the lung of the egg, transferring oxygen from the* chorion [9], *just under the shell, which has blood vessels filled with hemoglobin molecules to absorb oxygen.*

Connections: **Amphibian Reproduction** 164 **Animal Conservation** 344 **Bird Reproduction** 174 **Early Evolution** 92 **Reptiles** 130

B

C

D

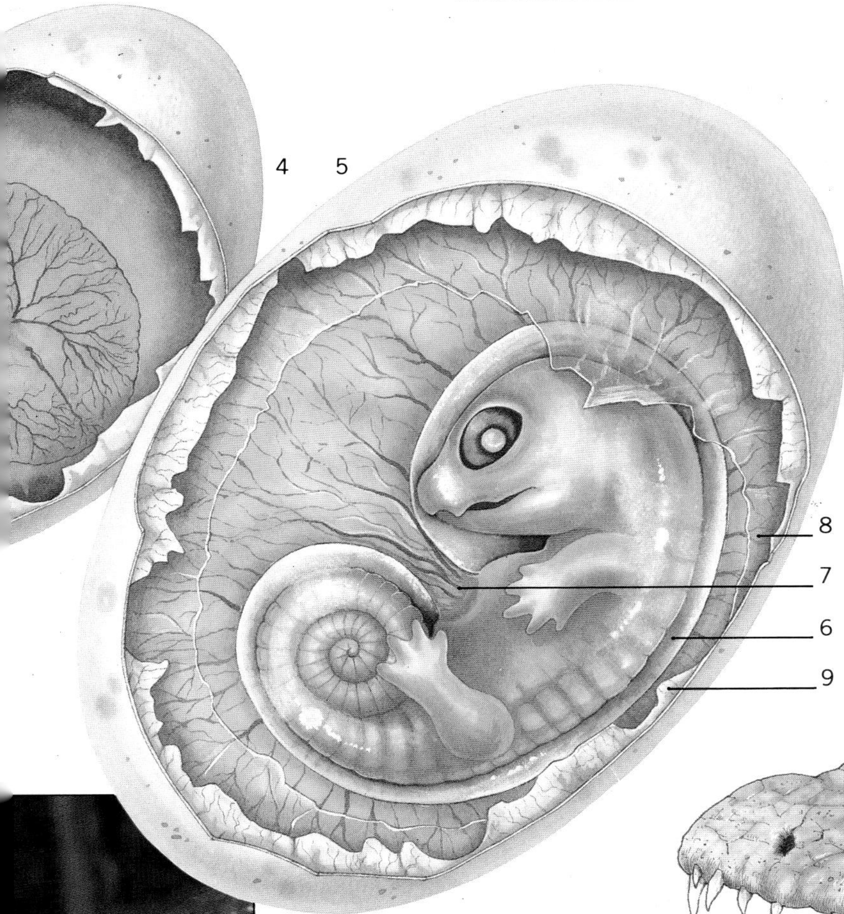

their eggs, protected by a leathery shell, for the Sun to incubate through the soil.

Similarly, most snakes simply lay their eggs and abandon them. But a minority of snakes, like the lizards, give birth to live young, either by hatching eggs inside the mother or, much more rarely, after direct nourishment of the growing young from the mother's bloodstream via the *oviduct*, which functions similarly to the placenta of the higher mammals (this is called *viviparity*). Live births, by whichever method, are less dependent on the temperature of their environments. Yet there are also snakes that show some signs of parental care for their eggs by, for example, coiling their bodies around them.

Both tortoises and turtles lay their eggs in sunny places on land, buried in large numbers in a nest that is then concealed. The mothers then display no further interest in either their eggs or their hatchlings.

E

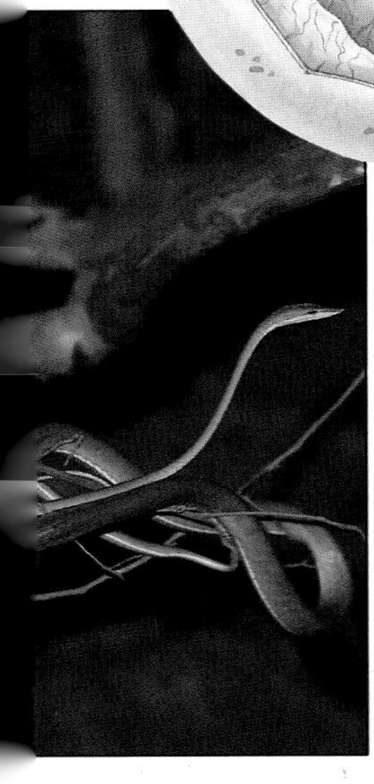

***Vine snakes** copulate en masse (left); like rattlesnakes and several other species of snake they hibernate in colonies and awaken at the same time in spring, after which group copulation occurs, with many males attempting to mate with a small number of females; this strategy helps ensure fertilization and the production of offspring.*

The Mating Ritual

How birds court their mates

The male East African widowbird has an enormously long black tail, which makes him visible to potential mates from over a half a mile away. Other birds, such as the blue bird of paradise, have evolved spectacularly colored feathers to attract their mates. Further forms of elaborate courtship display involve songs, aerial maneuvers and exhibiting prey. Courtship serves to cement the bond between the sexes, to coordinate the vital transfer of sperm from male to female, and to synchronize the behavior of the parent birds for the breeding season.

There are four main types of mating systems shown by birds, and of the 8,600 known species about 90 percent are seasonally *monogamous* – whereby an individual mates with only one partner each breeding season, but spends the rest of the year living an independent life. The next most common form is *promiscuity*, in which the pairing is casual. About 6 percent of birds, including many grouse and hummingbirds, show this form of sexual behavior. Other promiscuous species include the pectoral sandpiper, ruff, some species of pheasants, lyrebirds, manakins and most of the birds of paradise.

Rarer forms of mating systems include *polygyny* (about 2 percent), in which a male pairs with two or more females, and *polyandry* (about 1 percent), in which a female pairs with more than one male. Polygynous species, such as the savanna weaver and red bishop, tend to live in large groups in tropical areas, often occupying marshy habitats, whereas polyandry is found mostly in certain waders – notably the phalaropes, painted snipes, some sandpipers, such as the spotted sandpiper, and also in jacanas.

Display arenas

Certain species of birds tend to gather together in one spot to perform their courtship displays. The area where they gather is called a *lek*. Lek systems occur amongst promiscuous birds, and well-known lekking species are the black grouse, sage grouse, ruff, superb lyrebird and the birds of paradise, but many other birds also use leks, including some manakins and hummingbirds.

At a lek, male birds strut, clamber, or fly about displaying their often elaborate plumage, and enter into mainly visual competition for the attention of the drabber females, which are attracted to the performance. Some species of birds, such as the ruff, will also enter into mock combat displays. In this way the order of dominance can be decided within the group without the danger of serious injury. The successful males then go on to mate with the females. In black grouse leks, fewer than 10 percent of the males will be responsible for up to 80 percent of all the matings. Females tend therefore to mate with the most dominant males, thus ensuring that genes for these characteristics are passed onto their offspring.

***The great crested grebe* [A]** *has evolved an elaborate series of courtship displays, often called the "dance of the grebes." Many of the steps in the dance are ritualized versions of normal behavior, and serve to reinforce pair bonding. Great crested grebes are monogamous and the continued male presence after mating is essential if eggs are to hatch. Pairings occur toward the end of the winter, when sexually mature grebes gather in groups of about 100. There is little sexual dimorphism, and both males and females develop a brightly colored display plumage with distinctive crest and frill. Initially the pairings are fairly unstable and the dance rarely progresses beyond the "headshaking" display* [1], *in which prospective partners face each other and move their heads rapidly from side to side. Headshaking often ends with ritual preening* [2]. *As the breeding season advances, the dance reaches its greatest intensity in mated pairs. During the "discovery" display* [3] *one bird spreads its wings while the other dives and surfaces with its beak pointing downward. In the "retreat" display* [4] *one partner suddenly turns and dashes across the water away from its mate. In the "weed" dance* [5] *the pair rear up from the water pressed against each other.*

The House Builders

The 18 species of bowerbird are medium-sized birds found in rain forest and other woodland in northeastern Australia and New Guinea. They are unusual because the courtship rituals of many bowerbirds involve displays not only of their plumage but also of carefully built nestlike structures known as bowers.

The different species construct a whole variety of bowers, ranging from a simple display of leaves in a woodland clearing to complicated nestlike structures involving avenues and platforms. In general, those with the least ornate plumage make the most elaborate bowers. Perhaps the finest is that of the Vogelkop gardener bowerbird, a tent of sticks overarching a mossy stage, on which the bird displays colorful plants.

B

A

Connections: **Bird Reproduction** 174 **Birds** 132

The most colorful displays *are found in tropical rain forests. The elongated tail feathers of the male king bird of paradise [**F**] serve to attract a solitary female. The spectacular tail feather display of the male common peafowl [**G**] ensures that a successful male mates with a small harem of females.*

Birds of prey, *among other groups, typically have energetic courtship displays that exploit their aerial agility. The hen harrier [**B**] flies in a series of regular undulations. At great heights it glides with its wings at an angle, before tumbling, rising up and down, and occasionally looping the loop. The male may also pass food to the female on the wing. The display of tawny eagles [**C**] is a more extreme form of undulation flying, sometimes called "pothooks," in which the bird dives and swoops. Other species engage in solo flights. The Verreaux eagle [**D**] repeatedly flies in tight figure eight loops, this display is sometimes called a "pendulum." Pairs of African fish eagles [**E**] grapple talons in mid-air, tumbling and cartwheeling across the sky.*

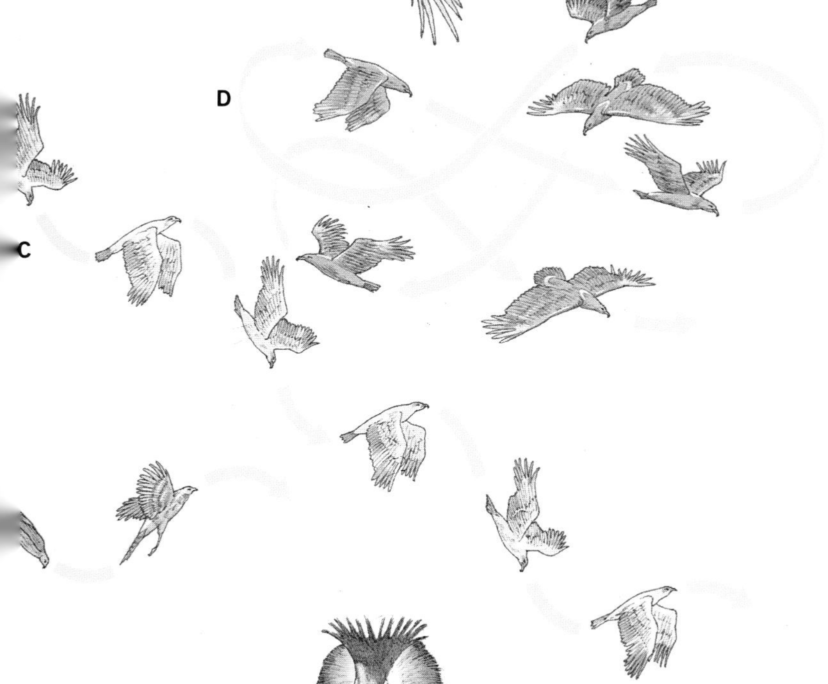

Female choice

Sexual selection is an evolutionary process seen in most polygamous species. The females' choice of mate leads to the elaboration of the males' plumage over time. With few exceptions, such as the phalaropes, it is the males that evolve the brighter plumage and the elaborate sexual courtship displays. Females tend to be more cryptically patterned to help them hide while on the nest. However, it is not always birds' plumages that are spectacularly colored. Puffins, for example, have brilliantly colored beaks that serve as their courtship display. Interestingly, their black, red and yellow beaks, colors which normally serve as nature's warning colors, have exactly the opposite effect for courting puffins.

The Chicken and the Egg

How birds reproduce

The common swift manages to mate while on the wing, but reproduction in many birds is inelegant and uncertain. One breed of penguin cannot, at first, even tell whether its prospective mate is male or female. The male roadrunner must bribe a would-be mate with food. All birds reproduce by laying hard-shelled eggs – sturdy life capsules for developing chicks. Laying eggs allows adult birds to follow an active, flying life-style throughout their reproductive cycle without having to support the extra weight of young developing inside their bodies.

Copulation in birds is very brief, the mating pair bringing their sex organs (*cloacae*) into contact in a rather clumsy-looking pairing session. However, this is normally sufficient for sperm transfer and, once inside the female, sperm swim to the upper end of the oviduct, where fertilization occurs, usually within a few days of mating. In some birds, including most waterfowl, hens, storks and the ostrich, the male cloaca is modified into an erectile penis.

After fertilization the eggs pass slowly down the oviduct and are laid one at a time in the nest. The hard, chalky shell of the bird's egg protects the embryo from microorganisms or invertebrates, while at the same time allowing gas exchange for respiration.

Incubation

Most birds begin to incubate their eggs only when the full clutch has been laid, with the effect that all the chicks hatch at about the same time. However, in birds of prey and owls incubation starts as soon as the first egg is laid, so that the young then hatch sequentially. The first nestling to hatch often dominates and is fed in preference to the younger, weaker ones. If prey is plentiful, more than one chick may survive, but if prey is scarce, the younger chicks usually die, or are killed by their siblings.

Birds incubate their eggs by using a special organ, the *brood patch*; an area of loose, bare skin that develops on the breast or belly. The brood patch is well supplied with small, muscular arteries that control the blood flow and therefore the temperature of the skin. Penguins, waterfowl and *pelecaniform* birds (pelicans, boobies, cormorants and relatives) lack a brood patch. This is why ducks pluck down from their undersides and incubate their eggs between the down and the bare skin. Penguins and pelecaniform birds use a different tactic of carrying their eggs on their feet to incubate them.

Hatching the eggs

For successful development and hatching, birds must keep their eggs at around 99.5°F (37.5°C). In the tropics and in desert habitats eggs must be saved from overheating. Birds achieve this by shielding them with their own bodies, or sometimes (as in some terns, gulls and shorebirds) by wetting them, usually with water transferred from their breast feathers. Brooding birds regularly turn their eggs to even out the temperature of the clutch. Incubation periods vary from between about 10 days (in some woodpeckers) to about 80 days in albatrosses and kiwis.

Waterfowl and gamebirds lay the largest clutches, of up to 20 eggs. Pelicans and shearwaters and their relatives lay just a single egg. Vultures and condors lay one or two eggs, and doves and hummingbirds invariably lay two. When young birds hatch, they may be naked and almost completely helpless, and need protecting and feeding within the nest. These are known as *altricial* young. Chicks that hatch fully feathered, able to run, and peck at food for themselves within a short time of hatching, are called *precocial*. Waterfowl, shorebirds and gamebirds tend to produce precocial young, whereas most perching birds have altricial young.

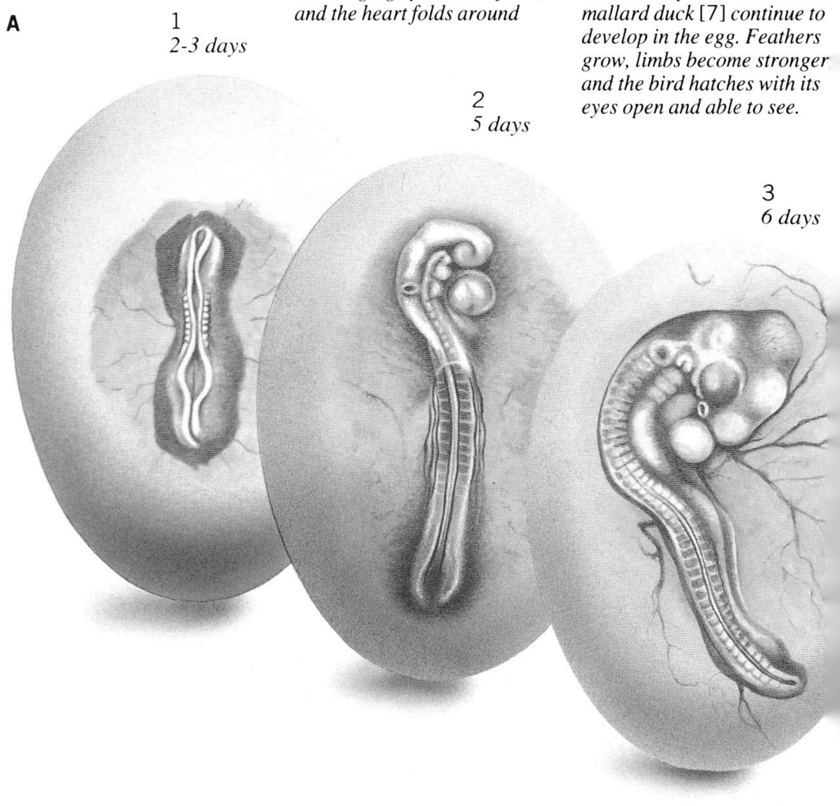

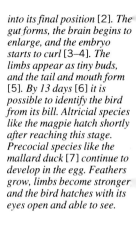

***A duck embryo* [A]** *grows from a patch of cells on the surface of the egg yolk. The yolk is its food store. First a network of tiny blood vessels spreads over the yolk and a simple heart develops. The developing embryo (enlarged here for clarity) begins to elongate and develops a vertebral column* [1]. *A head and bulging eye start to form, and the heart folds around into its final position* [2]. *The gut forms, the brain begins to enlarge, and the embryo starts to curl* [3–4]. *The limbs appear as tiny buds, and the tail and mouth form* [5]. *By 13 days* [6] *it is possible to identify the bird from its bill. Altricial species like the magpie hatch shortly after reaching this stage. Precocial species like the mallard duck* [7] *continue to develop in the egg. Feathers grow, limbs become stronger and the bird hatches with its eyes open and able to see.*

Connections: **Birds** 132 **Bird Courtship** 172 **Birds' Nests** 238 **Animal Growth and Development** 166

A hen's fertilized ovum [**B**] *moves from the ovary* [1] *into the enlarged head of the oviduct (the* infundibulum*)* [2]. *First it passes along the main section of oviduct, or* magnum [3], *whose walls add layers of egg white, or* albumen. *After about 3 or 4 hours it enters another section, called the* isthmus [4], *where the egg and shell membranes are deposited. The shell itself forms over the egg in a wider section of the oviduct just beyond the isthmus. This is called the* uterus [5], *or shell gland, and the process of shell formation takes about 20 hours. The egg is laid, through the cloaca* [6], *by contractions of the vagina.*

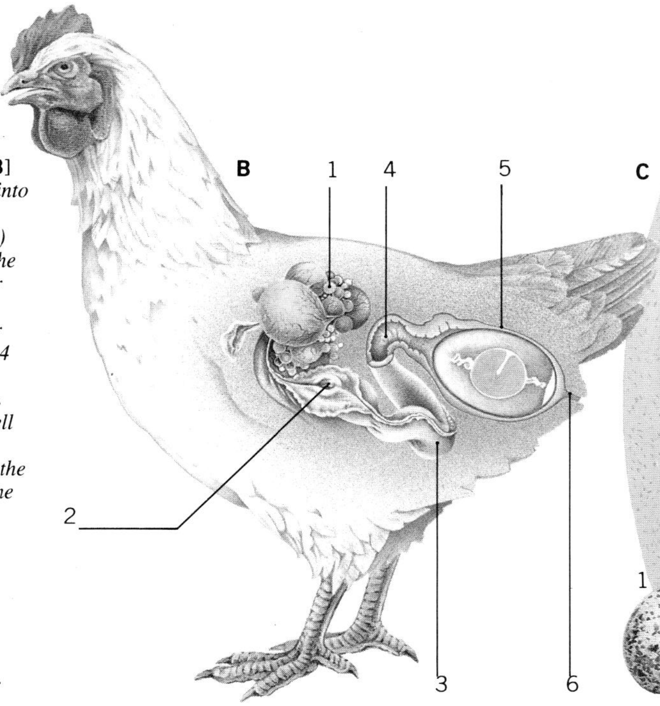

Egg sizes [**C**], *shapes and colorings vary widely. Many, like those of the wood warbler* [1], *are flecked to provide a rudimentary camouflage. Others, like those of the dunnock* [2], *are bright and conspicuous. Larger perching birds, like the blackbird, lay correspondingly larger eggs* [3]. *Eggs laid in safe places, such as holes, can be more spherical, like the tawny owl's* [4]. *Many birds of prey, such as the Egyptian vulture* [5], *lay eggs with dark, reddish-brown markings. Both emus* [6] *and ostriches* [7] *lay large eggs with heavy shells up to 0.08 in (2 mm) thick. The emu's egg changes color after it is laid from an initial dull green to a black sheen over a period of several days. The ostrich lays by far the largest egg of any living bird, weighing up to 3.3 lb (1.5 kg).*

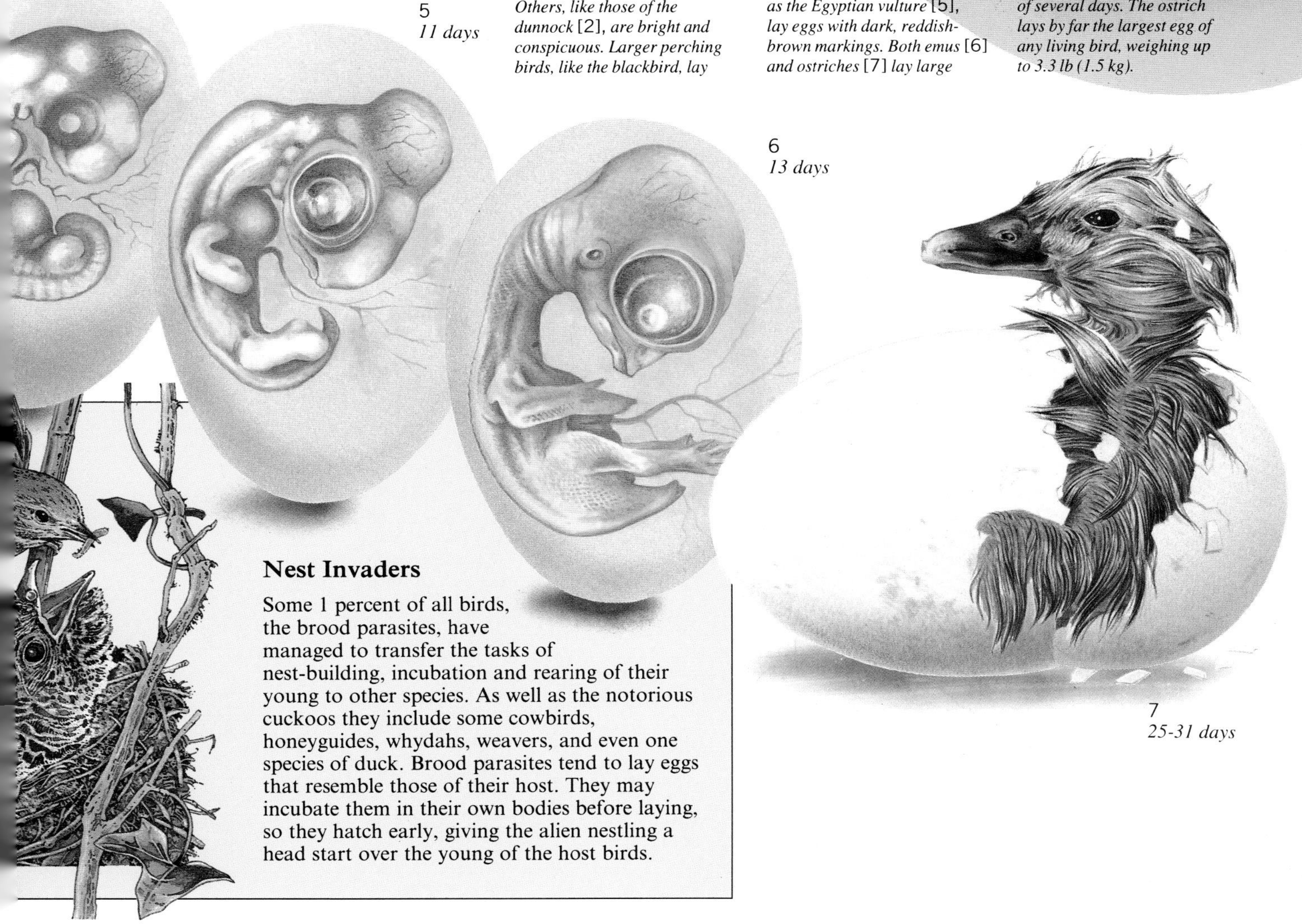

Nest Invaders

Some 1 percent of all birds, the brood parasites, have managed to transfer the tasks of nest-building, incubation and rearing of their young to other species. As well as the notorious cuckoos they include some cowbirds, honeyguides, whydahs, weavers, and even one species of duck. Brood parasites tend to lay eggs that resemble those of their host. They may incubate them in their own bodies before laying, so they hatch early, giving the alien nestling a head start over the young of the host birds.

Out of the Womb

How placental mammals reproduce

During its last two months in the womb the blue whale fetus increases its weight by 2 tons, at a rate of about 225 pounds (100 kilograms)/day. It is this ability to develop and grow in conditions of absolute security and comfort that characterizes the so-called placental mammals and explains their success. Blue whales are huge, and are born at a rate of one offspring per successful mating. Other mammals prefer to gamble on producing many young in the hope that some will survive. The tailless tenrec of Madagascar can produce 30 young in one go.

The placental mammals evolved about 90 million years ago. They are the most recent, and numerous, of the three major groups that make up the 4,000 or so living species of mammal. In this group the developing young, or fetus, is attached by means of a placenta to its mother's womb. The placenta is a remarkable structure in which the blood vessels of the fetus and the mother mingle so closely that food and oxygen can pass from the mother to the fetus, and waste and carbon dioxide can pass from fetus to mother for elimination without any interchange of blood taking place.

Powers of attraction

The first step in reproduction is to find a mate. Scent is the most common form of sexual attraction, but auditory and visual cues are also used. Many species gather in large groups during the breeding season. Whales, seals and sea lions migrate long distances to favored breeding beaches, while moose, which normally live in separate male and female groups, come together in the autumn. Such gatherings help to maintain the variation in the population, as animals from different ancestries mingle and mate.

Courtship is often essential to bring an animal into a state of receptivity for mating. Many male antelopes set about an elaborate courtship, in which they follow the female around stroking her with their legs and nudging her gently. But some harem-owning mammals, such as male elephant seals, forcibly impregnate the females. Species that live in large groups tend to be promiscuous. Often the males hold large harems, fighting each other for possession of the females. Sometimes only the very largest and most dominant males succeed in mating.

Other species form temporary or permanent pair bonds. This is particularly true of carnivores such as jackals, where the hunting skills of both parents may be needed to provide sufficient food to rear a family.

Cycles of reproduction

The timing of reproduction can decide between the death or survival of the young. Breeding in herbivores is usually timed so that the young are born in spring, when vegetation is in its first flush, while the breeding cycles of carnivores are often related to those of their prey. The young will also grow quicker if the temperature is warm, as they will have to expend less energy in shivering, and the mother will be able to devote more energy to milk production. In the moist tropics, where food is plentiful all year round, there are usually no marked breeding cycles.

In most mammals the females are receptive for mating only at certain times of year, or at certain regular intervals. Often the males undergo a similar fluctuation in sexual response. These reproductive cycles rely on an internal biological clock, which determines periods in which the animal may become sexually active. The clock causes certain hormones to be secreted that cause development of the sexual organs and promote the desire to mate. Sexual activity itself may be triggered by some environmental signal interacting with the underlying rhythm, such as the increasing day length in spring.

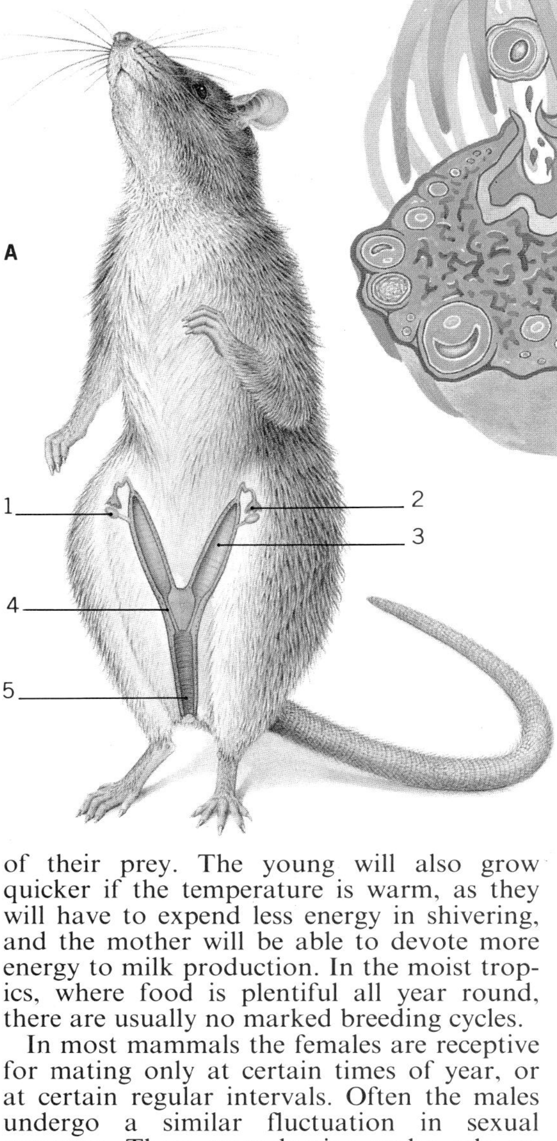

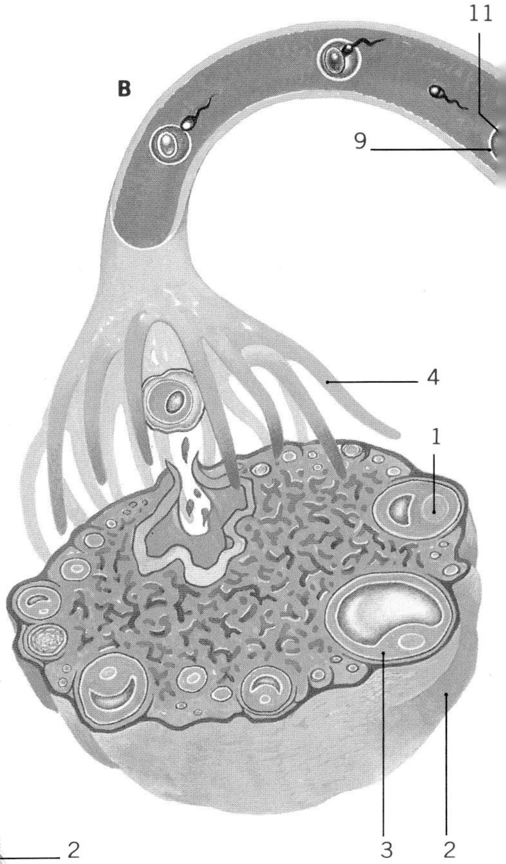

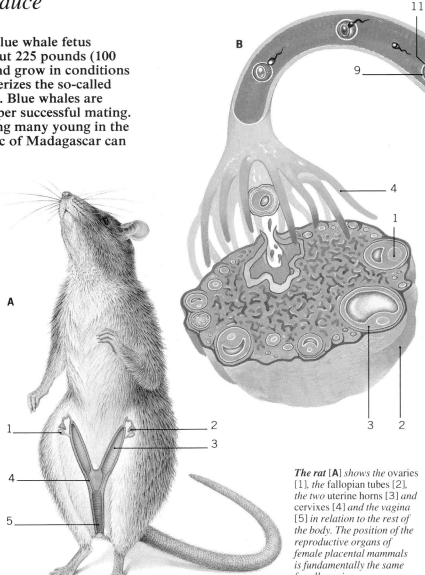

The rat **[A]** *shows the* ovaries [1], *the* fallopian tubes [2], *the two* uterine horns [3] *and* cervixes [4] *and the vagina* [5] *in relation to the rest of the body. The position of the reproductive organs of female placental mammals is fundamentally the same for all species.*

Fertilization **[B]** *begins with* ovulation, *whereby an* ovum *(egg)* [1] *develops in an ovary* [2] *into a* follicle [3]. *The follicle consists of the ovum, a sac of liquid and follicle cells. The pressure in the follicle increases until it bursts, releasing the ovum into the fallopian tube* [4]. *During ovulation the hormone* estrogen *is produced by the collapsed follicle. Estrogen causes the lining of the uterus to thicken and extend its network of blood vessels, from which the fertilized ovum will be nourished. Fertilization occurs when millions of sperm* [5] *are ejaculated from the male's penis during copulation. The sperms use their long tails or* flagella [6], *powered by*

Connections: Animal Cells 102 DNA 96 Heredity 98 Mammals 136 Mammal Societies 298 Primitive Mammal Reproduction 178

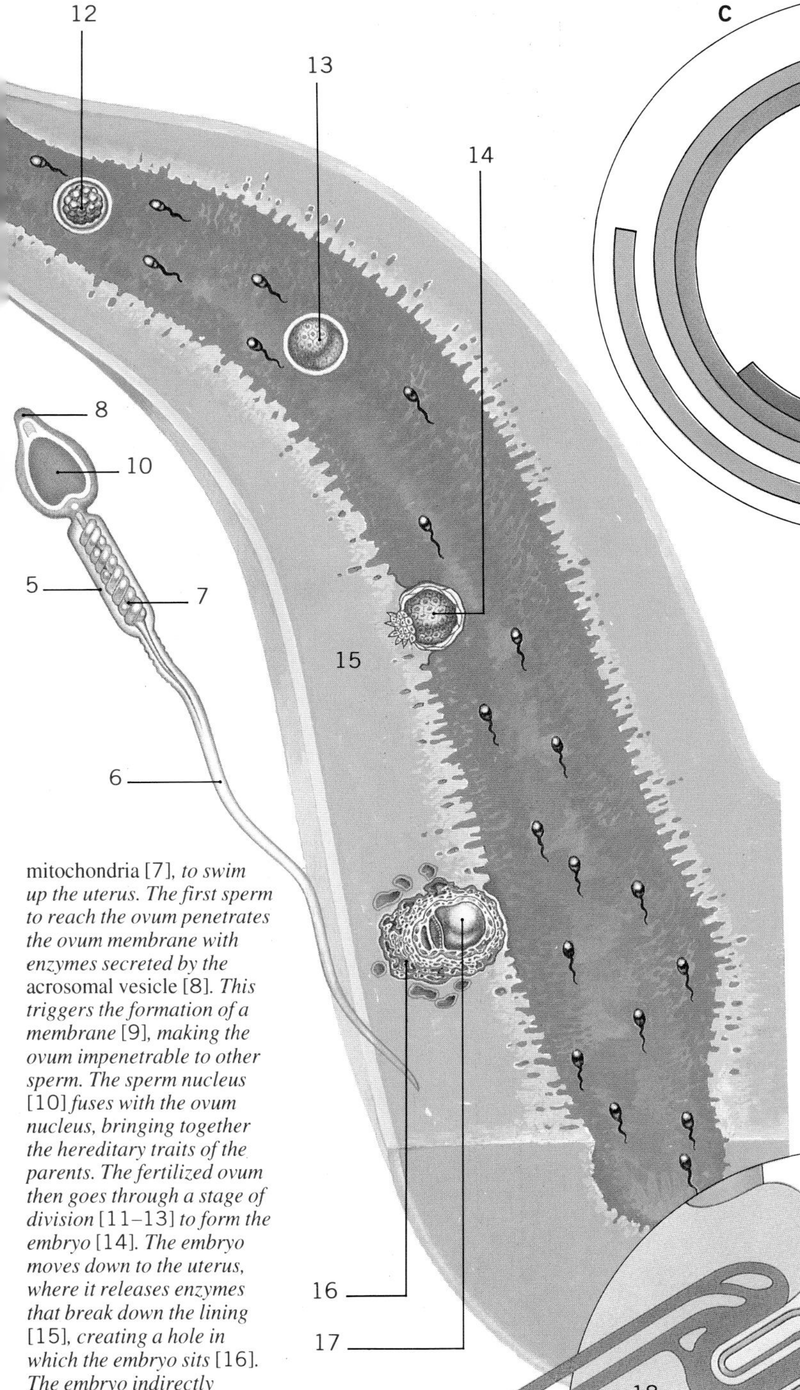

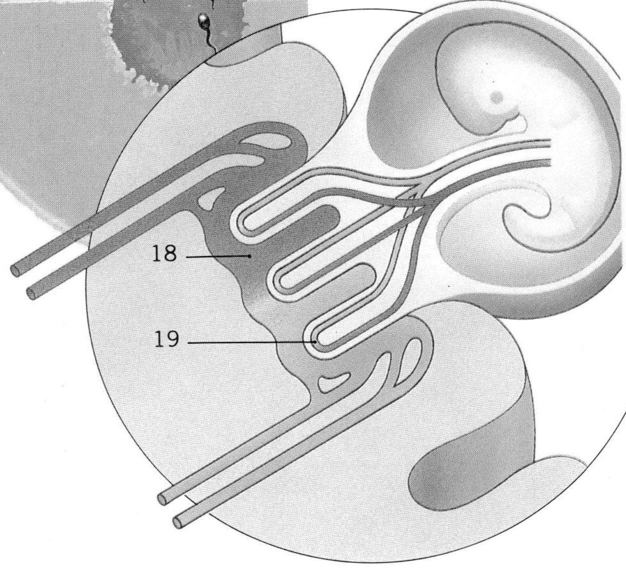

mitochondria [7], *to swim up the uterus. The first sperm to reach the ovum penetrates the ovum membrane with enzymes secreted by the* acrosomal vesicle [8]. *This triggers the formation of a membrane* [9], *making the ovum impenetrable to other sperm. The sperm nucleus* [10] *fuses with the ovum nucleus, bringing together the hereditary traits of the parents. The fertilized ovum then goes through a stage of division* [11–13] *to form the embryo* [14]. *The embryo moves down to the uterus, where it releases enzymes that break down the lining* [15], *creating a hole in which the embryo sits* [16]. *The embryo indirectly obtains its oxygen and food from the mother's blood as it flows through the uterus. The embryo* [17] *develops an organ, the* placenta [18], *which is comprised of millions of tiny appendages called* villi [19]. *The oxygen and food are absorbed from the mother's blood, via capillaries in the villi, into the embryo's blood.*

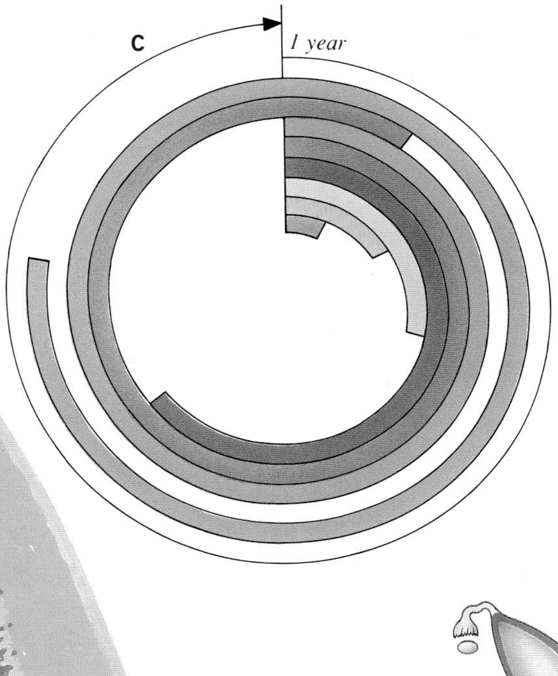

***The gestation periods** of mammals* [**C**] *vary greatly: from the rat, which has a 23–25-day gestation period, to the elephant, at 650–660 days. The deciding factors of the length of gestation are principally genetic and the animal's structural complexity, but the species' environments are relevant.*

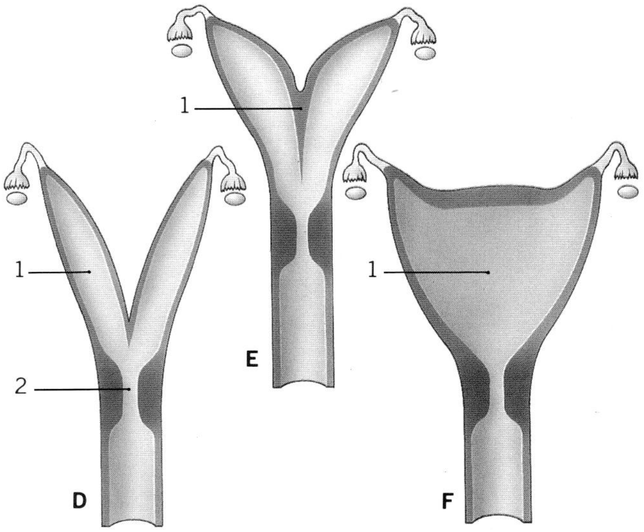

***Among the 4,000** or so species of placental mammals there are four basic types of uteri. The rat* [**A**] *has a* duplex uterus *composed of two cervixes, separate uterine horns, but no uterine body. Pigs and insectivores have a* bicornuate uterus [**D**], *which, like the rat's, has two uterine horns* [1] *but has only one cervix* [2]. *The* bipartite uterus [**E**] *is found in dogs, cats and cows. It has a more prominent uterine body but the horns are separated by a* septum [1]. *Primates, such as human beings, have a* simplex uterus [**F**]. *This type of uterus has no distinctive horns but a prominent uterine body* [1].

Eggs and Pouches

How primitive mammals reproduce

The platypus is one of Earth's oddest animals. It has a body like an otter but a bill and feet like a duck. The bill has an electric sensor for hunting and male hind feet bear poisonous spurs. Also, male testes are enclosed in the abdomen, a retained reptilian trait, and the female does not bear live young but lays eggs. The platypus is a rare surviving monotreme, or egg-laying mammal. The other primitive mammals are marsupials, which bear live young but have no placenta, preventing full embryonic development; instead they bear live young that complete fetal development in a pouch.

Although the monotremes and the marsupials are considered to be the most primitive mammals living today, it should not be concluded that they are the survivors of groups that were ancestral to the more advanced placental mammals. The reptiles gave rise to several lines of mammals, not all of which have survived. Monotremes are found only in Australasia; marsupials exist in South America, but far more variety exists in Australasia as a result of 130 million years of evolution without placental mammalian competition.

Platypus reproduction

The monotremes comprise the five species of echidna, or spiny anteater, as well as the platypus. The platypus is an aquatic mammal that lives along river banks, where it hunts underwater for invertebrates. About a month after mating the female lays two eggs in a nest at the end of a special breeding tunnel in the bank, out of sight of predators. She probably lies on her back, curling her tail to form a cup near the base of her belly into which the eggs roll. Platypus eggs are sticky and probably cling to her fur, as well as to each other. The female curls herself around the eggs to incubate them. She may remain in the burrow without feeding for up to 14 days until the eggs hatch. The baby platypuses, barely 0.6 inches (1.5 centimeters) long on hatching, have another feature unique to mammals but common in birds and reptiles: a temporary egg tooth with which to cut their way out of the egg. A few days after the eggs hatch milk begins to ooze from the mother's belly. Platypuses have no proper teats – the ducts from the mammary glands open into two longitudinal folds from which the young lap up milk. Their forelimbs are relatively large for clinging to the fur around the milk-secreting glands.

Parental pouches

The monotremes lay leathery eggs that resemble those of reptiles. The marsupials, on the other hand, give birth to extremely small, poorly developed young. They are named for the marsupium, or pouch, on their belly, which contains from 4 to 30 teats (according to the species). The pouch provides protection for the early days of the young's life. Not all pouches totally enclose the developing young. In some primitive opossums the young simply dangle from the teats, or cling to the fur of their mother's belly. Wombats and bandicoots and marsupial moles, which burrow for food, have large pouches that open to the rear of the animals to prevent dirt getting in. Kangaroos have front-opening pouches to prevent the young falling out as the animals hop along.

The early marsupial embryo relies on a yolk sac for nutriment, and on a substance called uterine milk that is secreted by glands in the wall of the womb. The yolk sac forms a series of extensions that act rather like a primitive placenta. Nutrients, waste and respiratory gases are exchanged between the maternal and fetal blood, but this exchange is only limited, since there is no close contact between the blood vessels as there is in the placenta of more advanced mammals. Only in marsupial martens and bandicoots does a more advanced placenta develop.

Red kangaroos *[**A**] tend to live in groups and solitary animals are rare. Most male kangaroos are promiscuous, with the strongest and largest one dominating the herd, monopolizing the females and fathering most of the young. Inevitably other males challenge for supremacy, engaging in ritual "sparring" matches that can often be prefaced by a stiff-legged walk as the two combatants size each other up [1]. Before the actual fight the males face each other and groom and scratch themselves [2]. The two kangaroos engage by locking forelimbs [3], with each of the opponents struggling to push and kick over the other [4].*

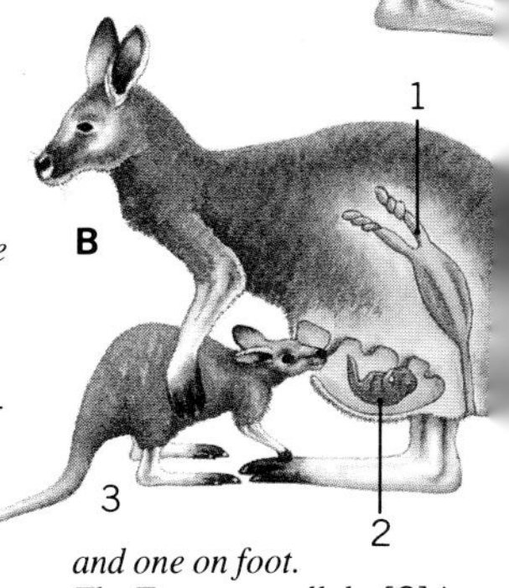

Kangaroos often inhabit *harsh environments, where infant mortalities can be high, especially in droughts. To counteract this, and the unavailability of males for mating, female reproductive cycles [**B**] allow two or three young to simultaneously develop at different stages: a fertilized egg in* diapause *(a state of suspended embryonic development that later continues in "normal" gestation) in the womb [1]; a suckling in the pouch [2], and another outside [3].*

*Kangaroo reproduction varies with species; the spiral diagrams [**C**] show the overlapping reproductive phases of three species. Starting at the top and ending at the inner arrows, each spiral represents a two-year breeding cycle (for reasons of clarity, each spiral is shown with a clear beginning and end, although it is only part of a continuous cycle). Not all species exhibit diapause; some give birth at any time of the year, others are seasonal.*

The Western gray kangaroo [1] breeds seasonally and has no diapause; after 30 days' gestation the "joey" is born, spending the next 320 days in pouch before leaving it. Before this happens, a new gestation begins, timed to finish just after the first joey leaves the pouch, leaving the mother with a joey in pouch and one on foot.

The Tammar wallaby [2] is also seasonal but has a diapause, after which a "normal" 27-day gestation occurs. A new-born joey spends the next 250 days in pouch, the fertilized egg remaining in diapause until the joey leaves the pouch.

The red kangaroo [3] is both nonseasonal and has a diapause. Its joey spends 235 days in pouch, leaving just before the next joey is born. For brief periods this species has a young at foot, one in pouch and a third gestating.

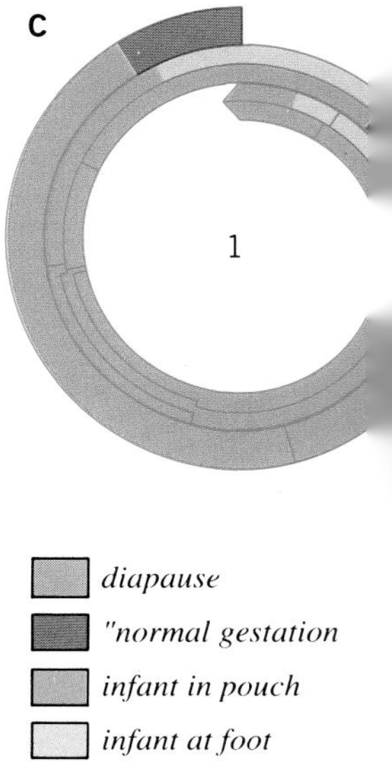

D

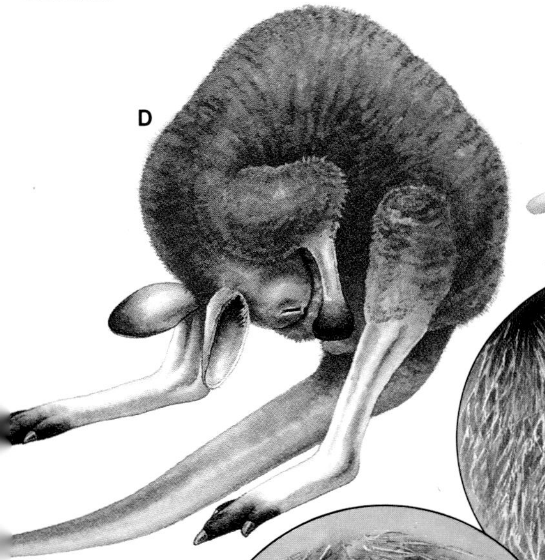

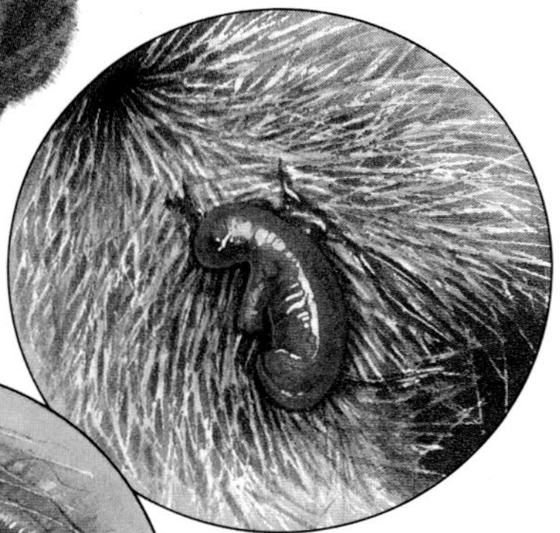

E

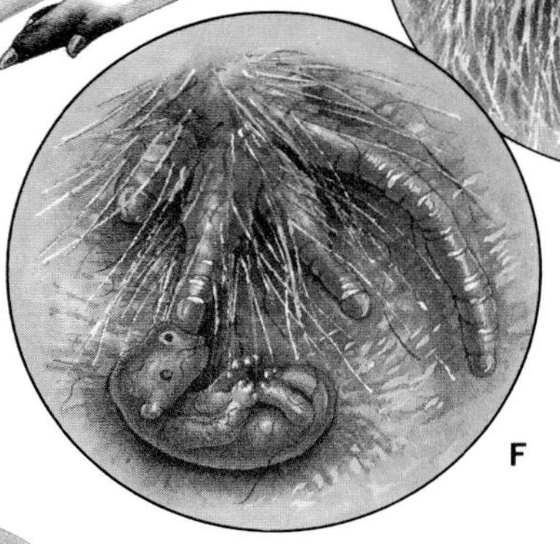

F

***Before giving birth** a kangaroo cleans her pouch for an hour or so by licking its inner lining [**D**]. A newborn joey [**E**] is 0.75 inch (2 cm) long; once free of its birth sac and umbilical cord it claws itself from the cloaca to its mother's pouch, a 6-in (15-cm) journey lasting three minutes. Inside the pouch [**F**] it sucks a nipple, which swells up so that the joey does not slip off during its six-month attachment.*

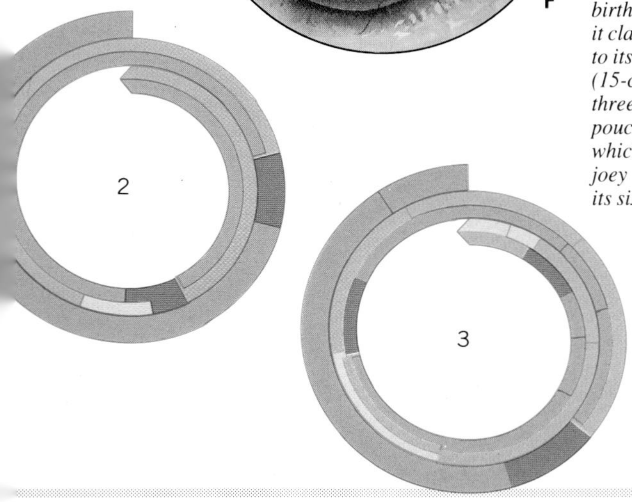

An Egg-laying Mammal

The echidna's primitive origins are shown in its reproductive methods [**G**]; it lays a soft-shelled, reptilelike egg [1] via the cloaca [2], which is then rolled by the hind limbs to a special incubation groove [3]. The minute hatchling is about 1.25 cm (0.5 inch) long. It suckles, but the milk comes from mammary glands that turn inward instead of forming nipples. Thus, young echidnas lick milk that seeps into the fur around the mother's glands.

G

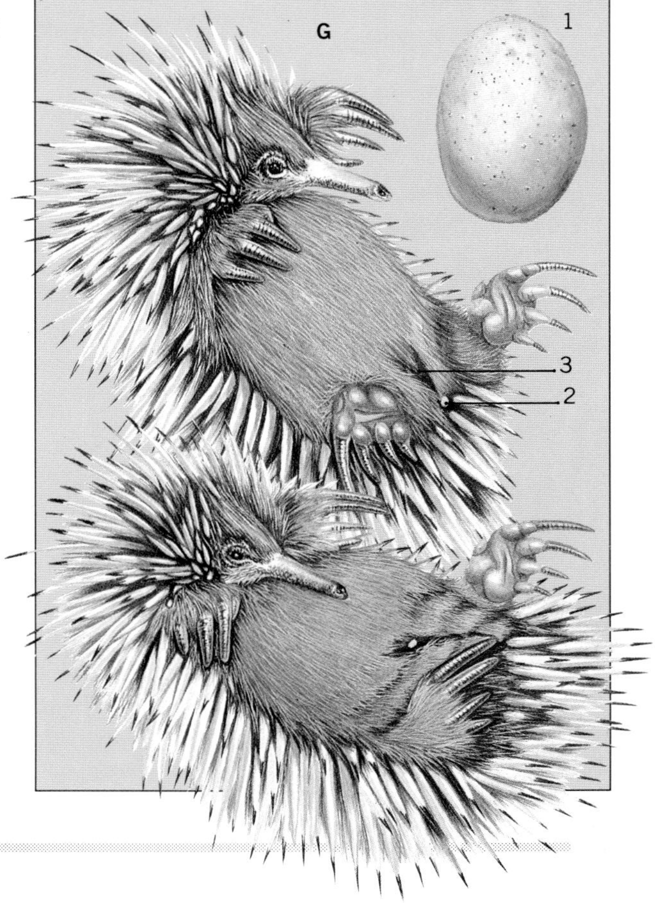

Photosynthesis

Plant Feeding

Parasitic Plants

Animal Parasites

5 The Search for Food

Internal Animal Parasites

Symbiosis

Fluid Feeders

Filter Feeders

Invertebrate Herbivores

Invertebrate Carnivores

Herbivores

Decomposers

Cycles of Life

Making Sunlight into Food

How plants photosynthesize

Virtually all life on Earth depends on the Sun's energy. Green plants, algae and some bacteria are able to convert this energy, through photosynthesis, into chemical energy that they can use. This energy is first stored in the form of simple sugars – and every year more than 150 billion tons of sugars are produced by photosynthesis. The evolution of photosynthesis was also instrumental in creating the Earth's atmosphere, since a by-product of the process is oxygen: in the last 2 billion years the concentration of this gas in the atmosphere has increased fiftyfold.

In green plants and algae photosynthesis takes place in the chloroplasts, miniature solar converters inside the plant's cells. Chloroplasts contain the green pigment chlorophyll, which absorbs light. The energy of light is transferred to the chlorophyll molecule, which then passes it on through a complicated chain of reactions and biochemical processes that finally result in the formation of simple organic (carbon-containing) compounds such as sugars. Carbon dioxide from the air provides the carbon atoms for these organic compounds. At the same time, water is split into its component atoms, producing oxygen gas: oxygen is thus a by-product of photosynthesis. Sugars are used by the plant as fuel for *respiration,* which generates chemical energy in *mitochondria* to power biochemical reactions essential for survival and growth. Respiration also produces carbon dioxide as a waste product, which can then be used again for photosynthesis.

The products of photosynthesis also represent the starting point for the formation of other simple organic molecules. These can then be combined into larger molecules such as proteins, nucleic acids, polysaccharides and lipids, from which all living material is made. Plants generally store food in the form of sucrose, a compound of the sugars glucose and fructose, and starch.

***Carbon dioxide** (CO_2) and water are the inorganic raw materials of photosynthesis. They arrive at a leaf's photosynthesizing cells by different routes* **[A]**: *CO_2 gas simply diffuses in via pores in the leaf* (stomata) [1] *and through the air spaces between the loosely packed cells of the leaf* mesophyll [2]; *water is drawn up from the roots through a system of woody xylem vessels* [3]. *The products of photosynthesis – simple water-soluble sugars – are loaded into phloem sieve tubes and distributed throughout the plant* [4].

Photosynthesis in a plant cell takes place within structures – organelles *– called* chloroplasts **[B]**. *Each chloroplast is bounded by a double membrane* [1] *that encloses a dense fluid known as the* stroma [2]. *A third system of membranes within the chloroplast forms an interconnected set of flat, disklike sacs called* thylakoids [3], *which are frequently stacked on top of one another to form structures called* grana. *The chloroplasts contain photosynthetic pigments, the most important of which is* chlorophyll. *This pigment absorbs light primarily in the blue, violet and red parts of the spectrum. Green light is not absorbed: it is because this light is reflected that leaves appear green. Photosynthesis involves a complex series of chemical reactions: for the sake of convenience these are usually divided into the* light-dependent reactions, *which occur on the thylakoid membrane, and* light-independent reactions, *which take place in the stroma.*

Minimizing energy loss

Energy is lost from the food chain at each upward step, that is when a plant is eaten by a herbivore, or a herbivore by a carnivore. Without photosynthesis to tap the Sun's virtually inexhaustible supply of energy, life would therefore rapidly run down. (The only survivors would be a small group of bacteria that can make use of the chemical energy locked into simple inorganic compounds.)

Photosynthesis is not a particularly efficient way of converting the Sun's energy into food. Only an average of about 1 percent of the light that hits a leaf is absorbed (up to 3 percent in the most productive cases), and even then the maximum possible level of photosynthesis is rarely achieved. Some desert plants, for example, that live in conditions of intense light, high temperature and low humidity, keep their stomata closed during the day to prevent water loss, and are thus unable to take up the carbon dioxide that is available. Many tropical and desert plants have developed successful variations on the usual pathways of photosynthesis in order to deal with such conditions. Net photosynthetic rates of the tropical crops maize, sugarcane and sorghum can be two to three times those of wheat and rice.

The overall efficiency of photosynthesis is also affected by a metabolic peculiarity of many plants. A substantial amount of the photosynthetically fixed carbon is almost immediately converted back into carbon dioxide by the process of *photorespiration,* especially when carbon dioxide levels are low. This apparently wasteful process resembles normal respiration in that it consumes oxygen and produces carbon dioxide, but occurs only in the light, takes place in structures called *peroxisomes*, not in the mitochondria, and does not generate any useful energy. It may, however, be involved in other functions of the plant, such as seed formation.

Connections: **Algae** 112 **Animal Cells** 102 **Atmosphere** 64 **Bacteria** 106 **Breathing** 126 **Cycles of Life** 206 **Early Evolution** 92

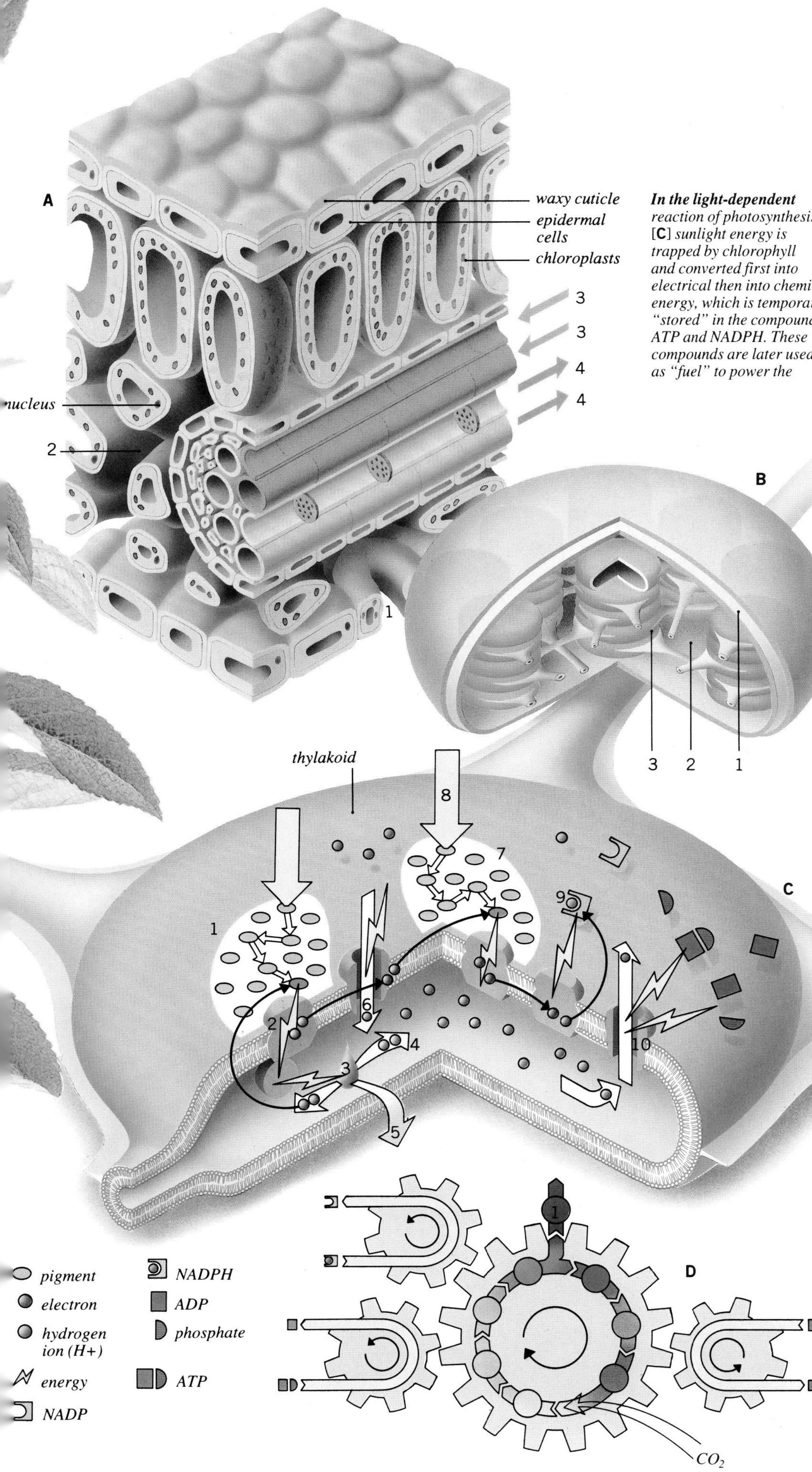

In the light-dependent *reaction of photosynthesis* **[C]** *sunlight energy is trapped by chlorophyll and converted first into electrical then into chemical energy, which is temporarily "stored" in the compounds ATP and NADPH. These compounds are later used as "fuel" to power the light-independent fixation of* CO_2 *into sugars. All the chemical equipment needed for the light-dependent reaction is located on the thylakoid membrane.*

Light-trapping pigments, including chlorophyll, are grouped together on the outer wall of the thylakoid sac into units called photosystems [1]. *When light strikes a pigment molecule, one of its electrons becomes "energized" and is passed through the photosystem to an* electron carrier *in the membrane* [2]. *Having lost an electron, the photosystem is left with a net positive charge: it is resupplied with electrons by the splitting of water* (H_2O) [3], *which also releases hydrogen ions* (H^+) *into the thylakoid sac* [4] *and liberates gaseous oxygen* (O_2) [5]. *The energized electron is passed to another carrier in the thylakoid membrane: in this process some of its energy is used to "pump" more* H^+ *into the thylakoid sac* [6]. *The electron passes through a second photosystem* [7]; *this absorbs more light* [8], *boosting the electron's energy level. The reenergized electron is now passed through other electron carriers, giving up some of its energy on the way to fuel the formation of NADPH from NADP and hydrogen ions* [9].

The result is that the H^+ *concentration within the thylakoid sac rises to over 1,000 times that in the stroma, generating a chemical pressure.* H^+ *can only "leak" back into the stroma through special membrane-spanning turbines – ATP synthetase enzymes* [10]. *As* H^+ *passes through these, it drives the synthesis of ATP from ADP and phosphate, as in cell mitochondria.*

The "energy-rich" compounds ATP and NADPH are then used to power the formation of sugar in the light-independent reactions in the stroma **[D]**. CO_2 *is bound to a series of intermediate compounds, using energy along the way, before finally being released as a sugar* [1].

Carnivorous Plants

Why some plants feed on flesh

The pale green butterwort looks harmless enough, but to the unwary insect these glistening leaves spell death. The butterwort has turned the tables on the animal kingdom and become a flesh eater. The deadly traps of Venus's-flytrap can snap shut on unwary insects in less than a second before slowly digesting them with a battery of enzymes. Other carnivorous plants, such as the pitcher plants found in Borneo and Malaysia, are large enough for birds and small mammals to get caught in. Unable to escape, they too will be digested by the plant.

All plants need a range of basic nutrients, including carbon, oxygen, nitrogen, phosphorus, potassium and other minerals and trace elements. These keep the plants alive and help them to build new tissues. Oxygen comes from the atmosphere, and all plants except parasitic plants also get their carbon from the air as carbon dioxide. Most plants take in the other nutrients, such as nitrogen, potassium and phosphorus, from the soil. But some plants live in soils poor in nitrates (plants' main source of nitrogen) and other minerals and have overcome this problem by becoming carnivorous. They supplement their nitrogen by digesting the protein in the bodies of insects and other small animals, which they catch in a variety of bizarre traps.

Swampy habitats

Carnivorous plants (sometimes called insectivorous) are usually found in boggy or swampy places. In these areas the waterlogged ground is usually very acidic and inhospitable to the soil bacteria that normally convert nitrogen, in the form of ammonia and nitrites, into the nitrates that plants prefer. Although most carnivorous plants are not absolutely dependent on their animal diet if living in good soil, in their normal habitats the nitrogen and other minerals they obtain from their prey could mean the difference between survival and extinction.

Deadly designs

Insectivorous plants have evolved several different strategies for trapping their prey. The leaves of sundews are covered with sticky tentacles. An insect alighting on the edge of the leaf is trapped by the tentacles, which are also triggered by touch to bend inward, moving the captured insect to the center of the leaf. Powerful digestive enzymes in the fluid released by glands at the tips of the tentacles rapidly liquefy the contents of the insect's body, which are then absorbed by the leaf.

Other carnivorous plants bear modified leaves in the shape of hollow pitchers that trap hapless insects. In some species of pitcher plants the mass of decaying bodies also provides food for the maggots of certain flies, which are resistant to the digestive enzymes. Some species of birds also make the most of an easy meal. They will often slit the sides of the pitcher to get at the maggots.

The size of pitchers can vary considerably in different species of plants. The vinelike tropical pitcher plants found in the forests of Malaysia and Borneo are the largest. After heavy rains the pitchers fill with water, and in some cases the pitchers are large enough for birds and even small mammals to drown in.

At the other end of the size scale is the aquatic bladderwort. Bladderworts float just under the surface of the water, where there is little nitrogen available. The bladderwort's stem carries tiny, translucent, air-filled, bladderlike traps. The mouth of each bladder is shut by a "door" carrying a few stiff bristles on the lower free edge, called the trigger hairs. When a tiny creature, such as a water flea or a mosquito larva, brushes against the trigger hairs, the door springs open and the victim is swept into the bladder as it fills with water. The door then snaps shut and the animal is trapped.

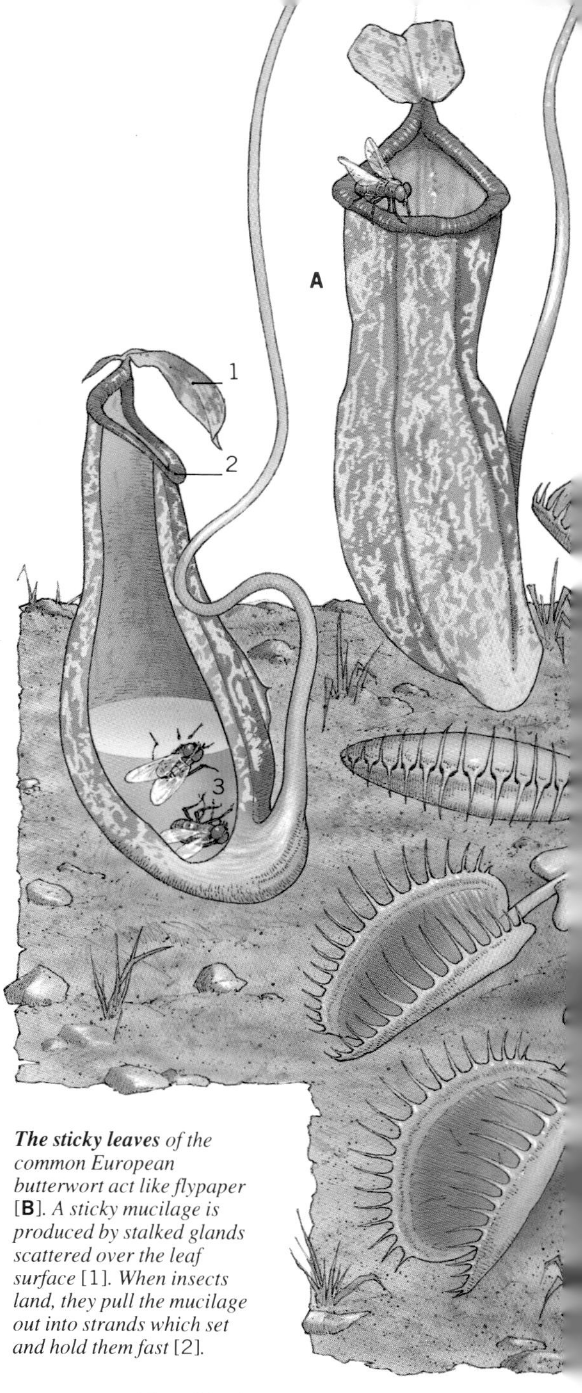

***The hollow, jug-shaped** traps of the* Nepenthes *pitcher plant hang at the end of its long leaves* [**A**]. *Each one has a lid at the top to keep out the heavy tropical rains* [1]. *Insects are attracted to the pitcher by its bright colors and by a sugary nectar that is produced by glands around the rim* [2]. *However, the surface of this rim is waxy and very slippery and most visiting insects rapidly lose their footing and fall in. Once inside, it is very difficult for them to escape: the upper areas of the inner surface are waxy; the lower regions are glassy smooth and covered with digestive glands; neither offer any foothold to an insect attempting to escape. They quickly tire of struggling and drown in the pool of water and digestive juices in the base of the trap* [3]. *The digestion and absorption of the insects' bodies is very efficient in* Nepenthes, *taking about two days for an average fly but only two hours for a small midge.*

***The sticky leaves** of the common European butterwort act like flypaper* [**B**]. *A sticky mucilage is produced by stalked glands scattered over the leaf surface* [1]. *When insects land, they pull the mucilage out into strands which set and hold them fast* [2].

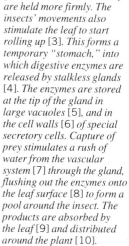

As they struggle, more glands are touched and they are held more firmly. The insects' movements also stimulate the leaf to start rolling up [3]. *This forms a temporary "stomach," into which digestive enzymes are released by stalkless glands* [4]. *The enzymes are stored at the tip of the gland in large vacuoles* [5], *and in the cell walls* [6] *of special secretory cells. Capture of prey stimulates a rush of water from the vascular system* [7] *through the gland, flushing out the enzymes onto the leaf surface* [8] *to form a pool around the insect. The products are absorbed by the leaf* [9] *and distributed around the plant* [10].

Connections: Cycles of Life 206 Parasitic Plants 186 Photosynthesis 182 Plant Architecture 114 Plant Defenses 244 Tropical Forest Plants 322

Venus's-flytrap [**C**] *is found in the bogs of North Carolina in the eastern United States of America. Ants, spiders and flies are its usual prey, though it may also capture other small animals such as snails and slugs.*

The leaf tips of Venus's-flytrap are modified into two kidney-shaped lobes [1] *hinged at the midrib* [2]. *Covering the inner surface of the lobes are two kinds of glands. The alluring glands, on the green outer margins of the trap, secrete a sugary nectar that attracts insects* [3]. *Farther in toward the midrib are the digestive-absorptive glands* [4], *which give the lobes their red color. Also on the inner surface of each lobe is a triangle of three tiny hairs that act as triggers* [5]. *If two or more of these hairs are touched in quick succession by an insect, they produce a tiny electrical current. This causes a change in the water retention of the membranes of the motor cells* [6] *in the midrib region. These cells quickly lose pressure and become limp, and the pressure from the cells in the outer epidermis forces the two lobes together* [7]. *Within two-fifths of a second, the trap has closed sufficiently to prevent the escape of larger insects* [8]. *As the prey struggles to escape, it further stimulates the trap to close completely. The insect's soft parts are slowly broken down by acids and enzymes released from the digestive glands and then absorbed. After the insect's soft parts have been digested, the leaves open again, releasing the indigestible skeleton.*

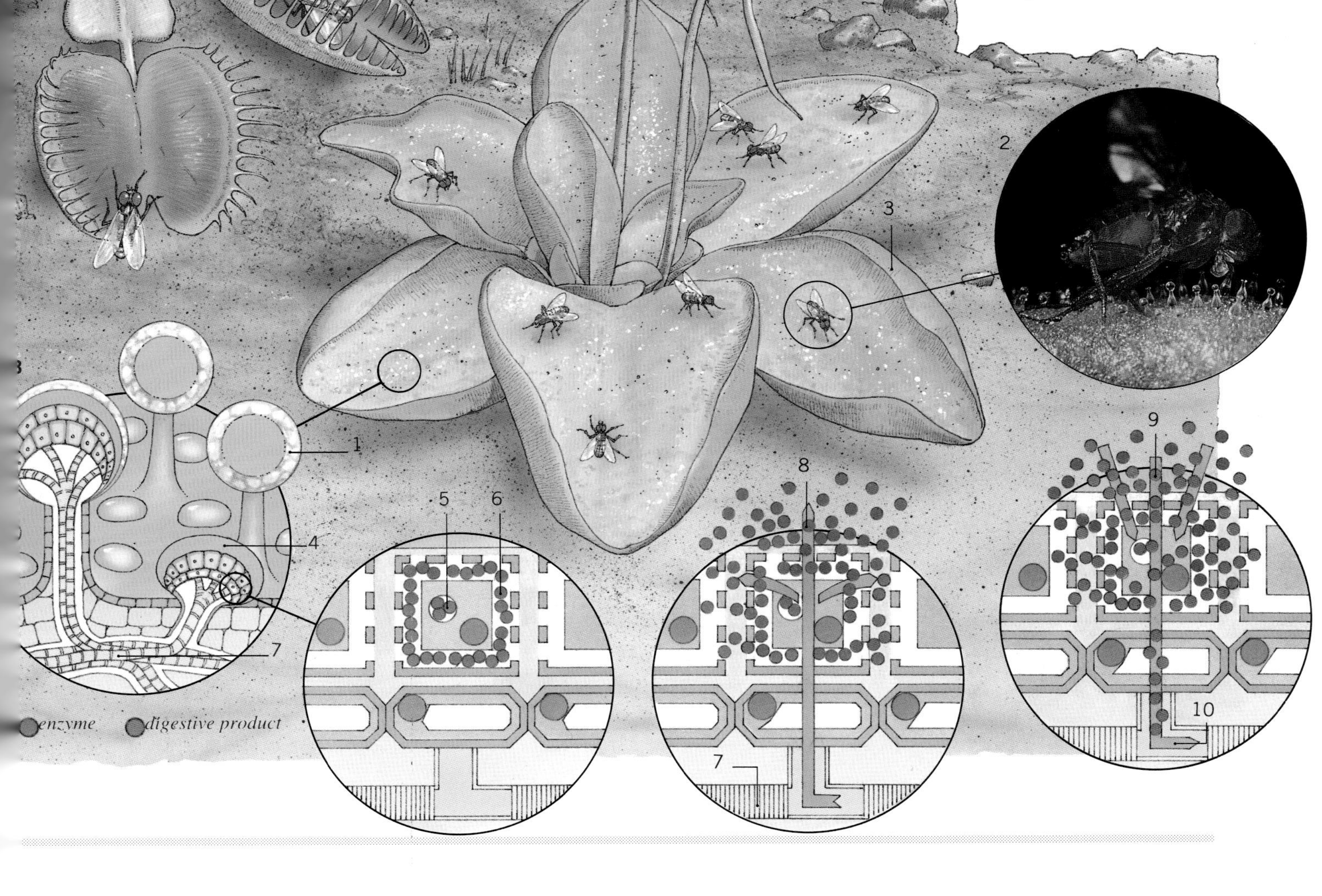

Plants on Plants

How some plants act as parasites on others

Around 1 percent of all the flowering plants – some 3,000 species – have evolved a bizarre, "vampirelike" way of life. They live as parasites on other plants, probably using characteristic chemical trails to "sniff out" their victim, and then tapping into its body and stealing its sap to fuel their own growth. This life-style has shaped some of the weirdest plant forms on earth, such as the giant flowers of *Rafflesia*. But more significantly these parasites are a threat to important tropical crop plants, and are hard to combat without damaging the crops.

The success of parasites, whether they are animals or plants, depends on their ability to find, and attach themselves to, suitable hosts. A parasitic animal, such as a flea, senses a potential host by sight or smell, moves toward it and actively anchors itself to its body. In contrast, a parasitic plant has far less control over its destiny. Its seeds must fall on, or very close to, an appropriate host. After this it relies on slow growth to approach and make contact with the host plant. In spite of these obstacles, parasitic flowering plants are a feature of most ecosystems from the Antarctic to the Equator.

No need to be green

Some parasitic plants "steal" from their hosts all they require to grow and reproduce. By plumbing in to the host's xylem they draw out water and minerals, while direct contact with the phloem gives them access to the sugars formed by the photosynthesizing host. Such parasites have no need to photosynthesize themselves and have consequently lost the pigments and structures typical of this process. Broomrape, for example, contains no chlorophyll and its leaves are no more than yellowish scales. Other parasites, such as mistletoe, do contain some chlorophyll and can, to an extent, photosynthesize, but their connections to their hosts also provide them with water, minerals and some sugars.

Tapping the sap

Some parasitic plants, for example dodder, successfully attach themselves to a wide variety of host species, but others are very particular in their choice of host. For such plants locating a new host is like finding a needle in a haystack. Some try to make the odds more favorable by releasing hundreds of thousands of tiny airborne seeds. Mistletoe, however, uses a different strategy. Its fruits are eaten by birds, and the mistletoe seeds pass through their bodies undigested to be excreted onto the bark of suitable host trees.

Once the seed has been successfully delivered, it germinates, and the young plant grows toward the host. In some species this germination and directional growth are thought to be stimulated by chemicals that naturally "leak" out of the host plant.

"Signature" chemicals on the host's surface also tell the parasite that it has reached its

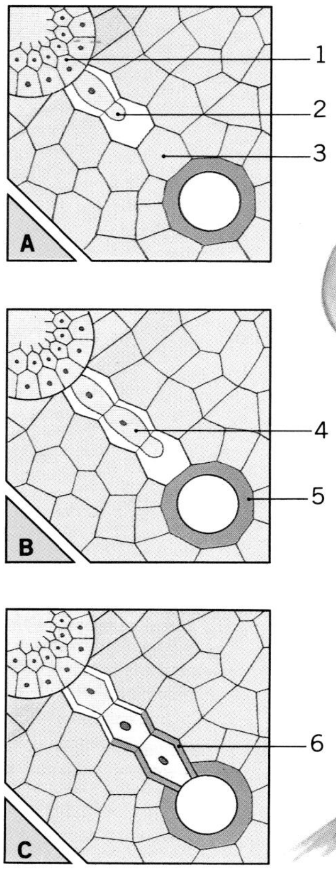

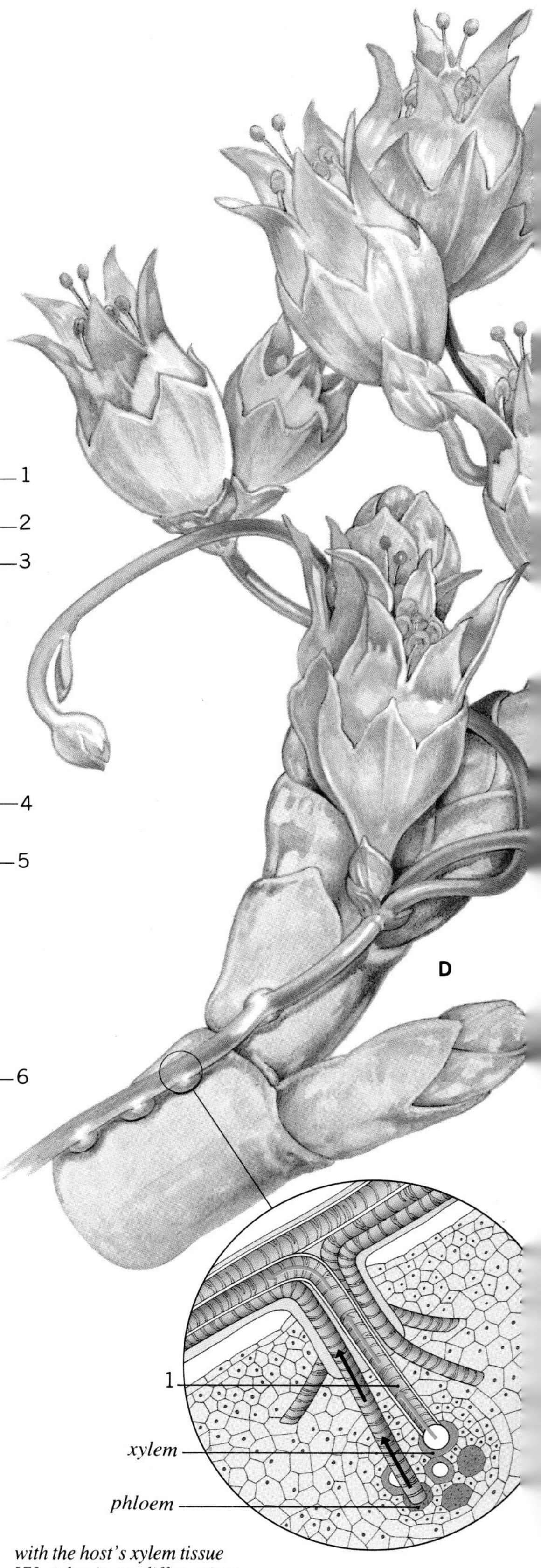

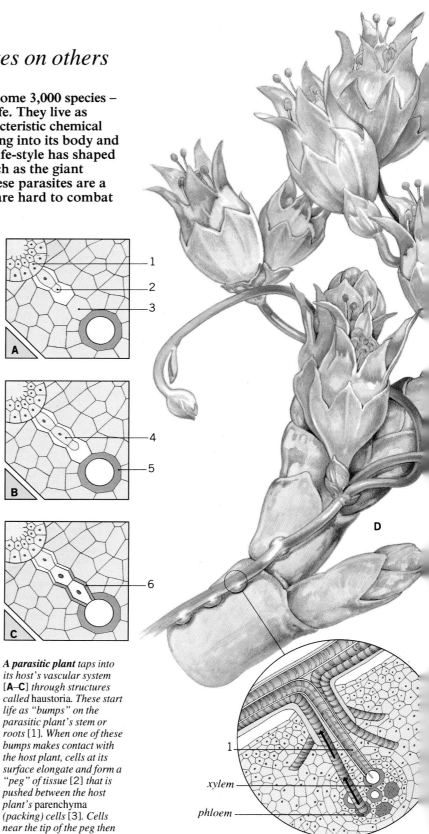

A parasitic plant *taps into its host's vascular system* [**A–C**] *through structures called* haustoria. *These start life as "bumps" on the parasitic plant's stem or roots* [1]. *When one of these bumps makes contact with the host plant, cells at its surface elongate and form a "peg" of tissue* [2] *that is pushed between the host plant's* parenchyma *(packing) cells* [3]. *Cells near the tip of the peg then begin to divide. They form thin filaments* [4], *which grow through the host's parenchyma cells, releasing special cellulose-dissolving enzymes which break down any cell walls in their path. If the filament makes contact with the host's xylem tissue* [5], *it begins to differentiate. It first expands, filling the space it has "carved out" from the host tissue. It then lays down a wall of woody* lignin [6]. *The filament cells die off and the lignified filament forms a water-conducting passage from the host to the parasite. Separate filaments make connections with the host plant's phloem to tap the sugary sap.*

Connections: **Animal Parasites** 188 **Decomposers** 204 **Flowering Plants** 120 **Fungi** 110 **Photosynthesis** 182 **Plant Architecture** 114

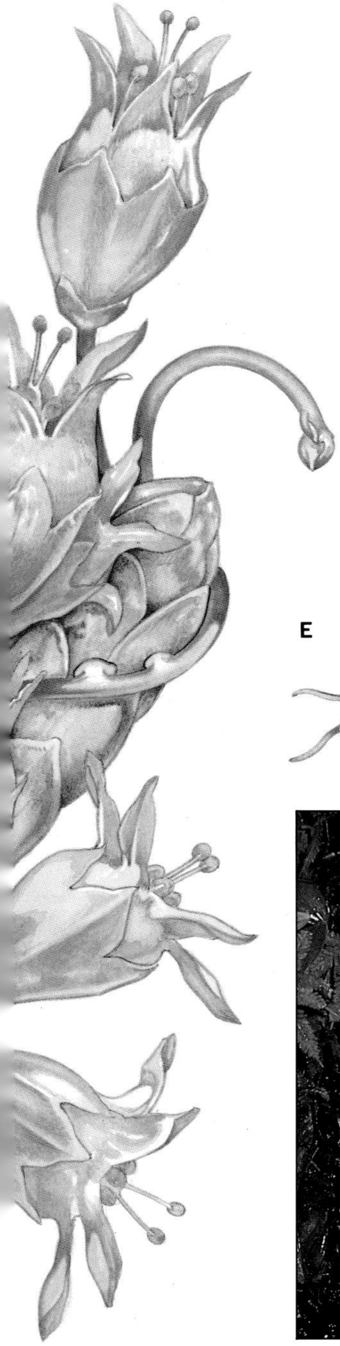

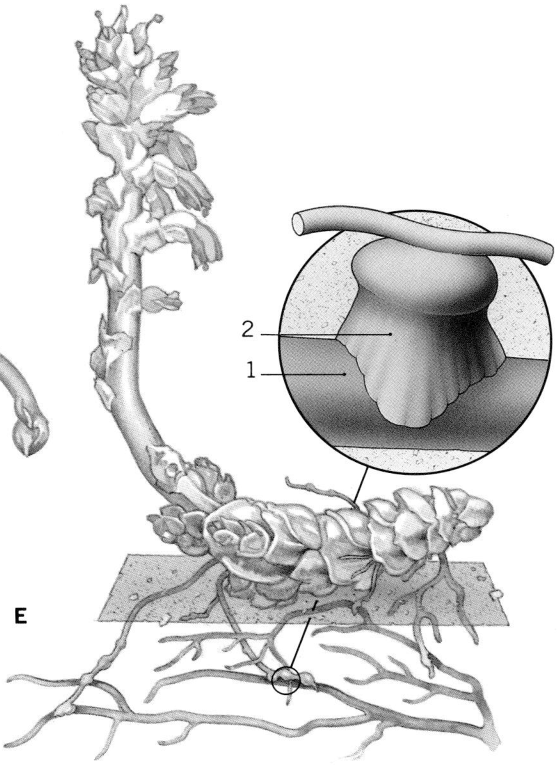

Lathraea **[E]** *forms attachments to the roots of flowering plants* [1]. *The haustorium develops into a large, cushionlike structure* [2] *that can completely encircle the host root. The penetrating power of the haustorium is great: it has even been known to break into electrical cables.*

Finding a new host *is one of the greatest challenges facing a plant parasite. Some, such as witchweed (below), increase their chances by producing up to 100,000 airborne seeds. The dwarf mistletoes, by contrast, form explosive fruits that can shoot seeds over distances of up to 50 ft (15 m).*

Rafflesia arnoldii *(left) grows in the jungles of Borneo and Sumatra. This specialized parasite has lost almost all the organs normally associated with flowering plants, such as roots and stem. It persists as a weblike system of filaments within the roots of* Vitis *species.* Rafflesia *produces giant flowers, which can grow to a diameter of up to 5 ft (1.5 m). The flowers produce a fetid odor to attract their pollinators – carrion-feeding flies. After pollination, berries containing sticky seeds are formed: these are eaten by fruit-eating rodents and thus carried to new hosts.*

odder **[D]**, *like many other arasitic plants, lacks hlorophyll and gains all the utrients it needs from its ost plant. Because it does ot photosynthesize, its leaves re reduced and resemble mall scales. Dodder seeds erminate in the soil and the oung plant forms a emporary anchoring root, hich sends up slender tems. When these make ontact with a host plant – in his case a glasswort – the oot rots away and the stem raps itself tightly around he host. This enables the arasite's haustoria* [1] *to enetrate the host's tissues nd make contact with the ylem and phloem, from hich it sucks nutrients.*

target. It then begins the process of tapping into the host's sap by producing structures known as *haustoria*. The parasite's tissue grows locally at those points where contact is made and then penetrates the host, establishing a direct link with its vascular system. In addition, the haustorium may grow around the host root or shoot, holding it in a vice-like grip that ensures that the contact cannot be broken.

Different parasitic species attach themselves to their hosts at different places. Some, for example witchweed, become attached to their host's roots through their own roots. Other parasites have perennial underground stems that form haustorial links with the host's roots. Such species may occasionally send up flowering shoots, but the main body of the parasite lives beneath the soil. In contrast, dodder can live independently in the ground until it finds a suitable host, which it may then tap into through haustoria on its stem. Dodder's host recognition system is not very advanced – it has been known to parasitize itself! Mistletoe taps into the stem of its host above ground and lives "rooted" into the stem and branches of the tree.

The devil's thread

In tropical regions parasitic plants are serious pests. In Africa witchweed can devastate crops of sorghum and maize, reducing yields by up to 90 percent. Mistletoe can cut the production of Australian Eucalyptus forest by 50 percent. And in the United States, the destructive effects of dodder are reflected in its familiar name – devil's thread. The control of plant parasites is very difficult because they are so intimately attached to their hosts. Weak applications of herbicide, which the crop can withstand but which kill the parasite, are sometimes effective.

lant Conservation 346 Plant Feeding 184 Seeds and Fruit 156 Symbiosis 192 Temperate Forests 318 Tropical Forest Plants 322

Unwelcome Guests

How parasites live on their hosts

In the Middle Ages people suffered horrendous deaths after being infected by itch mites. These tiny creatures, no larger than a full stop on this page, multiplied under the skin, creating boils the size of an egg. When mature, the mites would stream from the boils. Today, fortunately, itch mites are easily treated with special ointment. External parasites feed either by finding food on the host's skin, or by penetrating the skin and consuming whatever they find inside. This implies either a long-term scavenging operation, or a brief attacking raid and a quick getaway.

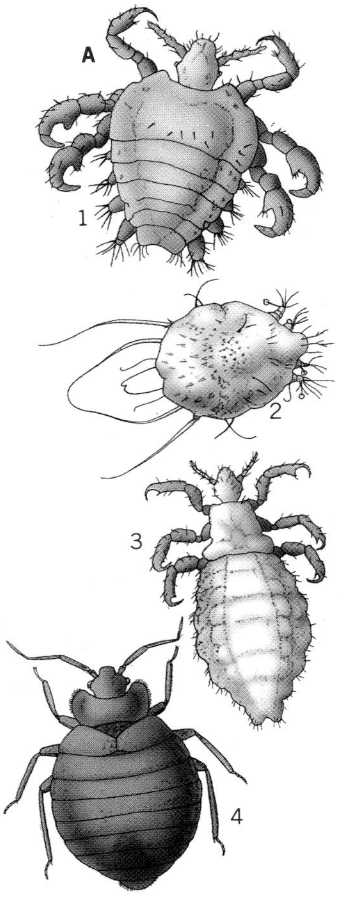

Some external parasites *[**A**] live either on, or nearby, humans. The crab louse [1] lives on areas of the body with widely spaced coarse hair. For this reason, unfortunately, one of its favorite sites is the pubic region. The acarus mite [2] is just about visible to the human eye. This mite is responsible for causing scabies. The body louse [3] is one of the more dangerous parasites. It is a carrier of epidemic typhus. The common bedbug [4] is found all over the world. During the day it is inactive, but at night it finds humans and sucks their blood.*

External parasites, *ectoparasites*, are for the most part species of insect or arachnid and tend to be less harmful than internal parasites, *endoparasites*. However, the blood-sucking feeding method of ticks and mites can spread disease. Viruses are transmitted in these parasites' saliva and can be responsible for causing various spotted fevers and even scrub-typhus, common in areas of Asia. Parasites are often host specific – so that one single parasite species generally lives on a particular host species. A cat flea usually lives only on a cat, for example, and a human flea lives and feeds only on humans.

Becoming attached to one's residence

For a parasite living on the outside of a host, the main problem is how to stay attached to the host. Not all parasites remain attached all the time; many, particularly the blood-sucking flying insects such as mosquitoes, live free-flying lives in the environment for much of their existence. But they must be able to attach themselves to their host and penetrate the skin or plant stem, at least while they are feeding. Some parasites, conversely, live their entire lives attached to their host.

The hairy surface of a mammal host is particularly accommodating to ectoparasites that are more or less permanent residents. Hair and fur can harbor a wide range of ticks, mites, fleas, lice and other small creatures. Strands of hair often act as good anchorage points for the mites' claws to cling onto. The scaly skin of reptiles would seem to be more hostile for ectoparasites – nonetheless, many insects have adapted to life on this rather less favorable environment. Ticks and other mites, for example, wedge themselves in between the scales of snakes and lizards.

Hosts of hosts

Insects can themselves act as hosts to smaller insect parasites. Meadow ants, for example, play host to a tiny mite of the *Antennophorus* species, a parasite that clings on under an ant's chin with six specially adapted hook-like legs. Its other two legs are much longer and are used to tap the ant's antennae. Such tapping corresponds to a natural signal to the ant for it to regurgitate a drop of food from which the mite can then sip.

For an ectoparasite, transferring to a new individual host is relatively easy, even for one that spends virtually all of its time resident on the host. The perfect opportunity arises whenever two individual hosts come into close physical contact. Those ectoparasites that live at large in the environment for some time can jump on and off their host, and fleas are the best-known example.

Pest removal

Just because they are more easily visible than endoparasites, however, ectoparasites are not necessarily easier to control. Spraying with pesticides is only of limited value and may cause damage to the environment. Moreover, because of the huge reproductive potential that insects have, new strains fairly readily evolve that are resistant to certain pesticides.

On the other hand, the spraying of fungal parasites that grow on the surfaces of animals and plants – to prevent their infestation – can often prove very effective.

The Helpful Leech

Before the advent of modern medicine, one of the most prescribed medical treatments was bloodletting. This treatment was carried out with the aid of leeches, and was used to cure almost any medical complaint. Bloodletting in such a way was once so common that the doctor applying the parasite was himself commonly known as a leech. Now rare in the world because of overcollection in former times, the medicinal leech, *Hirudo medicinalis*, is still sometimes used to cure hematoma – a clotting disorder that can form a solid swelling. The anticoagulant *hirudin* in the leech's saliva breaks down the blood clot, allowing the blood to flow freely once more. The leech attaches itself to its host quite painlessly and sucks blood without the host necessarily even noticing. When full, it drops off.

Connections: **Ants' Nests** 236 **Fluid Feeders** 194 **Fur and Feathers** 134 **Internal Animal Parasites** 190 **Parasitic Plants** 186 **Symbiosis** 192

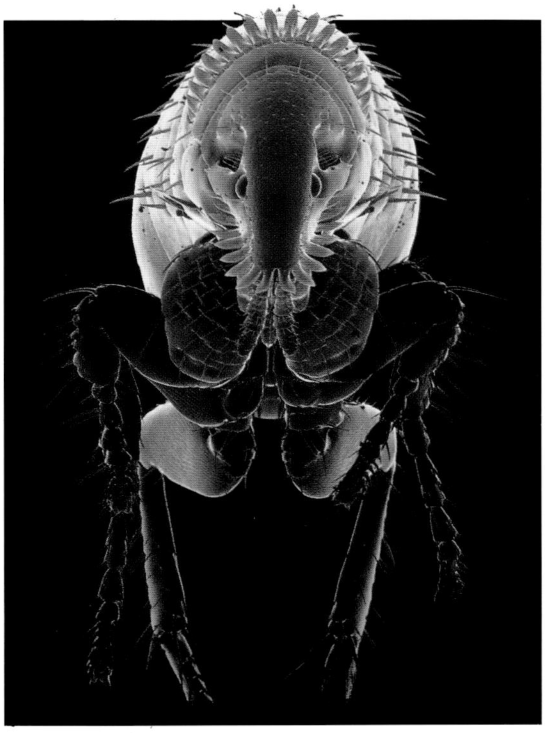

The cat flea (left) *has specialized hind limbs for jumping on and off its host. Its well-developed knee and hip joints mean that for its size the flea can jump vast distances, the equivalent of a man jumping a multistory building. The rounded shape of the head and the shape of the body, which is taller than it is wide, allows the flea to move easily through the host's thick fur. The hairs on the body and legs meanwhile stop the flea from falling out of the fur. Cat fleas are generally host specific – they rarely feed on any other species of animal – however, they will sometimes bite humans.*

***The three-host tick* [B] *is a member of the shield tick family. The female lays about 3,000 rounded eggs* [1], *which contain developing larvae. After a few weeks the tiny six-legged larvae emerge* [2]. *A larva attaches itself to a leg of the first host, where it feeds for four to five days* [3]. *When it has finished feeding, the larva drops to the ground, sheds its skin and molts into an eight-legged nymph* [4]. *When the molting process is complete, the nymph waits in the undergrowth before attaching itself to a second host* [5]. *The nymph again feeds for several days before dropping off and molting into an adult* [6]. *The adult waits on the ground* [7] *for the third host* [8] *on which to feed. Female ticks, once fully fed, drop from the host to the ground to lay their eggs. The choice of the victim varies from tick to tick. Some are host specific, others choose a variety of hosts such as cows, goats and sheep.*

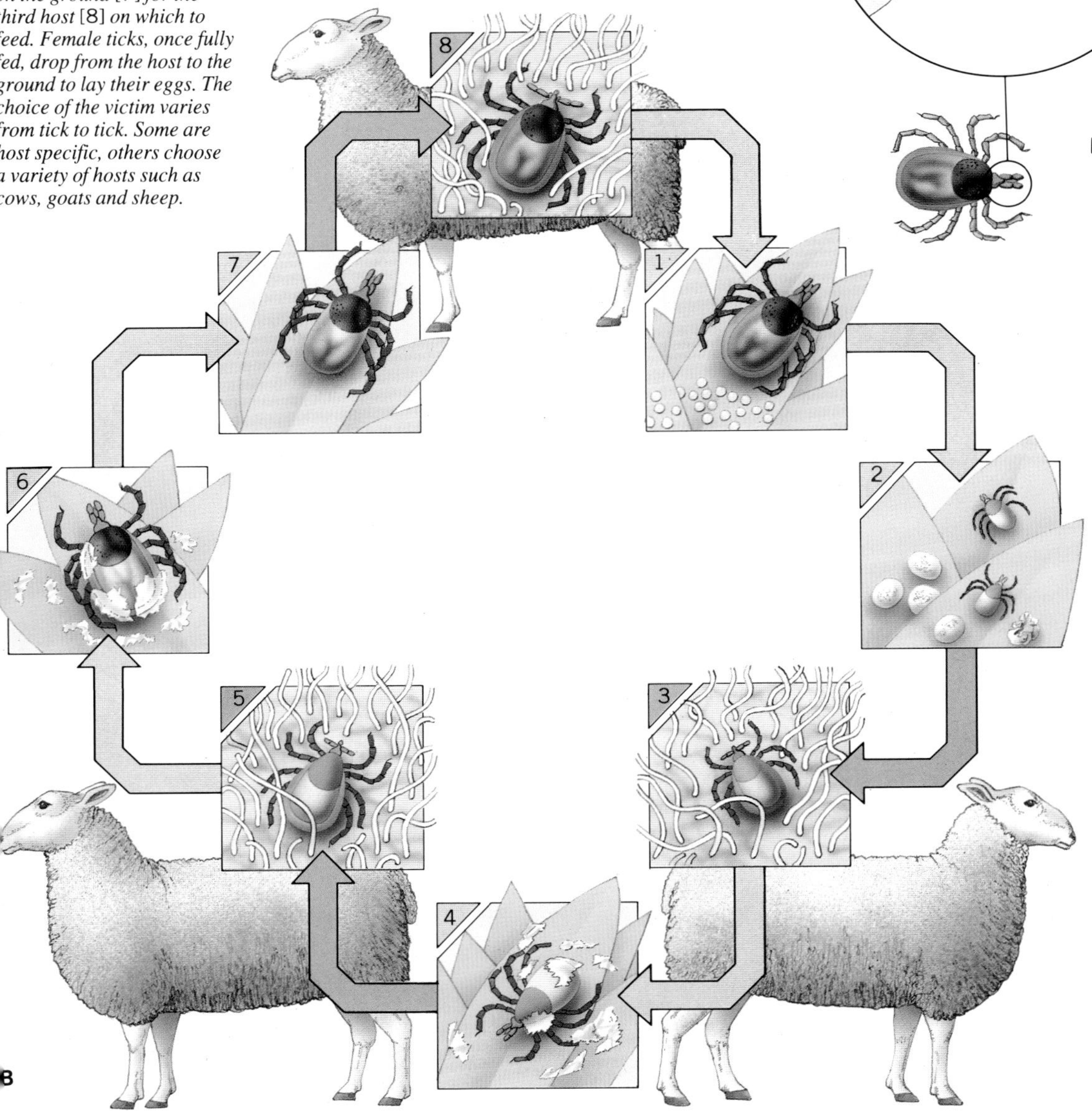

B

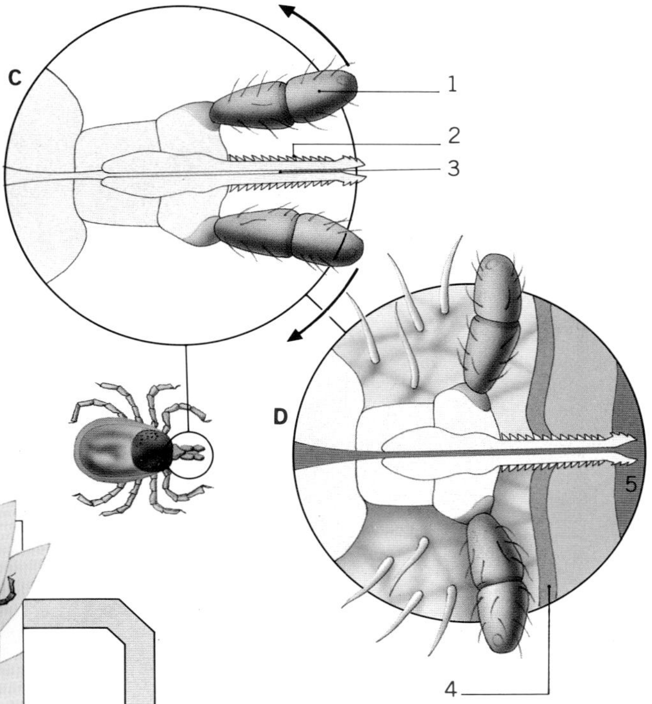

***The head and mouthparts** of the three-host tick are highly specialized. When the tick is searching for a suitable place to feed on the host's skin* [C], *the* pedipalps [1] *are kept alongside the* hypostome [2], *which surrounds the tick's feeding "tube," the* pharynx [3]. *This enables the tick to crawl through the host's fur. When feeding* [D], *the tick scratches at the host's skin* [4] *with tiny barbs on the hypostome. When the wound has reached a sufficient size, the tick inserts the hypostome by pushing with the pedipalps. The blood is pumped from the host, via the pincerlike* chelicera [5], *into the tick's body. The parasite's abdomen becomes greatly distended when feeding. It may grow from about 0.2 in (4 mm) to as much as 0.4 in (10 mm).*

The Enemy Within

How parasites live inside a host

The caterpillar in which the ichneumon wasp lays its eggs is an unwitting living larder for the wasp grubs. But although parasites only rarely kill their hosts, their effects can be wide-ranging – a male crab parasitized by a barnacle is likely to "change sex," losing more and more of its male characteristics on each successive molt. There are actually more parasites in the world than there are potential hosts, so most animals entertain more than one unwanted guest at any time. Even parasites are not immune from parasitism – there are protozoa that live off fleas.

Living inside another plant or animal is the ultimate luxury – all the parasite's needs are catered for. The host provides food, shelter and a stable environment, so the parasite can thrive and breed inside the body – as long as the host remains alive. Clearly it is not in the interests of the parasite to cause the death of the host, so in well-adapted parasitic relationships a "balanced parasitism" is established by which the parasite for a long time does no harm and only eventually causes the host to become diseased and inconvenienced.

Whether or not the parasite is the cause, there comes a time when the host does finally die. If the parasitic species is to survive, it must already have transferred to another individual host – even if it then has to find the correct part of the body or organ where it can grow and develop. Such transference may involve the parasite's adopting a different form for part of its life cycle – perhaps free-living in the environment, or even occupying a different host species.

The reproductive capabilities of parasites are frequently phenomenal, which enables parasites to maintain themselves despite the huge odds stacked against them of finding a new host species. Parasitic worms in particular seem to produce huge numbers of eggs. *Ascaris lumbricoides*, for example, which spends part of its life cycle in the human, can produce 200,000 eggs daily.

Microparasites

Microorganisms at large in the environment, generally in a state of dormancy – such as viruses or bacteria – can enter an animal host through the nose, mouth, or cuts in the skin, or a plant host through its sap system. Microorganisms that cannot survive at large for more than a short time must be passed from one host directly to another. Direct transference may occur through physical contact, for example, or through the use of carriers ("vectors") such as blood-sucking insects, which transfer the parasite as they feed themselves but are often not affected by them. Part of such a parasite's life cycle is thus spent in a mammal's bloodstream or cells; the other part is spent in the vectoring insect.

This sort of two-phased life cycle, in which a parasite has a primary and a secondary host of two different species, is common

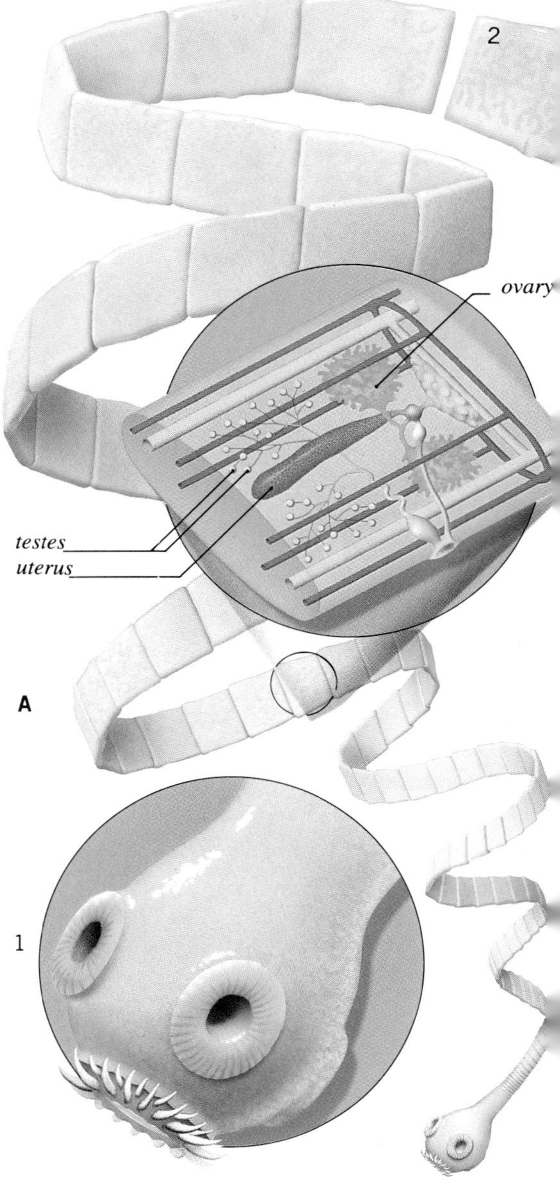

An adult tapeworm **[A]** *in the human gut has a "head" or* scolex [1], *which attaches it to the gut wall by means of hooks and suckers, and a "tail" of hundreds of identical flattened segments, or* proglottids [2]. *A tapeworm has no gut and absorbs nourishment through its surface. Mature segments bearing the tapeworm eggs break off from the "tail"* [3] *and are excreted in the feces. The individual segments are about 0.4 in (10 mm) long and 0.2 in (5 mm) wide. If infected feces are eaten by a pig, the eggs enter the pig's gut and the protective eggshell is broken down by gut enzymes* [4]. *The egg penetrates the gut wall* [5], *enters the bloodstream* [6] *and is carried to muscle, where it lodges and develops into a tapeworm larva* [7]. *Humans are reinfected by eating undercooked or uncooked pork or ham products* [8] *containing larvae. In the human gut the larva turns inside out, attaches itself to the gut wall* [9] *and the cycle resumes.*

Hard round marble *galls, often known as "oak apples" and common on oak trees, are the tree's response to infection by a parasitic insect – the tiny gall wasp. The female lays her eggs in the tissues of twigs. When the eggs hatch, the tissues swell up around the grubs to form a gall. Each "apple" contains a single grub, which eventually hatches from the gall leaving a neat round hole. Other species of gall wasp are responsible for the larger, spongy, oak apple galls on twigs, the round, red cherry galls and star-shaped spangle galls on the undersides of oak leaves, and the irregularly shaped galls on acorns.*

Connections: **Animal Cells** 102 **Animal Parasites** 188 **Bacteria** 106 **Parasitic Plants** 186 **Symbiosis** 192 **Viruses** 104

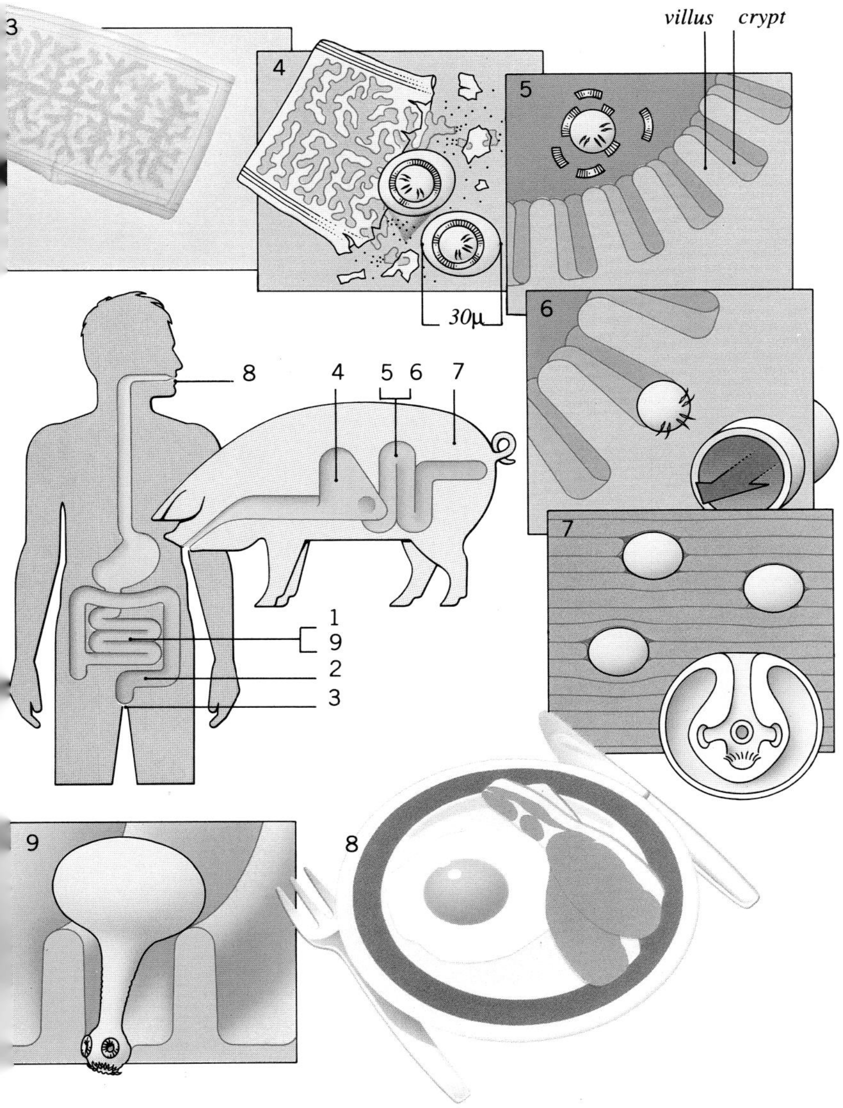

among larger parasites as well. Many are worms of various shapes, sizes and descriptions: the tapeworm is one typical example. Adult tapeworms parasitize many vertebrate species. The largest tapeworms live in sperm whales, and grow to a length of 100 ft (30 m).

Parasites and hosts

The parasite may enter through the mouth with infected food. Giardiasis, for example – a disease common in the Third World, where it is customary to use human feces as fertilizer – can be passed from human to human if infected feces are ingested. Others, such as the worm that causes bilharzia, live part of their life cycles in water and enter a swimming animal's (or human's) body by burrowing through the skin, using muscular burrowing movements and secretions that break the skin open. Once in the bloodstream, eggs are laid which then migrate to the intestinal wall; penetrating, they are passed out in the feces.

Parasites – especially worms – often lead to physical blockage of the intestine and other body ducts and cavities, which leads to disease and possibly to death. Elephantiasis is one such disease, resulting from blockage of the ducts of the lymphatic system by a threadlike nematode worm.

Very few vaccines are effective against parasites; none consistently protects humans. Correct diagnosis of parasitic diseases is also difficult, which only compounds the problem.

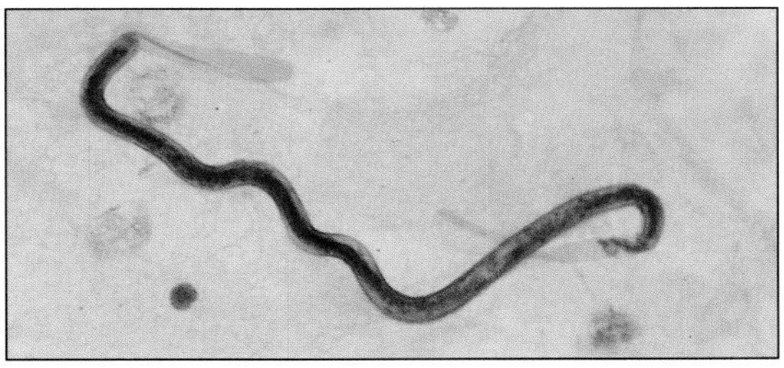

***The nematode worm** may be around 3 in (8 cm) long and only 0.01 in (0.3 mm) in diameter as an adult. It is the cause of elephantiasis, a particularly unpleasant tropical disease. Victims' legs and other parts of the body swell up to many times their normal size; the infected legs of a person with elephantiasis can measure 30 in (75 cm) around and the tissue of the scrotum around the testes can swell up into a mass weighing 50 lb (24 kg). The disease is transmitted between humans by mosquito. The worms are carried into the fine ducts of the lymph system, which they block, causing the tissues around them to swell.*

Banishing Malaria

Malaria is one of many diseases caused by microparasites injected by blood-sucking insects into human hosts. Prevention is mainly by controlling the vector: the *Anopheles* mosquito. One modern method destroys the mosquito in its larval stage so that it never reaches adulthood. In warm weather the larval stage lasts just a few days and the larva lives by clinging onto the film on the surface of standing water and absorbing oxygen there. They are attached to the film by their long breathing tubes, and use their mouth-brushes to filter food. If light oil or petrol is sprayed onto the water, the water's surface tension decreases and the film dissipates – and the larva sinks and drowns. This method has considerably reduced the incidence of malaria in tropical African countries.

The Unlikely Allies

How living things benefit from symbiosis

The intestines of Australian termites are host to thousands of tiny *Myxotricha* protozoa. They are essential to the termite because they help it break down the pulverized wood that is their food, and in return they are provided with all the nourishment they require. However, the protozoa themselves are in turn host to a further three species of bacteria, which obtain food from them: two on their outer membranes that help them to move around and one inside helping with the protozoa's digestion. All these relationships are mutually beneficial and are termed "symbiotic."

Every living thing on earth is engaged in relationships with other living things. There are those it relies on for food and, conversely, those that will eat it if they get the chance. Then there are those that benefit it indirectly in some way, such as the free-living soil fungi that unintentionally nourish plants by breaking down dead matter for their own sustenance, in turn releasing the nutrients.

For most living things these relationships are conducted with several species, not just one. But sometimes a closer and more specialized relationship evolves between two species that come to rely more and more upon each other. If this develops further still, and there is a physical link – one species lives inside or upon the other, perhaps – then biologists classify the relationship as one of several kinds of symbiosis.

Living together

There is a whole spectrum of relationships, from those in which both partners benefit (*mutualism*), to those where one gains food through the presence of, but without harming, the other (sometimes called *commensalism*, or "eating at the same table"), through to outright *parasitism*, in which one uses the other as a food source, sometimes actually causing harm to it. There are also relationships in which food is not the most important issue. The benefits may be protection, or transport, or simply support, as when climbing plants like ivy or clematis grow on a tree, or pseudoscorpions hitch a ride by grabbing the legs of a flying insect.

It is impossible to draw any sharp dividing line between all these different associations. Relationships that are mutually beneficial some of the time, for example, can under certain conditions slip into parasitism. Animals that are normally commensals – such as the bacteria and fungi in human intestines – may turn parasitic if a person's immune defenses are weakened. Occasionally even a parasite can provide some benefit to its host: at one time doctors used the malaria parasite to cure syphilis, because it provokes a fever of a temperature high enough to kill the potentially deadly syphilis bacteria. All such relationships are products of evolution and represent a precarious balance between two species, subject to change at any time should environmental conditions change.

Iridomyrmex ***ants*** *benefit from the sugary nectar of the ant plant* [**A**]. *This is produced in nectaries* [1] *that develop at the base of the flower* [2] *after the petals and sepals have fallen off. The plant benefits from vital minerals in the ants' defecation and waste materials* [3], *absorbed through the warty inner surface of the chambers* [4].

Measures of mutualism

Possibly the most interesting relationships are those in which both parties benefit. (Although these are best described as mutualism, some people use the term symbiosis just for these relationships.) In terms of freedom of movement, they range from the cleaner fish that swim the oceans freely, picking out food morsels from the teeth of sharks and other large fish, to single-celled plants that live out their whole lives within a coral animal. In terms of dependence, they range from those that rely completely on their partners to those that can get along without partners at all if they have to.

A remarkable number of mutualisms involve ants, probably because ants live in colonies, and their behavior patterns are complex yet controlled largely by outside forces. Their guard-dog mentality makes them useful to plants and to aphids.

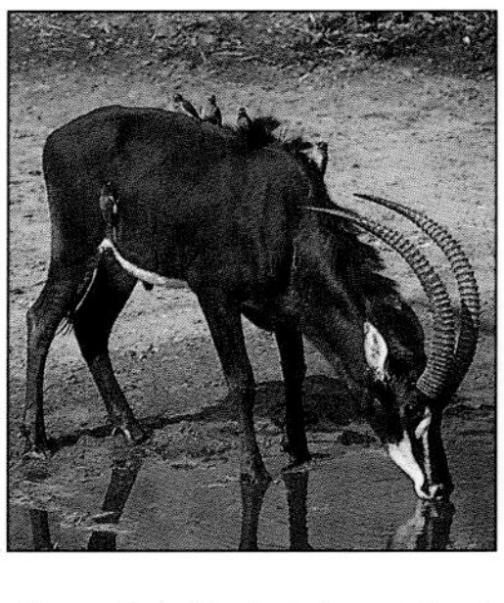

Oxpecker birds *(above) feed on the small ticks and insects found on the hides of various large African mammals like hippopotamuses and buffaloes. This is often known as a cleaning symbiosis. The bird gets an easy meal and the larger host is kept free from parasites.*

Connections: **Algae** 112 **Ants' Nests** 236 **Bacteria** 106 **Bioluminescence** 276 **Cycles of Life** 206 **Fungi** 110 **Herbivores** 202 **Pollination by Animals** 154

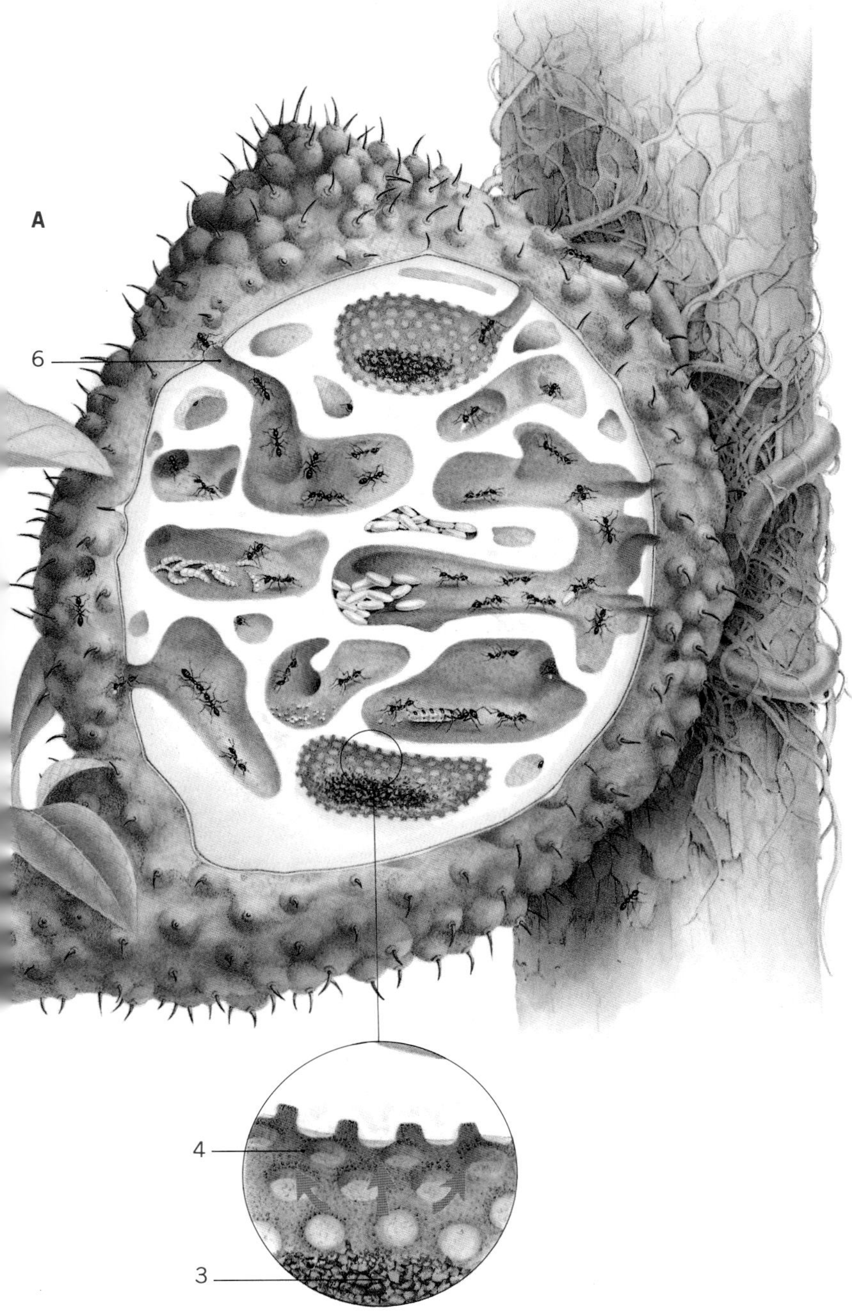

***Most of the plants** that house ants benefit from their aggressive response to leaf-eaters, but the ant plants,* Myrmecodia [**A**], *are different. They are epiphytic plants, growing suspended from trees in upland rain forests, where the soils are often lacking in nutrients. Therefore the mineral nutrients provided by the ants supplement the plants' poor diet. As the plants grow, their stems enlarge and they spontaneously develop cavities that are invaded by* Iridomyrmex ants [5]. *These chambers do not interconnect and have separate passages to the outside* [6]. *A complete ant colony soon becomes established in the plant.*

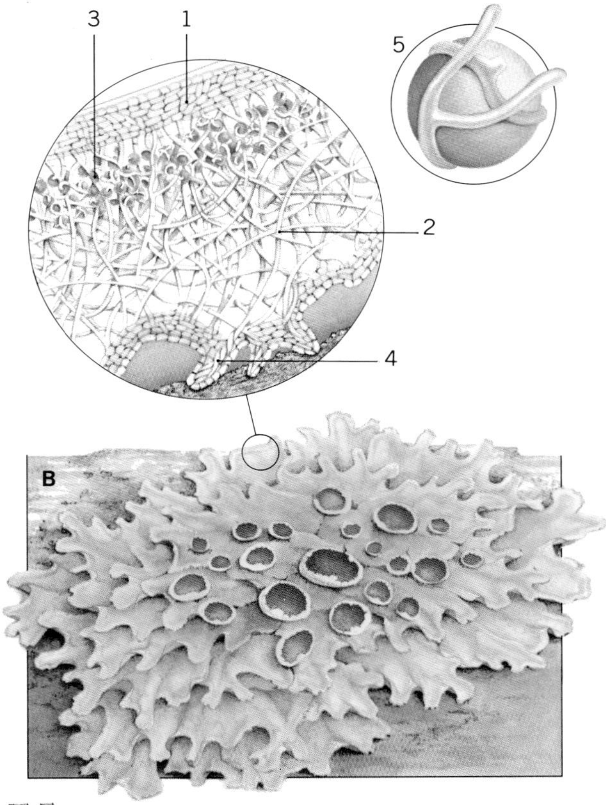

The lichen **Xanthoria** [**B**] *is found on coastline rocks. Lichens are the product of an intimate symbiosis between fungi and algae. The algae provide energy-rich sugars from photosynthesis; the fungi provides protection and water and minerals. The lichen's protective upper layer* [1] *is made up of matted fungal strands* (hyphae) [2]. *Below this is a layer of algal cells* [3] *closely associated with hyphae. The lichen clings on with rootlike structures at its base* [4]. *Vegetative asexual reproduction is achieved by* soredia [5]*: small clumps of algae and hyphae that disperse to form new colonies.*

Follow-The-Leader

The honeyguide – a bird related to the woodpecker – is distributed over a variety of habitats in southern and tropical Africa. It gets its name from its ability to attract the attention of a honey badger, also called a ratel, and lead it to a bees' nest that it has spotted. The honey badger then breaks open the nest with its powerful front claws and takes the honey. The honeyguide in turn eats the bee larvae and the wax-constructed nest itself. Its ability to digest bees' wax is unique among vertebrates, and is made possible by symbiotic bacteria in its intestines that provide the necessary enzymes. The bird initially gains the honey badger's notice by chattering loudly and fanning its tail. So instinctive is this behavior that honeyguides will readily accept humans as substitute honey badgers, and the humans happily reward them.

The Liquid Diet

How animals feed on fluids

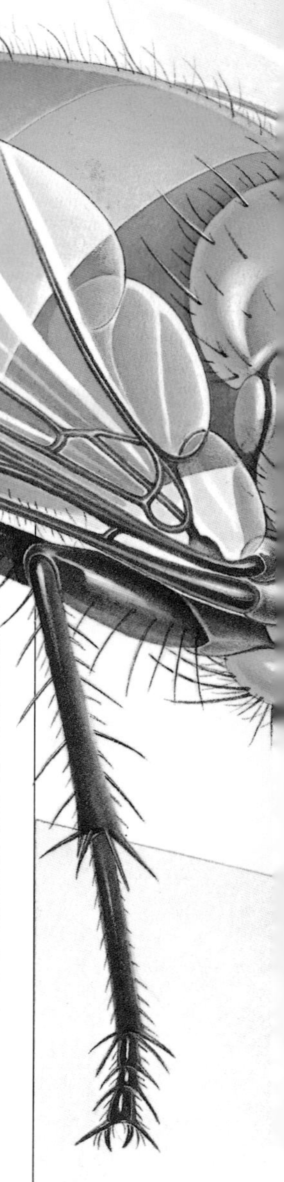

When the "bloodsucking" vampire bat bites its prey, its attentions are likely to go unnoticed, because this bat has evolved anesthetic saliva that stops the victim's blood from clotting. Many fluid-feeders have evolved yet another method of eating: they inject or dribble their digestive enzymes onto food, and only suck it up into their bodies after it has already been broken down. Many animals never touch solid food but live exclusively on nectar or sap, or more ghoulish liquids like blood or the tear fluid from animals' eyes. They are all intensively specialized for their way of life.

Nectar is a food source containing large amounts of energy-rich sugars. But the flowers that produce a supply of nectar tend to be few, widely scattered, and too fragile for many nectar eaters to be able to land on them. The consequence is that hummingbirds and some bats and moths – among others – have adapted to flying long distances in relation to their size, and have developed the ability to hover in front of the flowers once they find them, sucking out the nectar on the wing. Butterflies and moths feed on nectar solely in their adult stage – the one stage in their life cycle that requires energy for flight. In tropical forests nectar eaters that are not fussy about the flowers from which they get nectar can usually find food all the year round. In cooler climates flower production is seasonal and the butterflies and moths have a corresponding life cycle.

The sugars (*carbohydrates*) in nectar cannot provide a balanced diet. Nor can plant sap – which is 80 percent water and the rest mainly carbohydrate. Some small fluid feeders, such as whitefly, thrips and spider mites, cleverly avoid this potential problem by piercing and sucking out the nutritious contents of individual cells one at a time. Others have tiny organisms in their bodies that can synthesize the missing nutrients.

A taste for blood

Mosquitoes, leeches and other bloodsucking parasites have all had to find an answer to the potential problem of their food, clotting as they feed. On first beginning to feed they accordingly introduce into their victim's bloodstream an anticoagulant that prevents the blood from clotting. Many bloodsuckers in addition have very sharp skin-piercing mouthparts that enable them to feed unnoticed until some time into their meal.

Digestion from the outside in

The animals that secrete digestive enzymes onto their food in order to break it down into soluble compounds include many species of flies. The housefly, in particular, digests meat in this way. Some spiders inject enzymes into prey they have paralyzed and wait for the body contents to liquefy.

There are some advantages to this method of feeding. The animal does not require a large intestinal system in which to retain the food while digestion takes place, and is thus less heavy, and consumes less energy while flying and moving around. The major disadvantage, however, is that the animal has to wait until the enzymes have had their effect before it can consume its food.

The common vampire bat *(above) feeds primarily on domesticated livestock such as pigs, cattle, horses and donkeys – they rarely choose humans. They feed by making a tiny wound in the animal's hide with their razor-sharp incisors. Anticoagulants in the vampire's saliva keep the victim's blood flowing freely. One popular misconception of vampires is that they actively suck the blood from the victim, in reality they simply lap at the wounds with their tongues. Vampires go in search of their hosts at night, using their relatively good eyesight and well-developed sense of smell to locate their victims.*

A feast of tears

In the tropics moths from several families have become specialized tear-drinkers, sipping the eye-washing tears of grazing mammals such as cattle and deer. Some induce tear-shedding by sweeping their probosces across the animal's eye; others are less aggressive. Many of these moths feed on tears more during the dry season, and it is likely, therefore, that they are obtaining much-needed salt in this way. But tears also contain certain proteins, and at least some of the moths have been proved capable of digesting protein – a feature rare in nectar-feeding insects.

The greenbottle fly [**A**] *has highly adapted mouthparts. When it finds food, the greenbottle secretes digestive enzymes from its salivary gland* [1], *via the salivary channel* [2] *in the proboscis* [3], *onto the food. The saliva is squirted over the food through grooves* [4] *in the labella* [5], *situated on the end of the proboscis. The grooves keep their shape because of rings of chitin* [6]. *Once the food has been partly broken down to a liquid form by the enzymes, the greenbottle sucks up the liquid* [7] *through the grooves of the labella via the food channel* [8] *and into the midgut* [9], *where digestion is completed.*

Connections: Animal Parasites 188 Ants' Nests 236 Bird and Bat Flight 224 Invertebrate Carnivores 200 Invertebrate Herbivores 198

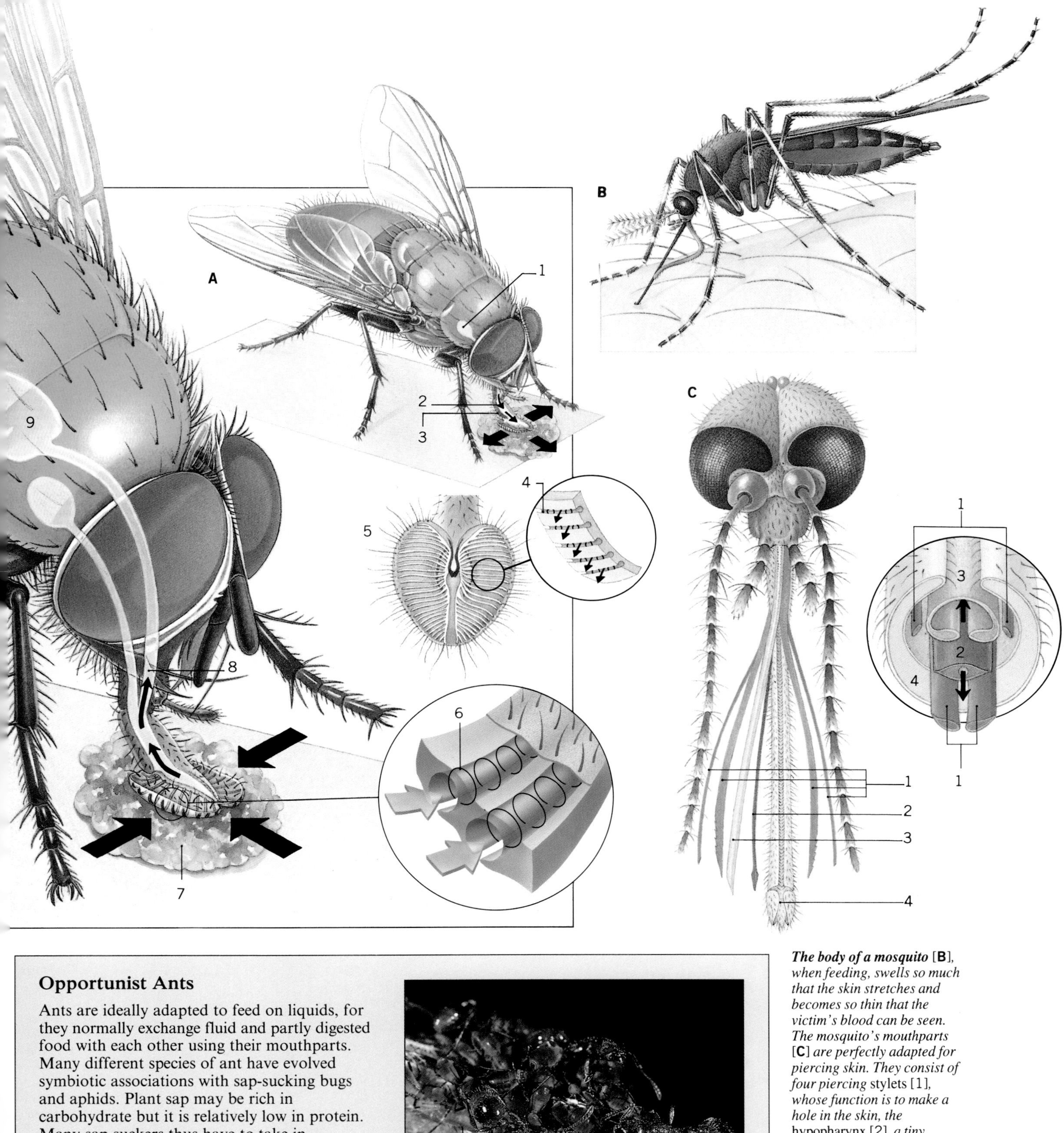

Opportunist Ants

Ants are ideally adapted to feed on liquids, for they normally exchange fluid and partly digested food with each other using their mouthparts. Many different species of ant have evolved symbiotic associations with sap-sucking bugs and aphids. Plant sap may be rich in carbohydrate but it is relatively low in protein. Many sap suckers thus have to take in considerable quantities of sap, with all its sugar, in order to obtain enough protein. Excess sugar is then exuded in the form of syrupy droplets.

By keeping the bugs and aphids well supplied with food, and by stroking them with their antennae, the ants can induce their "herds" to exude more droplets for the ants to drink. In this way a number of different species of ants actually stand guard over herds of bugs or aphids while they feed during the day.

The body of a mosquito [**B**], *when feeding, swells so much that the skin stretches and becomes so thin that the victim's blood can be seen. The mosquito's mouthparts* [**C**] *are perfectly adapted for piercing skin. They consist of four piercing* stylets [1], *whose function is to make a hole in the skin, the* hypopharynx [2], *a tiny tubelike structure down which is pumped anticoagulants to stop the victim's blood clotting, and finally the* labrum [3], *another tubelike organ up which the blood is sucked. These elements are protected by the* labium [4], *which guides the stylets into the skin rather than piercing it.*

Pollination by Animals 154 **Symbiosis** 192

Straining to Eat

How animals feed by filtering

In one day a blue whale may eat over four tons of food – not surprising, perhaps, for the largest animal that has ever lived. But this quantity is made up of the tiny, shrimplike krill and other diminutive creatures filtered through the whale's vast mouth. Filtration is a highly successful feeding method and is carried out in a variety of ways. The tongue of a flamingo shoots in and out like a piston up to 17 times a second, forcing water through a sieve in its beak. Oysters force water through their fixed filtration systems at up to 8 gallons/hour (37 liters/hour).

The essential requirement for a filter feeder is to pass large quantities of water containing high concentrations of food through its filtration system. The technique is thus limited to creatures that live in or around water.

Ram feeders, such as the enormous whales, herrings and other fishes, swim forward with their mouths open. This is the most common filter-feeding method employed by vertebrates. More than 20 species of fish are continuous ram feeders, including the basking shark, the manta ray, two species of freshwater paddlefishes and various herrings, sardines, anchovies and mackerels. Other equally mobile filter feeders, such as birds and many invertebrates, remain stationary when feeding. They have evolved many devices for creating a flow of water so that they need not remain dependent on water currents.

Beaks as sieves

The relatively few species of filter-feeding birds have evolved comblike straining devices in the shape of fine, horny plates, called *lamellae*, at the edges of their bills. Dabbling ducks such as shovelers and mallard feed by sifting food-containing water or mud. The bird moves its tongue, which is swollen at front and rear, back and forth extremely rapidly so that it acts as a piston and valve, drawing water into the mouth and then expelling it through the lamellae.

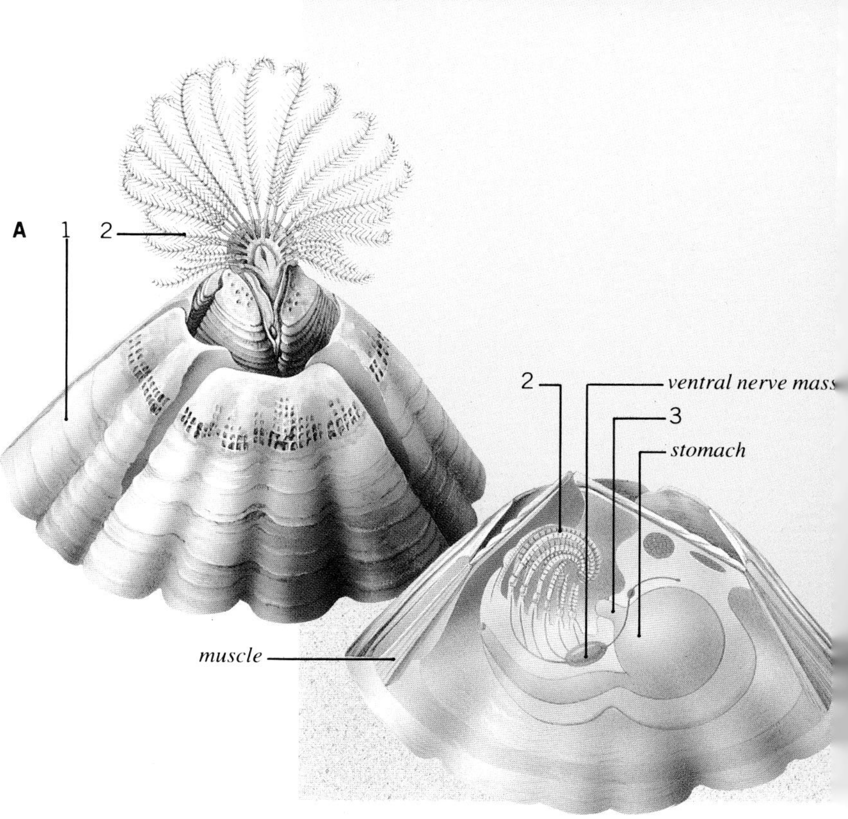

Barnacles **[A]** *lie on their backs, protected by a strong box of* calcareous *(calcium-based) plates* [1]. *When the tide is out and the barnacle needs to avoid drying, its limbs are folded away inside. When the tide is in, and they are ready to feed, the topmost, hinged, plates open and they kick their legs about in the water. Each limb divides near its base into two separate branches or* cirri [2], *each bearing a dense array of tiny bristles* (setae) *that filter food particles from the water. After being waved about briefly in the water, the limbs are drawn together like a fan being closed and pulled toward the mouth* [3].

Mass filtration

By far the greatest number of filter feeders are invertebrates. Many of them are stationary creatures that have to move the water through their fixed filters.

The most complex invertebrate filter-feeding systems are those of the bivalve mollusks, especially oysters. The filtering apparatus here is made up of modified gills, used for respiration as well as food gathering, and these are so large they fill the creature's body.

Large food particles are rejected and ejected with water as the oyster snaps its shell shut. The mucus strings on which smaller particles have been caught are wafted through the gut until they become wrapped around a unique rotating crystalline structure, called a *style*, found only in ciliary-feeding mollusks. This is the only rotating organ found in any animal. Here the stomach enzymes digest the food as it rotates on the style. In some cases an animal produces a disposable filter something like a vacuum-cleaner bag, rather than relying on part of its anatomy. The burrow-dwelling worm *Chaetopterus,* for example, secretes a mucous bag, through which it strains tiny planktonic food from the seawater. The minute pores in the bag soon become clogged up, so the worm regularly bundles up and eats the bag and produces a new one.

Infant filters

In addition to the huge numbers of stationary invertebrate filter feeders there are clouds of drifting planktonic animals that filter their food in the upper waters of the sea. These include both planktonic larvae of animals that live fixed to the seabed as adults – worms, mollusks, crustaceans, starfish, and so forth – and adult invertebrates, especially crustaceans, that are specialized filter feeders.

Connections: **Birds** 132 **Fish** 124 **Lake Life** 326 **Ocean Life** 332 **South Polar Life** 312

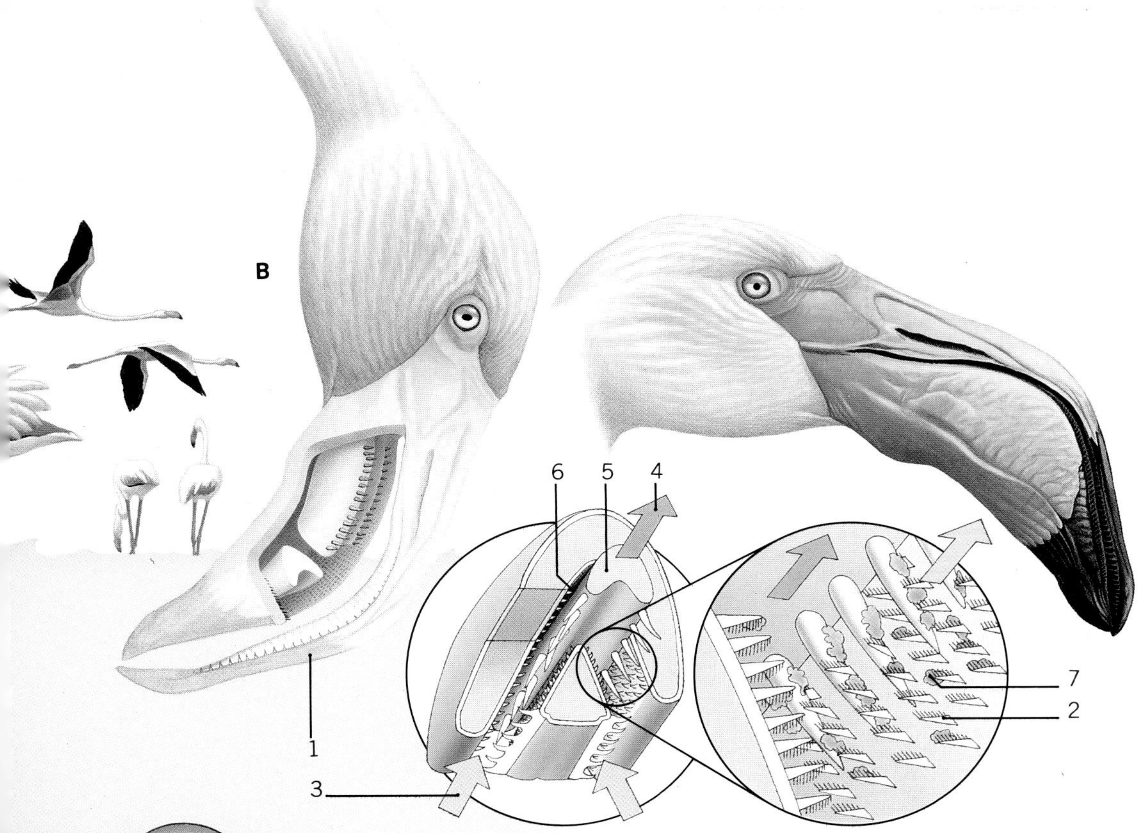

A flamingo *[**B**] feeds by wading into the water on its long legs, lowering its long neck, and holding its bent bill upside down in the water or mud, so that the bill's front half lies horizontally [1]. Sets of lamellae [2] strain out food as the water or mud is pumped through them [3] by the tongue action [4]. Different species eat different-sized foods, from minute algae to small invertebrates. Their lamellae vary accordingly. The large fleshy tongue [5] is narrow and cylindrical and lies within a deep bony trough in the lower mandible or jaw of the bill [6]. Protuberances on the tongue scrape food [7] off the lamellae for ingestion as the tongue moves back and forth.*

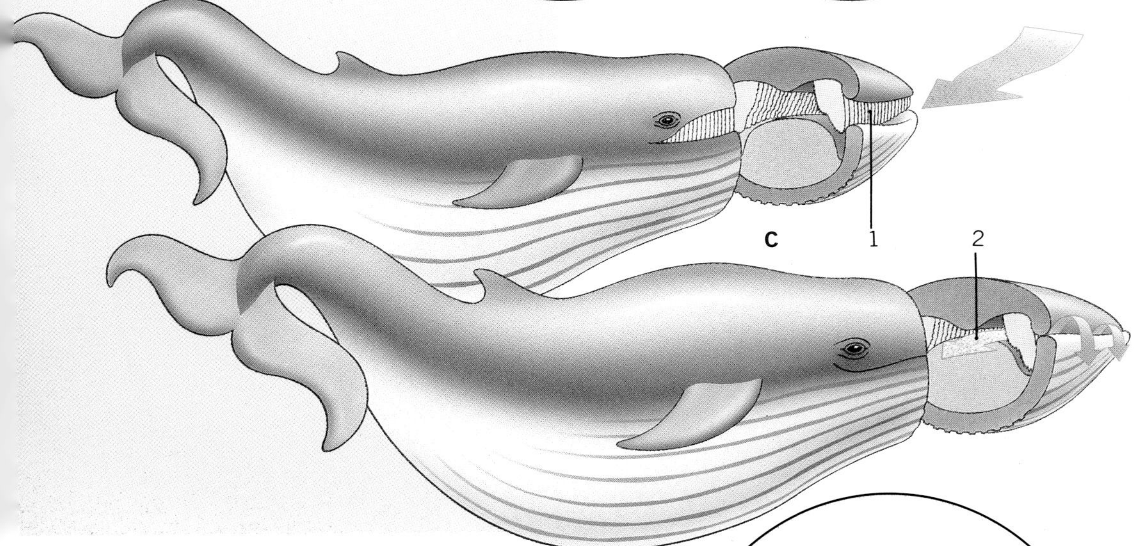

Baleen whales *[**C**] have no teeth, unlike killer whales or dolphins. Instead their great mouths contain hundreds of plates of* baleen *[1], or whalebone (*keratin*), which hang like a curtain from the upper jaw. Each baleen plate is less than 0.2 inches (0.5 centimeters) wide at the point of attachment, and tapers down to divide into a fringe of bristles on the inner margin at the bottom end. Adjacent fringes overlap one another to form a coarse mat that acts as an extremely effective filter. To feed, the whale forces water through the mat and between the baleen plates and then licks off the trapped food [2].*

Beyond Filtration

Many filter feeders, like the huge whale shark, can trap particles that are small enough to pass through the gaps in their "sieves." Three main mechanisms may be involved: the use of a sticky mucus-covered surface onto which particles adhere [1]; *interstitial impaction,* in which particles slightly denser than water end up trapped on the filtering structure while the water is diverted in two streams around them [2]; and *gravitational deposition* [3], in which dense particles sink and land on the filter. Many filter feeders employ filters with pores of varying sizes, enabling the sorting of particles into different categories. The number and arrangement of gill rakers varies in different species of the herring family living in the same seas, so that they take different-sized food items and thus avoid competition.

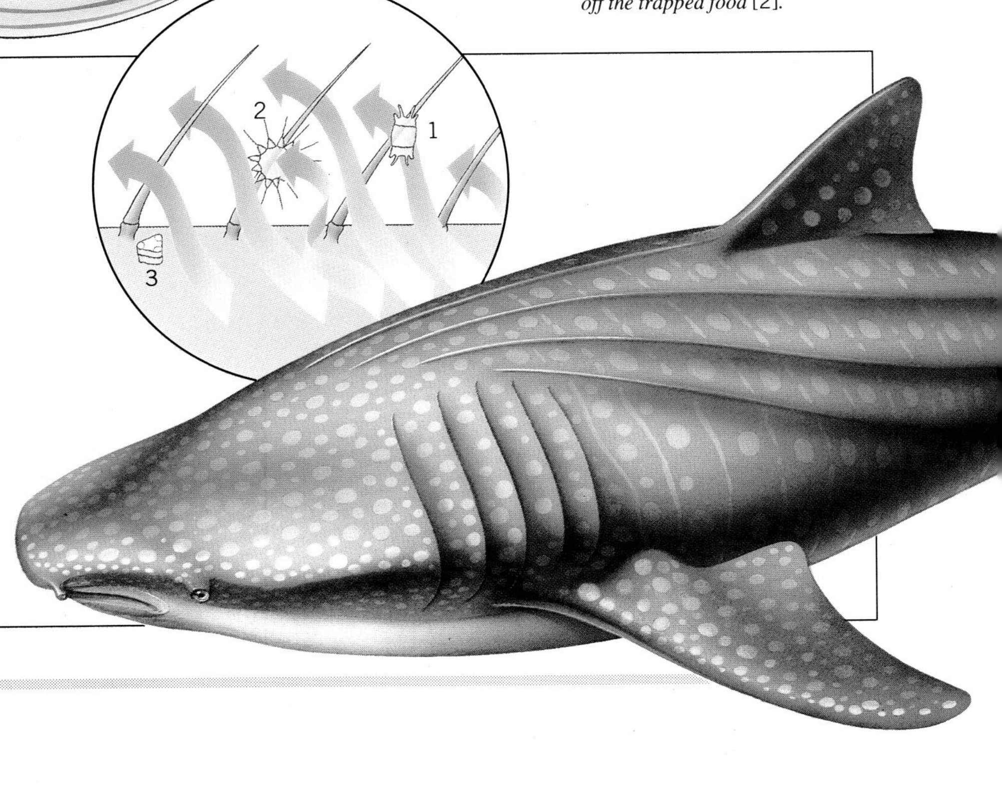

Green Feeders

How invertebrates eat plants

In order to get proper nourishment from leaves and other forms of vegetable matter, a single termite, just a tenth of an inch long, may contain up to 10 different species of microorganisms in its digestive tract; these break down the plant cellulose into substances that can then be absorbed by the termite. Ambrosia beetles use fungi instead: the fungus softens wood, making it easier for the beetle to chew. When they move to a new tree, the beetles carry with them a tiny piece of fungus. Only by such devices can many plant-eating invertebrates thrive.

Plant-eating (herbivorous) animals represent an important link in the food chain. They convert plant material into animal tissue, which in turn is eaten by predatory animals. In spite of their small size, invertebrates are among the most numerous and effective herbivores. Vast quantities of trees and shrubs are defoliated every year by caterpillars, crops are devastated by locusts, and household vegetable patches plagued by slugs.

Herbivores are described as primary consumers because they feed directly on plant material. But plant cell walls contain *cellulose* – a substance that cannot be digested by most animals – and may also contain many tough fibers. Many herbivores accordingly have strong mouthparts or teeth. Snails and slugs are an exception, producing within themselves their own cellulose-digesting enzymes. A few insects have alternatively evolved special relationships (symbioses) with cellulose-digesting microorganisms, which they harbor in their digestive tracts. Similarly, termites and some species of wood-boring beetles have symbiotic relationships with protozoa and bacteria.

Whereas some invertebrate herbivores feed on a wide range of plant material, others rely on more restricted diets, eating just sap, fruit, pollen or nectar, or the leaves of just one species of plant. A few highly specialized invertebrates feed on wood.

The larval forms of many insects are designed simply for eating vegetation. Caterpillars, for instance, have a very simple body form; expansion by molting is not impeded by elaborate appendages. Little energy is diverted into the production of complex structures, so more can be used for growth. The caterpillar's senses are also reduced to a minimum, and most species rely on camouflage or on a combination of warning coloration and toxic chemicals for defense.

Finding food

Invertebrates use an extensive range of senses to find their food. Visual signals are important for many insects. Bees recognize flowers by their color, and aphids have also been shown to be attracted to certain colors. So important are visual cues to bees that bee-pollinated flowers often have conspicuous patterns of guide lines to ensure that the insect is directed to the nectar.

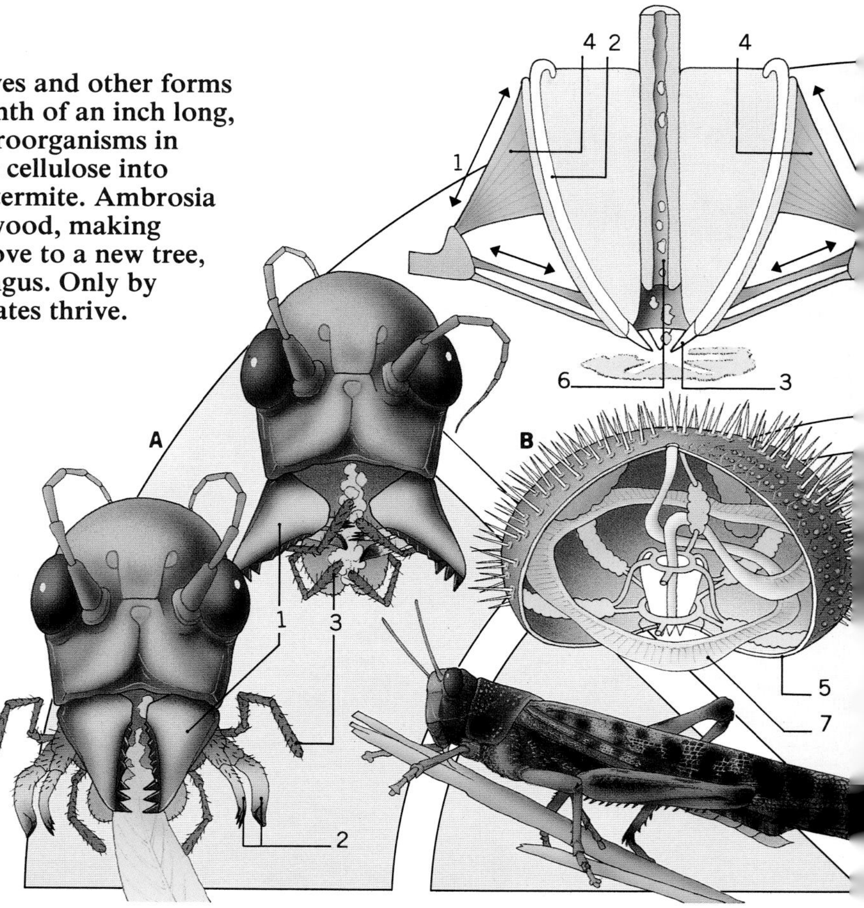

***The larva** of the spruce budworm moth (below) is one of the major pests affecting coniferous forests in the northern United States and Canada. Infestation densities as low as 15–20 caterpillars per square yard can lead to the rapid and severe defoliation of trees and their consequent death.*

***Like other insects**, the locust* **[A]** *has external mouthparts. Vegetation is cut and chewed by strong, toothed mandibles* [1]*, which open and close horizontally. The cut food is then held and pushed up into the mouth by a pair of* maxillae [2]*. The* palps [3] *have a sensory function and are thought to monitor food quality. Swarming locusts are the most notorious of insect pests. Within their migration zone, swarms of more than one billion individuals can strip the landscape bare of vegetation over hundreds of square miles. Swarming locusts can have disastrous effects on local human populations.*

Connections: Ants' Nests 236 Fluid Feeders 194 Grasslands 302 Herbivores 202 Invertebrate Carnivores 200 Invertebrates 122 Lake Life 326

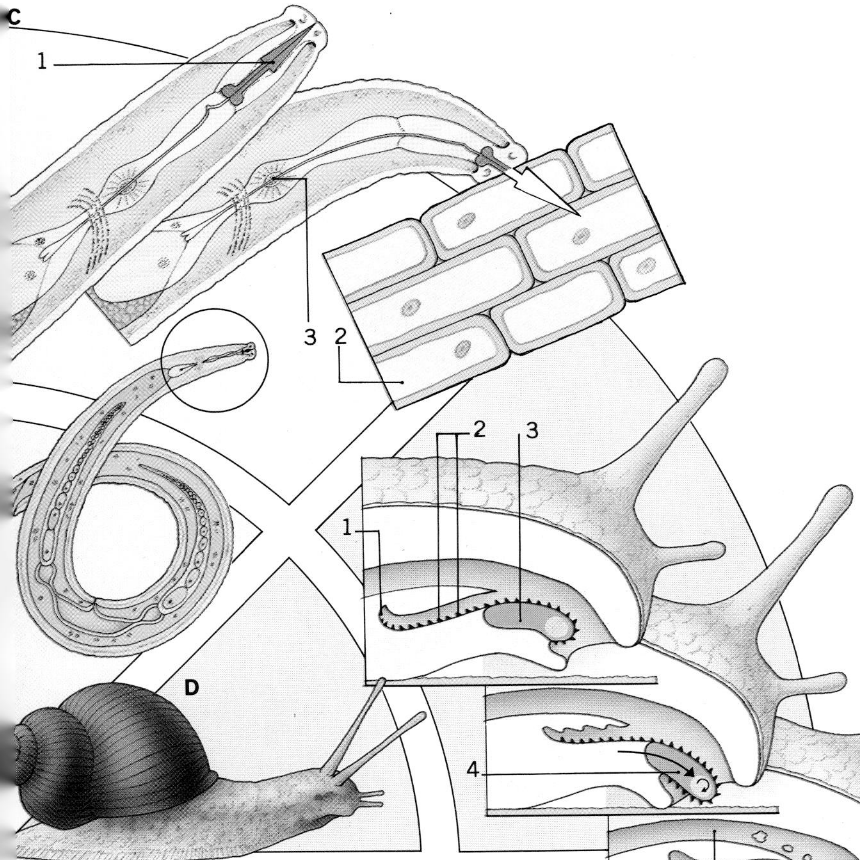

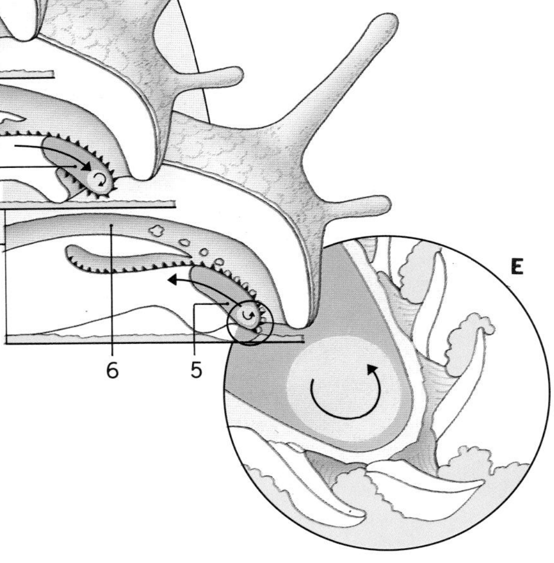

For other species smell is important. Many plants give off tiny traces of aromatic compounds or other chemicals that provide distinctive direction-finders. Taste is often used to check on the safety of food. Taste sensor cells are located in many parts of an insect's anatomy – butterflies, for example, taste with their feet.

Coping with food shortage

Some species of insects and plants manage to keep their populations roughly in balance, so that both plant and insect species thrive. But for many herbivorous invertebrates overgrazing or drought can cause a sudden crash in their food supply and hence their populations. Some – such as locusts – solve the problem by migrating, while others enter a period of dormancy. A few species of invertebrate have actually evolved the foresight to store food in their nests.

***The garden snail* [D]** *uses its radula* [1] *to rasp off vegetable food. The radula is a ribbon of flexible tissue, covered with rows of hard, angled teeth* [2]. *When feeding, the radula, which is attached to underlying cartilage known as the* odontophore [3], *extends* [4] *from the mouth and then retracts* [5], *in the process scraping up vegetable matter, which is passed along the esophagus* [6].

The radular teeth in closeup **[E]** *are seen to act like excavator shovels that cut and carry food. Constant wear erodes the teeth at the radula front, and these are replaced by new teeth formed at the radula rear.*

***The common sea urchin* [B]** *feeds on algae by means of a complex structure known as* Aristotle's lantern [1]. *Five calcium-based plates* [2] *tipped with hard, abrasive teeth* [3] *surround the mouth, which opens onto the underside of the body. A system of muscles* [4] *enables the teeth to be extruded below the* test *(shell)* [5] *and also permits a limited chewing action. The urchin's "bite" is powerful enough to tackle even the toughest seaweeds. Food particles suspended in water are drawn up into the* pharynx [6], *and excess water is internally siphoned off so that food is concentrated in the intestine* [7].

Some species *of soil-dwelling nematode worm* **[C]** *are equipped with a hollow, spearlike* stylet [1] *through which they feed on plant roots* [2], *attacking the zone of elongation. The nematode pierces the cell wall by a series of rapid thrusts of the stylet, which may then remain inserted in the cell for up to 30 minutes while fluids are sucked out via the bulb of the esophagus* [3]. *In some cases nematodes have been observed injecting a substance through the stylet and into the plant cell before feeding commences. This contains enzymes, which break down the food.*

Food Storage

In climates that feature a hot dry season or a cold winter the supply of food for small herbivores may dwindle at certain times of year. In response a few insects have evolved the ability to store food in their nests. Harvester ants and termites, which live in arid regions, bring home grass cuttings, seeds and other plant material and store it in special chambers in their underground nests. There may be large "drying rooms" in which the food is laid before being transported to the storerooms.

One of the most remarkable forms of storage is the "living larder" of the honeypot ants. A certain caste of ants remains in the nest and is solicitously fed regurgitated honey by other worker ants. Their abdomens distend as their stomachs fill with the liquid. They then regurgitate honey to other ants on demand.

Migration Patterns 228 Plant Defenses 244 Pollination by Animals 154 River Life 328 Termite Towers 234

Miniature Meat Eaters

How invertebrate carnivores feed

Of all today's carnivores the most aggressive and powerful for their size are not big cats or hunting dogs but small, active shellfish, snails, slugs, spiders and insects. The dragonflies that skim over ponds and streams on warm summer days are out hunting. These graceful insects have two pairs of wings, allowing them to reach speads of 18 miles an hour (30 km/h) as they search for smaller insects to eat. Fossil remains of dragonflies that flew 300 million years ago show that they were once even bigger, with wingspans as wide as 2.5 feet (75 centimeters).

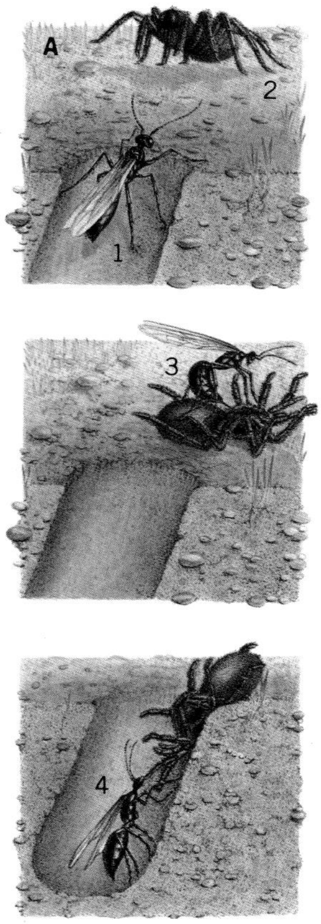

A female pepsis spider wasp eats nectar but provides her carnivorous larva with a living larder [**A**]. *Before laying eggs, the wasp* [1] *digs a hole to await a trapdoor spider* [2]. *Agility and a vicious sting enable her to overcome a venomous prey often larger than herself. She first dances round her victim, avoiding its venom fangs, which she may immobilize by stinging its head. The titanic battle ends in a paralyzing sting to the spider's nerve cord* [3]. *Dragging the inert body to the hole* [4], *the wasp lays an egg in it, food and home for the growing larva. Sealing the hole, she abandons the egg and finds new victims.*

One lurking predator is the sea slug, which moves steadily through rocky pools devouring sea anemones, quite immune to their stinging tentacles. Relatives of the sea slug, the dog whelks are carnivorous snails that rasp a round hole in the shells of other snails or bivalves, using filelike teeth, in order to feed on the soft inner parts. Other whelks use their own shells to wedge open the shells of oysters before consuming the oysters' flesh. Cuttlefish, squids and octopuses – all mollusk relatives of the whelks – are faster-moving hunters. Cuttlefish and octopuses seize crabs and prawns, squids take small fish.

Another marine hunter group is the starfish. Many feed on bivalve mollusks, gripping the two shells of mussels and similar creatures with strong, tubular feet on each "leg" of the star. A starfish keeps pulling until the bivalve is exhausted and its shell opens.

Many-legged hunters

Invertebrate land predators can be equally voracious. All centipedes are carnivorous, but the largest – the *Scolopendra* centipedes of Central America – can be as large as 10 inches (26 centimeters) long. They can even overcome mice, which they immobilize with poison claws. Despite their size, however, these centipedes are not as fast as the house centipede, which can travel at 20 inches (50 centimeters) a second to catch its insect prey.

A sticky end

Many spiders capture their prey using silken webs to entangle their victims. But web-spinning is not restricted to spiders. The freshwater caddis fly spins a net across fast-flowing water, anchoring it to stones on each side. Every so often the caddis fly checks the net for trapped water creatures. Conversely, many spiders do not spin webs. Big, hairy *mygalomorph* spiders will attack birds' nests but usually eat insects. They have a powerful bite but are rarely harmful to humans. The extremely keen-sighted and incredibly swift wolf spiders run down their prey. But jumping spiders, which leap on their victims, have the sharpest eyes of all the spiders. The European raft spider lives near streams and ponds and dives in pursuit of fish.

A sticky death also awaits the prey of the spitting spider. This tiny species makes no web but stalks small flies, waiting until just the right moment to spit a stream of poison glue, aimed with extreme precision, which fixes the fly to the ground.

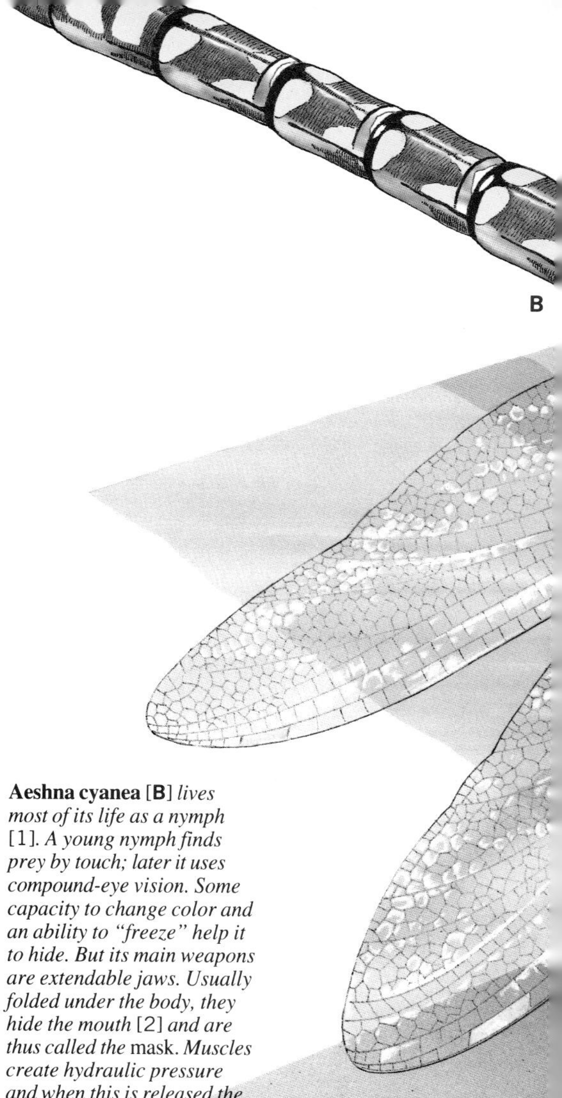

Aeshna cyanea [**B**] *lives most of its life as a nymph* [1]. *A young nymph finds prey by touch; later it uses compound-eye vision. Some capacity to change color and an ability to "freeze" help it to hide. But its main weapons are extendable jaws. Usually folded under the body, they hide the mouth* [2] *and are thus called the* mask. *Muscles create hydraulic pressure and when this is released the jaws* [3] *snap open to grab the prey.*

Maturity is reached when the final instar sloughs its old skin [4]. *Acute vision and speed make adult dragonflies* [5] *awesome hunters that catch prey in flight or at rest, holding prey in a leg "basket"* [6] *and taking it to the mouth with the front legs.*

Insect eyes [**C**] *resolve little detail, seeing only dark or light hexagonal cells. Each cell has one direct input to the brain, so the insect can detect rapid movement. A human eye sees beating wings as a blur* [1], *but an insect sees each stroke as changing cell patterns, like a slowed-down film* [2, 3].

Specialist tools

Long, sharp mouthparts, agility and aggressive behavior are characteristics of many predatory insects. Among the most active are ground beetles. They are usually black or brown, and conceal themselves beneath stones during daytime, emerging at night to hunt. Most use sharp mandibles to feed on passing insects, but one unusual snail-eating species has an elongated head, which it uses to reach right inside the shells of the snails.

Lacewing larvae – relatives of the antlions – are also predators, but they choose a more abundant source of food. They feed on aphids and other slow, soft-bodied arthropods that cannot defend themselves against the sharp, sucking mouthparts of the larvae.

Connections: Invertebrates 122 Lake Life 326 Poisonous Animals 250 Predatory Trappers 254 Spiders 252 Solitary Carnivores 246

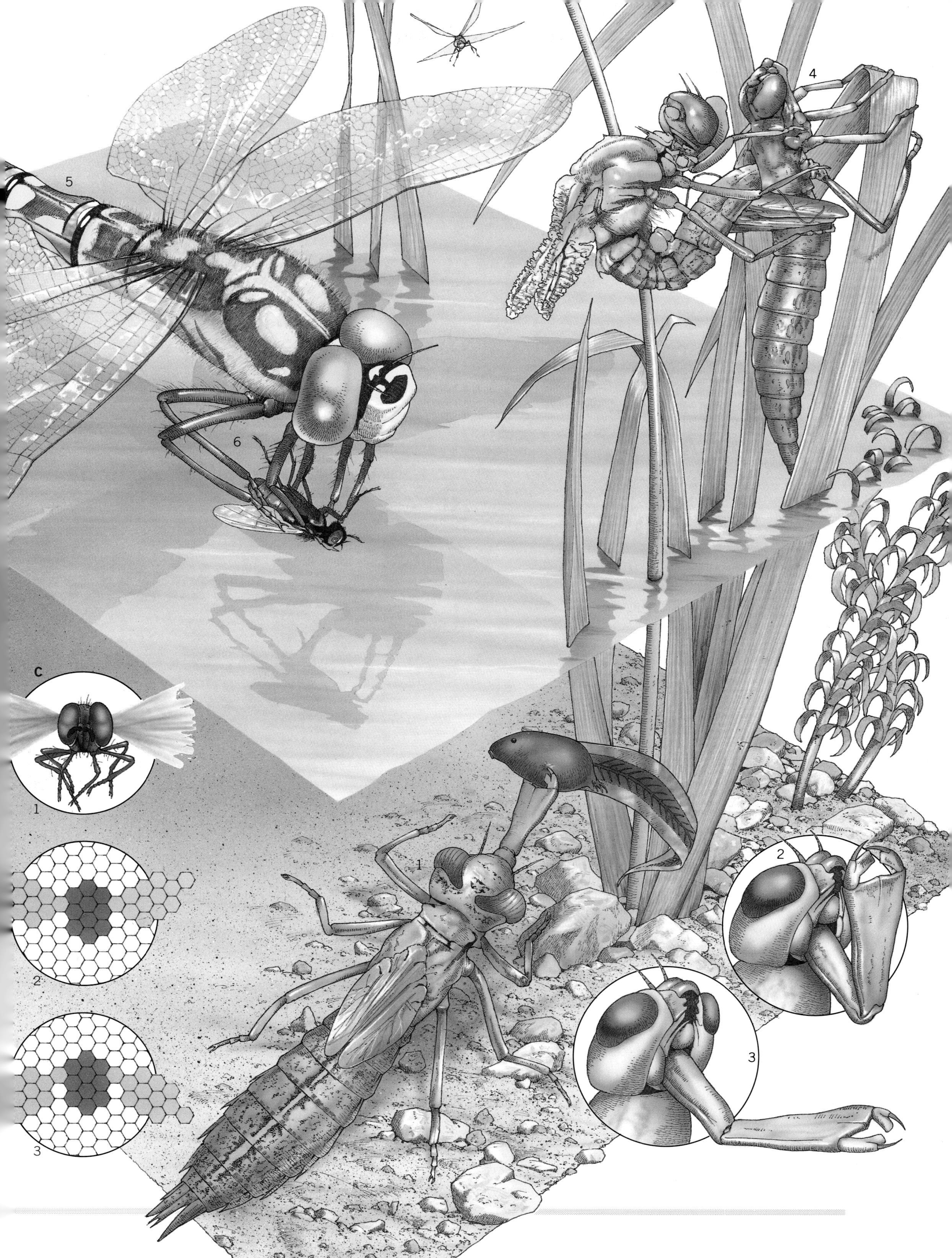

5
6
4
C
1
2
3
1
2
3

The Grazing Herds

How herbivores feed

Elephants digest less than half of everything they swallow. For this reason elephants have to consume several hundred pounds of vegetation in one day – often destroying large areas of savanna. But even without the problem of digestion – a problem various herbivores deal with in different ways – and despite the tremendous variety of plant food available (leaves, twigs, flowers, fruit, bark and roots), a plant diet is in any case low in protein and other nutrients. Herbivores must eat large quantities of food, and retain it for a long time, to extract energy from it.

As a foodstuff plants are not ideal. They often contain toxic chemicals that deter herbivores, and incorporate tough and woody fibers in their structure, which means they must be thoroughly chewed before digestion can proceed. In addition, plant cell walls are composed of the *polymer* (chain molecule) *cellulose* – potentially a good source of sugars – which cannot be broken down by mammalian enzymes. All these problems have been overcome by plant-eating mammals.

A herbivore's dentition is well suited to grinding plant matter, and its teeth grow throughout life to compensate for wear and tear. As food is being chewed, it is mixed with enzymes in the saliva that begin the digestive process and soften the food before it is swallowed. Food then passes slowly through a long and capacious digestive tract (gut), maximizing the amount of nutrients that can be extracted from it. Herbivore guts often have specialized areas containing bacteria, protozoa and fungi that can produce cellulose-splitting enzymes. These microorganisms break down cellulose, releasing soluble products that can be absorbed by the gut. The microorganisms are eventually digested by their mammal hosts, providing another source of nutrition. In rabbits the microbes are housed in an enlarged cecum, while in hippos, rhinos, elephants and horses they occupy the cecum and colon.

Ruminant mammals are able to regurgitate food in small amounts, once it has been partly digested, for chewing again ("chewing the cud"), reswallowing, and further digestion in special digestive chambers. This enables these animals to obtain a lot of food in a short time, then retreat to a sheltered place out of sight of predators to digest it. Cattle, sheep, goats, antelopes, giraffes, deer, hippos, camels, llamas, sloths and kangaroos are all ruminants. Because their food is softened internally before chewing, these animals have less developed jaws and facial muscles than nonruminants.

The quest for food

Most vertebrates find their food by sight, and to a lesser extent by smell. Flowers and fruit in particular are often conspicuous by their colors, which may also show whether a fruit is ripe. Smell can tell a herbivore whether a plant contains toxic chemicals.

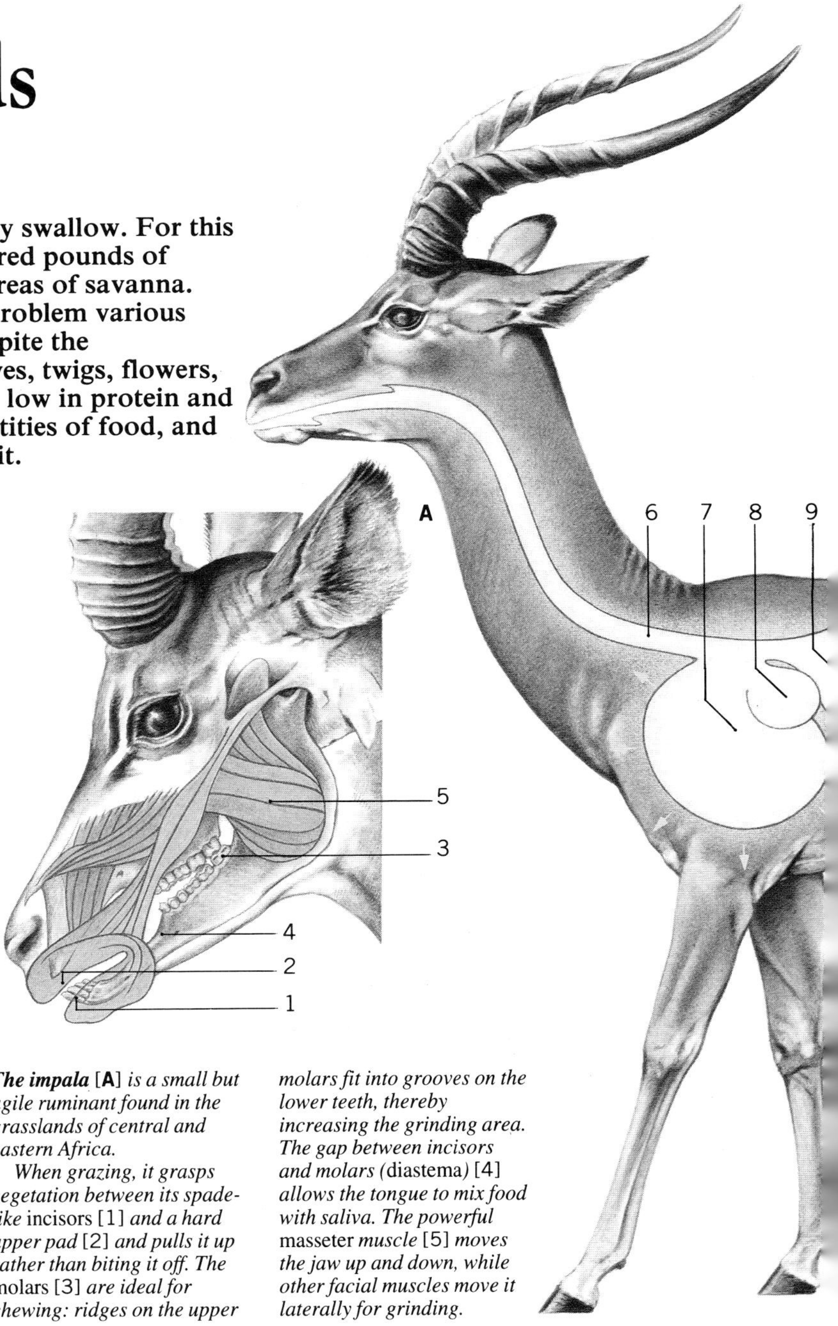

The impala **[A]** *is a small but agile ruminant found in the grasslands of central and eastern Africa.*

When grazing, it grasps vegetation between its spade-like incisors [1] *and a hard upper pad* [2] *and pulls it up rather than biting it off. The* molars [3] *are ideal for chewing: ridges on the upper molars fit into grooves on the lower teeth, thereby increasing the grinding area. The gap between incisors and molars* (diastema*)* [4] *allows the tongue to mix food with saliva. The powerful* masseter *muscle* [5] *moves the jaw up and down, while other facial muscles move it laterally for grinding.*

Many vertebrate herbivores live in large groups. Large flocks of birds have a better chance of finding food than single individuals. Large numbers of hoofed mammals serve a defensive purpose – if attacked by predators, each member of the herd has a slighter chance of being killed. Predators can also be avoided by remaining concealed during daylight hours and venturing out to feed at dawn and dusk, a habit that has been adopted by many deer and antelope.

In defense of food

If food is not particularly abundant, some species defend territories. Territoriality is particularly common during the breeding season, when more food is required for a growing family. For animals such as the antelopes and zebra of the African plains, however, a nomadic life style follows the sporadic rains and the new growth of grass.

After chewing, food passes down the esophagus [6] *to the largest chamber of the stomach – the* rumen [7]. *Enzymes produced by microbes in the rumen partially break down the food, which is then formed into small balls of cud. The cud is returned to the mouth and chewed at length, then reswallowed to let fermentation continue in the rumen and the* reticulum [8]. *Absorption occurs in the third chamber of the stomach – the* omasum [9]. *The remaining food particles proceed through the "true stomach"* [10], *small intestine* [11] *and* cecum [12], *where "normal" digestion takes place.*

Connections: **Bacteria** 106 **Plant Cells** 100 **Plant Defenses** 244 **Symbiosis** 192

B

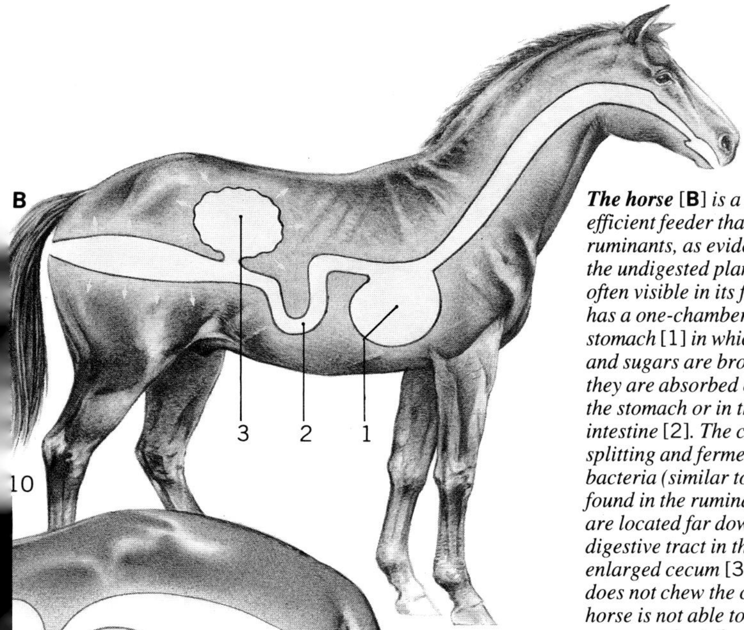

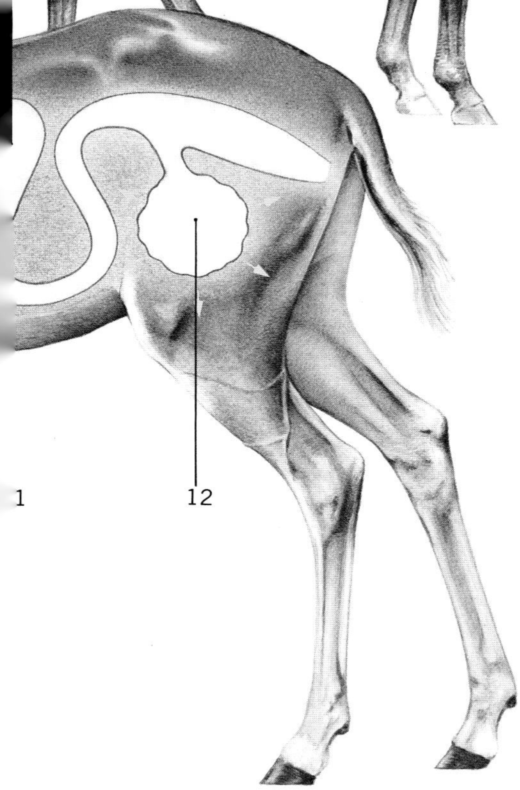

*__The horse__ [**B**] is a less efficient feeder than the ruminants, as evidenced by the undigested plant material often visible in its feces. It has a one-chambered stomach* [1] *in which proteins and sugars are broken down: they are absorbed either in the stomach or in the small intestine* [2]. *The cellulose-splitting and fermenting bacteria (similar to those found in the ruminant gut) are located far down the digestive tract in the enlarged cecum* [3]. *Since it does not chew the cud, the horse is not able to extract as many nutrients from its food as a ruminant, and consequently must take in more food to compensate.*

__Several herbivore__ species may conveniently share a habitat [**C**] *without competing for resources. On the African plains, giraffes* [1] *browse in branches up to 20 ft (6 m) above the ground. Elephants* [2] *too can browse tree canopies, using their trunks to pluck off vegetation. Eland* [3] *attack the middle branches with their horns, twisting twigs to break them off, while gerenuk* [4] *stand on their hind legs to reach higher branches. The black rhino* [5] *uses its hook-like upper lip to feed on bark, twigs and leaves (white rhinos have lengthened skulls and broad lips for grazing the short grasses that they favor). The wart hog* [6] *and dik-dik* [7] *eat buds and flowers, and will also dig up roots and tubers.*

Resource partitioning also occurs among the grazers: migrating zebra [8] *crop the taller, coarse grasses; wildebeest* [9] *feed on the leafy center layer, allowing small gazelles* [10] *to reach the tender new shoots.*

Slow Success

Some herbivores – such as sloths and koalas – survive on a particularly poor diet of tough leaves by means of a life style that involves very little expenditure of energy. Koalas, which feed only on tough eucalyptus leaves, sleep for 18 hours a day. Sloths also sleep a lot and move extremely slowly. In fact they have only half the musculature of most mammals and their food may take a whole week to pass through the digestive system. Thick fur reduces heat loss, saving more energy.

C

When the Rot Sets In

How dead matter is decomposed

More than 90 percent of all plant material is ultimately eaten by the "decomposers" – bacteria, insects and their larvae, worms and fungi – which break down the material, returning the basic vital nutrients back to the soil and to the ecosystems that rely on them. That leaves only approximately 10 percent of all plant material for the better-known herbivores, such as cows and sheep. Without the decomposers any nutrients would remain locked up in the dead bodies of plants and animals, which would quickly accumulate as a stifling debris on the ground.

The stench *of decaying flesh attracts female flies to a dead wood mouse* [**A**]. *Female flies deposit their eggs in moist, shaded parts of the carcass, such as the mouth, ears and anus. The larvae hatch after about 24 hours and release a cocktail of enzymes that liquefies the mouse's tissues.*

A

B

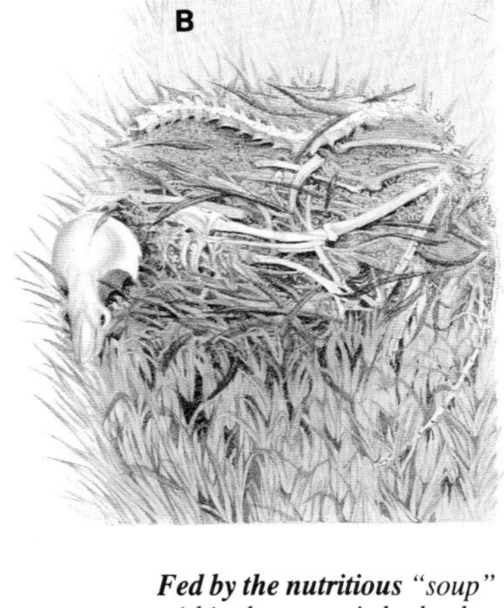

Fed by the nutritious *"soup" within the mouse's body, the larvae reach maturity in a matter of days* [**B**]. *About two weeks after hatching, they crawl underground to pupate and the adult flies emerge a few days later. Only the hard, calcified parts of the mouse are left behind to be broken down later.*

The body of an animal or plant is broken down by a combination of physical and chemical processes. Animal corpses are first attacked by scavengers, such as hyenas and vultures, beetles and ants. Material that is not completely digested is excreted as dung, which in turn is attacked by beetles and microorganisms. Fallen tree trunks are invaded by borer beetles and their larvae. Small creatures such as flatworms crush and consume tiny pieces of organic matter.

But the most numerous and important of the decomposers are the smallest – the bacteria and the fungi. They feed by secreting digestive enzymes onto their food and absorbing the soluble products directly through their own body surfaces. These organisms complete the breakdown process and release simple compounds back into the environment, where they can once again be taken up by green plants. Organic matter is composed primarily of carbon, hydrogen, oxygen and nitrogen: as it is broken down, carbon and hydrogen combine with oxygen in the process known as respiration. The products of this process – water and carbon dioxide gas – are also the raw materials of photosynthesis. Similarly, bacteria release nitrogen into the soil in the form of soluble nitrates that are taken up by plant roots.

Other organisms that contribute to the decomposition process include various types of beetle, maggots (fly larvae) and other larvae, ants and termites, slugs and snails, worms and woodlice, millipedes and mites. Other, larger, animals that help both to mix the surface layers of the soil and to enrich it with their excreta include mice, moles, badgers, rabbits and armadillos.

Fast food from the forests

The resources available within any ecosystem are limited and its productivity – the rate of new growth – therefore depends in part on how quickly and efficiently minerals and nutrients can be recycled. Many tropical rain forests – considered to be the most productive terrestrial ecosystems – thrive on relatively poor soils. The profusion of life in these forests depends on the rapid rate of decay in the warm, humid climate, which releases nutrients back into the soil. Such decay occurs more slowly on acidic, waterlogged soils into which oxygen cannot penetrate: many fungi and bacteria need oxygen to respire and are therefore unable to survive in these environments.

C

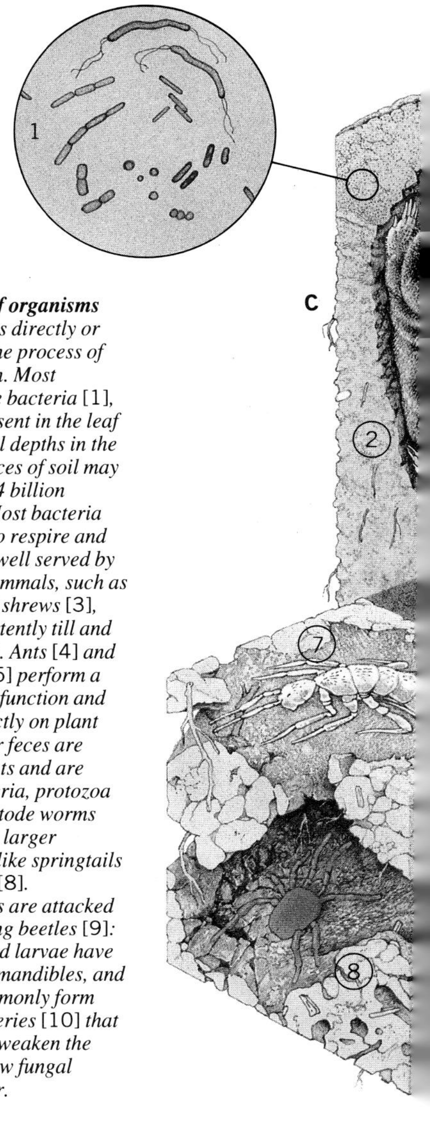

A multitude of organisms [**C**] *contributes directly or indirectly to the process of decomposition. Most significant are bacteria* [1], *which are present in the leaf litter and at all depths in the soil: 0.04 ounces of soil may contain up to 4 billion individuals. Most bacteria need oxygen to respire and are therefore well served by burrowing mammals, such as moles* [2] *and shrews* [3], *which inadvertently till and aerate the soil. Ants* [4] *and earthworms* [5] *perform a similar tilling function and also feed directly on plant material: their feces are rich in nutrients and are eaten by bacteria, protozoa and tiny nematode worms* [6], *as well as larger invertebrates like springtails* [7] *and mites* [8].

Tree stumps are attacked by wood-boring beetles [9]*: both adults and larvae have sharp cutting mandibles, and the grubs commonly form extensive galleries* [10] *that mechanically weaken the wood and allow fungal spores to enter.*

Decomposition in the sea

There are many sorts of scavenging fish, crabs, shrimps and other invertebrates. As tiny pieces of organic matter discarded by these and other scavengers float down through the water, they are consumed by myriad filter-feeding animals that trap the particles. These include sea anemones, barnacles, feather stars, soft corals, sea fans and a multitude of planktonic animals. Should the organic matter eventually reach the seabed, it becomes food for bottom-dwelling worms, sea cucumbers, sea slugs and snails. Even when whole carcasses of larger creatures sink to the sea floor, scavengers such as hagfish and crabs scurry to the attack, detecting their food by smell and touch in the total darkness.

Connections: **Bacteria** 106 **Crawling** 214 **Cycles of Life** 206 **Fungi** 110 **Protozoa** 108 **Soils** 30 **Termite Towers** 234

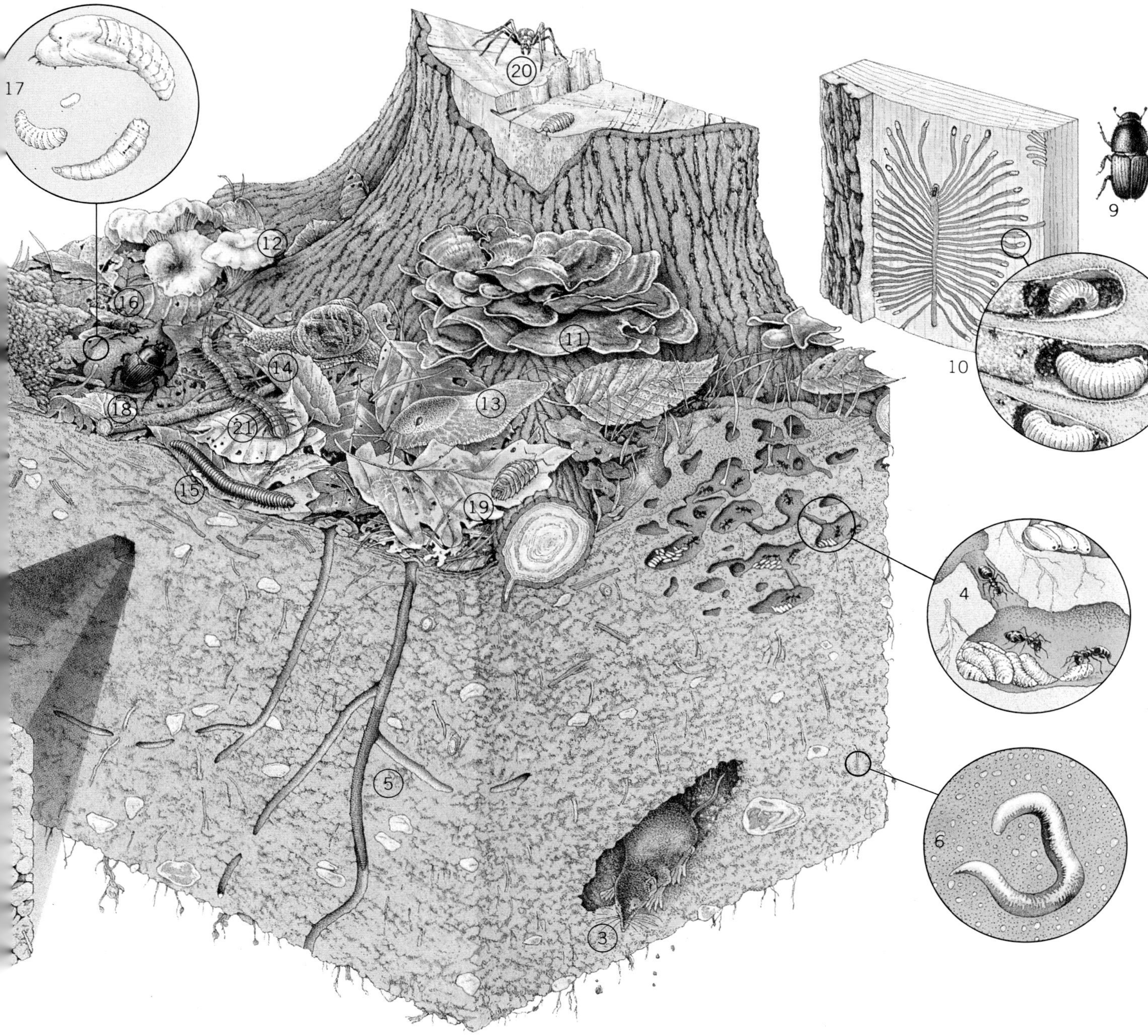

A network of fine fungal threads ramifies through the tree stump and secretes digestive juices into it: when conditions are right, some fungi produce large fruiting bodies – the familiar bracketlike structures [11] *and toadstools like the chanterelle* [12].

The leaf litter is made up of partially decomposed leaves, twigs, bark, animal corpses and feces. It supports fauna, such as slugs [13], *snails* [14], *millipedes* [15], *flies* [16], *fly larvae* [17] *and beetles* [18], *and scavengers such as woodlice* [19]. *Predators like spiders* [20] *and centipedes* [21] *exploit this abundant resource.*

Waste Disposal Units

Dung beetles are efficient recyclers of dung, carrying lumps down into the soil for their larvae to feed on and thus aiding its decomposition by soil microorganisms. In northern Australia the native dung beetles feed on dry, fibrous kangaroo dung. But since the late 18th century large numbers of cattle have been introduced, which produce a truly vast quantity of dung every day. The Australian dung beetles are unable to break down this huge amount and so, to combat the crisis, African dung beetles were introduced to Australia. So sensitive are these beetles to the odor of fecal gas that by the time buffalo dung first hits the ground most of the dung beetles within the area are already on the way toward it. With no native competitors, they multiplied rapidly and soon solved the problem.

The Elements of Life

How essential elements are recycled

Bacteria of the genus *Rhizobium* are reckoned to recycle around 200 billion tons of inert atmospheric nitrogen into ammonia every year. Most plants are dependent on this process of nitrogen fixation because they could not otherwise take up this essential element. But nitrogen is just one of the many chemical elements that flow through the bodies of all living organisms; others include carbon, oxygen and sulfur. These elements are continually being recycled through the environment as animals feed, respire, grow, and eventually die and decompose.

__Elemental carbon__ is in constant flux. Gaseous carbon dioxide (CO_2) is first incorporated into simple sugars by photosynthesis in green plants. These may be broken down (respired) by the plant to provide energy, a process that releases CO_2 back into the atmosphere. They may also be used to make larger molecules like starch and cellulose, which are ingested by herbivores that may themselves be eaten by carnivores. These animals also metabolize the sugars and release CO_2 in the process. Fungi and many bacteria release CO_2 as they decompose dead organic matter in the soil.

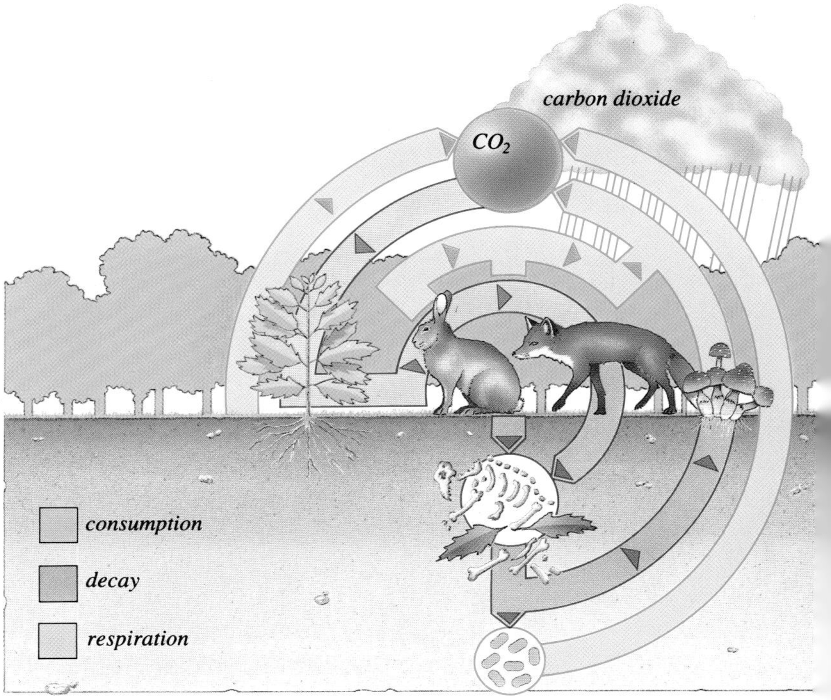

All living things are made up of molecules based on carbon. The carbon atoms come originally from the carbon dioxide in the atmosphere and enter the living world by photosynthesis in green plants, algae and some bacteria. These "primary producers" are the starting point of a cycle of consumption, decay and respiration in which carbon atoms are incorporated into complex organic compounds (with the net expenditure of energy) and then released from these compounds (liberating energy).

Geological processes also affect the Earth's carbon balance – carbon is effectively removed from the organic cycle when it is precipitated as limestone in the oceans, or forms fossil fuels. Conversely, large volumes of carbon dioxide are released whenever a volcano erupts.

The amount of carbon dioxide in the atmosphere, compared to that of the main constituents, nitrogen and oxygen, is very small – only 0.03 percent by volume – but it is increasing because human activities are unlocking the vast reserves of stored carbon in fossil fuels and long-lived forests, a process that is contributing to global warming.

Fixing nitrogen

Living organisms also need the element nitrogen, specifically to make proteins and the nucleic acids, DNA and RNA. Nitrogen makes up 78 percent of the atmosphere (oxygen accounts for 21 percent, and the remaining 1 percent is composed of traces of carbon dioxide, hydrogen and the "noble" inert gases such as argon, xenon and radon, along with various pollutants).

The vast majority of organisms, however, cannot use nitrogen in its gaseous form and can take up the element only after it has been "fixed" into water-soluble ammonia or nitrogen oxides. This fixation is carried out by bacteria, some of which live free in the soil (e.g. *Azotobacter*), but most of which form symbiotic associations with higher plants. The most important of these associations is between the bacterium *Rhizobium* and the roots of legumes, such as peas, beans and clover. After entering the plant through a root hair the bacteria multiply in the root cortex, which develops a bulbous swelling (nodule) made up of enlarged plant cells packed with *Rhizobium*. Drawing their energy from the legume's photosynthetic products, the bacteria are able to incorporate atmospheric nitrogen into ammonium compounds, some of which are then used by the legume to make proteins and nucleic acids. Leakage of ammonia and death and decay (or consumption) of the legumes releases nitrogen into the soil for other organisms to use.

Oxygen for life

All animals and plants, and many microorganisms, need a third component of the atmosphere – oxygen – to oxidize foodstuffs and generate energy during respiration. Oxygen is also combined with carbon, hydrogen and nitrogen to make biological molecules. All organisms that need it can use atmospheric oxygen directly. Oxygen gas is returned to the atmosphere from the living world by green plants and cyanobacteria, which give out oxygen during photosynthesis.

Atmospheric nitrogen *is "fixed" into ammonium compounds by specialized bacteria such as* Rhizobium, *which forms colonies within* nodules *on the roots of* leguminous *plants. Ammonium and nitrogen compounds are formed by the action of lightning in the atmosphere, and are also released into the soil when decomposers break down feces and organic remains. Other bacteria living free in the soil then convert ammonium compounds into* nitrites *(NO_2^-) and then* nitrates *(NO_3^-), which can be taken up by green plants. Some soil bacteria reverse the process, liberating N_2 gas from nitrates in the soil.*

Connections: Atmosphere 64 Bacteria 106 Breathing 126 Coal, Oil, and Gas 34 Decomposers 204 DNA 96 Global Warming 84 Minerals 32

Early, nitrogen-poor soils *were formed by the weathering of rocks on the surface of the Earth. The action of lightning in the turbulent early atmosphere brought about the formation of water-soluble ammonia and nitrogen oxides. These rained down onto the planet, enriching these soils with nitrogen. As life evolved, nitrogen was incorporated into more complex compounds, and today the element is constantly recycled in a series of biological and chemical processes. Free nitrogen (N_2) cannot be absorbed directly by plants or animals; its uptake depends on the action of bacteria in the soil.*

The Sulfur Cycle

Sulfur (S) is part of many proteins. It enters the proteins of plants (and subsequently of animals) when it is taken up from the soil as sulfate (SO_4^{--}). It is released by bacterial action on decomposing plant and animal remains as sulfide (S^{--}). Sulfides are oxidized to free sulfur by one group of soil and water bacteria, and then to sulfates by another. The sulfur cycle can be turned by bacteria alone, as there is yet another group of bacteria that can reduce sulfate to sulfide [1].

Communities of interdependent sulfur bacteria occur naturally in sulfur springs, living on each other's waste products. In certain conditions in the past the sulfur cycle seems locally to have stopped at the sulfur stage, resulting in sulfur lakes appearing in some warm, dry parts of the world.

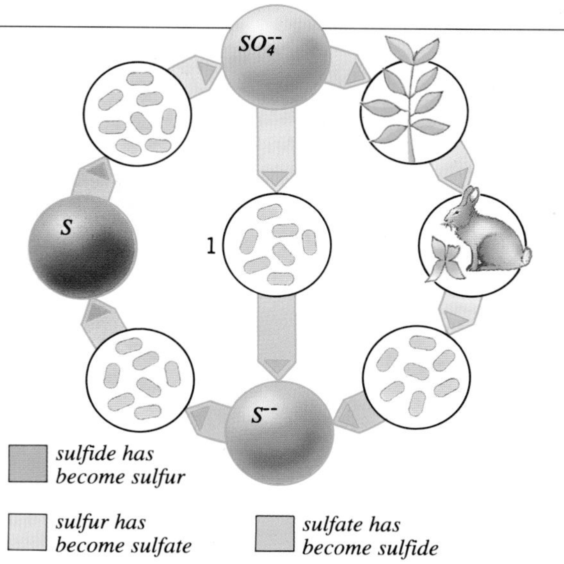

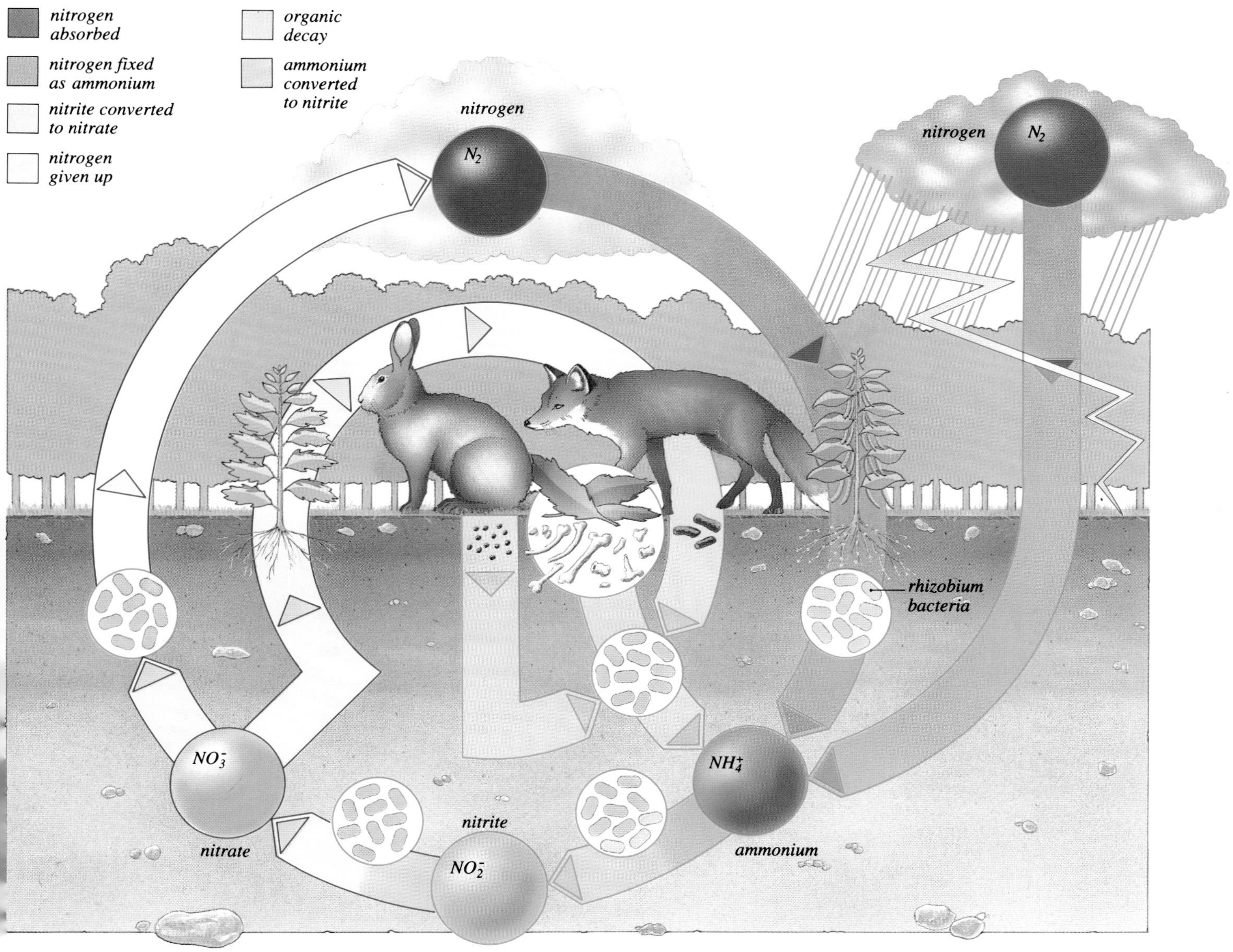

Photosynthesis 182 Soils 30

Swimming
Jet-propelled Animals
Crawling
Climbing
Leaping

6 Movement and Shelter

Walking and Running
Gliding and Soaring
Bird and Bat Flight
Insect Flight
Migration Patterns
Migration Techniques
Bees' and Wasps' Nests
Termite Towers
Ants' Nests
Birds' Nests
Beaver Lodges

Making Waves

How animals swim

Sloths and hedgehogs are fine swimmers. Unlikely as that seems, it is evidence that swimming is a skill possessed by a remarkable range of animals, from tiny amoebas to huge whales. Even a single drop of water placed under the microscope reveals myriad different swimming microorganisms. In all these creatures the basic principle of swimming is the same: forward propulsion is achieved by thrusting backward against the water. Some animals swim mainly by flexing their bodies; others use appendages such as legs, fins, flippers, or cilia – tiny whiplike hairs.

In order to reduce friction in the water, which would clearly slow forward movement, most animals that rely heavily on swimming have streamlined bodies – bodies shaped rather like a torpedo, a form that offers minimum resistance to water flowing over it. Fish have evolved streamlined bodies – the faster each type of fish can swim, the more closely its body shape approaches that of a torpedo. Very fast fish, such as tuna, are almost perfectly torpedo shaped.

Marine mammals evolved from four-legged land animals. Their arms and legs developed into flippers that can be laid flat against the body while moving forward, except for the brief time during which they are needed for propulsion. Whales and dolphins have lost their hind legs altogether, but those animals that do use their legs for swimming are in general equally well streamlined, following the thrust stroke, when the legs are flattened against the body. Even as the legs are brought forward for the next stroke they are usually kept folded close to the body to minimize resistance. Fish reduce resistance even further: they are coated in slimy mucus that permits water to slip freely over them.

Marine propellers

Fish have two main means of propulsion: the flexing of their bodies to thrust against the water and the sculling (or rippling) action of their fins. The undulating movements of their bodies enable most fish to swim rapidly, while the sculling action of the fins is primarily used for slow, precise movements, such as when a goldfish, for example, is feeding.

The fins in this case simply beat to and fro against the water. The tail fin in particular is used in this way, and by varying the degree to which it beats to one side or the other it also acts as a rudder. The dorsal (upper) fin in some fish is very long and wide; it ripples, achieving propulsion in much the same way as undulating the body from side to side. Very stiff-bodied fish, such as boxfish, pipefish and sea horses, propel themselves mainly with their paired pectoral fins, which beat in a sculling action that produces rather jerky movements.

In some fish the tail fin is operated by very powerful muscles and is capable of propelling the fish out of the water altogether. Flying fish, for example, can leap into the air and

A frog [**B**] *swims by a series of kicks from its powerful hind legs. The front legs are used only for steering. At the beginning of the swimming kick* [1] *the hind legs are drawn up to the body and are then thrown wide apart* [2], *at which point the soles of the feet are parallel with the spine. The toes then spread wide* [3], *opening the webs between them to present maximum surface area to the water. The legs are then straightened and brought together* [4], *propelling the frog forward. During the recovery kick* [5 *and* 6], *when the legs are drawn up to the body, the webs are closed to reduce drag.*

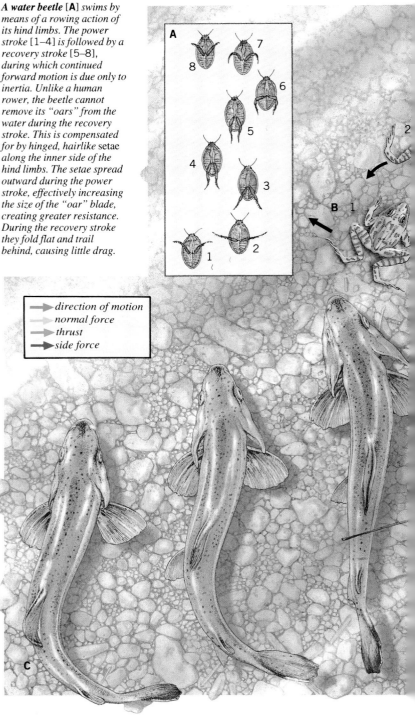

A water beetle [**A**] *swims by means of a rowing action of its hind limbs. The power stroke* [1–4] *is followed by a recovery stroke* [5–8], *during which continued forward motion is due only to inertia. Unlike a human rower, the beetle cannot remove its "oars" from the water during the recovery stroke. This is compensated for by hinged, hairlike* setae *along the inner side of the hind limbs. The setae spread outward during the power stroke, effectively increasing the size of the "oar" blade, creating greater resistance. During the recovery stroke they fold flat and trail behind, causing little drag.*

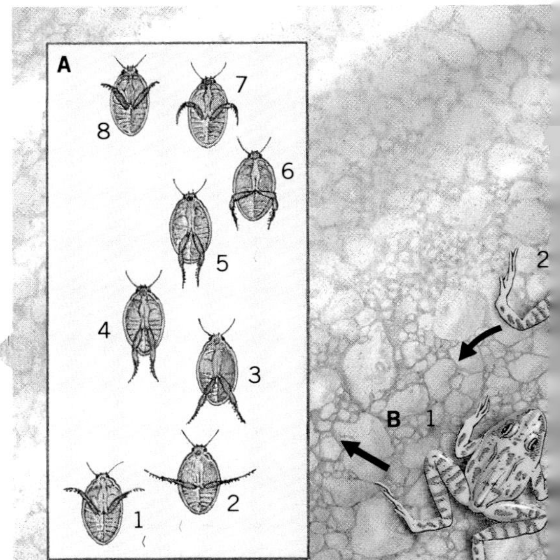

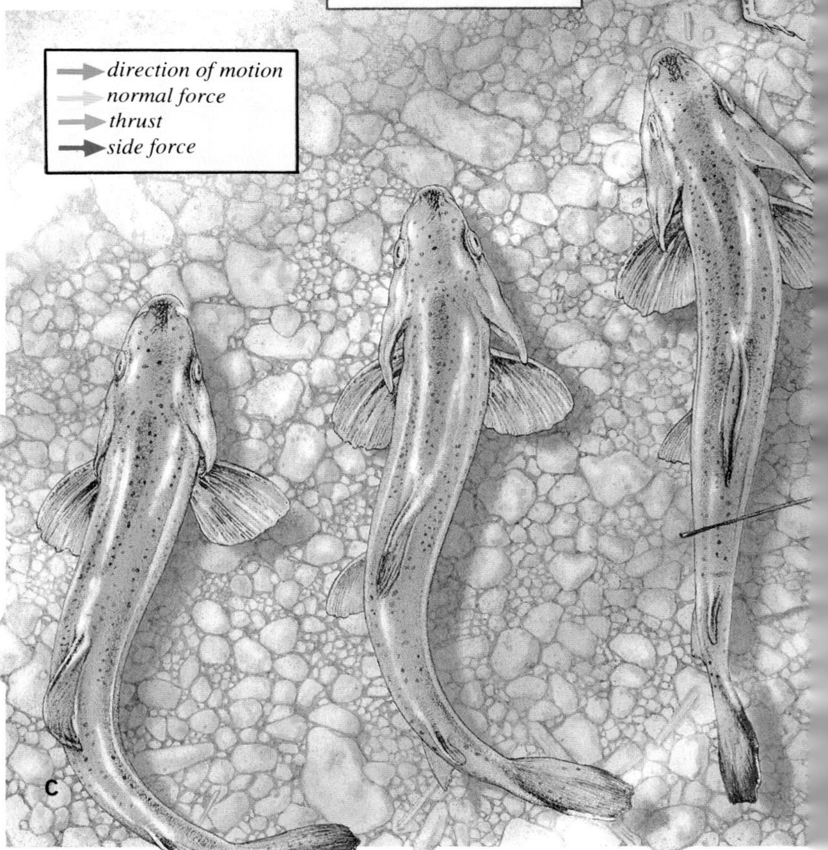

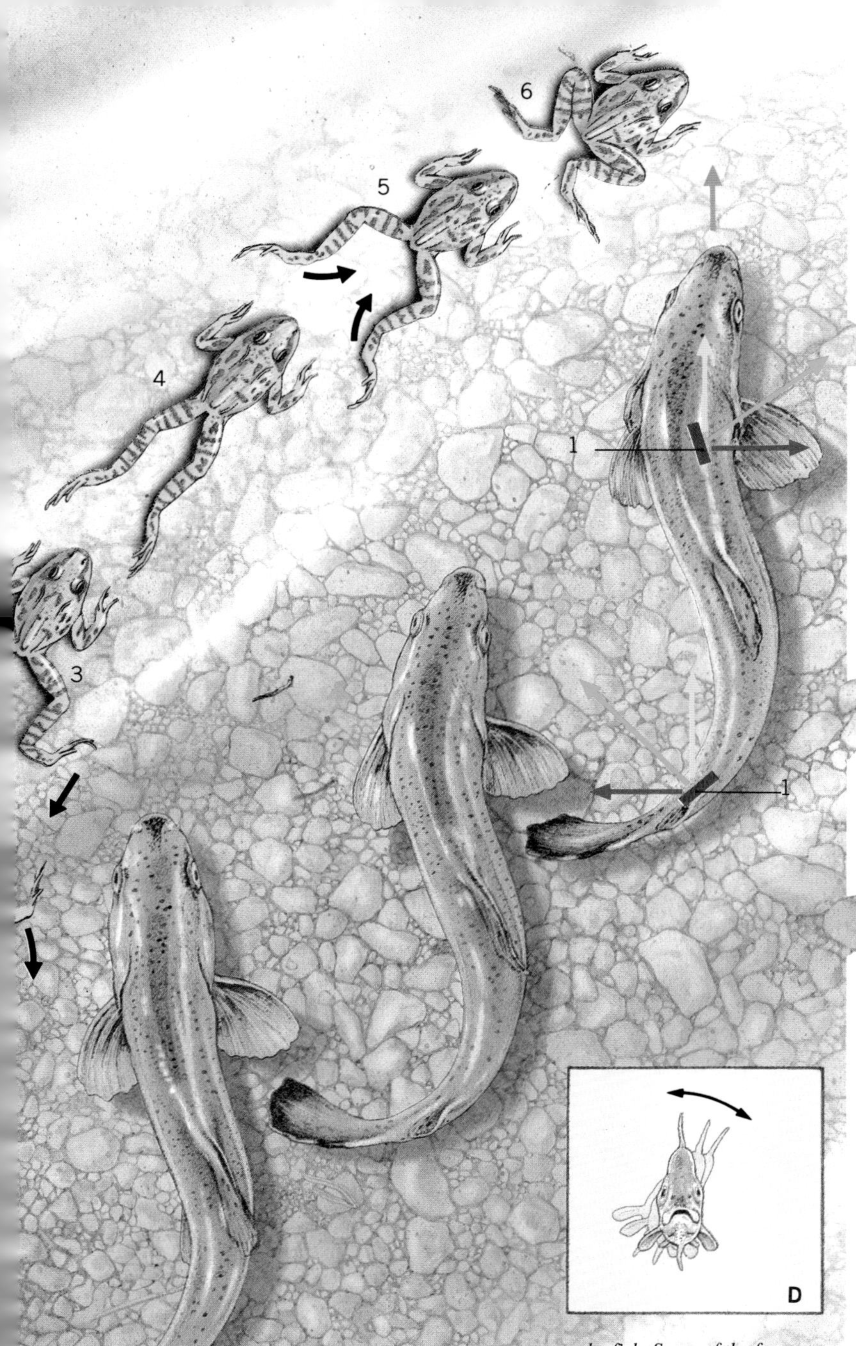

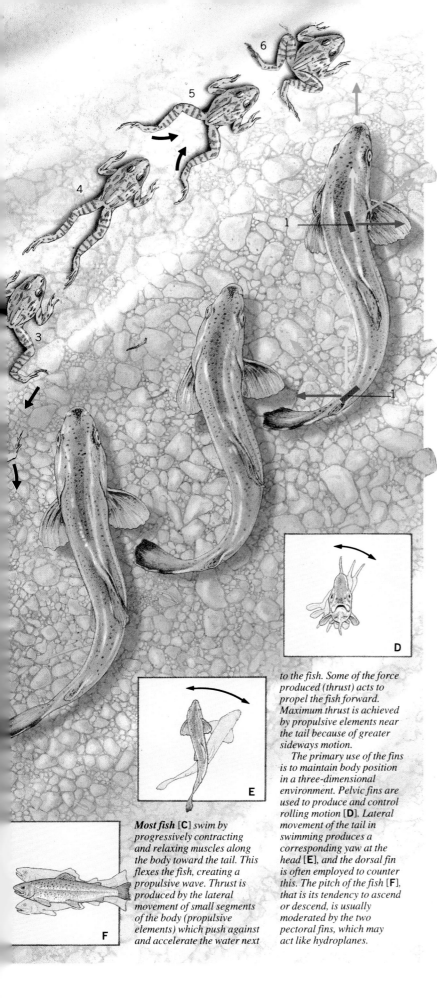

glide for extraordinary distances to escape predators, while salmon can leap large waterfalls in a desperate bid to reach their breeding grounds farther upstream.

Rippling along

Rays and skates, which are flattened for a life on the seabed, undulate their bodies and pectoral fins up and down instead of from side to side. The thrust is actually down and back at an angle, so the fish both propels itself forward and maintains its vertical position in the water. Flatfish appear to swim in a similar way, but in fact their bodies are flattened sideways and they lie on one side, so they are really undulating their bodies from side to side. Marine flatworms also swim with an up-and-down undulation of their greatly flattened bodies, and whales and dolphins use a similar motion, their huge flattened tail flukes providing the thrust.

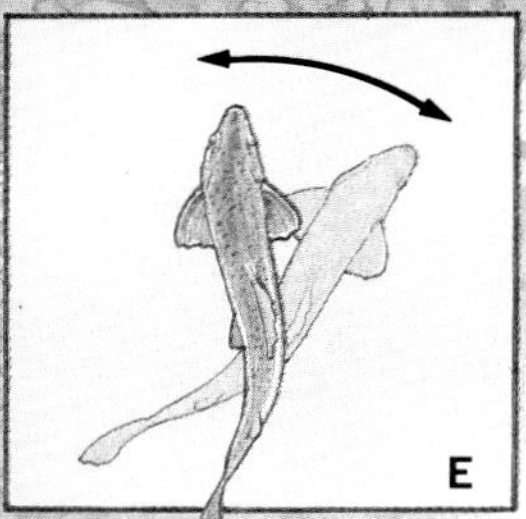

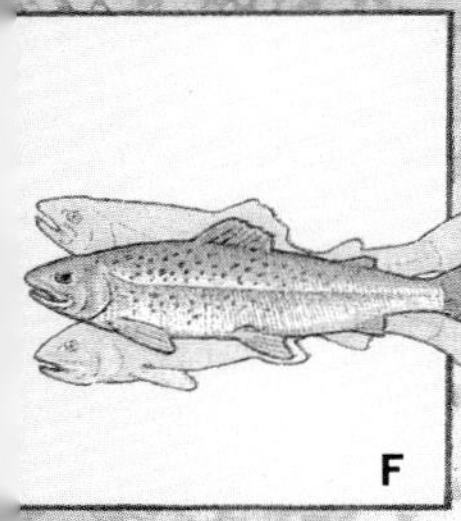

Most fish *[C] swim by progressively contracting and relaxing muscles along the body toward the tail. This flexes the fish, creating a propulsive wave. Thrust is produced by the lateral movement of small segments of the body (propulsive elements) which push against and accelerate the water next to the fish. Some of the force produced (thrust) acts to propel the fish forward. Maximum thrust is achieved by propulsive elements near the tail because of greater sideways motion.*

The primary use of the fins is to maintain body position in a three-dimensional environment. Pelvic fins are used to produce and control rolling motion [D]. Lateral movement of the tail in swimming produces a corresponding yaw at the head [E], and the dorsal fin is often employed to counter this. The pitch of the fish [F], that is its tendency to ascend or descend, is usually moderated by the two pectoral fins, which may act like hydroplanes.

In Buoyant Mood

To remain at a particular depth, a fish must withstand the water pressure, and its density must be roughly the same as that of the surrounding water – otherwise it drifts up or down. Animals of the upper layers of lakes and seas often have a large surface area to volume ratio to assist in floating, like the manta ray (below). Many marine larvae and larval fish have long spines that increase surface area. Sharks have oil-rich livers to assist buoyancy – oil being less dense than water, while marine mammals have a thick layer of fat under the skin, collapsible ribs and deflatable lungs. However, the shark's aid to buoyancy is not entirely successful – the fish will sink if it stops swimming. Cuttlefish have a spongy cuttlebone, and many fish have swim bladders filled with a volume of gas that can be voluntarily adjusted.

Lake Life 326 Mammals 136 Ocean Life 332 River Life 328

The Jet Set of the Sea

How marine animals move by jet propulsion

The so-called flying squid can propel itself right out of the water and has even been known to land on the decks of ships. It does this by forcefully expelling seawater in one direction, thus generating thrust in the opposite direction – jet propulsion. Larger squid sustain swimming speeds of up to 20 miles an hour (30 km/h). In the animal kingdom jet propulsion is used mainly by marine invertebrates, including the octopuses, cuttlefish and jellyfish. Even scallops can shoot themselves along at up to 12 inches a second (30 cm/sec) for short distances.

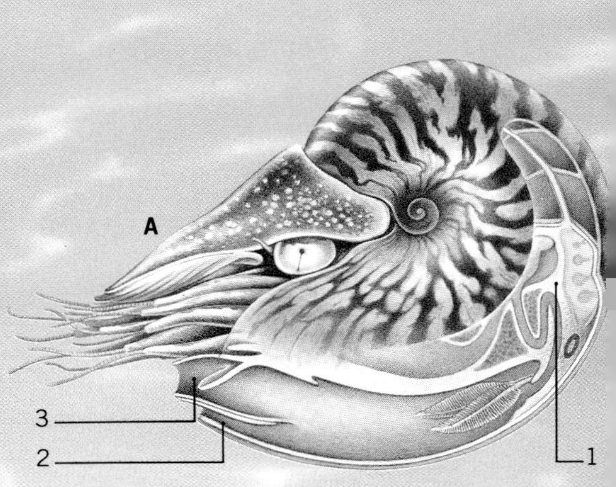

__The nautilus' shell__ [A] is divided into about 30 different compartments. But the body of the nautilus only occupies the first, and largest, chamber of its shell [1]. All the other chambers are self-contained buoyancy tanks filled with gas. The nautilus nevertheless is a poor shape for swimming. Water passes into the mantle [2] around its whole edge. It is expelled from the funnel [3] by the funnel muscles themselves and by the animal expanding its body in the shell. Unlike its relatives the squids, octopuses and cuttlefish, the nautilus cannot contract its mantle, which is attached to its shell.

Squid are the fastest jet-propelled animals in the oceans. The overall form of their body resembles a torpedo, the ideal streamlined shape for minimum drag. Squid and cuttlefish use a powerful combination of muscles that work in opposition to each other to produce the jet – by squeezing water out of the animal's body – and restore the body shape afterward. But whereas the squid has to keep swimming – or else it will sink – the cuttlefish has a spongy internal shell of "cuttlebone" that improves its buoyancy. The chambered cuttlebone contains gas at its front end and liquid toward its rear. On the other hand, the shell increases drag, so the cuttlefish is not such an efficient swimmer.

Water pumps

To achieve fast acceleration, the squid must pump out a large volume of water. This may be very effective when escaping from a predator, or when moving rapidly to seize prey, but it is very expensive in terms of the energy expended. In normal swimming the pressure generated by the squeezing of the squid's body is only one-tenth of that used for maximum acceleration. The squid directs the water out of its body through a funnel at the head end (and so swims backward) – but the funnel can be curved for swimming forward.

Jet propulsion serves other purposes for some animals. Cuttlefish direct a jet of water at the sandy seabed to flush out small invertebrate prey. Small cuttlefish and squid may use jet propulsion to bury themselves backward in the seabed.

Jellyfish

Although jellyfish might be said to have been ideally designed for jet propulsion, a jellyfish's weak body wall cannot produce very powerful contractions. Contractions of the rounded bell force out jets of water diagonally downward, generating both a backward and a downward thrust. The jellyfish is thus kept up and propelled forward. Between the outer and inner layers of the jellyfish's body wall is a layer of jellylike material, the mesoglea. As the muscles of the bell relax after pumping out a jet of water, the fluid pressure in the mesoglea restores the jellyfish's original shape, sucking in water ready for the next jet. The smooth shape of the jellyfish's bell helps to reduce drag.

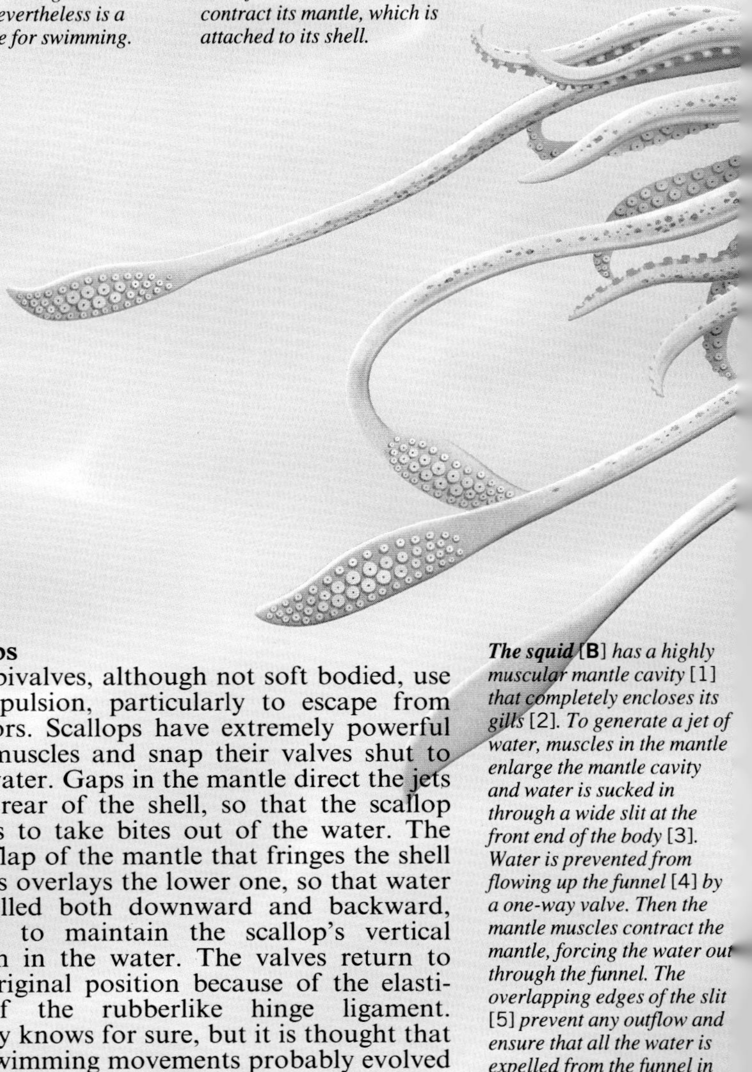

Scallops

Many bivalves, although not soft bodied, use jet propulsion, particularly to escape from predators. Scallops have extremely powerful hinge muscles and snap their valves shut to expel water. Gaps in the mantle direct the jets to the rear of the shell, so that the scallop appears to take bites out of the water. The upper flap of the mantle that fringes the shell margins overlays the lower one, so that water is expelled both downward and backward, helping to maintain the scallop's vertical position in the water. The valves return to their original position because of the elasticity of the rubberlike hinge ligament. Nobody knows for sure, but it is thought that these swimming movements probably evolved from the occasional cleaning contractions of the shells that the scallops perform in order to expel undigested food and other debris from the mantle cavity.

__The squid__ [B] has a highly muscular mantle cavity [1] that completely encloses its gills [2]. To generate a jet of water, muscles in the mantle enlarge the mantle cavity and water is sucked in through a wide slit at the front end of the body [3]. Water is prevented from flowing up the funnel [4] by a one-way valve. Then the mantle muscles contract the mantle, forcing the water out through the funnel. The overlapping edges of the slit [5] prevent any outflow and ensure that all the water is expelled from the funnel in a narrow jet for maximum thrust. The thrust also depends upon the whole mantle contracting

Connections: **Filter Feeders** 196 **Flight and Deception** 260 **Ocean Life** 332 **Predatory Trappers** 254 **Seashores** 330 **Swimming** 210

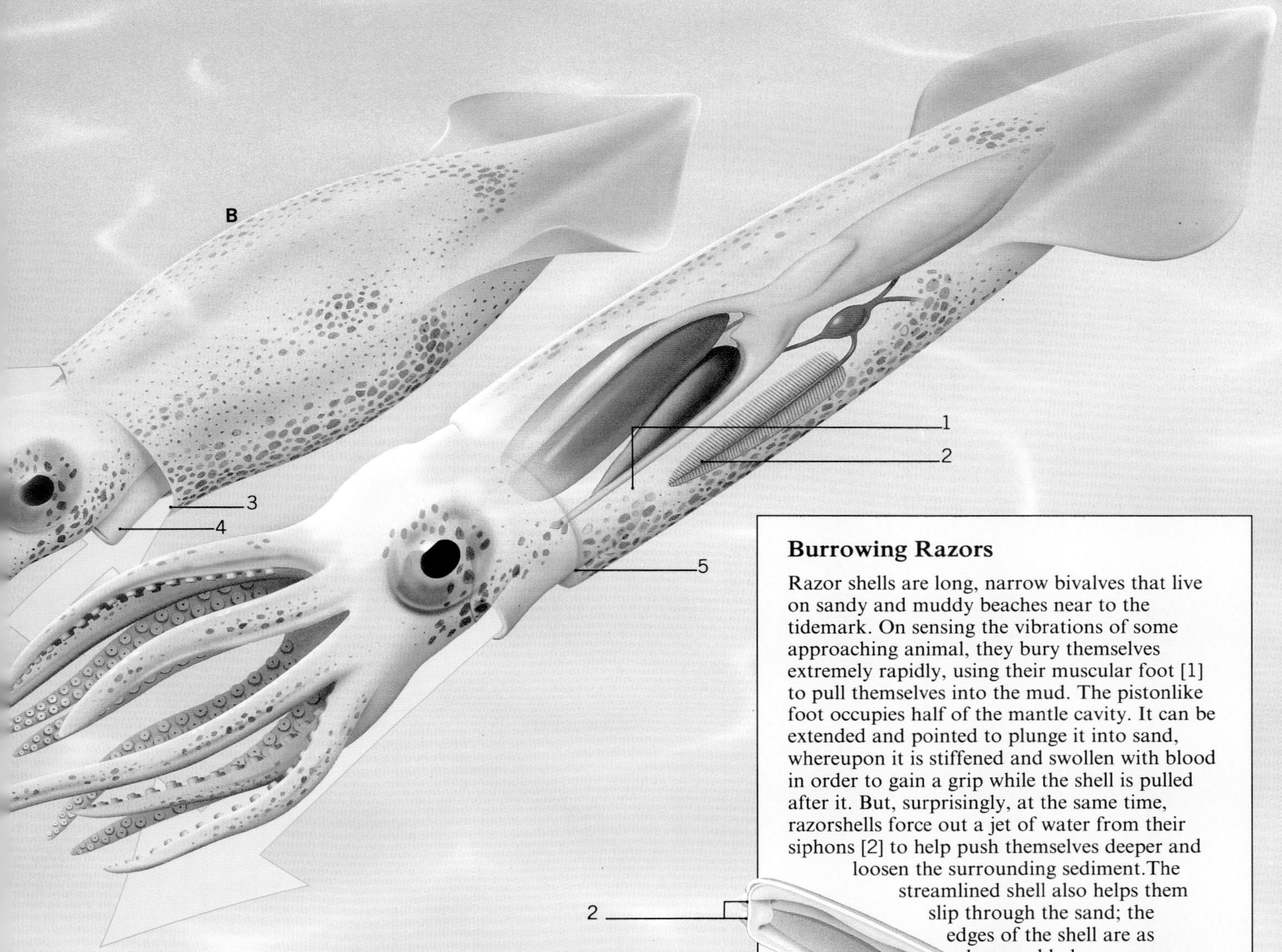

simultaneously. In order to achieve this, the squid has thicker nerve fibers going to the parts of the mantle farthest from the brain. This is because squid nerve fibers are not insulated by fatty sheaths like those of higher animals, and the fatter the nerve, the quicker it carries a signal from the brain. Gentler movements of the mantle muscles are used to circulate water over the gills so the squid can breathe.

***Squid, cuttlefish** and octopuses all have another use for jet propulsion – it is the mechanism they use to expel a cloud of ink, stored in their ink sacs, to confuse pursuing predators.*

Burrowing Razors

Razor shells are long, narrow bivalves that live on sandy and muddy beaches near to the tidemark. On sensing the vibrations of some approaching animal, they bury themselves extremely rapidly, using their muscular foot [1] to pull themselves into the mud. The pistonlike foot occupies half of the mantle cavity. It can be extended and pointed to plunge it into sand, whereupon it is stiffened and swollen with blood in order to gain a grip while the shell is pulled after it. But, surprisingly, at the same time, razorshells force out a jet of water from their siphons [2] to help push themselves deeper and loosen the surrounding sediment.The streamlined shell also helps them slip through the sand; the edges of the shell are as sharp as blades.

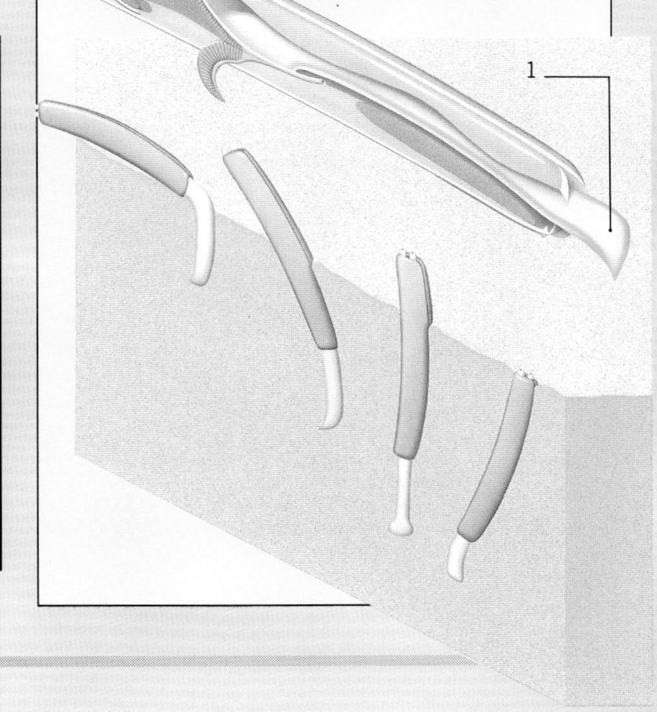

Slipping and Sliding

How animals crawl

An angered black mamba snake can pursue an animal that has disturbed it at speeds of over 6 miles an hour (10 km/h), with its head and the front part of its body held clear of the ground. Perhaps surprisingly, most groups of land-dwelling animals do include members that move without legs. At the other extreme to the black mamba, many invertebrates or their larvae crawl – literally – at a snail's pace. The microscopic amoeba is classified as a crawler, and even some large mammals, like the walrus, resort to crawling when they are out of the water.

Crawling uses the same principle as any other form of locomotion on land – the generation of a backward thrust against the ground. By pushing against a surface the animal is propelled forward. The body often has specialized appendages for gaining a purchase on the surface: the earthworm uses stiff bristles called *chaetae* that can be erected by special muscles, the snake uses the friction of its scales, while the snail uses sticky mucus on the sole of its "foot."

Stretching out

Typically a worm has a soft, tubular body enclosed in two layers of muscles. The worm's organs are suspended in a fluid – it is said to have a hydrostatic skeleton. Rings on the worm's body correspond to individual body segments, each separated from its neighbor by a partition or *septum*. When the worm contracts its muscles, the body fluid is compressed and the body segment changes shape. In this way the worm is actually able to alter the length of different parts of its body. By stretching and contracting its body segments in sequence, it is able to crawl and burrow through the earth.

Although they have legs, starfish, sea urchins and sea cucumbers move by using a similar mechanism to worms. They have hundreds – or even thousands – of tiny feet, each of which is like a tube filled with fluid. Muscle contractions extend and retract the tube feet, which cling to any surface with tiny suction tips and the aid of sticky mucus.

Skeletal shortcomings

Snakes do not have a hydrostatic skeleton like the earthworm, so they are incapable of significantly lengthening and shortening their total body length. But they achieve a similar effect by throwing the body into coils, thus effectively reducing the distance from head to tail. In the most familiar kind of movement – serpentine movement – the snake's body coils press against stones, roots, plant stems, ridges, or depressions in the ground, producing both a sideways and a backward thrust. The wider the coils, the greater the area of body wall in contact with the object and the faster the speed of movement. The process is enhanced by the use of muscles to erect the scales, thus increasing the friction on the object that the coils are being pushed against.

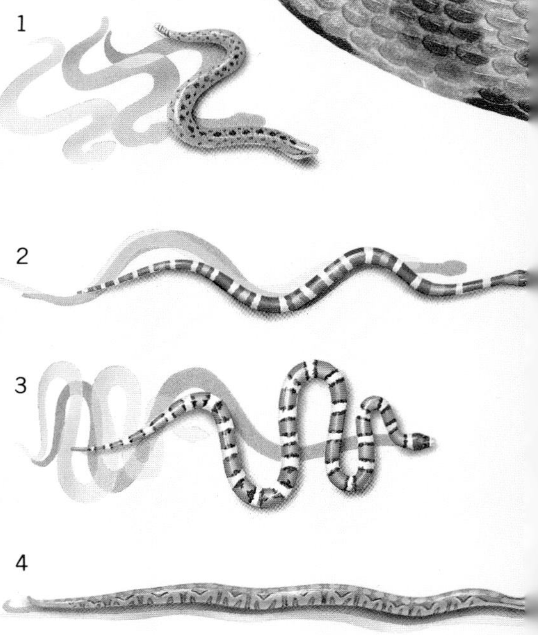

***Many snakes** in hot deserts* [A] *and on shifting surfaces move by "sidewinding"* [1]. *Their coils are formed in such a way that the main thrust is sideways instead of backward. They leave a "lazy S" trail in the sand. Only three parts of a sidewinder's body are in contact with the ground at any one time. This prevents any part of the belly resting on the desert sand long enough to become painfully hot. One modification of the common serpentine motion* [2] *is a concertina movement* [3], *often used in narrow tunnels or burrows. The snake's coils are wedged against the sides of the tunnel while the rest of its body is extended. Some very large boas, pythons and vipers, whose weight gives them grip, use "caterpillar movement"* [4]. *The body remains in a fairly straight position while successive groups of belly scales are erected, angled forward, then pressed back against the ground, levering the snake smoothly along.*

Gliding along

Terrestrial snails and slugs have a muscular "foot" along which waves of muscular contraction pass, pressing successive parts of the foot against the ground. Sticky mucus helps the foot to grip the surface. Flatworms and pond snails literally row themselves along the surface of their own mucus, using waves of tiny beating hairs to thrust back the mucus and propel themselves forward.

Somersaulting hydra

The hydra is related to corals. Normally it lives attached to submerged objects and feeds by trapping microscopic organisms. However, it is also capable of independent, if eccentric, movement. The hydra somersaults along by bending over until its tentacles touch the surface to which it is attached. It then stands upside down, mouth and tentacles on the surface, while it completes the somersault.

The earthworm's body [C] *is enclosed by two layers of muscle – the circular muscles running around the outside* [1] *and the longitudinal muscles on the inside* [2]. *The chaetae* [3] *are hairs that can be erected to provide a lever against the soil, or laid flat against the body. When at rest, the chaetae are extended and the longitudinal and circular muscles relaxed. The part of the worm that is extending has the chaetae withdrawn, circular muscles tense and longitudinal muscles relaxed. The part of the worm contracting has chaetae withdrawn, longitudinal mucles tense and circular muscles relaxed.*

Connections: **Amphibians** 128 **Invertebrates** 122 **Poisonous Animals** 250 **Protozoa** 108 **Reptiles** 130

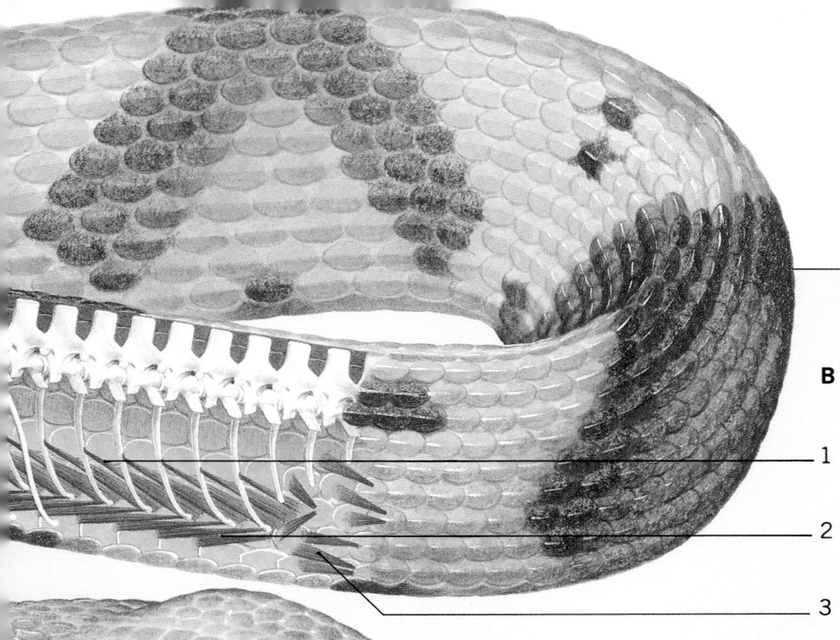

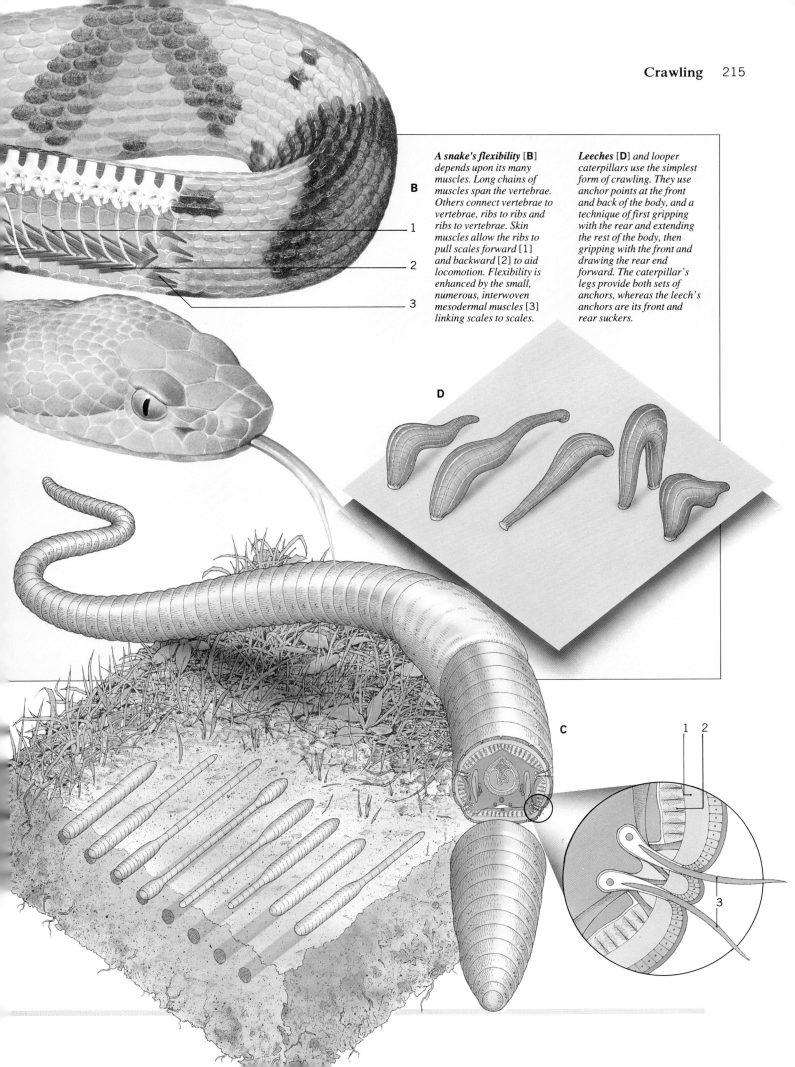

A snake's flexibility [**B**] *depends upon its many muscles. Long chains of muscles span the vertebrae. Others connect vertebrae to vertebrae, ribs to ribs and ribs to vertebrae. Skin muscles allow the ribs to pull scales forward* [1] *and backward* [2] *to aid locomotion. Flexibility is enhanced by the small, numerous, interwoven mesodermal muscles* [3] *linking scales to scales.*

Leeches [**D**] *and looper caterpillars use the simplest form of crawling. They use anchor points at the front and back of the body, and a technique of first gripping with the rear and extending the rest of the body, then gripping with the front and drawing the rear end forward. The caterpillar's legs provide both sets of anchors, whereas the leech's anchors are its front and rear suckers.*

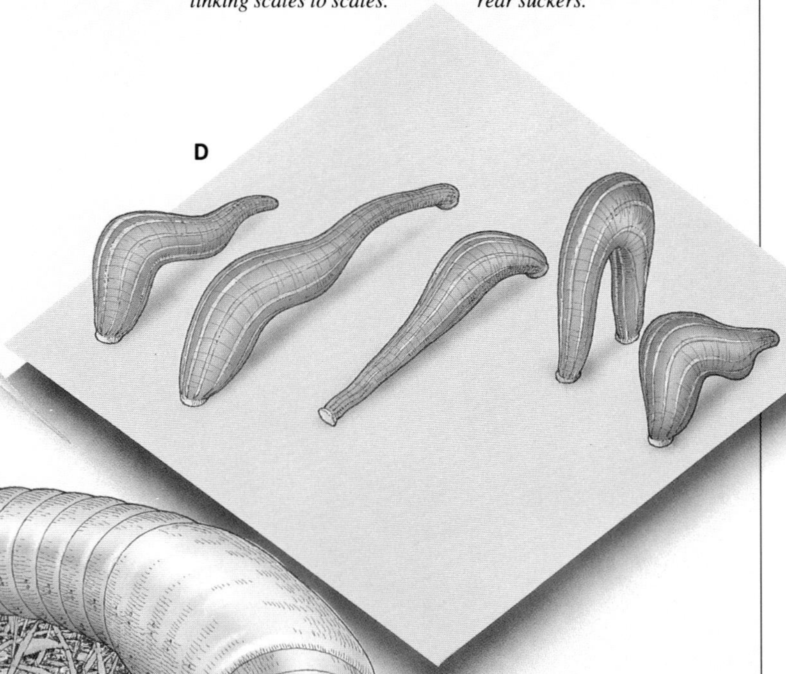

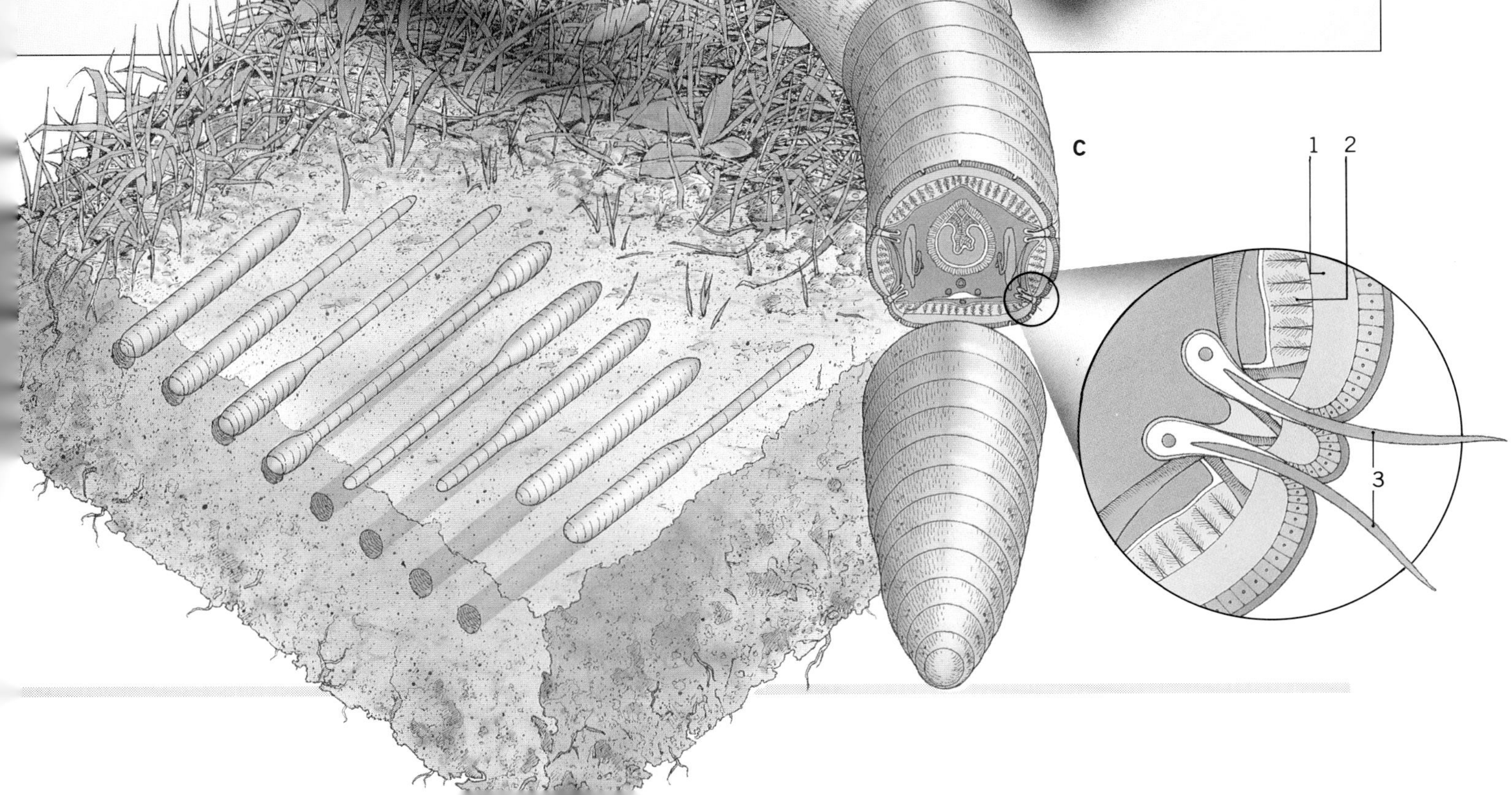

Taking to the Trees

How animals climb

If you tried to pull a gecko from a pane of glass you might well break the glass before the gecko lets go. Even dead geckos can sometimes be found still clinging to trees, so powerful is their grip. Other less obvious climbing animals include certain species of crab. The robber (or coconut) crabs found on some Pacific islands have sharp claws on their walking legs. They use these claws to shin up tall trees in order to escape from predators, or sometimes they use the shade of the tree's leaves to escape from the very hot Sun.

The ability to climb trees in search of food or shelter, or to escape enemies, has evolved separately in a host of different animals. Gibbons travel by swinging from branch to branch and need hardly ever come to the ground. The most important part of climbing is avoiding losing hold and balance and falling off the tree. Invertebrates are generally so small and light that they are little affected by the pull of gravity and show few structural adaptations for climbing: starfish, however, can climb vertical rocks using the suction power of their tube feet. Vertebrate climbers, are much larger and heavier bodied, and have to increase stability by keeping their center of gravity as low as possible. This is achieved by having short legs or, in animals such as bush babies and lemurs, which have long hind legs for leaping from tree to tree, by flexing them when climbing to keep the body pulled in close to the trunk or branch. Tree frogs have evolved horizontally flattened bodies to achieve the same end.

Gripping tails

Most climbing animals have specialized hands and fingers to make climbing easier. Many monkeys and apes, for example, have a thumb that is opposed to the rest of the digits on their hands and feet. It was from such animals that humans acquired great dexterity and manipulative skills. There are, however, also many animals, such as some anteaters, pangolins, opossums, some New World porcupines and New World monkeys, that have evolved a gripping tail that serves as a fifth limb for grasping. These types of tail are called *prehensile.* Such prehensile tails often have rough, naked pads where they grasp. The hairless tip of a spider monkey's tail, for example, has an extremely strong grip, yet at the same time it is sensitive and dexterous enough to pick up an object as small as a peanut. Spider monkeys can even hang by the tail to pluck food with their hands from the tips of otherwise unreachable slender branches. Chameleons, too, have evolved prehensile tails, which they can coil tightly around a branch for support.

Scaly climbers

Snakes can climb trees even though they are without legs or arms. Enlarged scales on their bellies can be tilted in and out to gain and relinquish a hold on tree bark. Many tree-climbing snakes have strongly prehensile tails. The tropical American tree boa uses its tail to climb trees concertina fashion. It coils its tail round the trunk, and by reaching up with its head it can hook its neck around the trunk higher up. It then loosens the grip of its tail and pulls the rear end of the body up to a level with the neck. By repeating this process it can climb even smooth bamboo trunks. Many snakes can extend their bodies across the space between two branches and use this ability to travel through trees, anchoring themselves alternately by their tails and necks. The vine snake, with its slender pencil-shaped body, has developed this ability to a remarkable degree. Its lightweight body and strengthened vertebrae enable it to bridge large gaps without its body sagging, as in other snakes. It can extend half its body length out into space without support.

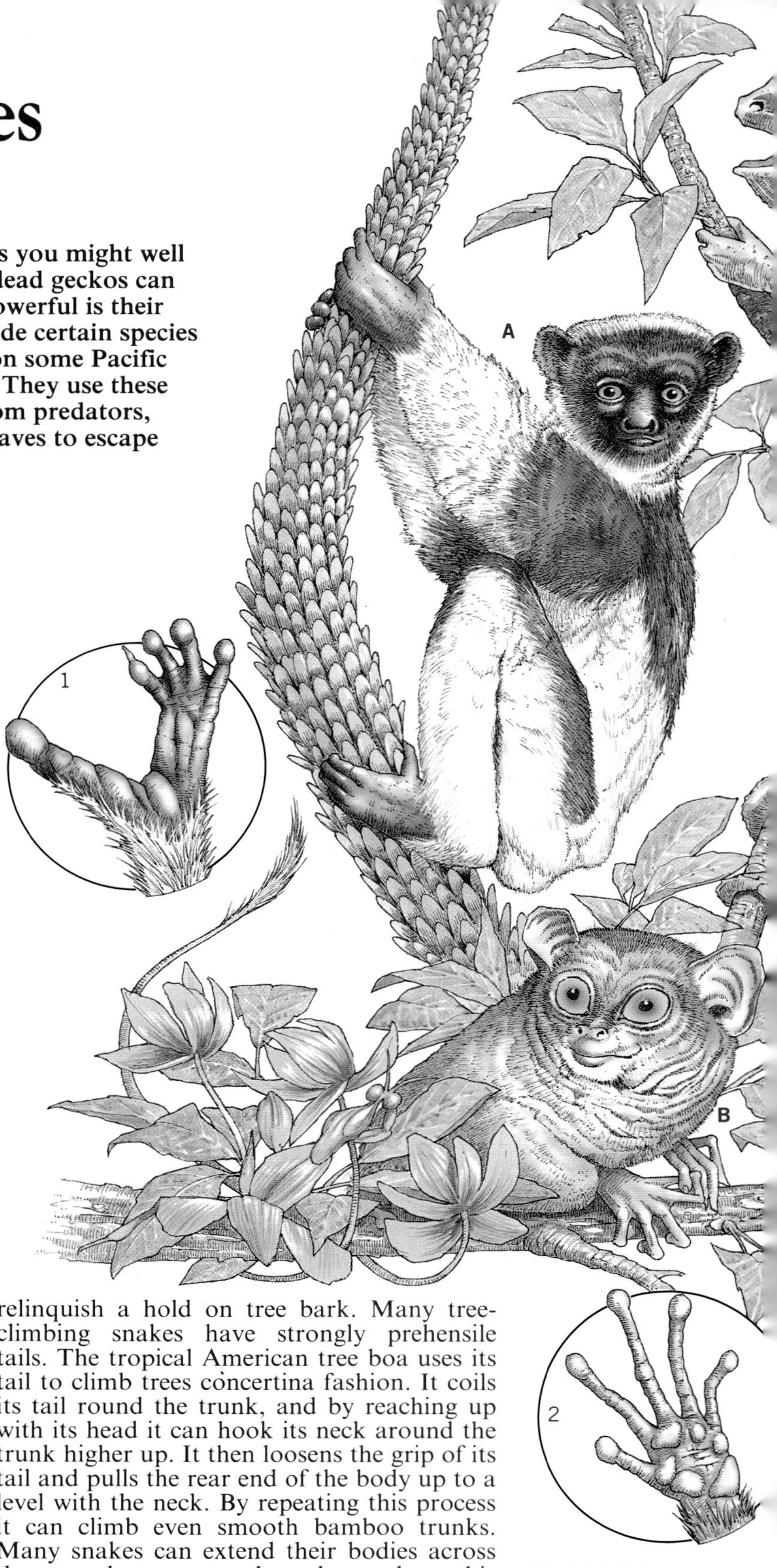

***The indri* [A]** *has almost no tail to balance with; despite this handicap, it is a superb forest acrobat, rarely leaving the trees. Much of its sure-footedness is owed to paws armed with four short toes and one extremely long first toe* [1] *that act together like powerful grasping callipers. The western tarsier's* **[B]**

Connections: **Amphibians** 128 **Crawling** 214 **Flight and Deception** 260 **Gliding and Soaring** 222 **Leaping** 218 **Mammals** 136 **Reptiles** 130

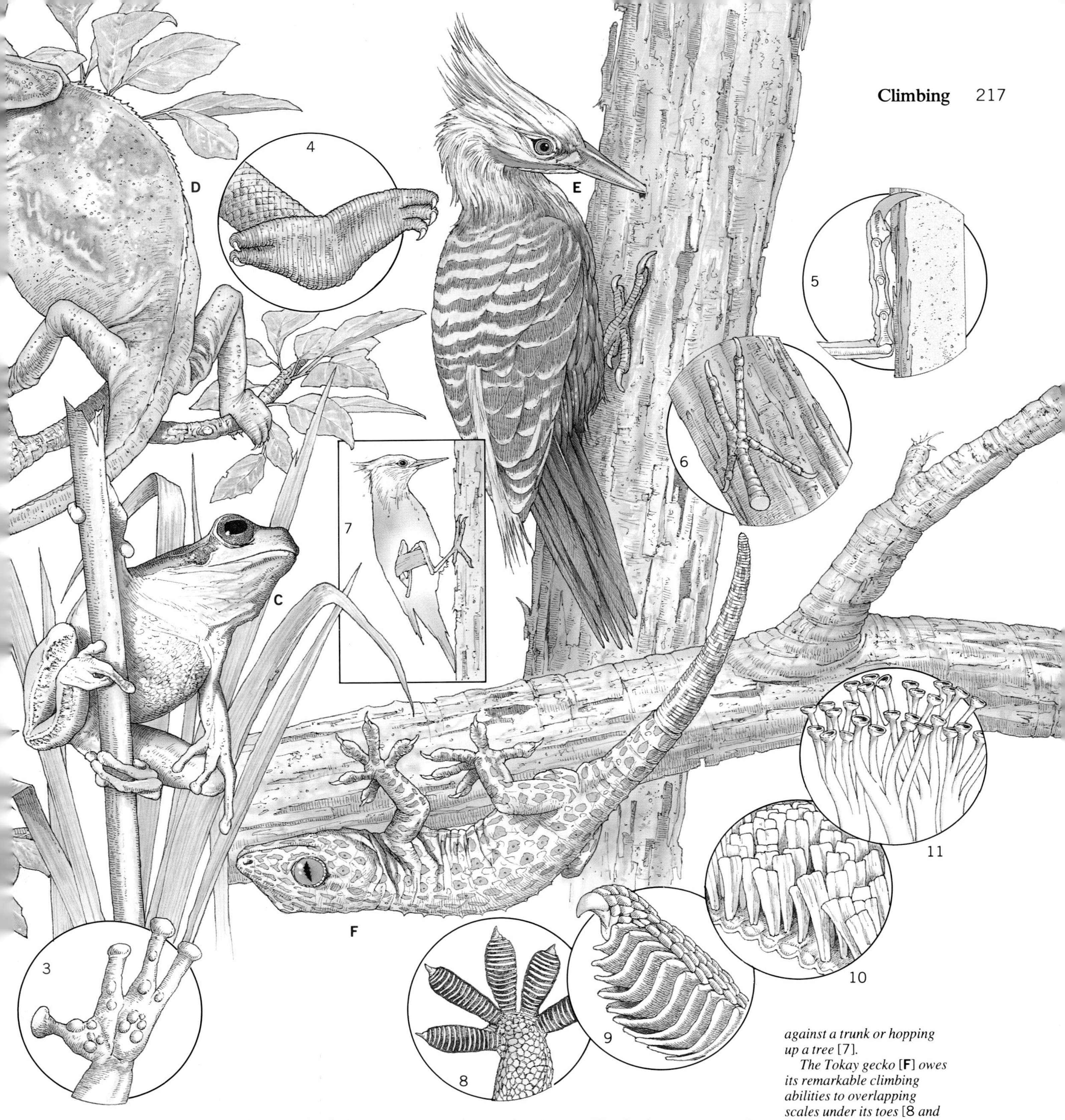

elongated fingers end in adhesive pads [2] that help it climb; it also has opposable first toes on its feet for excellent grasping (an opposable digit can touch the tips of all the other digits on a foot or hand).

*Most tree frogs [**C**] live high above land predators; circular pads on their digit tips [3] contain glands that secrete a substance that glues them to slippery leaves.*

*Other climbers have different arrangements of digits. Chameleons [**D**], for example, have what look like two opposable toes [4] on each limb that tightly grasp a stalk. In fact, each toe is made of fused digits – three on one and two on the other.*

*Like other climbing birds, the blond-crested woodpecker [**E**] has claws more curved than those of nonclimbing bird, and can pierce bark for grip [5]. Many woodpeckers can turn the fourth toe sideways or hold all toes forward for the best holding position [6]. Woodpeckers move up trunks in hops; their tails are specially modified for this, with a strengthened shaft acting as a brace. Broad feathers and broad vertebrae attached to powerful muscles help the bird push down its tail, balancing forces that would otherwise topple the bird backward when pecking against a trunk or hopping up a tree [7].*

*The Tokay gecko [**F**] owes its remarkable climbing abilities to overlapping scales under its toes [8 and 9]. Each scale has up to 150,000 hairlike setae [10], each dividing into up to 2,000 microscopic filaments ending in saucer-shaped plates [11]. Blood-filled sinuses push these plates into minute irregularities found on even the smoothest surface, creating about 100 million contact points.*

Hop, Skip and Jump

How animals move by leaping

Salmon, in their frantic efforts to get upstream, can leap waterfalls up to 10 feet (3 meters) high. Salmon, however, only need to leap occasionally, other animals have made hopping or leaping a major part of their life-style. In the dense rain forests it is easiest to cover long distances by swinging or jumping from branch to branch. And in the desert, hopping provides a fast getaway from predators in an environment with little cover. Some animals, such as kangaroos, are so dependent on hopping that they are unable to move their hind legs independently of each other.

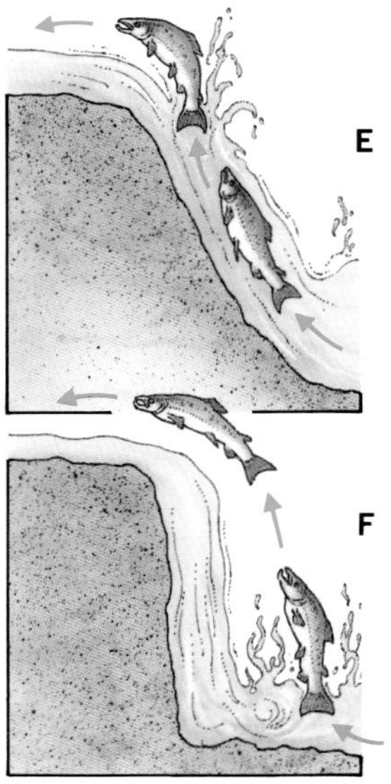

Salmon make long *migrations from the sea to river spawning grounds, leaping waterfalls to reach their goal. When possible, salmon take the easier option to swim up underwater* **[E]**. *In water a salmon need not work so hard to overcome gravity as when leaping out of water. Also, it can use its muscle power for the whole of the climb up the falls.*

Salmon jump best from deep water **[F]**, *which enables them to build up speed beforehand, and may make use of upcurrents. Many jumps fail and the fish drop back exhausted.*

Most hoppers and leapers have relatively long, slender legs. Just before takeoff, all the leg joints are tightly folded up beneath the body, for maximum thrust. The height of a jump is determined by the speed at which the animal leaves the ground. This takeoff speed does not depend on the creature's size and weight: for example, to jump 3 feet (1 meter) into the air, the takeoff speed would be the same for a frog or a flea as it is for a human.

Adaptations for leaping

Nevertheless, some animals can leap much higher than others. For instance, a bush baby can make a vertical standing jump of 8 feet (2.25 meters) – over three times the record for a top human athlete. This remarkable achievement does not mean that the bush baby has more efficient muscle in its hind limbs. It simply has much more muscle – about twice the amount found in a human leg in relation to its body weight.

Other structural adaptations that are needed by leapers include a strong framework to serve as an anchorage for the powerful leg muscles and to absorb the shock of landing. A typical frog, for instance, has a powerful pelvic girdle firmly attached to a stiffened vertebral column. The dorsal part of the pelvis, the *ilium*, is elongated and runs a long way forward, while the rear vertebrae are fused to form a solid rod, the *urostyle*. The vertebral column is very short, consisting of a few vertebrae that are braced to restrict any side-to-side bending. This arrangement ensures that the frog can launch itself into the air with the maximum possible efficiency. The pectoral girdle is flexible and the forelimbs are short and strong enough to absorb the relatively high impact as the frog lands.

Living catapults

Fleas are prodigious leapers. They are capable of making vertical jumps of up to 8 inches (20 centimeters) to enable them to reach a host that is standing above them. Such leaps are 150 times the flea's own body length, and are equivalent to a human jumping nearly a third of a mile (half a kilometre) from a standing start. A flea can reach a speed of over 40 inches (100 centimeters) per second in less than two-thousandths of a second, necessitating an acceleration over 20 times that produced when a typical space rocket blasts off the launch pad. This amazing feat depends partly on the flea's small size. Usually, the smaller the animal, the more powerful, relatively, are its muscles.

Muscles simply cannot contract this fast, so the flea uses a built-in catapult. It stores energy in a protein called *resilin* with elastic properties similar to those of rubber. This substance is situated at the base of the flea's hind legs and the muscles, when moving at normal speed, compress the resilin, which snaps back when a release mechanism is tripped and shoots the flea upward. As it hits the new host or, worse still, a floor or some other solid object, it experiences a deceleration equivalent to that experienced by a human being crashing an automobile into a brick wall at a speed of several hundred miles per hour. It is able to survive this astonishing impact only because it has a rigid external skeleton that protects the vital organs within.

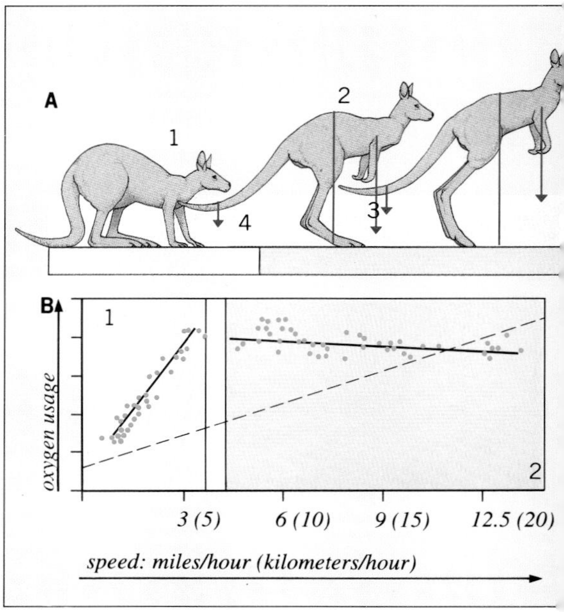

Almost all kangaroos **[A]** *hop at speed. But at low speeds they are clumsy, supporting themselves on forelegs and using the tail as a fifth leg as the hind legs slowly swing forward* [1]*; few species can move backward. This awkward gait results from the evolution of tiny forelegs, larger ones being unnecessary for the main locomotion – hopping. But high-speed hopping is very efficient, the animal balancing about a vertical fulcrum* [2]. *The body forward of this line* [3] *tends to pivot clockwise about the fulcrum, but the weight of the heavy tail* [4] *tends to pivot the opposite way, counterbalancing the head and shoulders as the body tilts forward for maximum speed.*

Plotting oxygen usage against speed on a graph **[B]** *shows the inefficiency of "five-legged" locomotion* [1] *compared to hopping* [2], *when oxygen usage is lower than at slower speeds. The dashed line shows oxygen usage for a 40-lb (18-kg) running quadruped mammal.*

Stored elastic energy *powers a wallaby leap* **[C]**. *The* gastrocnemius *muscle* [1] *tendon fits into the heel* [2] *and the* plantaris *muscle* [3] *tendon stretches over it and joins to the toes* [4]*; both exert powerful turning forces on the ankle. The graph* **[D]** *plots the length change of a hopping wallaby's plantaris muscle and tendon system (with feet in ground contact) against force stored in the system. At* [1] *the foot lands after a hop and the system is fully relaxed, having expended all stored energy. Between* [1] *and* [2] *the system is stretched as the ankle joint folds, reaching maximum stored energy at* [3], *when the body is nearest*

Connections: Climbing 216 Crawling 214 Gliding and Soaring 222 Swimming 210 Walking and Running 220

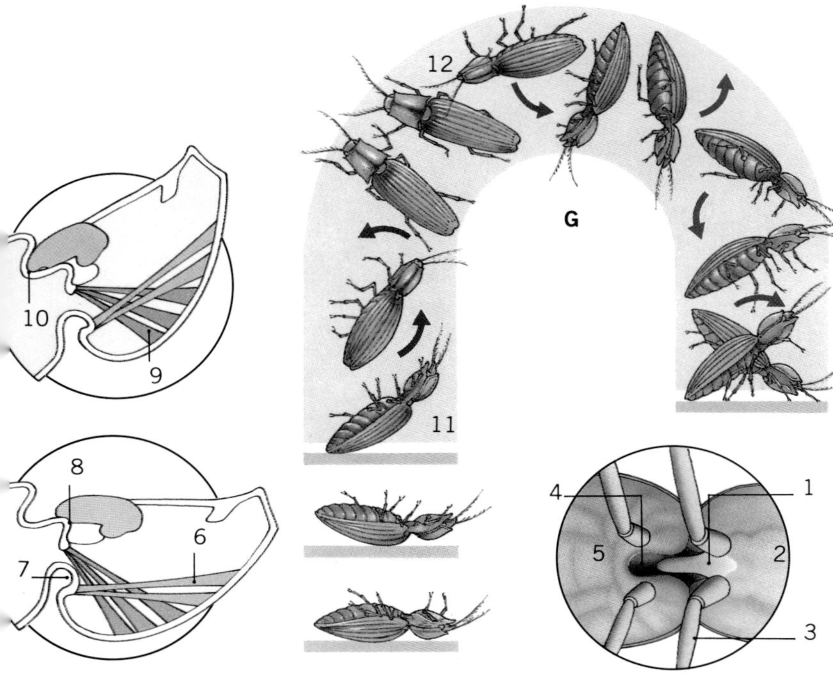

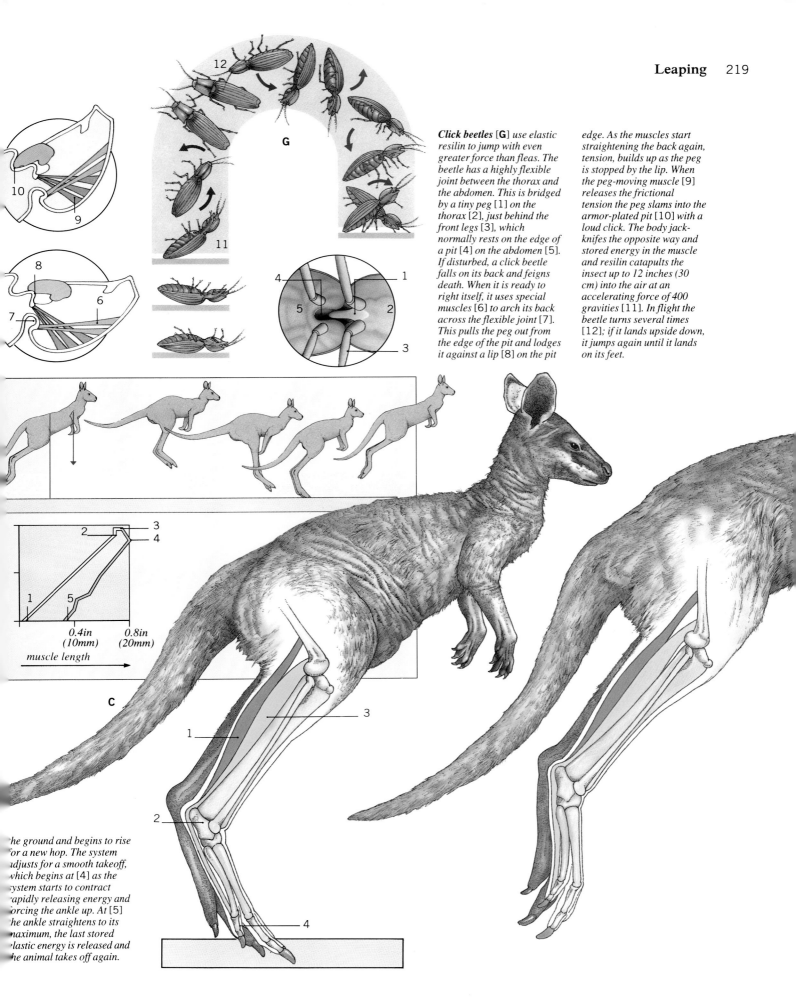

Click beetles **[G]** *use elastic resilin to jump with even greater force than fleas. The beetle has a highly flexible joint between the thorax and the abdomen. This is bridged by a tiny peg* [1] *on the thorax* [2], *just behind the front legs* [3], *which normally rests on the edge of a pit* [4] *on the abdomen* [5]. *If disturbed, a click beetle falls on its back and feigns death. When it is ready to right itself, it uses special muscles* [6] *to arch its back across the flexible joint* [7]. *This pulls the peg out from the edge of the pit and lodges it against a lip* [8] *on the pit edge. As the muscles start straightening the back again, tension, builds up as the peg is stopped by the lip. When the peg-moving muscle* [9] *releases the frictional tension the peg slams into the armor-plated pit* [10] *with a loud click. The body jack-knifes the opposite way and stored energy in the muscle and resilin catapults the insect up to 12 inches (30 cm) into the air at an accelerating force of 400 gravities* [11]. *In flight the beetle turns several times* [12]*; if it lands upside down, it jumps again until it lands on its feet.*

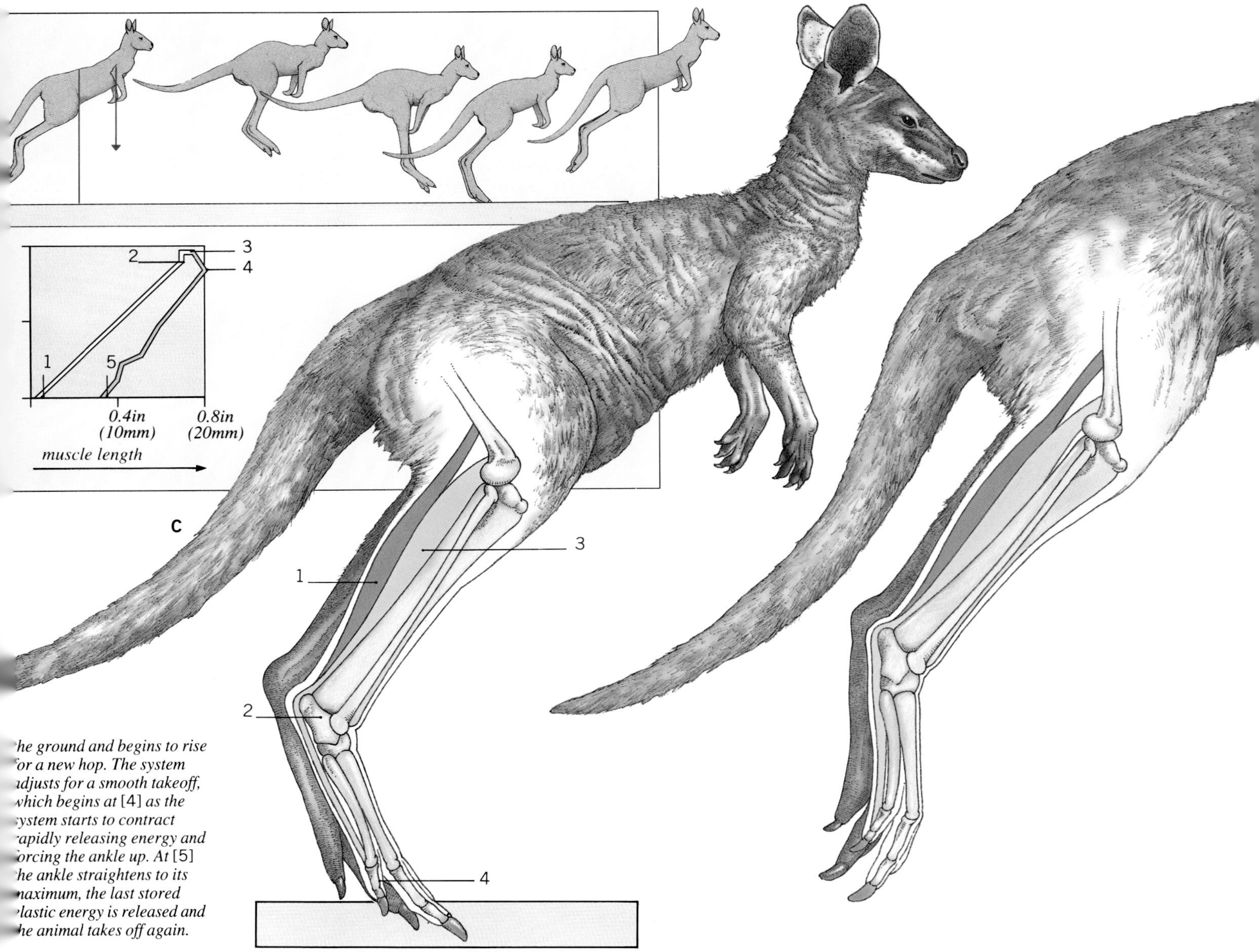

'he ground and begins to rise 'or a new hop. The system ıdjusts for a smooth takeoff, which begins at [4] *as the system starts to contract rapidly releasing energy and forcing the ankle up. At* [5] *the ankle straightens to its maximum, the last stored elastic energy is released and the animal takes off again.*

The Fleet of Foot

How animals walk and run

Some scientists believe *Tyrannosaurus rex*, "the most terrifying engine of destruction ever to walk the Earth," could accelerate its seven-ton bulk to speeds near 34 miles an hour (55 km/h), so powerful were its massive thigh muscles. Since the start of land-based animal life a variety of locomotive methods has evolved to take advantage of every possible ecological circumstance, enabling animals to colonize new areas, hunt prey and escape predators. Ultimately, evolution produced a versatile mammal that could walk, run, or climb on two legs – *Homo sapiens*.

The actions of walking and running both involve the transformation of chemical energy, produced by chemical reactions in the muscles, into mechanical energy and movement. Two different principles are involved: walking uses the pendulum principle, which requires very little energy, while running is a bouncing action that needs more energy but entails less friction with the ground (thus saving energy), as two or more feet may be off the ground for much of the time.

The swinging of the legs during walking can be compared to the oscillations of a pendulum. As the leg swings down, it loses potential energy by losing height but gains kinetic energy by increasing its speed; the reverse occurs as the leg swings up again. Just as the pendulum bob changes height as it swings, so do the bodies of animals – for example, human heads bob up and down as people walk along.

Running involves a different principle: the storage of elastic energy in tendons, especially in the heel (*Achilles*) tendons of vertebrates. Wedges of elastic cartilage perform the same function in leaping arthropods such as grasshoppers. As the heel lands on the ground, the tendon is compressed and its potential energy increases. As the foot pushes off again from the ground, this energy is converted into mechanical energy of movement, providing thrust against the ground to propel the body forward.

The evolution of walking

Walking and running rely on a system of levers – segments of the skeleton – pivoted at a series of joints. These levers are operated by muscles. Thus walking and running required the evolution of discrete paired (*antagonistic*) muscles, as opposed to the muscle blocks of fish. In the vertebrates these muscles are outside the body skeleton, whereas in arthropods they are inside the external skeleton.

Empirical evidence for the evolution of legged locomotion is scarce, most theories being hypothesized. The first stage was the evolution of firm supports to lift the body off the ground. The earliest terrestrial vertebrates evolved from fish that developed fleshy, muscular fins with bony supports. Amphibian limbs evolved from these supports. Amphibian and reptile locomotion is very inefficient because the legs are at the side of the body (as were the fins from which they evolved), and thus these animals have difficulty in lifting the body off the ground. The order in which these creatures' legs move during walking reflects their swimming ancestors' body undulations; a lizard such as an iguana twists its body from side to side as it walks, and rests with its body on the ground.

For mammalian locomotion the key evolution was of limbs positioned *underneath* the body. More elastic tendons, greater spinal flexibility and, for some, the ability to walk on tiptoes, or the evolution of the hoof all combined to give mammals superior locomotion. Among arthropods long legs increase the speed of animals like spiders; millipedes and centipedes, however, use large numbers of short legs not for speed but for power, as waves of motion pass along the rows of legs and enable the animal to force its way through loose soil or leaf mold.

An insect's knee* [A] *is where two sections of the hard, tubular exoskeleton* [1] *pivot about a peg-in-socket joint* [2] *covered by flexible membrane that stretches or folds* [3]. *Tendons* [4] *attached to flexor muscles* [5] *bend the joint and extensor muscles* [6] *straighten the joint.

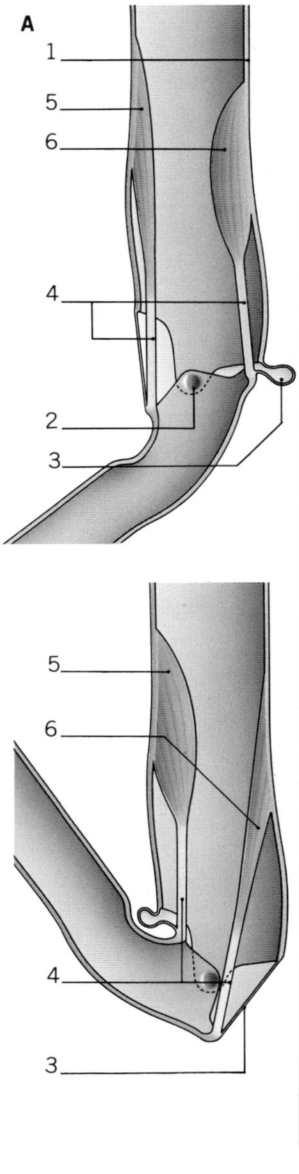

Walking on Two Legs

Certain primates, such as the siffaka (below) and, on occasion, the gorilla and chimpanzee, can walk on two legs for short distances, but it is only humans who use bipedal locomotion – a highly efficient form of locomotion that does not require four powerful limbs – all the time.

The evolution of a two-legged gait required a fundamental change in the orientation of the shoulder and hip girdles. It was also necessary for a hip girdle of greater strength to evolve, since it was now required to bear much of the animal's weight. The lower vertebrae of the spine became fused to give extra rigidity, thereby protecting them against the constant jarring impact of the legs as they hit the ground, and making them better able to take the weight of the rest of the vertebral column and the limbs and organs suspended from it.

Connections: Amphibians 128 Climbing 216 Crawling 214 Flight and Deception 260 Gliding and Soaring 222 Leaping 218 Mammals 136 Reptiles 130

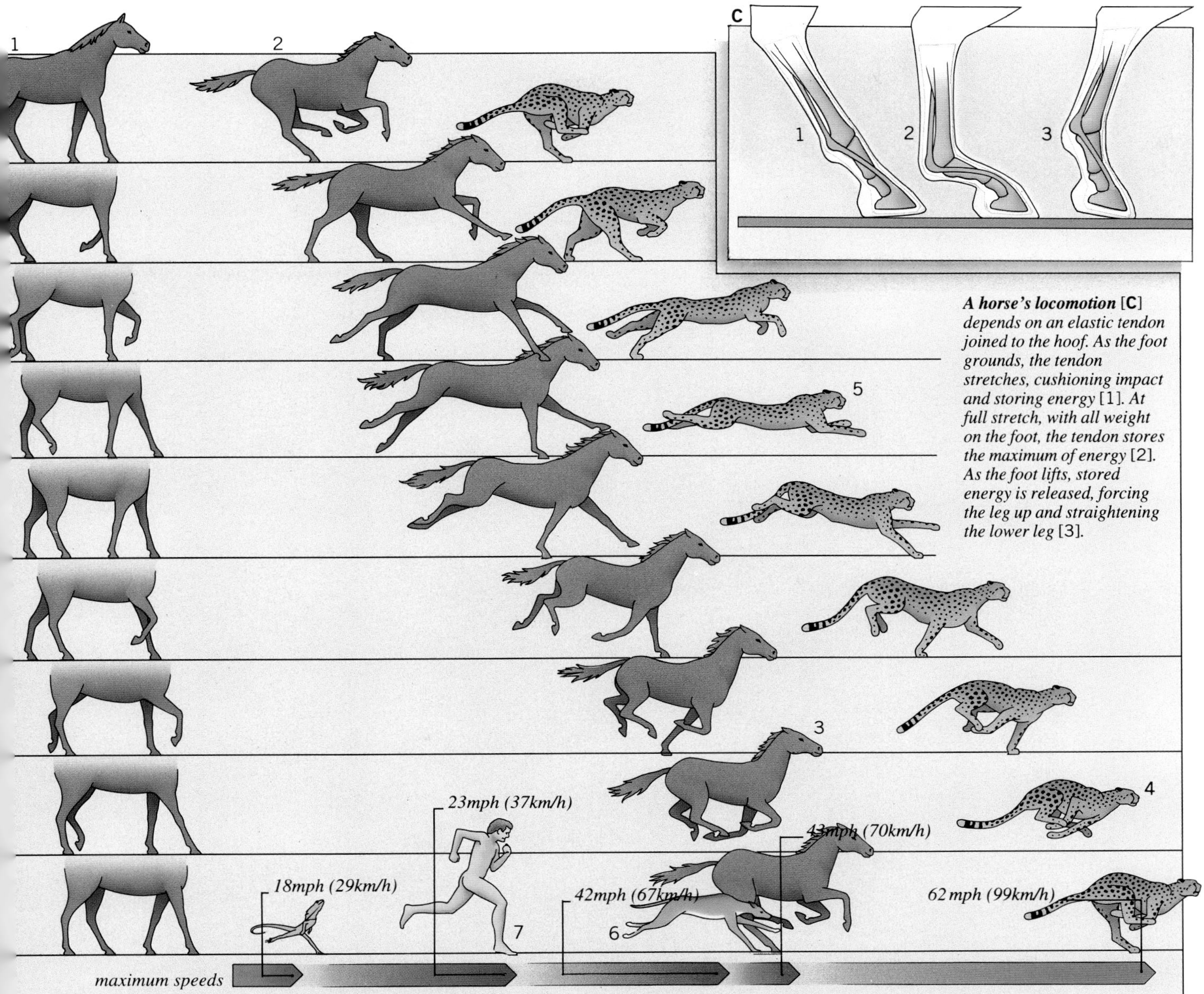

A horse's locomotion **[C]** *depends on an elastic tendon joined to the hoof. As the foot grounds, the tendon stretches, cushioning impact and storing energy* [1]. *At full stretch, with all weight on the foot, the tendon stores the maximum of energy* [2]. *As the foot lifts, stored energy is released, forcing the leg up and straightening the lower leg* [3].

Four-legged locomotion **[B]** *varies according to speed. A horse, for example, walks in a cycle of regular time-sequenced leg movements* [1], *balancing by keeping its center of gravity within the triangle of its three grounded feet. At maximum speed* [2], *regular sequence disappears; the horse is never supported by more than two legs and, at one stage, by none* [3]; *ground contact is so brief that balance is less important.*

The cheetah is airborne at the same stage as a horse [4] *but has an extra no-contact phase* [5]; *a whippet* [6] *has a similar cycle. Bipeds like human athletes* [7], *or the six-lined racerunner lizard do not rival a cheetah.*

The cheetah **[D]**, *the fastest mammal, derives its speed from an extremely flexible skeleton that allows its backbone to curve up and down. Its shoulder blades and hip joints permit maximum rotation* [1]. *These features extend its stride and give it faster leg movement cycles; relative to their heights, a cheetah's stride is about twice a horse's. The horse* **[E]**, *with its stiffer skeleton, is airborne at gallop for about a quarter of a cycle; the cheetah achieves double this. Although slower than a cheetah, a horse has much greater endurance, for the cheetah can only maintain maximum speed for bursts of about 15 seconds.*

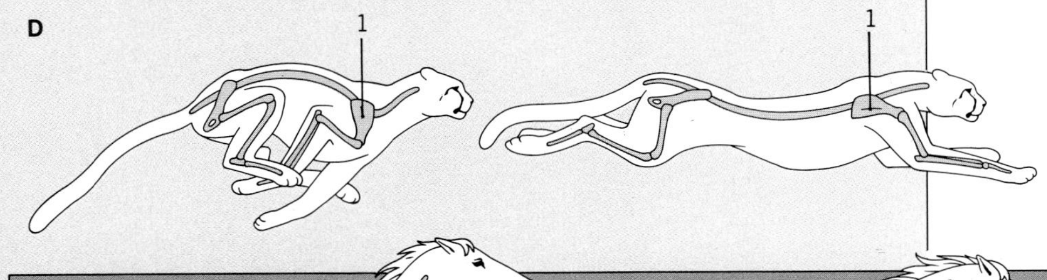

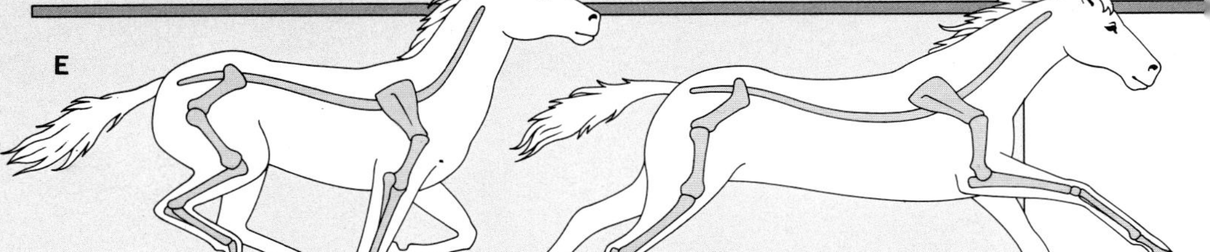

With the Greatest of Ease

How animals glide or soar

Some species of flying fish can glide as far as 295 feet (90 meters) at a height of 5 to 20 feet (1.5 to 6 meters), having built up speed in the water to some 32 miles an hour (50 km/h). Species of lizards, snakes, frogs, squirrels, colugos and one or two marsupials have – with more or less equal success – evolved the ability to glide on expanded flaps of skin or membrane stretched between their limbs. In each of these groups gliding has evolved independently, testifying to the success of this method of locomotion. The chief advantages of gliding seem to lie in saving energy and in escape.

Gliding requires a large, flattish surface area to support an animal's weight and generate lift. Any large flat surface – like a wing – will generate lift if it enters a flow of air at a tilted-up angle. This is because the wing splits the airstream in two, creating a circling movement of air around the wing – a vortex. The vortex hurries the air moving over the top of the wing, and slows the air over the lower surface. Because faster-moving air creates less pressure, the net effect is an upward push from under the wing. Of course this is counteracted by the weight of the falling creature. Animals generally create a large, flattish surface area by spreading out wings, fins, or skin membranes. Further anatomical adaptations may allow enlargement of these flight surfaces.

Gliding animals have also had to evolve muscular control of their gliding surfaces: they must be readily movable during takeoff and landing and to permit sudden maneuvres in flight yet held out rigidly during a steady glide. Sensory systems, particularly vision, have had likewise to evolve so that gliding creatures are more able to judge speed and distances accurately in flight.

Flying riverfish

Two groups of freshwater fish have evolved flight. The butterfly fish of West African rivers glide for short distances in pursuit of insect food. Their method of gliding is very similar to marine flying fish: they build up sufficient speed underwater before launching themselves into the air to chase their prey. The South American freshwater hatchetfish similarly catch insects, but to do so have actually evolved powered flight. Their huge breast muscles power large pectoral fins, producing enough lift to travel over the water for up to about 16 feet (5 meters).

Gliding reptiles

The 20 or so species of flying dragons, from the forests of southern India and Southeast Asia, are 10-inch (25-centimeter) long lizards that have evolved unique gliding membranes that are supported by between five and seven elongated ribs.

Flying geckos have skin flaps along the sides of the neck, a flattened body and tail and webs between the toes of all four feet, all of which help them glide from tree to tree.

__The flying squirrel__ [A], by making adjustments to the positions of its four limbs, can accurately control the shape and angle of its skin membrane, giving it great control in flight and the ability to make sudden maneuvers. The long, feathery tail is used as a rudder to stabilize the squirrel. Like other gliding animals, the flying squirrel has large, sharp claws with which to grasp the "target" tree. As it nears its landing point, the flying squirrel changes course by raising the tail. This causes the squirrel to travel upward and stall, thereby slowing the animal down.

A

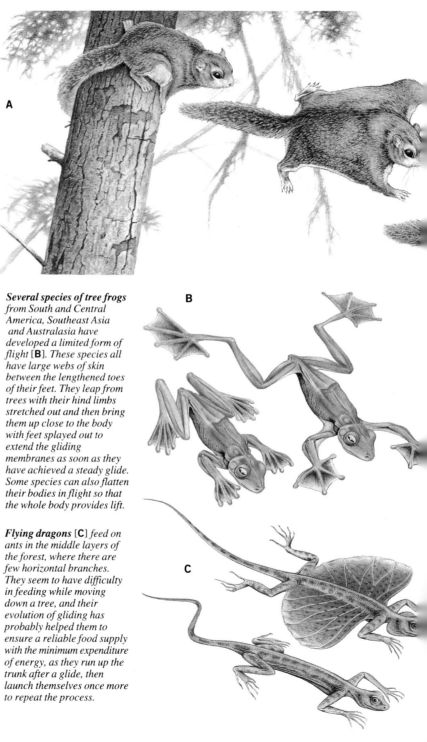

__Several species of tree frogs__ from South and Central America, Southeast Asia and Australasia have developed a limited form of flight [B]. These species all have large webs of skin between the lengthened toes of their feet. They leap from trees with their hind limbs stretched out and then bring them up close to the body with feet splayed out to extend the gliding membranes as soon as they have achieved a steady glide. Some species can also flatten their bodies in flight so that the whole body provides lift.

B

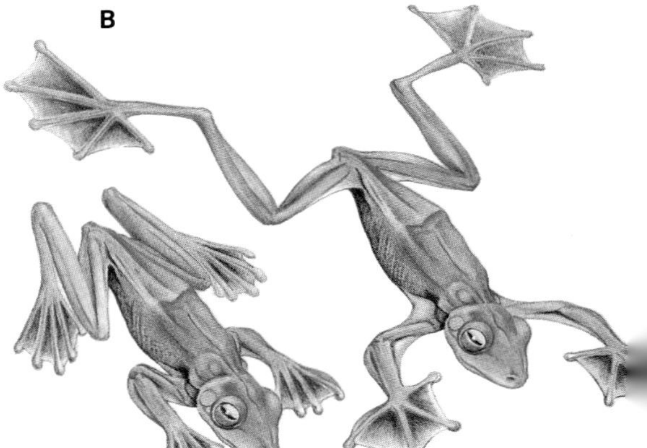

__Flying dragons__ [C] feed on ants in the middle layers of the forest, where there are few horizontal branches. They seem to have difficulty in feeding while moving down a tree, and their evolution of gliding has probably helped them to ensure a reliable food supply with the minimum expenditure of energy, as they run up the trunk after a glide, then launch themselves once more to repeat the process.

C

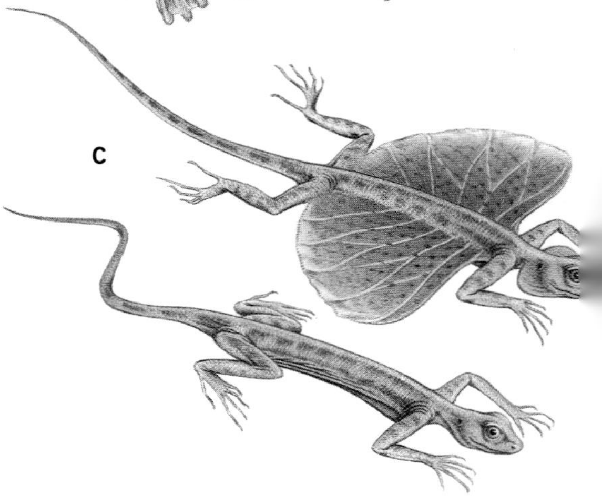

Connections: **Bird and Bat Flight** 224 **Early Evolution** 92 **Fur and Feathers** 134 **Insect Flight** 226

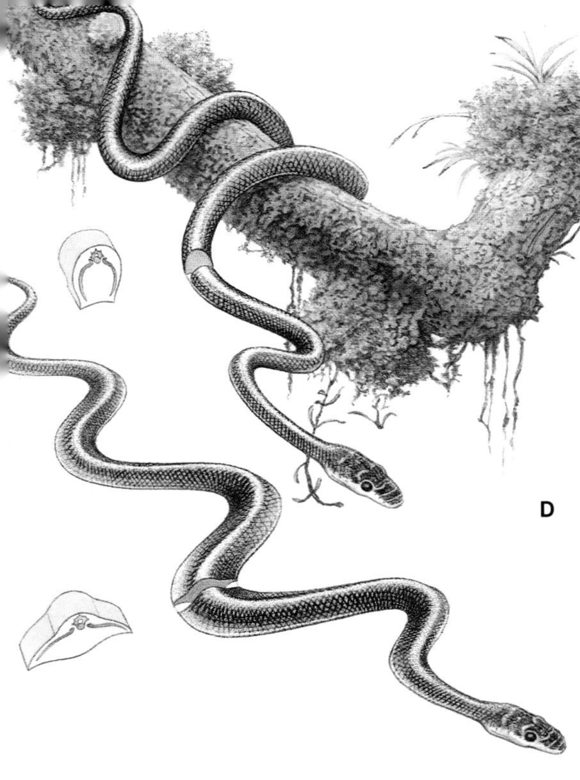

D

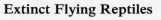

***The flying snakes* [D] *of the genus* Chrysopelea *are the most remarkable of present-day gliding reptiles. The two species have long, slender bodies and grooved fangs. They are well adapted to their habitat, moving quickly through the branches to catch their agile prey of geckos. It is likely that they evolved their gliding adaptation to help them hunt. When gliding, the snake launches itself from an appropriate branch into the air, where it makes a rapid series of S shapes while at the same time flattening its body to increase its surface area, thereby creating as much lift as possible.*

Extinct Flying Reptiles

Pterosaurs were the first flying vertebrates. They evolved about 220 million years ago – some 70 million years before the appearance of the first bird, *Archaeopteryx* – and diversified into many different types until they became extinct about 70 million years ago. Their fossilized remains, although scarce because of the fragility of their light, air-filled bones and their aerial life-style, have been found worldwide, except in Antarctica, mostly in marine deposits. Pterosaurs flew on wings of skin attached to the greatly elongated fourth "fingers" of each "hand" and to the thighs. Many of the smaller pterosaurs may have managed flapping flight, but the larger ones were probably gliders, using warm air currents over the oceans, sea cliffs, or land to keep them aloft. They included the largest of all flying vertebrates, *Pteranodon,* with a 23-foot (7-meter) wingspan.

Gliding mammals

Gliding flight occurs in the rodents (the flying squirrels of Eurasia and North and Central America and the unrelated African scaly tailed flying squirrels), in the marsupials (three families of gliders and possums in Australia and New Guinea) and in the colugos (an enigmatic order containing just one family of two species, in Malaysia and the Philippines). All of these are nocturnal forest dwellers that take off from the upper branches or trunk of a tall tree and glide by means of a flattish, broad, fur-covered flap of skin (called the *patagium*) that runs along both sides of their bodies.

When scampering about in trees, an animal like the flying squirrel tucks away its patagium to prevent it from snagging and being torn on twigs. When it wants to leave a tree, the squirrel holds out its limbs so that the patagium is extended into a taut membrane.

***The flying fish* [E], *by spreading its large, elongated pectoral fins, can make a flight lasting between 4 and 10 seconds. The fins are kept close to the fish's side when swimming through water to help reduce drag. When aided by flying into the wind, flying fish have been known to land on ocean-going ships, the decks of which are 30 feet (10 meters) above the sea surface. Often, particularly when pursued by a tenacious predator, these fish will make a number of consecutive glides. By beating their tails every time they hit the sea surface, they can get up enough speed to make another glide.*

E

On the Wing

How birds and bats fly

The swooping dive of the duck hawk takes it to speeds in excess of 180 miles an hour (290 km/h). The 24-pound (11-kilogram) Andean condor, one of the world's largest birds, can fly almost effortlessly, hardly flapping its wings, for hundreds of miles. Both birds demonstrate a spectacular command of the art of flight, but the principles behind the achievement are applied by every single bird that flies. For more than 8,500 species of birds – as well as the 950 or more species of bats – mastery of the air is the key to survival.

For a bird to fly, it has to solve two fundamental problems. It must overcome both the effects of gravity on its own body in order to rise into the air and stay there and also the resistance its body offers to the air, the *drag*, in order to propel itself forward.

To conquer gravity, the bird must generate lift. It does this by beating down on the air with its wings, so thrusting itself upward. On the downstroke the feathers lock together and spread out to give the wing the maximum resistance to the air. On the upstroke the wing flexes and rotates, and the primary feathers used for flying separate to offer the minimum possible air resistance.

Wings

The bird wing is made up of feathers attached to a modified forelimb to produce a large surface area of strong but flexible material. The bones of the "hand" are elongated and fused together to make a strong support for the long primary feathers. Viewed in section, the bird's wing is shaped like an aerofoil: its upper surface is convex and the wing tapers from front to back. When air flows over such a shape, the air speeds up as it flows over the upper surface and slows down over the concave lower surface. This creates a lower pressure above the wings, helping to lift the bird.

Aerodynamics

In free flight the whole wing is lifted up and down from the "shoulder." But most birds are perfectly capable of flying short distances by moving the wing only from the "wrist" down, that is, by flapping their "hands."

The bird steers by altering the angle of one or both wings, twisting its wings, and spreading and twisting its tail like a rudder. To slow down and land, it twists its wings so that they beat almost backward and forward rather than up and down; it also lowers its tail and spreads it out as a brake.

Taking off is more difficult, especially for large, heavy birds. Small birds, which do not need much lift, can simply leap off a branch and flap hard, but larger birds like swans may need a flapping takeoff run. As they run, the rush of air under their wings helps them off the ground. Birds have different-shaped wings for different purposes. The pheasant needs short wings to maneuver through undergrowth, and must flap them very quickly to stay aloft. Herons have large wings to allow them to make soft parachuting landings, which protect their delicate legs.

Soaring and gliding

If a bird flies toward a rising air current, it can remain airborne without flapping its wings as long as the air is rising faster than the bird is sinking. This kind of gliding flight is called soaring. Rising currents of air are formed where the wind is forced up over a hill or cliff. When land is heated up by the Sun, the air close to the ground becomes warm and rises in a column called a thermal. Vultures and large birds of prey use thermals to reach great heights, soaring in circles so as not to leave the thermal. When they reach the top of the thermal, where the air is no longer rising, they leave it and glide down until they reach another thermal.

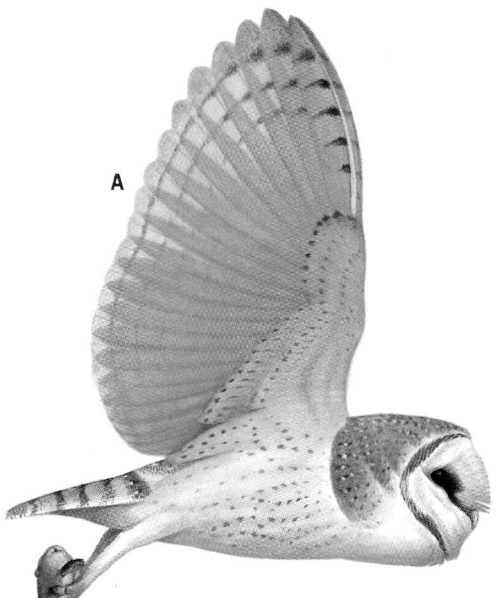

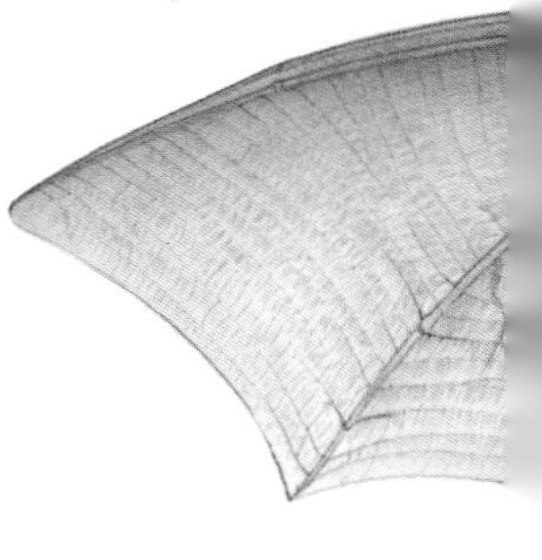

The barn owl *[***A***] is a skillful silent flyer. Like all birds the main muscles used for flight are located at the base of, and below, the wings. The positioning of these heavy flight muscles ensures that a bird's center of gravity is below the level of the wings, which makes the bird aerodynamically stable. There are two pairs of flight muscles, one pair for each wing – the powerful* depressor *or* pectorali, *muscles and the smaller, less powerful,* elevator *or* supracoracoideus *muscles – both pairs are firmly attached to the deep keel of the bird's sternum. During the upstroke* [**B**] *the pectoralis muscles* [1] *relax and the supracoracoideus muscles* [2], *which run from the sternum* [3] *over the top of the* coracoid [4] *and to the* humerus [5], *contract. This pulley system lifts the humerus even though the muscles that carry out this action are situated underneath it. The downstroke* [**C**], *which lifts the bird off the ground and propels it through the air, requires far more muscle power, which is why the pectoralis muscles are much bigger and stronger than the supracoracoideus. The pectoralis muscles are attached directly to the underneath of the humerus, making the downstroke more efficient than the upstroke. The downstroke is carried out simply by the strong pectoralis muscles contracting and the supracoracoideus relaxing. The* clavicles *or* wishbone [6] *support the coracoids and also help to brace the wings away from the bird's body.*

Connections: **Bird Courtship** 172 **Birds** 132 **Cave Life** 308 **Evolution to Date** 94 **Fur and Feathers** 134 **Gliding and Soaring** 222 **Insect Flight** 226

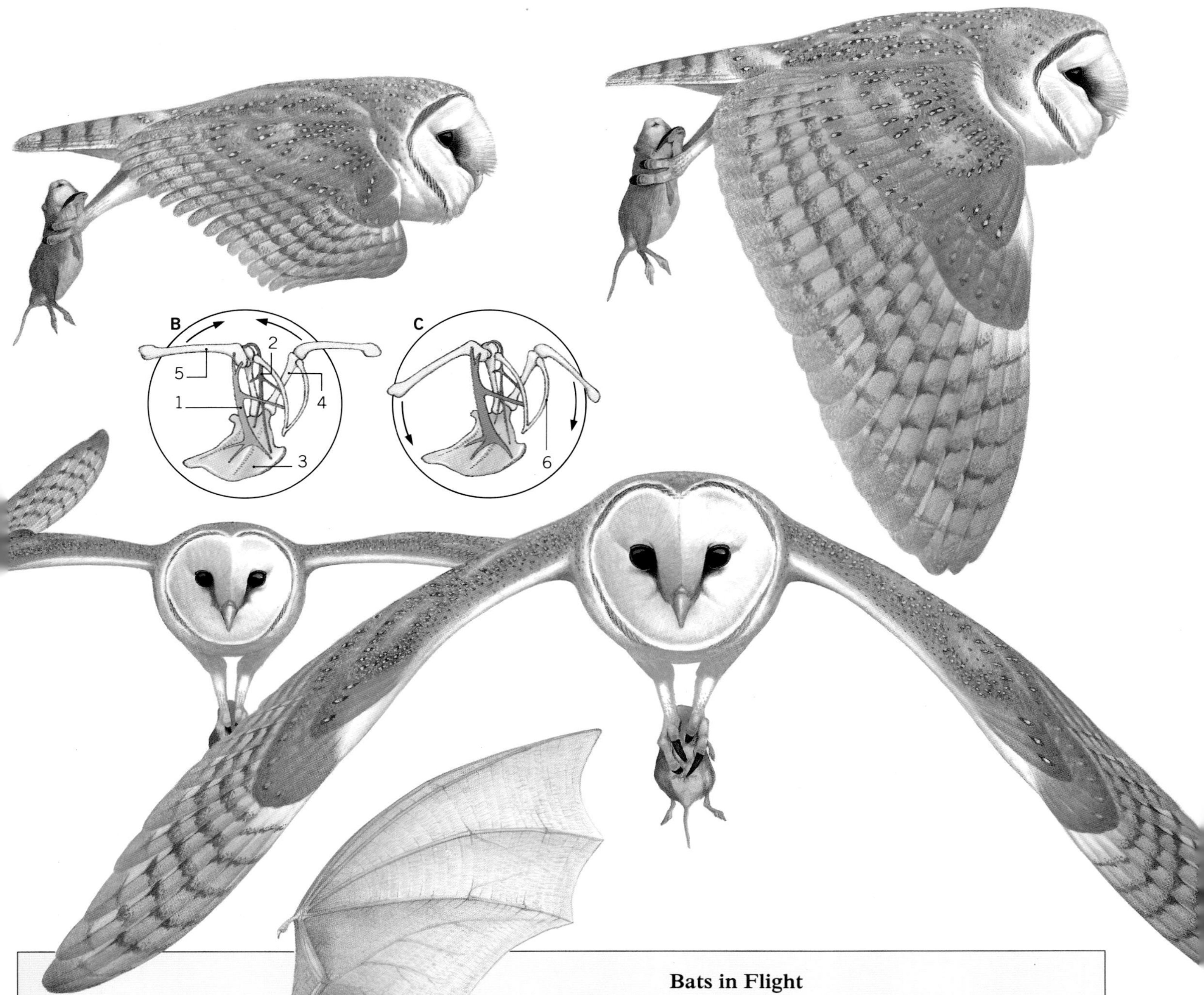

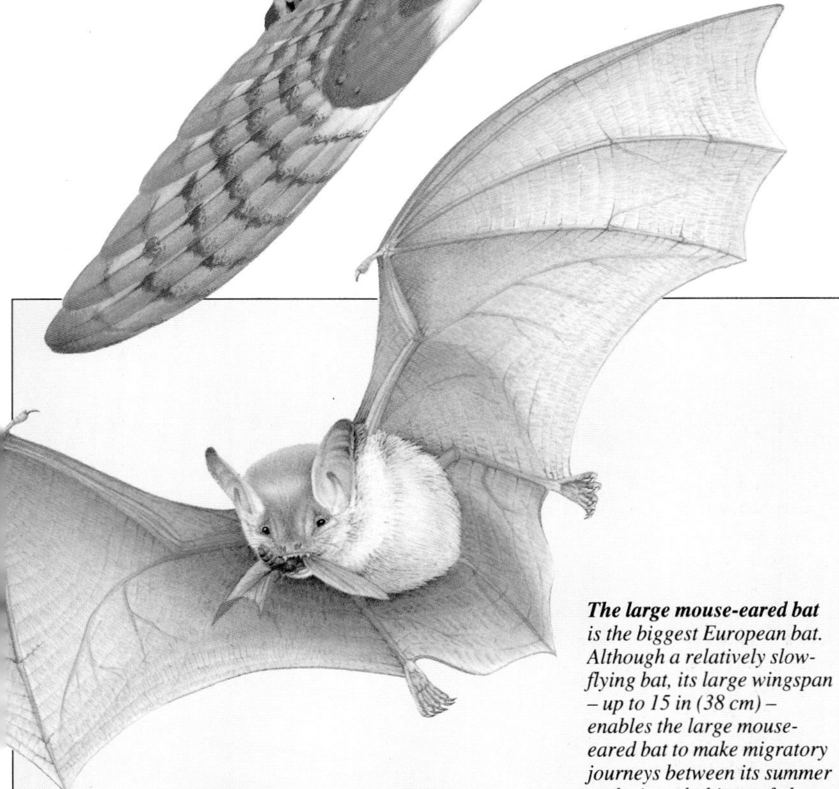

The large mouse-eared bat *is the biggest European bat. Although a relatively slow-flying bat, its large wingspan – up to 15 in (38 cm) – enables the large mouse-eared bat to make migratory journeys between its summer and winter habitats of about 125 miles (200 km).*

Bats in Flight

Bats, the only mammals that can truly fly, evolved long after the birds had taken to the air. They retain many orthodox mammalian features, including a bone structure that is strictly mammalian and is not honeycombed for lightness like that of birds. Instead of feathers bats are covered with skin and fur.

The wings are hugely extended webbed "arms" and "hands" with a thin layer of membranous skin stretched between thin and elongated finger bones. The "thumb" bears a small claw, for both grasping and clinging.

The major difference in flight technique between bats and birds is that birds use two large pairs of muscles to fly, but bats use four large pairs and several lesser pairs. Bat flight, therefore, is more acrobatic than that of birds, particularly in enclosed spaces such as caves.

Migration Techniques 230 **Solitary Carnivores** 246 **Sonar** 280 **Vision** 278

Lightweight Flight

How insects fly

Some hoverflies and midges beat their wings faster than 1,000 times a second when flying. Most insects average about 520 times a second – requiring muscle contraction and expansion every 0.0019 seconds or so. The evolution of flight in insects has given them many advantages. As a form of movement, it is thousands of times faster than crawling about on tiny legs, and it makes light of the mountainous obstacles that grass, rocks and bushes otherwise present. Of all the invertebrates only the insects have been able to develop true flight.

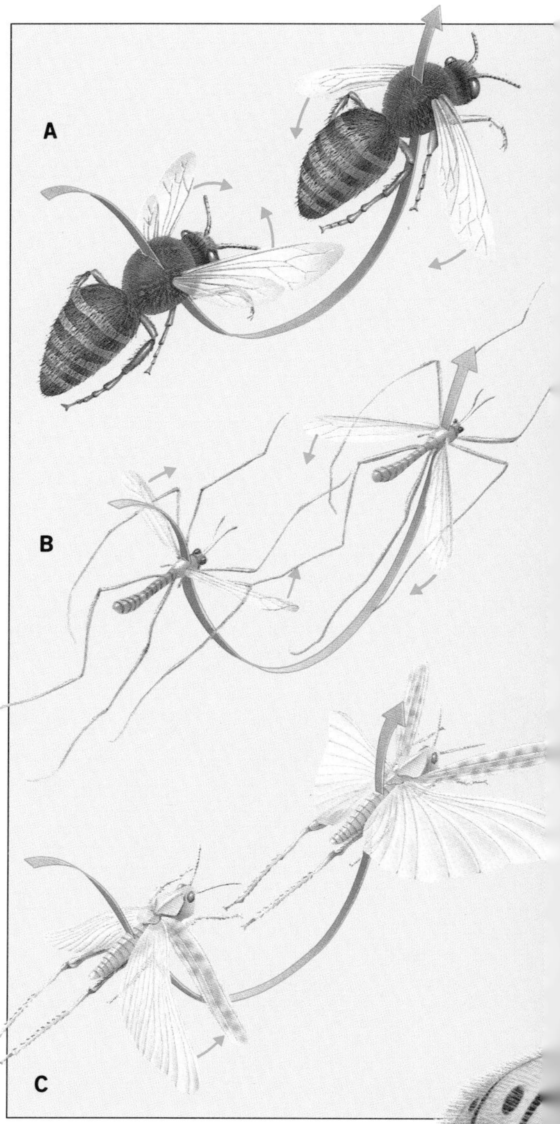

***Insects have evolved** many ways of reducing the surface area of their wings on the upstroke (the first insect in each diagram pair shows the upstroke). The more advanced fliers can all tilt their wings to slice like a knife through the air* [**A**]. *Other insects allow their wings to curve or camber* [**B**], *which not only reduces their surface area but also generates extra aerodynamic lift. Insects with large, gauzy hindwings often spread them on the downstroke but swivel them back on the upstroke so that they present a fraction of the surface area* [**C**]. *Like the spokes of an umbrella, the veins in these wings are a little too short for the extended membrane. When the wing is relaxed it forms pleats; when fanned out, the tightness of the membrane curves the veins downward, making the wing rigid and preventing the back edge from fluttering. Some four-winged species allow their fore- and hindwings to overlap on the upstroke, fanning them again for the downstroke* [**D**].

Many puzzles remain to be solved about the widely differing shapes of insects' wings. Unlike birds and bats, insects have no muscles within the wing, so they cannot directly change their wing shape while in flight. In both birds and bats the wings evolved from the forelimbs – they are basically legs that have, over millions of years, exchanged running movements for flying movements. Insect wings evolved as extensions of the exoskeleton, hard flaps that were originally there to protect the limb joints, but then found a new use as aids to gliding. As larger flaps evolved, they developed a network of veins supplying them with blood and a delicate membranous covering that grew lighter and more flexible.

Flapping wings

Insects rely on flapping flight – gliding may have suited their ancestors well enough as they evolved during the Carboniferous Period 300 million years ago, but that was before the advent of insect-eating birds and bats. A gliding insect today would be at the mercy of predators. For flapping flight to work the wings must change their surface area between the downstroke and the upstroke. The downstroke pushes the insect up, but if the wing is exactly the same shape on the upstroke, the effect will be cancelled out.

Wings with hinges

One straightforward solution is to have a wing that "collapses" automatically on the upstroke. This is found in a number of large-winged, slow-flying insects, such as the stone flies, scorpion flies and crickets, but also in some moths and most bugs. Their forewings are divided into a large front section and a smaller back section (the *clavus*), joined by a line known as the *claval furrow*. This acts as a one-way hinge, allowing the clavus to flip down when under pressure from above but not to flip up. On the downstroke the clavus works as hard as the front section of the wing, but on the upstroke it folds away to reduce the wing area. The most advanced, hovering fliers have narrow wings with complex aerodynamics. They can tilt their wings to slice through the air on the upstroke. The wing's leading spar cannot bend, but as it cuts through the air, aerodynamic forces make it twist. Running backward from this spar, and supporting the wing, are a number of veins, which are themselves twisted. When the spar twists, these veins curl downward, curving the wing into a shape like the wing of an airplane.

Mindless muscle power

Wings can only move as fast as the muscles that power them, and the muscles can only move as fast as the brain tells them to. The bottleneck in evolving faster wingbeats was the rate at which nerves could send out instructions to the muscles to contract. Some insects overcame this problem by developing special muscles attached to the flexible thorax instead of directly to the wings. The insect's nerves only have to send out a signal once for every forty muscle contractions to sustain a vibration of the thorax that keeps the wings flapping. This sort of flight muscle is found in the more advanced insects, such as flies, beetles, wasps and bees.

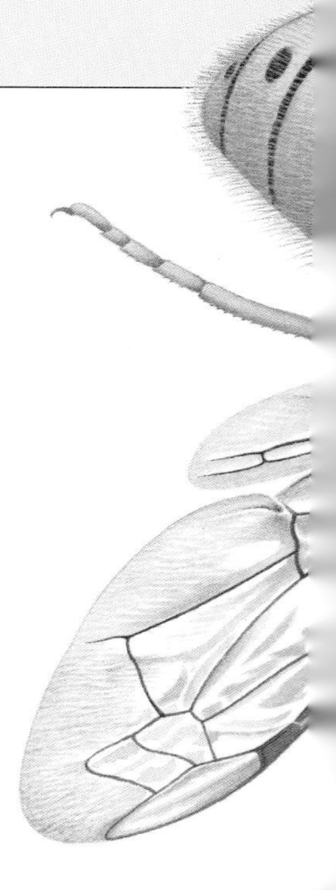

Connections: Bird and Bat Flight 224 Gliding and Soaring 222 Invertebrates 122 Pollination by Animals 154

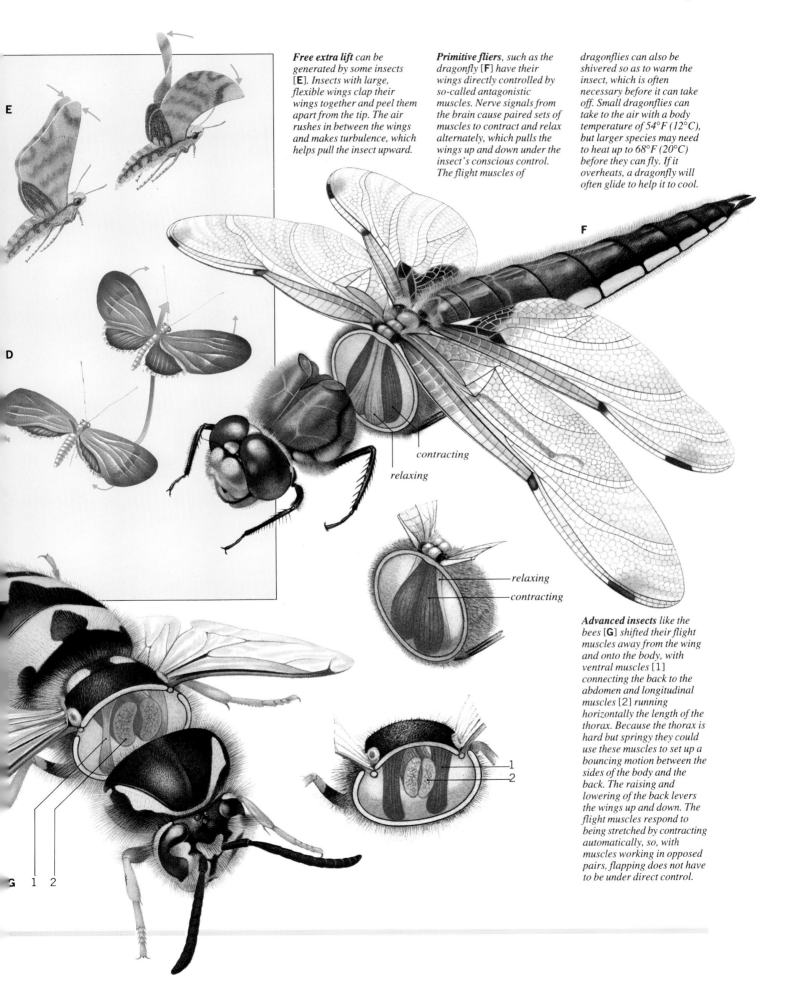

***Free extra lift** can be generated by some insects* [**E**]. *Insects with large, flexible wings clap their wings together and peel them apart from the tip. The air rushes in between the wings and makes turbulence, which helps pull the insect upward.*

***Primitive fliers**, such as the dragonfly* [**F**] *have their wings directly controlled by so-called antagonistic muscles. Nerve signals from the brain cause paired sets of muscles to contract and relax alternately, which pulls the wings up and down under the insect's conscious control. The flight muscles of dragonflies can also be shivered so as to warm the insect, which is often necessary before it can take off. Small dragonflies can take to the air with a body temperature of 54°F (12°C), but larger species may need to heat up to 68°F (20°C) before they can fly. If it overheats, a dragonfly will often glide to help it to cool.*

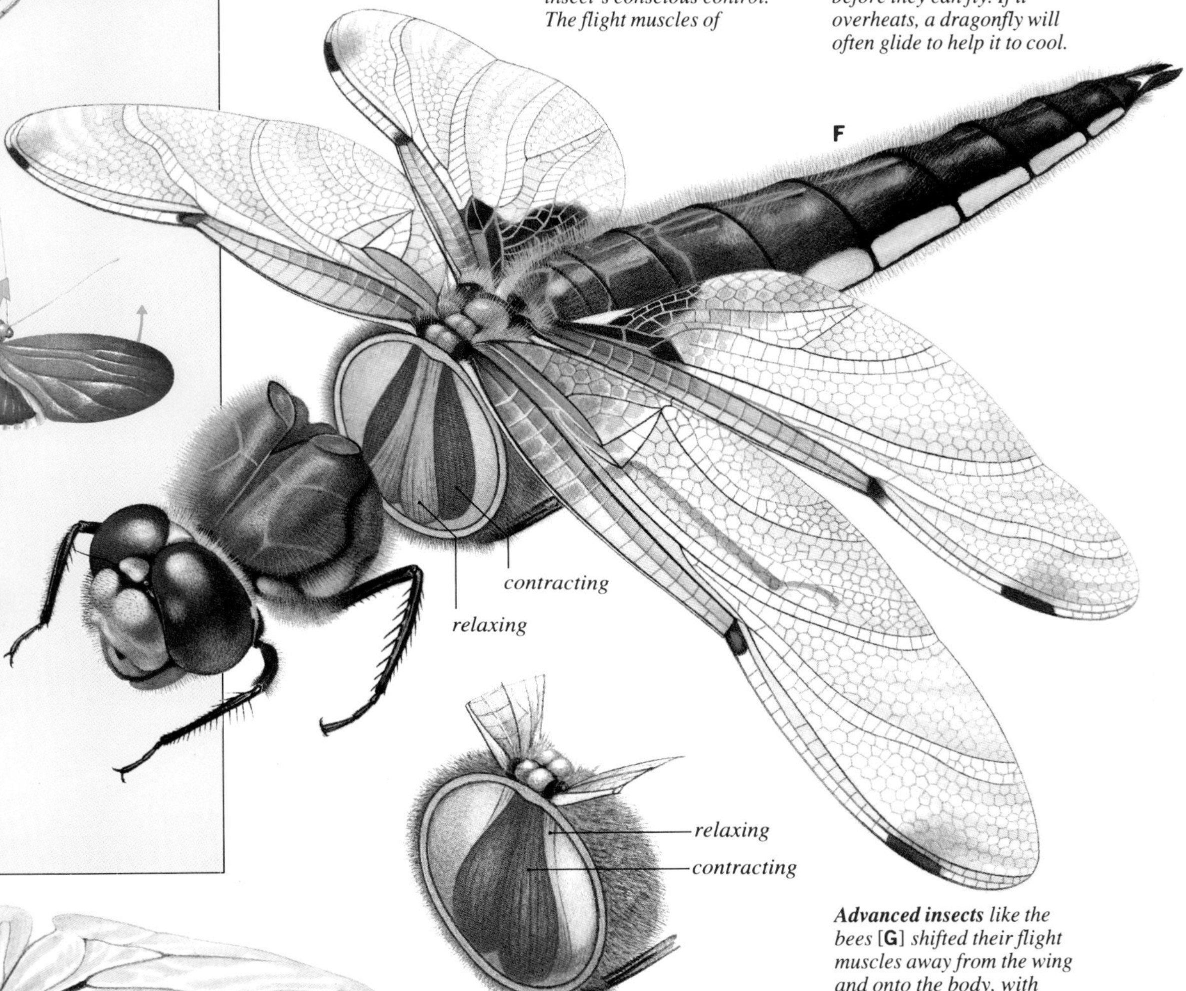

***Advanced insects** like the bees* [**G**] *shifted their flight muscles away from the wing and onto the body, with ventral muscles* [1] *connecting the back to the abdomen and longitudinal muscles* [2] *running horizontally the length of the thorax. Because the thorax is hard but springy they could use these muscles to set up a bouncing motion between the sides of the body and the back. The raising and lowering of the back levers the wings up and down. The flight muscles respond to being stretched by contracting automatically, so, with muscles working in opposed pairs, flapping does not have to be under direct control.*

The Pathfinders

Why some animals migrate

The tiny blackpoll warbler travels from North America to South America, taking only 80–90 hours to make the journey. This is the equivalent of a human running a succession of four-minute miles for 80 hours. If the warbler were dependent on petrol as a fuel instead of its body fat, it could maintain a fuel consumption of 720,000 miles/gallon (253,000 kilometers/liter). Animals migrate for different reasons: some make irregular movements in response to climatic changes or food shortages, while others are true migrants, making regular journeys tied to the seasons.

In autumn, birds living in Arctic or north temperate regions start to experience food shortages at the very time they need more energy for metabolism as the weather becomes colder. To compound the problem there are also fewer hours of daylight in which to find food. The solution for these birds is to fly south to warmer regions for winter, returning north the following spring to breed. Despite the huge risks involved in long journeys, from storms, exhaustion, hunger and predators, on balance the journey is beneficial and the migrants stand a better chance of breeding the following year.

It is easy, therefore, to understand why birds migrate south for winter. But once in the warm, food-rich southern countries, what is the benefit of returning to the north to breed in the more unpredictable spring of northern regions? There are two main reasons. First, there is a great increase in the availability of food, particularly insects, in the temperate and northern regions during spring and early summer, enabling them to rear more young than in the less seasonal conditions of the tropics. Second, because the tropics are densely populated with non-migratory animals, competition for food and nesting sites is greater there and predators of eggs and young are much more common.

A variety of migrants

Although birds include the most migratory species, other groups of animals also make regular migrations. As with birds, these are mostly concerned with food supply and reproduction, such as the movements of whales, which find abundant food in the productive cold polar waters but migrate to warm tropical and subtropical seas to breed. Many fish make long migrations between feeding and breeding areas, either entirely

__The deep warm waters__ of the Sargasso Sea [**A**] *are spawning grounds for American eels and their close relatives the European eels. Slowly the larval hatchlings make their way back to fresh water in North America or Europe; for an American eel this journey takes about a year, but for the European eel the journey lasts about three years. Newly hatched larvae* [1] *are about 0.25 in (7 mm) long; fully grown eels* [6] *are over 39 in (1 m) long. Between these extremes the transparent flattened larvae* [1–4] *grow until they reach a length of about 3 in (80 mm), by which stage they are near the coast and metamorphose into transparent eels,* elvers, *about 2.5 in (65 mm)* [5]. *(The map colors correspond to colors for each larval stage and show distances drifted at each stage.) Reaching river estuaries, the elvers change from drifters to strong swimmers, many battling against the current to move upstream. A mature American eel passes through the same growth stages as its European cousin, but matures in 6–10 years, whereas the European eel takes up to 15 years to develop. The mature eels are now ready to adapt back to a deep-sea life, for the mating urge beckons, and they migrate back to the Sargasso Sea, where they mate and die.*

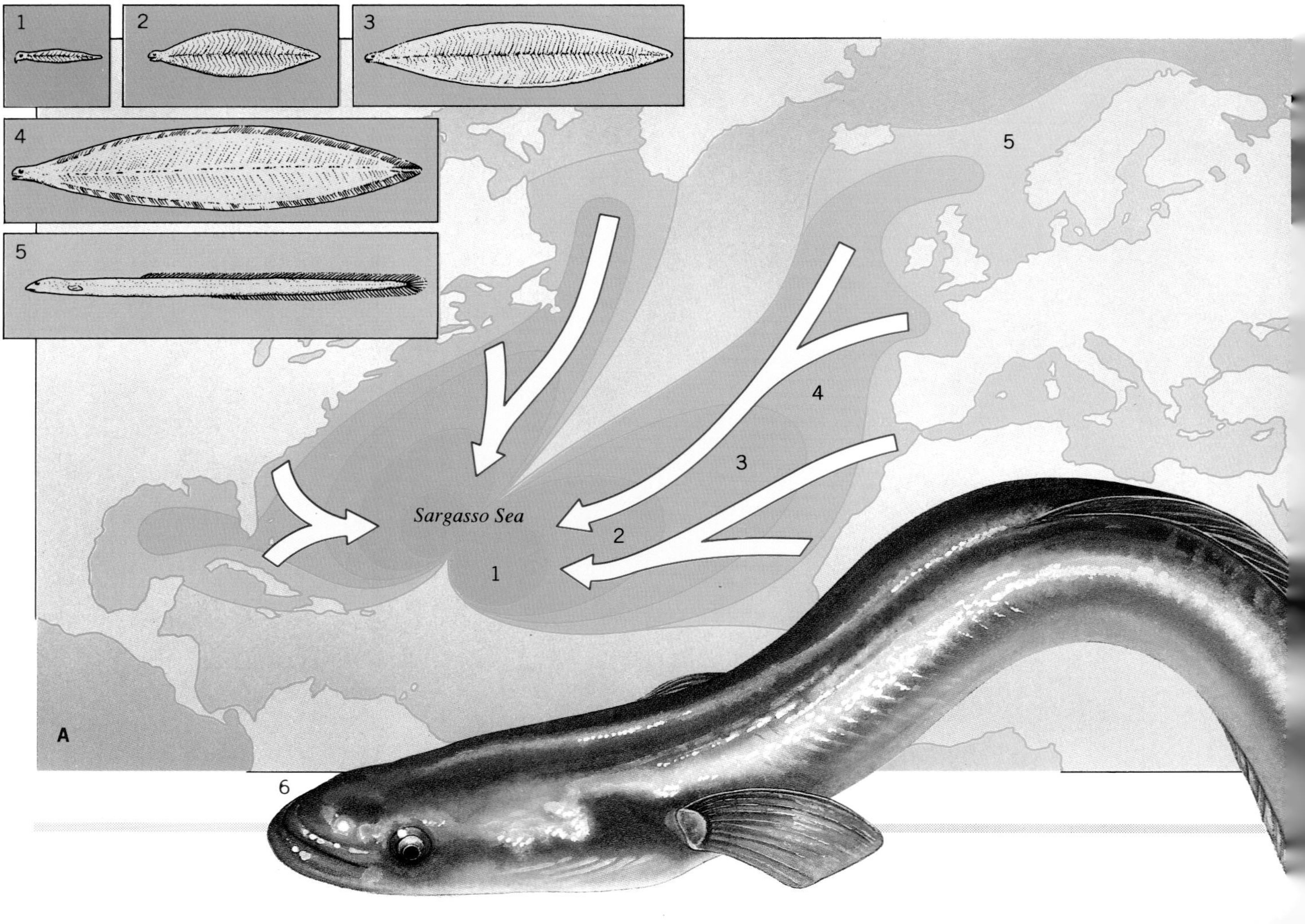

B

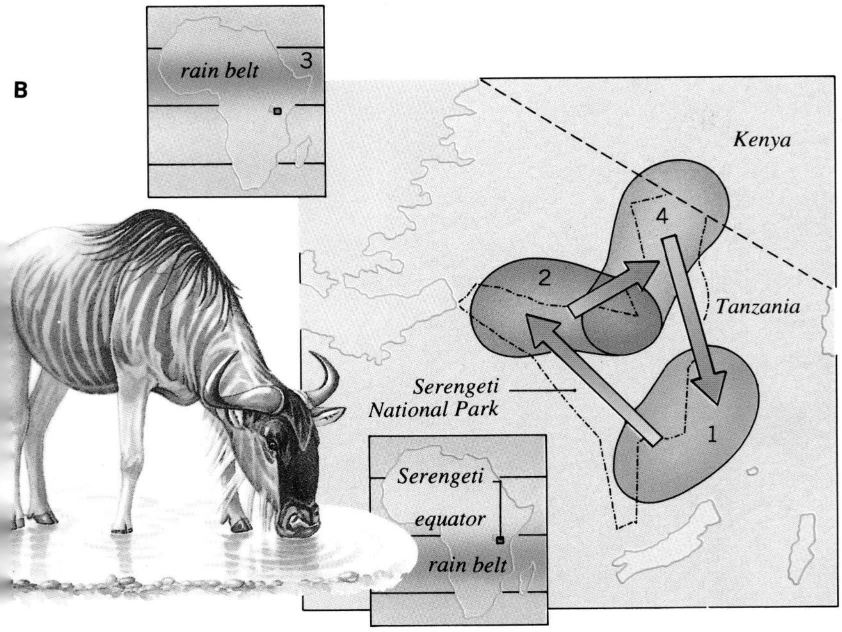

within the ocean or between sea and freshwaters, as in the case of salmon. Sea turtles, too, travel immense distances to reach sandy breeding beaches free from predators on isolated islands and coasts. Amphibians, such as frogs and toads, return from the land every spring to ponds and streams to spawn.

Misleading directions

Heredity probably has an important role in migration, especially in long-distance migrators. Starlings, for example, migrate in a southwesterly direction from the eastern Baltic, briefly resting in the Netherlands, to winter in England. To study the role of instinct, 11,000 birds were ringed in the Netherlands and, to simulate being blown off course by a gale, were taken to Switzerland. Most starlings migrate in groups, with adult, experienced birds guiding the young migrants. The birds were re-released in Switzerland, but the adults were separated from the young. The adults reoriented toward their normal winter quarters, some stopping in northern France, some reaching England. The immature birds headed off in the same direction in which they would have left the Baltic for England, but failed to correct for their displacement, ending up in southern France or Spain. The experiment showed that starling navigation combines instinct with experience gained from adult guides.

Migrant Hunters

The Elenora's falcon is an elegant bird of prey that breeds in the Mediterranean. Instead of breeding in spring and summer like European birds, it times its breeding cycle so that it has its young in autumn [1], migrating south to winter in Madagascar [2] to coincide with the huge flocks of small migrant birds that cross the Mediterranean [3] on their way to their African winter quarters; these birds are the staple diet of the young falcons.

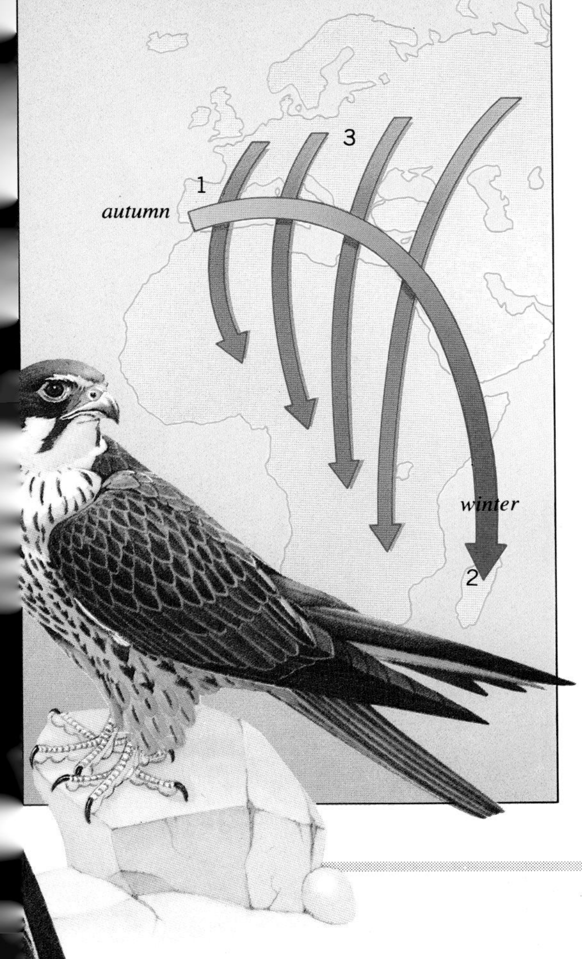

Wildebeest migrate **[B]** *to find grass, which is determined by movements of the equatorial rainbelt. They spend November to April in the Serengeti Plain, when the rain belt is in the south* [1] *and grass growth is greatest. Mating ends before the rains stop and grass stops growing; migration to the western Serengeti* [2] *follows just before May. Although the rains have moved north* [3] *and grass growth here is not at a maximum, more grass is available than on the plains. August and September are spent in the northern Serengeti* [4]*, where the "dry" season is wetter than dry seasons in the other areas and grass is plentiful.*

Desert locust nymphs *will daily march 1,000 ft (300 m) to find food, but adults* **[C]** *fly in huge swarms and may cover 2,000 miles (3,200 km) a year, devastating vegetation en route. Locusts usually migrate downwind to low-pressure zones, where rain is most likely (though winds may blow them out to sea, where they drown), following the seasonally shifting rain belts shown on the map* [1]*. Only if they reach a wet zone, with plentiful food for the nymphs, do locusts sexually mature and breed. If conditions are too dry, locusts change their appearance, growing longer wings and becoming gregarious – ready to swarm.*

C

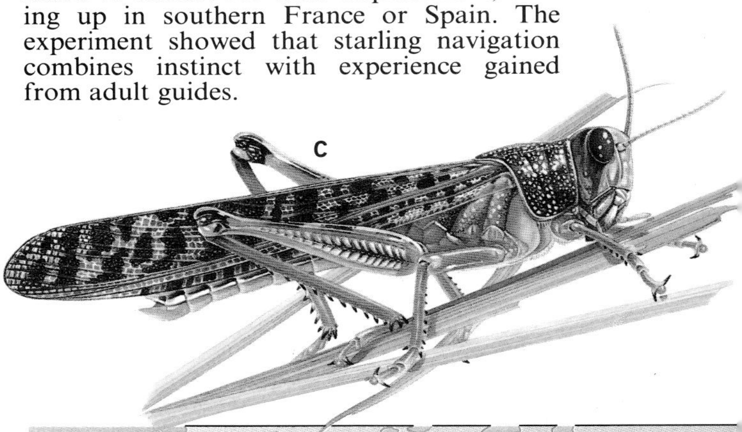

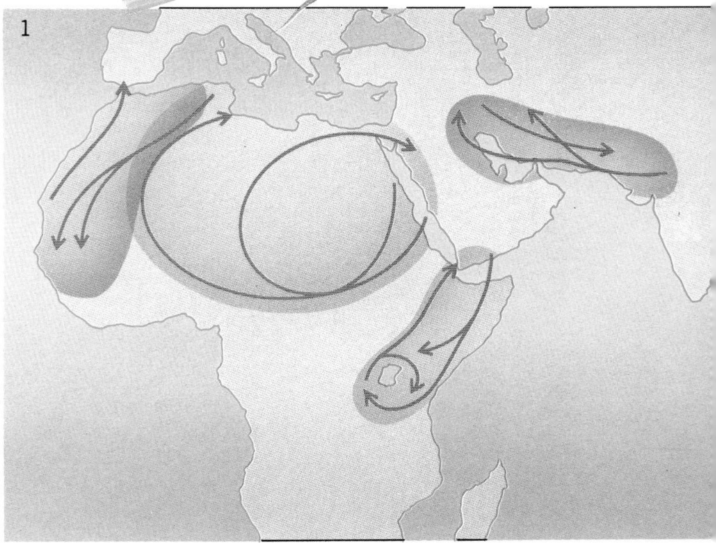

Connections: **Birds** 132 **Herbivores** 202 **Migration Techniques** 230 **Seasons** 86

Nature's Navigators

How animals migrate

In one famous experiment a Manx shearwater was taken from its nest burrow on the island of Skokholm off the coast of Wales to Boston, almost 3,000 miles (5,000 kilometers) away on the opposite side of the Atlantic, and released. It flew home in just 12 days, arriving the day before the letter confirming its release. Researchers are only beginning to understand how animals use a variety of techniques – from mapping visible landmarks to sensing patterns of sound or magnetism – to make such long, hazardous journeys and reach precise targets with such amazing accuracy.

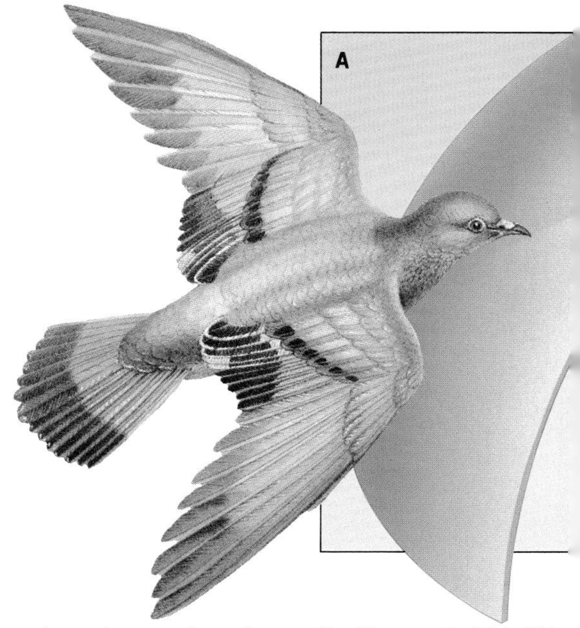

Some animal migrants have a magnetic sense to help them navigate. It seems that an animal like the barnacle goose can orient itself by determining the angle the Earth's lines of magnetic force make with its body as it travels through them. Birds find navigation hard in times of high sunspot activity, when magnetic storms rage across the globe. Whales, too, can suffer from false information. The previously inexplicable mass strandings of whales have been found to occur where natural magnetic fields are distorted by geological formations. It appears that whales build up an internal map of magnetic contors and use this to navigate along a system of invisible undersea magnetic "roads" – an advantage if traveling through thousands of miles of dark ocean.

Olfactory maps

Various mammals move about over short distances with the help of scent trails. Scent also guides salamanders and other amphibians to their breeding ponds, and may help marine turtles navigate thousands of miles across the oceans to their nesting beaches. Salmon also cross the seas, to spawn in precisely the same stream where they hatched. They probably make the bulk of their journey guided by the position of the Sun, by water currents and by

Homing pigeons **[A]** *successfully navigate home over hundreds of miles from sites never previously visited; these astonishing feats have inspired many experiments, some of which are shown below. Evidence suggests visual landmarks are fairly unimportant guides, so the obvious alternative is solar navigation; but, as 16th-century sailors were only too well aware, accurate solar navigation needs an accurate clock. Bird behavior is influenced by light and darkness, and laboratory birds can be tricked by artificial lighting to make their "body clocks" run fast or slow. On release, a homing pigeon (shown here in the Southern Hemisphere) with an unaltered clock flies in the correct direction relative to the Sun to reach its home loft when released* [1] *(in all diagrams red arrows show the correct direction to the pigeon loft and black arrows show the observed scatter of actual flight directions). If a pigeon with a clock "set" six hours fast is released at noon, it assumes the time is 6.00 pm and the Sun is in a 6.00 pm position* [2]*, much farther west than in reality. The pigeon therefore flies too far east in its attempt to reach home* [3].

But pigeons can also navigate if the Sun is hidden by clouds [4]*, when birds with altered clocks navigate just as successfully as birds with normal clocks* [5]*. In fact, they have a "backup" navigational aid using the Earth's magnetic fields. This can be tested by attaching a small bar magnet to a pigeon's neck to distort the magnetic field around it; on an overcast day the "homing" directions flown by such a pigeon are completely random* [6]*. Confirming the magnet's effect, a pigeon with a similar, nonmagnetic, brass bar is not disoriented. On sunny days, however, a pigeon with a magnet and a normal clock reverts to solar navigation and is not disoriented* [7] *by its disturbed magnetic field.*

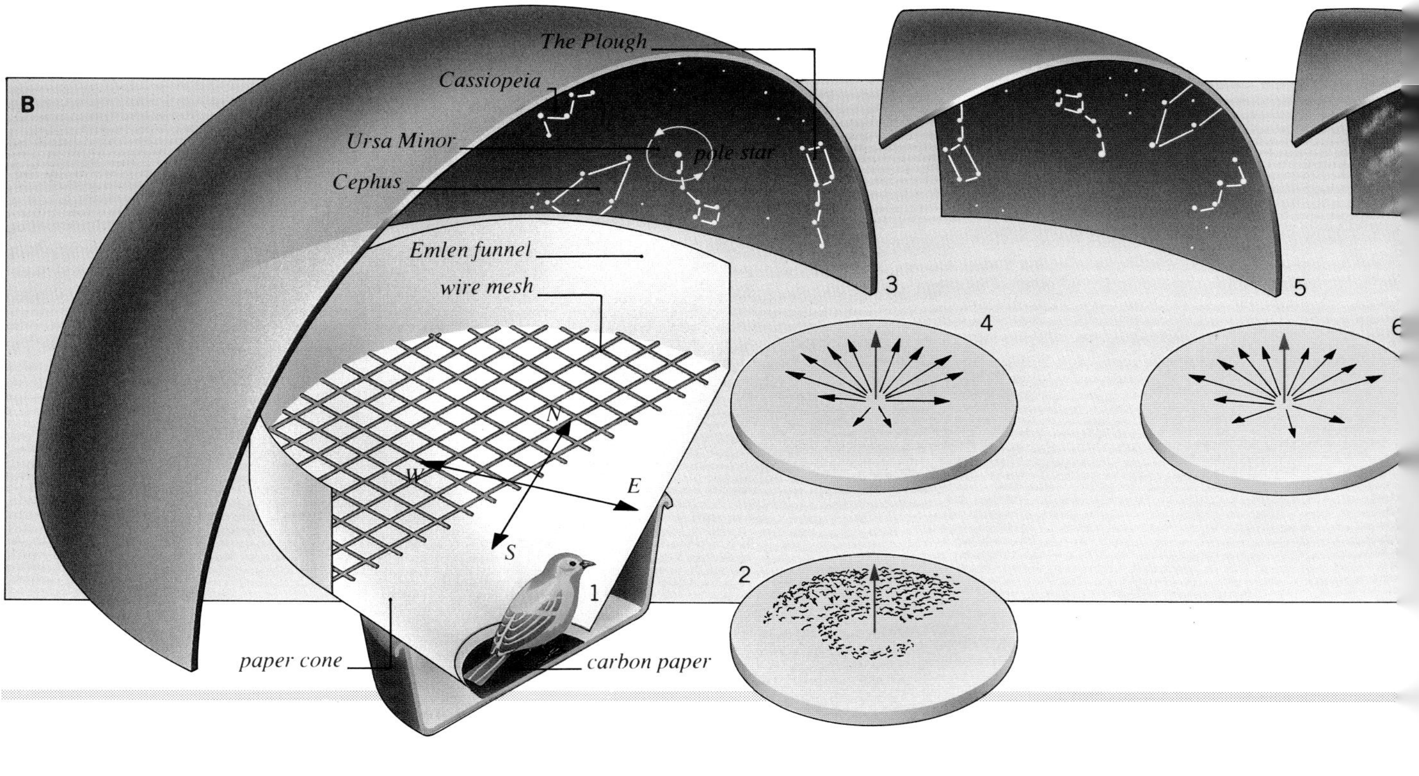

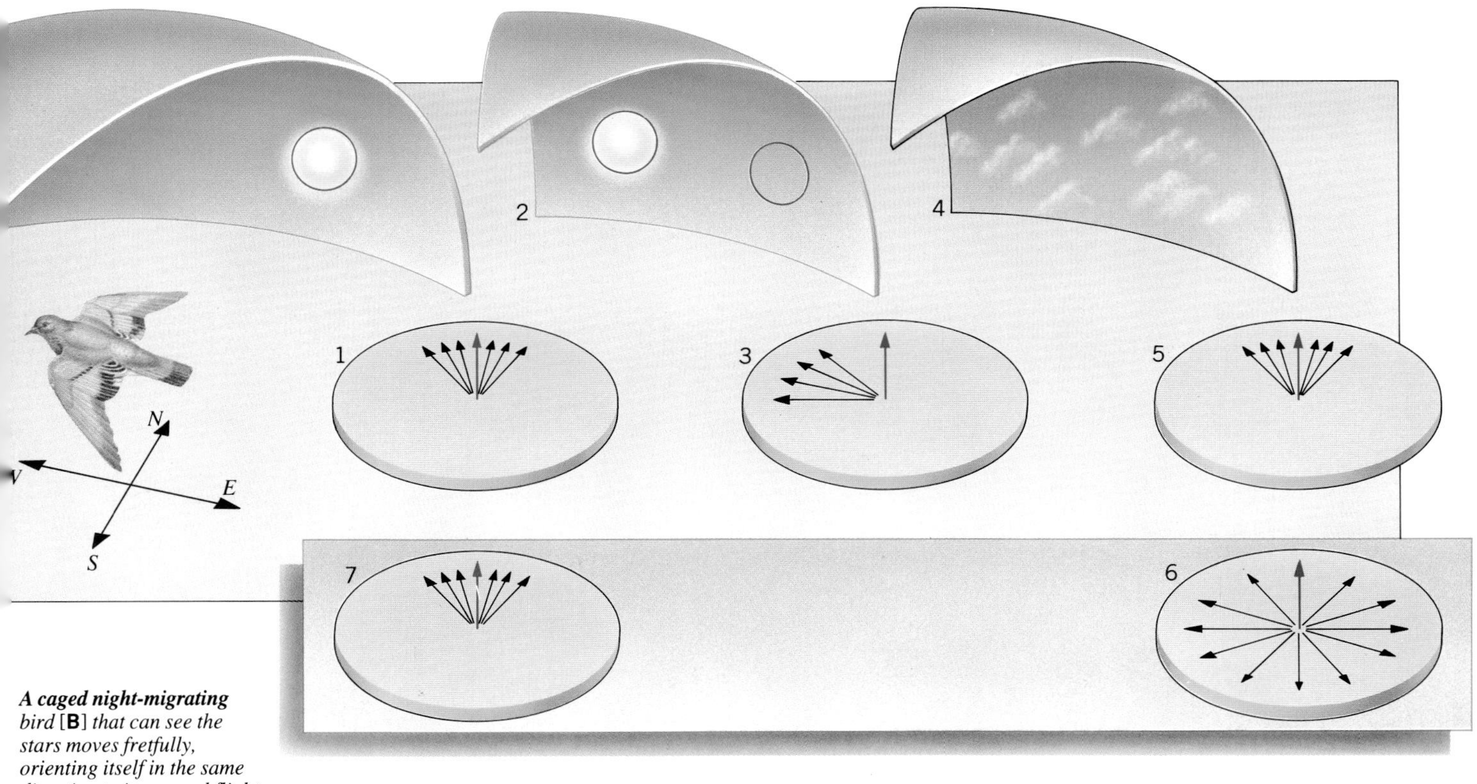

***A caged night-migrating** bird* [**B**] *that can see the stars moves fretfully, orienting itself in the same direction to its normal flight – north in the spring and south in the autumn.*

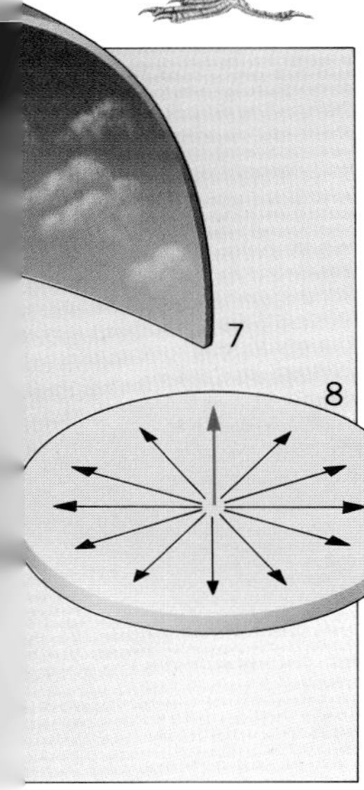

This led to experiments to see how indigo buntings [1] *navigated by stars (Moon and planet motions were ruled out as too complex) – did they use an internal body clock like day navigators? When caged in an Emlen funnel placed under stars projected in a planetarium, the bunting could see the sky and tried to escape, leaving carbon paper footprints on the paper funnel* [2], *which could then be interpreted to show the direction of orientation. The projected stars could be repositioned to change the apparent time and alter the bunting's clock. As the Earth revolves every 24 hours and its axis of revolution points north to the Pole Star, the stars appear to revolve around the Pole Star, a constant reference point from which to orient. Under stars set at normal time* [3], *the bird oriented. successfully northward* [4] *(in all diagrams red arrows show the correct migratory direction and black arrows show the scatter of actual observed orientations). When the stars were shifted 12 hours forward* [5], *the bird still oriented in the correct direction* [6], *showing that the bunting made direct use of the stars but did not compensate with an internal clock. But if the stars were eliminated and the planetarium diffusely lit to simulate stars obscured by clouds* [7], *the bunting's orientation attempts were entirely random* [8]. *Further experiments confirmed that the birds navigate by observing the rotation of the stars closest to the Pole Star, which could be eliminated from the sky yet not impede correct orientation.*

their magnetic senses. But when they near fresh water they can distinguish the exact scent of their stream of birth, which they carry in their memories.

Most birds have a poor sense of smell, but some, like shearwaters and petrels, use scent detection to navigate home across the sea to their nesting burrows. Until recently birds were also thought to have poor low-frequency hearing, but research shows that some birds can hear infrasounds – very low or muffled sounds – as low as 0.1 vibration per second: far more sensitive than humans. There are many natural infrasound sources – ocean waves, atmospheric jet streams, wind blowing through mountains or over sand dunes – which birds use to orient themselves.

Judging angles

The waves that make up sunlight vibrate at random angles. As sunlight passes through the atmosphere, part of it is *polarized* so that it vibrates in one plane only. Some migrants use their ability to detect polarized light to orient and navigate. The pattern of polarization follows the Sun around the sky and can be used to determine the Sun's position, even when it is hidden by clouds, mountains, or the horizon. Experiments with pigeons show that they can detect when the plane of polarization of light alters.

Insect navigators

Bees have long been known to use their sensitivity to polarized light to navigate by using the Sun's position in the sky. Small areas of cells in the top of the bee's eyes detect the pattern of polarization, and even if the bee can see only a tiny area of clear sky, it can compare part of the pattern with a map of the whole pattern stored in its brain and locate the Sun's position accurately.

Connections: **Communication** 292 **Earth's Magnetic Field** 14 **Hearing** 282 **Migration Patterns** 228 **Special Senses** 286

Made to Measure

How bees and wasps create their nests

Some bees' nests can contain as many as 80,000 individuals. The physical organization of the colony's structure is reflected in the distinctive castes of the inhabitants – the queen, the workers and the drones – each of which has not only set tasks but a well-defined body shape. Yet there are other species of bees and wasps for whom a nest may contain far fewer insects – sometimes only a dozen or so. And then there are the solitary species, hunting alone, and retreating to a nest or burrow underground. Even for these, the structure of their home is all-important.

A

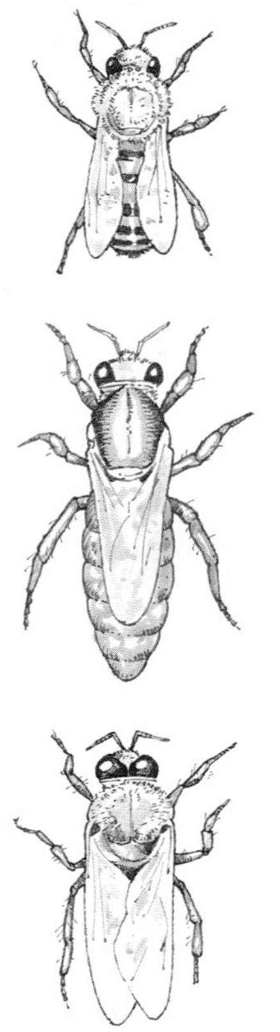

Honeybee societies *are organized into three castes* [**A**]. *Smallest and most numerous are the workers* [1]. *These sterile females build, maintain and defend the nest, care for the queen and drones, and rear the brood. Combs on their front legs and "baskets" on their hind legs are used to collect pollen. The queen* [2] *is the largest bee in the hive and the mother of all the workers. Her main task is laying eggs. She also produces pheromones that maintain the sterility of the workers. The male drone* [3] *is distinguished by his large compound eyes. His sole function is to fertilize a female on the wing.*

The feature central to the nests of all bees and wasps, whether solitary or social, is the cell – the structure in which a single grub develops from egg to pupa. Only in the most social species, such as honeybees, are the cells grouped together into combs. Honeybee combs are complex wax edifices with certain areas set aside for rearing the young, and other parts devoted to the storage of honey and pollen. The cells, which slant upward toward the opening so as to be able to hold liquid honey, are usually arranged into a vertical comb. Each cell has a characteristic hexagonal shape. This is no accident of design, as such a structure makes for a strong comb that can be constructed with the minimum of building material and the smallest possible expenditure of energy.

Building in wax

Wax is produced as thin flakes from glands on the underside of the worker bee's abdomen. The bee kneads the wax between her front legs and her mouthparts, and mixes it with saliva until it is flexible enough to work with. Temperature in the hive is maintained at a fairly constant 95°F (35°C) to keep the wax workable. If it falls below this level, the workers generate heat through muscle spasms: if it rises too high, the bees fan their wings, and sometimes bring water back to the hive and allow it to evaporate, thereby increasing cooling.

In addition to wax, bees also use a special resinous glue – *propolis* – for plugging gaps and cracks in the hive (and for coating the bodies of dead companions to mummify them). Propolis is scraped from the sticky buds of certain trees with the mandibles, then transferred to the pollen baskets for transport. The greatest demand for propolis comes in the autumn, when the bees are endeavoring to eliminate cold drafts from the nest.

Nesting underground

Many species of wasps and bees build their nests underground in specially excavated chambers, or in abandoned mouse or vole burrows. Social vespid wasps build surprisingly strong combs from paper. Wood scraped from twigs or posts is mixed with saliva to form pellets that are bonded together as they dry into a paperlike mass, which is then fashioned into tiers of hexagonal cells.

B

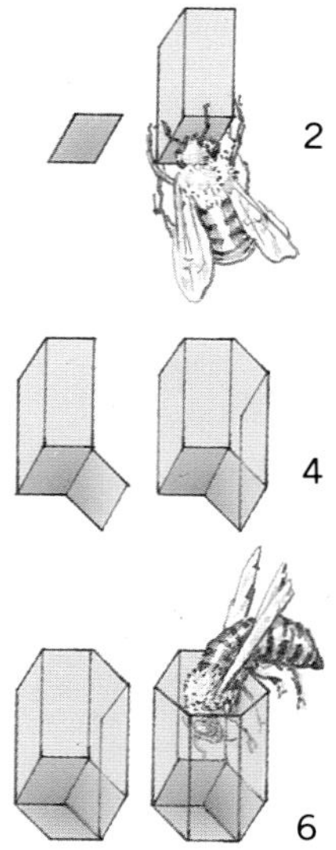

D

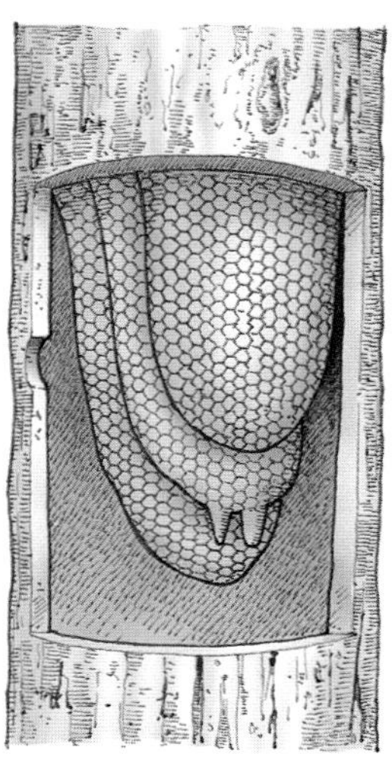

C

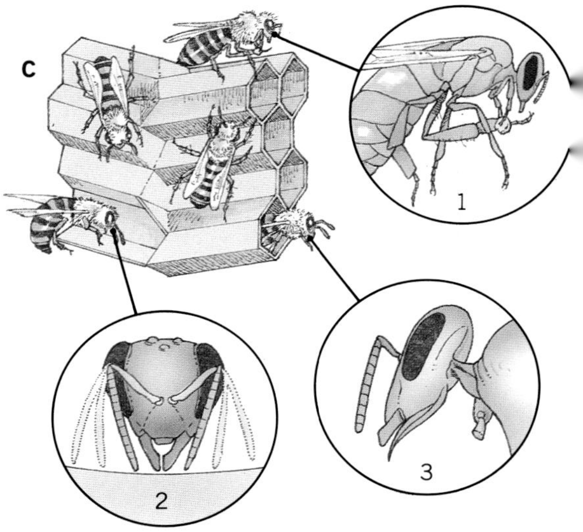

Each honeybee cell *is the work of many individuals, and is built upward from the base* [**B**]. *A rhomb-shaped section of the base is made first* [1], *followed by two adjoining cell walls* [2]: *a second rhomb is added at the base* [3] *and two more cell walls are erected* [4]. *The third rhombic section* [5] *and two walls* [6] *complete the hexagon. Honeybee combs hang vertically, the cells on both sides separated by a wall down the middle* [**C**]. *Workers* [1] *knead and soften the wax, fashioning it into walls that vary in thickness by no more than 0.0008 in. Bees achieve this astonishing accuracy by "prodding" the wall with their antennae and using its elasticity (or "give") as a measure of thickness* [2]. *The bee's own head serves as a "plummet," allowing it to tell up from down* [3]. *There are sensory bristles on the back of the bee's neck, which help it judge with enormous accuracy the movements of its own head.*

Bees and wasps *show a remarkable diversity of techniques for making their nests. Potter wasps (right) fashion tiny clay pots, in which they lay their eggs and provision their young with paralysed grubs. The mason bee molds sand and dust into pellets with her saliva to build her*

Connections: Ants' Nests 236 Beaver Lodges 240 Birds' Nests 238 Communication 292 Invertebrate Carnivores 200 Invertebrate Herbivores 198

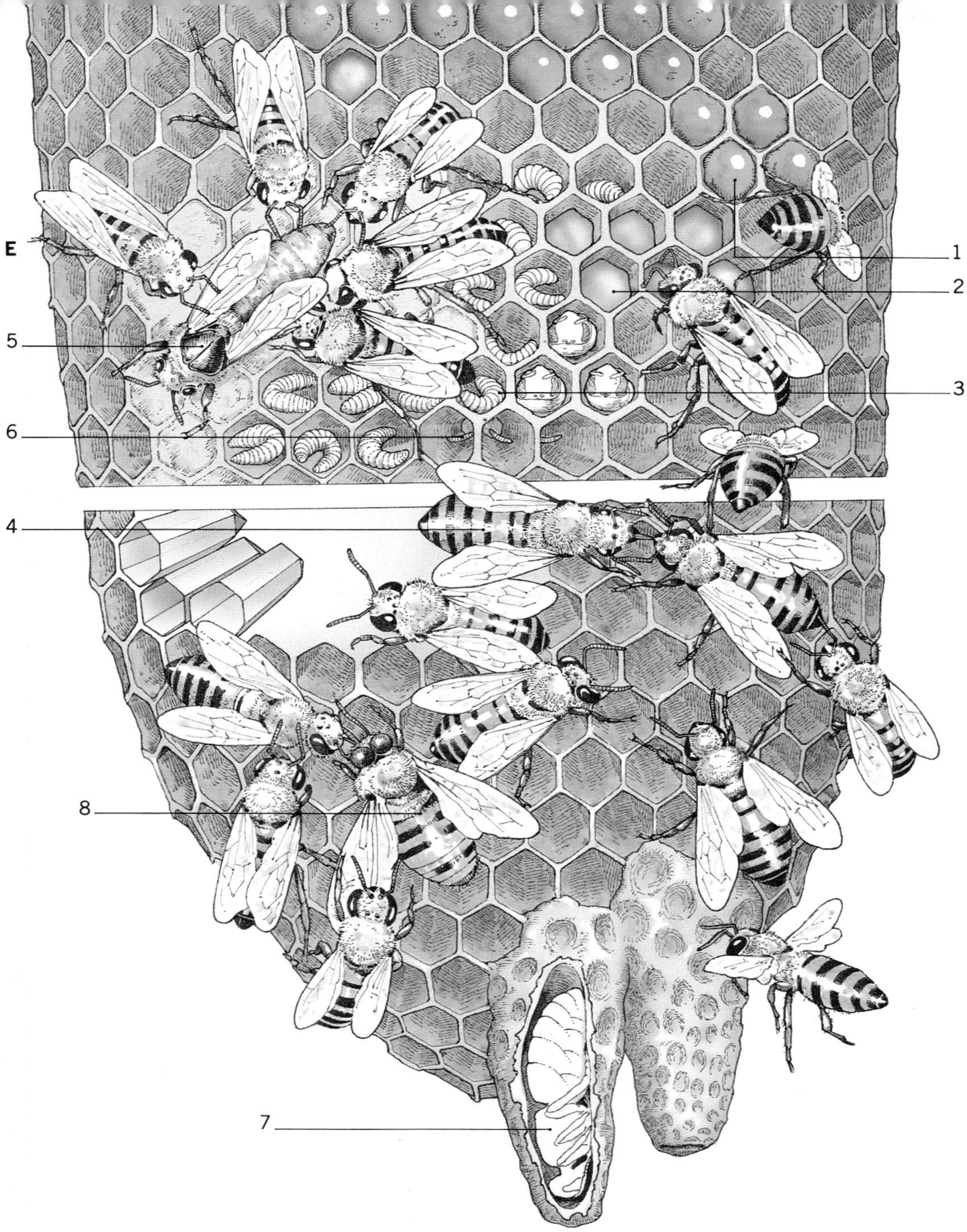

The honeybees' nest consists of a number of wax combs suspended in a shelter – in this case an old tree [**D**]. *The cells* [**E**] *that are situated at the periphery of the comb* [1] *contain nectar mixed with saliva, which the workers fan with their wings to evaporate excess water before capping the cell with wax: this mixture is gradually converted into honey. Other cells are used to contain reserves of pollen* [2]. *Developing larvae* [3] *in uncapped cells are carefully groomed by workers to prevent fungal infestations. They are fed by young workers with regurgitated food, and with honey and pollen from the storage cells. On emerging from their cells, workers are fed regurgitated nectar and pollen by their sisters* [4]. *The newly vacated cell is cleaned thoroughly by the workers and reused. The queen* [5], *surrounded by a retinue of workers, here rests on a group of capped cells, each of which contains a worker pupa. Eggs* [6] *are laid by the queen in the center of the comb. When new queens are needed, the workers construct extra-large cells on the edge of the comb in order to hold them. In the illustration, one queen cell has been cut open to reveal the pupa within* [7]. *Drones* [8] *do not perform "housekeeping" duties, and are driven away in times of food scarcity.*

camouflaged pots and honey stores. Each pot is provisioned with "bee bread" – a mixture of nectar and pollen – upon which a single egg is laid. Mining bees dig into hard clay soil, excavating a group of underground cells and impregnating their walls with a secretion that hardens and helps to stop the cells from collapsing. Leaf-cutter bees use narrow crevices or old beetle borings in wood as nest sites. They cut oval pieces of leaf and roll them to make their cells, which they then cap with carefully cut circles of leaf. The nest hole or crevice itself is subsequently plugged with a pile of leaf disks.

Because wasps do not store food in their cells (grubs are fed on the remains of insect prey), the combs usually lie horizontally, with the open end of each cell facing downward.

Lone hunters

Though social species are more prominent in terms of sheer numbers, most bees and wasps are in fact solitary, with no sterile worker caste. In these species, nest building is carried out by the mother. Solitary bees provision their chamber – which may be made of soil or wax – with pollen and nectar. The hunting wasps pounce upon caterpillars or other insects, inject a paralysing poison into them, then drag them to the wasps' underground nests, where they form a living larder upon which an egg is laid. The wasp carefully closes the entrance to the burrow with a stone, and in some species may even scatter other stones around the entrance to camouflage it.

Invertebrate Reproduction 160 Pollination by Animals 154 Termite Towers 234

Colonies and Castles

How termites build and live in their towers

In tropical zones there may be as many as 10,000 termites/square yard of soil surface. Found mainly in the warmer parts of the world, they play an extremely important role in decomposing plant remains, because they have bacteria in their stomachs that are able to digest woody fibers. Termites are social insects; they live in colonies. Their nests range from small colonies inhabiting tunnels in rotting tree stumps to colonies of several million in huge mounds so tall that were humans to build on the same scale their homes would be over half a mile high.

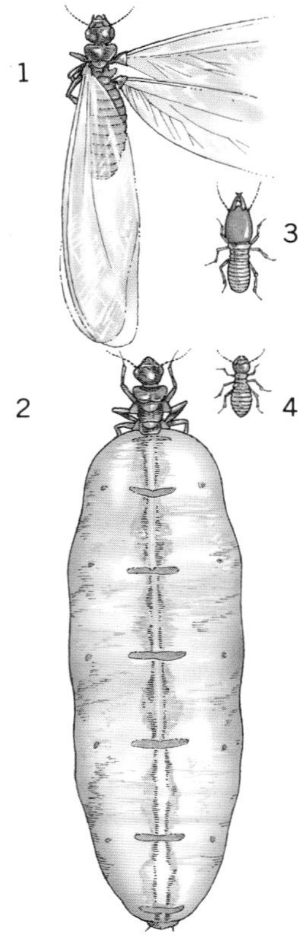

***Most species of termite** have a variety of castes or types. There are the temporarily winged reproductives (male and female), called* alates *[1], responsible for setting up new colonies, the queen termite [2], the enlarged abdomen of which produces thousands of eggs, the soldier termites [3], which protect the colony, and the workers [4], which collect food, care for the queen and serve as builders.*

Most of the 2,000 or so species of termite have two basic building materials. The principal structural building material, particularly with species that build larger structures, is made from a combination of soil particles and saliva. The second material, known as *carton*, is made from a combination of saliva and fecal pellets, and is usually used to build the "partitions" within the nest. To protect themselves from predators and from the drying effect of the atmosphere, termites often construct covered runways between their nests and their foraging grounds.

The social order

In contrast to the colonies of bees and ants, termite colonies contain more or less equal numbers of males and females, most of which are sterile. As in other social insects, there are several distinct castes, with characteristic roles and physical features. There is usually a single royal couple in each colony. Although larger than the other termites, the king is not as large as the queen, whose abdomen increases in size dramatically as she embarks on a life of egg-laying.

There are usually only two sterile castes, the ordinary workers and the soldiers. Soldiers, reproductives and young nymphs do not forage or feed themselves, but are instead fed on regurgitated food and anal secretions by the older workers.

Differentiation into the various castes appears to be controlled by hormones. Special hormones – pheromones – produced by the reproductive pair and by the soldiers suppress the metamorphosis of immature workers into reproductive adults. On the death of the king or queen, certain nymphs – "secondary reproductives," which have reduced compound eyes and sometimes small wing pads – develop into replacement reproductives.

The balance of the different castes in a colony is carefully maintained, so that there is a specific ratio of soldiers to workers and nymphs. If the balance is upset, it may be restored by the differentiation of immature nymphs into the required caste, or by cannibalistic workers.

Tasks for life

The division of labor in the colony varies with both caste and age. Young workers nurse juveniles, build and repair the nest, and

***The mounds of termites** [**A**] dominate the African savanna. Most termites prefer to eat dead plant material that has been partly softened by fungus. This food supply is limited in dry conditions because fungi need moisture. For this reason* Macrotermes *termites create fungus chambers [1]. These are combs of carton that provide a large surface area on which the fungus grows. The fungus flourishes in the humid atmosphere of the nest as it breaks down the feces in the carton walls. Some termite species dig deep tunnels [2] to find underground water to ensure that the nest is moist enough for the fungus to thrive.*

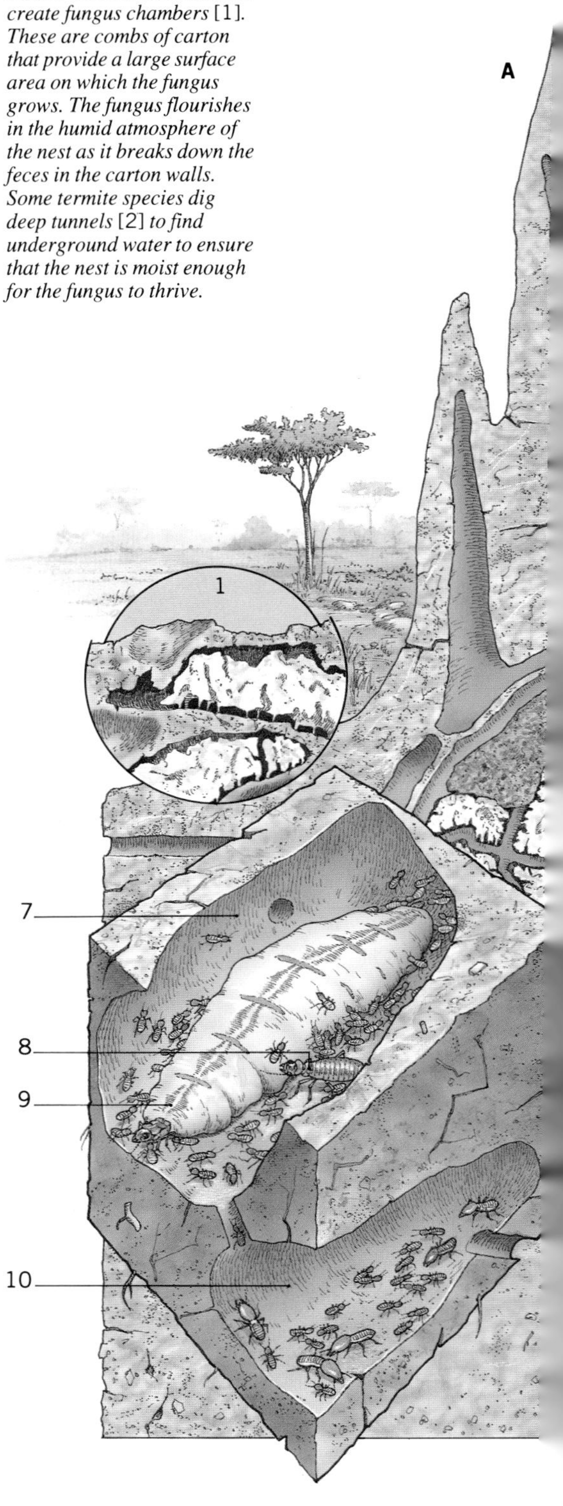

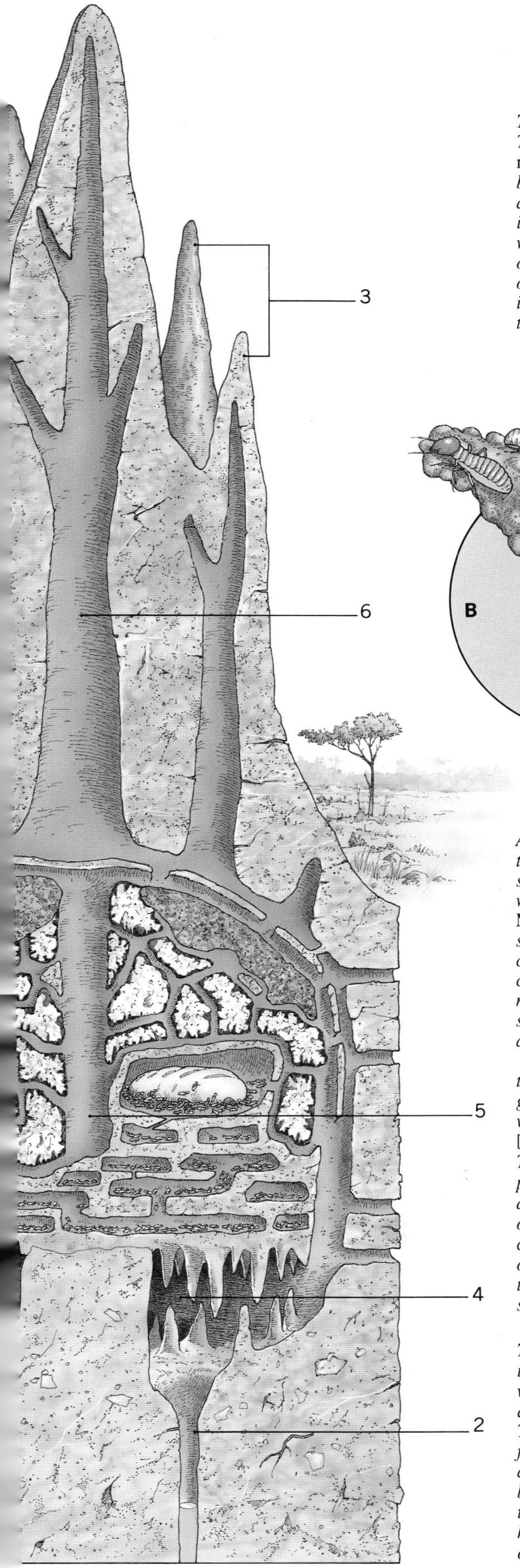

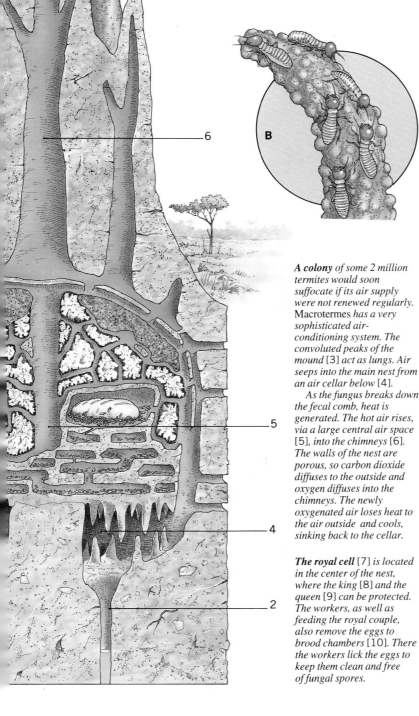

Termites are adept builders. The species Macrotermes natalensis *is capable of building structures such as arches* [**B**]. *The construction is carried out in the usual way, by using excrement or saliva to bind particles of soil. The termites instinctively know when to start angling the arch.*

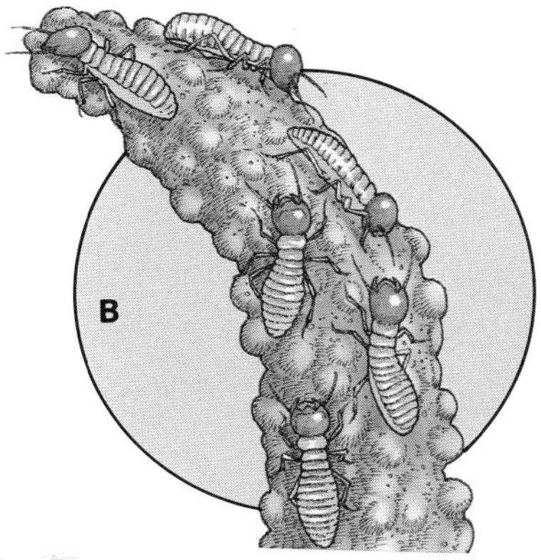

A colony *of some 2 million termites would soon suffocate if its air supply were not renewed regularly.* Macrotermes *has a very sophisticated air-conditioning system. The convoluted peaks of the mound* [3] *act as lungs. Air seeps into the main nest from an air cellar below* [4].

As the fungus breaks down the fecal comb, heat is generated. The hot air rises, via a large central air space [5], *into the chimneys* [6]. *The walls of the nest are porous, so carbon dioxide diffuses to the outside and oxygen diffuses into the chimneys. The newly oxygenated air loses heat to the air outside and cools, sinking back to the cellar.*

The royal cell [7] *is located in the center of the nest, where the king* [8] *and the queen* [9] *can be protected. The workers, as well as feeding the royal couple, also remove the eggs to brood chambers* [10]. *There the workers lick the eggs to keep them clean and free of fungal spores.*

feed and groom the royal couple and other members of the colony, while the larger workers are sent out to forage. This ensures that the maximum work has been obtained from a worker before it is exposed to dangers outside the nest.

Mushrooms and pagodas

Termite nests vary enormously in size and shape. Some rise like chimneys from the ground, others hang from branches high in the forest canopy. Some South American termites plaster their nests to trees, and then build up to 40 water-shedding mud barriers in layers above the nest to keep off the heavy tropical rain. Other forest species build tall nest mounds with pagodalike eaves to intercept the rain. *Cubitermes* termites are well known for their mushroom-shaped nests, which often have several "caps" piled one on top of the other.

Young Settlers

New termite colonies are usually formed by the swarming of newly produced, winged reproductive termites (*alates*). Alates can be either male or female. The male is attracted to the female by an odor secreted from a gland located on the underside of the female. The swarming occurs in certain weather conditions, or at particular times of the year, depending on the species. Colonies of the same species in the same area will often produce alates at the same time, allowing for breeding between individuals from different nests. Unlike bees, termites do not mate in flight. The swarming flight simply serves to disperse the alates and encourage breeding between colonies. After descending to the ground, they shed their wings and pair, then dig a small nest and seal themselves in before mating and starting a new colony.

Building a Society

How ants work together to make a nest

In some areas of rain forest the population of ants is so dense that their combined weight is thought to exceed that of any other species of animal. Ants are exceptionally successful – they have adapted to fill a wide variety of habitats, from the rain forests to deserts. Living in colonies of up to a million individuals, ants cooperate closely to build nests in which eggs and larvae are carefully nurtured. Some underground nests are extremely large and elaborate – for example, those excavated by North American leaf-cutter ants may be up to 20 feet (6 meters) deep and 49 feet (15 meters) in diameter.

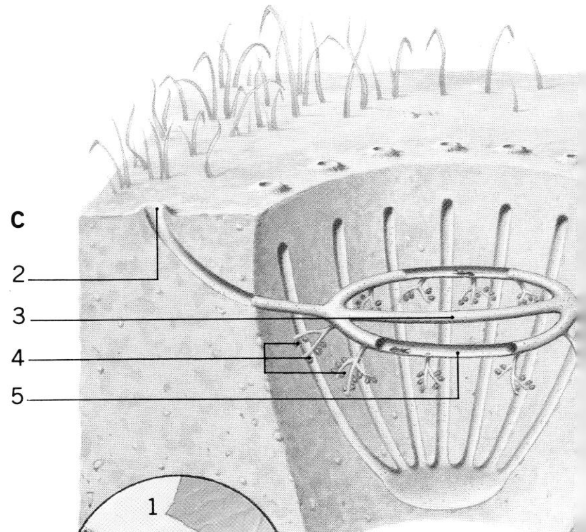

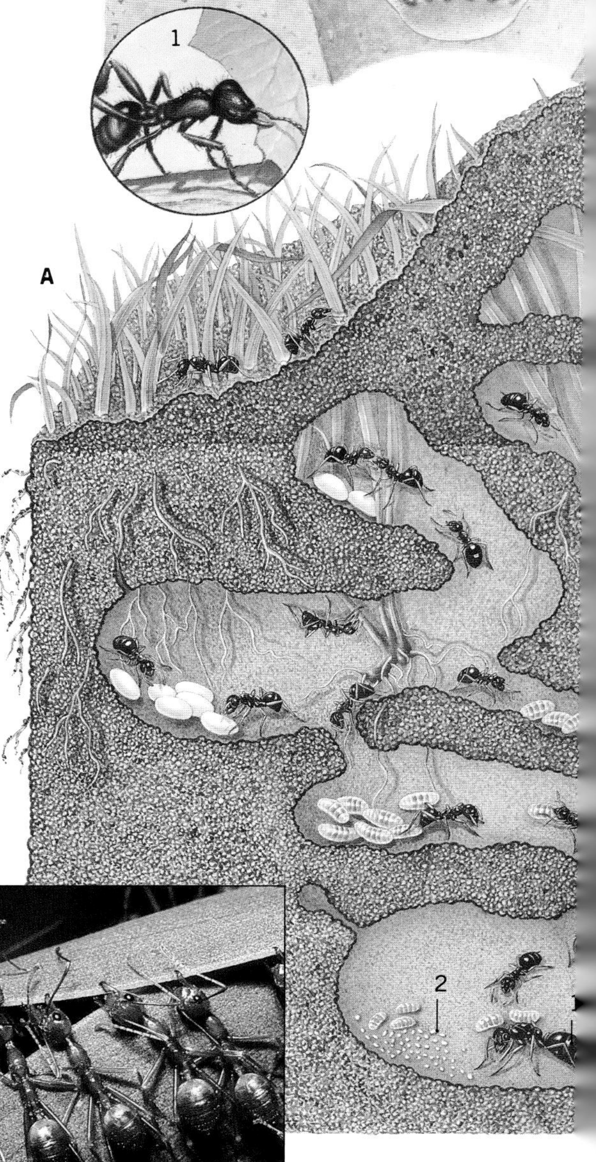

__The European garden ant__ nests in a wide variety of sites – in rotting tree stumps, under stones, or in soft soil [**A**]. *The main nest extends deep into the ground: excavated soil is piled up to form a conspicuous mound. A series of chambers is linked by galleries, which lead to entrance holes. In the heart of the nest is the queen* [1]*: fed by her attendant workers on regurgitated liquid food, and constantly groomed to keep her body free of parasites and fungal growths, she devotes her whole life to laying eggs* [2]. *The eggs, and subsequent larvae* [3] *and pupae* [4], *are sorted by the workers* [5] *and are tended to in separate chambers of the nest. The mound above the nest serves as an important temperature regulator: when the Sun warms one side of the mound, the eggs and larvae are carried near the heated surface; in cold weather, the brood is transferred to the interior of the nest.*

Ant colonies vary in size from about a dozen individuals to a million or more. They are highly structured societies where the role of each individual is dictated by the caste it is born into. The community is dominated by one or more queens, fertile females whose sole function is to lay the eggs from which the other members of the colony are derived. Most of the eggs develop into worker ants – sterile females. There may be more than one size of worker, each having a different task to perform. For example, soldier ants often have large heads with powerful mandibles for defending the colony against attack from neighboring ant colonies.

The tasks performed by worker ants vary with their age. Young, small workers attend the queen, and look after the eggs and smallest larvae. Middle-aged workers tend to the larger larvae and pupae, while the oldest workers guard the nest entrances and forage for food.

Setting up a new home

At certain times of year, male ants and new queens are produced. These winged adults gather in the nest until outside conditions are just right, then they leave in a swarm. All the colonies of one species in a given area swarm at the same time in response to the same environmental cues, thereby increasing the chance that ants from different parents will mate. A few days after mating, the males die, and the queen travels on alone to find a suitable nest site. She pulls off her wings and seals herself in a chamber in the ground or in a rotting tree, and spends several months in this hideaway, laying eggs and tending her brood. The first workers to emerge take over care of the brood and the queen, and begin nest building.

Builders, farmers and gardeners

Some ants use their mandibles as trowels to excavate the earth, piling it around the nest entrance to form volcanolike craters; other species carry the loose earth away from the nest so that the entrances are almost invisible. Others still build conspicuous mounds above ground, covering them with loose vegetation. These mounds may help to regulate the temperature within the nest. Unlike termites, which build air-conditioning "chimneys," ants have no sophisticated temperature regu-

__Tailor ants__ live high in the branches of Asian forests, and build their nests from living leaves. Several large workers cooperate to fold over a suitable leaf, pinning it together with their mandibles and feet. Small workers then pick up a larva and work it about, using its silk to stitch the leaf.

Connections: Beaver Lodges 240 Bees' and Wasps' Nests 232 Birds' Nests 238 Decomposers 204 Fungi 110 Mimicry 266 Scent Communication 294

lation systems in their nests. Instead, they move eggs, larvae and themselves to where the temperature and humidity are suitable.

The structure of an ant nest reflects its occupants' diet. Harvester ants live in areas prone to seasonal drought. They safeguard their food supply by constructing vast underground chambers in which seeds are dried and stored. Other ant species "farm" aphids: by stroking the insects with their antennae, the ants can induce them to release a sugary liquid from their abdomens. Some ant species even "herd" their flock into the nest at night.

Ants play an important role in the recycling of nutrients in most ecosystems by feeding on, and hence breaking down, small animals – especially invertebrates – and plants. And in the process of building their extensive underground nests they "till" the soil, bringing deeper layers to the surface, where the nutrients become available to plants.

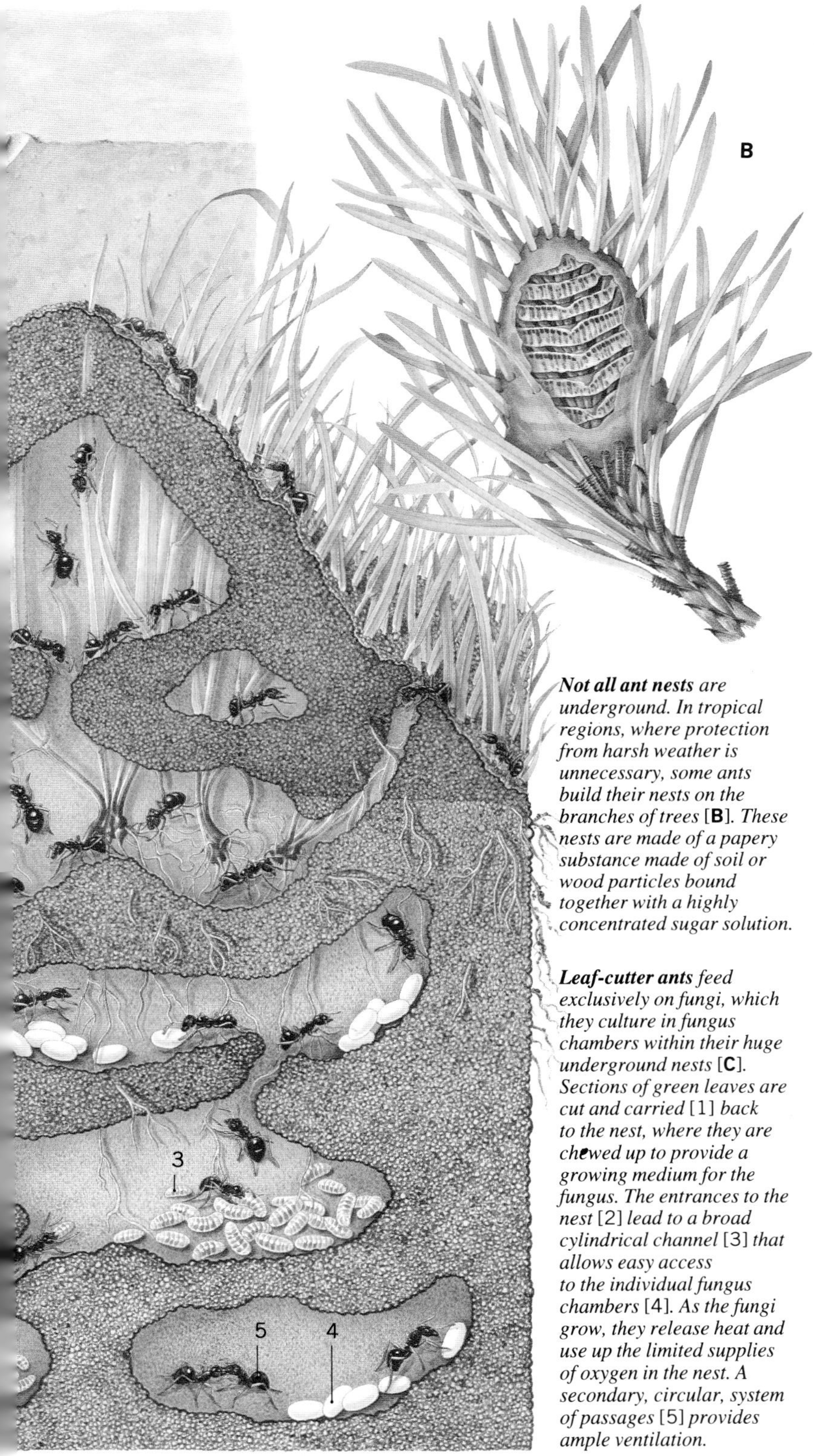

***Not all ant nests** are underground. In tropical regions, where protection from harsh weather is unnecessary, some ants build their nests on the branches of trees* [**B**]. *These nests are made of a papery substance made of soil or wood particles bound together with a highly concentrated sugar solution.*

***Leaf-cutter ants** feed exclusively on fungi, which they culture in fungus chambers within their huge underground nests* [**C**]. *Sections of green leaves are cut and carried* [1] *back to the nest, where they are chewed up to provide a growing medium for the fungus. The entrances to the nest* [2] *lead to a broad cylindrical channel* [3] *that allows easy access to the individual fungus chambers* [4]. *As the fungi grow, they release heat and use up the limited supplies of oxygen in the nest. A secondary, circular, system of passages* [5] *provides ample ventilation.*

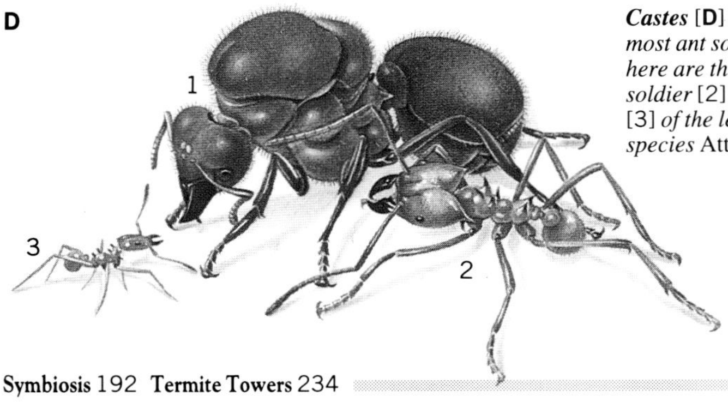

***Castes** [**D**] are a feature of most ant societies. Shown here are the queen* [1], *soldier* [2] *and small worker* [3] *of the leaf-cutting species* Atta caphalotes.

House Guests

Ant nests attract many unwelcome visitors. Among the commonest are woodlice, which steal honeydew from the ants' herded aphids. Some species of beetle lurk in the less frequented galleries of the nest, pouncing on disabled or solitary ants. When threatened, they emit foul-smelling secretions and beat a hasty retreat. Other intruders mimic the appearance and odor of their ant hosts and effectively trick them into accepting them as members of the colony. The caterpillars of some blue butterflies make their way into ant nests and release a sugary secretion, which the ants collect. In return, the ants tolerate the caterpillars' consumption of their eggs and larvae. When the adult butterfly emerges, it must run the gauntlet of hostile ants, protected by a coat of scales on its wings, and leave the nest.

Symbiosis 192 Termite Towers 234

The Avian Artisan

How birds build their nests

The large, untidy nest of the stork is reused year after year. Perched on a chimneystack or even, these days, on a specially prepared platform, its layers of twigs and branches are so capacious that tree sparrows may nest in its base. Unruly as it appears, however, the stork's nest is constructed with care. Different birds build nests that fulfill similar needs: to protect eggs and young from predators; to provide a site for incubation; and to provide a roost for the adults. But whereas some birds produce marvels of architectural intricacy, others may nest in a simple scrape in the ground.

***A swallow's nest* [A]**, *sheltered high under the eaves of a roof, is built by the male and female, and keeps the hungry chicks warm and dry and safe from the attention of birds of prey.*

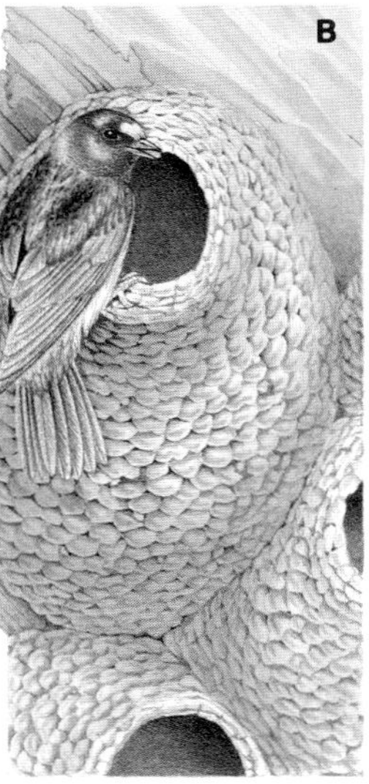

***Cliff swallow nests* [B]** *are built by both male and female birds, originally on cliff or rock faces. Many birds, however, now build tightly packed nest colonies inside buildings such as barns. The nest is made of overlapping mud pellets that dry to an extremely durable protective structure.*

Pigeons and doves make modest platforms of thin twigs, just sufficient to support their clutch of eggs. Thrushes and many other medium-sized songbirds build cup-shaped nests from twigs, grasses, hair and mud, in the fork of a tree or bush, partly secured on the surrounding twigs. More complicated in construction are the deep cup nests of the reed-nesting warblers, which are woven around parallel reed stems so that they move with the stems and remain intact as the wind blows. Swifts, martins and some swallows site their nests well away from predators by sticking them to cliff overhangs or on the vertical walls of buildings.

The *megapodes* of Australasia – including the brush-turkeys and the mallee fowl – build large mounds that contain layers of rotting vegetation. Laid in these mounds, their eggs are then warmed by the heat produced by the decaying matter.

Among tropical birds, suspended nests are relatively common. Such nests may well have evolved as protection against tree-climbing predators such as snakes and lizards. Some of the most impressive hanging nests are built by the oropendolas and caciques of Central and South America. Crested oropendolas, for example, breed in colonies: up to 100 nests dangle down from a single tree. Each nest is a 3-foot (1-meter) long flask of plant fibers, firmly woven into the twigs of the tree. Perhaps the most complicated of all nests is the huge communal nest of the sociable weaver bird of southern Africa. Such colonies may contain up to 300 birds, and the nest is normally suspended in the crown of a tall tree. Below the domed roof each pair of weavers has its own separate nest and entrance tunnel. Their relatives the village weavers also breed in colonies, but build separate nests.

Nesting without suspense

Many birds nest in holes in trees, and some – such as woodpeckers and barbets – excavate their own nest tunnels in tree trunks. Sand martins and kingfishers dig out long tunnels in earth or sand banks. Probably the strangest hole-nesters are some of the hornbills, which seal over the entrance to the nesting hole, leaving a small gap through which the male delivers food to the female for the whole period of egg incubation. The female remains walled up in her nest-prison until the

***The reed warbler's nest* [C]** *is built round dry reed stems that safely secure it above the wet reed bed. The initial construction stages, linking the dry stems, demand considerable acrobatic abilities, for the reeds may constantly sway in the wind. The "handles" that lock the nest to the reeds are the basis for the cock and hen to add grass, feathers or flowers to build up the nest*

A redstart's nest [D] *is insulated with other birds' feathers. The female makes the main nest structure from dried leaves, moss, hair and other natural materials, building it in locations that include tree holes, gaps in walls, nest boxes or old swallows' nests.*

A song thrush nest [E] *is built so soundly that it will survive bad weather months after its builders have gone. The robust grass and twig nest is "plastered" on the inside with a mixture of mud, animal droppings and saliva which, when hard, provides a secure home for the young in a tree or bush.*

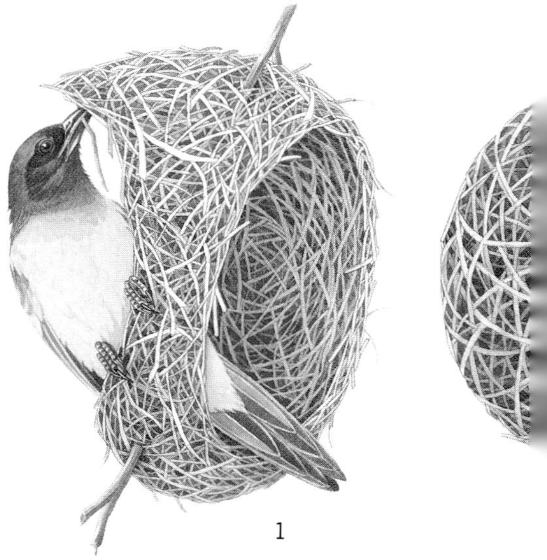

***The weaver bird's nest* [F]** *is built by the male. Using no adhesive, he loops, twists and knots leaf strips to make an enclosed hanging structure. Starting with a ring attached to a forked twig* [1], *he gradually adds a roof and entrance* [2]. *When the finished nest is accepted* [3] *by the female, she inserts soft, feathery grass tops or feathers to make a thick, soft lining around the egg chamber base.*

"Weaving" the nest is a little misleading: since the green leaves used are rarely over three times the bird's length, knotting forms most of the nest's construction. Young male birds' first nests

Connections: Beaver Lodges 240 Bird Courtship 172 Bird Reproduction 174 Birds 132 Fur and Feathers 134 Tropical Forest Animals 320 Wetlands 324

D

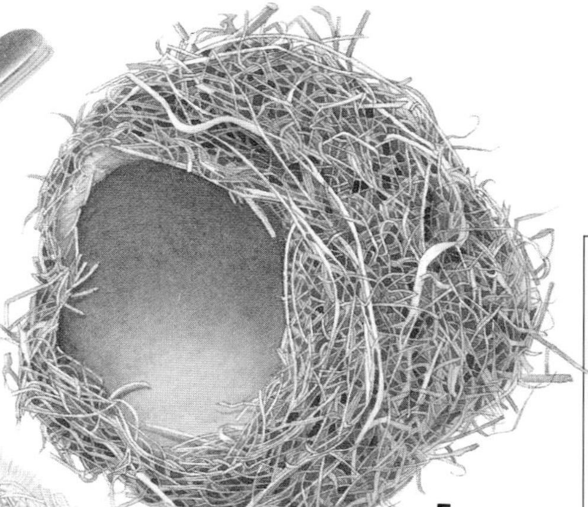

E

Edible Nests

The cave-dwelling swiftlet of Malaysia and Indonesia has dispensed with the mud and twiglets of which its relatives, the swifts and swallows, make their nests and instead creates a small, cup-shaped nest almost entirely from its own saliva. Many of the swiftlets, however, are obliged to build a nest up to three times between February and May because the first ones are removed by local villagers – at some risk in the dark and slippery caves – to eat or to sell as a delicacy to restaurateurs in the Far East, particularly Hong Kong. The nests, black or white depending on which of the two species of swiftlet made them, may be eaten raw (although they are rather rubbery), but are generally boiled in a soup; the white is the more nutritious – and more expensive – variety.

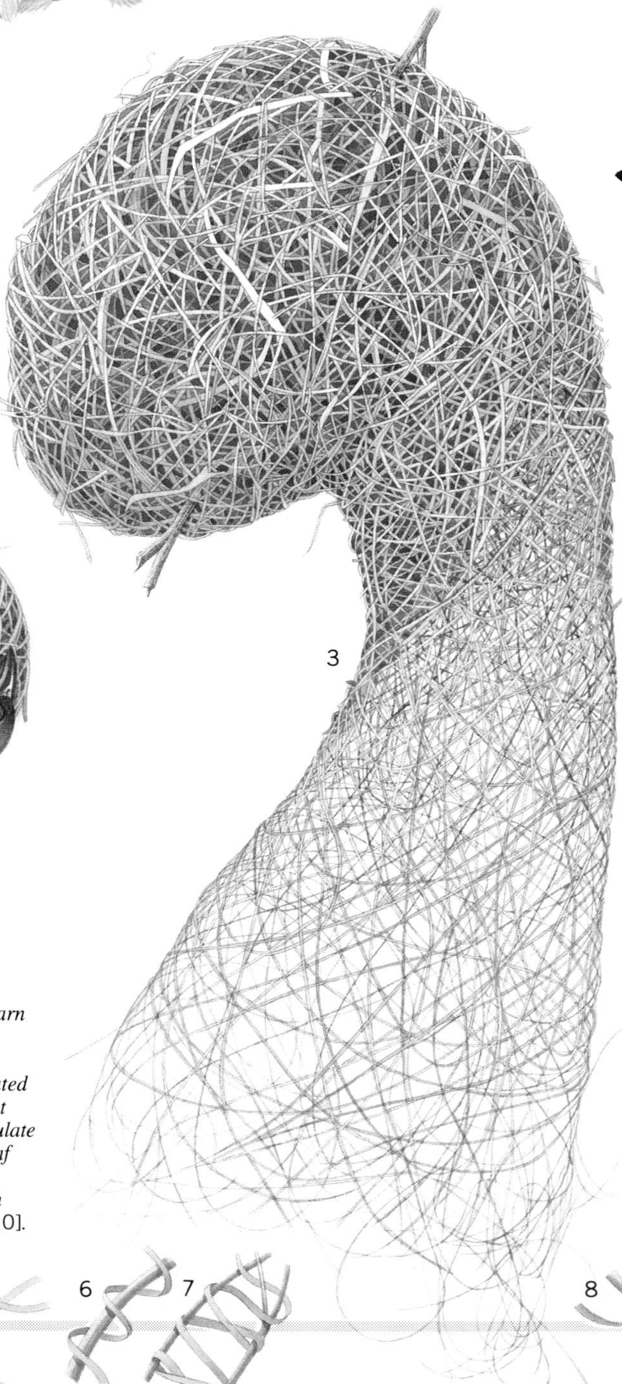

3

2

chicks are half-grown. The smallest nests of all are those of the vervain and bee hummingbirds. They build tiny cup-shaped nests not quite 2 inches (5 centimeters) across, from moss, plant fibers and lichens.

Grebes keep their eggs safe from land predators by building floating nests of water-sodden vegetation. If scared, the mother covers her eggs with more vegetation before retreating to a safe place. Floating nests also have another advantage in that they move up or down with the water level.

Some birds make no obvious nest at all. Cliff-nesting guillemots, for example, lay their eggs directly onto a rocky ledge or crevice amid the sheer precipices. Many waders dig a slight scrape in shingle, relying largely on the camouflaged patterns of their eggs for concealment. The tropical fairy tern simply wedges its single egg into the fork of a tree and incubates it there.

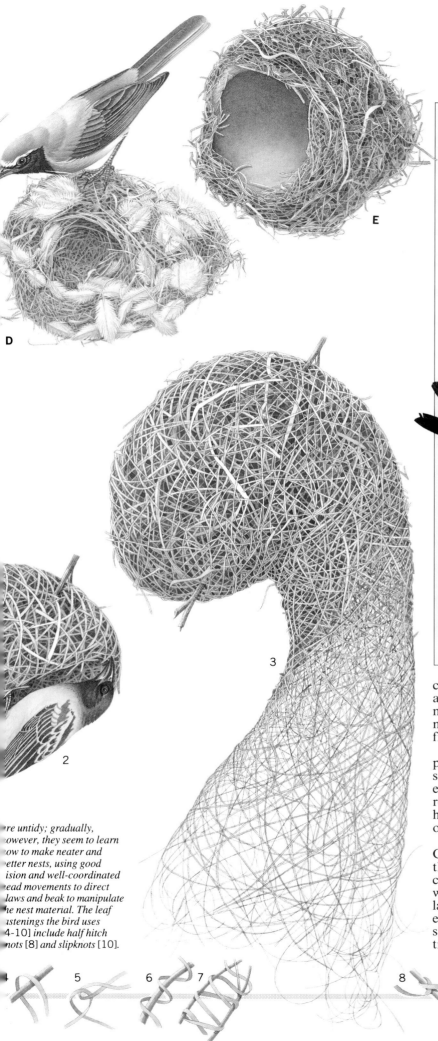

re untidy; gradually, owever, they seem to learn ow to make neater and etter nests, using good ision and well-coordinated ead movements to direct laws and beak to manipulate he nest material. The leaf istenings the bird uses [4-10] *include half hitch nots* [8] *and slipknots* [10].

Nature's Lumberjacks

How beavers build their lodges

Complex wooden fortresses surrounded by lakes, dams up to half a mile in length and intricate webs of canals testify to the beaver's engineering prowess. Beavers are among the greatest natural architects. A single animal can fell a tree 5 inches (12 centimeters) in diameter in less than half an hour, while a whole family can clear several hundred trees and saplings in the course of a single winter. Beavers rapidly reshape the environment to their own specifications, disturbing whole communities of plants and animals, and even changing the course of human history.

Through their prolific forest clearance, beavers have been transforming northern landscapes for the past five million years. But their impact on the environment is undoubtedly less now than it once was. As recently as 10,000 years ago, our present-day species coexisted with giant beavers such as the North American, which was over 6.5 feet (2 meters) long and weighed up to 700 pounds (320 kilograms). The two species alive today have suffered from overhunting and habitat destruction: the European beaver disappeared from much of Europe by the end of the 19th century, and now persists only in isolated indigenous populations. The North American beaver is still found over much of North America, though in most places it has been reintroduced following heavy hunting.

At up to 66 pounds (30 kilograms) in weight and 4 feet (1.2 meters) in length, beavers are the second largest rodents in the world: only the South American capybara is bigger. But beavers are distinguished from other rodents, not so much by their size but by their well-organized social system, which, together with their adaptations for tree felling, swimming and diving, is the key to their success. They live and work in small family units, usually consisting of an adult male and female, the young of the current year (a single litter of between two and four), young born the previous year, and perhaps one or more nonbreeding individuals.

*__The beaver's dentition__ [**A**] is well adapted to the animal's unique way of life. Like all other rodents, including the squirrels, to which they are closely related, beavers have huge incisors that regrow as fast as they are worn down. These razor-sharp teeth – once used as knife blades by the American Indians – are matched by powerful jaw muscles that give the animal a thickset appearance. A beaver can block off the back of its throat with its tongue and close its lips behind its incisors so that it can gnaw wood without choking on the sawdust. This mechanism also enables the animal to carry twigs and branches underwater in its open mouth without drowning.*

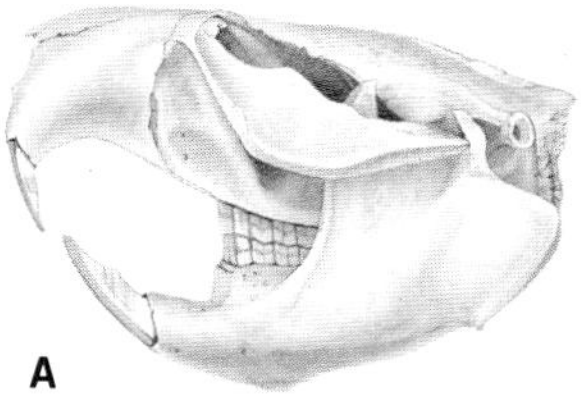

A

A home to call your own

A beaver family invests considerable time and effort in building and maintaining its lodge, dam and canals. Through these feats of civil engineering, the family radically alters its habitat in a way that improves its chances of survival. The lodge provides formidable protection against predators, while the dam floods a large area around the lodge, thereby increasing the beavers' range and enabling them to transport heavy building materials. In addition, the dam ensures that the water around the lodge is deep enough for all its entrances to be permanently submerged, thus creating a moat around the beaver fortress that can only be breached from underwater. It is important to the beavers that underwater access to the lodge is maintained throughout the winter, when the lake is frozen over. The animals store logs and twigs at the bottom of their lake, and in the winter this forms a refrigerated foodstore, which can only be reached via the lodge.

*__Beavers begin lodge__ construction [**B**] by tunneling into the bank or bed of a lake. Branches, sticks and stones, often mixed with grass and leaves, are piled over the tunnel entrance and reinforced with mud: the resulting dome-shaped structure [1] can reach a height of up to 6 ft (1.8 m) and a diameter of up to 40 ft (12 m). The beavers remove or chew away some of the sticks and other material to hollow out a living chamber [2], which they line with wood shavings. Leading off the chamber are several tunnels, each with its entrance opening below water level [3].*

Changing the face of the forest

Extensive damage can result from the activities of beavers. Apart from the trees felled for building materials, flooding caused by beaver dams drowns many trees' roots, as well as inundating farmland and roads. But this destruction is balanced by a number of beneficial effects: flooding creates new habitats for waterfowl, moose and of course fish. And over the years, as silt is deposited in the dammed lake, aquatic plants proliferate: as these die and decay, a layer of peat slowly builds up. Eventually, grasses and other land plants colonize the area, producing a damp and fertile meadow. Some of the finest agricultural land in the northern United States owes its existence to the activities of beavers.

In the absence of a suitable habitat, beavers dam a shallow stretch of water [4] to create their own lake. Construction varies with the particulars of the river or stream: sometimes a large tree is felled and anchored to the bank with heavy vertical timbers and stones. Alternatively, stout stakes buried in the river bed make up the main structural elements of the dam. Using their dexterous, unwebbed forepaws [5], the beavers fill gaps in the dam with sticks, stones, vegetation and a layer of mud. Their webbed hind feet [6] and long, flattened tails [7] make them excellent swimmers.

Connections: Ants' Nests 236 Bees' and Wasps' Nests 232 Birds' Nests 238 Boreal Forests 316 Termite Towers 234

B
1
2
3
4
5
6
7

7 Attack and Defense

Plant Defenses
Solitary Carnivores
Group Carnivores
Poisonous Animals
Spiders
Predatory Trappers
Claws, Jaws and Horns
Armor and Spines
Flight and
Deception
Color
Camouflage
Mimicry
Chemical Defenses
Termite Defenses

Pugnacious Plants

How plants compete and defend themselves

Aphids wandering onto wild potato plants are held fast by the sticky secretions from hairs on the leaves and, unable to move, the aphids starve to death. In an arms race that has been going on for millions of years, plants have evolved an armory of defenses to protect themselves against hungry herbivores. Thorns and spines, for example, such as those found on rose or cactus stems, deter plant-eating animals, while many other plants produce unpleasant-tasting or toxic chemicals. Herbivores, however, are not the only hazard, plants also face competition from other plants.

All plants have to compete for the essential resources of light, water and minerals from the soil, not only with other members of their own species but also with other plant species that share their immediate habitat.

Two plant species with similar requirements growing together will compete to such an extent that one may eventually supplant the other completely. Ecologists have formalized this rule as the "competitive exclusion principle," which states that two such species of plants (or any other organism) cannot indefinitely occupy exactly the same habitat. In reality, natural habitats are usually very varied. Their many microhabitats, or *niches*, provide footholds for many different species.

Living together

Competition for light and available water determines the makeup of most plant communities. Even in a small area of grassland, different species will exploit wetter or drier areas, or prefer to grow in the shade of a stone or on the sunny face of a slope. Variation in height and arrangements of leaves allow different plants to make the most of the available light.

When they are competing with many other plants in a natural habitat, the individuals of any particular species are rarely as large or as numerous as they might be if they were growing alone. Relatively infertile soils generally provide a chance for more different species to grow together, whereas richer soils are often taken over by one dominant species. A typical flower-rich downland or alpine meadow can only develop on fairly infertile soil, where the growth of grasses and rampant weeds such as docks are suppressed, both by soil conditions and by grazing animals. Such habitats are composed of a patchwork of scattered individuals of different species.

As well as competing for light and nutrients, some plants produce chemicals that prevent seedlings of other species becoming established near them.

Chemical defenses

Plants can make a wide range of chemicals that appear to have no particular role in their everyday life-support processes, and which are often unpalatable or toxic to animals but do not harm the plant. These secondary products of metabolism – like the mustard oil

In temperate countryside *[**A**], many of the most successful weed species are plants with well-developed defenses against large herbivores. In addition to preventing the plants from being eaten, these defenses also inhibit animal movement. The common stinging nettle [1] employs sophisticated stinging hairs that act like a hypodermic syringe and inject an irritant liquid into the skin of any animal that gets too close. Each stinging hair consists of a single cell with glasslike silicated walls enclosing the irritant liquid.*

B

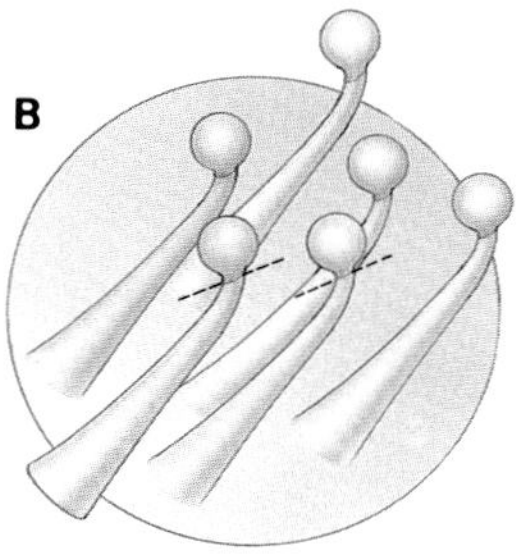

C

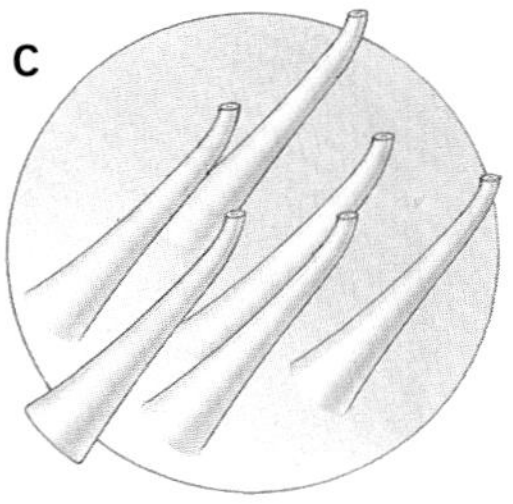

The stinging hair *of a nettle [**B**] has a rounded bulbous end that breaks off along a predetermined fracture line (dotted) at which the slightest pressure will form a hollow, needle-sharp point [**C**]. The active ingredients of the irritant liquid include acetylcholine, formic acid and histamine.*

Connections: **Herbivores** 202 **Invertebrate Herbivores** 198 **Parasitic Plants** 186

***Many plants** eschew chemicals and rely on purely physical defenses. Holly [2] has hard spines formed at the margins of the leaves. Often, spines are found only on the lower leaves of a tree; higher up, the leaves lack spines. The thistle [3] is heavily protected with sharp spines on the leaves and stem, and the unopened flowers are protected by spiny bracts while they develop. The blackberry [4] and rose [5] have curved thorns along the stem. However, defense is only a secondary function. The primary function of the thorns is to enable the plants to climb. They are formed as emergences of surface tissue, and are only superficially attached. If they become embedded in flesh, they break off before the plant stem is broken.*

Deadly Leaves

Sorghum, or Indian millet, is grown in many tropical and subtropical countries for its grain. It is one of approximately 1,000 plant species that produce a variety of cyanide-based compounds to deter herbivores. Sorghum stores its poison in the epidermal cell layer of its leaves [1] as *durrhin.* This is a *cyanogenic glycoside,* a compound made of sugars bound to cyanide complexes, and in this form it is harmless. However, when the leaf blade is damaged by a herbivore, two special enzymes present in the *mesophyll* layer of the leaf [2] are released that break down the cyanogenic glycoside to eventually liberate cyanide. Once ingested, cyanide disables respiratory enzymes in the *mitochondria,* necessary for respiration, with frequently lethal consequences.

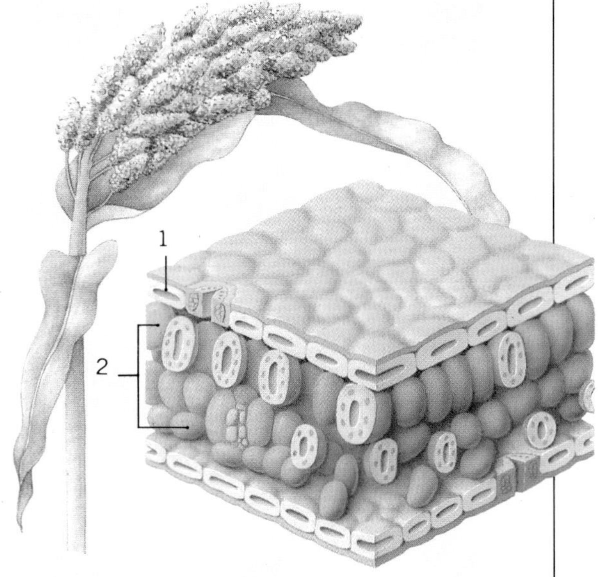

substances produced by members of the Brassica (cabbage) family, the poisonous *atropine* from deadly nightshade, and *digitalin* from foxgloves – have probably evolved as chemical defenses against insects.

The mustard oils in Brassicas break down to give these plants their strong distinctive smells and bitter taste. Although this deters many insects, some, such as the cabbage-white butterfly and its relatives, have become adapted to depend on Brassicas as food-plants for their caterpillars.

Plants are also threatened by attack from fungal pests, and produce chemical defenses against them. *Phytoalexins* are chemicals produced by plants in response to fungal attack. Unlike a plant's inbuilt chemical and mechanical defenses, phytoalexins, such as the *sesquiterpenoid rishitin,* produced against potato blight, are only made when needed and are not present in the plant all the time.

Plant Architecture 114 Plant Feeding 184

The Lone Hunter

How solitary vertebrates hunt

The enormous sperm whale is the largest predator in the sea – average males reach a length of at least 50 feet (15 meters) and weigh around 33 tons. Fast-swimming sperm whales hunt alone, seeking prey as large as giant squid from the depths of the oceans. One whale was found entangled in a submarine cable at a depth of 4,100 feet (1,240 meters). Another was observed diving, for nearly two hours, in water known to be 11,200 feet (3,400 meters) deep. Sperm whales are thought to use echolocation in these pitch-black depths to track down their prey.

Most predators live solitary lives, using their wits and skill to find food. Hunting animals come in all sizes – and so does their prey. Some, such as the jaguar, which stalks tapir in the forest, choose their large victims carefully; others select smaller prey. The osprey soars 100 feet (30 meters) above lakes and rivers before swooping down with pinpoint accuracy to snatch a single salmon. Some birds are less selective. Nightjars operate rather like flying vacuum cleaners, gathering large quantities of tiny insects as they fly with mouths agape through the twilight sky. Whatever their method of hunting, all predators need acute senses to locate their prey. Once it has been found, some use speed and strength to capture it; others must rely on patience and stealth.

Sharpened senses

Finding food is a matter of life and death, and many predators have senses that are more highly developed than those of humans. The eyes of eagles and owls can detect small movements of mice and voles many yards below them. A good sense of smell is particularly valuable to nocturnal predators. Night-hunting cats have a keen sense of smell, and so too do moles, which hunt for worms in total darkness beneath the ground. The conspicuous pits of the pit vipers enable these snakes to detect infrared radiation and therefore hunt warm-blooded animals at night. One group of birds with an unexpectedly acute sense of smell is the kiwis. They have nostrils at the tips of their beaks that allow them to sniff grubs in the soil.

Acute hearing is the most vital sense of the bat-eared foxes. Their huge ears enable them to find the termites, other insects and small rodents on which they feed. Like many predators, they can also rotate their ears to pick up the smallest sound, and by comparing signals received in both ears, they can locate the direction of the sound precisely. Perhaps the most sensitive ears of all are the East African honey badger's, which can hear the sounds of tiny dung beetle larvae in their cocoons beneath the ground.

The thrill of the chase

Some predators chase their victims. To catch them, they must run, swim or fly faster than their prey. The fastest fish are the hunters of the open ocean. Tuna can swim at speeds of up to 43 miles an hour (70 km/h), and the fastest sharks at up to 31 miles an hour (50 km/h). On land, the cheetah is said to be the animal kingdom's fastest runner, but other big cats – such as leopards and tigers – can manage 43 miles an hour (70 km/h) over short distances. In addition the tiger can also leap more than 16 feet (5 meters) high to pull prey from trees.

Because meat is a more nourishing food than vegetation, hunting animals tend to spend less time feeding than do herbivorous creatures. Big cats, crocodiles and foxes can all be seen lounging or dozing between meals, while most herbivores graze all day. Leopards will often catch herbivores too large to eat at one time. For this reason they may drag a dead antelope up into the branches of a tree, where it is safe from other predators and can provide food for several days.

The talons of birds of prey **[A]** *are sharp and strong to ensure that these hunting birds can effectively grab their prey first time, as a second attacking dive may not be possible due to the prey escaping down its burrow. Birds of prey have forward-facing eyes. Although this restricts their field of vision when compared to other nonhunting birds that have eyes on the sides of their heads, the binocular eyesight of birds of prey gives extraordinarily accurate information on the exact location of prey. Big cats, such as the ocelot* **[B]**, *have retractable claws. In this way a highly efficient hunting weapon converts into a tough, padded paw for running without wearing down the claws. The claws are retracted* [1] *and extended* [2] *by muscles in the paw, that control the outer toe bones. The boa constrictor* **[C]** *kills its prey by first striking out and grasping the prey in its sharp, backward-facing teeth, making it almost impossible for the prey to escape. The snake then coils itself around the victim and squeezes. The victim, unable to breathe, is asphyxiated. The elasticated hinges of its jaws enable the boa to swallow its victims whole. A meal the size of an ocelot will last a boa several weeks.*

Connections: **Claws, Jaws and Horns** 256 **Fur and Feathers** 134 **Group Carnivores** 248 **Invertebrate Carnivores** 200 **Poisonous Animals** 250

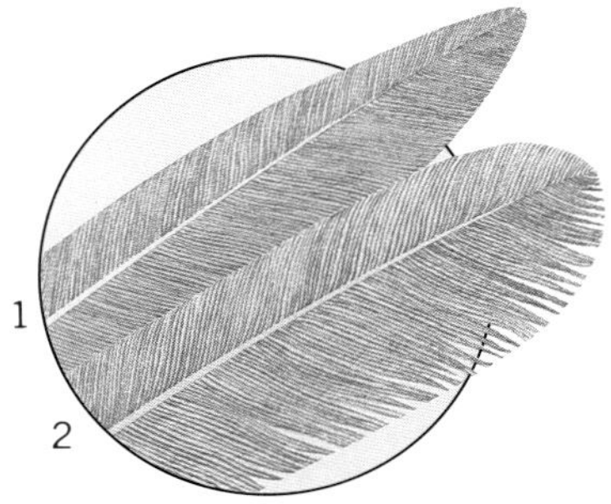

The feathers of an eagle *[1] are much narrower and more pointed than those of an owl [2]. The owl's larger, more rounded feathers have a greater surface area. This enables the owl to glide easily, thereby dispensing with the need for noisy wing beats. The frayed edges of the feathers ease the flow of air over them, also helping to reduce noise. These adaptations for silent flight were thought to have evolved to allow owls to swoop down silently and take their prey by surprise, but scientists now think that the adaptations developed so that owls can listen for the quiet squeaks of their prey.*

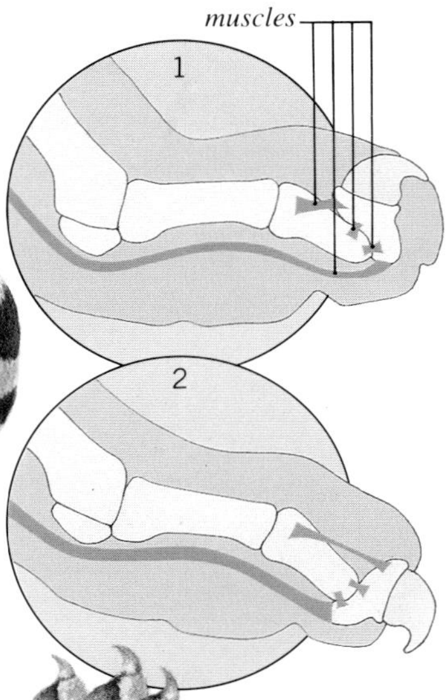

Razor teeth

Sharks are among nature's most formidable predators. They have an acute sense of smell, which helps them locate injured creatures. Most are fast swimmers and all have razor-sharp teeth. Once a shark locates its victim, it moves in, grasping the prey in its mouth. Strong muscles give a powerful, biting grip as the teeth lock into position. Once it has hold of its prey, the shark moves its head and body in a series of twisting movements that help the jaws and teeth shear the food into pieces.

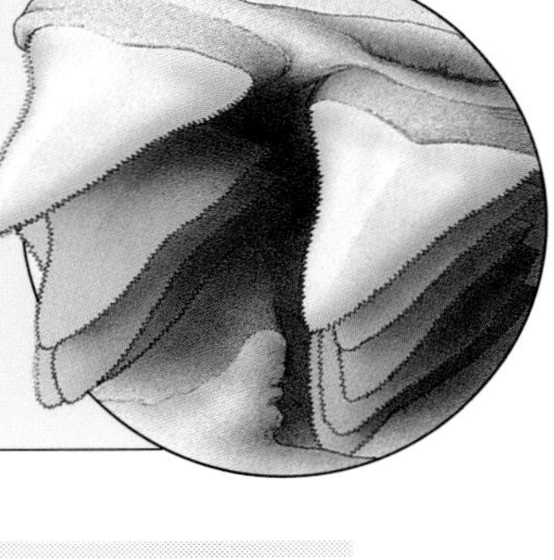

A shark's teeth *are arranged in parallel rows along the line of the jaw. Because its teeth are not firmly embedded in the jaw bone, as the shark grasps its prey, it may lose several teeth. These, however, are continually replaced by new teeth, which grow from grooves in the shark's jaws.*

Predatory Trappers 254 Sonar 280 Spiders 252 Vision 278

The Predatory Pack

How carnivores hunt in groups

The bloodstained faces of chimpanzees feeding on the flesh of a freshly killed animal must be one of the least expected sights in nature. Chimpanzees form one of the best-organized groups of hunters. Teams of strong, experienced males ruthlessly pursue red colobus monkeys through the treetops of the West African forest. One or two work as drivers, chasing one colobus monkey ahead of the other chimpanzees and isolating it. Two or three block its escape routes on each side, while one of the strongest chimpanzees rushes ahead to ambush it and move in for the kill.

Successful hunters must not only possess the guile and speed to capture their prey but also the strength to overpower and kill it. Lone predators must be faster, cleverer or larger than their prey; but hunters who work in groups can develop team tactics that enable them to tackle animals larger than themselves.

Some animals, such as the army ants of the South American forest, or lions of the African plains, form permanent social groups for hunting. Others, such as the wolves of the Arctic tundra, hunt alone for small animals, birds or fish during the summer, but form hunting packs during the autumn and winter. The advantage of doing so is obvious. A wolf pack can attack and kill caribou (reindeer) and may even take on a moose (elk). Working together, a group of lions can pull down and kill a buffalo that weighs three times as much as each lion. In the rivers and lakes of South America, large groups of piranha fish, each one no more than 1 foot (30 centimeters) long, can kill large animals, such as capybaras, which may weigh as much as 88 pounds (40 kilograms).

The hunters in most groups are usually the strongest and fittest animals. Often the younger and weaker members are left behind in a protected place, but when the hunters return with food, it is shared among them all. This may well mean that some animals receive less food than they might have had they hunted alone. Nonetheless, when large prey is available, teamwork gives every member of the group the best chance of survival. In addition, the group can defend the kill from scavengers tempted to steal it.

The rules of association

Hunters are commonly closely related to each other. A pack of wolves may include an adult male and female, up to five cubs and youngsters, two or three yearlings, and uncles and aunts with no territories of their own. Such groups are bound together by complex hierarchies and patterns of behavior.

Hyenas hunt in well-coordinated teams. They communicate with a rich vocabulary of sounds and signs. Their tails, normally carried in a downward-pointing direction, can be raised to indicate aggression or held over the back to show excitement. To reinforce the bonds between the group, members may growl, yelp, or whine together, or join in a chorus of cackling laughter. Before a hunt, team members sniff the mouths and necks of each other, and the hyenas stand head to tail to sniff and lick each other. Only when these procedures are completed does the group move off to hunt.

Pack-hunting techniques

Hunting techniques vary from species to species. Teams of piranhas travel long distances together, searching for shoals of fish to eat. A single "scout" piranha selects and seizes a victim. The blood leaking into the water attracts the other piranhas, who race in to seize small pieces of flesh from the prey with their sharp teeth. On land, lions are heavy and cannot run at great speed: they have to rely on their strength to overpower their prey. On the other hand, long-legged dogs and wolves have sufficient stamina for a chase of several miles.

Wolf in the Sheep-pen

Bred to obey the commands of the shepherd when rounding up sheep, the sheepdog nonetheless has many of the hunting characteristics of its wolf ancestors of many thousands of years ago. Running left and right behind a group of sheep, crouching low to the ground, the sheepdog keeps the flock together in a similar way to a hunting wolf seeking to isolate a victim. Any sheep that strays is chased back to the flock with great precision. The highly disciplined behavior of the sheepdog is similar to the discipline learned by every member in the hierarchy of a wolf pack.

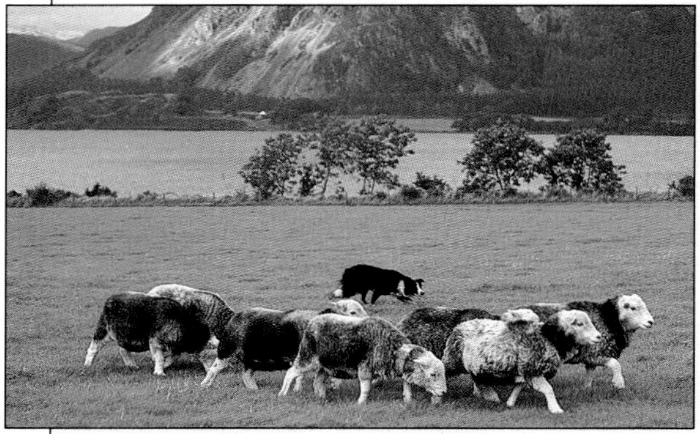

Connections: **Camouflage** 264 **Grasslands** 302 **Mammal Societies** 298 **Solitary Carnivores** 246 **Tundra** 314

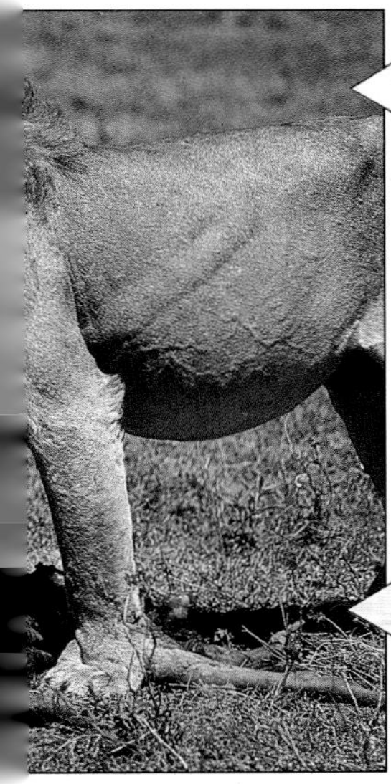

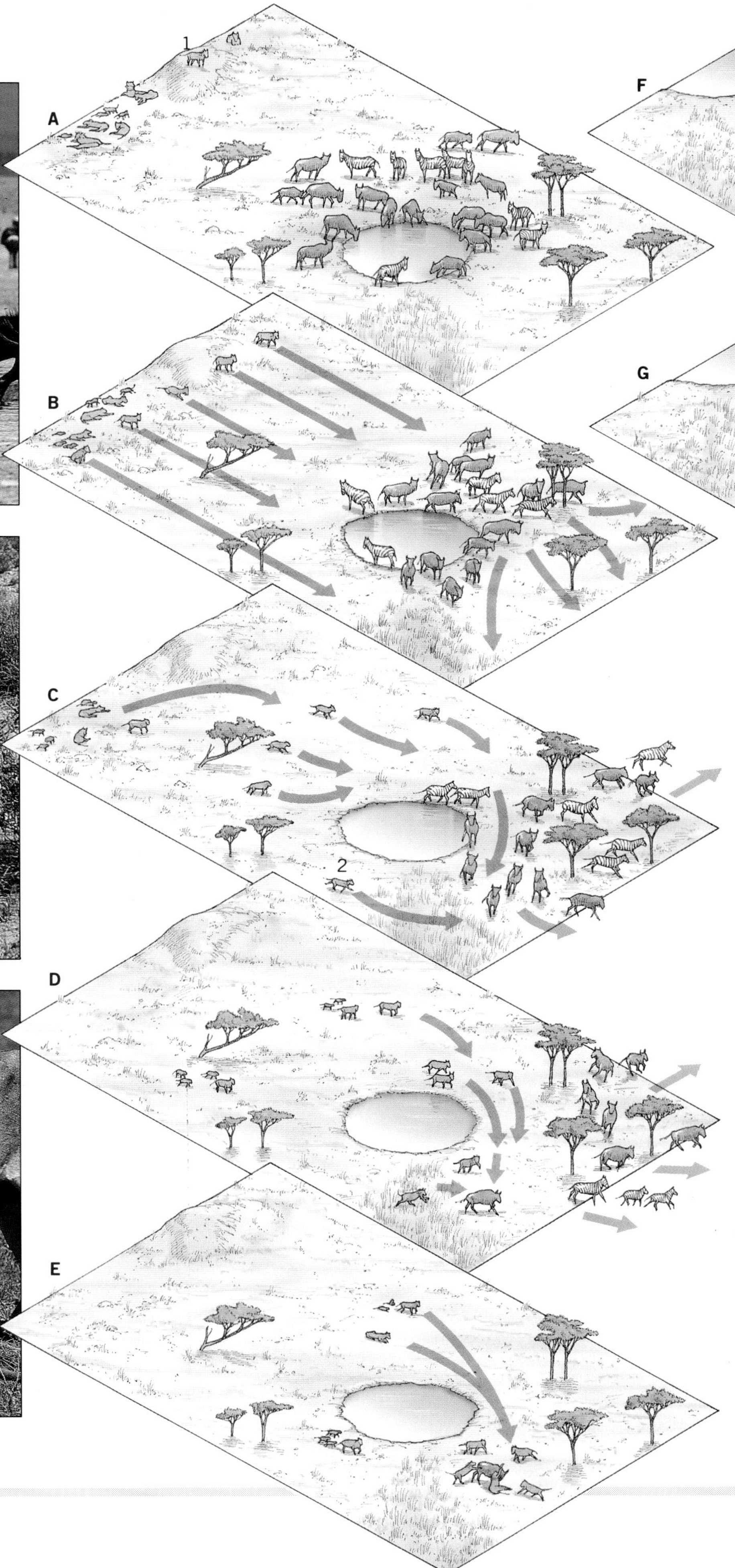

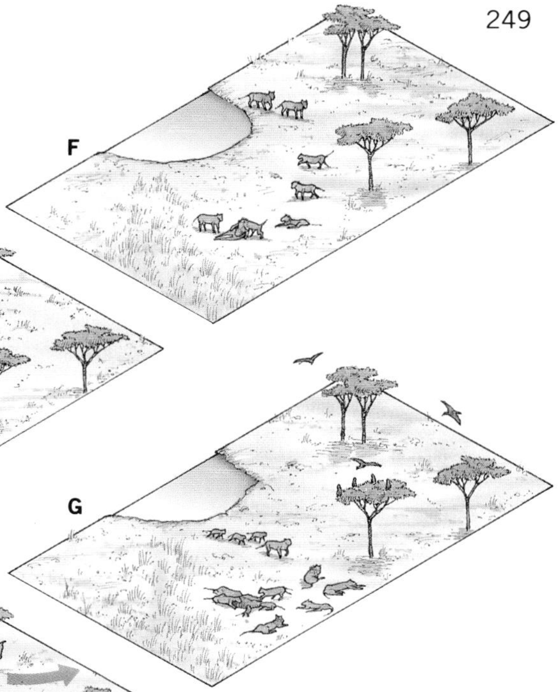

Lions *are the only species of cat that hunt in groups. The advantages of this method of hunting are that the lions can kill prey larger than themselves, such as wildebeest and zebra (thereby increasing the number of available prey); and by hunting together each individual expends less energy. The hunting technique usually follows a standard pattern. As the family group rests on the plain* [**A**]*, one female often stands on a raised position* [1] *looking out at a herd of wildebeest or zebra. At a given moment, the young and strong females in the pride will line up and advance on the herd* [**B**]*, while the male(s) protect the rest of the pride from hyenas or other potential attackers. The lionesses then rush the herd* [**C** *and top left*] *toward a hidden lioness* [2] *that has assumed a camouflaged position downwind of the herd. The lioness often positions herself close to a well-worn track, or a gap between trees, where the prey is likely to pass. This female then singles out a weak individual* [**D**] *for the kill* [**E** *and middle left*]*, which is usually achieved by the lioness knocking the prey to the ground and then strangling it by biting its neck. The actual feeding follows a strict social order, with the male(s) feeding first* [**F** *and bottom left*]*, followed by the lionesses and cubs* [**G**].

Lethal Injection

How animals use poison

The poison-arrow frogs of Central and South America have glands that produce the deadliest animal toxin. The toxin of some species is so powerful that a mere 0.0000004 ounces can kill a human being. This does not prevent native Indian hunters from using the lethal poison to hunt game. The poison from a single golden poison-arrow frog is enough to tip 40 arrows – a sufficient amount to paralyze game, but not to render it inedible. Venomous animals, such as snakes, can eat their victims for a different reason – their poison is protein based, and can be digested harmlessly in the gut.

Animals that use toxins to kill or subdue prey and deter predators fall into two main groups. Venomous species, such as vipers, cobras, sting rays and wasps, are capable of injecting toxins into their victims by means of fangs, spines, stings or other weapons. To be effective, a viper's venom must reach its victim's bloodstream. Death may then follow quickly. Yet the venom of even a rattlesnake is relatively harmless if applied to the skin or swallowed. Indeed it would be necessary to drink 750,000 times the amount of rattlesnake venom that is injected in a typical bite to kill a human.

Poisonous animals, such as puffer fish and various species of caterpillar, on the other hand, have poisons that act more as a deterrent to potential predators rather than as a weapon that can be used to kill prey. Such animals must be eaten before the toxins present on their spines or accumulated in their tissues take effect.

The deadliest snakes

The species of snake that contain the most venom are the cobras. A single cobra may have as much as 0.1 ounces (350 milligrams) of stored venom. Although this may not sound a great deal, scientists estimate that just 0.04 ounces of dried venom can kill over 160,000 mice – or the equivalent of 165 humans. The king cobra of Asia, which feeds mainly on other snakes, rather than on small mammals such as rats as most cobras do, grows up to nearly 20 feet (6 meters) long, and is the largest of the world's poisonous snakes. Despite having fangs only about 0.6 inches (1.5 centimeters) long, this species of snake can deliver more venom than any other species – sufficient to kill an elephant – but causes few human deaths, because it is not particularly aggressive, making it less likely to strike out.

The snake that poses the greatest overall threat to human life is the saw-scaled viper. The danger lies in the fact that this species of snake is abundant and widespread (occurring in West Africa, the Middle East, India and Sri Lanka). It tends to live in heavily populated areas, and is aggressive, alert, fast-moving and well camouflaged, with an exceptionally potent venom. A single adult male or female of this species contains enough venom to kill eight adult humans.

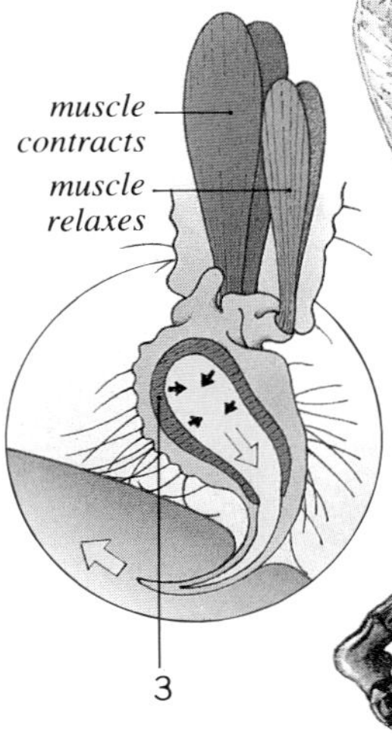

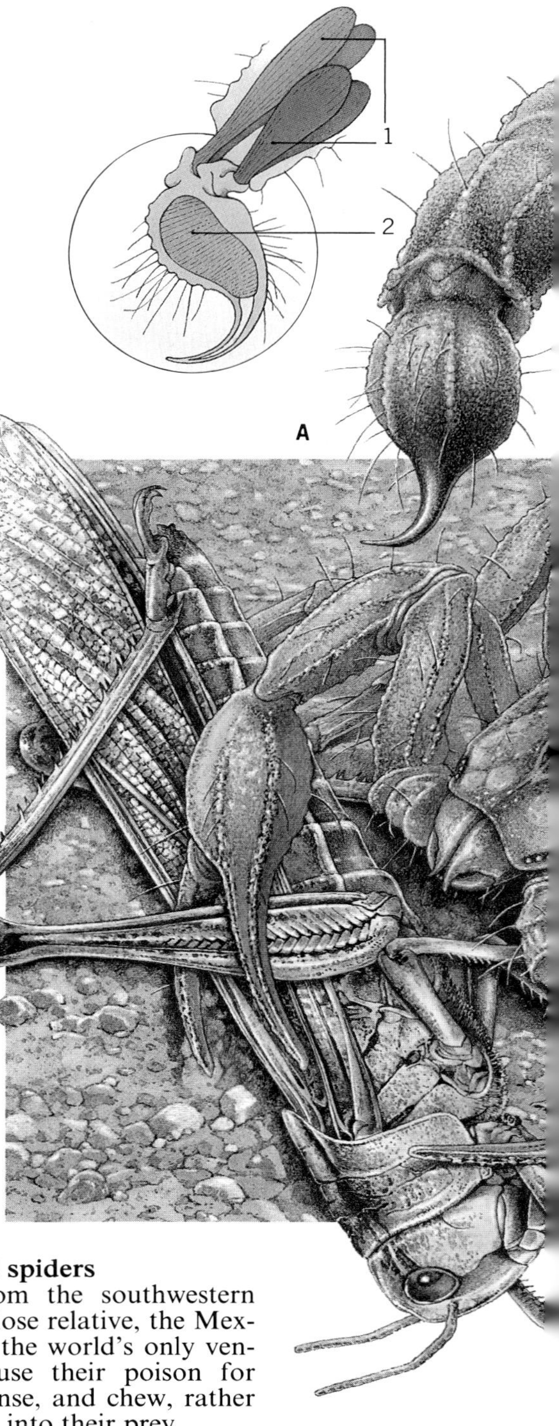

***The scorpion's sting** is worked by opposing muscles [1], fixed to the base of the sting, which contract and relax, forcing the sharp tip into the victim's tissue. The poison, stored in the poison gland [2], is forced down and out of the tip of the sting by muscles [3] located around the gland.*

Venomous lizards and spiders

The Gila monster, from the southwestern United States, and its close relative, the Mexican beaded lizard, are the world's only venomous lizards. They use their poison for attack rather than defense, and chew, rather than inject, their venom into their prey.

Among the world's most venomous spiders are several species of black widows, with a vast range that includes southern Europe, Africa, Asia, Australia, New Zealand and much of the American continent, and the three species of Australian funnel-web spiders. They prey almost exclusively on insects, injecting them with poison, then sucking out their body contents. The toxins of the female black widow spider may be up to 15 times as toxic as rattlesnake venom: humans bitten by the spider may suffer nausea, vomiting and paralysis, but rarely die, because only a small amount of poison is delivered.

***Of the 600 or so species** of scorpion, only a few have venom powerful enough to endanger human life; scorpions of the* Buthidae *family* **[A]** *are an example. Most scorpions are active at night and hunt mainly insects, although some of the larger species, which grow up to 7.5 in (19 cm) long, will attack rodents and small lizards. Most scorpions kill by holding their prey with their pincers and injecting poison via the sting.*

Connections: **Chemical Defenses** 268 **Invertebrates** 122 **Plant Defenses** 244 **Predatory Trappers** 254 **Reptiles** 130 **Spiders** 252

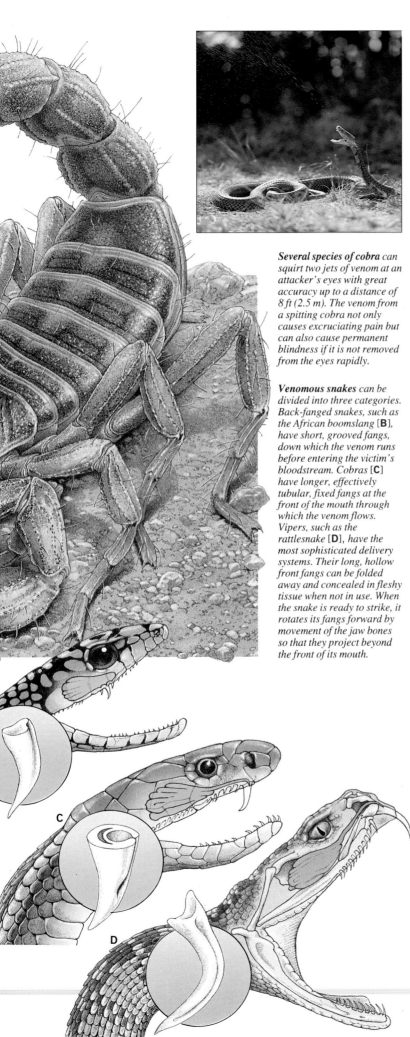

Several species of cobra *can squirt two jets of venom at an attacker's eyes with great accuracy up to a distance of 8 ft (2.5 m). The venom from a spitting cobra not only causes excruciating pain but can also cause permanent blindness if it is not removed from the eyes rapidly.*

Venomous snakes *can be divided into three categories. Back-fanged snakes, such as the African boomslang* **[B]**, *have short, grooved fangs, down which the venom runs before entering the victim's bloodstream. Cobras* **[C]** *have longer, effectively tubular, fixed fangs at the front of the mouth through which the venom flows. Vipers, such as the rattlesnake* **[D]**, *have the most sophisticated delivery systems. Their long, hollow front fangs can be folded away and concealed in fleshy tissue when not in use. When the snake is ready to strike, it rotates its fangs forward by movement of the jaw bones so that they project beyond the front of its mouth.*

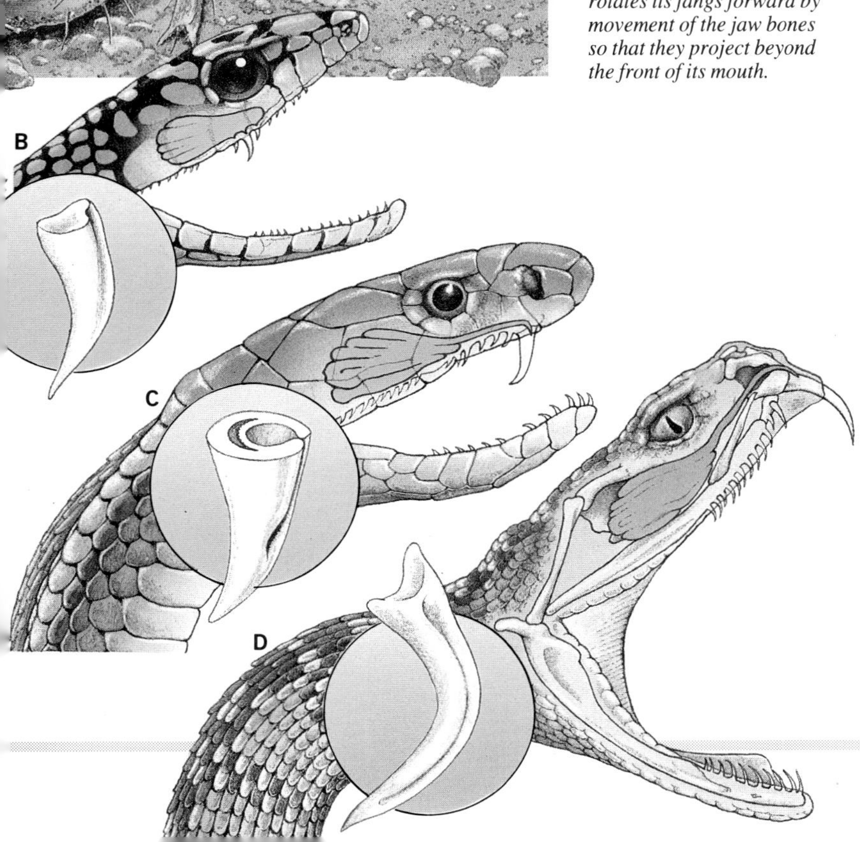

The Deadly Cocktail

Animal venoms can be divided into two basic types: *neurotoxic*, which act on the central nervous system of the victim, killing it by stopping the heart and lungs but causing little tissue damage; and *hemotoxic*, which kill by breaking down the victim's tissues. Every venom contains elements of both types, although each is predominantly neurotoxic or hemotoxic. Hemotoxic constituents, apart from acids that simply destroy tissues, include other substances that have more subtle effects: *hemolysins* break down and destroy red blood cells; *cytolysins* break down white blood cells and other defensive cells; *anticoagulants* prevent blood clotting; *thrombins*, conversely, promote blood clotting; *antibacterials* prevent bacterial invasion at the puncture site. These venoms also contain enzymes, such as *hyaluronidase* (which helps venom to spread quickly through the victim's body), *phospholipase* (which breaks down cell membranes) and *protease* (which liquefies the victim's tissues).

Cobra (Africa/Asia/India)

venom: primarily neurotoxic
plus anticoagulant and hemolysin
effects: paralysis, then respiratory failure

Viper (Eurasia/Africa)

venom: primarily neurotoxic plus enzymes
effects: stinging/burning pain, tissue damage
heart failure, blood poisoning

Brown spider (Eurasia/Americas)

venom: primarily hemotoxic
hemolysin, cytolysin, hyaluronidase
effects: stinging/burning pain
bleeding, blistering/ulceration
fever, vomiting
possible convulsions/heart attack

Short-tailed shrew (North America)

venom: present in saliva
primarily neurotoxic plus enzymes
effects: localized pain/discomfort
reddening at wound

Puffer fish (Indian/Pacific Oceans)

poison: primarily neurotoxic
effects: poisonous only when eaten
tingling lips/tongue
salivation, vomiting
numbness and muscular paralysis
mental confusion, convulsions, death

European earth salamander (Europe)

venom: primarily neurotoxic alkaloids
effects: (unknown on humans)
cardiac arrhythmia, convulsions
paralysis, death

Architects in Silk

How spiders spin and use webs

Spider silk is one of the strongest natural substances known. A single thread of it can be stretched by nearly a third without snapping, and would have to be about 50 miles (80 kilometers) long to break purely under its own weight. With the silk secreted by their glands, spiders build an enormous variety of webs, from flimsy hammocks to delicate spirals within spirals, sticky sheets and thick funnels. They also use their silk to wrap up prey, to cocoon eggs, and even in mating, when the male enfolds its sperm in a miniweb, which it places inside the female.

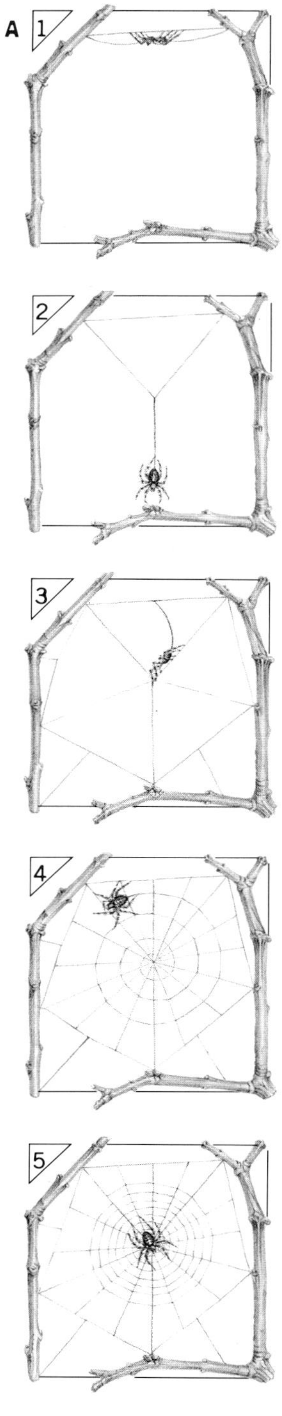

Spiders produce their silk as a fluid containing a protein called *fibroin*. This solidifies into an insoluble thread when the proteins rearrange themselves under tension as the silk is drawn out of the spider's body. Spiders have several glands to make silks for different uses – as draglines attached to a fixed point by silken suction disks, as prey-catching spirals, and as wrapping for prey, cocoons, eggs, or sperm. The silk used to make structures such as cocoons and draglines is dry, but threads used to trap prey are often sticky. The Australian bolas spider spins a special strand of silk that it uses in a unique way. The thread ends in a small ball (the *bolas*), which is coated with female moth pheromones. Male moths are attracted to the ball, but end up trapped on the sticky thread.

Weaving webs

Spiders' webs range in complexity from apparently chaotic tangles of threads to dense sheets, tubes, and the intricate, highly organized orb web of species such as the familiar garden spiders.

Webs spun above the ground require some sort of scaffolding: a framework of threads anchored to rocks, vegetation, or other solid structures. This framework may be surprisingly large in relation to the size of the spider. The spider selects a high perch, and allows a line of silk to float on the air until it touches

When spinning its web *[**A**], the spider first casts out a thread of silk to form a horizontal strut [1], which it then reinforces to form a bridge-line. A second, drooping thread is trailed across below the bridge-line. Halfway along the second thread, the spider drops down on a vertical thread until it reaches a fixed object [2]. It pulls the silk taut and anchors it, forming a "Y" shape, the center of which forms the hub of the web. The spider then spins the framework threads and the radials, which are linked together at the hub [3]. After spinning the remainder of the radials, a wide temporary spiral of dry silk is laid down, working from the inside of the web outward [4]. This holds the web together while the spider lays down the sticky spiral. This is laid down starting from the outside, and is attached successively to each radial thread [5]. The spider eats the remains of the dry spiral as it proceeds. The central dry spirals are left as a platform for the spider, which will often lie in wait there during the night, but will retreat to a nearby silk shelter during the day. Spiders tend to spin their webs at night, when they are less likely to attract the attention of birds. Most orb spiders need to spin a new web every night, and they eat the old one to save the protein.*

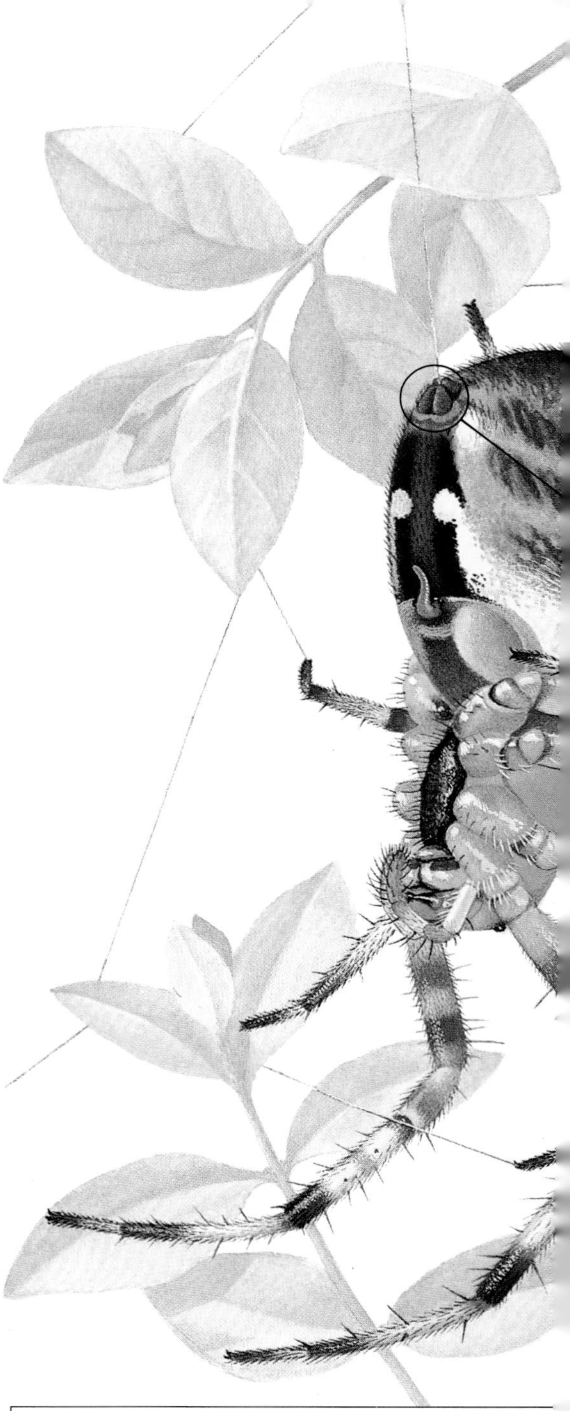

The Web-casting Spider

Most webs are used as passive traps, but a few spiders are more active hunters. The web-casting spider, of the genus Dinopsis – sometimes called the ogre-faced spider because of its huge eyes – spins a small rectangular web [1] about the size of a postage stamp. It fluffs up the strands of silk into thousands of tiny loops using hairs on its hind legs. This traps insects by getting tangled in their hairs and scales. The spider hangs upside down [2], holding the web by four threads, one at each corner, in the two front pairs of legs. As soon as an insect passes below [3], it opens the net and spreads it over the prey. Using its free legs, the spider spins the prey round and round until it is covered in silk, then delivers a fatal bite. Several attempts are often required before a capture is made by a web-casting spider.

Connections: **Breathing** 126 **Invertebrate Carnivores** 200 **Invertebrates** 122 **Invertebrate Reproduction** 160 **Poisonous Animals** 250 **Predatory Trappers** 254

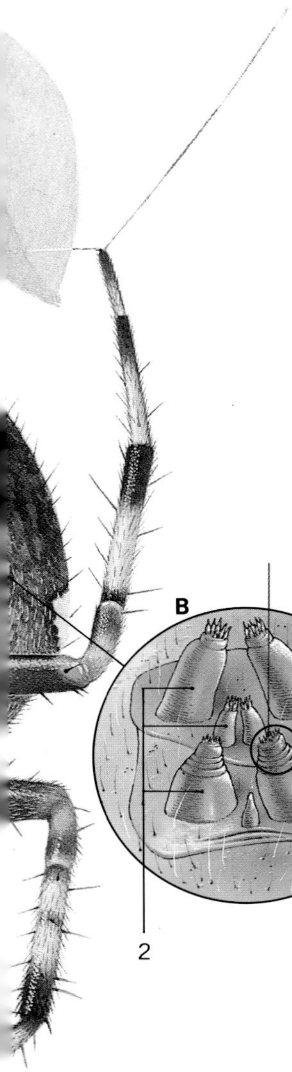

Spider silk *is produced by special silk glands inside the spider's abdomen* [**B**]. *Each gland* [1] *is connected by a small tube to an organ called a spinneret* [2], *which opens to the outside through a number of tiny nipples called spigots* [3]. *Most spiders have three pairs of spinnerets at the tips of their abdomens which can be moved by muscles* [4] *in various directions while spinning. The silk is not squeezed out of the spinnerets by the action of the muscles but is pulled out, either using the claws on the spider's hind legs, or by attaching the silk to a fixed object and then walking away from it.*

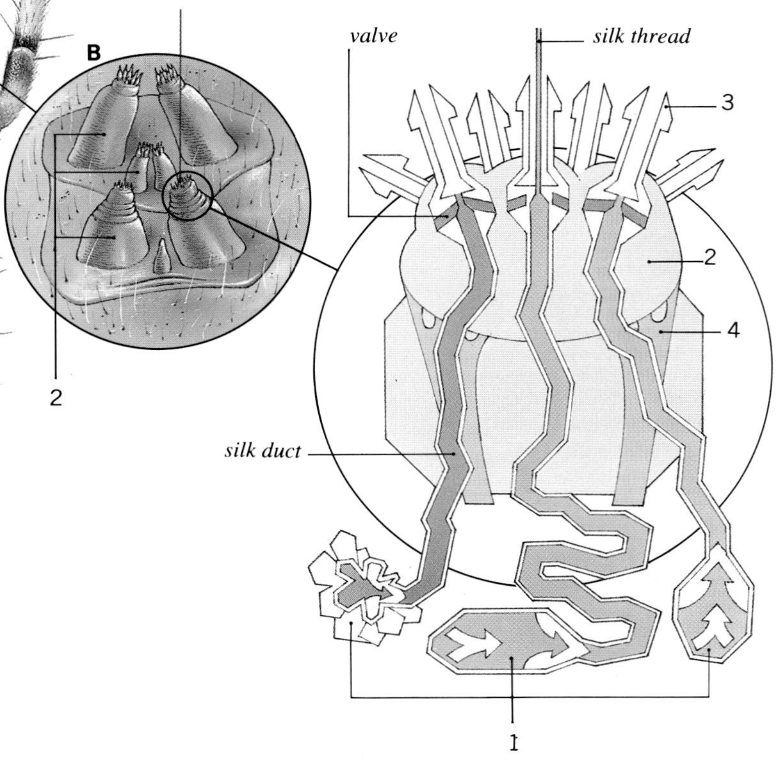

The successful hunter *(above) spins a straitjacket of silk around a trapped insect to prevent its escape. Spiders frequently enfold relatively large prey in this way while they attempt to bite them and inflict a paralyzing poison. Once paralyzed, the prey can be unwrapped and eaten, or saved for later. Spiders usually walk on the underside of their own webs, dangling by their claws, to keep their bodies from coming into contact with the sticky spirals of thread.*

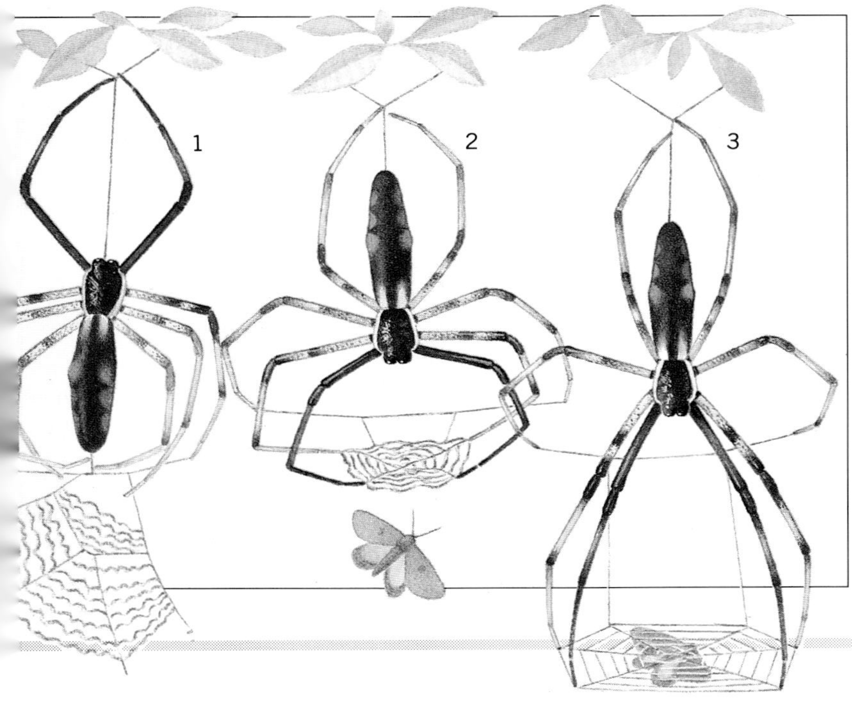

and sticks to some object. Other threads can be spun from this base line. Within the framework, a denser meshwork of threads is spun to trap the prey. In the orb web, this takes the form of a spiral of sticky silk. When the spider has laid down a section of sticky silk, it jerks the thread sharply, causing the glue to form a series of blobs along the thread, which are very effective in trapping prey. Although the sticky thread traps insects, the spider is able to walk around its web without difficulty. It walks only on the dry threads, and uses special brushes on its claws to grip the fine silk threads. An oily coating on its feet helps to prevent them from sticking if it accidentally encounters the glue.

Web wonders

Experiments have shown that construction of the web is purely instinctive, controlled by touch. Each successive step is triggered by the position of existing threads and the degree of tension on them. Newly hatched spiderlings can spin a perfect orb web first time.

Spiders are extremely sensitive to touch and vibration. Once an insect blunders into the web, the spider can tell exactly where it is by the tension and vibrations of the various threads that it disturbs. Courting male spiders, which are usually much smaller than females, vibrate the web in a special way to signal they are prospective mates, not meals.

Lying in Wait

How animals trap their prey

A small fish drifts unwittingly closer to the tentacles of a Portuguese man-of-war. Suddenly there is contact. Instantly a venomous barb explodes from one of the stinging cells arranged in batteries along the tentacle. The fish, helplessly paralyzed, is slowly drawn into the Portuguese man-of-war's mouth. In this way, a fearsome marine predator need not actually pursue its prey. Such sit-and-wait predators include animals that exhibit some of the most complex camouflage of the animal world, and have remarkable adaptations for seizing their prey, or luring it to them.

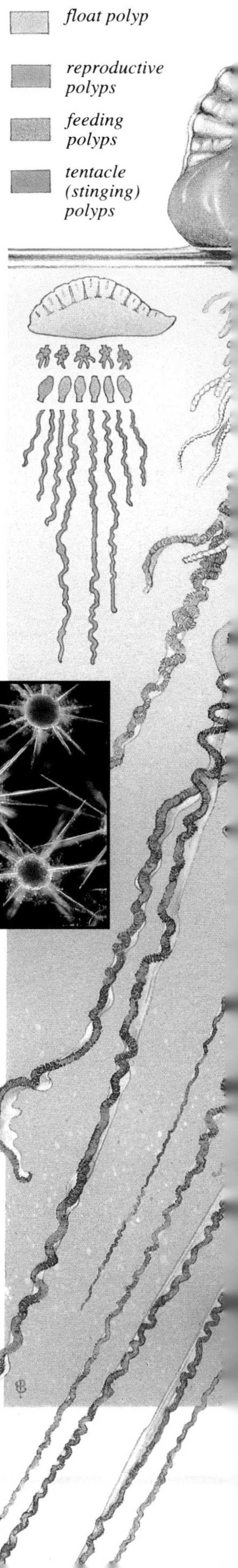

The Portuguese man-of-war *(*Physalia *sp.) is found in warm Atlantic waters. Though it appears to be a single organism, it is in fact a colony of many hundred hydralike individuals (*polyps*). The individuals are not all alike, but are specialized to perform individual functions.*

If a lie-in-wait predator is not to give away its presence, it must be able to seize its prey very rapidly. Chameleons, frogs and toads have long, sticky tongues that they can extend at lightning speed to catch a passing insect. Mantids rapidly unfold their long "praying" front legs, and dragonfly larvae unleash a similarly folded "mask," derived from their mouthparts. The octopus, lying in wait unseen in a crevice or under an overhang, can quickly extend tentacles armed with suckers to cling onto its prey.

For such sudden, snatching moves to be successful, the predator must have good – preferably stereoscopic – vision. Frogs and toads have big, bulging eyes that are positioned well forward on the head.

Many animals that catch their prey by lying in wait prefer to hide in burrows or cover themselves in sand. Tiger beetle larvae wait at the entrance to their burrows. Ghost crabs lie in wait on the beach, with only their stalked eyes protruding above the sand. The antlion – the larva of a lacewing fly – digs a small pit in the sand and waits at the bottom to seize any small creature that tumbles into it. Sometimes it gets impatient and bombards its prey with sand to prevent it from escaping. Cuttlefish and rays flick sand over their backs to conceal themselves as they wait on the sea bed. Crocodiles and alligators, conversely, leave the sand of the shore to hide just below the water, just their nostrils and eyes showing above the surface.

Perhaps the ultimate master of camouflage is the chameleon, which is able to change its color to match its surroundings. The octopus, too, can change color. Praying mantids cannot change color, but they have evolved into an astonishing range of different shapes and colors to resemble the plants on which they habitually lie in wait.

The Asian flower mantis *uses petallike camouflage and the ability to keep perfectly still in order to trap its prey.*

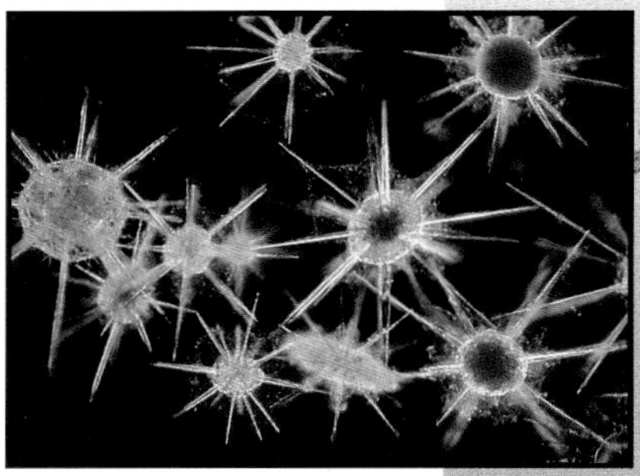

Radiolaria are tiny marine animals. Their delicate glasslike shells are drawn out into long spines. Streaming along these spines is a sticky mucus that traps food particles, and carries them to the animal's mouth.

Waiting with baited breath

A more effective way of catching prey than simply sitting and waiting is the use of lures to attract the prey within range. The best-known example of this is the anglerfish, the female of which has a long, stiff, threadlike lure, often with a fleshy tip, that it wiggles just in front of its mouth. When a small fish is attracted by the quivering lure and approaches, the anglerfish opens its huge mouth and the fish is sucked in with the inrushing water. In the deep sea, where visibility is close to zero, other species of anglerfish have adapted by using luminous lures.

The alligator snapping turtle of North America has a fleshy pink, wormlike lure on the floor of its mouth. It opens its mouth and wiggles the lure to attract fish.

The simplest lure, however, is that of the fishing spider, which vibrates a leg in the water to attract passing fish and tadpoles.

Using camouflage as a form of lure, the sabre-toothed blenny – a small fish of coral reefs – is a mimic. It has the same striking blue and black stripes as the cleaner wrasse, a small fish that feeds by removing parasites from the skin and gills of larger fish. The blenny mimics the cleaner wrasse's dance, which attracts fish who want to be cleaned. But instead of performing a beneficial service, the blenny darts in and bites a chunk of flesh out of its "customers."

Connections: **Bioluminescence** 276 **Camouflage** 264 **Coral Reefs** 336 **Mimicry** 266

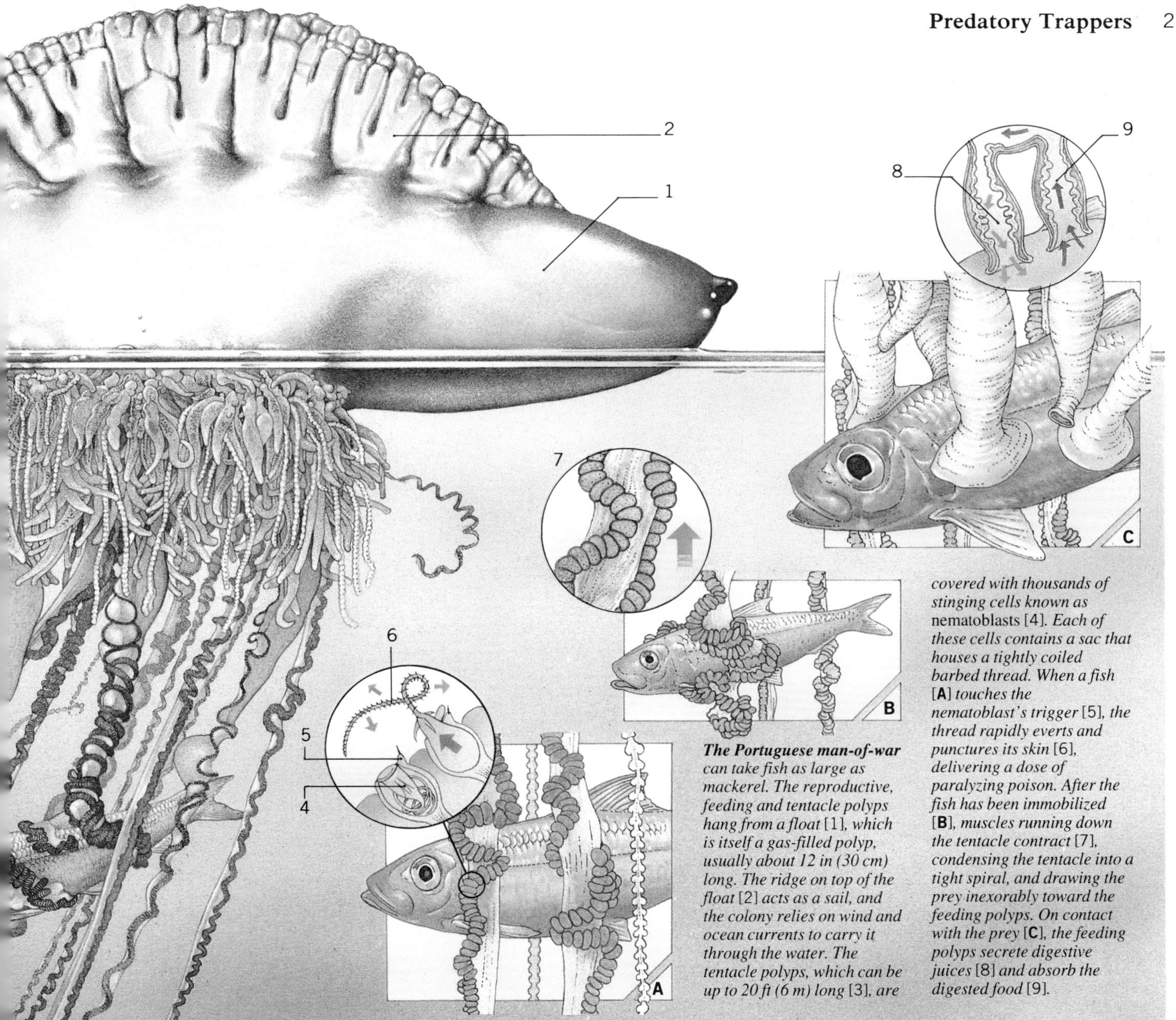

The Portuguese man-of-war *can take fish as large as mackerel. The reproductive, feeding and tentacle polyps hang from a float* [1]*, which is itself a gas-filled polyp, usually about 12 in (30 cm) long. The ridge on top of the float* [2] *acts as a sail, and the colony relies on wind and ocean currents to carry it through the water. The tentacle polyps, which can be up to 20 ft (6 m) long* [3]*, are covered with thousands of stinging cells known as* nematoblasts [4]. *Each of these cells contains a sac that houses a tightly coiled barbed thread. When a fish* [**A**] *touches the nematoblast's trigger* [5]*, the thread rapidly everts and punctures its skin* [6]*, delivering a dose of paralyzing poison. After the fish has been immobilized* [**B**]*, muscles running down the tentacle contract* [7]*, condensing the tentacle into a tight spiral, and drawing the prey inexorably toward the feeding polyps. On contact with the prey* [**C**]*, the feeding polyps secrete digestive juices* [8] *and absorb the digested food* [9].

Twist in the Tail

The copperhead, *Agkistrodon contortrix,* is the most common venomous snake in easterly areas of the United States. Also called the highland moccasin, it is a pit viper that in adulthood reaches a length of less than 3 feet (1 meter). Although many attacks on man are reported, the venom is weak and rarely fatal. Younger members of the species have a sulfur-yellow tip to their tail, which they wave to mesmerize and lure small animals, such as frogs, within reach of their striking fangs. In older snakes the tail can take on a wormlike appearance.

Using some form of lure in this way saves a great deal of energy compared with what would have to be expended during a chase; so fewer, smaller meals are necessary. With no need to pursue their prey, small predators also reduce their own exposure to larger predators.

The Best Form of Defense

How animals use claws, jaws and horns

The flightless cassowaries of Australia and New Guinea can be very dangerous birds when cornered – and have killed many people with the long, straight spikes that grow from their modified inner toes. Many other creatures can turn the tables on an attacker and fight back using horns, antlers, teeth, legs, or claws. Usually these weapons were evolved not for beating off predators but for use in jousts between members of the same species. Although the weapons can be fearsome – some bull elephants' tusks are up to 11 feet (3.5 meters) long – such battles are rarely fatal.

The earliest stage in the development of horns and antlers may have been the evolution of small horns used to establish dominance between rivals of the same species. In the next stage, horns became larger, and were used to pierce or bruise the body of a rival and also to serve as a defensive shield, blocking the opponent's attack. Bighorn sheep and bison use their horns as battering rams in head-on fights between rival males. Within a particular species of wild sheep, horn size relates to the animal's social position. Some horns have become further modified for locking rivals together during wrestling and pushing tournaments. In extreme cases the horns are used for ritualized displays, removing the animals even further from any real fighting, with its attendant risk of damage.

Different methods may be used against predators from those used in fights between rivals of the same species. The giraffe, for instance, has small, blunt horns used only in disputes between males, and defends itself instead by kicking out with its long front and powerful back legs.

The North America caribou (reindeer) are unique among all deer in that antlers are found in females as well as males. The reason is not clear, but it may be so that females are in a position to compete successfully with males in winter, when accessible food is limited by snowdrifts.

***The fighting style** of pig species* [**A**] *is reflected in the form and placement of their offensive and defensive capabilities. Wild boars* [1] *fight shoulder to shoulder, each slashing at the side of the other's body with their tusks (extended teeth). The vulnerable hindquarters are pivoted away from the opponent, and the shoulders are protected by thickened skin covered with coarse, matted hair.*

Warthogs [2] *fight head to head, and the frontal skull is thickened to protect the brain from the shock of impact. There are two sets of well-developed tusks. The impressive upper tusks, up to 16 in (40 cm) in length, are in fact the less effective weapons. When fighting, most of the damage is done by the sharper lower tusks. The prominent facial warts that give the animal its name have developed as defensive features that protect vulnerable eyes from the incurving tusks.*

Connections: **Armor and Spines** 258 **Camouflage** 264 **Chemical Defenses** 268 **Flight and Deception** 260 **Group Carnivores** 248 **Herbivores** 202 **Mammals** 136

__Musk oxen__ do not use their horns to just fight each other, or for individual defense. When threatened by potential attackers they use teamwork, and quickly form a defensive circle, or "hedgehog position," with their heads and horns turned out to make an intimidating barrier.

*__Animal horns__ [**B**] consist of either keratin or bone, or in some cases, keratin-covered bone. A rhinoceros horn [1] consists of only keratin in the form of compacted hairlike fibers and is thus not a true horn. Yak horns [2] are extensions of the frontal bone sheathed in keratin. A pronghorn [3] has bone horns covered by annually renewed keratin. Caribou antlers [4], like those of other deer, are shed and regrown annually. Antelope horns, like those of the ibex [5], have a bone core and a thick keratin layer that may be ridged or spiral in form.*

Teeth to beware of

The tusks of elephants are greatly elongated upper incisor teeth. They have a wide range of uses, including moving branches, prizing off bark and digging for roots, as well as in defense against predators and fights between rivals, most of which are harmless.

The long tusks of walruses are greatly enlarged upper canine teeth. Although they have various other uses, such as levering the animals up onto ice floes, they evolved primarily to indicate an individual's status within the walrus society. The walrus with the largest tusks is generally the most dominant animal. The tusks are also used in defense against predators, chiefly polar bears.

Birds that fight back

Various groups of birds have weapons that can inflict damage on rivals or predators. Male pheasants and turkeys have sharp spurs on the backs of their legs for fighting: these are made of a bony core covered with a sharply pointed horny sheath. According to species, there may be one, two or three spurs. Some of the peacock-pheasants may have as many as four of these dangerous instruments on each leg.

The long, dagger-shaped beaks of birds such as gannets and herons, though primarily used for catching fish, can serve as formidable weapons against an attacker.

Some birds use their wings for offensive purposes: mute swans will protect their eggs or young by battering an attacker with their wings. The heavy "wrist joint," which bears a bony knob, can be a powerful weapon, capable of deterring most predators and bruising a human badly. However, the potential threat to humans has been greatly exaggerated, and there are hardly any authenticated records of broken human arm or rib bones.

Mimicry 266 Poisonous Animals 250 Solitary Carnivores 246 Termite Defenses 270

Swords and Shields

How animals use armor or spines in defense

The quills of a porcupine, erected in defense, are capable of deterring predators as formidable as leopards. The shell of a tortoise is both home and armor. The spines of the stickleback fish make it virtually unswallowable. Many creatures carry with them an effective means of withstanding unwelcome attentions from would-be predators. Faster-moving animals can often make do with spines, quills, or barbs, but slower creatures like tortoises and pangolins rely on thick, heavy armor-plating proportionate to that of a battle tank.

Many different groups of invertebrates protect themselves with armor of one sort or another. One of the most efficient designs of invertebrate armor is the mollusk shell. As well as providing protection against predators and mechanical injury, the shells of land- and shore-dwelling mollusks help prevent loss of vital body water. Species that live on the shore often grow thick shells to withstand the battering of waves, whereas shells of land-dwelling species tend to be thinner.

Arthropod armor

On land, the jointed-limbed invertebrates (arthropods), such as insects, crabs and spiders, protect themselves with a tough outer surface (cuticle) made up of several layers. Its innermost layer is made of the strong, resilient, celluloselike polysaccharide called *chitin*, which forms a fibrous framework: this is strengthened by other materials.

In insects, this strengthening material is a protein called *sclerotin*. After being secreted by the epidermal cells, this is modified by types of protein to produce a tough, waterproof layer. In crabs and other crustaceans, by contrast, the chitin is strengthened by chalky substances, creating a strong but heavy and relatively inflexible armor.

Like armadillos and hedgehogs, pill woodlice and pill millipedes combine their protection with the ability to roll themselves up, so that their vulnerable vital organs are tucked away inside, and water loss is reduced to a minimum. Rolled up, these creatures become larger and thus more difficult for a small predator to swallow. Also, their tough outer coats are too slippery for most predators to grip onto.

Armored fish

The scales of various fish, such as boxfish, cowfish and trunkfish (or cofferfish), are joined edge to edge, forming a strong, ridged suit of armor. The mouth emerges at one end, the tail at the other, and the fins from the sides, top and bottom.

The spines of sticklebacks have a locking device that keeps them erect when the fish is in danger. Because the stickleback swims directly away from a predator, the latter is forced to seize it tail first, so that the spines jam in the mouth, making the fish almost impossible to swallow.

***The spines** of the European hedgehog [**A**] are actually hairs modified into hollow tubes with reinforcing ridges on the inside walls, making for a strong but light structure. They taper to a sharp point at one end, and their bulbous bases [1] are buried in the skin, allowing the spine to pivot when erected or pulled flat by a pair of antagonistic muscles [2]. The spines are raised when the animal is threatened: at the same time, a thick band of muscle [3] running around the hedgehog's flanks contracts, drawing the animal into a tight ball, difficult for a predator to penetrate.*

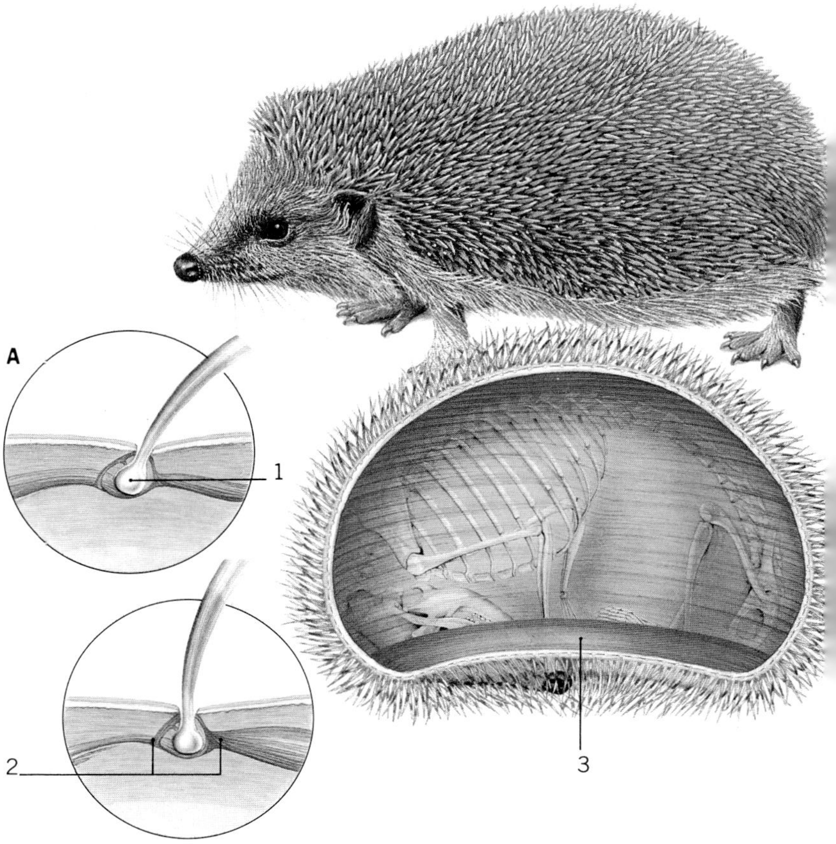

Borrowed Armor

The hermit crab adopts an empty whelk or other marine snail shell to provide it with an armorplated home, moving to a bigger one as it grows. It has a modified abdomen with a twisted shape to fit the spiral snail shell, and its last two pairs of legs and hind appendages (*uropods*) are specialized for gripping the shell.

When the hermit crab has grown too large for its shell, it waits until it has found a suitable larger shell before risking the changeover. The crab explores its potential new home, and if the fit, weight or ability to move about in the shell is poor, it will reject it, just like a human rejecting an ill-fitting pair of shoes, and return to its old shell until it can find a suitable new one.

The hermit crab uses empty shells: it never kills or ejects the occupant. Two crabs may fight fiercely for possession of an empty shell.

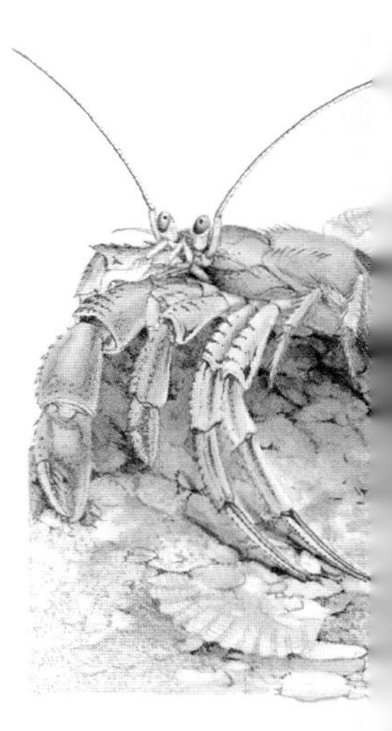

Connections: Chemical Defenses 268 Claws, Jaws and Horns 256 Flight and Deception 260 Invertebrates 122 Reptiles 130

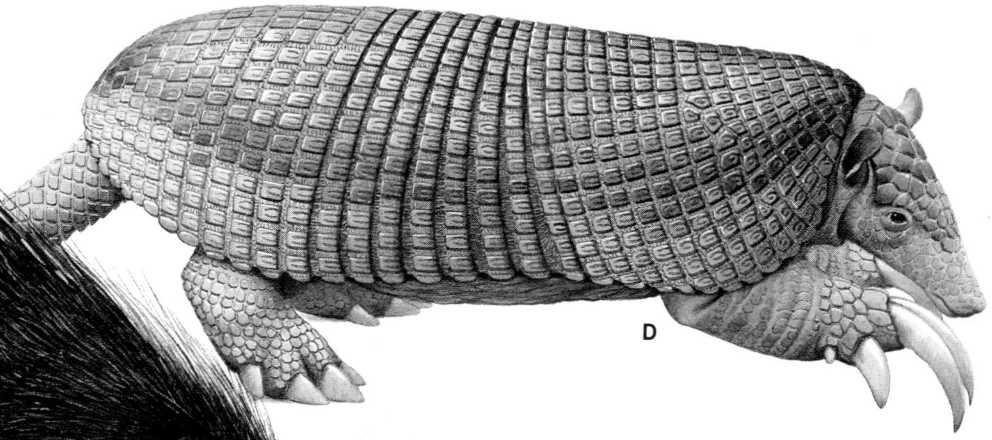

D

C

B

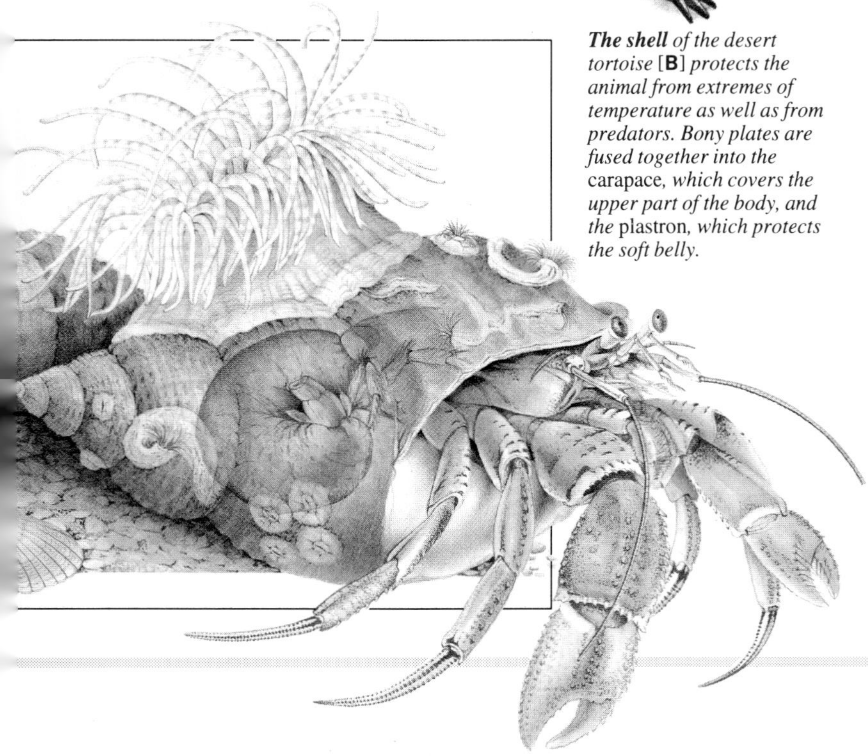

The giant armadillo *[**D**] of South America is the largest member of its family. The biggest examples can measure 5 ft (1.5 m) from head to tail, and can weigh up to 132 lb (60 kg). Its protective armor consists of thick, horny plates reinforced beneath the skin by areas of bone. Broad shields cover the shoulders and pelvis and a number of articulated half-rings protect the back and give a little flexibility. Even with all its protection, the giant armadillo is surprisingly agile. When threatened, this species of armadillo can partly roll itself up to deter its attacker. However, it is more likely to run away. Its relative, the three-banded armadillo, on the other hand, is capable of more spectacular feats: it can roll itself into a tight ball, leaving no chink in its armor for a predator to exploit.*

The Indian porcupine *[**C**] is a ground-dwelling creature that feeds on roots and bulbs. The rear part of its body is surrounded by a formidable array of sharp, stout, cylindrical quills approximately 20 in (50 cm) long. When threatened, the animal performs a series of warning signals. It grunts, stamps its hind feet, and erects and rattles its tail quills. If the predator persists, the porcupine runs at it backward so that quills pierce the enemy's skin and break off, often resulting in a painful or occasionally even fatal septic wound.*

The shell *of the desert tortoise [**B**] protects the animal from extremes of temperature as well as from predators. Bony plates are fused together into the* carapace, *which covers the upper part of the body, and the* plastron, *which protects the soft belly.*

Reptile armor

The shell of a tortoise or turtle is an inflexible structure made up of fused bony plates attached to the backbone and covered with thin dermal (skin) plates. Some species have a hinge or hinges on the shell. The hinges allow the front and back of the shell to close tightly, like a drawbridge, protecting the animals' soft body parts. A fast-closing shell can give a painful pinch to potential predators.

Armored mammals

The pangolin's scales develop from the animal's thick skin, probably from modified hairs (like the horn of a rhinoceros), and overlap like tiles on a roof. They are shed and replaced regularly. They cover all parts of its body except for the underside and the inner surfaces of the limbs. If attacked, the pangolin curls itself into a ball, rendering it safe from all but the largest cats and hyenas.

Evasive Action

How animals run, hide, or bluff

When some species of sea cucumber are being chased by a predator, they expel their entrails through the mouth and anus. This has the effect of distracting the attacker so that the sea cucumber can escape to safety. The entrails grow back in a few weeks. Other animals may use startling or distracting tactics to avoid capture, although few methods are as drastic as that of the sea cucumber. Startle displays usually rely on using warning colors generally recognized throughout the animal world – reds, oranges and yellows – that signal danger or poison.

Animals that do not have natural armor or warning colors with which to deter the attacks of predators usually run away or hide. These are popular defense strategies because, in the long run, they use up less energy than developing survival weapons. Animals such as antelopes and gazelles, for example, are already adapted for traveling long distances over the plains, and an extra turn of speed to improve the chances of survival can be achieved, in evolutionary terms, by simply increasing leg length and developing more powerful leg muscles.

As well as individual flight, some animals also send warning signals to their fellows – ghost crabs and kangaroo rats, for example, thump the floor of their burrows to warn of danger. Birds use cries and shrieks, and some species even have different cries for aerial and terrestrial predators.

Hiding or running?

Animals that are really fleet of foot, such as gazelles, may start to run as soon as a predator appears on the scene, but other animals freeze, hoping their camouflage or inactivity will conceal them. Many of these animals, if they realize this tactic is not succeeding, will take flight in a burst of speed that startles and confuses the predator. Female pheasants, for example, will remain on their eggs until almost stepped on, then burst out of the undergrowth with a loud flapping of wings. Rheas (large flightless birds of South America) will dodge about when pursued, often doubling back on their tracks and sometimes even leaping right over the head of the pursuer as they run in the opposite direction. They then duck down among the long grass and test their camouflage once again.

Some animals, such as frogs and the South American capybara, when threatened will leap into water to escape attack, as most terrestrial predators do not willingly follow prey into water. Moorhens not only take to the water but also sink and wait almost submerged until the danger has passed.

Startle displays

There are three main kinds of startle display. There is the sudden revelation of large false eyes, which hopefully deceive the predator into thinking it is attacking a much larger animal. Alternatively, flash colors (generally

B

D

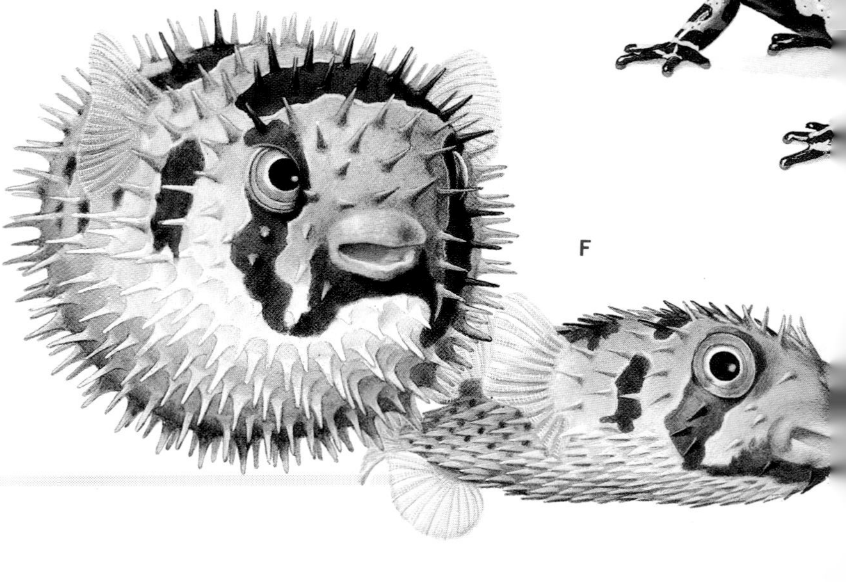

F

The Australian frilled lizard *[**A**], when threatened, hisses violently, lashes its tail, and fans out its black and red frill to startle predators away. Many butterflies, such as the peacock butterfly [**B**], use eyespots to trick predators into thinking that they have disturbed a larger animal. Persistent hunters will attack the spots, not the vulnerable body. The hognosed snake [**C**] feigns death if threatened. It rolls on to its back, hangs open its mouth and emits a smell similar to decaying flesh. The five-lined skink [**D**] wriggles its tail to distract predators. The vertebrae of the tail have special fracture points so that, if pulled, the tail breaks off, allowing the skink to escape. The tail eventually grows back. The fire-bellied toad [**E**] warns off predators by rolling onto its back and revealing its black and red underside, demonstrating it is poisonous. The porcupine fish [**F**] can inflate its body so that the spines in its skin stick out and deter larger fish from biting.*

A

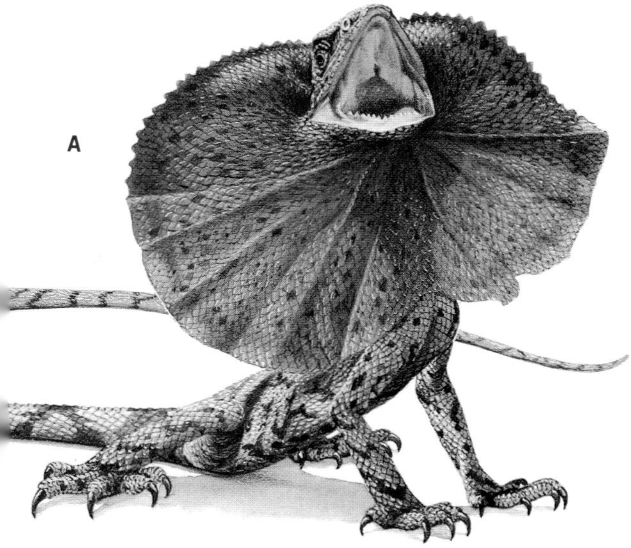

G

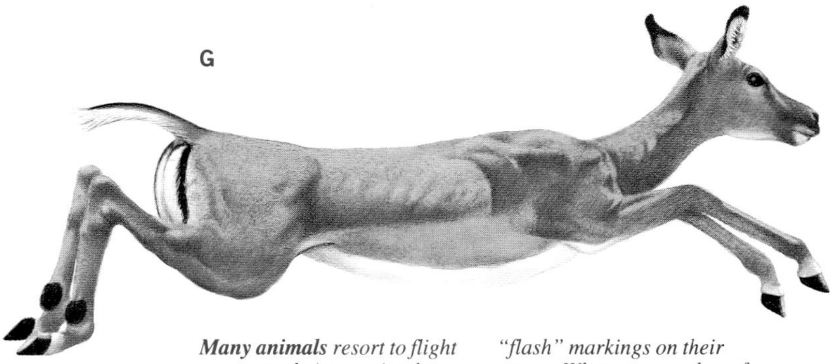

Many animals *resort to flight to escape their enemies, but for some, additional physical adaptations can greatly improve the chance of survival, not just for the individual but also for the whole group. Impala* [**G**] *are just one of many species of grazing antelope that have conspicuous black and white "flash" markings on their rumps. When one member of the group senses danger and runs away, its flash markings are quickly spotted by other members of the group grazing nearby, which are then also alerted to the danger. Other species of animals, such as rabbits, have a similar survival technique.*

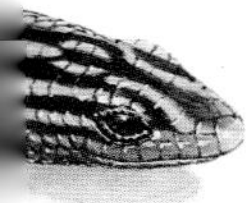

C

red, orange or yellow), unexpectedly displayed in underwings, throats or other parts of the body, will scare most predators. A sudden increase in size is also a very effective deterrent – as in the case of great horned owls, which fluff their feathers and spread their wings when in danger, or toads which puff themselves up and stand on tiptoe. Sudden noises – like hissing – are also deterrents.

Concealment devices

Small animals have to rely on escaping to a familiar bolt hole in order to escape from attackers. Rabbits, mice and other small mammals often build a series of tunnels with several exits and entrances through which they can escape from enemies.

Beach crabs often construct a turret of sand or mud around their burrow entrances. The turret gives them a lookout point above which only their stalked eyes protrude.

Distraction Displays

Some defensive displays are actually designed to draw attention to the prey. These types of display are usually performed by adults to protect their young. Adult plovers, for example, are a species of bird that often employs this technique. An adult pretends to be injured, trailing an apparently broken wing along the ground to lure the predator away from its nest or young, in the hope that it appears a better catch for the predator. The display is reinforced by a plaintive distress cry. Before starting the display, the bird surreptitiously creeps a short distance away from the nest. Even if the predator does not follow the bird, by looking up to see what the fuss is about it has lost sight of the nest, and may not be able to find it again. When the parent bird judges the predator to be sufficiently far from the nest, it flies off to safety.

E

Connections: Armor and Spines 258 Camouflage 264 Chemical Defenses 268 Claws, Jaws and Horns 256 Mimicry 266

Shades of Difference

How color is created in nature

Squid, cuttlefish and octopuses have evolved a whole language based on colors. Waves of changing hues and patterns wash across their bodies. Their colors also betray their moods. A male octopus, for example, will blush bright red if it sees a female, or if it is annoyed. Colors can be used by flowers to attract pollinators, or by animals to attract mates. They can warn off rivals, advertise the presence of a deadly poison or provide concealment. They can even be used to adjust body temperature, by differentially absorbing or reflecting the heat of the Sun.

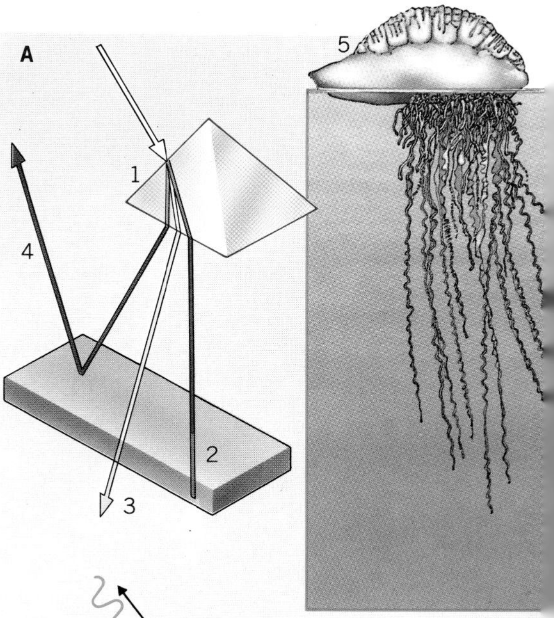

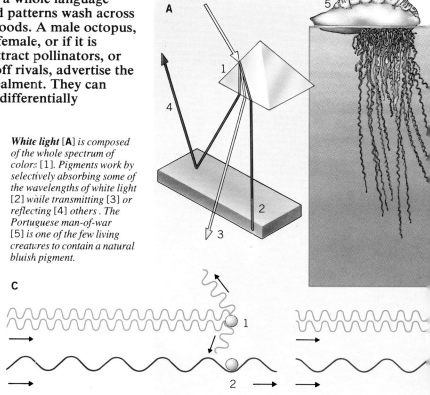

White light *[A] is composed of the whole spectrum of colors [1]. Pigments work by selectively absorbing some of the wavelengths of white light [2] while transmitting [3] or reflecting [4] others. The Portuguese man-of-war [5] is one of the few living creatures to contain a natural bluish pigment.*

Light is made up of pulses of energy that behave like waves. And color is the way in which the brain interprets the action of different wavelengths of light on certain sensitive cells in the eye. Different wavelengths – the distance between successive crests in a series of waves – produce different colors. But visible light is only part of the much broader electromagnetic spectrum. The shorter visible wavelengths produce blue light, which becomes invisible ultraviolet as the wavelengths decrease further. The longer wavelengths give red light, which slowly grades into infrared radiation (heat).

How colors arise

Many colors arise because peculiar structures in the outer coverings of plants and animals interfere with and scatter light in special ways. But most colors have a chemical basis, often depending on substances that lie just below the transparent surface of plants and animals. Oils and fats that are naturally contained in living cells reflect distinct wavelengths, usually red, orange or yellow. *Guanine* deposited in fish scales produces mirror-like reflections. In areas of bare skin, such as the human face, or a domestic cock's comb, the red light reflected by red blood cells shows through where blood vessels run close to the surface.

Local color

Pigments are complex chemicals that reflect specific colors. They are often produced in special cells. In the green parts of plants, most cells contain the green pigment chlorophyll. Plant pigments may be located in the cell sap, in special membrane-bounded bags (*plastids*), in the cytoplasm, or in the cell wall. The leaves of many deciduous plants turn brilliant shades of red, orange and gold in autumn. This is due to the breakdown of some pigments into others as the nutrients locked in the leaves are converted to soluble forms to be stored in the trunk. The pigments responsible are usually *carotenoids*, colorful compounds that are also found in animal groups, from the protozoa through to the highest vertebrates.

Some animals also change color, either seasonally or, like the squid, under conscious control. In these latter species, pigments are linked to muscle fibers and are free to move around in the cells, expand and contract, and even to respond to nerve signals. The changing position or size of pigment areas leads to a change in the pattern or intensity of color. Because plant color change is achieved mainly by synthesis or breakdown of pigments, it is much slower than the color changes in animals. Certain flowers that produce nectar only for one day become paler as the day goes on. By changing color when they no longer have nectar, they save pollinators a wasted visit and increase the chances of unpollinated flowers being visited.

Borrowed colors

Not all creatures generate their own colors. The brilliantly colored mantles of giant clams, for example, owe their colors to the presence of symbiotic algae inside their tissues. Some corals, too, contain colorful symbiotic algae. The sloth, on the other hand, carries its algae on the outside. Algae living in the sloth's fur contribute to its camouflage. In wet weather they turn green, camouflaging the sloth against the green tree trunks. In dry weather the algae dry out and become brown, and so does the bark to which the sloth clings. In a few cases the pigmentation of an animal depends on its diet. The pink of certain species of flamingos, for instance, is derived from pigments in the crustaceans that the flamingos eat.

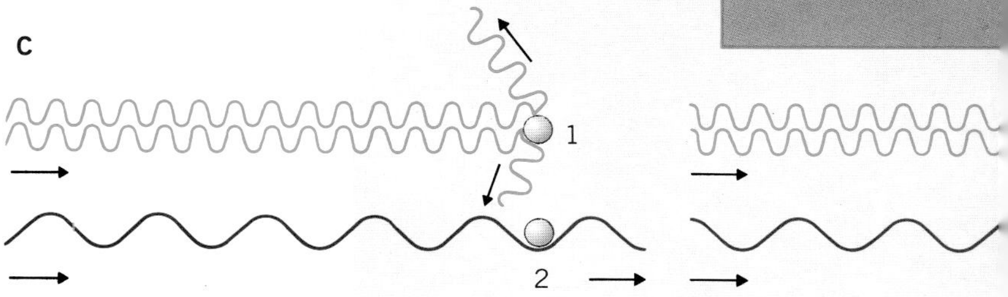

The wing *of the Morpho butterfly [E] has laminated scales of clear film on its surface [1]. Light is reflected from each of the boundaries between these films, or* lamella. *The light that is reflected from a lower boundary has to travel farther than light reflected from one nearer the surface. If this extra distance – which depends on the angle between the light rays and the layers – allows the peaks and troughs of two waves to match up when they recombine, then the wave is intensified [2]. Other wavelengths – other colors – will not be reinforced, or may even be cancelled out if the peaks of one wave match up with the troughs of the other. Then that color will disappear. The interference creates an iridescent shimmer. Different colors will be reinforced at different angles, so the butterfly's wing appears to change color as it tilts [3]. Many other animals have iridescent colors, such as the green and pink of the wavy-top shell.*

Connections: Algae 112 Bioluminescence 276 Camouflage 264 Communication 292 Coral Reefs 336 Flowering Plants 120 Mimicry 266 Vision 278

***Most bluish animals,** like the hyacinth macaw [**B**], rely on structural coloration, such as the selective scattering of wavelengths [**C**], for their hue. A particle, or even a bubble of air or liquid – which itself has no pigment – will scatter light whose wavelength is approximately equal to its own diameter [1]. Light of a longer wavelength is far more likely to simply pass the particle by [2]. Therefore, when white light falls on a substance containing air bubbles about 0.5 μ (10^{-6}m) in diameter – roughly the wavelength of blue light – the component of the light that has that wavelength will be scattered [3], giving the object a blue coloration. Conversely, its shadow will appear reddish hued. The barbs of a hyacinth macaw's contour feathers [**D**] have a clear cuticle [1] surrounding alveolar pith cells [2], which contain air cavities that scatter blue light. The cells are backed by the dark pigment melanin [3], which increases the color's intensity. The apparent white color of, for example, polar bear hair – or the hair of creatures with albinism – is similarly caused by air cavities – in this case about 1 μ across – inside the hair. Cavities this size scatter all light completely, yielding white.*

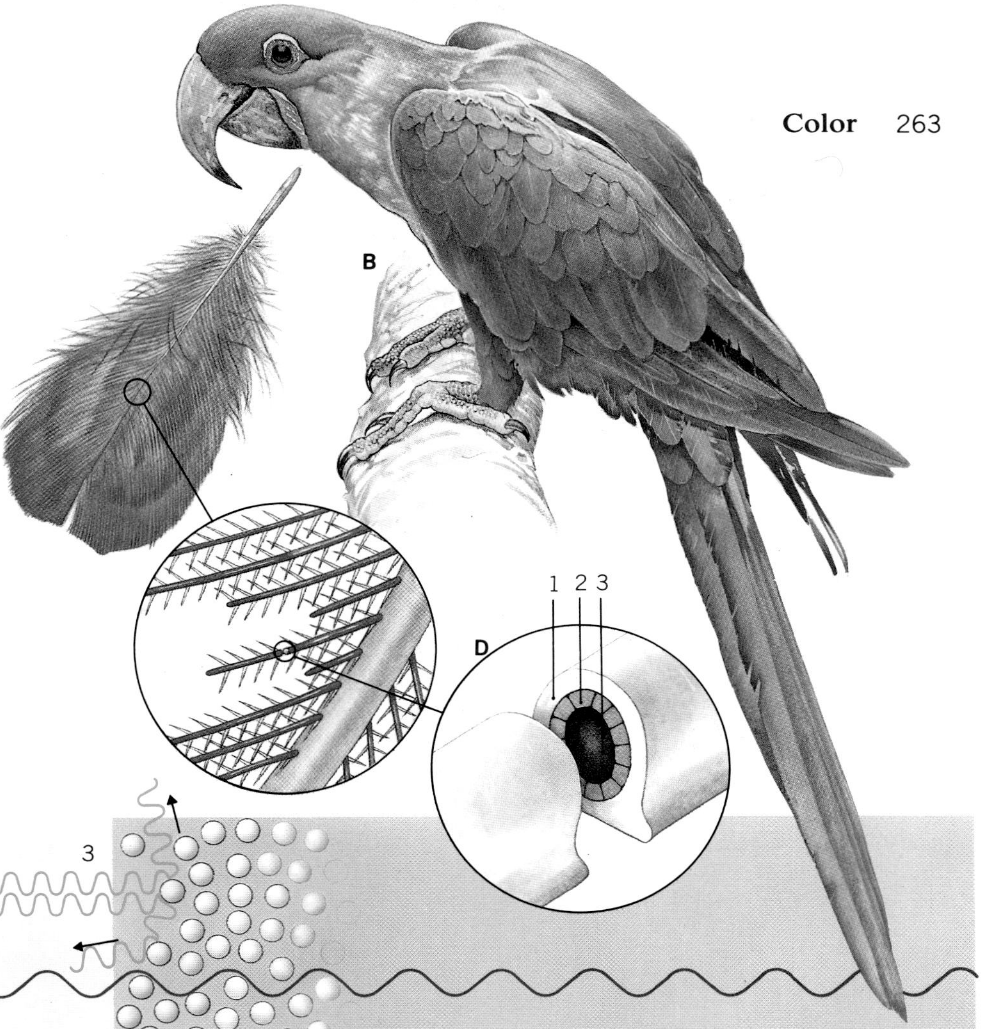

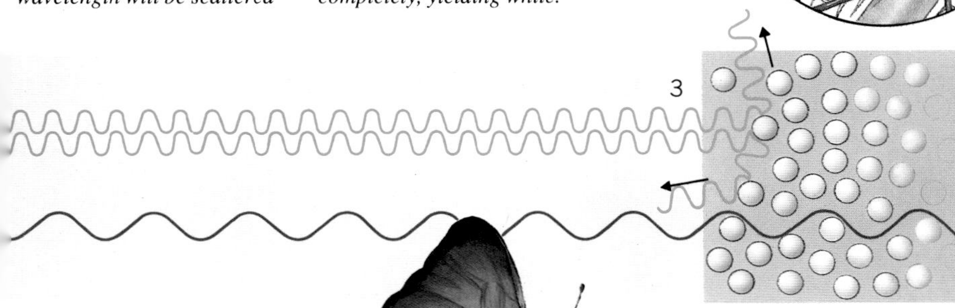

The Cloak of Invisibility

How animals use camouflage

Military camouflage is an essential part of modern warfare. But for millions of years, animals have been using a far more impressive array of cryptic patterns and colors in their battle for survival. The deception is practiced by both the predator – to approach its quarry unseen – and the prey – to remain unobtrusive. Some animals change color gradually, mirroring the changing seasons; while others, such as the cuttlefish, can undergo dramatic color changes in less than a second. In some cases, the disguise is so convincing as to confound the most experienced naturalist.

***Chameleons** can change color in under two minutes. They change color not only for camouflage but also to regulate their body temperatures, to threaten rivals, and to court their prospective mates. Color changes are accomplished under the animal's direct nervous control.*

The most obvious way for an animal to hide itself while remaining in full view is by looking like something else. Clearly the imitated object should not itself provoke alarm or attack. A whole host of creatures disguise themselves as leaves, twigs, bird droppings and other inanimate objects. But there are, broadly, four subtler ways to disappear.

Vanishing tricks

The first is represented by those animals that simply match the background color of their habitat exactly. So despite its seemingly gaudy coloration, a brilliant green tree frog is invisible against a background of lush rain forest leaves. Similarly, an Arctic hare or stoat in its white winter coat can move undetected against the blanket snow of its habitat.

In the second category, animals use the trick of disruptive coloration to break up their outline as seen against a background. A tiger's stripes admirably serve this purpose as it stalks its prey through the long grass. The eggs of birds such as terns and plovers that are laid on bare ground have effective disruptive camouflage that helps them to merge with pebbles or lumps of soil.

Light and shade

Species of the third category display a striking form of shading in which the animal's upper surface is darker than the lower. The purpose of this coloration is to obscure the shadow cast on the lower part of the body. Animals that normally live upside down, such as sloths and some species of catfish, have reverse countershading. This strategy has been refined to perfection by various deep-sea fish, squid and crustaceans. These animals have silvery sides, so when they are viewed from any angle below the horizontal, they reflect light from above and are therefore not visible as a silhouette. Even when viewed from directly beneath, these creatures have no silhouette, since light-producing organs on their undersurface closely match the light output of the distant sky above.

The fourth major camouflage strategy, often used with one or several of the other techniques, is to appear less like a solid body by eliminating shadows cast from the side. Many animals achieve this by having flattened bodies, or by pressing themselves into hollows in the ground.

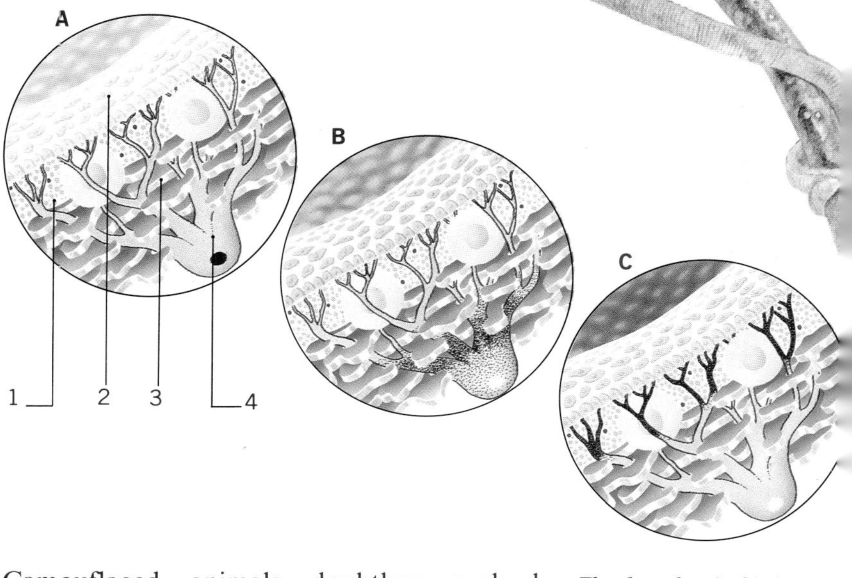

***The chameleon's skin** has a layer of yellow oil droplets distributed among yellow cells [1] just beneath the epidermis [2]. Under this layer is a layer of* guanophores [3], *which are colorless when backed by a clear surface. Then the skin is the color of the light reflected from the yellow droplets* [**A**]. *Guanophores reflect blue light when they have a dark backing, when the* melanophores [4] *let their dark granular pigment spread out* [**B**]. *The chameleon's green color is due to the mixing of reflected yellow and blue light. It gets darker as the melanophore pigment spreads farther into the yellow* [**C**].

Camouflaged animals doubtless evolved through a combination of genetic mutation and natural selection. If a chance mutation helped an animal to avoid capture, or to catch more prey by altering its appearance, then that animal would have been more likely to survive than its fellows. The camouflage trait would then be passed to its offspring.

Quick change artists

Cuttlefish can change color in less than a second. They do so by means of very complex structures, with an elastic, pigment-filled cell surrounded by muscle cells. The pigment cell is tiny and spherical when the muscle cells are relaxed, but when they contract it is drawn out into a flat disk so that the pigment is visible. These *chromatophores* have different colored pigment cells that can expand or contract independently, giving a huge range of color patterns.

Connections: Animal Cells 102 Color 262 Communication 292 Flight and Deception 260 Fur and Feathers 134 Mimicry 266

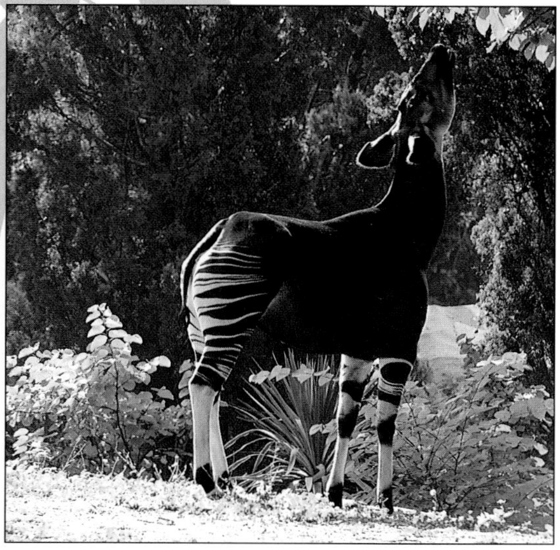

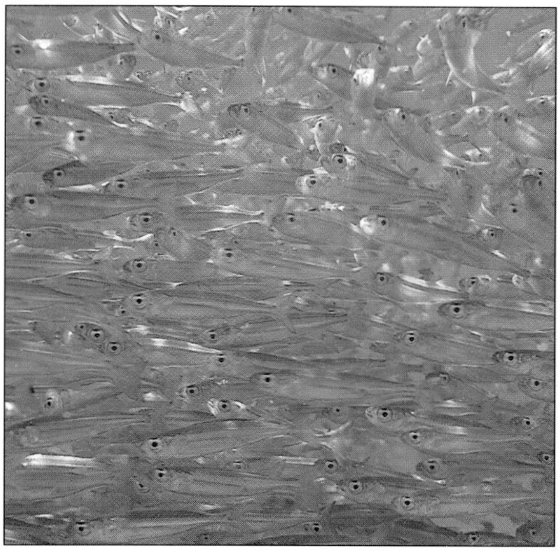

The four basic camouflage *strategies are used by a wide range of animals. A combination of techniques provides a subtler, more impenetrable disguise. The okapi (top left) uses a mixture of disruptive markings on its rump and legs with the characteristic dark back and lighter belly that make it harder for a predator to spot it by its shadow. Herring (bottom right) have a similar dark/light coloration. In addition, they tend to swim in large shoals, and when they are disturbed the sudden glint of sunlight off hundreds of silver sides can confuse an attacker and make it unsure of where to strike. Although the polar bear (top right) is the perfect color to disappear against its snowy Arctic background, it still flattens its body to the ground to eliminate shadows when creeping up on unsuspecting seals. The leopard (bottom left) will similarly flatten itself against the branch of a tree when it wishes to be unobserved. The jaguar of South America has similar coloring to the leopard. The "spots" of both cats are really dark rings, but the jaguar's have a dot at the center of each, while the leopard's are unoccupied. The cheetah, which hunts in more open country, has smaller, solid spots, perhaps because it does not need to suggest dappled shadows.*

Do-it-yourself Disguise

Many species of spider crab actively camouflage themselves by attaching fragments of marine detritus to their bodies. One of the prime exponents of this form of home decoration is *Oregonia gracilis*, a crab found on the northwestern coast of the United States. Using its pincerlike front legs, the crab picks up small pieces of algae, tubes secreted by burrowing worms, wood chips – or almost any other available material. The fragments are manipulated in the crab's mouthparts in order to "roughen" their edges, and then attached to the body by means of tiny hooked bristles (called *setae*). The setae on *Oregonia* are present on the upper shell, the walking legs, and on a ridge above the eyes. In daylight hours, the crabs remain immobile, their bodies pressed into the sea floor.

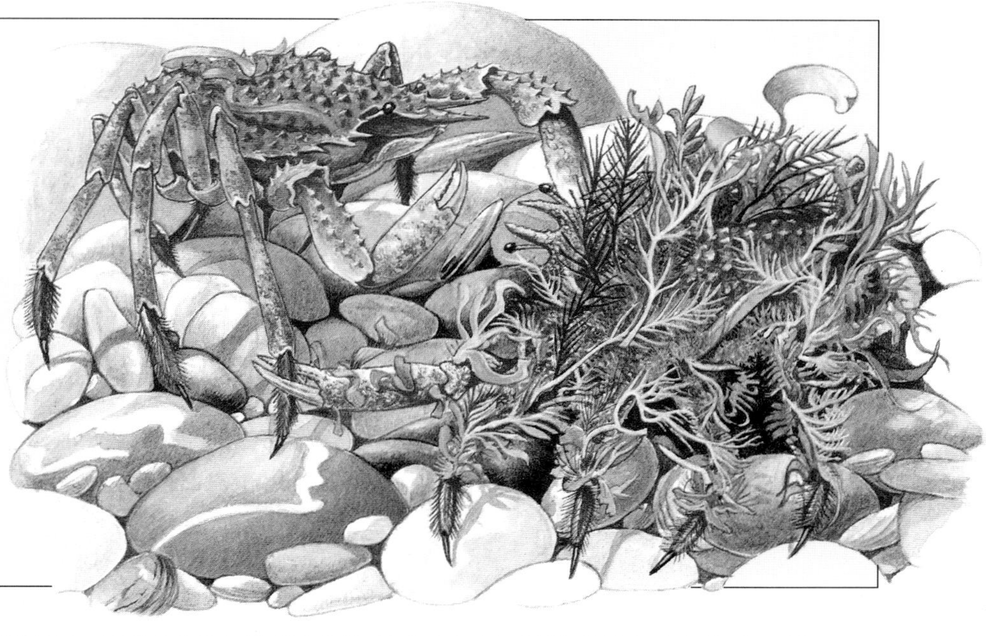

Mistaken Identity

How living things use mimicry

Macleay's specter, an Australian stick insect, must be one of the most successful lifelong mimics anywhere. It imitates ants when very young. Eventually, however, it grows too big for this ploy to succeed, so the juveniles curl their abdomens over their backs and pretend to be baby scorpions. The adults in their turn are too large for this trick, and instead resemble a cluster of dry, curled-up leaves. There are even "bird droppings" that move, "leaves" that walk, "seaweeds" that swim away, "scorpions" with no stings, and "tree stumps" that fly – nature is full of mimics.

Many animals escape predation by looking like an inanimate object. But it is not just structural and surface appearances that have evolved, for fascinating behavioral strategies have also developed to reinforce deception. When at rest, stick caterpillars hold their bodies away from the bark of trees, rigid and unmoving like a branching twig. The potoo, a bark-colored bird related to nightjars, perches atop a broken branch and points its bill skyward, so that its head appears beveled like a continuation of the stump. When threatened, the African leaf fish, a highly flattened, thin-bodied fish, drifts gently to the sea bed and lies there on its side.

Mimicry can also benefit predators. A species of African jumping spider both looks and moves like an ant. This means that – often aggressive – ants will mistake it for one of their own, and relax their guard.

False alarms

Instead of camouflaging themselves from predators, some animals use false signals to trick predators that they are other, more dangerous, animals that the predator would not consider attacking. A very common kind of mimicry is known as *Batesian* mimicry, where a nonpoisonous species imitates a poisonous one. This enables it to enjoy the same reduction in predation pressure, without having to expend energy synthesizing poisons. Such mimicry will work only if the animal being mimicked – the model – is more common then the mimic. Only if a predator usually gets a bad experience from attacking such an animal will it learn to avoid it. Poisonous animals usually bear bright yellow, red and black patterns, almost universally recognized as danger signals by animals. It pays them to be distinctive, because they want to be recognized and remembered.

Flying mimics

Batesian mimicry is very common among butterflies, where certain species render themselves poisonous or distasteful by taking in toxic compounds from the plants upon which they feed as caterpillars. Certain groups of butterfly species in South America and Africa are all moderately poisonous, and have similar patterns of warning coloration, even though they are made up of several different species. This is known as *Mullerian* mimicry. The butterflies benefit because the predator has only to learn one pattern in order to learn to avoid several species, thus reducing the predation on each.

Plant deceptions

Animals are not the only mimics. Some plants mimic animals in order to attract pollinators. Carrion plant flowers – with their mottled red surface and fetid smell – lure flies which come to lay their eggs on the "food." Defensive mimicry also occurs in plants. The gaily colored *Heliconia* butterflies of the South and Central American rain forests lay their eggs on passionflower vines, so their caterpillars can feed on the vine leaves. However, to ensure an ample food supply for her offspring, the female butterfly will not lay her eggs on stems and leaves already occupied by eggs. The passion vine produces mock eggs on its leaves to deter this invasion.

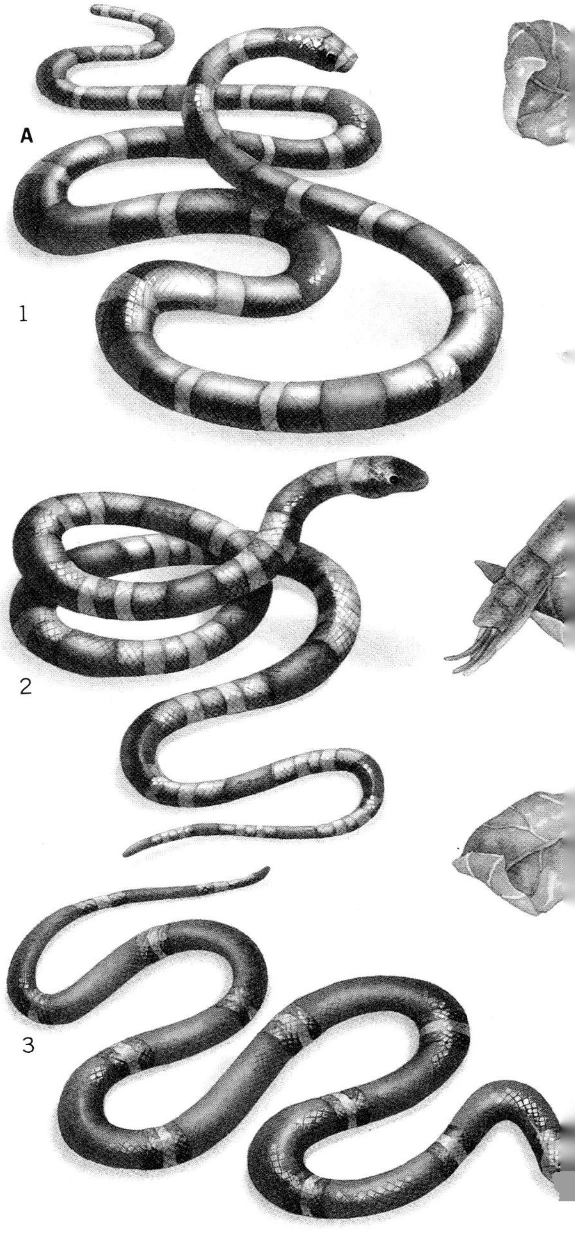

***Many coral snakes** of South America* [**A**] *have black, red, and yellow warning stripes. Some have no poison, some, like* Micrurus lemniscatus [1], *have lethal venom, and some, like* Oxyrhopus trigeminus [2], *are mildly poisonous: all inhabit the same areas. A mildly poisonous snake gives a predator an unpleasant taste that it remembers when it sees a similar snake. A harmless species has no poison, but is protected by mimicking the poisonous species. Lethal species also benefit – if attacked, their predators would not live to learn by experience. So both harmless and lethal species mimic moderately poisonous species (which must be the most numerous locally for a statistical probability that an attacker's experience will be unpleasant). Disputed by some scientists,* Mertensian *mimicry, as this is known, may be used by king snakes* [3], *which, although unrelated to coral snakes, have very similar color banding but no venom.*

***Many different insects** [**C**] mimic wasps or bees that deter predators with unpleasant stings. Key recognition points of the common wasp* [1] *are its size, a black and yellow body, transparent wings and a tiny waist. One "impostor" is the wasp-mimic moth* [2], *whose transparent wings are shaped like a hornet's; the moth even mimics its role model's flight patterns. The wasp beetle* [3] *convinces at a distance, and also imitates the jerkiness of a wasp's flight. A hoverfly* [4] *has a black thorax, which gives the impression that the abdomen ends abruptly in a narrow waist, even though this is only an optical illusion.*

Connections: **Camouflage** 264 **Color** 262 **Flight and Deception** 260 **Plant Defenses** 244 **Pollination by Animals** 154 **Poisonous Animals** 250

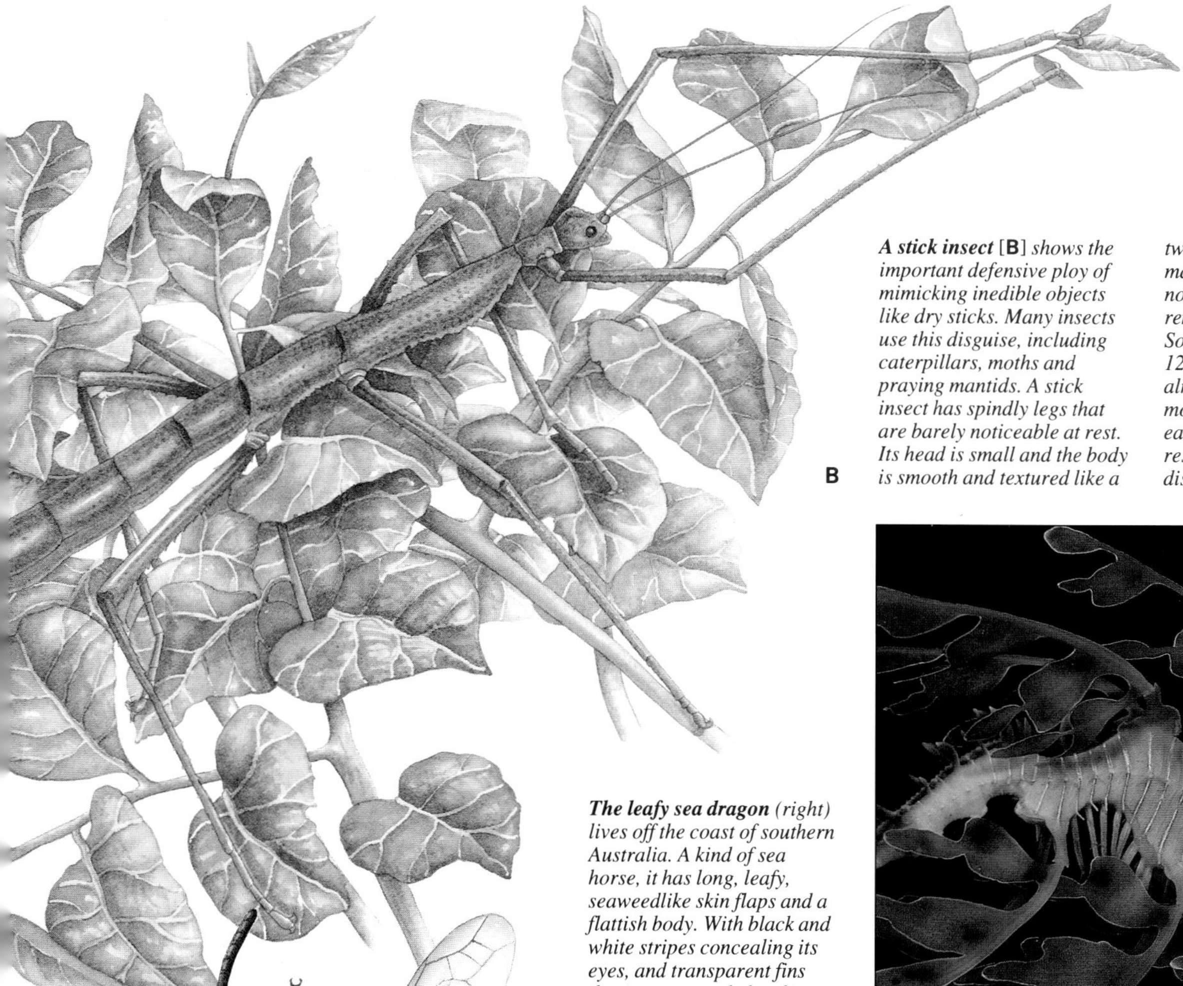

B

A stick insect [**B**] *shows the important defensive ploy of mimicking inedible objects like dry sticks. Many insects use this disguise, including caterpillars, moths and praying mantids. A stick insect has spindly legs that are barely noticeable at rest. Its head is small and the body is smooth and textured like a twig, with slight ridges and markings mimicking twig nodes. Stick insects may remain motionless for hours. Some stick insects grow up to 12 in (30 cm) long, yet are almost invisible until they move; even then, they move each leg very slowly, responding to any disturbance by "freezing."*

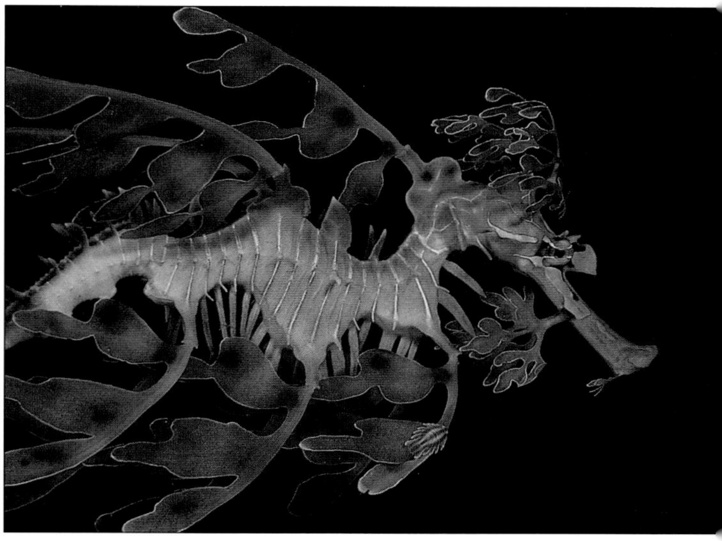

The leafy sea dragon *(right) lives off the coast of southern Australia. A kind of sea horse, it has long, leafy, seaweedlike skin flaps and a flattish body. With black and white stripes concealing its eyes, and transparent fins that are extremely hard to detect, the fish resembles drifting seaweed.*

C

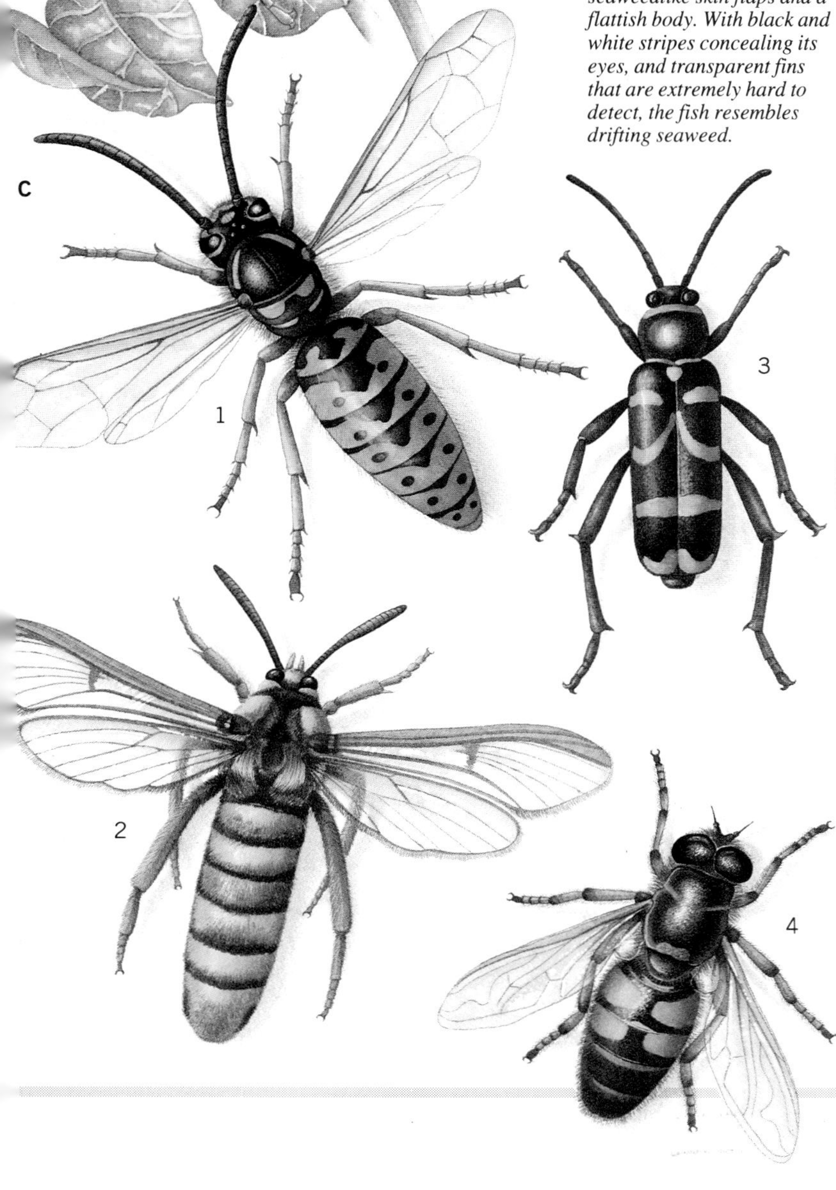

Stone plants *(above) are desert plants with leaves that have a waxy surface to minimize transpiration (water loss from leaves). Like other plants that can survive extreme aridity, this allows them to store large amounts of water in their leaves. However, this adaptation for a desert existence also means that they are potentially a precious source of liquid refreshment for desert animals. To avoid being eaten, these South African desert plants, which only just protrude above the ground, have evolved a shape and coloring that enable them to blend in with the surrounding stones.*

Chemical Warfare

How animals use chemicals in defense

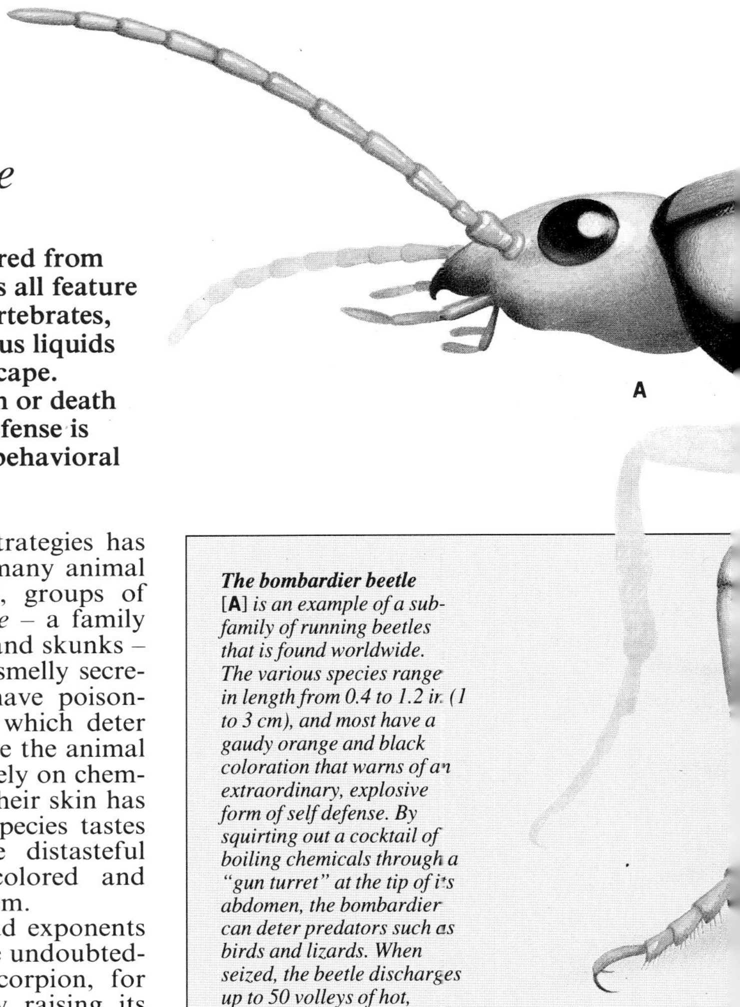
A

Jets of boiling, caustic liquid, paralyzing poisons delivered from explosive syringes and foul-smelling or stupefying fluids all feature in the chemical armory of the animal world. Some invertebrates, such as octopuses and squids, discharge inks or luminous liquids that confuse their predators while they make a hasty escape. Other species manufacture potent toxins that cause pain or death in any predator foolish enough to eat them. Chemical defense is commonly accompanied by a distinctive coloration or behavioral ritual that serves to warn off a potential foe.

***When threatened**, the darkling beetle (top) virtually stands on its head. If its enemy is not deterred by this display, it is sprayed with a smelly skin irritant, discharged from glands in the beetle's abdomen. The bush cricket (above) exudes drops of distasteful fluid from its joints.*

A variety of chemical defense strategies has evolved quite independently in many animal groups. Among the vertebrates, groups of mammals such as the *Mustelidae* – a family that includes the weasels, stoats and skunks – are well known for their use of smelly secretions in defense. Many toads have poison-producing glands in their skin, which deter any predator rash enough to seize the animal in its jaws. Though birds rarely rely on chemical secretions for defense since their skin has few glands, the flesh of many species tastes foul or induces nausea. These distasteful species are usually brightly colored and predators soon learn to avoid them.

The finest and most widespread exponents of chemical defense, however, are undoubtedly the invertebrates. A whip scorpion, for example, responds to attack by raising its abdomen and spraying its aggressor with a fine mist of highly acidic fluid from a pair of large anal glands.

Some beetles, when seized by predators, react by reflex bleeding, often from their leg joints. The blood of blister beetles contains a substance called *cantharidin*, a powerful blistering agent, very effective in deterring ants and other predators.

Weapons on the wing

Caterpillars of swallowtail and Parnassus butterflies have a brilliant red or orange two-pronged defensive weapon (called the *osmeterium*) just behind the head. This is normally hidden in a pouch, but can be displayed suddenly if the insect is disturbed or attacked. This warning display may itself deter the predator, but if the attack continues, the extended "horns" extrude a highly odorous secretion of fatty acids, which the caterpillar wipes against its enemy to repel it. There is evidence that swallowtail butterflies cooperate to enhance the effectiveness of their chemical defense. Disturbing one individual on the edge of a group makes surrounding butterflies erect their osmeteria: their combined action may produce a protective chemical "fog" around the group. Chemical defense, however, carries a cost, for an animal must invest energy in producing toxins or stinging structures. But some species of butterfly cut down on this investment by extracting defensive chemicals from the plants on which they (or their caterpillars) feed. For example, monarch caterpillars eat milkweed plants, which contain toxic compounds called cardenolides. The toxins accumulate in the bodies of the caterpillars, protecting them, and subsequently their butterflies, from predators.

***The bombardier beetle** [**A**] is an example of a subfamily of running beetles that is found worldwide. The various species range in length from 0.4 to 1.2 in (1 to 3 cm), and most have a gaudy orange and black coloration that warns of an extraordinary, explosive form of self defense. By squirting out a cocktail of boiling chemicals through a "gun turret" at the tip of its abdomen, the bombardier can deter predators such as birds and lizards. When seized, the beetle discharges up to 50 volleys of hot, irritant solution. The beetle's aim is surprisingly accurate. Just by swiveling the tip of its abdomen, it can direct the spray to either side and backward or forward. Each discharge is accompanied by a loud, startling click, and the emission of a cloud of caustic vapor that may distract or even temporarily blind any would-be predator.*

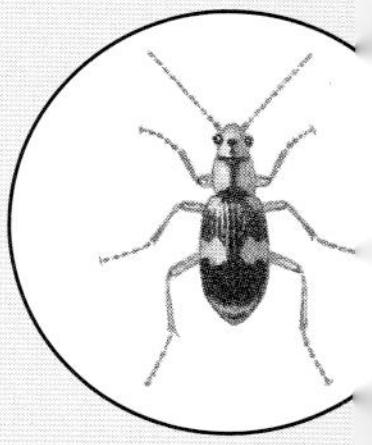

Glued to the floor

Onychophorans, caterpillarlike invertebrates that live in humid habitats in the tropics, discharge an odorless fluid from glands on the ends of projections on their heads. This fluid, which can be squirted over distances of up to 6 inches (15 centimeters), hardens almost immediately to a rubbery consistency, entangling and immobilizing potential predators.

When attacked, aphids exude droplets of liquid wax from the tips of peglike structures (or cornicles) on their abdomens. The wax solidifies on contact with the attacker, preventing it from moving; it may also serve to protect the aphids' own bodies.

Connections: Ants' Nests 236 Bees' and Wasps' Nests 232 Flight and Deception 260 Invertebrates 122 Jet-propelled Animals 212 Poisonous Animals 250

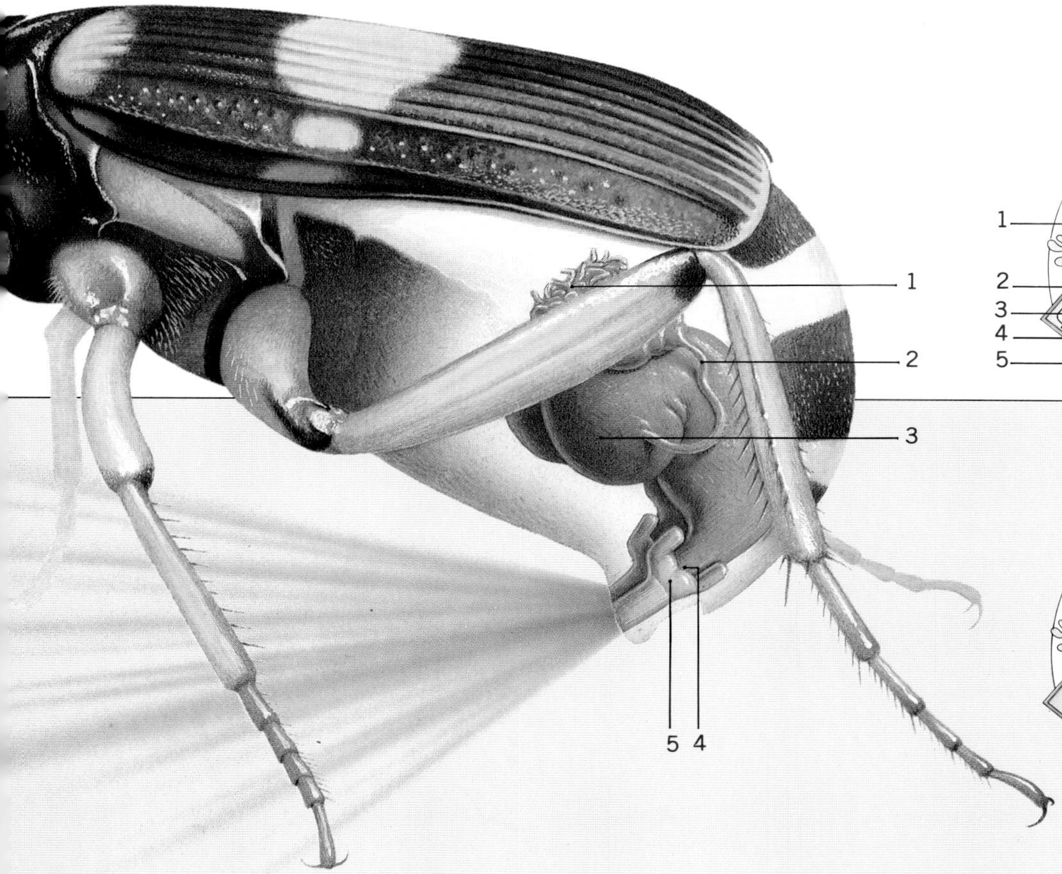

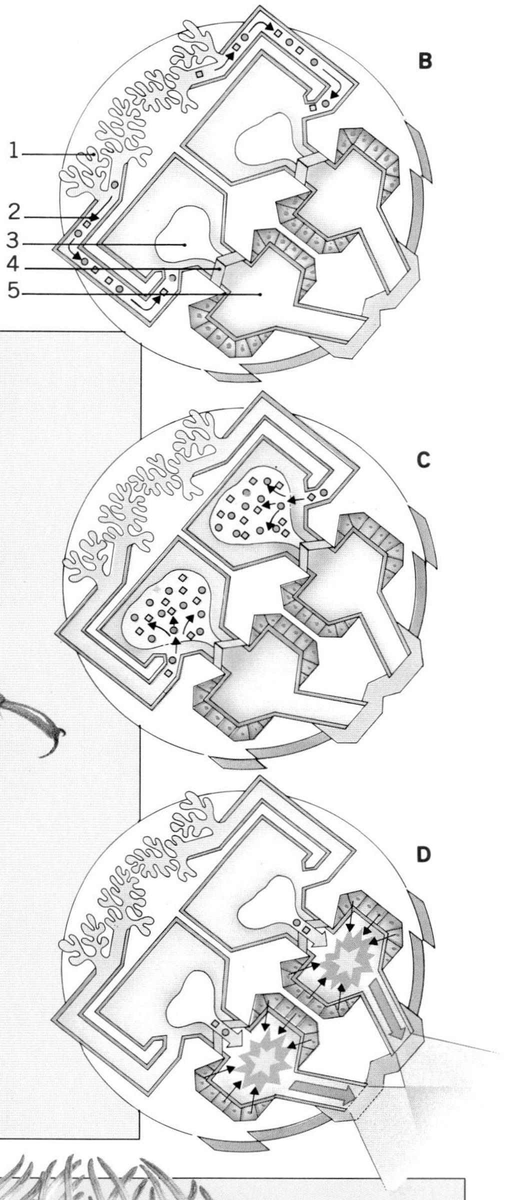

***Two pygidial glands** in the bombardier beetle's abdomen produce the irritant chemicals and generate the explosive force needed to expel them. Each gland [**A** and **B**] consists of a battery of secretory cells [1] arrayed around a duct [2] that drains into a "reservoir" [3]. A muscular valve [4] connects the reservoir to a thick-walled, heat-resistant "combustion chamber" [5]. The secretory cells produce hydroquinone compounds and hydrogen peroxide, which are stored in the reservoir until needed [**C**]. Each reservoir contains enough chemicals to produce around 50 blasts. If the beetle is attacked, the valve between the reservoir and combustion chamber opens: enzymes are released into the combustion chamber, where they mix with the hydroquinone and hydrogen peroxide in a violent heat-generating reaction [**D**]. The pressure within the combustion chamber builds up and forces the caustic liquid through an opening beside the beetle's anus.*

Invertebrate Defense

Sea slugs are among the best protected of all marine organisms. Despite their apparent lack of armor – shells, claws, or spines – they are rarely troubled by hungry fish. The secret of the sea slug's success lies in its cerata [1] – the long projections that cover its body. Many of these projections contain *cnidocytes*, or stinging cells [2], which discharge a barbed, poison-tipped thread into anything that touches them.

The cnidocytes are not produced by the sea slug itself, but stolen from its prey – the sea anemone. When an anemone is eaten, the stinging cells from its tentacles are not digested but are transported through the sea slug's body into the ends of its cerata. In this way the sea slug obtains its defense "for free," thereby saving vital energy. Its spectacular coloration acts as a warning to potential predators.

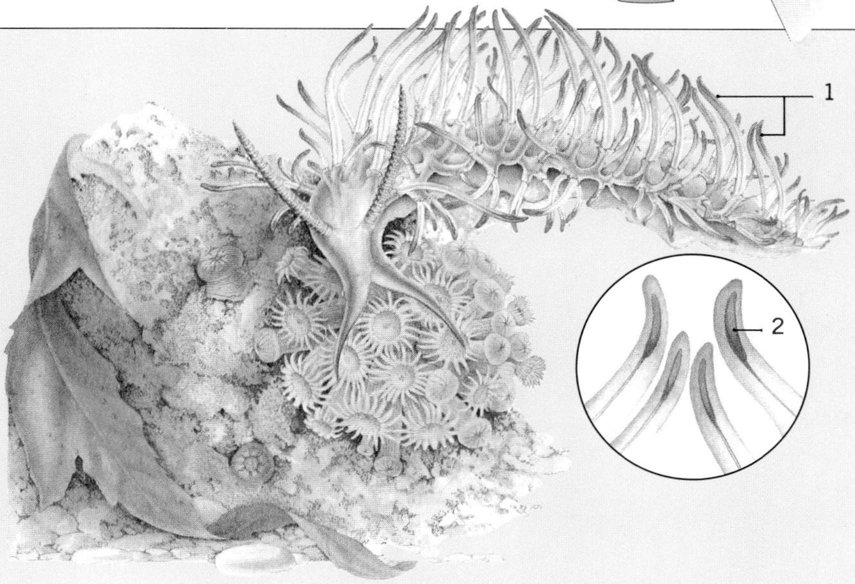

Predatory Trappers 254 Scent Communication 294 Smell and Taste 284

Insect Repellants

How a termite colony defends itself

Some termite workers defend the colony by sacrificing their own bodies in an acid explosion of gut contents, but, blind and soft-bodied, most termites are vulnerable to attack. Soldier termites, whose sole function is to defend the colony, have evolved a whole battery of substances, from irritants and toxins to anticoagulants and glues: a formidable armory of chemical weapons with which to kill, incapacitate, or repel their main enemies, the ants. No other animals have developed such a varied arsenal of chemical weapons and delivery systems.

Various mammals, including the aardvark and aardwolf of Africa, the pangolin of Africa and Asia and the South American anteaters, have evolved specializations to break open termites' nests and eat them. But the termites' greatest enemies are ants. It is against these formidable adversaries that they have evolved their sophisticated weaponry. Soldiers of many termite species are protected with armored heads and large biting jaws, with which they slash, seize or bludgeon their foes. But others employ a remarkable array of chemical weapons in addition.

The chemical armory

Termites have evolved three main methods of chemical defense. In the first, the soldier bites its enemy and simultaneously releases a toxic, oily chemical into the wound. This method has evolved independently several times in different families of termites.

In one genus of biter-injectors, the African mound-building *Macrotermes,* the chemical is a greasy, paraffin waxlike mixture of long-chain hydrocarbons called *alkanes,* which are stored in a frontal gland in the head. Each time the soldier bites, it covers the ant's punctured cuticle with anticoagulant oil, with the result that the ant bleeds to death.

Termites of the genus *Cubitermes* secrete more complex hydrocarbons called *diterpenes.* Others, of the genus *Armitermes,* have a nozzlelike protrusion on their foreheads from which they drip toxic, oily, modified fatty acids into the wounds made by their sharp-pointed, tonglike mandibles.

The second method of chemical defense, found in the subfamily *Rhinotermitinae,* involves daubing oily contact poisons onto the attacker's body from a bristly "paint brush" formed from a modified "upper lip" or *labrum.* The poison is stored not just in the head as in the biter-injectors but also in a reservoir in the abdomen, and may exceed 35 percent of the soldier's dry weight.

The third method, glue-squirting, is found in the most diverse and abundant group of termites, the subfamily *Nasutitermitinae,* which has more than 500 of the world's 2,200 species throughout the tropics. This takes its name from the *nasus,* a long snoutlike elongation of the forehead, from which the glue is squirted at the enemies. It is usually effective against termite-eating mammals as well as

Most primitive termites lack chemical defenses, but have formidable mandibles with which they can bite and crush their enemies. Soldier termites of more advanced families rely more upon their chemical armory, and consequently their mandibles are less well developed.

Macrotermes *soldiers* **[A]**, *for example, have long, slender mandibles with which they slash at their foe to break through their hard cuticles, following up their thrusts with the delivery of toxic anticoagulants produced by three glands located in or on the head.*

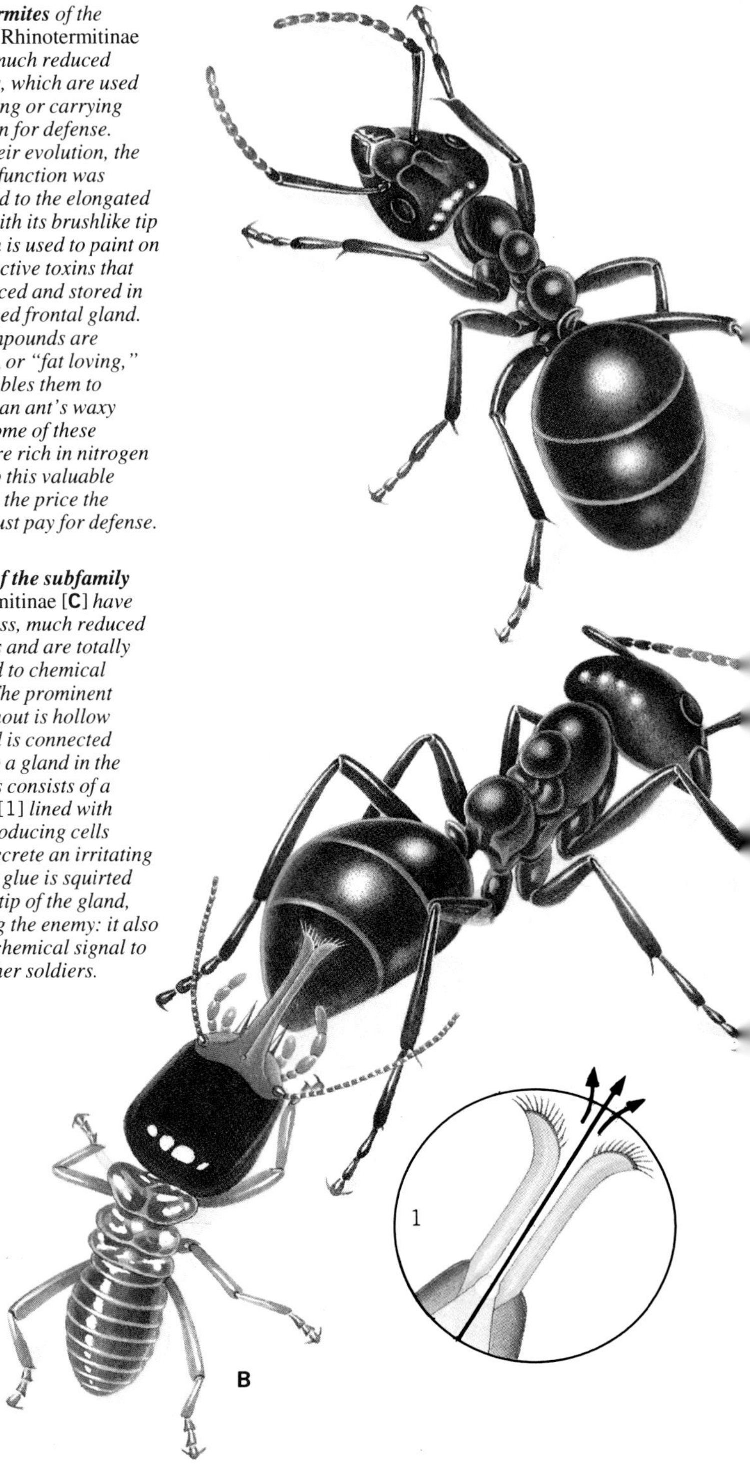

***Soldier termites** of the subfamily* Rhinotermitinae **[B]** *have much reduced mandibles, which are used for grabbing or carrying rather than for defense. During their evolution, the defensive function was transferred to the elongated labrum, with its brushlike tip* [1], *which is used to paint on highly reactive toxins that are produced and stored in the enlarged frontal gland. These compounds are* lipophilic, *or "fat loving," which enables them to penetrate an ant's waxy cuticle. Some of these poisons are rich in nitrogen – using up this valuable element is the price the termite must pay for defense.*

Soldiers of the subfamily Nasutitermitinae **[C]** *have functionless, much reduced mandibles and are totally committed to chemical defense. The prominent pointed snout is hollow inside and is connected directly to a gland in the head. This consists of a reservoir* [1] *lined with poison-producing cells* [2] *that secrete an irritating glue. This glue is squirted out of the tip of the gland, entangling the enemy: it also acts as a chemical signal to attract other soldiers.*

Connections: Chemical Defenses 268 Poisonous Animals 250 Termite Towers 234

A

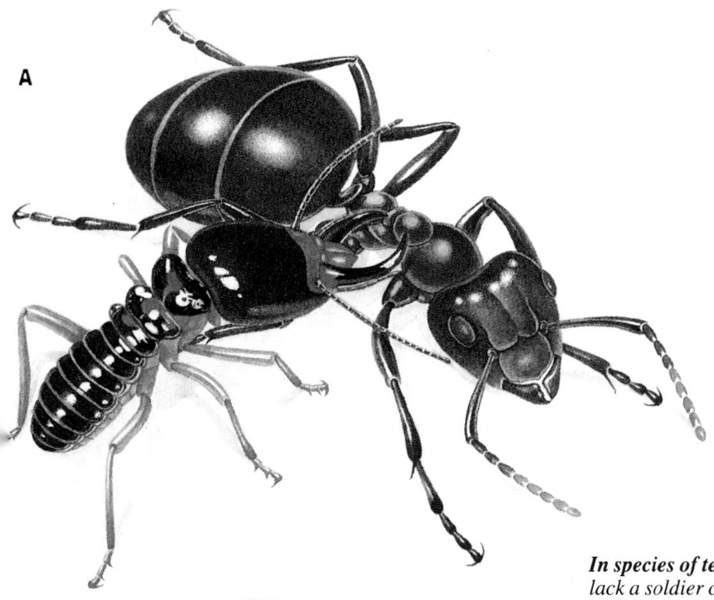

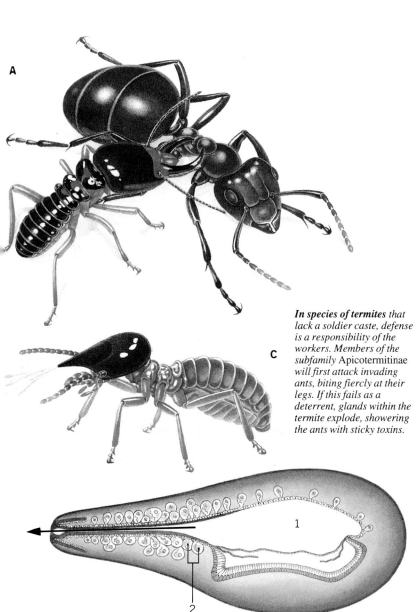

ants and other insects. The glue is similar to pine resin and acts as an entangling agent, an irritant and a contact poison.

To avoid being poisoned by their own toxins, both when they are being synthesized and stored and when they are used in battle, termites detoxify their chemical weapons using enzymes that catalyze their reduction to harmless products.

Detonating in defense

Not all termites use biting jaws or chemicals in defense. Some of the soil-feeding higher termites even lack soldiers. Instead, special workers defend the colony against marauding ants by sacrificing themselves in spectacular kamikaze fashion, showering their attackers with feces and other acidic gut contents.

C

In species of termites that lack a soldier caste, defense is a responsibility of the workers. Members of the subfamily Apicotermitinae *will first attack invading ants, biting fiercly at their legs. If this fails as a deterrent, glands within the termite explode, showering the ants with sticky toxins.*

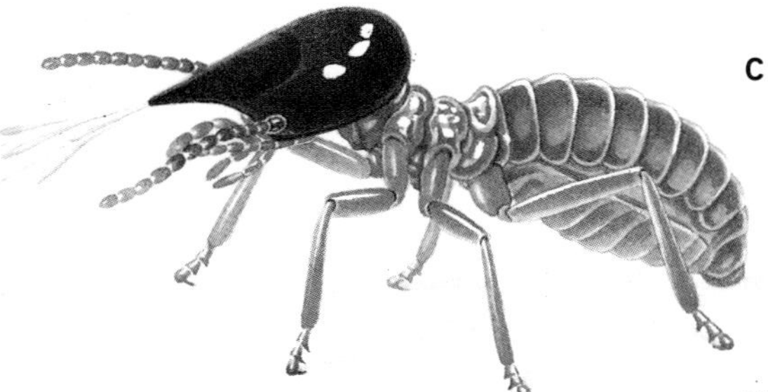

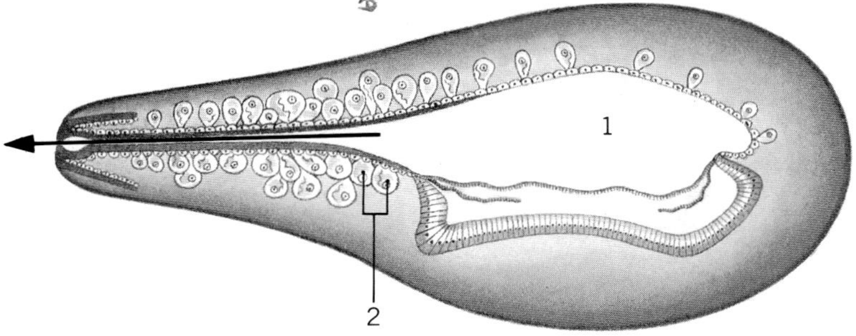

Termite History

The seven termite families come from a common cockroachlike ancestral stock. The six families of lower termites [**A**–**F**] depend for their defense mainly on their soldiers' jaws, although the chemically defended *Rhinotermitidae* also belong in this group. The seventh family – the *Termitidae* [**G**] – comprises four subfamilies of higher termites, all of which use chemical defenses. Included here are the soldierless *Apicotermitinae,* whose workers are explosive.

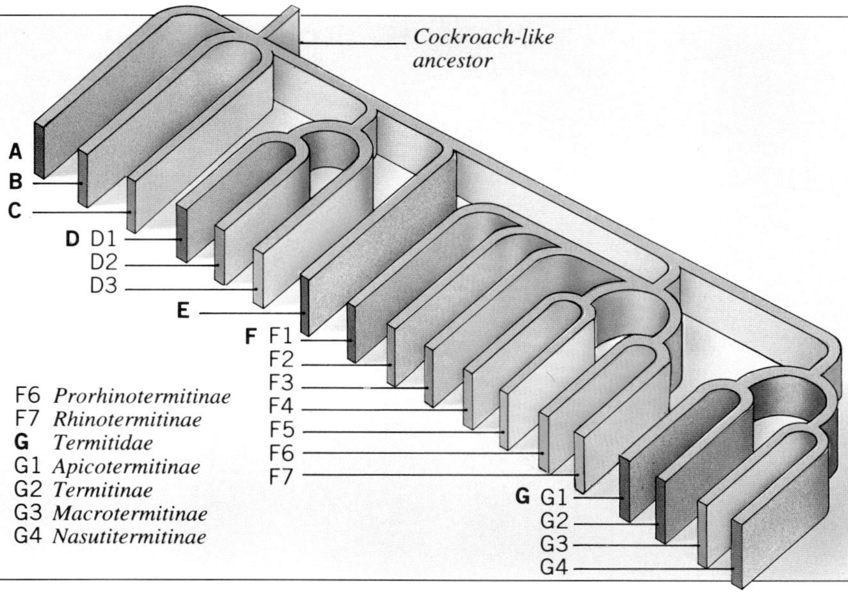

A *Mastotermitidae*
B *Kalotermitidae*
C *Hodotermitidae*
D *Termopsidae*
D1 *Termopsinae*
D2 *Porotermitinae*
D3 *Stolotermitinae*
E *Serritermitidae*
F *Rhinotermitidae*
F1 *Stylotermitinae*
F2 *Termitogetoninae*
F3 *Psammotermitinae*
F4 *Heterotermitinae*
F5 *Coptotermitinae*
F6 *Prorhinotermitinae*
F7 *Rhinotermitinae*
G *Termitidae*
G1 *Apicotermitinae*
G2 *Termitinae*
G3 *Macrotermitinae*
G4 *Nasutitermitinae*

8 Senses and Communication

Biological Clocks
Bioluminescence
Vision
Sonar
Hearing
Smell and Taste
Special Senses
Nervous Systems
Intelligence
Communication
Scent Communication
Sound Communication
Mammal Societies

Keeping Time

How animals use biological clocks

Sheep that are moved from Australia to North America will eventually change their breeding season by six months. Sheep that live in the Southern Hemisphere tend to give birth in October, when it is spring, but when they are brought across the equator, their breeding patterns change and they will then give birth in May, when it is spring in the Northern Hemisphere. Precisely how and why the breeding cycle changes no one is sure, but we do know that all kinds of organisms, from the tiniest alga to the largest whale, have biological clocks that run their lives.

Biological clocks are as old as life itself. From the earliest days of life on this planet, living creatures have had to contend with a regularly changing environment, dictated by the daily fluctuations in light and temperature due to the rotation of the Earth, and by the seasonal changes due to the movement of the Earth around the Sun and the different phases of the Moon, with their powerful effect on tides. So it is not surprising to find that even simple cells, as well as the bodies of multicellular organisms, have their own internal clocks, which regulate many of the biochemical reactions of the cells.

Daily and seasonal routines

Diurnal rhythms are very common – many flowers open in the morning and close at night, and most animals are either diurnal or nocturnal in their activity. By confining their activities to a particular time of day, animals avoid competition with one another, so more species can occupy the same habitat. Internal clocks allow animals that live in darkened burrows to emerge at the right time of day without needing to surface to check the light.

The changing seasons bring new problems: cold or dry seasons bring a reduction in food and water, and hostile weather conditions. Animals and plants need some means of preparing themselves in advance of these changes in time to ensure their survival.

Seasonal cycles are also important for reproduction, so that the young are born and seeds produced when conditions are suitable for their growth. For larger animals this may mean mating in autumn in order to give birth in spring, or migrating to breeding grounds where food is abundant and the climate kinder to their young.

Moon and sea

The regular environmental changes on the seashore are among the most violent on Earth. At low tide seashore creatures are exposed to solar radiation, dramatically increasing salinity as the remaining sea water evaporates, and the possibility of attack from birds and other predators; while at high tide they are buffeted by the waves and inundated with salt water. Many seashore animals, therefore, need advance warning of the ebb and flow of the tides in order to burrow or seek refuge in crevices in time.

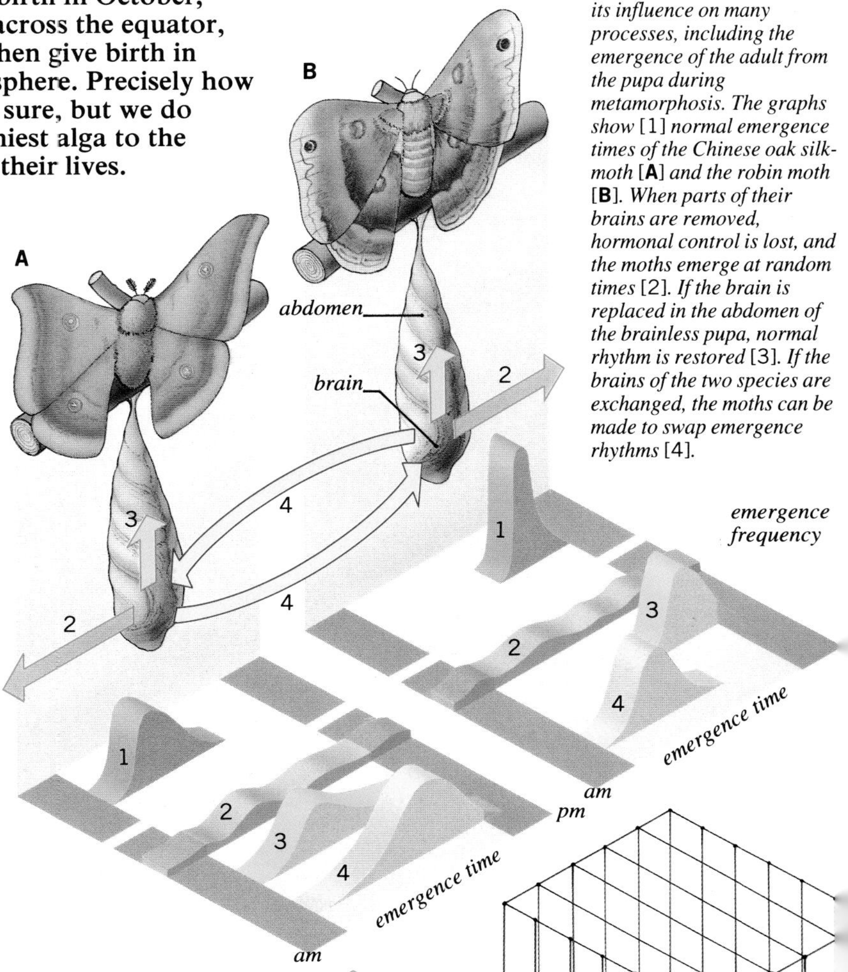

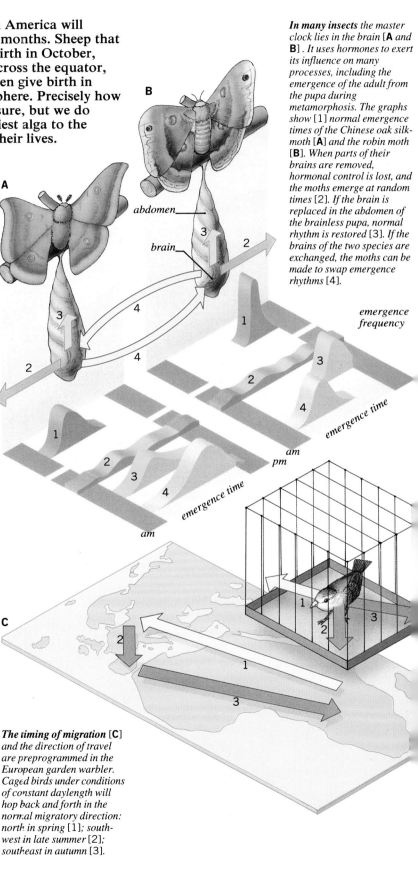

***In many insects** the master clock lies in the brain* [**A** *and* **B**]. *It uses hormones to exert its influence on many processes, including the emergence of the adult from the pupa during metamorphosis. The graphs show* [1] *normal emergence times of the Chinese oak silk-moth* [**A**] *and the robin moth* [**B**]. *When parts of their brains are removed, hormonal control is lost, and the moths emerge at random times* [2]. *If the brain is replaced in the abdomen of the brainless pupa, normal rhythm is restored* [3]. *If the brains of the two species are exchanged, the moths can be made to swap emergence rhythms* [4].

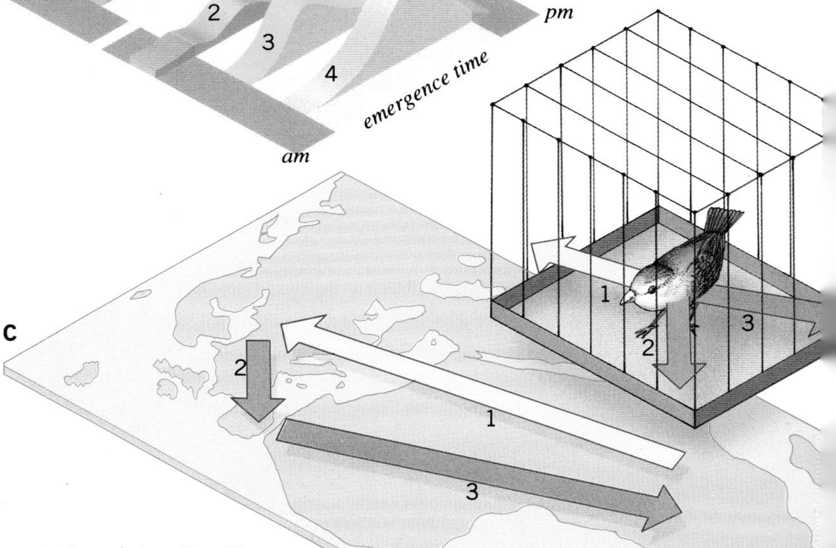

The timing of migration [**C**] *and the direction of travel are preprogrammed in the European garden warbler. Caged birds under conditions of constant daylength will hop back and forth in the normal migratory direction: north in spring* [1]; *southwest in late summer* [2]; *southeast in autumn* [3].

Humans have a daily, *or "circadian," rhythm of sleep, in which they are generally awake by day and asleep by night. This graph* **[D]** *shows the changing behavior of a man who stayed in a cave away from normal cycles of day and night. The cycle of waking and sleeping gradually changes to a 25-hour cycle, interspersed with erratic long cycles. However, the basic pattern persists for many months, showing that it is controlled by an internal clock. Under normal conditions, the clock is tuned to the 24-hour day by the normal cycle of light and dark or, in modern city dwellers, by other signals connected with the daily routine (including mechanical clocks). When people fly from one side of the world to the other, a new routine is imposed. Furthermore, the natural circadian clocks governing various bodily functions readjust at different rates, so the body's metabolic synchronization is lost. During this time the person experiences jet lag until the body readjusts itself to the new daily routine. This pattern is typical of biological clocks – a basic rhythm of approximately the right period is "trained" to match the environmental changes by external signals.*

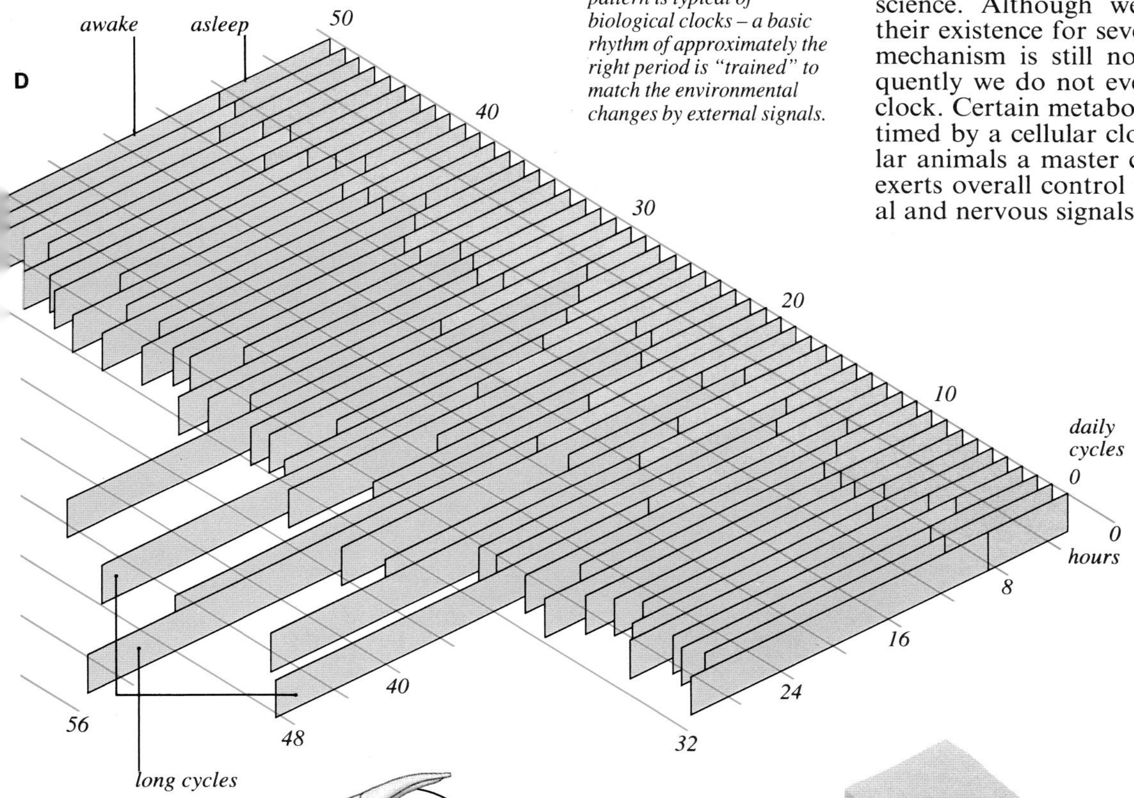

Many marine animals also use these cycles for reproduction. Some deep-water fish spawn only at certain states of the tide. They are responding to changes in temperature, pressure and salinity associated with tides, rather than to moonlight. The grunion, a small silvery fish, gathers in huge numbers on Californian beaches to spawn at the peak of the spring tides.

The continuing mystery

Biological clocks present a great challenge to science. Although we have been aware of their existence for several centuries, the exact mechanism is still not understood, and frequently we do not even know the site of the clock. Certain metabolic functions seem to be timed by a cellular clock, while in multicellular animals a master clock in the brain often exerts overall control via a range of hormonal and nervous signals.

Many shore crabs, *like the European shore crab* **[G]**, *use a tidal clock to regulate their daily activity pattern. They emerge to feed as the tide comes in, but retreat to a burrow or crevice as the tide recedes to avoid being seen by predators – especially shore birds like the crab plover. Other scavenging crabs that feed at low tide may use a double timer, emerging to feed at low tide only if this occurs during the hours of darkness, when they are less visible to predators. Since the tides are determined by the phases of the Moon, many of these animals use a lunar clock, responding to changes in the intensity of moonlight.*

Fiddler crabs **[E]** *show an unusual daily rhythm for shore animals. They emerge to feed at low tide and retreat into their burrows at high tide. It is risky for them to be exposed because they are then vulnerable to predators, but they must come out at low tide in order to perform their mating displays. The flatworm* Convoluta **[F]** *has both a tidal and a daily clock. It derives essential carbohydrates from symbiotic green algae that live in its tissues. However, these must have light to photosynthesize, so it moves to the surface of the sand only at low tide during the day. At high tide and at night it remains buried in the sand.*

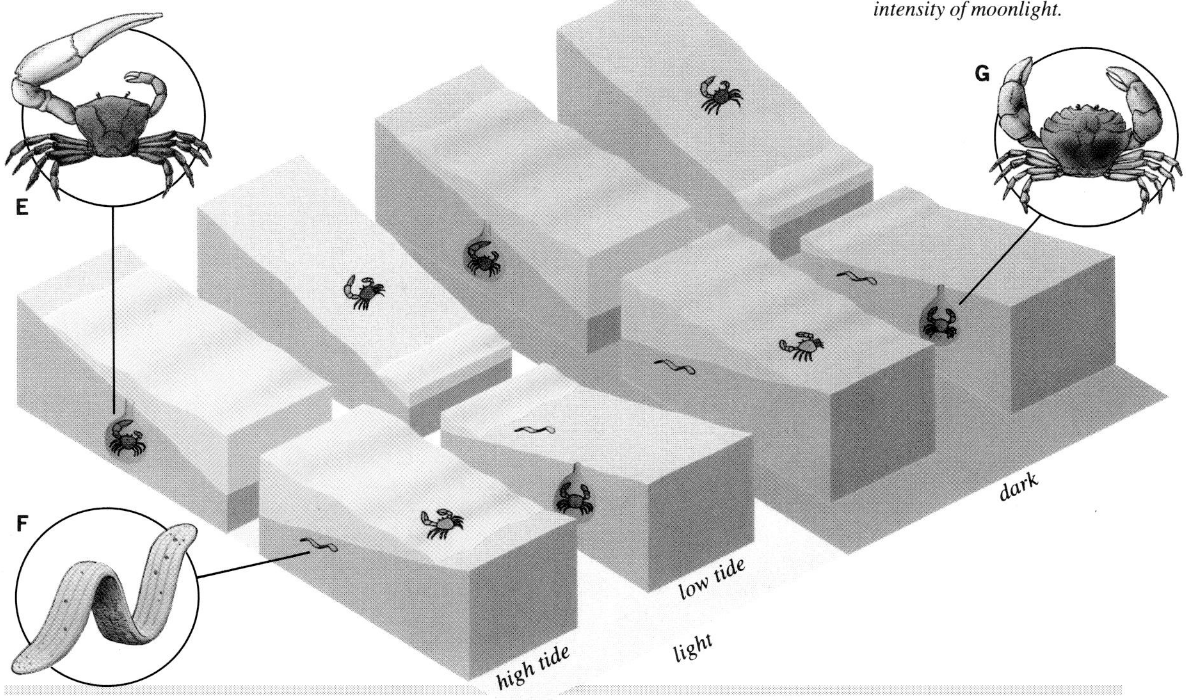

Special Senses 286 Seasons 86

Glowing in the Dark

Why some organisms are luminescent

The light of the flashlight fish can be seen from 100 feet (30 meters) away, even in the dark depths of the oceans – it is the brightest light produced by any living organism. For an organism to produce light visible over a distance, and to do so without at the same time producing heat, is hard to imagine – but species of fish, shrimp, squid, centipede, beetle, fungus and bacterium can all do it. Such light – bioluminescence – may be produced as a continuous glow, or as a series of flashes, and in some species can be switched off and on at will to produce spectacular patterns.

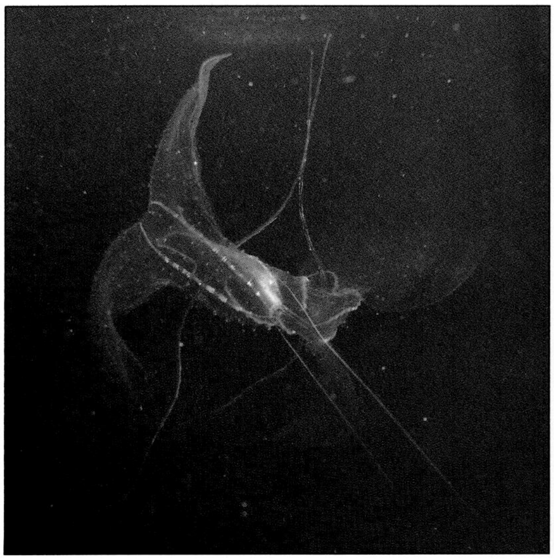

Comb jellies *of the genus* Beroe *have a yellow tint to them due to the algae in their body tissue. However, chemical reactions in the walls of their eight digestive cavities also generate brightly colored spots of bioluminescent light, which flash through the animal's transparent skin when it is disturbed, and which may also help to attract prey. Unlike those species that have tentacles, which feed on plankton by catching them in their sticky, threadlike tentacles, comb jellies of the genus* Beroe *feed on other comb jellies by sucking them into their large mouths.*

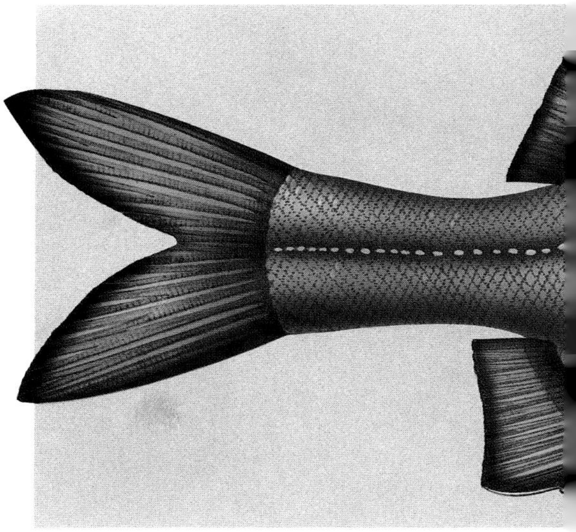

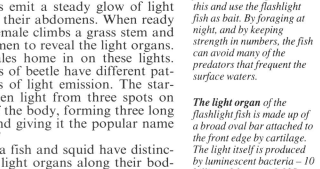

The flashlight fish *is a small fish that grows to a length of about 3.5 in (9 cm). It lives in dark crevices in the coral reefs of the Red Sea and Indian Ocean. On dark nights, large numbers of these fish move to the surface to feed. The light organs under each eye are thought to act as lures for the fish's prey – in parts of the eastern Indian Ocean the native farmers have long realized this and use the flashlight fish as bait. By foraging at night, and by keeping strength in numbers, the fish can avoid many of the predators that frequent the surface waters.*

The light organ *of the flashlight fish is made up of a broad oval bar attached to the front edge by cartilage. The light itself is produced by luminescent bacteria – 10 billion of them per 0.035 fluid oz – which are held inside special polygonal, tubelike compartments. Loops of fine blood capillaries provide nutrients and oxygen for the bacteria, essential if they are to emit*

The light produced during bioluminescence is generated during chemical reactions inside living cells and tissues. Such reactions generally involve the breakdown of a protein called *luciferin* in the presence of a biological catalyst or enzyme, *luciferase.* In most species the reaction requires oxygen. During this breakdown, energy is released in the form of light. A different reaction – one not requiring an enzyme – occurs in some species of jellyfish, shrimps and marine worms; this involves a luminescent protein that emits light when mixed with calcium or iron ions, or with oxygen.

The fact that most bioluminescent reactions in this way involve oxygen has led to the suggestion that the ability to luminesce may originally have arisen as a means of getting rid of the gas. This would have been important to primitive organisms that had evolved in an atmosphere largely devoid of oxygen, a gas they could not use. Once plants evolved and oxygen began to accumulate in the atmosphere, these organisms would then have had to find ways to adjust their metabolism.

The source of the light

In most species the light-generating reaction occurs in their own tissues and cells. This includes squid, deep-sea fish, some marine worms and some marine algae (*dinoflagellates*), which flash when disturbed, producing the so-called phosphorescent seas of tropical and subtropical regions. In some of the more advanced species the reaction is confined to special light organs, in which the light is focused by a lens and reflector, and colored by passing through certain pigments. Light emission by some of these animals may be controlled by the nervous system.

An exception to the general rule, however, are the few fish of both deep-sea and shallow waters – and some shrimps – that use light emitted by symbiotic bioluminescent bacteria, which live in special pouches. Some light producers store up energy that they gain from external light sources in order to reemit it later. These animals, which are said to phosphoresce, store the light energy as electronic energy in their bodies.

The light of love

Many animals use light displays to attract mates by night, or in the darkness of the ocean depths. Glowworms, which are really beetles, are some of the most familiar. The wingless females emit a steady glow of light from the tips of their abdomens. When ready to mate, each female climbs a grass stem and twists her abdomen to reveal the light organs. The winged males home in on these lights. Different species of beetle have different patterns and colors of light emission. The starworm emits green light from three spots on each segment of the body, forming three long rows of light, and giving it the popular name of "night train."

Many deep-sea fish and squid have distinctive patterns of light organs along their bodies. These patterns differ even in closely related species, and are thought to aid species recognition during courtship. Some female deep-sea anglerfish flash their lures to attract males as well as prey. Unsuspecting smaller fish mistake the modified luminous structure for a worm, but are then themselves eaten.

Flashing Flies

Both male and female fireflies flash lights to attract mates. The males fly close to the ground and emit flashes of light while performing aerobatics, which produces distinctive light patterns in the air. Each species has its own characteristic pattern. However, species recognition is more specific than this – the female (who responds only to a male of the same species) must respond with pulses of light at exactly the correct time interval after the male's flashes before the male will fly over to investigate. But things are not always what they appear to be. In North America the signal of a female *Photinus* firefly is mimicked by the female of a larger species, *Photuris.* When a *Photinus* male flies past flashing, the female *Photuris* responds with the correct signal, but as he flies in to mate she seizes him and eats him.

***North American* Photinus** *fireflies advertise themselves with specific light patterns:* P. brimleyi [1], P. collustrans [2], P. ignitus [3], P. granulatus [4]. *The light organs of fireflies are located on the underside of the abdomen and thorax, and arranged in different patterns according to species.*

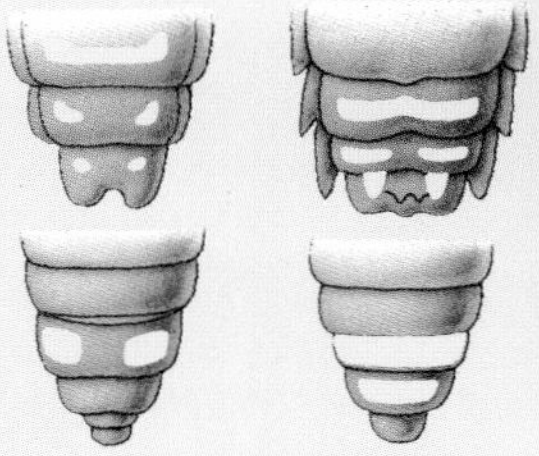

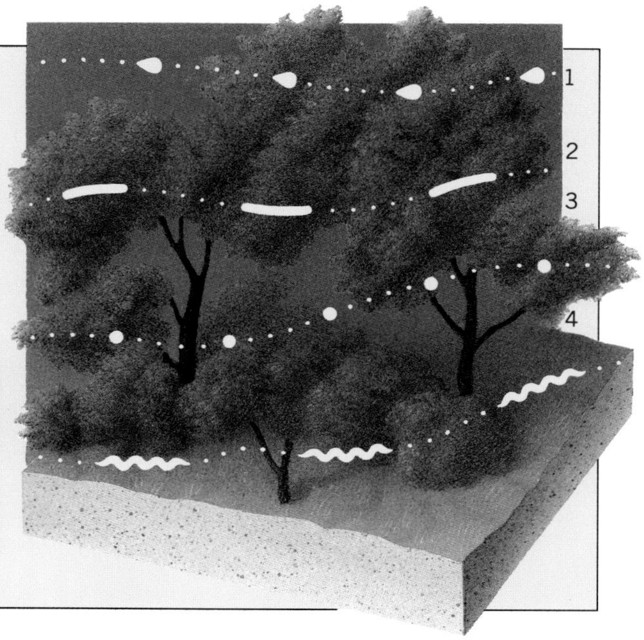

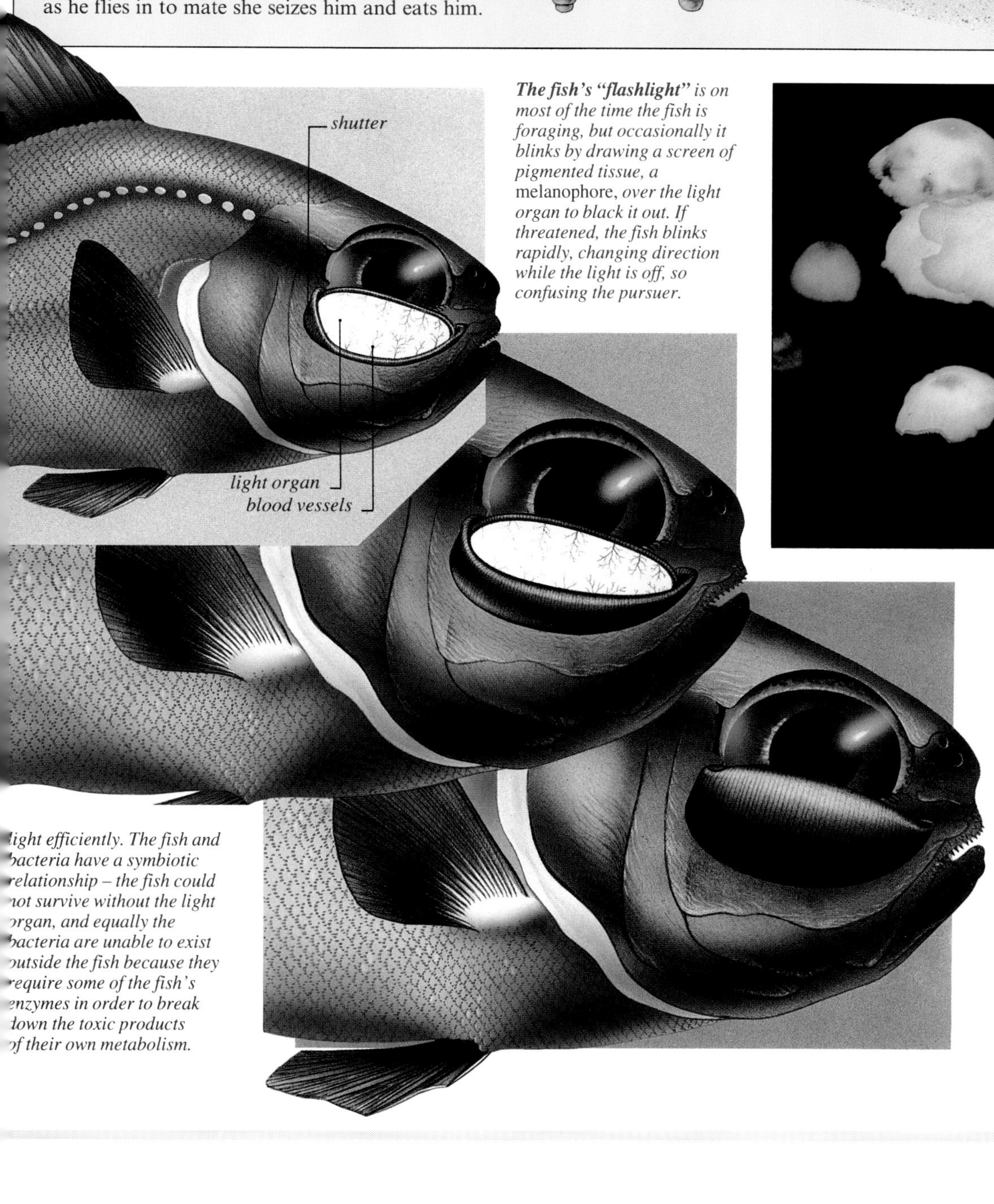

The fish's "flashlight" *is on most of the time the fish is foraging, but occasionally it blinks by drawing a screen of pigmented tissue, a* melanophore, *over the light organ to black it out. If threatened, the fish blinks rapidly, changing direction while the light is off, so confusing the pursuer.*

ight efficiently. The fish and bacteria have a symbiotic relationship – the fish could not survive without the light organ, and equally the bacteria are unable to exist outside the fish because they require some of the fish's enzymes in order to break down the toxic products of their own metabolism.

Luminous fungi *are a common feature in tropical forests, where they are found growing on rotting trees. These fungi come from Costa Rica, but they are also frequently found in temperate woodlands. Their luminosity is probably an adaptational hangover from the time when fungi had to get rid of excess oxygen in the atmosphere that they could not use. Some scientists, however, also believe that their luminosity attracts insects at night, which helps spore dispersal. Young females of some tropical peoples use glowing fungi as a body paint in the hope that it will attract prospective partners.*

Getting the Picture

How animal vision works

The chameleon has a unique way of spotting potential attackers or prey – its eyes, situated on small turrets, can move independently of each other, so that one can look forward while the other looks back. In some birds of prey the eyes fill the orbits so completely that they can scarcely move them around, and the birds have to turn their heads instead, yet they can still see in detail a mouse moving in the grass hundreds of yards below them. At the other end of the scale, even some of the tiniest creatures – the single-celled protozoans – are able to detect light and swim toward it.

In most animals, the light-sensitive cells are grouped into organs called eyes, which have specialized reflecting surfaces and lenses for maximizing the amount of light received. Light sensitivity is achieved using special pigments – complex chemicals that absorb packets of light energy (*photons*) and convert them into electrical nerve impulses. The subtlety of color detectable by the human eye relies on analyzing the information received from a combination of three different color pigments together with the intensity of light absorbed by each.

Eye variations

Not all animals see as we do. The simple but numerous eyes of the scallop, for example, cannot relay a large, detailed image, but are extremely sensitive to movements, perhaps of a potential predator. The compound eyes of insects and crustaceans are made up of many individual eyes called *ommatidia*. Compound eyes were once thought to produce a mosaic image rather like magnified newsprint, but in fact the image is probably more blurred due to the slight overlap of fields of view between adjacent ommatidia. Many such eyes together produce "flicker vision," for perceiving small movements in their field of vision.

Aquatic and marine animals need specially adapted eyes, because the *refraction* (bending) of light at the eye's surface is less in water than in air, and a more spherical eye shape is needed. A few species, such as water beetles like the water boatman and the four-eyed fish of South America, have eyes in two parts: one set of eyes is adapted for underwater vision, the other for above the water.

Judging speed and distance

More complex animals, including insects, crustaceans and vertebrates, have paired eyes. This gives them a wide field of vision, while at the same time allowing the two eyes to converge on an object. The two views of the object seen by the two eyes can be put together by the brain to produce a three-dimensional image, which is important in judging distance and the speed of moving objects. Animals for which such skills are important, particularly predators, like birds of prey and big cats, tend to have the eyes placed well forward on the head to give a large overlap between the two fields of view. Prey species such as

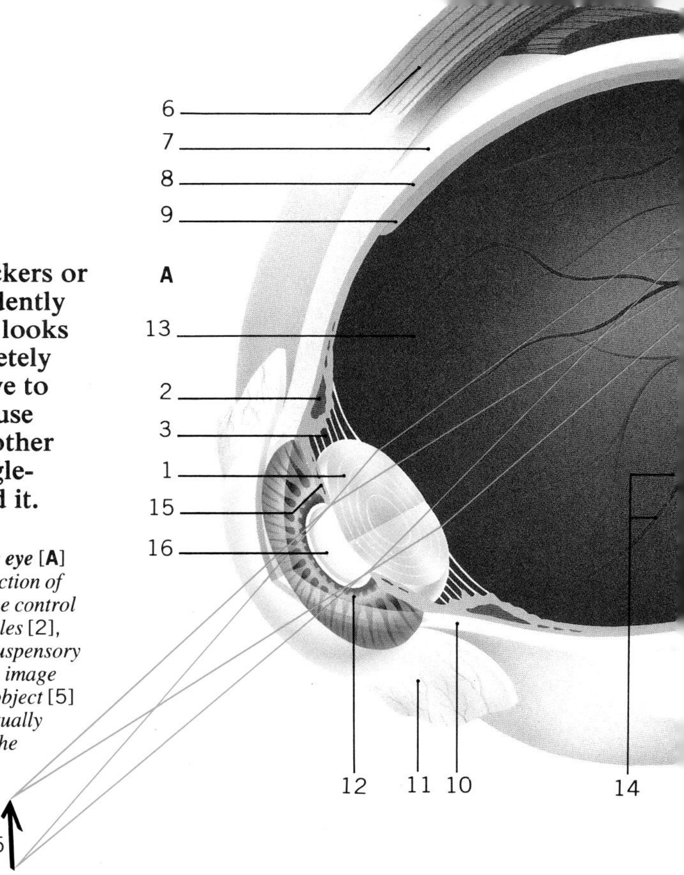

***Light entering the eye* [A]** *is focused by the action of a lens* [1] *under the control of the ciliary muscles* [2], *which act on the suspensory ligaments* [3]. *The image* [4] *formed by an object* [5] *on the retina is actually upside down, but the brain is able to correct this.*

***The eyeball* [A]** *is held in place in the orbit by muscles* [6] *that also allow it to move. It is covered in three layers of tissue: the* sclera [7], *a tough, fibrous coat; the* choroid [8], *which supplies nutrients, and is pigmented to reduce internal reflection; and the* retina [9], *where the light-sensitive cells are located. The front of the eye is protected by the transparent* cornea [10] *and* conjunctiva [11]. *The* aqueous [12] *(behind the cornea) and* vitreous humors [13] *help keep the shape of the eye, and also contain* blood vessels [14]. *The* iris [15] *can be dilated or contracted by muscles to control the amount of light entering the eye through the* pupil [16]. *The* optic nerve [17] *carries the visual information to the brain. Where it exits the retina there are no receptors and there is a blind spot* [18]. *Most of the time, however, the brain can compensate.*

***The visible spectrum* [C]** *of deep-sea fish is limited to a little blue light. Other fish have a broad range. Many snakes can see far-red, using special pit organs, as well as ultraviolet. Birds and insects may also see into the ultraviolet. The primate range, including humans, lies between red and blue.*

The light-sensing cells *in the vertebrate retina* **[B]** *are the* rods [1] *and* cones [2] – *highly specialized nerve cells. The outer segment is comprised of membranous disks* [3] *containing a light-sensitive pigment. The inner segment has a branched base* [4] *that links to nerve fibers. If sufficient light* [5] *is absorbed by the pigments, an electrical signal is produced by the adjacent nerve fiber. Before reaching the brain via the optic nerve* [6], *messages pass through a series of neurones in the retina –* horizontal *cells* [7], bipolar *cells* [8], amacrine *cells* [9], ganglion *cells* [10] *– which organize the sensory*

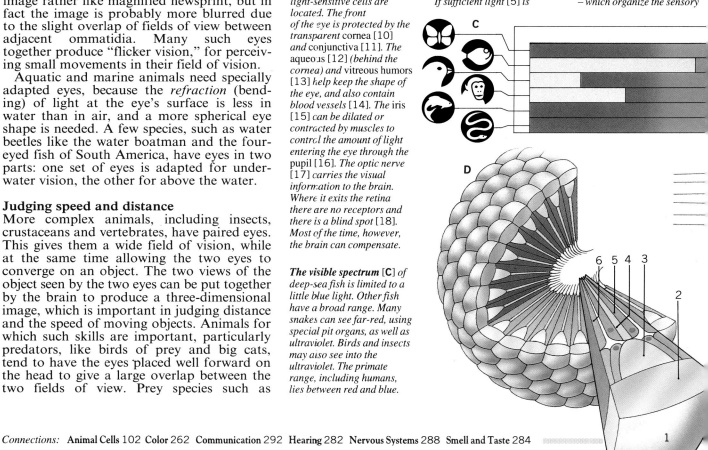

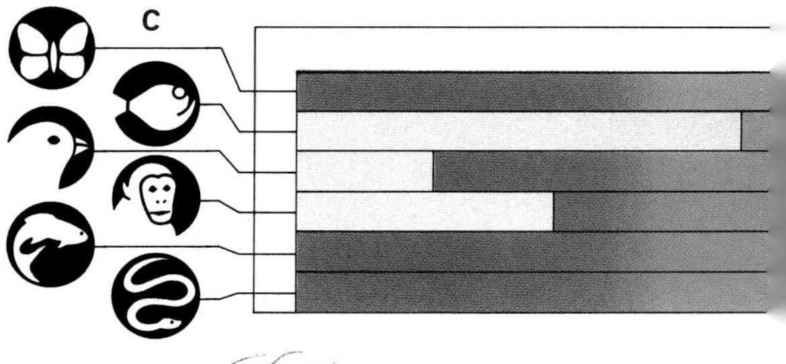

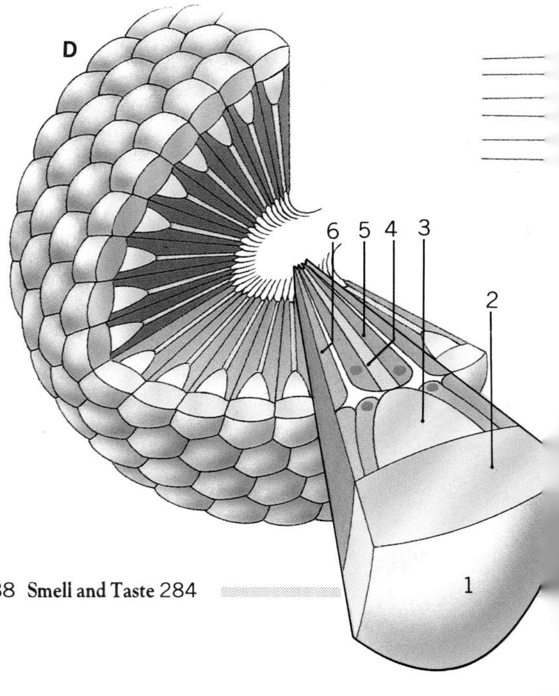

Connections: Animal Cells 102 Color 262 Communication 292 Hearing 282 Nervous Systems 288 Smell and Taste 284

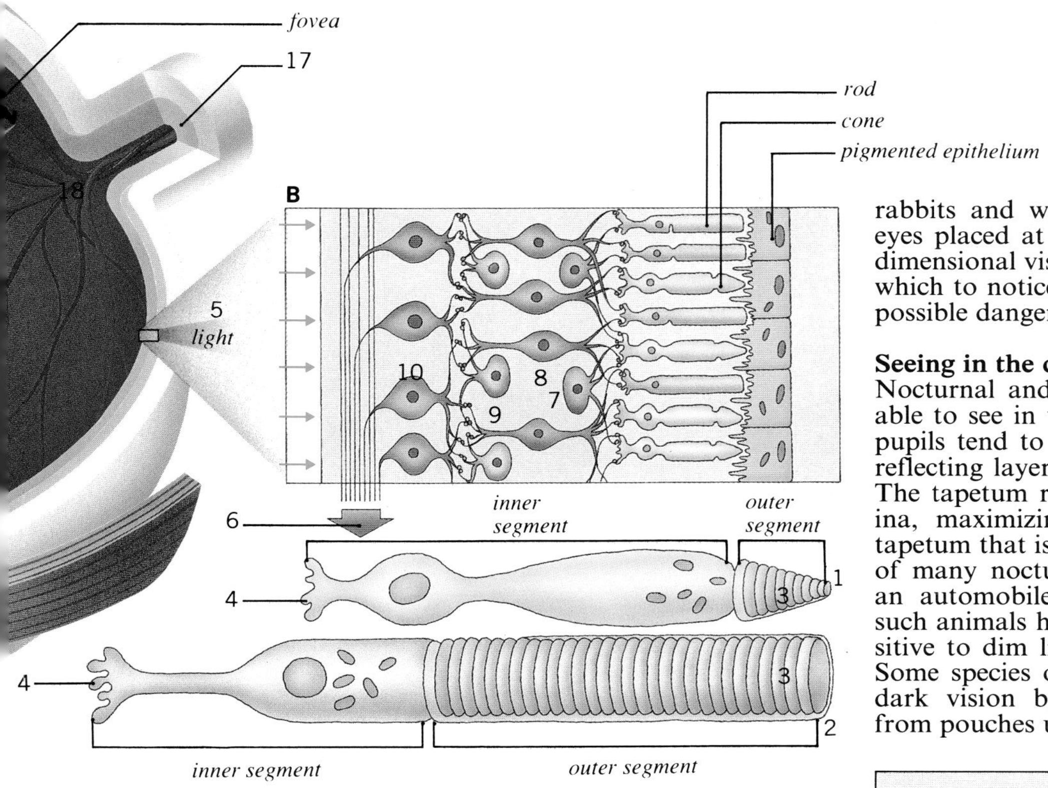

rabbits and wading birds tend to have the eyes placed at the side, giving limited three-dimensional vision, but a wide field of view in which to notice movements that might signal possible danger.

Seeing in the dark

Nocturnal and deep-sea animals need to be able to see in very dim light. Their eyes and pupils tend to be large, and there is often a reflecting layer of *tapetum* behind the retina. The tapetum reflects light back into the retina, maximizing absorption, and it is the tapetum that is responsible for the "eyeshine" of many nocturnal animals when caught in an automobile's headlights. The retinas of such animals have many rods, which are sensitive to dim light, and relatively few cones. Some species of deep-sea fish augment their dark vision by producing light themselves from pouches under the eyes.

information. Rods are found throughout the retina, except at the fovea, *and are sensitive to dim light. They contain only one type of pigment, so do not detect color. There are three types of cones, all stimulated by bright light. Each contains a different pigment and is sensitive to different wavelengths of light. The degree to which different cones are stimulated gives the brain information about color. Cones are highly concentrated at the fovea, giving great detail here, where most images are focused. Each cone has its own connection to the brain, so that very detailed information is received.*

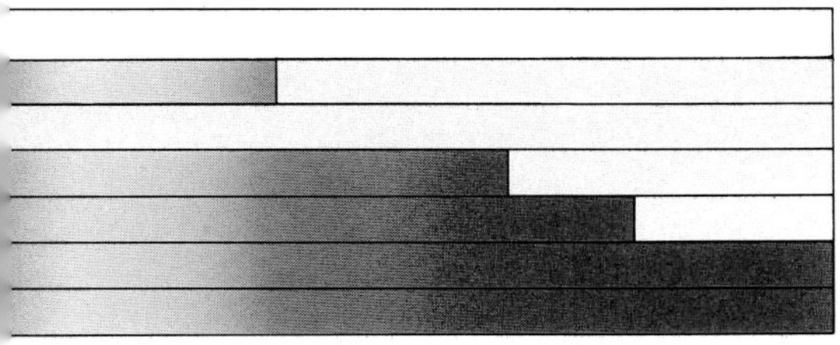

Insect compound eyes **[D** *and* **E]** *are made up of many individual units (from one up to 28,000) called ommatidia, which are arranged in a hemispherical fashion. Each ommatidium* [1] *has its own cornea* [2] *and lens* [3]. *A light-sensitive region called the* rhabdom [4] *contains the visual pigment. This is surrounded by retinal cells* [5] *that transmit the electrical stimulus from the excited visual pigment to the brain. Cells containing screening pigment* [6] *prevent light entering one ommatidium from entering its neighbor. In the* apposition *eye of daytime insects* **[D]** *this means each ommatidium can only receive light from a small part of the whole field of view. Thus the whole image formed in the brain consists of the overlap of many adjacent spots of light. Nocturnal insects have a* superposition *eye* **[E]**, *which is constructed in a very similar way to the apposition eye, and acts in the same way during the day* **[F]**. *In dim light, however, the pigment withdraws toward the outer surface of the eye* **[G]**, *thus allowing diffracted light to reach the rhabdom from adjacent ommatidia as well. This produces an image that is brighter than it otherwise would be, though it may be less distinct.*

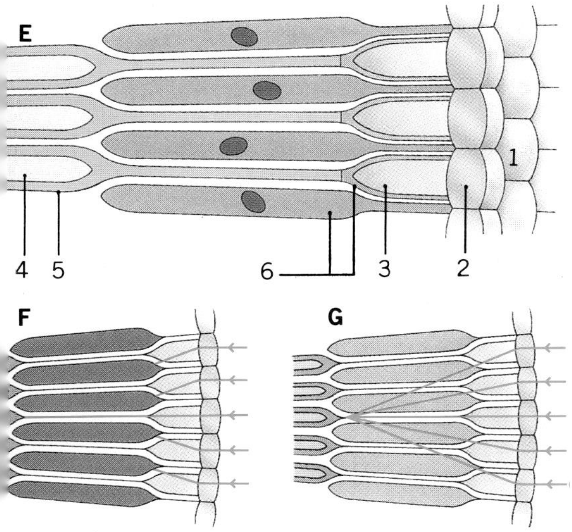

Eye Designs

The simplest eyes, like those of flatworms [1], are just cups lined with a light-sensitive retina. In tube-worm eyes [2] each light receptor lies at the bottom of a pigmented tube. It receives only light from a particular angle and functions as a basic compound eye. The mirror eyes of scallops [3] form an image by reflection of incident light, in a similar way to a reflecting telescope. Shrimps and lobsters have a *superposition* eye [4], in which mirrors channel light to form a single, particularly bright, image.

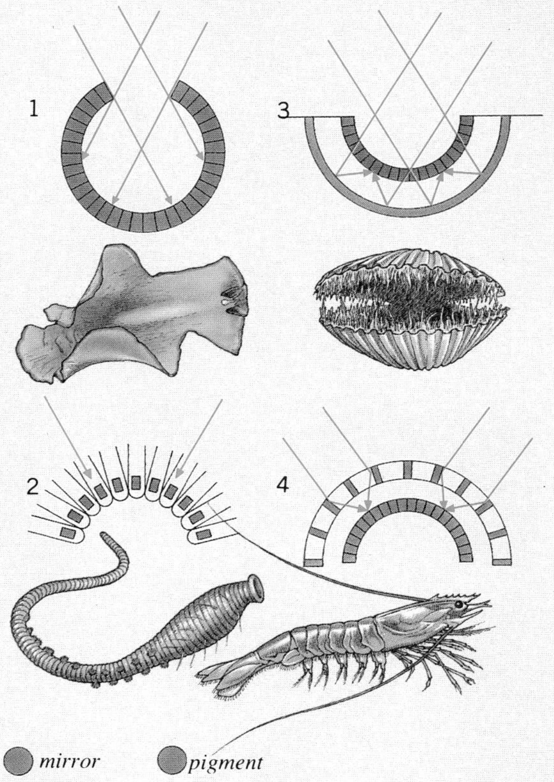

Seeing With Sound

How animals use sonar and sound waves

Divers who swim in front of a school of dolphins describe sensing an intense vibration through the body. Each dolphin is "looking" at the diver using a narrow, focused beam of high-energy ultrasound. The sound bounces off the diver and sends back echoes that yield a remarkable amount of information. Bats and some species of seal have a similar technique, while less sophisticated versions of echolocation, in which the sounds are audible to human ears, can be used by cave-dwelling birds to find their nests in the dark. Even shrews and tenrecs may use sonar.

__One feature of bats__ that is particularly variable is their faces. The many differences are often explained by the way in which the ultrasound systems bats employ is generated. Some bats, such as the Mediterranean horseshoe bat **[A]**, *emit their ultrasound through their noses. To enable them to do this they have an elaborate nose, which incorporates a fold of skin forming an organ called a "nose leaf"* [1]. *Others, such as the mouse-eared bat* **[B]**, *"screech" their ultrasound through open mouths. In these species the nose is simple and resembles that of a normal insectivore.*

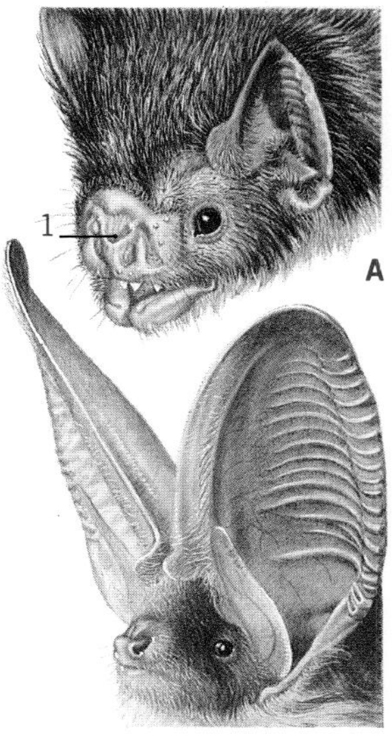

A

B

In many ways ultrasound is a better sense than vision – it literally penetrates the body. One dolphin "looking" at another, for example, can probably tell if it has eaten recently, or if it is unwell. More importantly, echolocation can be used where vision is impossible, particularly in the murky but fertile seas close to the coasts, which are often the richest hunting grounds for large fish-eaters.

In the muddy waters of rivers such as the Indus and the Amazon, river dolphins have come to depend so heavily on echolocation that they are virtually blind, only distinguishing light from dark. And the largest of the toothed whales, the sperm whale, puts echolocation to work in the dark waters below 3,300 feet (1,000 meters), where it pursues deep-sea squid. For bats, echolocation gives access to moths and other night-flying insects, which originally became nocturnal to escape from insect-eating birds.

Hunting for the right frequency

Both bats and dolphins use high-frequency sound outside human hearing range because it gives a more precise echo. (If the wavelength of the sound is larger than the reflecting object, the echo will be weak or nonexistent.) A human voice has a wavelength of about 14 inches (35 centimeters), which would be useless to a bat catching moths.

Dolphins pursue far larger prey, and can manage with a minimum wavelength of half an inch or so. But to achieve this, they have to work much harder, because sound travels five times faster in water than in air. This means that for the same frequency of sound emitted, the wavelength is five times longer. So dolphins emit at frequencies up to 270 kHz (270,000 cycles per second) compared to a maximum of 160 kHz for bats. Yet they also emit quite low frequencies, down to 0.25 kHz, which gives a wavelength of 22 feet 4 inches (6.8 meters). These sounds are used to survey the seabed and are probably involved in navigation.

Navigating with sound

Apart from the whales, dolphins and bats that can navigate using ultrasound, there is also one species of bird that uses a very simple form of echolocation with which to navigate. The oilbirds of South America live in caves. At sunset they emerge to seek out fruit, before returning to their caves to feed. To navigate in the darkness of the caves, which are sometimes half a mile deep, they emit clicking sounds, audible to the human ear, from which they can build a mental picture of their surroundings, and so avoid flying into the cave walls and each other.

Bat trackers

The echolocation signals of bats differ widely, depending on the sort of prey they take, and their habitat. Bats that hunt in the open produce pulses of sound that fluctuate from a high frequency down to a lower one. Horseshoe bats, however, hunt moths in dense foliage. These bats have a two-part sound pulse. The first part is a constant frequency. When a flying moth is caught in this constant frequency, the flapping of its wings sets up a characteristic echo that is distinguishable from that of rustling leaves.

__A dolphin__ **[D]** *generates clicks and whistles by trapping air, by means of the blowhole, when it surfaces to breathe. The trapped air can then be squeezed through a complex series of valves located in the* nasal sac, *found beneath the blowhole to create sound. It is thought that the sounds generated by the valves are then focused by the* melon, *a large, fatty organ located in the dolphin's forehead, which acts rather like a "sound lens." The returning sounds, once they have bounced off the target, are picked up by a fatty channel in the hollow lower jaw and transferred, via an unusually thin section of bone, the* panbone, *and on to the middle ear.*

A dolphin's sonar is highly advanced. For example, a dolphin can not only single out a specific fish from a shoal [1] *by interpreting the reflected waves from the fish's swimbladder, but it is also capable of deciphering information from more than just one fish* [2], *enabling it to quantify the shoal.*

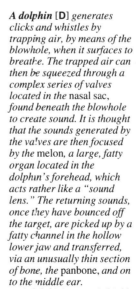

D

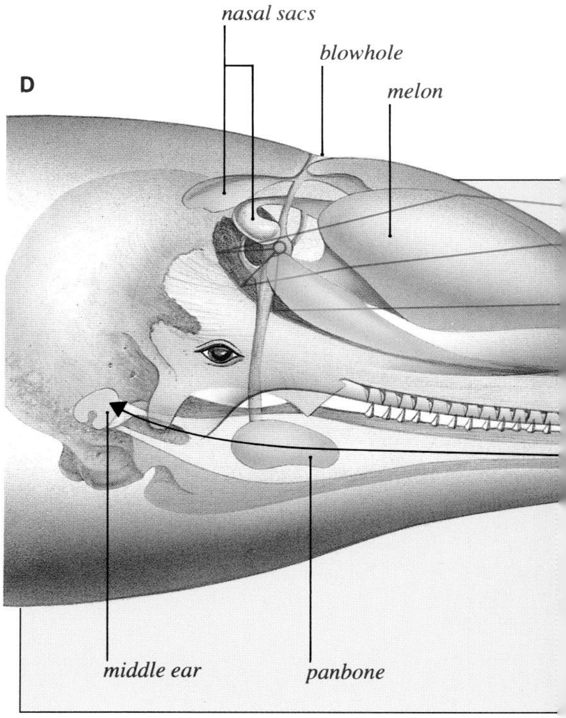

Connections: **Bird and Bat Flight** 224 **Birds** 132 **Cave Life** 308 **Mammals** 136 **Migration Techniques** 230 **Sound Communication** 296 **Special Senses** 286

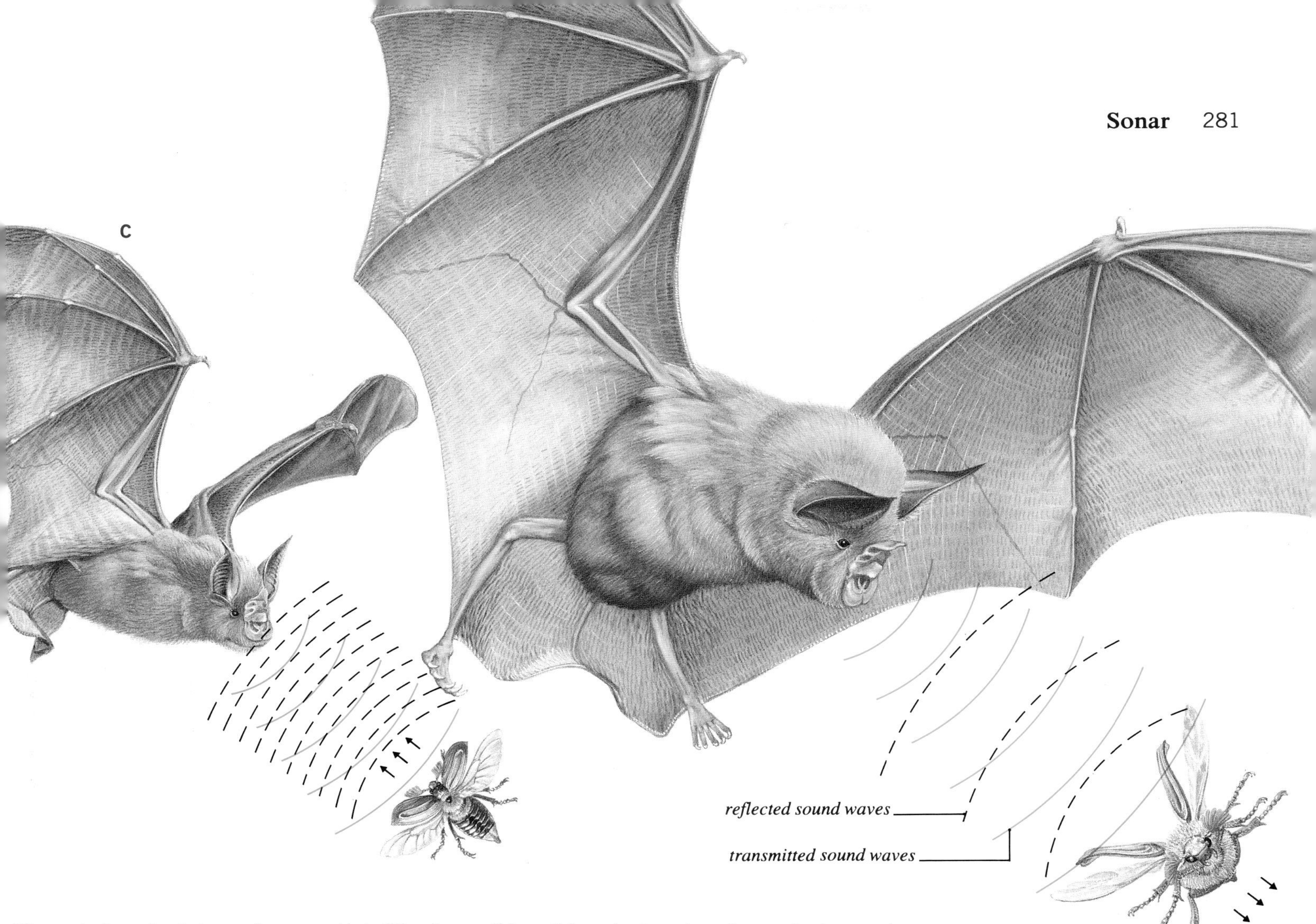

The greater horseshoe bat [**C**] *is an efficient hunter, capable of catching moths on the wing with its sharp teeth or in its claws. The bat's echolocation signals (sound pulses) are produced in its larynx. The bat can find out information about its prey's relative position and velocity by emitting a constant high-frequency whistle. When the prey, such as a moth, is caught in this constant frequency of sound, an echo bounces back off the moth's body, returning to the bat. If the pulse bounces back from a moth that is moving away from the bat, the note drops. This is known as downward* Doppler shift. *If the moth is moving toward the bat, then the note rises (upward Doppler shift). The Doppler effect is the same aural sensation experienced when a patrol car with its siren on drives past a standing figure. The note of the siren rises as it approaches the figure and then falls as the car passes.*

Some moths, however, have developed ways of evading a bat's attacking run. Certain species can hear the bat's pulses and can take evasive action; other species can actually send out a high-pitched squeek that resembles that of the bat and which seems to jam the bat's sonar equipment.

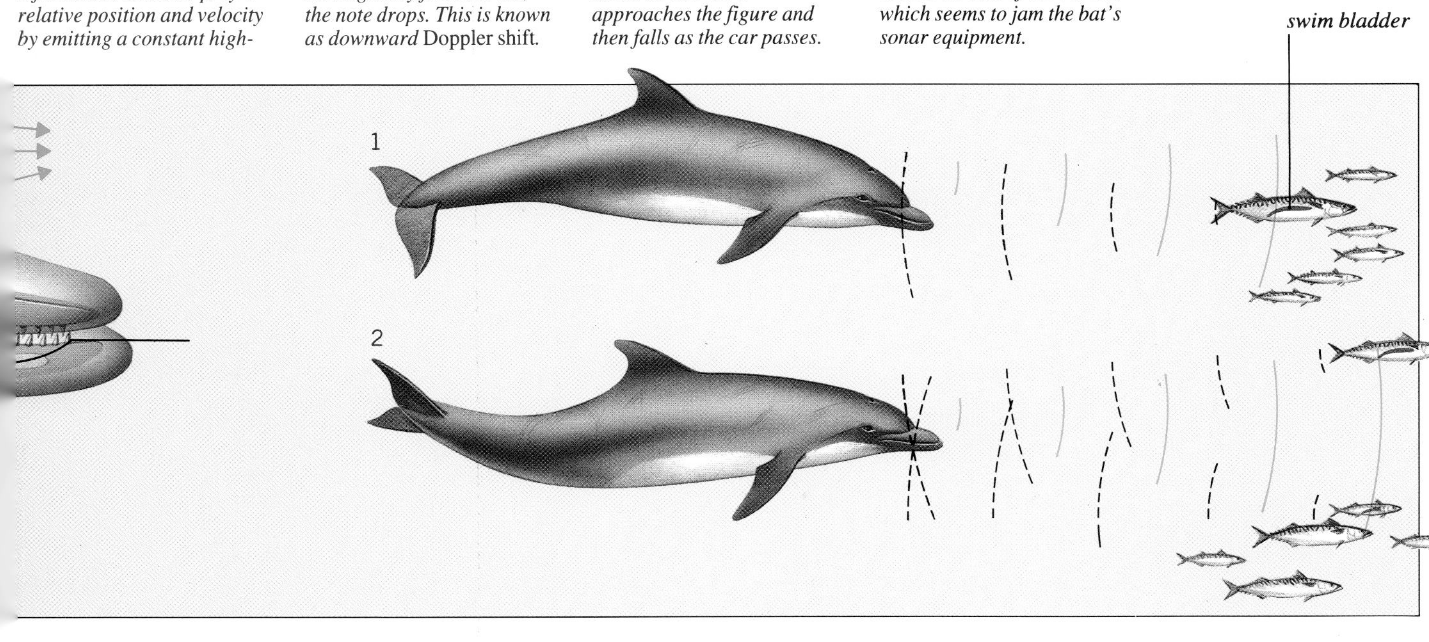

Tuning In

How animals hear

Elephants can communicate with their stomachs. They emit low growling noises that travel great distances across vast African plains. The sounds are of such a low frequency that the human ear cannot detect them. Our ears have a certain frequency range, and the thousands of animal noises outside that range are inaudible to us. Animals "hear" in a variety of ways. Some animals, for example, have "ears" on their knees, others on their feet. Animals that hunt at night rely heavily on hearing in their search for prey, while some underwater creatures use loud sounds to disorient prey.

The most highly developed ear in the animal kingdom belongs to the mammals. The mammalian ear is divided into three basic sections: the outer, middle and inner ear. The outer ear consists essentially of the *pinna,* which funnels sound into the ear, and the ear canal, along which the sound waves travel until they reach the eardrum. The size and shape of the pinna, however, is not a reliable guide to the sharpness of an animal's hearing.

An amplifier system

From the eardrum the sound waves are transmitted to, and amplified by, a series of small bones, called the *ossicles,* situated in the air-filled middle ear. It is the ossicles that amplify the sounds we hear. In humans sound is only amplified 18 times. This is quite sufficient for our needs, but the kangaroo rat, a small mammal hunted by rattlesnakes, needs much sharper hearing to survive. The sounds the kangaroo rat hears are amplified 100 times – sharp enough to hear the rustle of a snake's scales. The ossicles cause a membrane, the oval window, to vibrate. This sets the internal fluids of the inner ear in motion, which stimulates hair cells in the complex structures of the inner ear, the cochlea and the semicircular canals, and these in turn generate nerve impulses that are interpreted by the brain as sound.

Underwater hearing

Sound is an important way of communicating underwater as well as on land. Although some aquatic bugs "listen to," or sense, vibrations in the surface of the water, surface vibrations distort quickly, and a three-dimensional wave is essential for good hearing. Under the surface, for instance, whale song can be heard for hundreds of miles. Many species of fish have calling sounds to attract mates. Fish hear in much the same way as we do, except the fish ear lacks the air-filled middle ear (and therefore the ossicles) common to the mammalian ear. Fish also lack the cochlea; the hair-bearing cells are instead located in a simple chamberlike organ called the *lagena.* Most fish have three semicircular canals, but lampreys have two and hagfish just one. However, fish also have a *lateral line* running along their bodies that contains hair-bearing cells that give the fish information about water currents.

The amphibian ear

Unlike most mammals, amphibians have no outer ear, or pinna. This makes their bodies much more streamlined when traveling through water. In many amphibians the eardrum is clearly visible on either side of the head. The eardrum is connected to the oval window by two small bones and, like fish, amphibians have no cochlea – only the small chamberlike lagena. The lack of the cochlea in both fish and amphibians means that these groups of animals cannot detect a wide range of different pitches. Their range of pitch, however, is sufficient for their needs. They hear only sounds within the range of those produced by members of the same group or species, such as mating or other communicating calls. This means that their brains do not have to decipher those sounds that are relevant to them and those that are not, and so valuable "brain power" is saved.

***Invertebrate animals** have diverse ways of detecting sound. Most species of spider* **[A]**, *for example, have hairs on their eight legs and feet that are attached to sensory nerve fibers. These run directly to the brain, and when sound waves travel through the air, they disturb the hairs, which then send signals straight to the brain via the nerve fibers. The length of the hairs determines the frequencies that can be detected. Earwigs* **[B]** *detect sound through the pincers on their abdomens. The pincers are very sensitive to the vibrations in the air created by sound waves.*

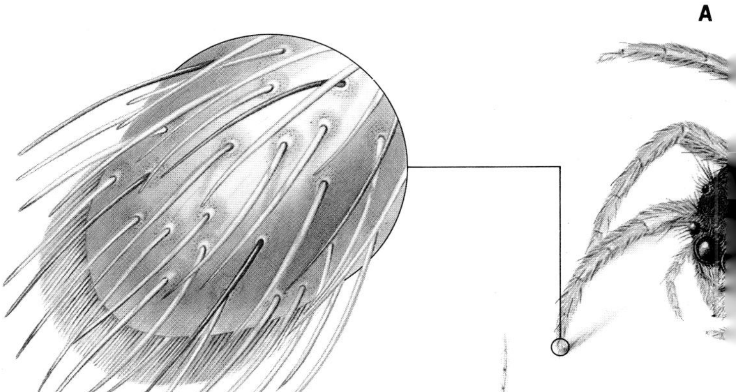

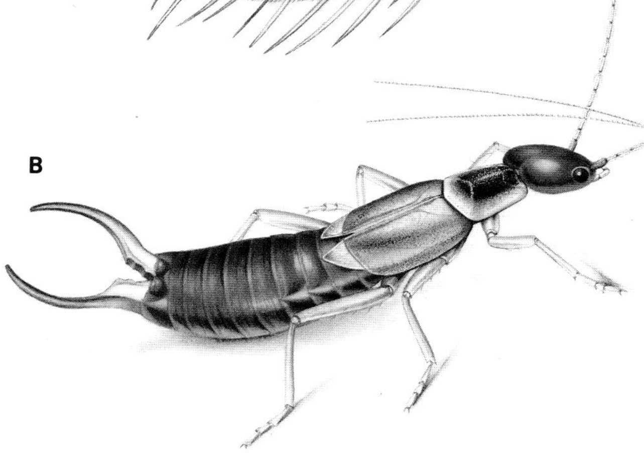

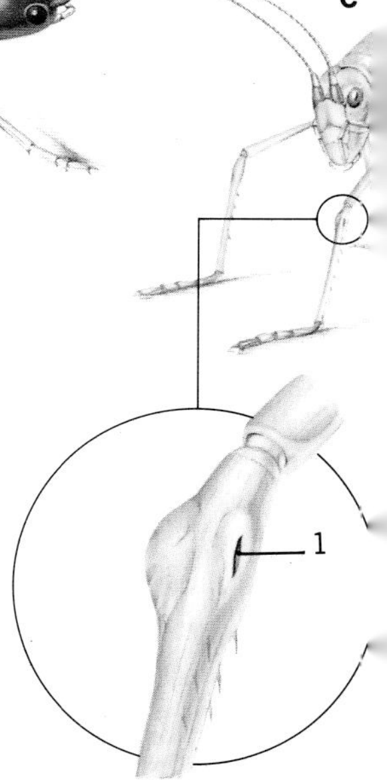

***Crickets** [C] have a stretched membrane of skin, the* tympanum [1], *located on the "knees" of the front legs. These detect sound vibrations, which are then sent to the brain. Grasshoppers have similar membranes, but in most species they are located on the underneath of the abdomen.*

Connections: Mammal Societies 298 Migration Techniques 230 Nervous Systems 288 Smell and Taste 284 Solitary Carnivores 246 Sonar 280

***The caracal* [D]** *is found in Africa and Asia. To help track and catch its nocturnal prey, the caracal has large pinnae that can be directed to help the animal pinpoint its prey. The hearing range of the caracal (in Hz, or cycles per second) is compared with that of other creatures in the graph (below).*

D

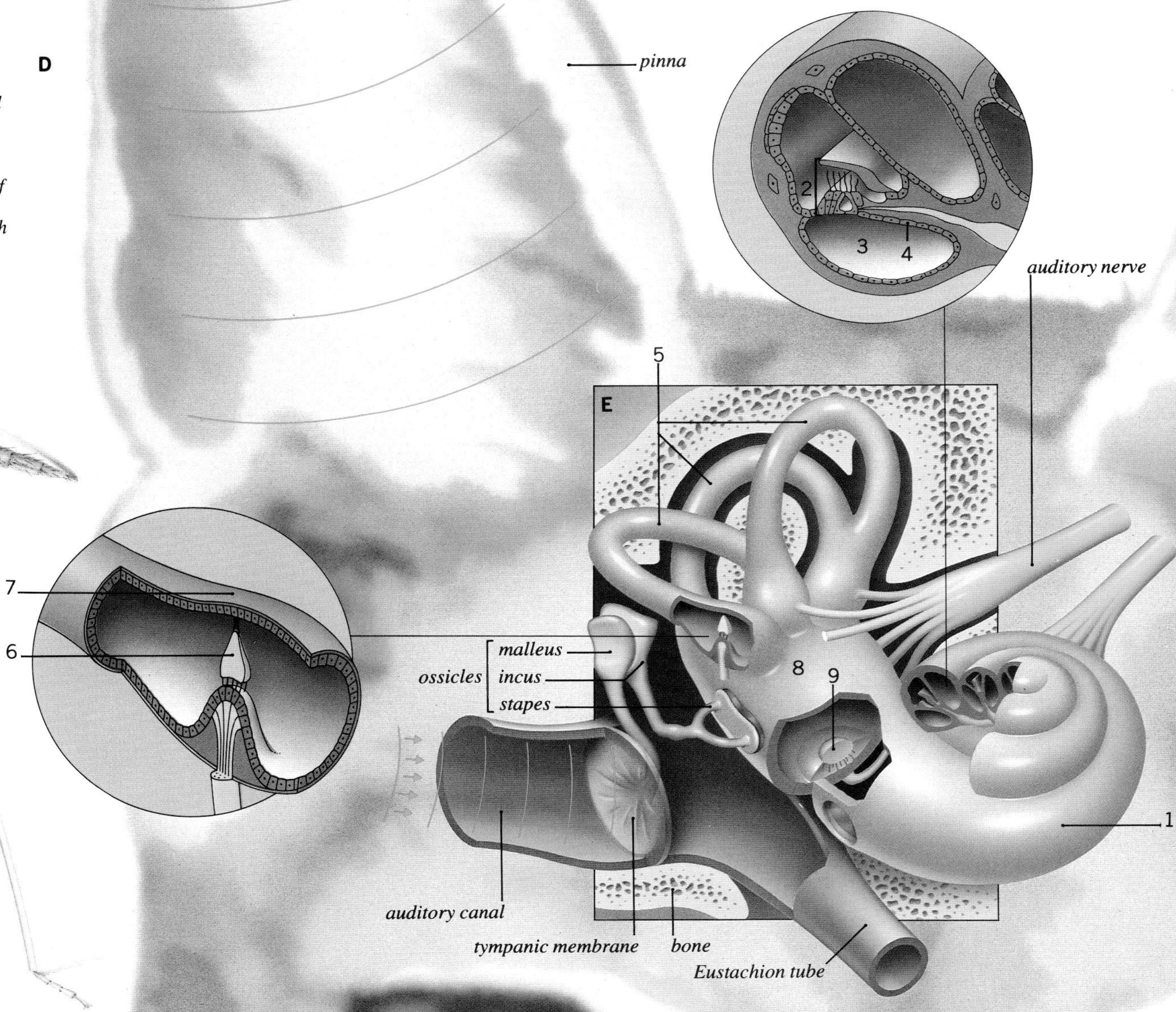

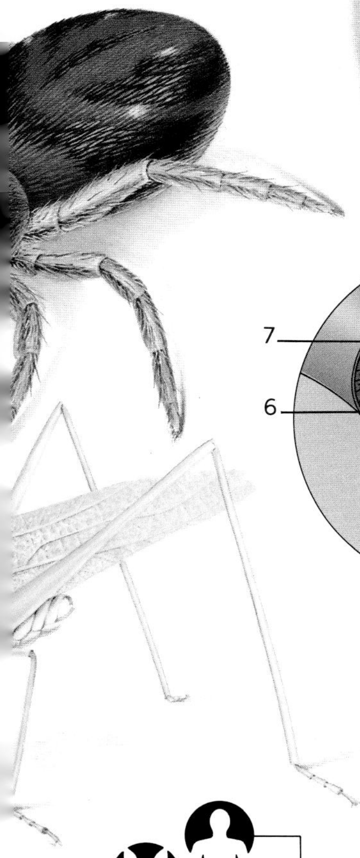

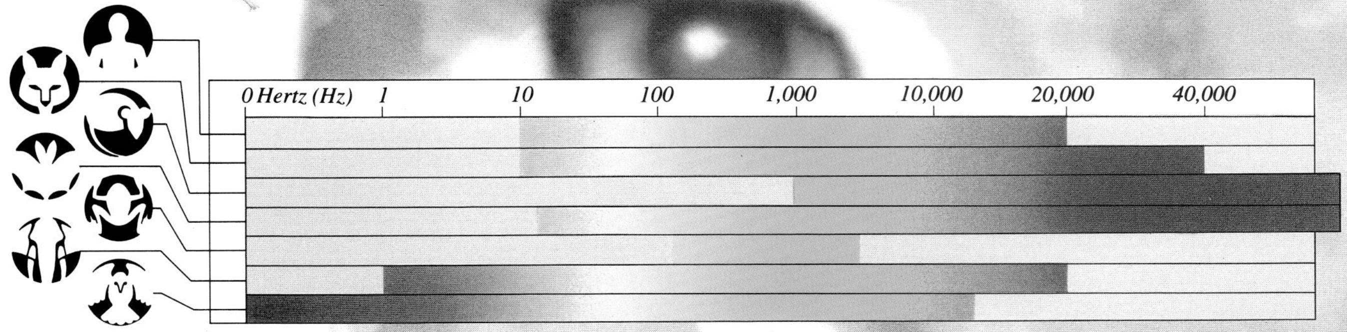

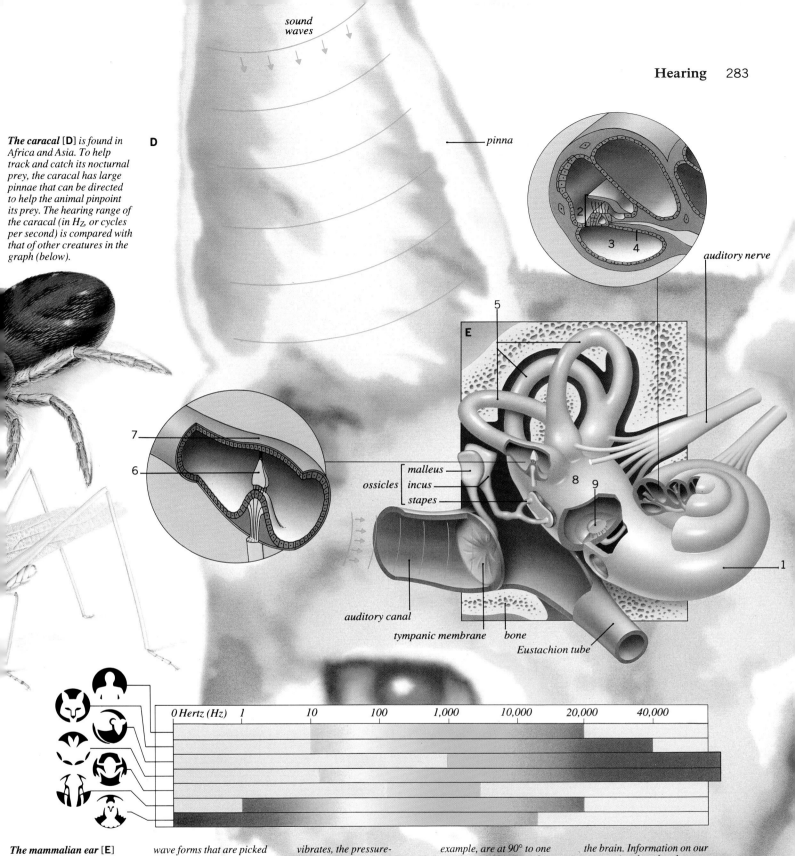

***The mammalian ear* [E]** *is a complex structure that enables mammals to hear a wide range of frequencies. Frequencies are separated in the cochlea* [1] *by the* organ of Corti [2], *part of which is a layer of pressure-sensitive cells. When sound enters the inner ear, the fluid in the* scala tympani [3] *creates wave forms that are picked up by the* basilar membrane [4]. *Sound waves of different frequencies create different wave forms. High-frequency sound vibrates the basilar membrane located at the end of the cochlea, while low-frequency sound vibrates the membrane at the apex. Wherever the membrane vibrates, the pressure-sensitive cells of the organ of Corti are stimulated, sending impulses to the brain via the auditory nerve.*

The mammalian ear is, however, more than just an organ of hearing. The inner ear also helps mammals to keep their balance. The three semicircular canals [5], *for example, are at 90° to one another, and between them cover the horizontal and vertical planes. When we move our heads, the fluid in the semicircular canals is displaced, which causes the* cupula [6] *in the* ampulla [7] *to flex, which in turn pulls on nerve fibers in the ampulla that send electrical signals to the brain. Information on our posture is relayed to the brain via organs in the* utricle [8]. *The utricle is lined with sensitive nerve cells that are triggered by* otoliths [9] *(tiny crystals of calcium carbonate). The otoliths press on the sensitive nerve cells because of the pull of gravity.*

Sound Communication 296 **Special Senses** 286 **Vision** 278

Traces on the Wind

How animals smell and taste

The most highly developed sense of "smell" belongs to the European freshwater eel, which is able to detect one part in three million million million of certain chemicals. Such powers enable migrating eels to travel thousands of miles across the Atlantic Ocean to find their spawning grounds in the Sargasso Sea. Other fish can detect the secretions from the skin of a predator long before it gets dangerously close. Some mammals also have considerable powers of smell – the polar bear, for example, can detect a dead seal from a distance of 12 miles (20 kilometers).

The distinction between taste and smell is not always clear cut. In terrestrial animals, smell is the ability to detect the presence of traces of airborne chemicals, while taste is the detection of substances dissolved in water. But since smell sensors are analyzing airborne chemicals after they have dissolved in the thin films of water that line the *olfactory* (smell) organs, even this distinction is somewhat dubious. When we humans have a cold, we often find it difficult to taste food. This is because our noses are blocked and the scent-laden particles from our mouths are unable to reach the smell sensors at the back of the nasal chambers: what we think of as taste is really a combination of taste and smell.

In fish, which do not sample the air, the distinction is even less obvious – smell is used for sensing chemicals at a distance or in very small concentrations, while taste is used to judge the quality and nature of substances in higher concentrations at close range.

The mystery of smell

Humans, who are not noted for having greatly sensitive noses, are nonetheless able to distinguish between about 10,000 different chemical compounds by their odors. This is possible not because the human nose contains 10,000 different types of smell sensors, rather it is thought that smells can be classed into about seven groups – camphorlike, musky, floral, pepperminty, etherlike, pungent and putrid. In each group, the chemicals that produce these odors are somewhat similar in molecular shape and size. There are probably at least seven different types of sensors, each of which is receptive to more or less one group of chemicals. Chemicals that are a good fit in the sensor's receptor site produce greater stimulation than those that only loosely fit the site, and some intermediate chemicals may stimulate more than one kind of receptor. This provides a kind of smell code – the brain interprets the smells according to the degree of stimulation of all the different sensors.

Sensitive nose

The acuity of these senses depends to a large extent on the numbers and density of the sensory cells involved. Some male moths have over 150,000 sensory cells on their antennae, many specifically responding to the female sex scent (*pheromone*). The sensory tissue at the back of the human nose covers an area of about 2 square inches (14 square centimeters), whereas that of a dog, which has a much greater capacity for detecting odors, may cover 23 square inches (150 square centimeters). The olfactory membranes of the bloodhound detect very weak smells, as well as recognizing many more different smells.

Taste operates in a similar way. Among the mammals, herbivores have the most taste buds. Insect-eating bats have about 800 taste buds, humans 9,000, and other omnivores such as pigs have about 15,000. Birds have a relatively poor sense of smell, relying more on sight and sound, but fulmars and petrels can detect carrion many miles away, as can turkey vultures, and kiwis use their noses to find underground food. Reptiles rely more on smell than on taste, while amphibians, on the other hand, use taste more than smell.

__The monitor lizard__ [A] is just one example of the many lizards, amphibians and snakes that possess a highly developed sensory structure known as Jacobson's organ. *It is used in conjunction with the animal's darting tongue. Jacobson's organ [1] consists of paired cavities above the upper palate. These cavities are lined with receptors that are linked directly to the brain, and are totally separate from the receptors at the back of the nasal passage [2]. The tongue collects sample molecules [3] from the air and transfers them to the duct (opening) of the organ. Inside, minute* ciliae *(hairs) [4] circulate the molecules.*

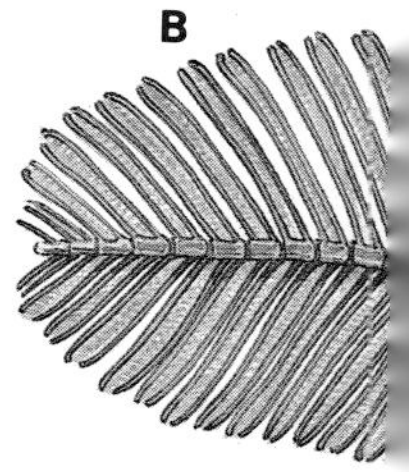

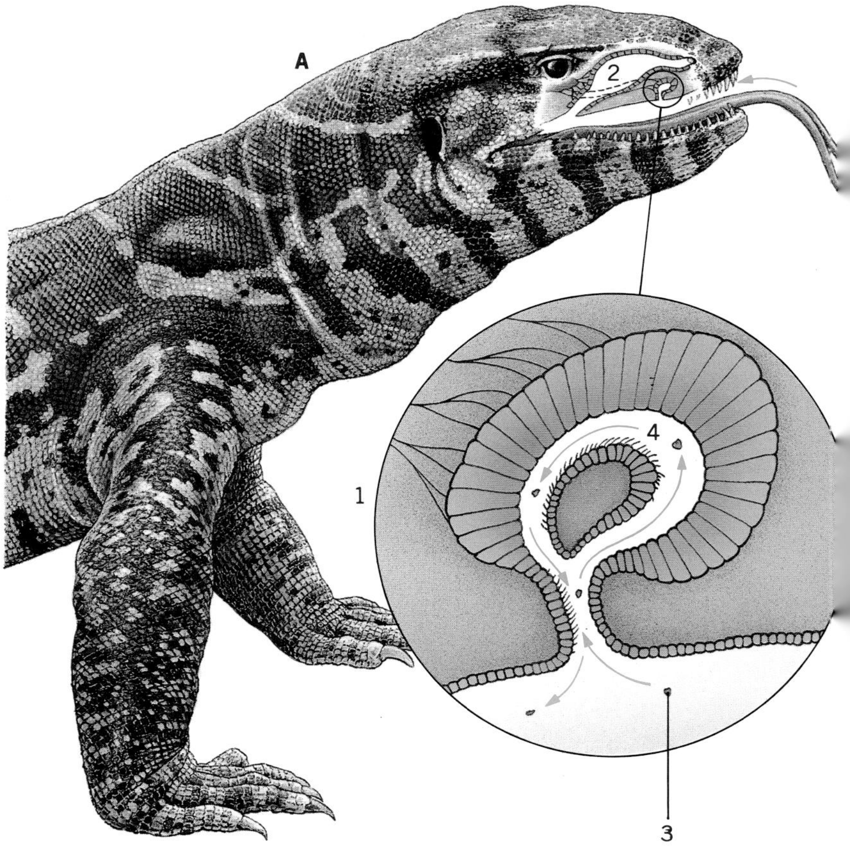

Connections: Hearing 282 Invertebrate Reproduction 160 Migration Patterns 228 Migration Techniques 230 Nervous Systems 288 Scent Communication 294

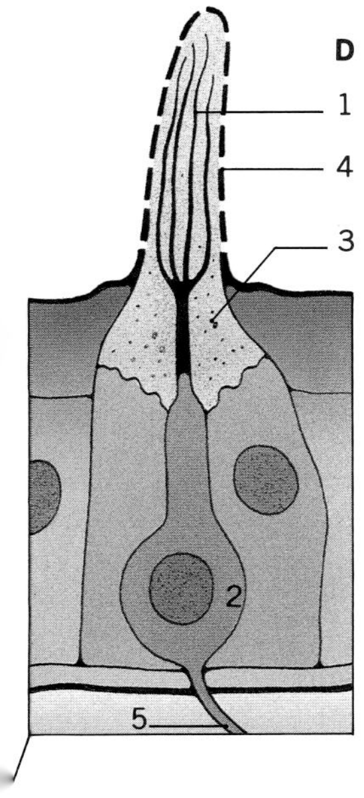

Many insects, *such as the male African moon moth* [**B**], *have smell receptors located on their antennae, while vertebrates, such as the German shepherd* [**C**], *have functionally similar internal receptors lining the nasal passages. In both cases, the reception surfaces* [**D**] *are the modified* dendrites [1] *of receptor cells* [2]. *The dendrite is bathed in an aqueous solution* [3], *which mediates receptor contact. Among insects, each receptor is contained within a chitin-covered peg pierced by pores* [4], *allowing air to enter. In vertebrates the aqueous solution forms a mucous layer. Information is passed to the brain via the axon* [5].

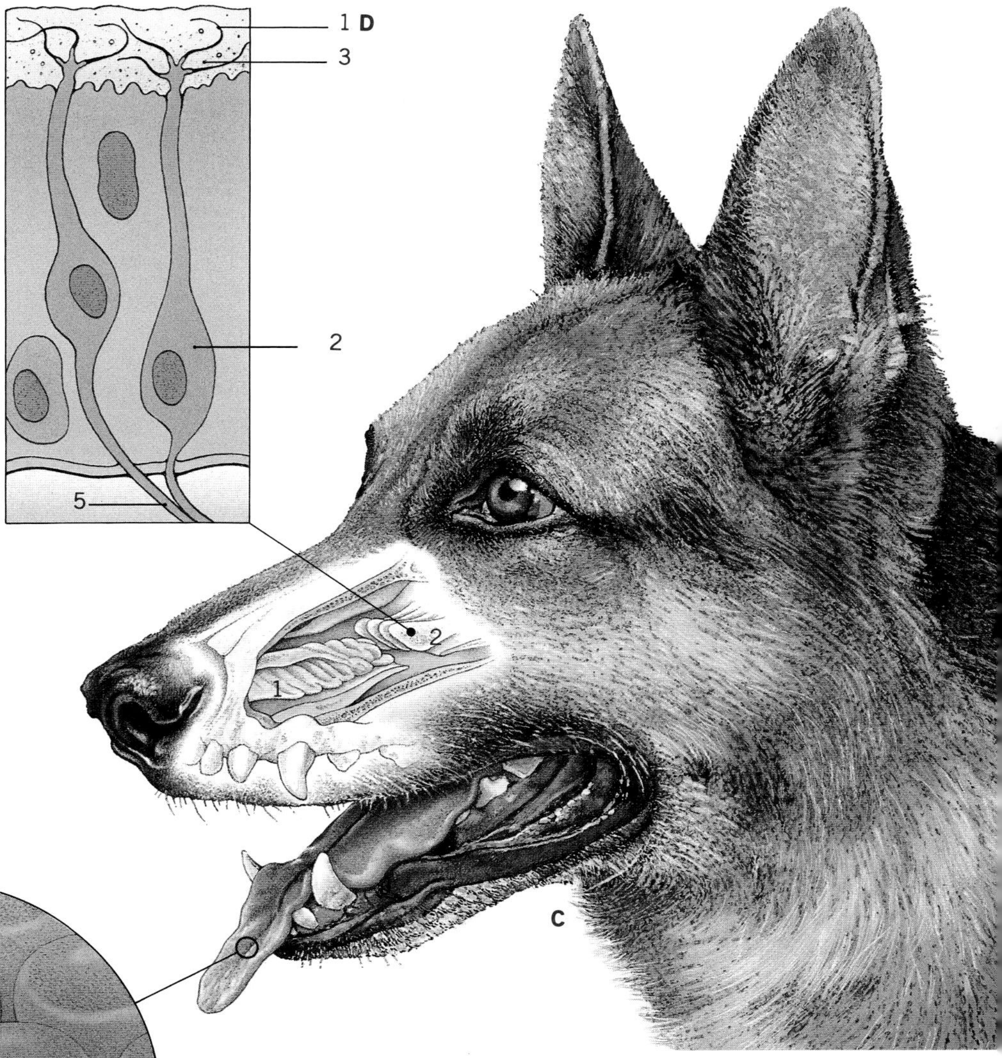

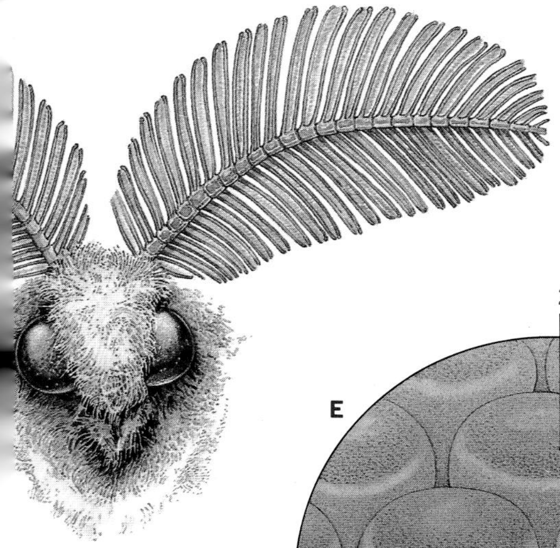

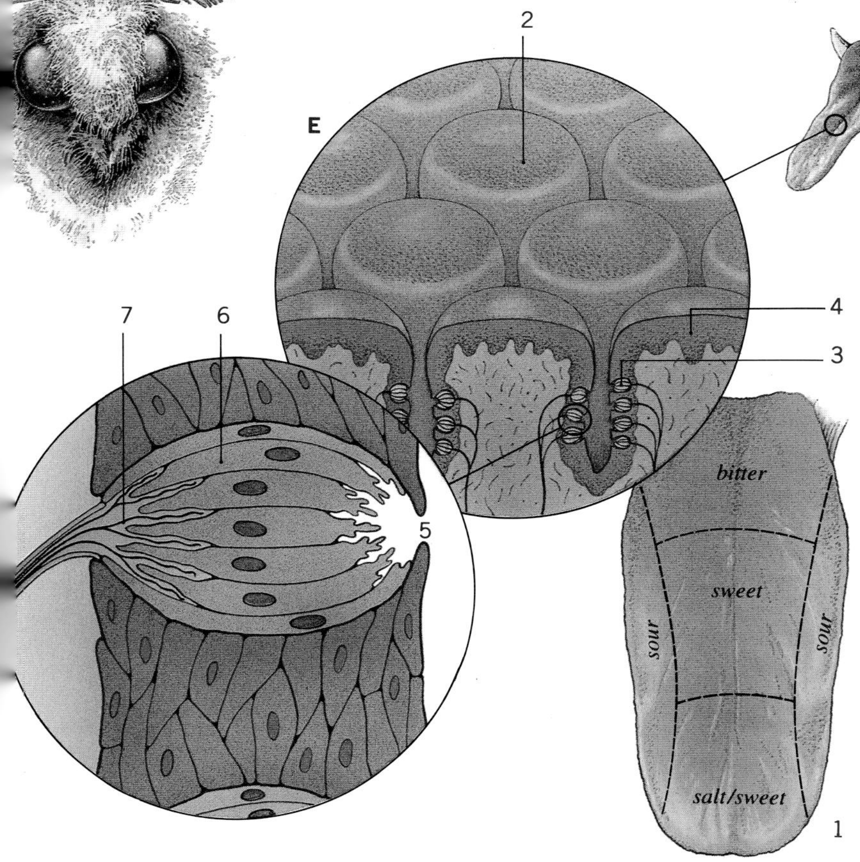

The keen sense of smell *displayed by many mammals owes much to the internal structure of the nasal passages, and these are most highly developed in predators such as dogs* [**C**]. *Air entering the nostrils first passes over the turbinal bones* [1], *where it is warmed, moistened and filtered by a layer of mucus secreted by covering cells. Processed air then passes over the smell receptors, which are located in the olfactory epithelium* [2].

Mammalian taste receptors, or taste buds, are concentrated on the surface of the tongue [**E**] *in zones that respond to different flavor sensations. On the human tongue* [1], *sour is detected at the edges, sweet in the center, bitter at the rear and sweet/salt at the tip. These zones broadly correspond to variations in the structure of the raised papillae* [2] *that form the surface of much of the tongue. The taste buds* [3] *are mainly located in the grooves between papillae, embedded in the epithelium* [4], *and are open at the apical aperture* [5]. *Each bud contains a number of individual receptor cells* [6] *with elongated receptor surfaces. As with the sense of smell, molecular contact is also mediated through solution in water (mucus). The receptor cells in a taste bud are served by separate nerve cells* [7], *which pass the sensory information to the brain's olfactory centers.*

Sixth Senses

How some animals use acute special senses

A freshwater shrimp flicks its tail to propel itself forward – and the predatory duck-billed platypus instantly moves in to snap it up. Both are in murky river water, and the platypus has poor eyesight, but it "sees" the shrimp as clearly as a flashing light, thanks to tiny pores around its "beak" that sense the electric fields produced by the muscular movements. Sharks have similar sense organs. If two such distantly related animals possess the same sensory ability, then it may be also present in other aquatic animals: at the moment we just don't know.

The blue shark *[**A**], like other sharks, has two special sensory systems, the* ampullae of Lorenzini *and the lateral line, to help it find prey. The lateral line [**B**] detects distant, low-frequency vibrations created by the swimming movements of fish or aquatic mammals. The* lateralis *system consists of skin openings [1] that allow seawater through tubes [2] into the lateral canal [3], which contains bunches of sensory hairs [4], each of which is encased by a gelatinous cap, the* cupula *[**C**]. Vibrations in the water are transmitted to the cupula by the tubes, causing the cupula and the hairs inside it*

Modern scientists still know little about electrical senses. Investigations of sensory powers that involve electricity, magnetism, air or water movements, or infrared (heat) are all relatively recent because human beings are so heavily biased toward sight and sound. Biologists were slow to realize how different the senses of other creatures might be.

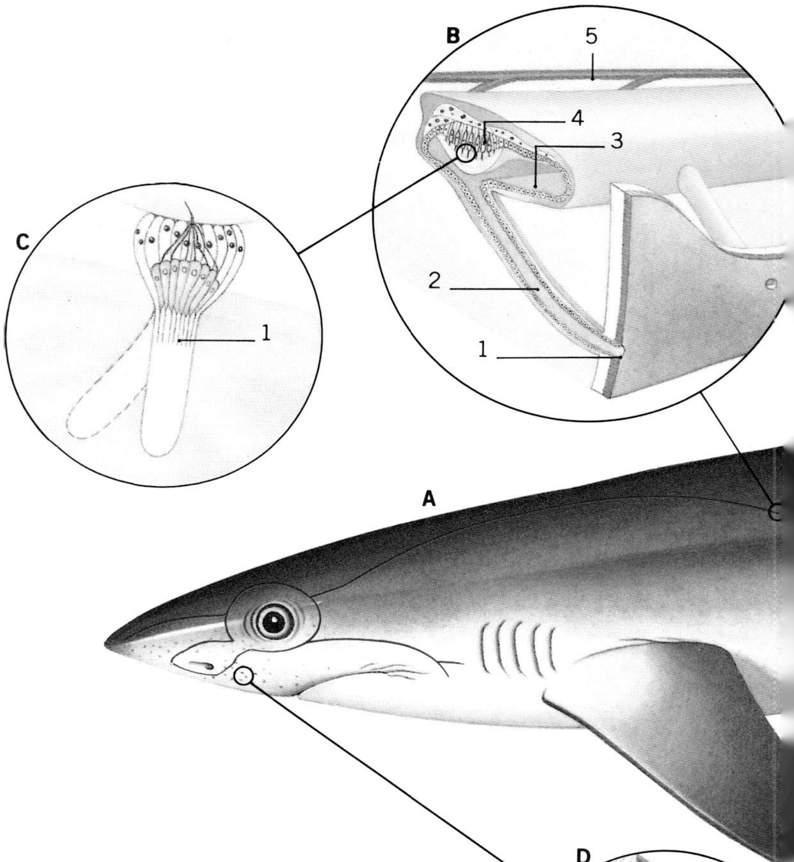

Electrical echoes

An electrical sense could be useful to any predator hunting in seas, rivers, or lakes, for water transmits electric current well. Several fish are thought to possess the faculty. A few, such as the elephant-snout fish of Africa, even use electrical fields as a form of echolocation. The elephant-snout produces pulses of electricity from modified muscles at the base of the tail, so setting up an electric field. Pores on its head then "read" the electricity, sensing any disturbances produced in the field by rocks or the riverbank. This enables the fish to navigate even in murky, silt-laden water. The fish can also detect its prey with its electricity-reading pores, either directly or by echolocation. The electric eel, which has electrical organs of such power it can stun its prey with them, also generates weaker signals for echolocation.

Touching tactics

Like most fish, the elephant-snout and electric eel have a line along the side of the body that corresponds to another important sense organ. This is the lateral line, which picks up water movements and low-frequency vibrations, and helps with hunting and safety.

The lateral line is just one aspect of a complex and varied sensory system: that of touch. This is a sense that humans share, but we cannot imagine the intense and detailed touch sensations that a mouse foraging at night receives through its whiskers, or that a catfish experiences through the fleshy barbels that hang down around its mouth. In some animals, touch is the predominant sense, and yields a surprising amount of information. In a flying bird, whiskerlike feathers relay information about the air currents around the body to the brain, which uses this information to regulate flight movements. Parts of insects, mainly their feet and antennae ends, are covered with sensory hairs – called *sensilla* – that fulfil the same purpose.

A sense of attraction

Another sense that humans have – although we use it relatively little – is the ability to detect magnetism. Some animals can detect the Earth's magnetic fields, and some are thought to rely heavily on this sense, particularly those that migrate over long distances. Birds were among the first animals thought to use a magnetic sense during migration, and there was unexpected confirmation of this when flocks of migrating birds were observed over an area of Sweden that has deposits of iron so rich that they distort the Earth's magnetic fields. Flocks flying in a straight line to the southwest suddenly became disoriented as they crossed the area, and lost their bearings for several minutes before resuming their steady course. Honeybees may also use a magnetic sense in order to orient a new comb in the same direction as their last one.

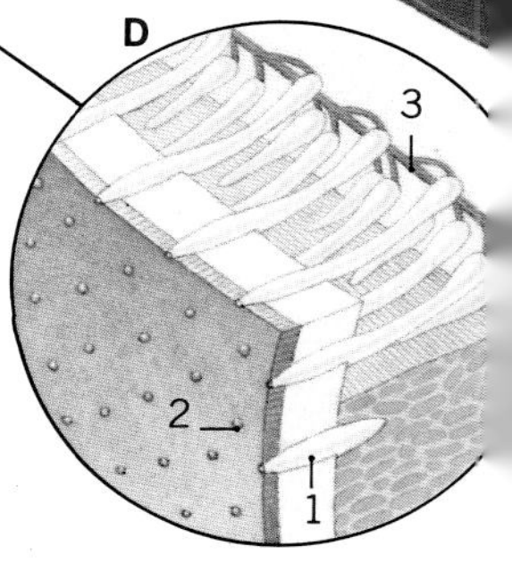

The ampullae of Lorenzini *[**D**] are bulblike electrical receptors [1] that open to the water through pores [2] on the skin of the shark's upper jaw. The ampullae connect to nerves [3] linked to the brain, which conduct signals of electrical activity generated by potential prey.*

Connections: **Biological Clocks** 274 **Bioluminescence** 276 **Communication** 292 **Hearing** 282 **Earth's Magnetic Field** 14 **Migration Techniques** 230

[1] *to bend. Any movement of the hairs is detected by the sensory cells, indicating the vibrations' strength and direction. The signals* [**B**] *reach the brain via the lateral-line nerve* [5]. *Coupled with its other senses, these systems equip the shark to be a formidable hunter.*

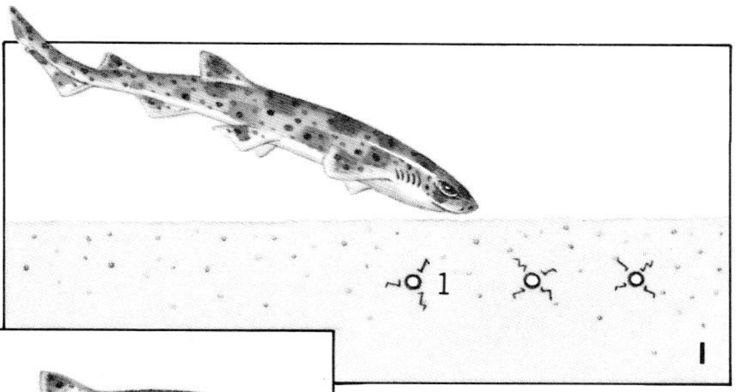

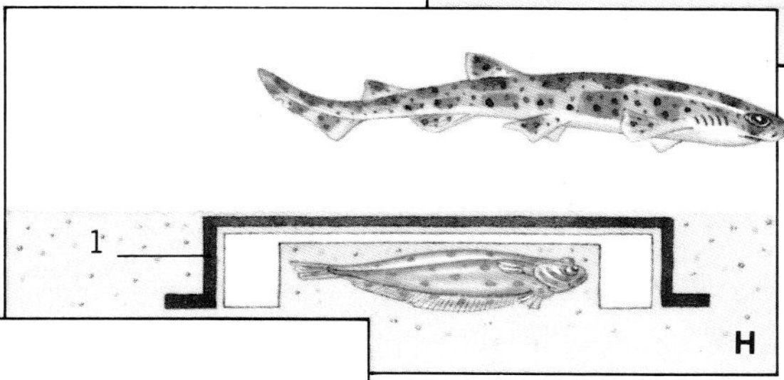

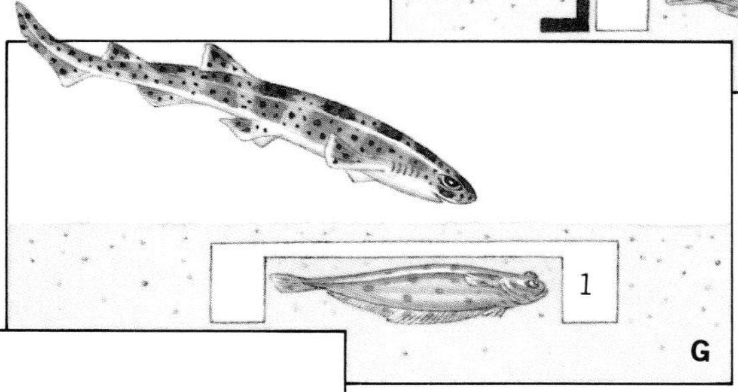

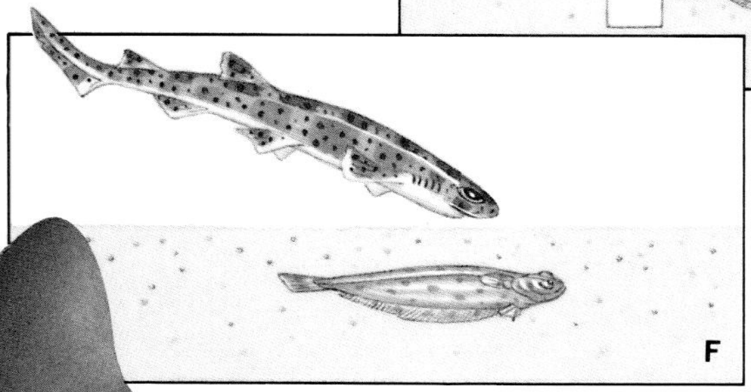

A small shark *(a greater spotted dogfish) detects a flatfish on the seabed, even when the flatfish is covered in sand and invisible to the human eye* [**F**]. *The fish is betrayed by electrical fields generated by the slightest muscle movement, which the shark senses with its ampullae of Lorenzini, tiny pores around its snout. Close-range experiments suggest that sharks can detect voltage gradients equivalent to a 1.5-volt battery linked to underwater electrodes 940 miles (1,500 kilometers) apart.*

The experiment [**G**] *showed that a shark can detect fish enclosed in a container of agar jelly* [1], *which allows electrical fields through but keeps in smells, and deadens sound and water vibration – the other indicators that attract sharks.*

The next stage of the experiment [**H**] *was to cover the agar container with nonconducting material that eliminated the electrical emissions* [1]. *The shark ignored the fish underneath, which was "hidden" from its predator's senses.*

As final proof, [**I**] *the researchers buried electrodes in sand and generated weak electric fields by battery. The shark homed onto the electrodes as if to a fish* [1]. *Varying the current allowed the shark's sensitivity to be measured.*

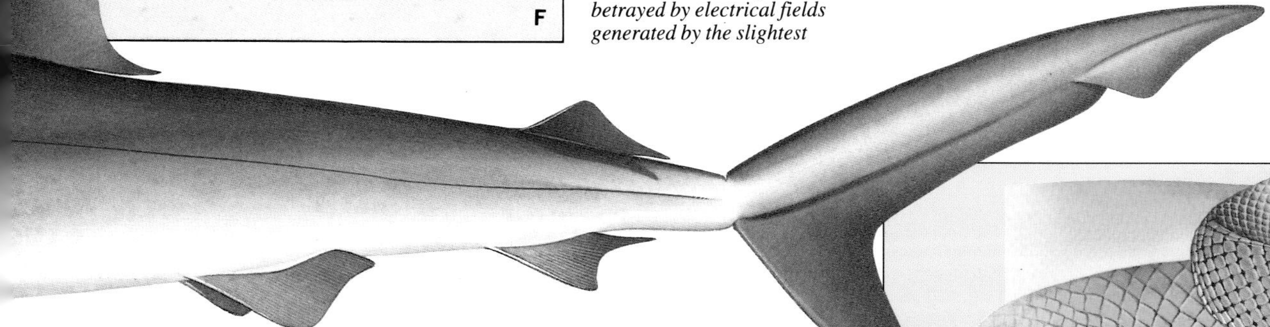

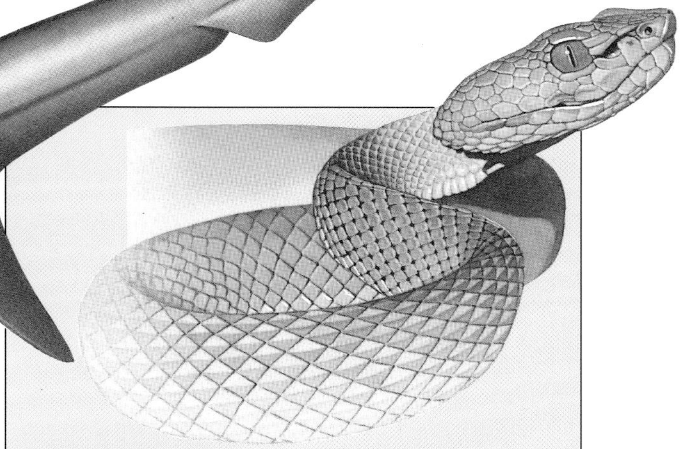

Feeling the Heat

Many snakes feed on small nocturnal mammals and must find their prey in semidarkness. They do so with the aid of heat-sensitive organs located in pits on the upper jaw. By using electrodes implanted in the brain, scientists have discovered that these heat sensors feed their information into the same area of the brain as the eyes do. So a snake must actually see a thermal (infrared) image of the animal superimposed on its visual image. At dusk, or by moonlight, it can make the most of both sensory powers by combining the two images.

When a snake opens its mouth to deal the death blow, the heat-sensing pits are no longer directed at the prey. However, research has shown that the snake has a second set of heat sensors located inside the mouth, which fix on the target as it strikes.

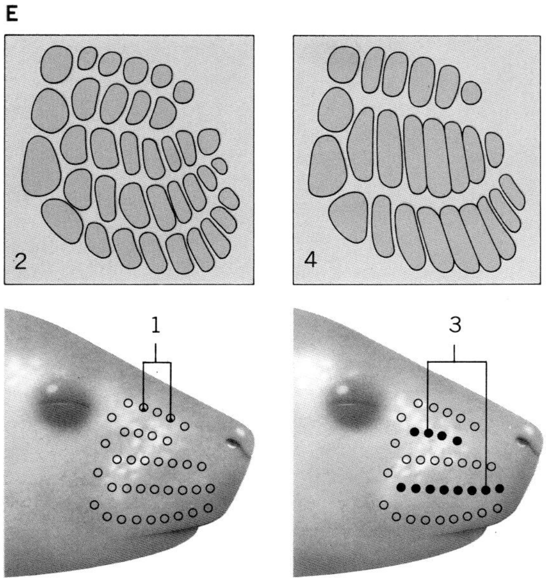

This young mouse [**E**] *has sensory hairs, or whiskers, around its snout* [1], *each of which is extremely sensitive. Each hair connects to a specialized bunch of neurons (cells specialized to carry nerve impulses) that will only react to the whisker with which it is connected. The neuron bunches have specific locations in the brain* [2], *and it can be shown that removing whiskers* [3, *solid black dots*] *will cause the corresponding bunches of neurons to cease to grow and eventually disappear* [4], *giving some indication of the extent to which the animal's brain and sensory inputs are organized for precision data gathering.*

Nervous Systems 288 **Smell and Taste** 284 **Solitary Carnivores** 246 **Sonar** 280 **Vision** 278

The Electric Circuit

How nervous systems work

An animal's nervous system is like a computer, employing exactly the same means of transmitting information: electronics. What's more, an animal's nervous system has memory banks, information processors and output devices that deliver the nerve signals to their destinations just like computers. The nervous system is the body's communications network. Without it an animal cannot function, since it cannot coordinate itself: it cannot move, digest its food, or breathe, since all these activities are triggered and controlled by the nerve network.

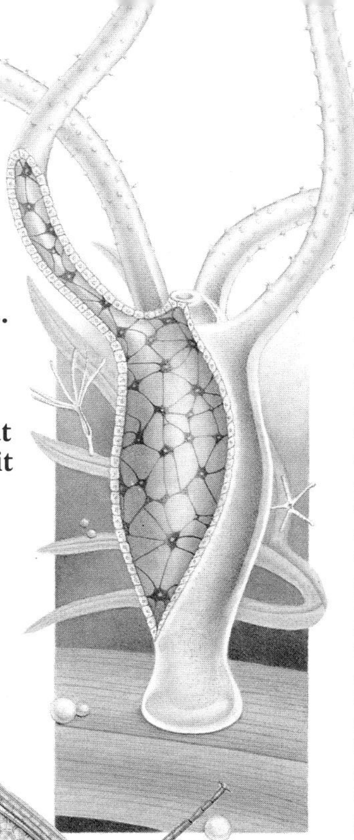

The hydra [**A**], *a close relative of the sea anemone, has an extremely primitive nervous system that is often called a* nerve net. *Individual neurons, some of which act as receptors, are distributed throughout the animal's body and are linked into a nondirectional network. A stimulus, such as physical contact, detected at a particular place on the body, triggers a nerve impulse that diffuses through the nerve net to all parts of the animal. As the impulse spreads, it triggers a generalized contractionof all the hydra's muscular cells. This results in the animal shrinking into itself, away from any potential danger.*

All animals are made up of cells: tiny packages of living tissue. A microscopic animal may consist of a single multipurpose cell, but the bodies of larger animals are constructed from clusters and chains of cells acting together. Each cell has its own function, and a complex animal may be built up from over 200 distinct cell types. Some of the most specialized of these are the cells that carry information through the animal's body to coordinate its actions: the nerve cells or *neurons*.

A typical neuron resembles a tiny tree, with branches, roots and a thick stem. The roots of one neuron connect with the branches of others to form extensive chains and networks, allowing a signal generated within any one neuron to be passed on through the system. Each neuron has a very weak electrical charge, like a miniature battery. Normally the inside of the neuron is negative while the outside is positive, but if part of the cell is stimulated – by a sensory mechanism or by another nerve signal – the inside of the cell in the immediate vicinity becomes briefly positive. This effect ripples along the neuron like a wave. It travels the length of the cell's stem – known as the *axon* – and along the roots and branches, or *dendrites*. It carries on until it reaches the very tip of each dendrite, where it jumps a tiny gap, a *synapse*, over to the next neuron.

The centralized system

The nervous system of an extremely simple animal, such as a sea anemone, consists of a network of basically identical neurons extending throughout its body. In more sophisticated creatures the nervous system is based on a central core, and the neurons are of three basic types. *Sensory neurons* collect information that has been gathered by the senses, while *motor neurons* deliver messages to the animal's muscles. The sensory and motor neurons are linked by the interneurons of the central nervous system, which process the sensory information and coordinate the motor neurons to produce an appropriate response.

Automatic functions

Most bodily functions, such as the beating of the heart, digestion and heat regulation, are performed without conscious thought. Such functions are governed by the *autonomic* nervous system, itself composed of two systems – the *sympathetic* and the *parasympathetic*. The sympathetic nerves, when stimulated, prepare the body for emergency action – such as fight or flight – by increasing heart and respiration rates and slowing down less immediately vital activities such as digestion. The parasympathetic nerves are most active during periods of rest. They slow the heart and breathing, but increase rates of activities such as digestion.

The seat of intelligence

Sending signals is fairly straightforward, but analyzing those signals is complex. As a result, the central nervous system of an advanced animal like a mammal has become highly sophisticated, with millions of interneurons linked together in complex electronic circuits, creating a central information processor, or brain. In many animals the potential of the neuron has been exploited still further, and their brains have the capacity to store information, recall it and use it: the basis of intelligence.

The evolutionary advance *marked by the arthropod exoskeleton is mirrored inside the body by the development of the nervous system. The grasshopper* [**B**] *has a well-defined brain* [1] *located in the head which is connected to the main nerve pathway* [2]. *The brain is largely concerned with control of the mouthparts and collating information from the eyes and antennae. Movements of the jointed limbs, and other basic body functions, are largely controlled by a series of* ganglia *(neuron clusters)* [3] *situated along the nerve pathway. The largest ganglion is that corresponding to the grasshopper's hind limbs.*

Connections: **Animal Cells** 102 **Hearing** 282 **Intelligence** 290 **Mammals** 136 **Smell and Taste** 284 **Special Senses** 286 **Vision** 278

The cat, a vertebrate, **[C]** *has a central nervous system consisting of a brain and spinal cord connected to a branching network of sensory and motor nerves. Sensory nerves pass information to the brain, and motor nerves convey instructions to muscles. Although all nerves are ultimately interconnected with the brain, a considerable amount of processing is carried out within the spinal cord, as in the case of the reflex action shown. When receptors are stimulated, a pulse passes along sensory neurons* [1] *into the spinal column* [2] *via a bundle of nerve fibers* [3]. *Inside the gray matter* [4] *of the spinal cord, the impulse triggers an immediate motor response, via the motor neuron* [5], *to the relevant muscles* [6].

Nerve impulses are passed along motor neurons **[D]** *by an electrochemical process. A ripple of potential travels down an axon until it reaches the tangle of dendrites* [1] *at the synapse* [2] *with another neuron. Each dendrite ends in synaptic knobs* **[E]**. *Within each synaptic knob are small* presynaptic vesicles [3] *that contain chemicals known as* neurotransmitters. *The electrical impulse causes the vesicles to move to the surface of the synaptic knob* [4], *releasing neurotransmitters, which travel across the synaptic gap* [5] *and are absorbed at specialized sites* [6] *on the receiving cell* [7].

Inside a motor neuron, the neurotransmitters provoke a change in potential in the cell body [8] *around the nucleus* [9], *and an impulse is "fired" down the axon* [10]. *The axon is insulated by a fatty myelin sheath* [11] *secreted by a series of individual* Schwann cells [12]. *An axon connects to muscle fibers through an end-plate structure* [13]. *The terminal synapses* [14] *can only make contact through* synaptic clefts [15], *which extend into the muscle* [16]. *Specialized neurotransmitters contained in synaptic vesicles* [17] *are released across these clefts.*

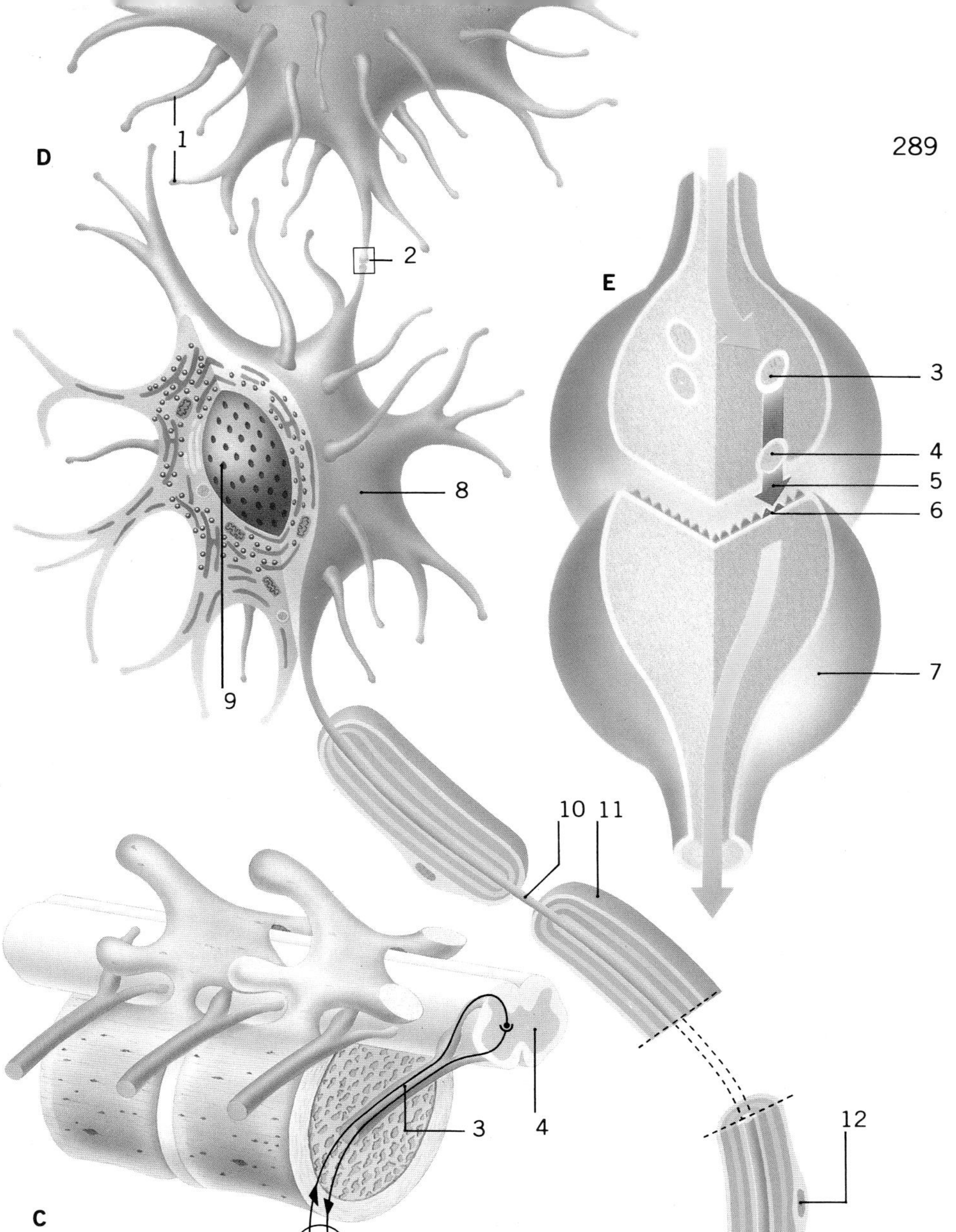

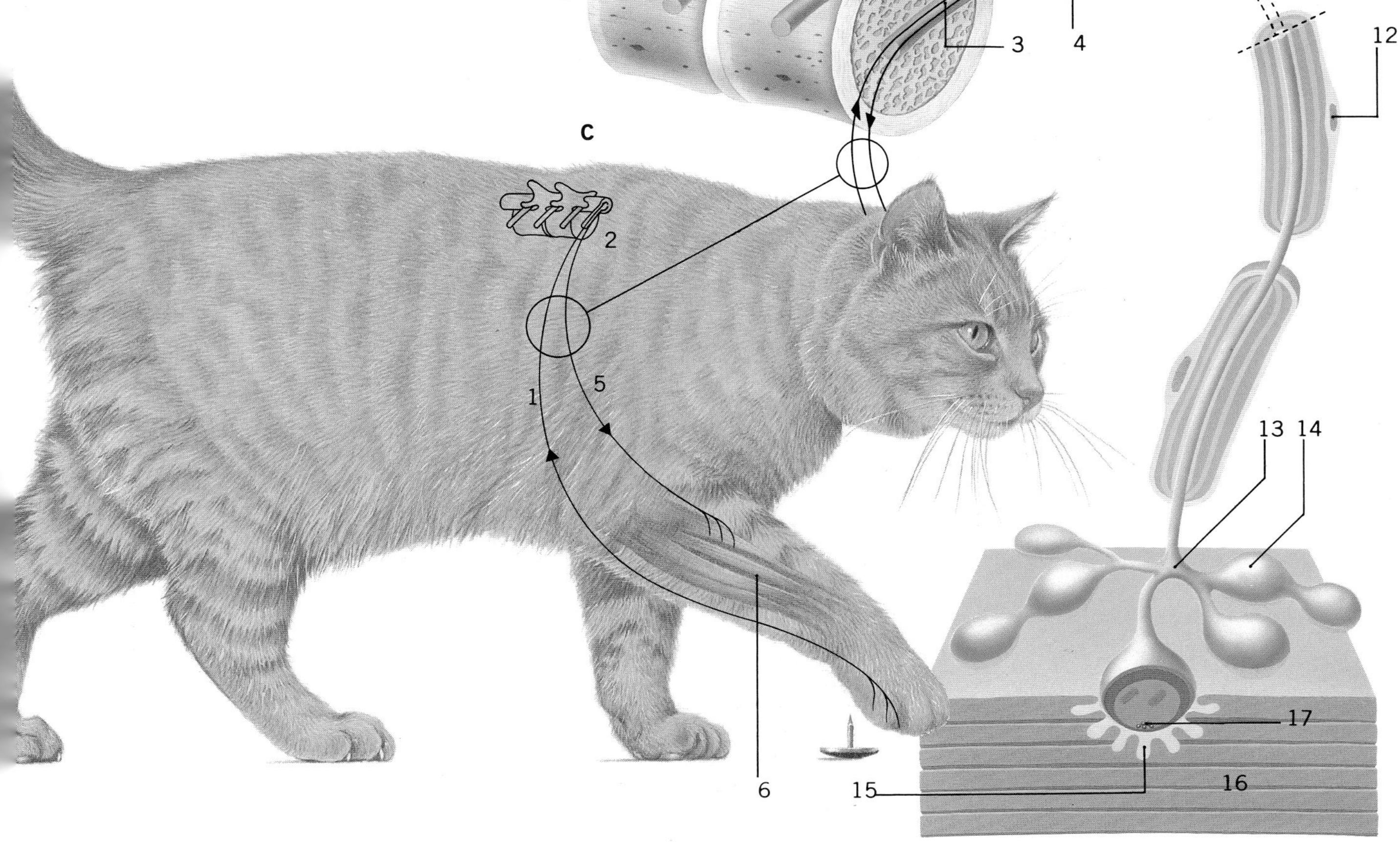

The Power of Thought

How animals developed coordination and intelligence

There are many thousands of different chemical reactions taking place every minute in the average human brain. Yet humans are commonly thought to use only a minute fraction of their brain's potential. All animals make reflex movements, which bypass the analytical processes of the brain, many of them as a major mode of everyday life. But most animals learn from experience, and can learn to make judgments of one kind or another. And humans have evolved the even greater abilities and capacities that stem from reflective thought – they know that they know.

One of the most striking characteristics of any animal is its ability to move in a purposeful, directed fashion. It may twitch, slither, swim, walk, or even fly, but unless it is badly injured, its movements are never aimless. They are coordinated by the nervous system.

Coordination is partly a question of straightforward mechanical efficiency. A millipede, for example, has a vast number of legs that must work cooperatively: in a millipede on the move the nerve impulses at work are almost visible as the waves of movement ripple along the fringing limbs.

But where is it going? It may be making its way toward food or a mate, or moving away from danger, but either way, it is moving in response to sensory stimuli. Its sense organs obtain information about its surroundings, and the information is transmitted down sensory nerves to its brain, which issues appropriate commands to the animal's muscles.

Some stimuli short-circuit the system, bypassing the interpretation stage to generate an immediate reflex response. This is not only an important survival aid but also provides less obvious services: a human's ability to stand upright, for example, relies entirely on a reflex response to signals that are generated in the inner ear.

Instincts and learning

Many aspects of animal behavior have a lot in common with reflexes. When a breeding herring gull returns to the nest, its chicks automatically peck at its bill to beg for food. A crude model of the parent gull would inspire the same response, as long as it had the right coloration: a yellow bill with an orange spot near the tip. The chick reacts to these colors on instinct.

Such instincts may be connected within the young bird's nervous system in exactly the same way as our balance reflex, bypassing the brain, but they may be refined by experience. Experiments suggest that although the nestling has an inbuilt pecking reflex, it has to learn how to use it. For some days after hatching, it pecks at virtually anything, and it only registers the benefits of pecking at the spot on its parent's bill after trial and error.

Learning from experience

To profit from experience in this way an animal has to have a good memory. Storing information is one of the main functions of the central nervous system. Even quite primitive animals, such as insects, can memorize details such as the routes to food sources. In more advanced animals, memory is central to their lives, for it allows them to learn from their own experience and, equally important, to benefit from the experience of others.

A young chimpanzee, for example, may attempt to eat the most unlikely things, gradually discovering – and learning – which are edible and which are not. But it is not alone: it lives in an extended family group, and the adults in the group already know what is good to eat and where to find it.

By copying the behavior of its elders, the young chimpanzee can dispense with the inefficient process of trial and error. It can learn complex skills that have been developed and refined over hundreds of years and passed down through the generations.

A

***Chimpanzees* [A]** *demonstrate high intelligence in their use of tools. They have been observed using sticks to "fish" for termites: a suitable twig is selected, wetted with saliva at one end, and inserted into the termite mound. After a while the stick is withdrawn and the termites clinging to it are eaten. Young chimpanzees are thought to learn this complex behavior by observing their parents: aged between two and three years, they manipulate and prepare sticks as a form of play; by their fourth year of life, they have mastered the efficient use of this tool.*

Connections: Animal Growth and Development 166 Communication 292 Mammal Societies 298 Nervous Systems 288 Vision 278

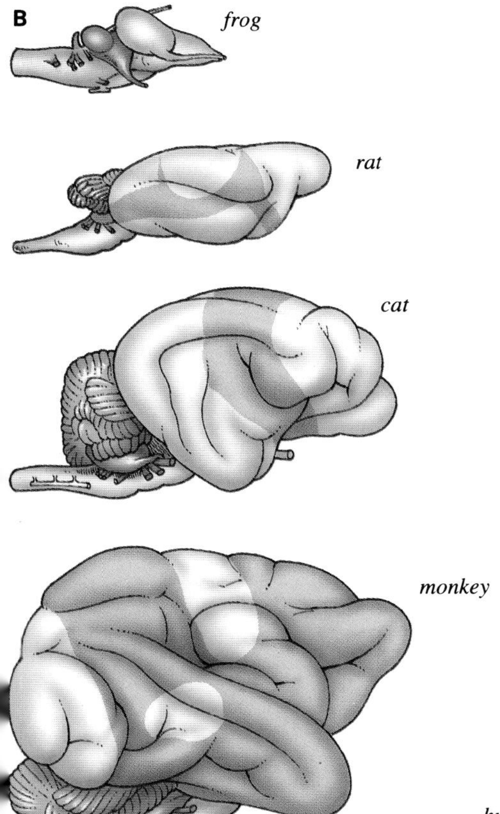

The vertebrate brain [**B**] *has three major functional and structural regions, known as the* forebrain, midbrain *and* hindbrain. *In primitive animals, each of these parts was concerned with a specific sense – the forebrain with smell, the midbrain with vision, and the hindbrain with balance and hearing. This primitive arrangement is still possessed by fish and amphibians, but in higher animals different parts of the brain have expanded or adapted their function to meet the changing needs of the organism. Most notably, part of the forebrain, the* cerebrum [1], *developed into a complex, deeply fissured structure. Its outer layer, the* cortex, *contains areas that coordinate movement and sensory information, and large regions (shaded gray) concerned with association, reasoning and judgment. It houses the "conscious" mind. Unconscious activities, such as breathing, are under the control of the deeper, primitive brain.*

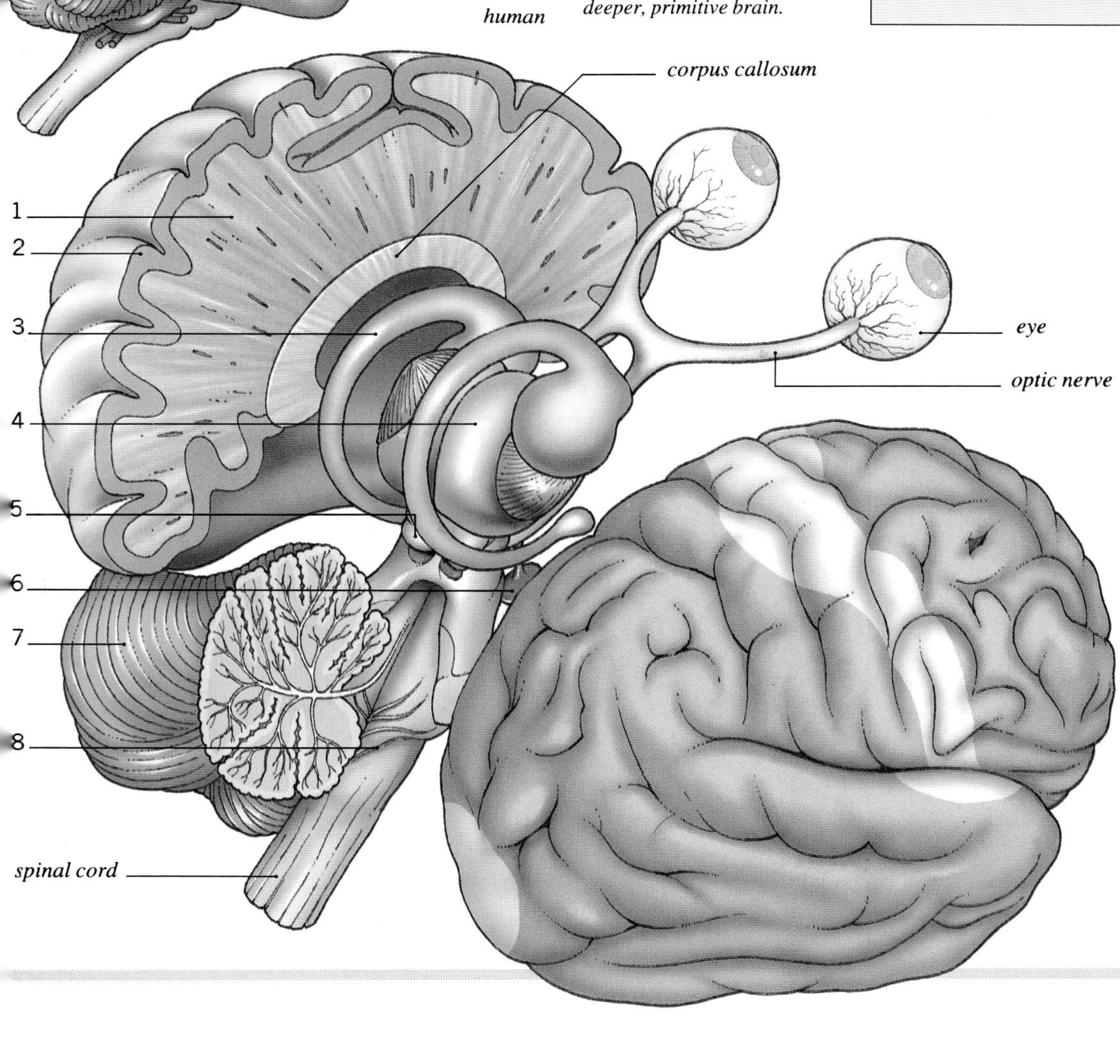

First Impressions

Chickens, ducks and geese – and a number of other birds, fish and mammals – all have another evolutionary specialty that represents an improved chance of survival for the young. At birth, and for a number of hours afterward (perhaps up to 30), the young animal somehow fixes its total attention on the first large object that holds it by moving or by being close enough, and thereafter follows it as if it was a parent figure. The first such object is generally a parent, but young animals may instead adopt other animals (like the ducklings with the hen, below), or even inanimate objects.

Most cerebral functions *depend on the coordinated activity of* neurons *in different regions of the brain. However, certain regions appear to be more closely related to particular activities. In the forebrain, the* cerebrum [1] *and the overlying* cerebral cortex [2] *act as coordinating centers and contain memory circuits. The* limbic system [3] *controls emotional responses, such as fear and aggression. The* thalamus [4] *coordinates sensory and motor signals, and relays them to the cerebrum: the* hypothalamus [5] *and* pituitary *glands control the body's hormonal system. Visual, tactile and auditory inputs are coordinated by the* tectum [6] *– part of the midbrain. In the hindbrain, the* cerebellum [7] *controls the muscle activity needed for refined limb movements and maintaining posture. The* medulla [8] *contains reflex centers that are involved in respiration, heartbeat regulation and gastric function.*

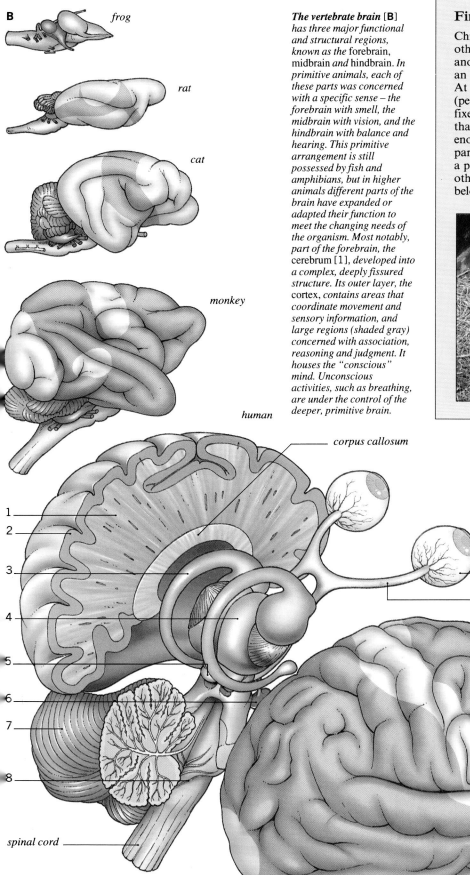

Breaking the Code

How living things use signals to communicate

A female chimp who is ready to mate develops a bright pink swelling around her genitals that is so conspicuous it may be seen by an interested male over half a mile away. The meaning of the signal is fairly obvious. However, a chimpanzee that seems to be laughing at a joke is in fact grinning because it is frightened. Many other animals make visual signals whose real meaning can often only be worked out after long and careful study. Animals signal each other for courtship, for defense, for warning, or, like the honeybee, to tell others where to find food.

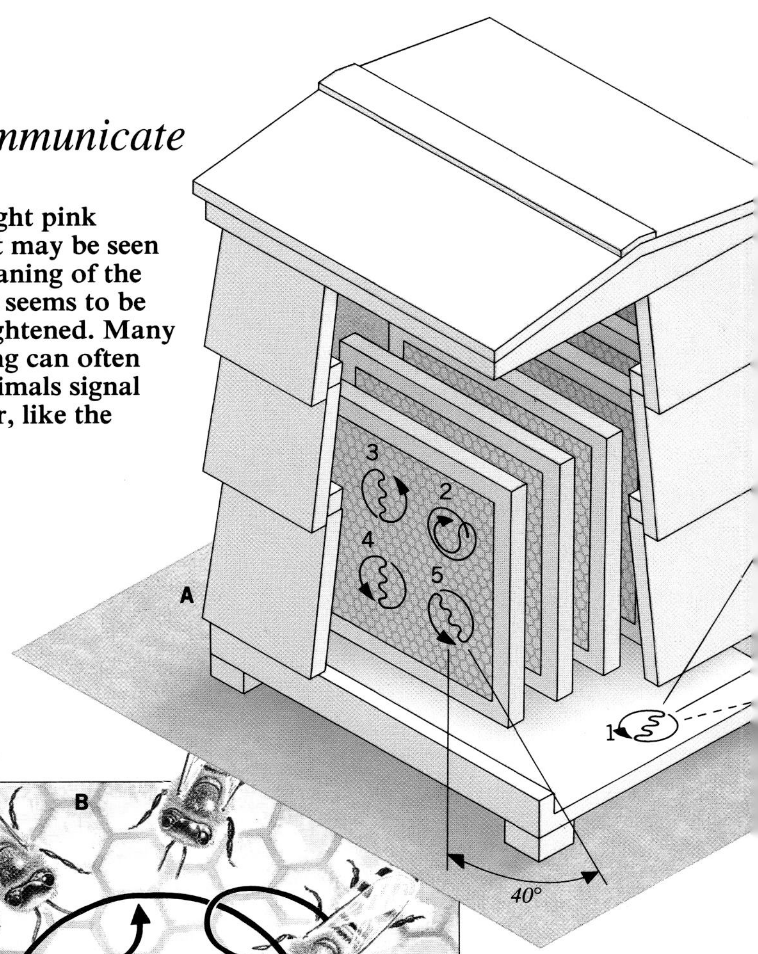

One of the most spectacular sights of the natural world may be seen in the mangrove swamps of Malaysia, where fireflies gather to mate. The mating assemblies form along stretches of open water and are so large that they can extend for as much as 330 feet (100 meters), with a firefly on almost every leaf of every tree. They flash about 90 times a minute, but in perfect synchronization. Even those on the outermost trees flash in time with each other. It is the females that flash to find a mate, and the advantages of flashing en masse may be that males are attracted from a wider area, while potential predators are deterred by the dazzling pulses of light, or simply glutted by so much available prey.

Squid semaphore

Fireflies are not alone in generating light for displays. In the deep ocean, where no light penetrates, live species of squid whose tentacles are lit up with rows of light organs, looking like strings of fairy lights. But it is by the light of day that visual communication can be at its most varied and informative. The most complex invertebrate displays occur in the cuttlefish, squids and octopuses because their vision is the most highly developed, with eyes as sophisticated as those of birds and mammals. Most of these mollusks can change color at will. When angry, frightened, or sexually aroused, many species pulse with vivid bands of color, one startling hue sweeping over the animal, to be replaced by another that is equally brilliant. An octopus supplements the color changes with bizarre textural changes to its rubbery skin, raising small bumps all over it, or suddenly sprouting large fingerlike projections.

Shades of meaning

Visual communication is relatively unimportant only in the lower invertebrates, whose simple eyes can do little more than detect light and shade. The external skeletons of insects, spiders and crustaceans make them less versatile in visual display, and rapid color changes are impossible. Nevertheless, butterflies use their colorful wings to good effect in courtship displays, while male fiddler crabs have one hugely enlarged claw, which apparently has an irresistible charm when waved at a female of the species. A great many insects, such as grasshoppers, moths and mantids,

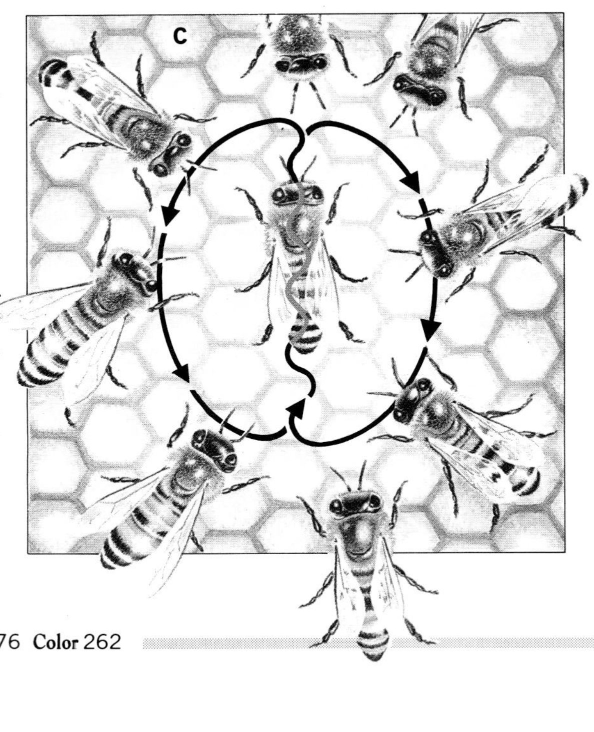

***Inside a dark hive,** foraging worker honeybees perform complex dances on combs, using touch (conveyed to bee "dance-followers" via their antennae), air vibrations and smell to communicate nectar and pollen locations to others. Nectar and pollen provide carbohydrates and protein – both vital to a colony. The round dance* **[B]** *indicates food about 150–300 ft (50–100 m) from the hive, the dance's length and vigor showing the source's richness. A waggle dance* **[C]** *consists of a figure eight with a waggle in the middle, and indicates more distant sources (waggle orientation shows direction and waggle frequency shows distance). The dancing bee's smell may indicate the flower types. To communicate and utilize the information requires a knowledge of the Sun's position (even if hidden, the Sun's position can be detected by light-polarization levels), a sense of time and wind speed, the use of visual landmarks and perhaps some sort of magnetic sensory device.*

Connections: Bees' and Wasps' Nests 232 Bioluminescence 276 Color 262

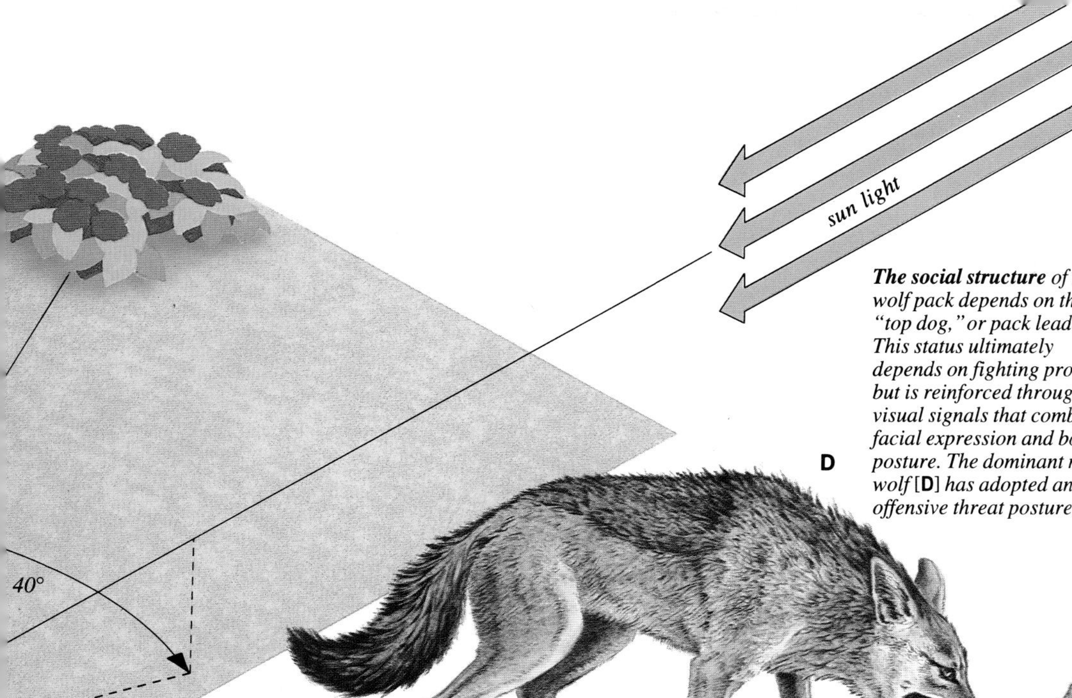

The social structure *of the wolf pack depends on the "top dog," or pack leader. This status ultimately depends on fighting prowess, but is reinforced through visual signals that combine facial expression and body posture. The dominant male wolf* [**D**] *has adopted an offensive threat posture, with arched back, erect hair and open jaws, all indicating that aggression is barely controlled. The other wolf* [**E**] *stands in the defensive threat posture. The subservient crouch and the positioning of the tail hung between the legs both acknowledge the superiority of the dominant male.*

Bee dances [**A**] *usually take place in the darkness of the hive on the comb sides, and are interpreted mainly by touch and smell. However, the bees are essentially communicating visual directions based on the Sun's position. This is apparent when the waggle dance is performed on the horizontal entrance board of a hive* [1]. *The angle of the waggle indicates a nectar source 40° to the left of the Sun. Inside the hive, the vertical surfaces do not affect the round dance* [2], *but a downward* [3] *waggle indicates a direction away from the Sun, and an upward one* [4] *toward the Sun. The dance* [5] *is a vertical representation of* [1].

often use their wings to present startling threat displays whenever they are approached by predators.

Higher animals

The equivalent threat display among birds is to fluff out the feathers, fan out the tail, or raise a crest of specialized feathers on the head. Such displays may be involved in attracting a mate, deterring a rival or seeing off a predator.

Mammals that are threatened by predators can make their fur stand on end to increase their apparent size, and male hooded seals have strange, inflatable membranes that balloon out from their nostrils to intimidate other males and impress sexually receptive females. Elephants may attempt to see off potentially dangerous intruders by mounting a mock charge, trumpeting and waving their ears away from their bodies like flags.

The male three-spined *stickleback (above) defends its nest or territory against another male by adopting a threatening head-down position. This posture carries an unambiguous visual signal by displaying both the red belly and the iridescent blue head at the same time. The response is instinctive, and is triggered purely by the visual stimulus of another male stickleback. This has been shown to be the case in a classic experiment of animal behavior. In the laboratory, the head-down response has been triggered by exposure to suitably painted bits of wood, and even by the fish's own reflection in a mirror.*

Floral Signposts

To help the pollinator find its nectar reward, flowers frequently have visual signposts. Sometimes these are patterns visible to us, as with the contrasting yellow and blue of the forget-me-not. But often it is an ultraviolet pattern only seen by insects. Mountain dryads [1] and alpine arnicas [2] look yellow to us, but insects see dark veins leading to a darker center, where nectar is found.

Mammal Societies 298 **Pollination by Animals** 154 **Scent Communication** 294 **Sound Communication** 296 **Vision** 278

Chemical Conversation

How scent is used to communicate

Camels swish their tails when urinating to spray themselves with the powerful scents the urine contains. White rhinos stamp in their feces to spread their own smell with each step they take. Nothing symbolizes the human break with its animal ancestry more powerfully than a stick of deodorant. For all other mammals, personal smell is one of the joys of being alive, a source of reassurance and comfort, a way of establishing bonds with close relatives, finding the way home or claiming territory. Most importantly, it is a vital element in sex and reproduction.

Anyone who "owns" a cat is an unwitting participant in their chemical conversation. Cats rub themselves against the legs of people in their home in order to leave a trace of their personal perfume, produced by scent glands on the cheek. When stroked, they often turn and lick their fur vigorously to pick up traces of the stroker's scent. This behavior is a way of establishing a "shared scent" that is part of group identity.

In the wild, animals do this only with members of their own species. European badgers "musk" other members of the set, raising the tail and backing toward another animal, or straddling it, so anointing it with the pungent secretion from a gland next to the anus. By exchanging scents regularly, the colony establishes a common identity, but one that is flexible enough to admit newcomers and change with the composition of the group.

Scent identification

The group smell undoubtedly enables badgers to distinguish their own kin from strangers. If an alien badger feeds within the territory, its feeding spots will quickly be noted and scent-marked by the possessive residents, apparently as a warning. Hamsters, finding the scent-mark of an intruder within their territory, grind their teeth in a display of aggression. But the old idea that scent-marking is all to do with territory is far too simplistic. Individuals without territories often scent-mark, and those with territories scent-mark their entire home patch, and not just the boundaries. Some dabs of scent are clearly there to mark out trails, which can be followed even on the darkest night. Badger scent-trails, laid on traditional pathways over farmland, are so strong that they survive ploughing. Bushbabies urinate on their hands before setting off to feed, and can follow the track of smelly paw-prints home.

In the mating season, scents take on another, far more obvious, role. Dominant males usually make more numerous scent-marks at this time, and there may be constituents in the scent that affect the behavior of females or subordinate males. Among rats, substances in the males' urine can accelerate puberty in young females, or block pregnancy in females that have mated with another male. In crowded conditions, the urine of adult females can delay puberty in young

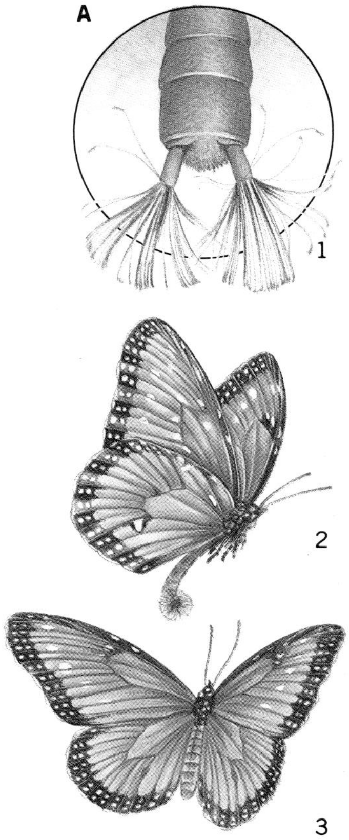

Odor plays an important *role in insect courtship* [**A**], *both in attracting the sexes to one another and in influencing their subsequent behavior. The male queen butterfly courts the female using scent produced by an abdominal gland and dispersed by two retractable "hair pencils"* [1] *at the tip of the abdomen. The male* [2] *flies around the female* [3], *occasionally bumping into her, at which point he liberally douses her head and antennae with his scent. The pheromones released by the male induce the female to alight on the ground and remain in one place long enough for copulation to take place.*

Scent glands *are present in nearly all mammals, but perhaps the prime exponent of scent communication is the oribi* [**B**] – *a dwarf antelope found throughout central Africa. This animal has no less than six scent glands: one, visible as a dark patch* [1], *is found beneath each ear. Another is situated to the front of each eye and appears as a groove, covered with a fleshy lid, leading to a dark, glandular mass* [2]. *Other scent glands are found on the feet* [3], *on the knees* [4], *and between the false hooves on the hind legs* [5]: *male oribis also have tufted saclike glands* [6] *situated near the scrotum.*

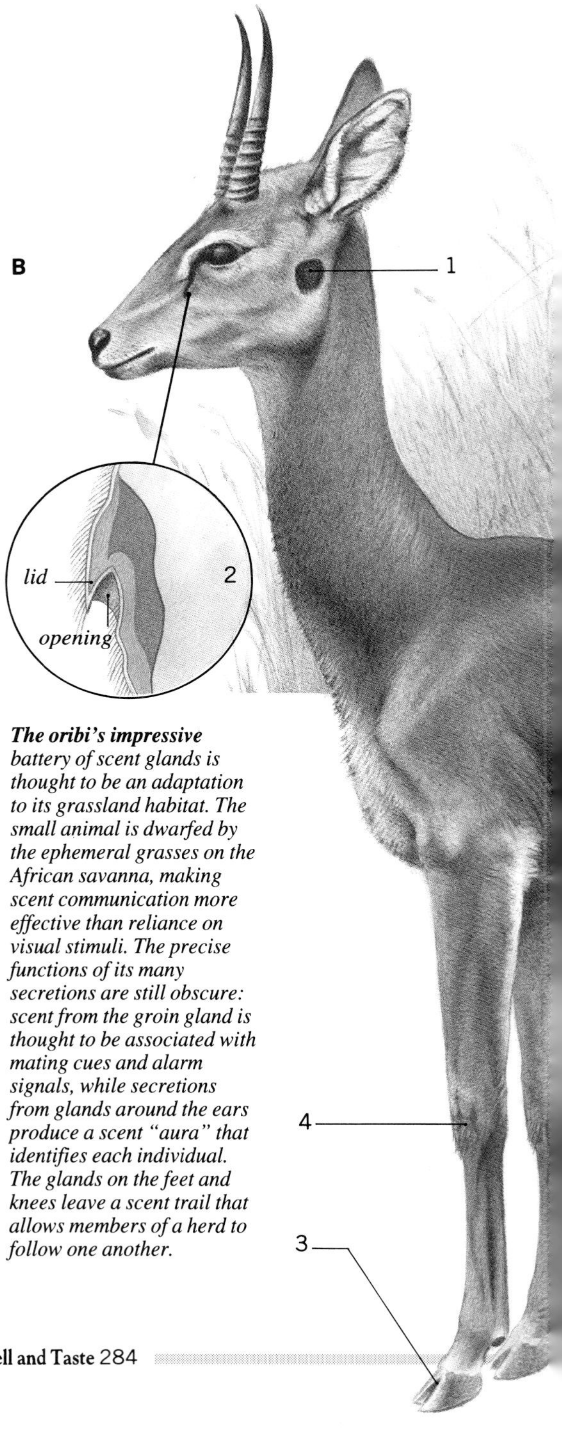

The oribi's impressive *battery of scent glands is thought to be an adaptation to its grassland habitat. The small animal is dwarfed by the ephemeral grasses on the African savanna, making scent communication more effective than reliance on visual stimuli. The precise functions of its many secretions are still obscure: scent from the groin gland is thought to be associated with mating cues and alarm signals, while secretions from glands around the ears produce a scent "aura" that identifies each individual. The glands on the feet and knees leave a scent trail that allows members of a herd to follow one another.*

Connections: **Invertebrate Reproduction** 160 **Plant Defenses** 244 **Pollination by Animals** 154 **Smell and Taste** 284

D

Coke's hartebeest *[D] uses secretions from its facial gland to mark itself: by rubbing its gland on its own forequarters with a sideways swing of the head, it creates a prominent dark patch on its coat. It then rubs its body against its rivals, mates and inanimate objects to mark them indirectly.*

C

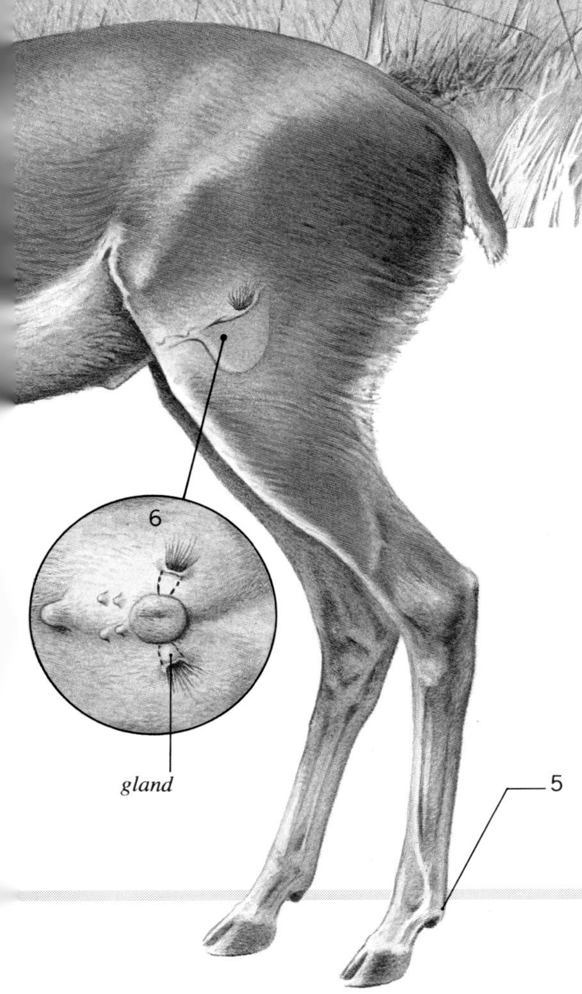

The oribi's facial glands *are used for marking out territory [C]. A number of antelopes, such as the dik-dik and impala, stake out their territory rather crudely with feces and urine. The oribi's marking behavior, however, is more systematic. The male first bites off a grass stem to head height [1], then pushes his scent gland over the cut stem [2], covering it with a sticky black secretion [3]. Moreover, he periodically re-marks the same stem to keep the scent fresh [4]. Such marking is most common along the male's territorial boundaries, but is also done at random throughout his range, and at times of stress, mating and conflict.*

females. Hormonal substances that control behavior in this way are known as *pheromones*. Those that affect the timing of physical change, such as puberty, are called priming pheromones. Others that simply attract a mate or warn of danger are called signaling pheromones.

Perhaps most surprising of all is the fact that some plants produce alarm pheromones when attacked by insects. Trees in particular seem to respond to such signals from others by packing their leaves with more tannins – bitter chemicals that interfere with the digestive process and thus deter herbivorous insects. Maize plants produce an alarm pheromone in response to aphid attack; this alarm pheromone is in turn used as a signal by parasitic wasps that lay their eggs inside aphids. The parasitic wasps can thus target the aphids for their own purposes through the information given by a third party.

Spreading Fear

Most social insects produce alarm pheromones when their colony is attacked. The alarm pheromones emitted by soldier ants attract other soldiers and rouse them to aggressive behavior. Some species of ants have subverted the message of the alarm substance to their own ends. Slave-maker ants attack other ant colonies and carry off the pupae to rear as slaves. Their attack is facilitated by the huge quantities of alarm pheromone they produce from their enlarged Dufour's gland [1] – a comparison of the size of the gland in the slave [2] and the slave-maker [3] reveals their pheromone-generating capacity. The pheromones cause panic among the colony's defenders, leaving it defenseless. An alarm pheromone used in this fashion is known as a "propaganda substance."

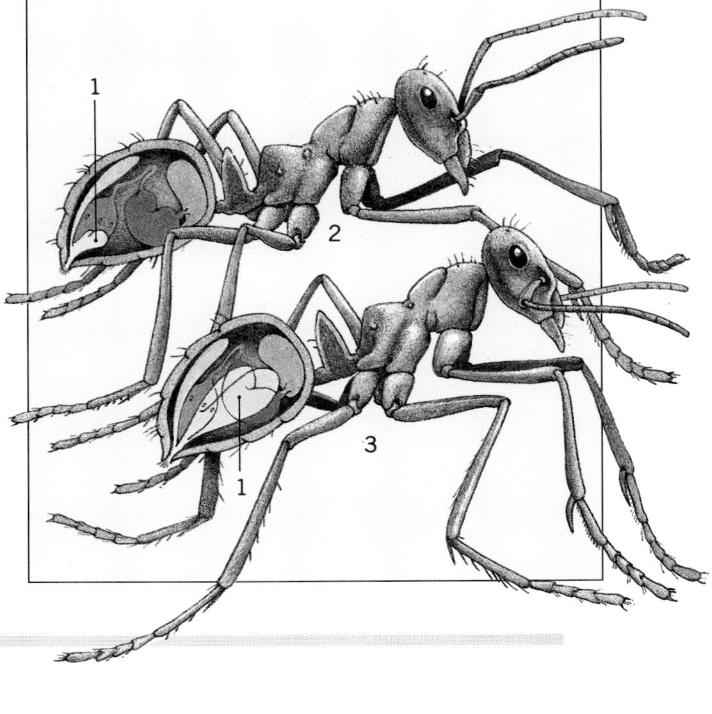

Loud and Clear

How animals communicate with sound

One deep-sea baleen whale can hear another's booming groans from at least 50 miles (80 kilometers) away. Variations in temperature, salinity and density of the sea create invisible "pipes" that trap and transmit sound so that it loses little energy over long distances. Bird audio signals vary from mute storks that clack their beaks to "mastersingers" with repertoires of several thousand songs; like humans, many birds have regional accents passed on from parent to young. Fruit flies and other insects beat their wings to signal, with thousands of different "songs" to distinguish species.

The male European buzzing spider is as adept as the baleen whale in using the special features of its environment to transmit a sound message. Sitting on a leaf, it moves its abdomen back and forth very rapidly, so that the leaf vibrates like the skin of a drum. The buzzing sound of the leaf acts like a mating call to attract female spiders of the species. Male mole crickets excavate special burrows for the same purpose: they dig two trumpet-shaped openings that act as megaphones to amplify the cricket's own mating call. Measurements of the sound taken just over a yard above the "megaphone" have been recorded at levels as high as 92 decibels – as loud as traffic on a main city street.

Not all animal communication is audible to the human ear; indeed, studies are often hampered by our own insensitivity to certain frequencies of sound. Small animals – shrews, mice and voles, for example – generally communicate in high-pitched sounds way beyond our hearing range. Such frequencies are also beyond the range of predatory owls. Nonetheless, owls do possess superb hearing sensitivity, and so do many other birds. A recording made of an American whippoorwill's song will only reveal that it actually contains an extra note normally inaudible to the human ear when replayed at an eighth of its normal speed. A second recording of a mockingbird actually imitating the whippoorwill's call, played back at the slower speed, also revealed the extra note, suggesting that birds in general can distinguish more sounds per second than humans can.

Sending signals

Sound is ordinarily transmitted by the vibration of air or water molecules. But these are not the only forms of vibration that convey information. The ground transmits vibrations too. The species of mole rat that mostly lives alone rather than in colonies hammers its head on the roof of its burrow to announce its presence to other mole rats, so avoiding a tunnel confrontation. Yet in the mating season, it is by hammering out a different rhythm characteristic to its own species that the same mole rat finds a mate.

Many insects have vibration detectors in their legs. The legs of a cockroach carry sound sensors so sensitive that they can detect the footfall of another cockroach.

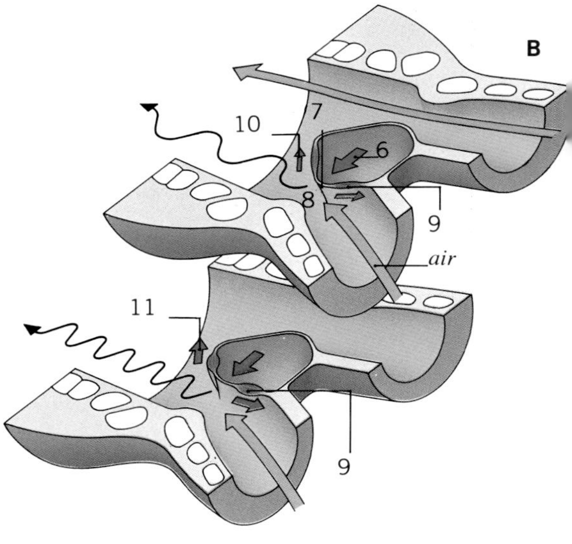

Many birds **[A]** *show astonishing range, quality and agility in their singing.*

A bird's respiratory system is very complex; its relatively small lungs [1] *are supplemented by connected unmuscled air sacs* [2] *filled or emptied by the chest muscles. To sing, a bird first closes a valve in one of the two bronchi* [3] *between the lung and* syrinx [4, *and enlarged* **B**], *or voice box, allowing it to compress air in the sacs. Air pressure in the clavicular sac* [5] *surrounding the syrinx* [**B**] *forces* [6] *very fine tympanic membranes* [7] *into the bronchial passage* [8], *briefly closing it. Muscles in the syrinx* [9] *then tense* [10 *and* 11], *opposing air pressure in the sac and pulling the membrane back to reopen the bronchial passage. Air rushes across the tensed membrane, vibrating it in song. As tension grows, so the song's pitch rises, just as a drum's note rises as its skin gets tenser – though the bird's song is more versatile. (Each pair of syrinx muscles of songbirds works independently, allowing different notes to be sung in each passage, or a note in one and nothing in the other.) The highly tensed muscles at* [11] *produce a higher note than at* [10], *where the muscles create less tension across the membrane. If the membrane is too taut, it stops vibrating, letting air pass through silently.*

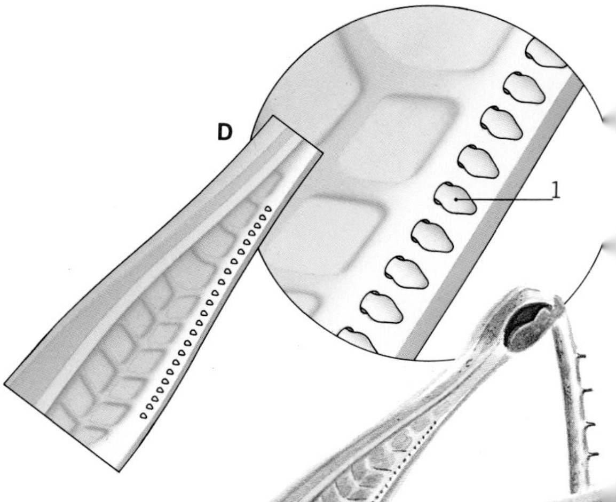

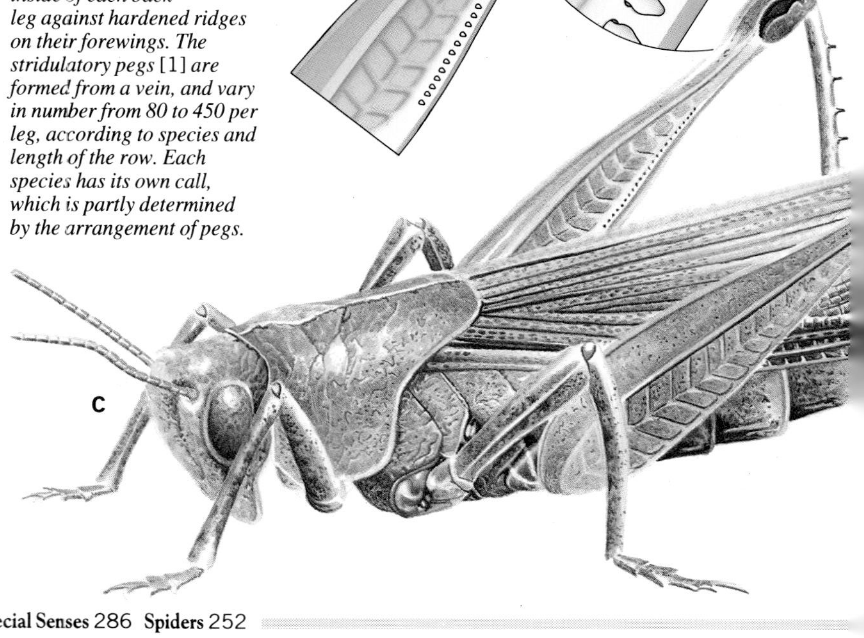

Grasshoppers **[C]** *have problems visually attracting mates in long grass, so they seek partners using sound signals; many other animals that are nocturnal, or live in dense vegetation use sound to communicate.*

Grasshoppers make their rasping mating calls by stridulating*: scraping a row of protruding pegs* [**D**] *on the inside of each back leg against hardened ridges on their forewings. The stridulatory pegs* [1] *are formed from a vein, and vary in number from 80 to 450 per leg, according to species and length of the row. Each species has its own call, which is partly determined by the arrangement of pegs.*

Connections: **Communication** 292 **Hearing** 282 **Sonar** 280 **Special Senses** 286 **Spiders** 252

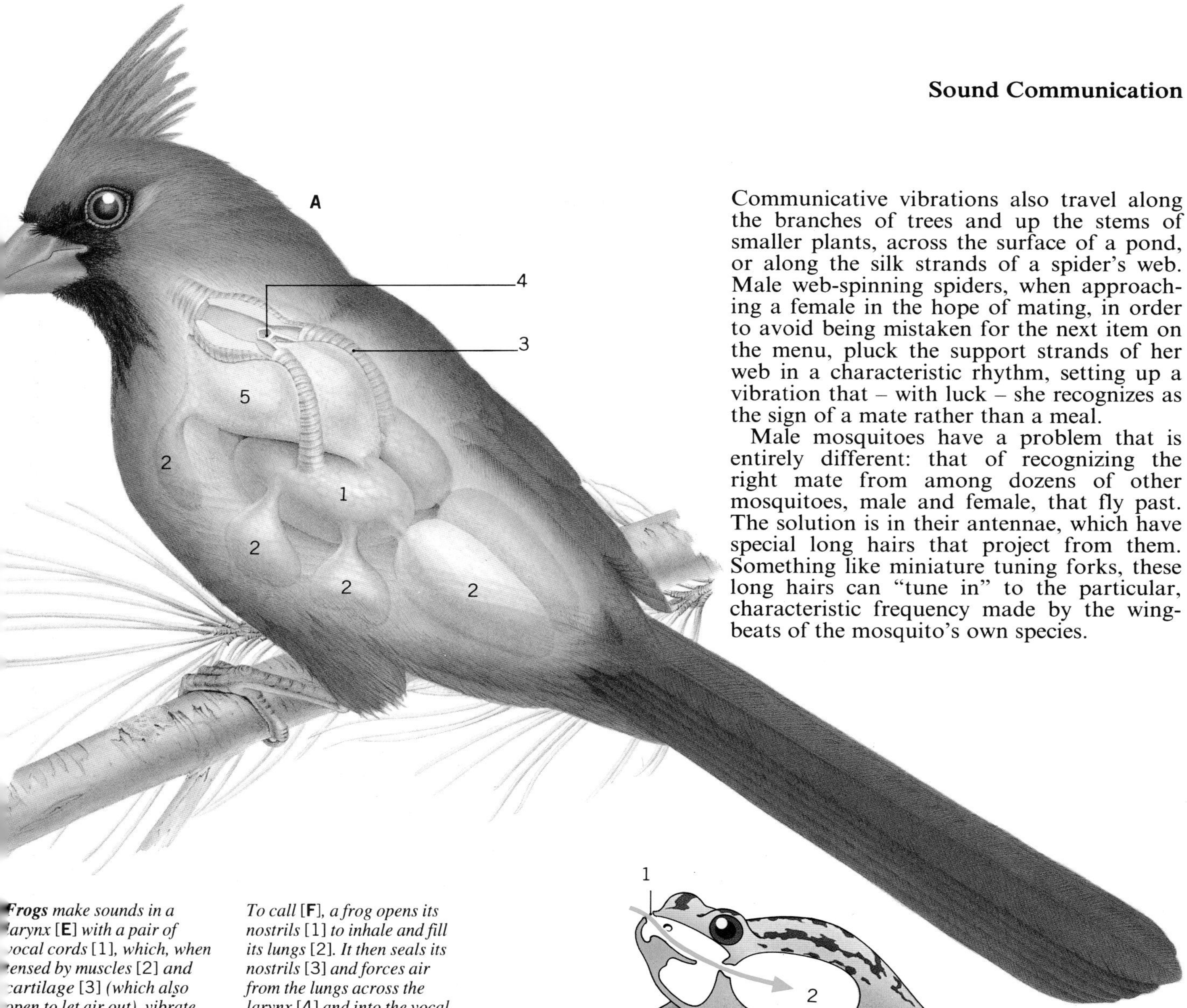

Communicative vibrations also travel along the branches of trees and up the stems of smaller plants, across the surface of a pond, or along the silk strands of a spider's web. Male web-spinning spiders, when approaching a female in the hope of mating, in order to avoid being mistaken for the next item on the menu, pluck the support strands of her web in a characteristic rhythm, setting up a vibration that – with luck – she recognizes as the sign of a mate rather than a meal.

Male mosquitoes have a problem that is entirely different: that of recognizing the right mate from among dozens of other mosquitoes, male and female, that fly past. The solution is in their antennae, which have special long hairs that project from them. Something like miniature tuning forks, these long hairs can "tune in" to the particular, characteristic frequency made by the wingbeats of the mosquito's own species.

***Frogs** make sounds in a larynx [**E**] with a pair of vocal cords [1], which, when tensed by muscles [2] and cartilage [3] (which also open to let air out), vibrate as air crosses them; greater tension produces higher notes, and faster air makes louder notes. Mating calls vary for species, but the basic process is the same.*

*To call [**F**], a frog opens its nostrils [1] to inhale and fill its lungs [2]. It then seals its nostrils [3] and forces air from the lungs across the larynx [4] and into the vocal sac [5]. Air is shunted back and forth from lung to sac to produce croaking. Vocal sacs [**G**] amplify the sound up to a hundred times, and may be single [1] or paired [2].*

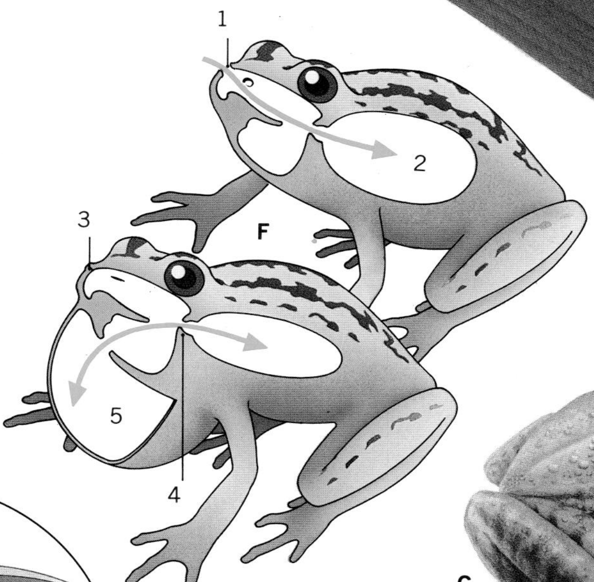

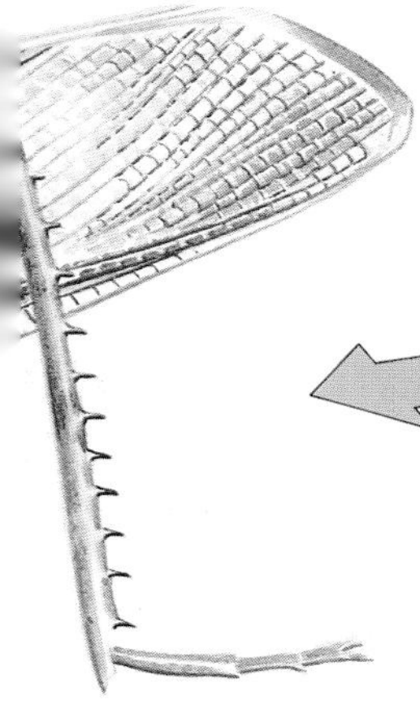

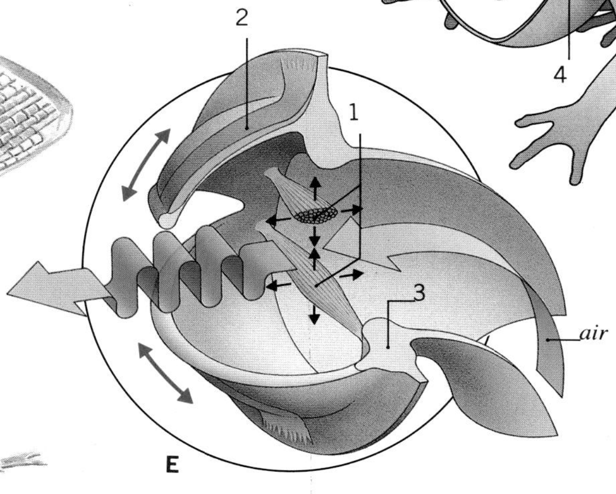

Community Life

How mammal societies work

Rank is such an integral part of Rhesus monkey society that it even affects the sex of offspring. Low-ranking mothers, for example, will generally give birth to more sons than daughters, whereas high-ranking mothers give birth to more daughters. Scientists believe that this sex determination occurs because males leave their group, and even if they are low-ranking they still have a chance to advance their status in a different group. Daughters, however, tend to remain in the same group, and will therefore always retain the status they were born with.

Not all mammals are social. Orangutans of the rain forests of Sumatra and Borneo are largely solitary, and the adults interact very little, except when they mate, or fight over territories. The young leave their mothers once they reach maturity and there is no further contact. Similar life-styles are found among moose and polar bears, but these are exceptional for large mammals. In general, as mammals have evolved, becoming larger and more intelligent, they have developed more elaborate social systems.

The binding factor

The major factor that keeps mammalian social systems together is milk – the liquid food that sustains all young mammals from the moment they are born up until they are weaned. It creates a bond between the mother and her young that often lasts way beyond weaning. Among birds, the males can contribute to hatching and feeding of the chicks as much as the females – they can even take over, as happens in a few species. Such role-reversal is impossible in mammals, and the scope for sharing parental responsibility is far more limited. For this reason, monogamous pairs (animals that only have one partner during their lifetimes), which are relatively common in birds, are unusual among mammals. Given that the prime objective is to pass on his genes to as many young as possible, a male's energies are better spent in trying to mate with other females than in caring for the young he has already sired.

Male competition

Females, with their prolonged pregnancy followed by milk production, are the limiting factor in mammalian reproduction – a female mammal can only bear so many young a year. Inevitably, where there is a limited resource, there is competition, and most males have a lifelong preoccupation with mating – it is by far their most important objective. Good examples of competitive systems can be seen in the Ugandan kob (a type of antelope) and the hammerhead bat, where a traditional breeding site or *lek* is the scene of intense displays, or ritualized battles, which lead to the most dominant males securing the most favored sites within the lek. The females then make their way to those sites and mating takes place.

***Meerkats** (above) are extremely social creatures, and at any one time several meerkats will stand guard on their hind limbs looking for approaching danger. In this way they are always prepared to warn other members of the group. Meerkats live in close-knit extended family groups.*

***Chimpanzees** have a highly complex social structure compared to other apes such as gorillas and orangutans. The number of individuals in a chimp community varies between 20 and 120. The community is made up of subgroups known as parties, each consisting of between three and six individuals.*

The parties, instead of being based on a family group, can be made up of either all males, all females, young, or adults, or a combination of sexes and ages. Moreover, individuals will often leave one party in preference for another. In this way there is a wide association within the whole community.

A more common mammalian system is for males and females to live in groups based loosely on a mother and her daughters. Such systems are seen in animals as diverse as mice and elephants, foxes and dolphins. In these matrilineal groups, young males are either driven out as they mature, as happens with lions, or they are tolerated but kept in subordinate positions, as with chimpanzees. The dominant male may secure all the matings, or there may be occasional chances for other, lower-ranking males to mate as well. Dominance hierarchies are established by occasional fights, frequent aggressive displays, and other mechanisms, such as grooming.

Some scientists also believe that there may well be hormonal factors at work in maintaining hierarchies. Dominant males and females, through aggressive behavior, are thought to suppress the sexual development of their subordinates.

***The relationship** between mother and offspring [**H**] is very close and long lasting. Grooming is one of the most important social interactions, and serves to bond the relationship. Large male chimps [**I**] protect the community by patrolling its borders and chasing away males from other communities.*

Connections: **Ants' Nests** 236 **Communication** 292 **Group Carnivores** 248 **Herbivores** 202 **Intelligence** 290 **Mammals** 136 **Mammal Reproduction** 176

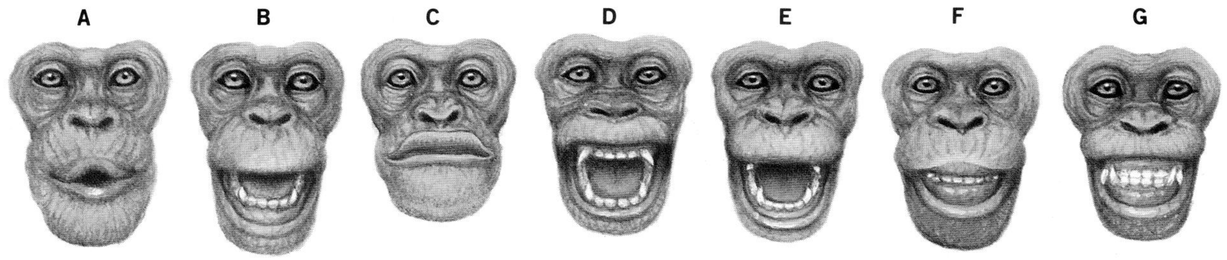

Facial expressions *can often reveal a chimpanzee's mood at a particular time. Each facial expression shown carries the same letter* **[A – G]** *as the activity it accompanies. When calling to one another, chimps alternate between the two "calling faces"* **[A** *and* **B]**. *The calling sounds, known as "pant-hoots," are made in a variety of contexts, such as alerting other members of a group to a food source, or when two different groups meet. The "display face"* **[C]** *is made by an aggressive chimp, particularly during charging displays, or when attacking another chimp. The subordinate chimp may well respond with a full, open grin* **[D]**. *This usually indicates fear, in deference to the superior chimp, but may also be used in times of great excitement. The "play face"* **[E]** *is shown, particularly by young chimps, during periods of play, when young chimps learn adult skills, such as "termite fishing" – using a stick to obtain termites from nests. Pouting* **[F]**, *usually accompanied by a whimpering sound, shows dissatisfaction. Young chimps who want to be fed or groomed will pout. The closed grin* **[G]**, *like the full, open grin, is indicative of subservience. Often a low-ranking chimp when approaching a superior will make this expression to show respect.*

Naked Mole Rats

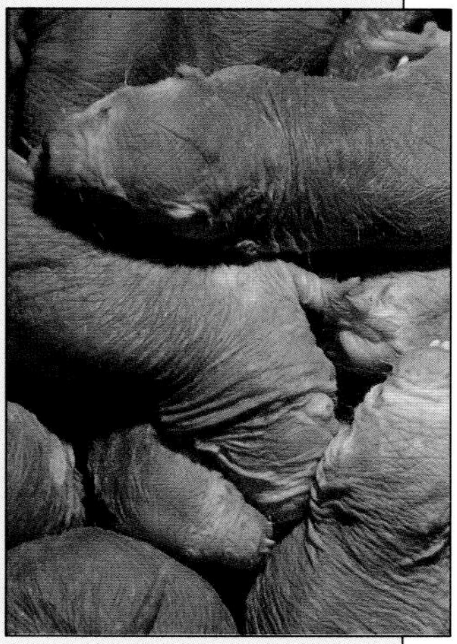

The naked mole rat of eastern Africa is probably unique among mammals, in that it has a social structure akin to that of bees, wasps, ants and termites. The large, isolated colonies contain only one breeding female, who does very little apart from produce young. Most of the hard work – excavating tunnels, defense, collecting food and feeding the young – is performed by a caste of small animals called "frequent workers." Another caste, known as "infrequent workers," spends most of its time lounging in the nest chamber with the queen and her pups, and thus helping to keep them warm. When the queen dies, one of the infrequent workers takes over her role, but the frequent workers never have the chance to breed.

Grasslands
Desert Life
Mountain Life
Cave Life
North Polar Life
South Polar Life
Tundra

9 The Living Environments

Boreal Forests
Temperate Forests
Tropical Forest Animals
Tropical Forest Plants
Wetlands
Lake Life
River Life
Seashores
Ocean Life
Estuaries
Coral Reefs
Land Pollution
Water Pollution
Air Pollution
Animal Conservation
Plant Conservation

The Fragile Balance

How grassland ecosystems work

Grasslands are some of the most valuable places on Earth, both ecologically and economically, because grasslands are not just the sweeping prairies of America, or the savannas of Africa: they can be completely man-made, like the vast grain-producing areas of the United States and the Ukraine. Wild grasslands provide a varied and productive habitat. Among the grasses many other flowering plants grow. Grazing animals and their predators roam the surface, while the ground beneath their feet is honeycombed with the tunnels of burrowing animals.

Grasslands first appeared about 25 million years ago, when the world's climate changed. Extensive grasslands developed on plains and rolling uplands in tropical and temperate regions where there was insufficient rainfall, or the soil was too poor in nutrients to support continuous tree cover. In Europe, where the climate and geography led to forest as the natural cover, most grassland is completely man-made, having been reclaimed from the forest for agricultural purposes. High-altitude grassland, above the treeline in mountains such as the Alps, is one of the few natural European grassland habitats.

The great natural grasslands of the world, especially the African savannas, are the home of many of the larger species of land mammals, both herbivores and their predators. Predominant among the grazing animals are the swift-running hoofed mammals – the deer, antelope, cattle and horse families – which often live in large herds, partly as protection from predators. The herds of grazing animals provide a well-stocked larder for carnivores. Lions, cheetahs, leopards, hyenas and hunting dogs in Africa prey on hoofed animals, and coyotes in North America hunt prairie dogs and other smaller prey.

A delicate equilibrium

In grasslands, the grasses and other herbaceous plants are the primary producers of basic foodstuff. Grassland can support so many relatively large grazing animals because, unlike many other herbaceous plants, grasses thrive on being eaten. The leaves of grass grow from the base and can survive and regrow even if their tips are nibbled off. The permanent shoot, from which the whole plant grows, is kept safe close to the ground, at the base of the plant, or even underneath the soil.

Little of the world's grassland has remained untouched by man. For thousands of years, nomadic herdsmen have grazed their cattle, sheep and goats on the grasslands of Asia and Africa, and farmers grow wheat on much of the former North American tall-grass prairie. Some of the larger animals, such as the North American bison, have been hunted to near extinction. A combination of overgrazing and many severe droughts now threaten to destroy large areas of the dry grasslands around the Sahara in Africa.

***The African savanna** is home to large numbers of animals. The savanna's plains provide abundant vegetation for herbivores. In wetter areas, trees such as the baobab* [1], *umbrella* [2] *and* Euphorbia candelabrum [3] *grow alongside the whistling thorn* [4], *home of the weaverbird* [5].

***Many of the kills** attributed to lions* [6], *which do indeed prey on herbivores such as wildebeest* [7], *zebra* [8] *and waterbuck* [9], *are actually often carried out by hyenas* [10] *– the lions then scavenge at the kill. Another hunter, the leopard* [11], *often lies in trees overlooking waterholes. If it makes a kill, it may drag the victim up a tree away from scavengers such as the marabou stork* [12]. *The numerous species of herbivores, including elephants* [13], *cape buffalo* [14], *giraffe* [15] *and impala* [16], *can all survive on the savanna because they feed on different types of vegetation, and competition for food is therefore reduced. The hides of these large grazers, particularly the buffalo, provide food in the form of ticks and sometimes flesh and blood for the tiny oxpeckers* [17]. *The banded mongooses* [18] *also have their own niche, feeding on animals such as the African striped skink* [19]. *Dung beetles* [20] *collect dung on which to lay their eggs.*

Connections: **Decomposers** 204 **Flowering Plants** 120 **Group Carnivores** 248 **Herbivores** 202 **Termite Towers** 234

11
3
15
13
4
8
17
6
16
14
18
20

The Arid Wastelands

How wildlife survives in the desert

In the creosote bush of North American desert regions, the water content of the leaves drops to 50 percent of the leaves' dry weight – in comparison, the leaves of temperate forest trees have a water content of up to 300 percent of their dry weight. Water is a life-giving element: it has the power to turn vast regions of barren desert into areas thick with colorful vegetation. Water also affects the extent of deserts. During summer the Sahara is 40 percent larger in area than in the winter, when the more frequent rains force back the desert boundaries.

When we think of desert landscapes, we think of vast stretches of rolling sand dunes, interspersed with the occasional oasis. But most deserts are not like this – only relatively small areas of certain deserts, such as the Sahara, have endless miles of sand. More often, desert landscapes are rocky, barren and sometimes mountainous.

To survive in such harsh conditions, plants have evolved a variety of adaptations. Many, such as the African grass, have accelerated life cycles. Its seeds can lay dormant for many years, but with the advent of rain, the seeds germinate and new plants grow, flower and set seed in just over two weeks. Other plants, such as the cacti of North America, are highly specialized water storers – such plants are called succulents. One of the most remarkable desert survival techniques belongs to "resurrection plants" such as *Xerophyta*. In times of drought the leaves of these plants shrivel and turn brown, and reduce their water content to a bare minimum. As soon as rain falls, however, the leaves rapidly absorb water, turn green and start to photosynthesize. This adaptation allows resurrection plants to start photosynthesizing much quicker than those plants that have to generate new leaves.

Desert diets

Desert food chains begin with plants. Bugs, grasshoppers and crickets feed on the sparse vegetation. After the rains, swarms of millions of locusts may invade.

Many desert plant-eaters get the moisture they need from their food. Desert carnivores such as grasshopper mice, which eat invertebrates, also obtain the water they need from their food. Other carnivores range from the antlion to the puma. In the American deserts, bobcats are at the top of the food chain, killing and eating such herbivores as mule deer and jack rabbits.

The main conservation threats to desert wildlife are human activities: hunting animals and collecting plants. The Arabian oryx was hunted to extinction in the wild, but has since been reintroduced. The African wild ass has also been hunted, and has suffered from incessant military activity almost throughout its range. Cacti have suffered merciless exploitation: the "living rock" cacti of Mexico are sought out by collectors.

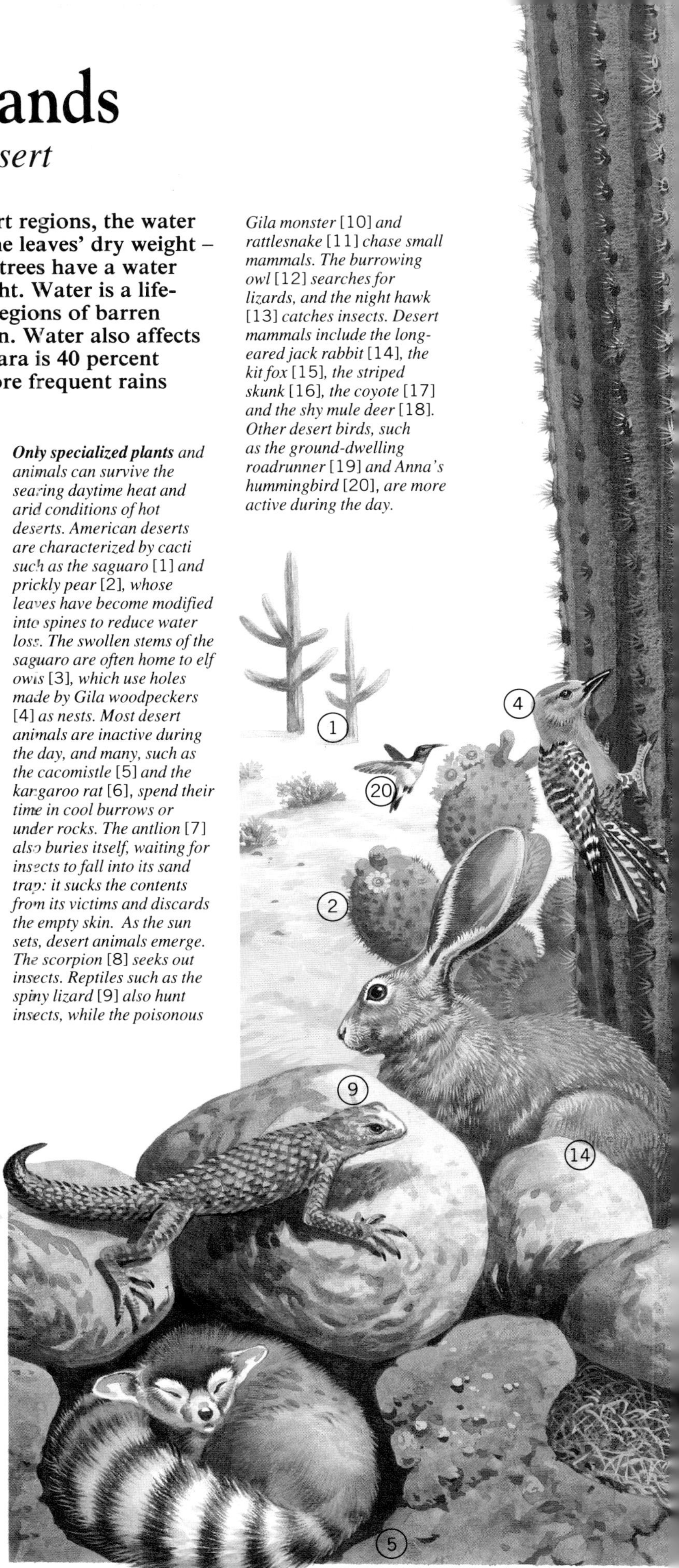

***Only specialized plants** and animals can survive the searing daytime heat and arid conditions of hot deserts. American deserts are characterized by cacti such as the saguaro* [1] *and prickly pear* [2], *whose leaves have become modified into spines to reduce water loss. The swollen stems of the saguaro are often home to elf owls* [3], *which use holes made by Gila woodpeckers* [4] *as nests. Most desert animals are inactive during the day, and many, such as the cacomistle* [5] *and the kangaroo rat* [6], *spend their time in cool burrows or under rocks. The antlion* [7] *also buries itself, waiting for insects to fall into its sand trap: it sucks the contents from its victims and discards the empty skin. As the sun sets, desert animals emerge. The scorpion* [8] *seeks out insects. Reptiles such as the spiny lizard* [9] *also hunt insects, while the poisonous Gila monster* [10] *and rattlesnake* [11] *chase small mammals. The burrowing owl* [12] *searches for lizards, and the night hawk* [13] *catches insects. Desert mammals include the long-eared jack rabbit* [14], *the kit fox* [15], *the striped skunk* [16], *the coyote* [17] *and the shy mule deer* [18]. *Other desert birds, such as the ground-dwelling roadrunner* [19] *and Anna's hummingbird* [20], *are more active during the day.*

Connections: **Animals in Heat and Drought** 142 **Deserts** 54 **Germination** 158 **Grasslands** 302 **Migration Patterns** 228 **Plant Defenses** 244

Plants in Heat and Drought 138 Reptiles 130

Scaling the Peaks

How plants and animals live on the mountain slopes

The temperature on a mountain slope falls about 0.9°F (0.5°C) every 330 feet (100 meters). Tropical mountains are oases of cold in a landscape of unremitting heat. Wildlife adapted to one mountain environment is thus unable to migrate to other mountains, so ecosystems develop in isolation, often with strange, unique consequences. Mountaintop hazards include intense sunlight (which some single-celled plants protect themselves against by turning red, often staining whole snowfields with this color) and forceful winds, as well as the freezing temperatures.

Red algae are the only plants living on the snowfields above the snowline, but not the only source of food for animals. Mountains act as blocking walls to moving masses of air, forcing them to divert or rise rapidly, thus creating violent winds. The winds that sweep upward from the lower slopes bring with them pollen, tiny seeds, flying insects and other nutritious debris. Waiting to feast on it are hordes of springtails that live among the snowfields and are so well adapted to mountain life that they can survive being frozen solid within a glacier for up to three years. Preying on the springtails in turn are a variety of mites, beetles and flies.

Clinging for life

On the mountain peaks the ecosystem is relatively impoverished. Farther down, however, lichens and mosses appear on the rocks, and flowering plants among the scree slopes. Such plants are highly specialized, many have developed thick, succulent leaves that improve their ability to store water because rainfall is sparse and quickly drains away. Two plant adaptations are particularly common: a compound mound shape that retains warm daytime air after nightfall, and a rosette in which the leaves are pressed close to the ground so that the *stomata* (pores) are shielded from the Sun and drying winds. As a protection against the intense sunlight, the upper surface of the leaves is usually thickened. Some plants, such as the alpine snowbell of Europe, can actually generate heat to melt the snow layer above them in spring.

High climbers

Feeding on the plant seeds are mice and voles, while insects such as grasshoppers and thrips feed on the plants themselves. All mountain insects stay close to the ground to avoid being blown away. Many shelter within mound-forming plants, and some species have lost the power of flight altogether. In Europe, butterflies and moths take nectar from flowering plants, while in Africa this job is also performed by sunbirds, and in the New World by hummingbirds. The scarcity of these pollinators forces the mountain flowers to evolve exceptionally eye-catching displays, creating a blaze of color in spring. Birds such as snow buntings and accentors feed on the insects.

Vertical migration

Many animals take advantage of the different vegetation zones created by mountains by moving up and down the mountainside. Most mountain migrators feed high up during the summer, then move down to escape the severe winter cold. The snowcocks of central Asia, for example, move with the snowline, feeding largely on onions and crocus bulbs. They follow wild goats, relying on them to dig up the bulbs during their search for food.

Not all animals migrate, and some of those that stay on the mountains all year, such as the ptarmigan and the mountain hare, change color with the seasons, growing white feathers or fur for winter camouflage, then reverting to brown for the summer. The camouflage is needed to avoid the attentions of birds of prey, which use the air currents and strong winds to assist their flight among the mountain peaks.

***Most animals** of the Rocky Mountains do not roam freely from base to summit, instead there are strict zones of life dependent on the vegetation. Above the treeline [**A**], where only low-growing plants survive, few animals are resident in winter. During the warmer months, however, bighorn sheep [1] and the sure-footed mountain goat [2] scratch a living on the sparse vegetation. Smaller mammals include the pika [3] and hoary marmot [4]. Birds such as the endangered bald eagle [5], which hunts the white-tailed ptarmigan [6], and the water pipit [7], which spends most of the time on the ground searching for insects, are also temporary visitors. Moving down the mountainside [**B**], the wildlife is more active. Predators such as the wolf [1], the cougar [2] and the American marmot [3] hunt, while the black bear [4] searches for berries and the moose [5] browses on trees. Birds also are more apparent. Clark's nutcracker [6], the red crossbill [7] and the northern three-toed woodpecker [8] are all found here. The increase of vegetation in the mixed-forest zone [**C**] creates abundant food for animals, such as the white-tailed deer [1], the white-tailed jackrabbit [2], the red fox [3], the pine squirrel [4], the North American porcupine [5] and the short-tailed weasel [6]. Birds of the mixed-forest zone include Stellar's jay [7], the hermit thrush [8] and the blue grouse [9].*

***An equatorial mountain** has several vegetation zones. At its summit [**A**] it is too cold for anything to grow, while lower down lies the Afro-alpine region [**B**], where plants such as giant lobelia [1] and senecio [2] can survive the cold. Beneath this layer is the subalpine moorland region [**C**], where larger plants such as shrublike tree heaths [1] can take root. The bamboo belt [**D**], as it approaches its lower reaches at about 8,250 ft (2,500 m), gives way gradually to montane forest [**E**]. Beneath the montane belt, savanna [**F**], usually characterized by acacia trees [1], red oat [2] and Bermuda grasses [3], begins.*

***The zones** of a typical African mountain are governed by temperature, wind, exposure to sunlight and rainfall, as well as height. Many mountains have a dry and a wet side because as warm, moist air is propelled up the mountain, it cools, releasing moisture in the form of relief rainfall or snow.*

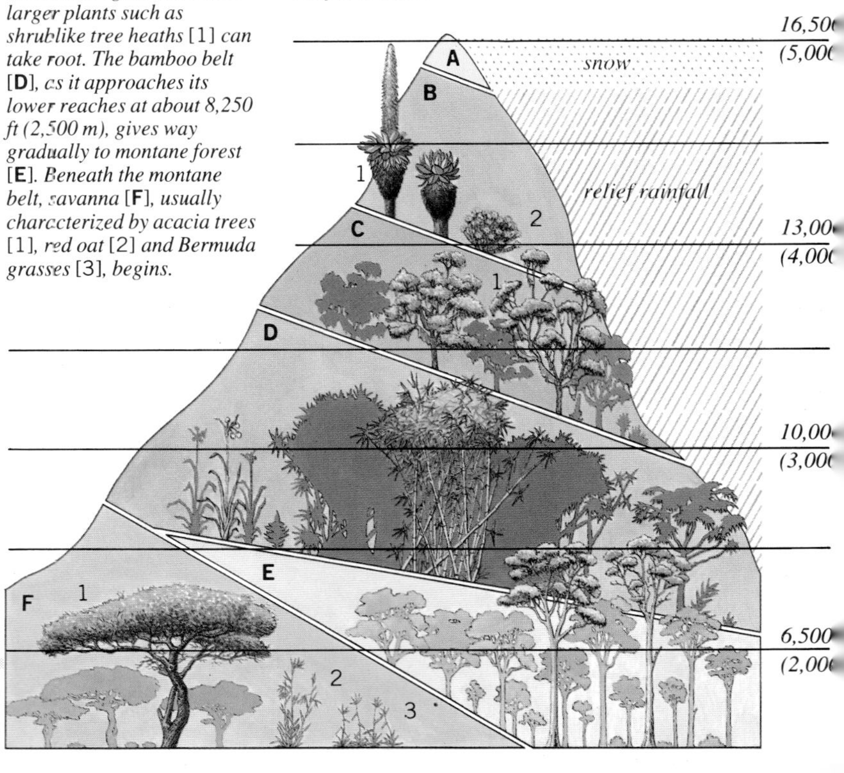

Connections: Animal Adaptation to Pressure 146

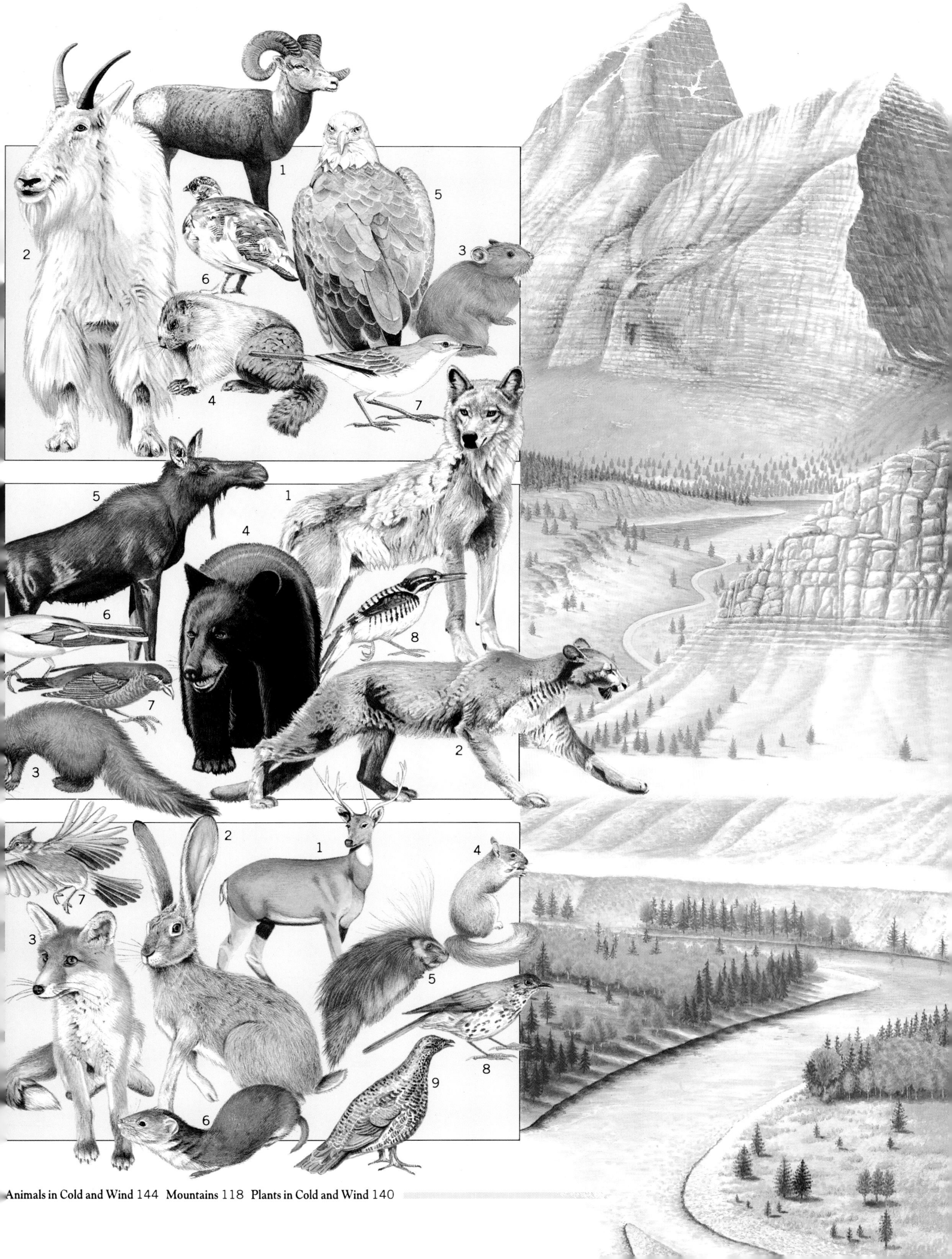

Animals in Cold and Wind 144 Mountains 118 Plants in Cold and Wind 140

Living in the Dark

How animals have adapted to caves

In tropical caves, cockroaches the size of frogs compete for food with millipedes as thick as your fingers. Most seemingly deserted caves are in fact teeming with life. Caves are like evolutionary islands, harboring animals that the outside world has long forgotten. Some animals are too specialized even to move from one cave to another. Some of the cave predators, for example, are directly descended from animals that have inhabited caves since the end of the last ice age, when the warming of the global climate forced them to retreat underground.

A cave is a barren environment, devoid of the light that sustains plant growth. Without plants and algae there is no food production, and caves would be entirely lifeless if they did not receive food from the outside world.

Dead leaves, which nourish fungi and some insects, may be blown into the cave by the wind. Other nutritious debris is washed in by streams that trickle down through sinkholes from the world above. But such supplies are meager, and the cave ecosystem really depends on animals that feed in the outside world but come to the caves for shelter.

Rich pickings

The droppings from cave-roosting creatures, particularly bats, are the key to life in most caves – a food source that sustains a whole community of specialized cave dwellers. The droppings form huge piles beneath the roost sites, piles that seem to be strangely alive by the light of a caver's lamp, for they are, in fact, seething with beetles, millipedes, springtails and, in tropical caves, cockroaches.

In some tropical caves, cave-nesting birds are another source of droppings. In parts of Trinidad, Venezuela and the Peruvian Amazon, the oilbird raises its chicks in precarious nests built on the narrow ledges of caves. The birds feed almost entirely on fruit, a poor diet for a growing nestling as it is low in protein. The chicks grow slowly, so the vulnerable nesting period is far longer than in other birds. By nesting in the relative safety of caves, the oilbirds have overcome the problems that their poor diet entails. To find their way in the darkness, they have evolved a system of echolocation based on clicking noises made with the tongue.

Underground hunters

Predators are as important in the cave ecosystem as anywhere else. Snakes may enter tropical caves, or even live there all year round, preying on bats and cave-nesting birds. Other predators are cave specialists and never emerge into the daylight. Among these are many species of spider. Some of the tropical spiders can measure 18 inches (45 centimeters) across and prey on young bats. Other predators include the blind salamanders and blind cave fish, whose acute senses of touch and smell, and an uncanny sensitivity to vibrations, guide them to their prey.

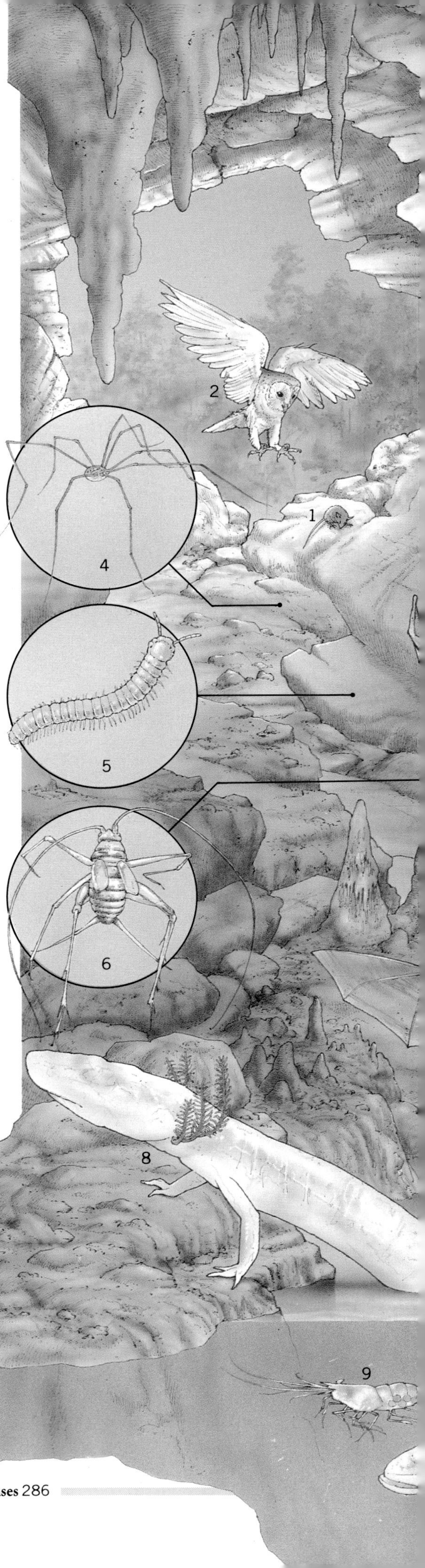

***Although each** cave ecosystem is unique, it is possible to identify the sorts of animals that might be found in a North American or European cave* [**A**]. *The most important distinction is between those animals that shelter in the cave and venture out to feed and those that permanently inhabit the cool damp darkness. Small mammals, such as the mouse* [1], *often nest around the entrance to caves. When they venture out at night, they attract outside predators such as the barn owl* [2]. *Deeper inside the cave, large colonies of bats, such as the greater horseshoe bat* [3], *may roost during the day, collectively leaving at night.*

The permanent residents, which are known as troglodytes, tend to be much smaller in size. The harvestman [4] *is a predator that feeds on other invertebrates such as the dung-eating millipede* [5] *and the scavenging cave cricket* [6]. *The spider and cricket both have elongated limbs for maximum agility over rock. Underground pools and lakes often support the largest troglodytes. Blind cave fish* [7] *and amphibians such as the Texas blind salamander* [8] *feed on invertebrates such as the blind cave shrimp* [9].

Connections: Caves 36 Decomposers 204 Invertebrate Carnivores 200 Sonar 280 Special Senses 286

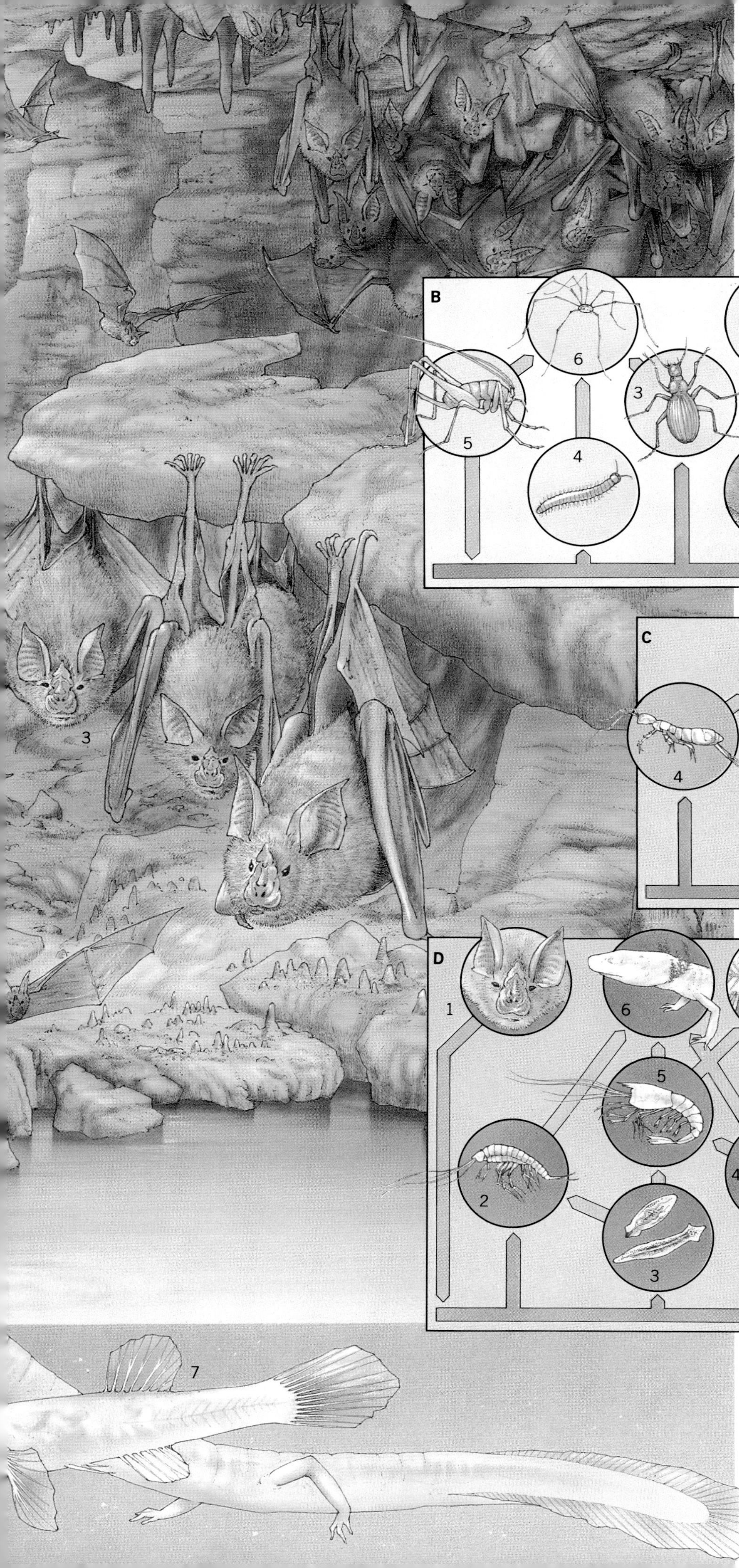

primary production

primary consumption

secondary consumption

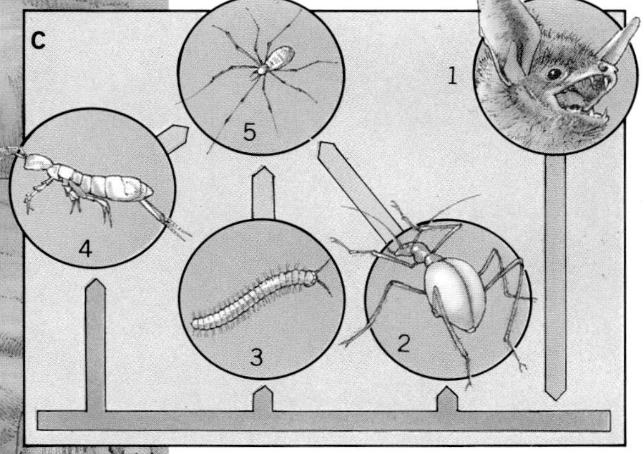

The food webs *found in caves are extremely fragile, and are based on very meager resources.*

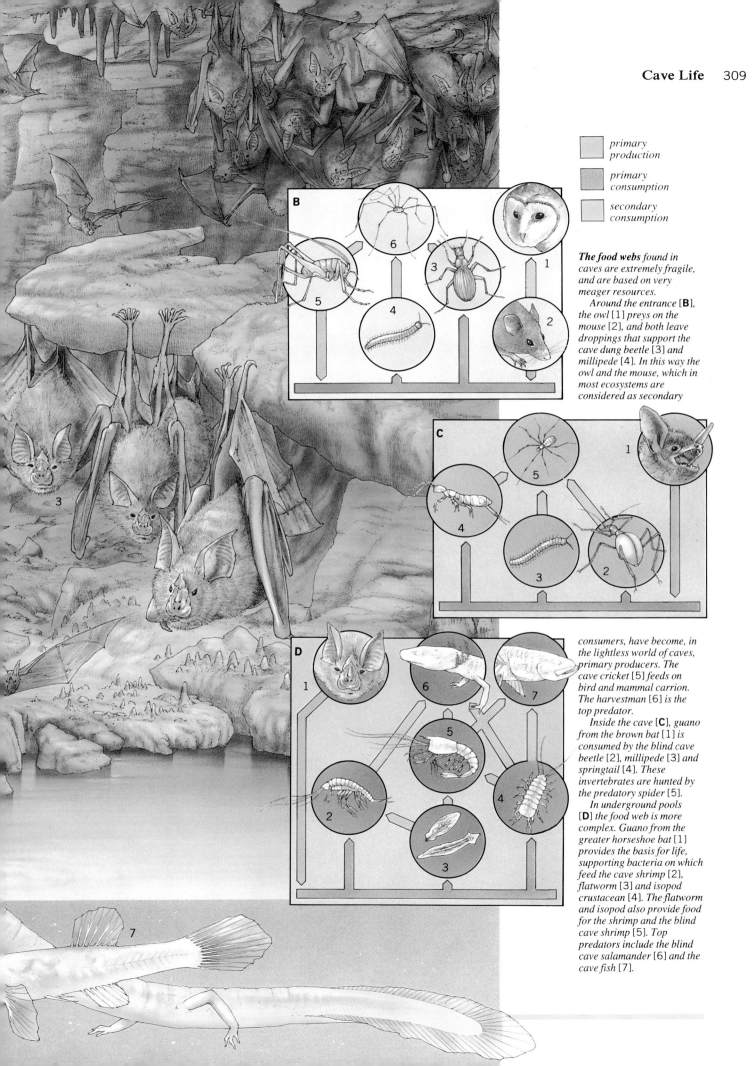

*Around the entrance [**B**], the owl [1] preys on the mouse [2], and both leave droppings that support the cave dung beetle [3] and millipede [4]. In this way the owl and the mouse, which in most ecosystems are considered as secondary consumers, have become, in the lightless world of caves, primary producers. The cave cricket [5] feeds on bird and mammal carrion. The harvestman [6] is the top predator.*

*Inside the cave [**C**], guano from the brown bat [1] is consumed by the blind cave beetle [2], millipede [3] and springtail [4]. These invertebrates are hunted by the predatory spider [5].*

*In underground pools [**D**] the food web is more complex. Guano from the greater horseshoe bat [1] provides the basis for life, supporting bacteria on which feed the cave shrimp [2], flatworm [3] and isopod crustacean [4]. The flatworm and isopod also provide food for the shrimp and the blind cave shrimp [5]. Top predators include the blind cave salamander [6] and the cave fish [7].*

The Frozen North

How plants and animals survive near the North Pole

Life during an Arctic winter is a constant struggle for survival. Intense cold, and some of the harshest conditions on Earth, present the ultimate challenge to the plants and animals of the region – that of simply staying alive until the next spring. Many birds and some animals leave the Arctic altogether for the winter months. Other animals and many plants take advantage of the insulation afforded by the snow, and spend the winter beneath it. But for many of those who live on the surface, life is hard indeed, and the modern world has brought a new threat – humans.

In the Arctic the air is very dry, very cold, and cannot hold much moisture – so snow is not as frequent or as heavy as is often supposed. For all that, most of the plants of the far north prefer snowy conditions. Those that get covered by a blanket of snow are insulated from the worst of the cold, for temperatures beneath the snow can be up to 45°F (25°C) higher than those above. Plants not covered by snow drifts have to endure frequent icy blasts and a dehydrating wind. The ground-hugging cushion forms adopted by many plants help to decrease snow abrasion, while drought-resistant features like small leaves and abundant hairs cut down water loss.

At the bottom of frozen pools, among the vegetation, or buried in the soil, insects survive the winter in a state of inactivity, most of them in the form of eggs or larvae. Surprisingly, virtually no Arctic mammals hibernate, perhaps because the risk of freezing to death is too great; although grizzly bears hide up and go into a deep winter sleep. Elsewhere, however, life goes on busily.

The Arctic fox is so well adapted to the cold that it can sleep out on open snow at –112°F (–80°C) for an hour without ill effect. In winter the foxes depend as much on scavenging as on hunting. They trail polar bears across the ice packs, hoping for scraps of the bears' seal victims. They also eat voles and lemmings, which they dig out of the snow.

The human threat

Several Arctic mammals, including the stoats and Arctic foxes, molt to white in the winter. This is a survival tactic, for the white hairs are hollow, and probably help to retain heat as well as providing camouflage. Until recently, white fur had long attracted the attentions of the fur hunters, however.

The presence of humans presents further survival hazards to the plants and animals of the far north. Oil and mineral exploitation seriously threatens the established ecosystem. More insidious threats are high levels of toxic substances, including pesticide residues, found in Arctic wildlife. Such residues emphasize how pollutants affect ecosystems far from their origin. Worst of all, the warming of the atmosphere not only threatens the arctic environment but also could release more carbon dioxide from the frozen tundra, in turn accelerating the warming process.

Connections: Animals in Cold and Wind 144 Camouflage 264 Fur and Feathers 134 Migration Patterns 228

Plants and animals *of the Arctic show a variety of adaptations to extreme cold. The dwarfed size of the plants minimizes the damage caused by snow blasting, and ensures that most species are completely covered by the limited snowfall. Insects lie dormant in the winter, having eliminated water from their bodies to prevent the damage caused by freezing.*

Most Arctic mammals and birds must maintain their body temperatures at between 98° and 102°F (37° and 39°C) throughout the winter. Larger predators, such as the gray wolf [1], *the Arctic fox* [2] *and the ermine* [3] *– a member of the weasel family – lay down a layer of insulating fat, and develop a thick, white coat: the Arctic fox even has fur on the underside of its paws. The largest Arctic predator, the polar bear* [4], *may travel over 13 miles (20 km) searching for its main prey – the harp seal* [5]. *The gregarious walrus* [6] *has a tough hide and a thick fat layer, which helps protect against cold and against the tusks of other walruses. Smaller mammals, such as the Norway lemming* [7] *and the Arctic hare* [8], *cannot "afford" to invest energy producing extra layers of insulating fatty tissue: they shelter from the cold in burrows in the snow.*

Overwintering birds, such as the ptarmigan [9], *increase their feather density to ward off the cold; they may also burrow into the snow when resting overnight. Puffins* [10] *may excavate nests up to 6 ft (2 m) deep. Graylag* [11] *and snow geese* [12] *avoid the winter conditions, journeying over 2,000 miles (3,500 km) to reach the Gulf of Mexico.*

Plants in Cold and Wind 140 Polar Regions 42 South Polar Life 312 Tundra 314

The Antarctic Wilderness

How life survives near the south pole

If the ice cap of Antarctica were to melt, the world's sea level would rise by more than 164 feet (50 meters). Antarctica is a massive continent, as big as China and India combined, and only between 2 and 3 percent of it is ice free. Although the coldest place on Earth, life exists on Antarctica, albeit mostly at its edges. The major part of the ecosystem is based not on the land but in the sea. Plankton teems offshore and supports one of the richest marine systems in the world. Antarctica's riches have not gone unnoticed. Overfishing and mineral exploitation threaten the world's last great wilderness.

Virtually nothing lives in the cold heart of Antarctica. In the few places where bare rock pierces the ice, a few lichens and mosses may cling on, but otherwise it is lifeless. Only at its fringes, where seabirds nest and seals haul themselves onto the shore, does Antarctica support much life. Even so, there is little by way of vegetation.

Antarctic waters are cold, and are enriched by nutrient-bearing currents and upwellings that sustain millions and millions of plankton. These tiny plants and animals are the basis of the major food chain. Among the animal plankton that harvest the plant plankton lives an animal that is among the most abundant in the sea: the krill. Part of the shrimp family, the krill forms the basic diet of fish, seabirds and marine mammals.

Antarctic animals deal with the cold in various ways. Many are well insulated. Under their dense feathers, penguins have a layer of blubber; seals too have blubber. Some of the fish, such as the icefish, even use a type of "antifreeze": their blood contains *glycoprotein* cooling stabilizers, highly effective inhibitors of ice formation.

Exploited wilderness

With its rich marine wildlife and its desolate interior, Antarctica is the world's last great wilderness, but now it is threatened. The slaughter of seals and whales by human predators in the last century may have been halted, but our exploitation of the southern ocean continues. Huge fleets of fishing boats trawl the subantarctic waters for fish, krill and squid. Overfishing could seriously affect the Antarctic ecosystem by leaving too little food for its wildlife. Finally, the hole in the Earth's ozone shield in the atmosphere over Antarctica exposes Antarctic wildlife to potentially dangerous ultraviolet radiation from the Sun.

Beneath its icy surface, the seventh continent is believed to hold abundant mineral wealth. Mines, oil wells and pipelines, as well as tourist development, could pose a massive environmental threat if the resources were exploited. To ensure that the potential threat does not become an environmentally disastrous reality, several countries have signed the Antarctic Treaty. The Treaty basically ensures that the preservation and conservation of the environment will be upheld.

Antarctica supports *five species of penguin. The emperor penguin* [1] *is the largest of all the penguins. The parents produce a single egg in the autumn which is incubated by the male. When the egg hatches, the female returns to protect her young from predators such as the Antarctic skua* [2]. *The long breeding cycle of the king penguin* [3] *restricts the parents to producing only one egg every two years. Adélie penguins* [4], *however, the most common species of penguin found on Antarctica, lay two eggs, in early summer, and the subsequent chicks take only 4 months to reach independence. The chinstrap penguin* [5] *gets its name from its facial markings. Chinstraps often take over the nesting sites of adélies to rear their young. Gentoo penguins* [6], *like all the other species of Antarctic penguin, set up crèches to protect the chicks, brown in color, from predators while the parents hunt for food. Antarctic fur seals* [7] *feed mainly on the large swarms of krill found in the region. Killer whales* [8], *sometimes known as orca whales, are highly successful predators. They feed on fish, squid, seals, seabirds and even other whales. To improve their chances of a good catch, killer whales often hunt in packs. The wandering albatross* [9] *is the largest of the albatross species. It is a resident of Antarctica, where it breeds only once every two years. These birds mature late in their lives, and only start to reproduce when they are about 12 to 14 years old. Unlike the albatross, the Arctic tern* [10] *is a migratory bird. During the Antarctic summer these birds fly from the north pole to the south to feed.*

Connections: Algae 112 Animals in Cold and Wind 144 Fur and Feathers 134 North Polar Life 310 Ocean Circulations 46 Ocean Life 332 Ozone Hole 68

10
3
6
4
1
5

Land of the Midnight Sun

How the tundra flourishes in summer

For most of the year the barren, wind-blown tundra is an eerie moonlit landscape. But for about 50 hectic summer days it enjoys lukewarm 24-hour sunshine that partially thaws the permafrost. The thaw depth determines the vegetation growth and, thus, the animals it can feed. As well as caribou and migratory birds, the tundra supports lemmings, rodents that breed all year. Eating mainly grass, their numbers explode in years of plenty; finally they overcrop the grass and are forced into mass migrations in search of food in which thousands die crossing the icy rivers.

As the winter fades in the far north, the melting of the snow reveals an open, rolling landscape – tundra. This barren habitat lies north of the Arctic Circle, surrounding the Arctic Ocean and stretching across northern North America, Greenland, northern Scandinavia and the former Soviet Union. The land and its occupants – as well as the occupants of the coastal waters – prepare for a mad rush to grow, and to reproduce, as quickly as they can during the short summer season.

Permanent residents

Tundra vegetation is often patchy and low, rarely growing more than 3 feet high. There are no trees. Even woody species such as Arctic willow hug the ground, where temperatures are higher and the wind less vicious.

Some animals are resident in the tundra throughout the year, notably voles, shrews, lemmings and Arctic foxes. The four-year-long cycle that culminates in a population explosion of lemmings remains so far unexplained. The phenomenon is, however, common among many other small rodents, and is probably connected with the presence or absence of predators or food. As summer draws near, other birds and mammals move in. Herds of caribou (or reindeer) up to 100,000 strong wander north with the summer, returning to their traditional calving grounds. Butterflies, bumblebees, hoverflies and mosquitoes are all on the wing. The mosquitoes, flying in massive swarms when the air is still, are the scourge of mammals.

Offshore, open water appears as the sea ice melts. There, abundant plankton support whales – including the beluga – and also feed many fish and other marine life forms, such as squid. These in turn are preyed upon by animals higher in the food chain.

In summer the tundra predators thrive. The wolves hunt in the drier land areas; the polar bears make a good living from fish and seals on the cold shores. Birds pick on smaller mammals, or sometimes even other birds, and the Arctic fox feeds its young on lemmings, birds and other rodents.

Despite the fox's attentions, plenty of birds survive to make the flight back south. As the summer ends, the birds take to the air, the caribou start tracking south again, and the fox and its fellow residents are left to scratch a living in the tougher winter conditions.

***Tundra plants grow** very slowly, but once the snow melts, they flower rapidly. In the summer, plants like the saxifrage* [1]*, arctic poppy* [2] *and mountain aven* [3] *use up resources stored from the previous summer in bursts of flower. The sun also warms the ground, but only to a depth of about 3 feet (1meter), beyond which it is permanently frozen. Meltwater cannot drain away; the waterlogged surface is ideal for mosquito* [4] *eggs, which are frozen until they hatch during the following spring.*

Migratory birds, like willow grouse [5]*, golden plovers* [6] *and flocks of wetland birds, pour in to feed on the short-lived greenery and insects; many pair first – time is too precious for courtship rituals.*

Some small mammals, like the red-backed vole [7]*, survive the cold by living in the insulating snow. The larger musk oxen* [8] *and caribou* [9] *are protected by thick coats. The oxen gorge themselves in the summer, storing surplus energy as fat. Both animals use the herd's collective warmth to minimize heat loss.*

Oxen protect themselves and their young from wolves [10] *by bunching together to form a circle and lowering their horns; high above soars another hunter, a white-tailed sea eagle* [11].

The Northern Forests

How wildlife lives in a boreal forest

Boreal forests – the great, dark, conifer woods of the north – stretch for more than 7,500 miles (12,000 kilometers) around the top of the world. They sweep across the top of Siberia, Eastern Europe, Scandinavia, Canada and Alaska, forming a ring sometimes 1,250 miles (2,000 kilometers) wide. They are young ecosystems, having developed only since the last ice age retreated about 10,000 years ago. Although they are less diverse than older ecosystems, they do have a range of wildlife that is well adapted to the long, hard winters and brief summers.

The dark, pointed profiles of conifer trees dominate the boreal forest. Species such as pine, fir and spruce are ideally suited to the challenges of the environment: the shape of their branches allows them to shed weights of snow rather than break, and their slender trunks are flexible in the face of the strong winds. The root systems of these trees tend to be shallow, enabling them to obtain water from the surface soils; deeper roots would strike the unyielding permafrost beneath.

The soils beneath the conifer canopy are poor, acid, and often thickly covered with needles. In these infertile conditions, the conifers' growth is assisted by fungi, which help to break down the needles and supply the trees with nutrients. In return, they probably gain carbohydrates from the tree.

Deciduous trees such as birch and aspen also occur in the boreal forests, but the early spring growth of the evergreens allows them to overtop and dominate the deciduous trees. Where enough light reaches the ground, plants like bilberry and sphagnum moss grow.

The great providers

The conifer trees are fundamental to the boreal food chain. Moth and sawfly larvae eat the needles. Other insect larvae, including those of the wood wasp and pine sawfly, attack the wood itself. Capercaillies eat the conifer needles during winter, when food is short. To crossbills and red squirrels, conifer seeds are important dietary elements. Caribou, which migrate south from the tundra to spend the winter in the shelter of the forest, depend upon scraping away the snow beneath the trees to find lichens and other vegetation. Small rodents such as voles feed on bark, buds, fungi, seeds and berries. During spring and summer in the boreal forests, insects are abundant, and they attract large numbers of migratory birds to the habitat to breed. The abundant voles and other small mammals are the prey of the red fox and the great gray owl. Other hunters include wolverines, wolves and brown bears.

The boreal forests are one of the least disturbed habitats in the world, but even so they have long been harvested for lumber and, on a local scale, berries and fungi. Large areas have been cleared for farmland. But many boreal forests are now at least protected nature reserves and national parks.

__The structure of__ a boreal forest is relatively simple compared to other forest types. The canopy of evergreen conifers forms a dense barrier to light, below which a thick layer of needles accumulates: the chemistry of the needles, the low temperature and wetness of the forest floor (which leads to poor aeration) all prevent the action of soil-enriching decomposers. Deep shade and low soil fertility mean that few shrubs can grow under the canopy, and the vegetation here is restricted to patches of mosses, lichens and grasses.

During the harsh winters, temperatures plummet and food becomes scarce. Some animals, such as the brown bear [1], *avoid adversity by hibernating: others, like the goshawk* [2], *may move south. Remaining grazers, such as the elk* [3] *and bison* [4], *and carnivores like the lynx* [5], *the wolverine* [6] *and the great gray owl* [7] *are protected by a covering of thick fur or feathers.*

Conifer seeds take several years to mature within cones, and therefore constitute a reliable source of winter food for birds and small mammals. The nutcracker [8] *and red crossbill* [9] *have developed the powerful bills needed to extract these seeds from their protective scales. Spruce grouse* [10] *survive winter on an austere diet of conifer needles, while the goldcrest* [11] *picks out grubs and insects buried within the bark of a tree.*

Plants and animals of boreal forests are well prepared to take full advantage of the short summer growing season. Evergreen needles can begin photosynthesis as soon as light levels and temperature increase, and swarms of herbivorous insects take immediate advantage of this new resource.

Some insects are themselves parasitized by ichneumon flies [12]. *Mosquitoes and gnats rise from the surface of ponds and marshes in search of blood, and wood-boring bark beetles* [13] *can attack either dead or living trees.*

Connections: Air Pollution 342 Conifers and Cone Bearers 118 Mountain Life 306 Plants in Cold and Wind 140 Soils 30 Temperate Forests 318 Tundra 314

7
2
5
4
9
12
13

The Deciduous Woodlands

How seasonal change affects temperate forest life

The changing seasons dominate deciduous woodlands. Woodland animals and plants live and breed in conditions that vary widely throughout the year. The key to their success is an ability to survive the winter and then take full advantage of the spring and summer. Deciduous woodland is the natural vegetation of much of temperate Europe and North America, and also occurs in the limited temperate lands of the Southern Hemisphere. But its water-retentive soils make excellent farmland, and in many places woodlands have been replaced by agricultural development.

In the depths of winter there seems to be little life in a deciduous woodland. The trees are bare, snow may lie along the branches, and there is little birdsong to be heard. Yet even so, there is activity. On the woodland floor shrews hunt for invertebrates such as woodlice and earthworms. Also active are resident birds like the jay in Europe and the hairy woodpecker in North America. The jay survives the winter by eating the acorns it stored in the autumn. The hairy woodpecker changes diet, eating invertebrates in the warmer months and seeds in the winter.

Many species of birds – and a few bats – simply avoid the winter: they head for warmer climates after breeding. Mammals like the dormouse and the striped skunk stay put but become less active, or hibernate.

Bursts of color

As winter retreats, the days lengthen, the temperature rises and the snow melts; greenery and life return to the woods. First come the spring-flowering plants, which expand their leaves and produce their flowers before the canopy closes over them and shades them from the Sun's light. Woodland trees – oak, hickory, maple, ash, beech and many others – burst into leaf, and no sooner are their leaves expanded than they are being eaten by insects and their larvae.

Migrating birds such as warblers return in spring to build their nests and raise their broods. The plentiful harvest of insects provides breeding birds with a protein-rich diet for their young. Down on the woodland floor deer browse the vegetation, and in the dense understory mice are busy searching out seeds, buds and invertebrates.

All parts of the wood – from the woodland floor to the top of the canopy – have their predators. Hunting spiders chase small invertebrates across the woodland floor, and far above, superbly adapted woodland hawks hunt small birds among the branches.

By late summer the migratory birds have raised their young and begin to leave. Trees are preparing for the dormancy of winter by withdrawing the nutrients from their leaves and shedding them. The leaves turn brown, red and gold as they die, and the jays hunt once more for acorns. The cycle begins again.

***With spring**, deciduous forests come to life. The hedgehog [1] emerges from its winter sleep, foraging, along with the blackbird [2], for insects hidden among the bluebells [3], primroses [4], wood anemones [5] and cuckoopints [6]. The forest floor provides cover for the woodcock [7] incubating her eggs. As the days lengthen, roe deer [8] battle for mates, while the purple emperor [9] searches for aphids' honeydew, preferring this to*

Connections: Biological Clocks 274 Birds 132 Boreal Forests 316 Decomposers 204 Flowering Plants 120 Fungi 110 Mammals 136 Seasons 86

the nectar of the honeysuckle [10]. The predatory sparrowhawk [11] hunts birds such as chiffchaffs [12], wood warblers [13] and blue tits [14], which search the branches of the oak [15] for insects. The green woodpecker [16] hunts on the ground, but also looks for insects in tree bark. As the days become wetter, fungi such as fly agaric [17], wood blewits [18] and shaggy parasols [19] produce fruiting bodies, and wood mice [20] search for food, often finding a meal from cast-off deer antlers. The badger [21] may mate as late as October, but delayed implantation of the egg allows the adult to rear its young in the spring.

***As winter sets in** and the days grow shorter, jays [22] begin to search for the food stores they made in the autumn and great tits [23] forage for berries. The tawny owl [24], however, preys on small living mammals, its silent flight allowing it to listen out for the squeaks of its prey. On the ground, the grey squirrel [25] continues to search for nuts, while below ground, the dormouse [26] hibernates, but voles [27] remain active, hiding below the snow cover with their food stores. The fox [28], like the tawny owl, is a twilight hunter that emerges from its burrow at dusk to hunt small mammals.*

The Treetop Zoo

How animals live in the rain forest

An area of forest covering nearly 40 square miles (100 square kilometers) may contain up to 400 different types of bird, 125 types of mammal, 150 types of reptile and amphibian and 150 types of butterfly and moth. Every possible niche is inhabited, from the dead wood on the forest floor to the branches way above. The greatest abundance and variety of life is up in the canopy, where the trees thrust their branches, leaves and flowers toward the sunlight. The animals that live in the canopy are adapted to life up there – many never touch the ground.

Life is richest in the canopy, where the bulk of the foliage is and where most of the plant-eaters live. Caterpillars of moths and butterflies feed on the leaves, and are in turn eaten by birds and frogs. Larger herbivores like howler monkeys have large, specialized intestines that enable them to derive enough goodness from leaves that contain few nutrients. Even the epiphytic plants have their own mini-ecosystems: tree frog tadpoles and aquatic insect larvae develop in the trapped water, snakes lurk to catch small mammals that come for a drink, and small animals that fall in and drown help provide the plant with nutrients.

Life in the canopy is very noisy, with animals calling to keep in contact, to pass on alarms, or to issue territorial threats. Canopy animals are adapted to life in their treetop habitat. Parrots use strong claws and beaks to clamber among the twigs. Fruit bats move among the trees by flying, but other mammals – far from solid ground – have to cling on and climb around. Sloths hang from branches by their claws as they feed on leaves. Spider monkeys' tails are prehensile, and function as fifth limbs as the animals swing through the branches.

Floor dwellers

On the forest floor, small animals like ants, beetles, earthworms and termites feed on fallen leaves, dead wood and animal corpses. They are vital to the forest's ecology, for together with fungi and bacteria they break down dead material into substances plants can take up and use. This recycling of the forest's nutrients is particularly important because many of the rain forest soils are very poor in the minerals required for plant growth. These plants and decomposers are fed upon by animals further up the food chain. Giant boars graze in clearings; pheasants scratch around for insects, seeds or fallen fruit; and anteaters use their long, sticky tongues to capture termites and ants.

At the top end of the food chain are the large predators. Leopards hunt birds and monkeys in the lower canopy and deer on the forest floor. Above the canopy float keen-eyed forest eagles, fierce predators of the air. The rain forest ecosystem is linked together by a complex food chain in which all the forest organisms depend upon each other.

Bromeliad plants *[**A**] can store up to 2 gallons (10 liters) of water in their funnel of leaves, providing an oasis for a host of insects and small animals, such as poison arrow frogs, which bring their tadpoles to bromeliads to complete their metamorphosis. The highly evolved camouflage of thornbugs [**B**] makes them difficult to distinguish from genuine thorns; close up, their bright colors are a warning to predators. The floor-dwelling agouti [**C**] is a seedeater. Its habit of stripping and burying the seeds of the black palm is essential to the tree's life cycle. Species of morpho butterflies are among the world's most stunning. The iridescent colors of this blue morpho [**D**] come from millions of scales on its wings. The large spiders of the rain forest prey on small animals as well as insects. The wandering spider [**E**] injects its prey with venom before macerating the body with powerful mouthparts and sucking out the liquid.*

A

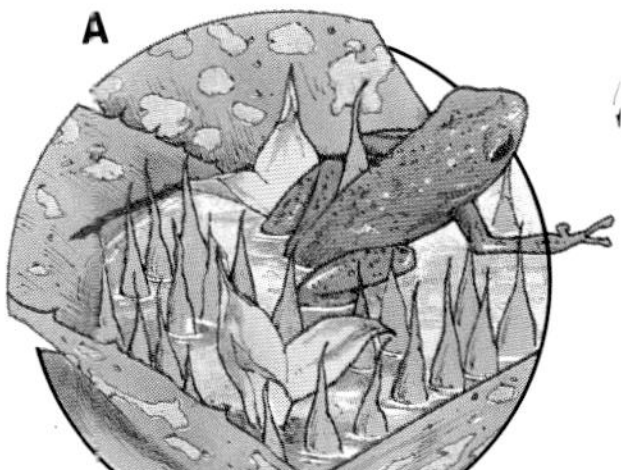

B

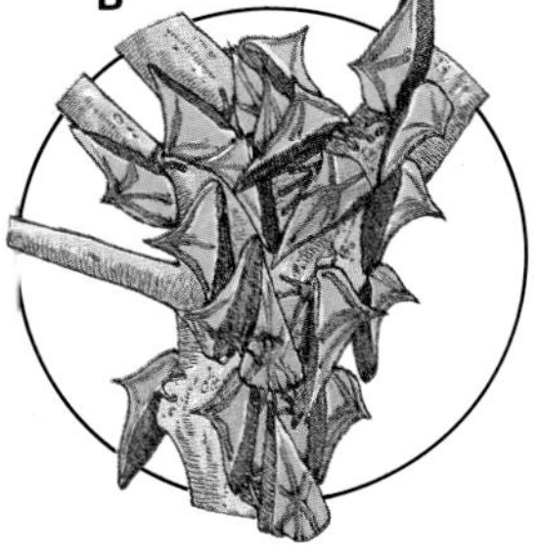

C

Connections: **Climbing** 216 **Cycles of Life** 206 **Decomposers** 204 **Gliding and Soaring** 222 **Mimicry** 266 **Poisonous Animals** 250 **Spiders** 252

F

The steamy jungles *[F] of Central and South America are the most species rich of the world's ecosystems. The upper canopies of these forests, where the light levels are highest, are very productive, and many animals spend their whole lives in the treetops. Monkeys, like the spider monkey [1] and squirrel monkey [2] feed principally on the abundant fruit and nuts of the forest, but will also eat insects, spiders, eggs and small vertebrates. The territorial cries of the howler monkeys [3] can be heard for miles, but the most raucous animals are the screeching parrots and macaws, such as the scarlet macaw [4], the blue and yellow macaw [5] and the blue-headed parrot [6]. These colorful birds use immensely strong beaks to crunch up seeds and nuts. The long bill of the keel-billed toucan [7] is useful for reaching fruit, and is probably also involved in sexual displays. The purple honeycreeper [8] and the ruby-throated hummingbird [9] have long, curved beaks to probe inside flowers for their nectar. The ithomid and tiger butterflies [10 and 11] are also nectar feeders. The broad-billed motmot [12] and the rufous-tailed jacamar [13] are both skillful hunters of spiders and of insects like leafcutter ants [14]. Harpy eagles [15] are awesome predators of the treetops, seizing arboreal mammals with razor sharp talons. Such mammals include the coati [16] and the opossum [17], which feed on insects and fruit, the ant-eating tamandua [18] and the two-toed sloth [19], which hangs upside down. Other predators are the jaguar [20] on the ground and lower branches and the blunt-headed tree snake [21], which resembles a twisting liana. The vivid coloration of the eyelash viper [22] attracts tree frogs [23], iguanas [24] and other reptiles to their death.*

D

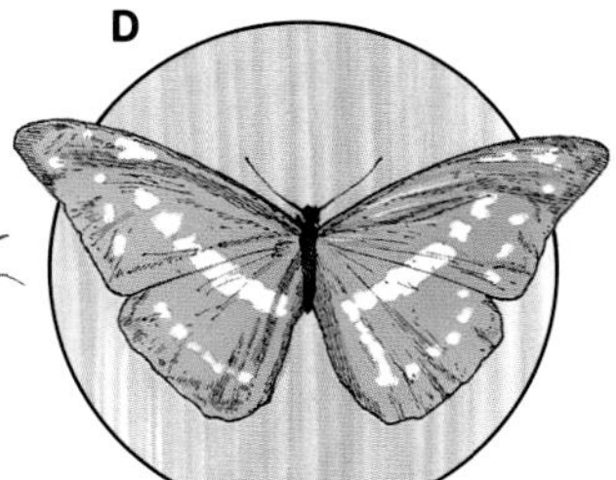

E

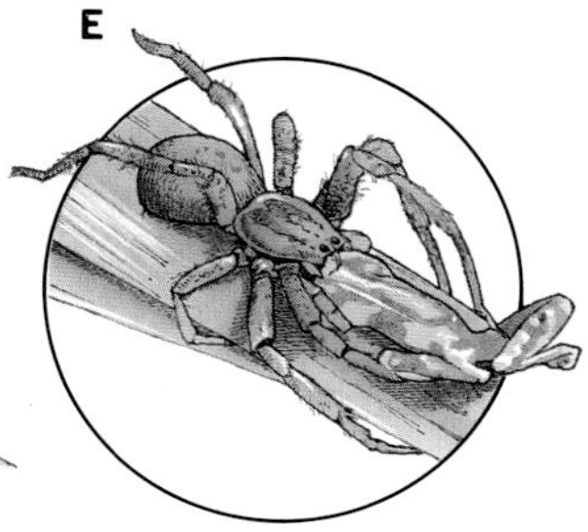

The Great Green Canopy

How tropical rain forest plants live

In two acres of rain forest there may be as many as 200 species of tree, whereas in the same area of the richest temperate forest there might at best be about 25. The tops of the rain forest trees together form the vast, green canopy – a mass of branches, leaves, fruit and flowers creating an aerial world in which most of the forest animals live. Above the canopy tower the forest giants, the emergents, some of them as tall as a twenty-story building. Way below are shrubs and climbers, the understory, and on the shady floor are minute algae and delicate fungi.

Some rain forests may well have existed largely undisturbed for 60 million years. In these stable conditions the forests have become the richest habitats in the world. They support half the total number of species of animals and plants on the planet.

Despite its diversity, the rain forest canopy looks from the air like a vast green carpet, spotted with the colors of flowers and birds, and frequently interrupted by the tall emergents. With its epiphytes – "air plants" that grow on other plants – and its climbers, a rain forest canopy tree may support over 30 other species of plants.

Beneath the canopy, the forest is shady, with sparse undergrowth. To reach for the sunlight, some plants climb toward the light, using established trees for support. Others are better adapted to survive in the shade. When a forest giant falls, the trees that have survived below race for the gap, fighting to outgrow the rest and fill the vacancy.

The forest's plant life supports its animals, providing shelter and sleeping places, as well as food in the form of fruit, flowers and foliage. Plants are also responsible for driving the forest ecosystem, especially in helping to recycle vital water and nutrients.

Rapid recycling

On the forest floor fungi thread their way through the thin soil, helping to break down plant and animal remains into nutrients that can then be used again by plants. One sixteenth of a cubic inch of forest soil may hold several yards of fungal threads (*hyphae*) and as many bacteria as there are people on Earth. Decomposition is so rapid on the floor that there is little accumulation of leaf litter. On this shallow soil, many trees grow buttress roots, which spread out around the base of the trunk and help hold it up.

Every tree dies and falls sooner or later, but in the last few decades the forests have come under a much more serious threat: destruction by humans. Forest destruction does not just harm a rich ecosystem, it has huge environmental effects. Locally the removal of large areas of vegetation can lead to runoff and flooding. On the global scale, forest burning puts high additional amounts of carbon dioxide into the atmosphere, increasing the greenhouse gases and contributing to global warming.

A

B

C

D

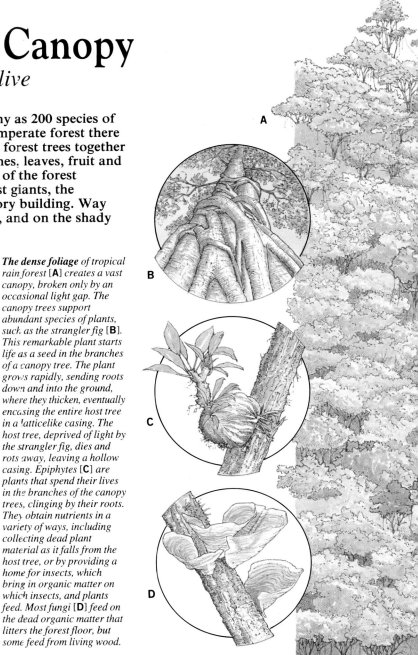

__The dense foliage__ of tropical rain forest **[A]** *creates a vast canopy, broken only by an occasional light gap. The canopy trees support abundant species of plants, such as the strangler fig* **[B]**. *This remarkable plant starts life as a seed in the branches of a canopy tree. The plant grows rapidly, sending roots down and into the ground, where they thicken, eventually encasing the entire host tree in a latticelike casing. The host tree, deprived of light by the strangler fig, dies and rots away, leaving a hollow casing. Epiphytes* **[C]** *are plants that spend their lives in the branches of the canopy trees, clinging by their roots. They obtain nutrients in a variety of ways, including collecting dead plant material as it falls from the host tree, or by providing a home for insects, which bring in organic matter on which insects, and plants feed. Most fungi* **[D]** *feed on the dead organic matter that litters the forest floor, but some feed from living wood.*

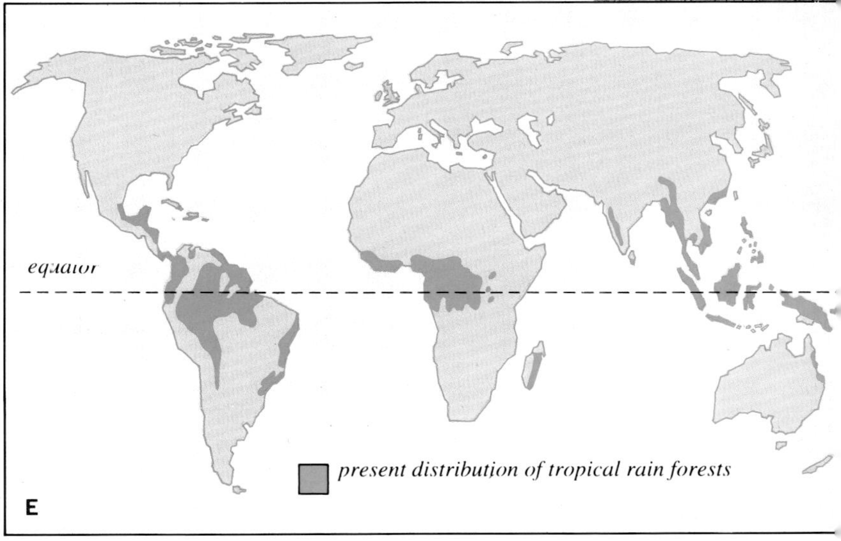

Connections: Cycles of Life 206 Decomposers 204 Fungi 110 Global Warming 84 Parasitic Plants 186 Plant Conservation 346 Plant Feeding 184

Today's rain forests **[E]** *are being destroyed at an alarming rate. It has been estimated that, at the present rate of deforestation, all the world's rain forest will have disappeared within 50 years. Moreover, the slash-and-burn method of clearing means that the poor-quality soil is leeched and plants can never grow back.*

Tropical rain forests *have distinct layers. The tallest trees, the ones that break through the canopy, are the emergents. Beneath these lie the trees that form the canopy. This is the richest part of the rain forest. It is where most photosynthesis takes place, contains most of the branches, flowers and fruits, and therefore attracts the majority of the wildlife. Beneath the canopy little sunlight penetrates, and for this reason the vegetation is relatively sparse. However, in lighter areas younger trees, shrubs and lianas grow, forming the understory. The forest floor is dark and humid. In this gloomy environment only shade-tolerant plants can survive.*

The roots **[F]** *of rain forest trees rarely grow deeper than 18 in (45 cm). This is because the soil nutrients are found only near the surface, and because rainwater drains quickly. To stop the trees falling over, they have supporting buttresses, which grow 17 ft (5 m) up the tree's trunk.*

The rainfall water cycle **[G]** *works when water, evaporated from the sea, is carried to the forest by the prevailing wind, where it condenses and falls, only to be reevaporated. In* **[H]**, *water runs off* [1] *an area that is cleared* [2]. *The cycle fails and the forest becomes increasingly desertified* [3].

Plants in Heat and Drought 138 **Soils** 30 **Symbiosis** 192 **Tropical Forest Animals** 320

Among the Reeds

How plants and animals live in wetlands

Some swamps produce up to eight times as much plant matter as an average wheatfield. Wetlands, wild half-worlds between land and water, occupy some 6 percent of the world's land surface, their ever-changing patterns of floating vegetation, solid ground and open water providing plenty of opportunities for a wide range of plants and animals. Wetlands are found throughout the world, in a wide range of climates, and vary from swaying reedswamps and steamy forests to desolate peat bogs. They are among the most threatened habitats in the world.

Wetlands form where water gathers. They occur at the edges of lakes, for example, and where rivers reach the sea. Plants grow out into the open water; once established, they slow the flow of water and trap silt; the silt is then invaded by more vegetation, and the open water is itself gradually taken over. The type of wetland formed by this process is swamp or marsh, and is often dominated by species such as reeds, papyrus or swamp cypress. Another important type of wetland is bog, which is typically dominated by sphagnum moss and cotton grass: such vegetation usually develops in closed inland basins, where rainfall is high but evaporation of surface water is relatively low.

Temperate reed swamps are one of the most familiar types of wetland. Reeds are very well adapted to the wetland environment. They can tolerate a high water table, and their new shoots grow quickly upward to reach the light above the water. There they grow tall and dense, and although other plants grow beneath them, few can overtop them. The reeds provide a habitat for wildlife throughout the year. Together with other wetland plants, the reeds also provide food for the wetland animals. Moth larvae eat the leaves of the reeds, and other herbivorous insects bore their way into the reed stems.

Wetland hunters

The smaller flying insects fall prey to such hunters as dragonflies. Dragonfly larvae develop in the water, spending up to a year submerged as nymphs; they then climb reed stems to emerge from their larval skins as adults, and find plenty of food among the vegetation. Amphibians such as newts and frogs also feed on insects. Small birds feed on the insects too, and when the insect abundance ends with the passing of the summer, these birds leave.

The open water provides food for many species that shelter in the reed swamp itself; vegetation, water invertebrates and fish provide food for ducks, water shrews, water voles and otters. But wetlands are easily damaged and destroyed by drainage and pollution. Swampy areas, so vital to many types of wildlife, are widely regarded as wastelands, and are always under threat of reclamation. Bogs suffer a different type of damage – they are exploited for their peat.

Connections: Estuaries 334 Lake Life 326 Lakes 60 Rivers 58 River Life 328 Water Pollution 340

The term "wetlands" *is applied to a diverse range of environments that share certain features. All are characterized by an indistinct and ever-changing boundary between water and land: and though their waters are usually poor in oxygen, swamps and marshes present a variety of habitats in which a great number of organisms can thrive without competing directly with one another.*

A typical reed swamp in a temperate region is dominated by shallow-water plants such as bulrushes [1] *and common reeds* [2], *which provide excellent breeding sites for birds: the moorhen* [3] *builds its reed nest on a mudbank or tussock, while the reed warbler* [4] *weaves a nest around a few reed stems. In deeper waters, plants like duckweed* [5] *and Canadian pondweed* [6], *which can survive total immersion, provide plentiful food for invertebrates. The showy, solitary flowers of the water lily* [7] *float on the water surface, carried aloft on long stems that contain air spaces for buoyancy. Water snails* [8] *feeding on the submerged vegetation breathe through lungs rather than gills, which allows them to survive periods of drought. Pond skaters* [9] *exploit surface tension to glide over the water and to detect their prey – fallen flying insects. Dragonflies* [10] *spend up to a year underwater as nymphs, the adults emerging for a short period to mate.*

The fish found in reed swamps, such as the stickleback [11], *tench* [12], *pike* [13] *and bream* [14], *are usually deep bodied, and thus better able to weave between thick-growing vegetation: most are tolerant of reduced oxygen levels. Amphibians, such as the common frog* [15] *and great crested newt* [16], *find the conditions ideal for breeding and development.*

Equally agile on land and in water, otters [17] *are able to prey on small mammals, such as water voles* [18] *and harvest mice* [19], *as well as fish and amphibians. Predatory birds include gray herons* [20], *kingfishers* [21] *and marsh harriers* [22].

The Living Lake

How a lake supports life

California's Mono Lake is nearly three times saltier than the Atlantic Ocean, yet it still supports a bustling ecosystem of algae, brine shrimps, brine flies and a variety of bird species – but no fish. Lakes are complex and delicate ecosystems, where balances change with the seasons, climatic variations and water levels. Pollution can destroy these fragile balances all too easily. Sewage and fertilizers artificially boost nutrient levels, thus encouraging blue-green algae growth, ultimately leading to oxygen starvation and the death of many life-forms.

A lake has three main zones – a *pelagic* zone (deep, open water in the center), a *littoral* zone (the area on the gently sloping sides), and a *benthic* zone (the bottom below the littoral zone). Each zone has a different variety of life-forms, often overlapping each other. Ultimately, all depend on photosynthetic algae growth.

Large algae and plankton colonies grow in the still, pelagic water. Most plankton eat algae, including some that rapidly multiply in sunny weather, when algae is abundant. Plankton are a vital food link between algae and larger creatures: growth rates of perch closely correlate to sunshine and plentiful plankton – their food. A lake's waters, as well as providing birds and fish with all their food, are also a night refuge for birds from land predators such as otters.

Life at the bottom

Benthic animals living on or in the mud, mostly eat organic debris from above, though there are also benthic algae. Larger animals include worms, larvae and sometimes mollusks. Phantom midges live by day on the bottom, but feed on the surface at night, when they are invisible in the dark. Animals at great depths have breathing systems to cope with low oxygen levels. As a lake ages, it grows more fertile and its ecosystem changes, favoring species that can cope with less oxygen and more congested conditions.

Large plants (*macrophytes*) in the littoral zone help create the lake's most complex animal community. Reeds give shelter from wind and waves, and trap sediment, home for fragile mud organisms. Slime made of algae and other microorganisms adheres to underground plants, and is a major food source of pond snails. Underwater plants give cover to crayfish and other crustacea.

The water surface hosts many insects, including pond skaters, which use surface tension to avoid sinking. Some air-breathing insects store air to dive in the water. Further down, various nymphs and larvae extract oxygen from the water.

The roach is one of many littoral fish; its feeding creates plant debris subsequently eaten by invertebrates. The camouflaged pike ambushes prey, starting with larvae in its first year and graduating to tadpoles, young fish and, finally, large fish, birds or water voles.

***Lake food chains** begin with algae [1], prey to plankton like* Daphnia *[2], in turn prey to carnivorous plankton like cyclops [3]. Minnows [4] reduce the chance of individuals being preyed upon by swimming in shoals. Perch [5], normally pelagic residents, breed in littoral waters, trailing mucus-covered eggs around plants [6]. Carp [7], however, are benthic, using whiskerlike barbels [8] to find food, such as caddis fly larvae [9], which build "armor" from plant fragments. Midge larvae [10] are prey for many ducks, fish and birds, particularly when they hatch into vast insect swarms.* Tubifex *worms [11] have oxygen-carrying hemoglobin in their blood, and are a rich food source for fish.*

Underwater plants such as wormwort [12] and Canadian pondweed [13] are both bottom-rooted and photosynthesize in minimal light. Slime on their surfaces is food for the great pond snail [14]. The great diving beetle [15] attacks anything from worms to tadpoles [16]. Dragonflies [17] eat smaller insects; living underwater as carnivorous nymphs, they reach a last molt stage [18] before emerging as adults.

Some lakeland birds, such as the great crested grebe [19] and tufted ducks [20], feed by diving for fish, others, such as the mallard [21], "dabble" for plant matter. The osprey [22] catches fish in its talons.

Plants like bulrushes [23] grow near the water's edge and help the slow colonization of the lake by the land. Long stems of lilies [24] convey oxygen to roots at the bottom. Floating plants, like the water soldier [25], have roots taking nutrients directly from the water. A water shrew [26] dives for beetles.

Connections: Algae 112 Decomposers 204 Fish 124 Invertebrate Carnivores 200 Lakes 60

23
17
22
20
19
24
3
2
1
25
26
5
4
13
7
14
12
15
11
10
9
8

Streams of Life

How rivers provide many different environments

Less than 1 percent of the world's water is fresh river water – yet in this tiny portion of the biosphere lives a great variety of animals and plants. One scientist has calculated that 64 million microscopic animals, weighing in total some 450 pounds (200 kilograms), drift under one bridge on the Missouri River every 24 hours. The flowing water continuously replenishes oxygen, carbon dioxide and nutrients, providing river plants and animals of all sizes with the basics of life. Along their lengths, rivers present wildlife with many opportunities and challenges.

Rivers *support a huge variety of wildlife. Some species thrive in the upper reaches, while others prefer the lower. Birds treat the upper river mainly as a source of food. The dipper* [1] *dives underwater to hunt for insect larvae and other invertebrates. The gray wagtail* [2] *feeds on similar aquatic prey, but also takes insects, such as the damselfly* [3], *on the wing. Only strong- swimming fish occupy the upper river. Young brown trout* [4] *hold themselves motionless head-on to the current, alert to any passing morsel. The bullhead* [5] *hunts among the stones on the bed. Insect larvae make up the majority of the population, and some species, such as the damselfly nymph* [6], *have predatory mobile forms. Crustaceans are represented by the freshwater shrimp* [7], *and mollusks by the freshwater limpet* [8], *which feeds on algae. Insect larvae employ a variety of feeding strategies, and not all are active predators. The blackfly larva* [9] *anchors itself to a stone, and uses head bristles to filter food from the water. Some species of caddisfly larvae* [10] *spin a net to catch food. Birds are more numerous around the lower river. Swallows* [11] *swoop overhead, attracted*

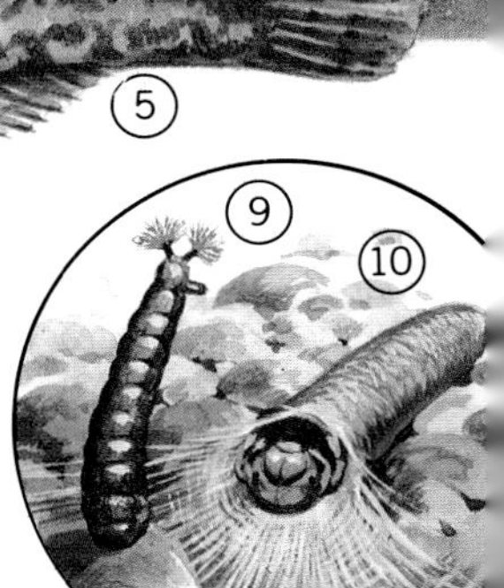

Connections: **Algae** 112 **Estuaries** 334 **Invertebrate Carnivores** 200 **Fish** 124 **Lake Life** 326 **Rivers** 58

Near its source, a river is steep, shallow and turbulent. There is little in the way of food. Plankton cannot survive in the rough waters, and there is too little sediment for larger plants to root in. Some algae grow on the rocks, grazed on by invertebrates such as mayfly nymphs. The food chain is mainly supported by detritus – organic material that washes in from the river banks, falls in from overhanging vegetation, or comes from the dead inhabitants of the stream.

Right next to the riverbed, friction slows the current and there is a thin, still band of water. Flattened invertebrates, such as water pennies, colonize this zone.

Putting down roots

Flowing downstream, along with the detritus, are tiny worms, snails and insect larvae, carried along by the current. But as the river widens and deepens, the current slows sufficiently for plants to take root. On their leaves grow algae, which are eaten by snails and other invertebrates. Plants provide not only food but also shelter for river animals. They add to the ecological niches available and to the complexity of the ecosystem. Along the riverbanks, vegetation provides another niche, inhabited by birds such as moorhens, and mammals such as the herbivorous water voles and carnivorous otters.

Detritus-eating and herbivorous invertebrates are hunted by fish and carnivorous insect larvae, and these hunters themselves fall prey to predators: larger fish, otters, turtles, crocodiles and birds. Down on the mud – or in it – are worms, snails and crustaceans, feeding on organic material, and in turn providing food for bottom-feeding fish.

In its later stages river waters are often very muddy with the sediment they carry. It becomes more and more difficult for animals to find their way through the water. River dolphins have solved this problem by using echolocation to navigate and find their prey. A number of fish – including the electric eel – have evolved specialized organs that use electricity also for detecting prey. Rivers are vital for wildlife – and for people. We must treat these ecosystems with respect so they retain their fascination for us, and their value.

by the annual swarms of adult mayflies [12]. *Among reeds* [13] *at the river's edge, a pair of swans* [14] *make their nest. The water vole* [15] *is one of the few mammals to have adopted an aquatic life-style. Fish occupy all levels of the lower river, searching for food among the stems of yellow water lilies* [16]. *Adult brown trout* [17] *take food at the surface, and the roach* [18] *patrols the middle waters. At the bottom, the bream* [19] *swims in a near vertical, head-down position as it feeds. Amphibians such as the smooth newt* [20] *are also present. Invertebrates of the slow waters such as mayfly larvae* [21], *which emerge in sychronized swarms of adults, and* tubifex *worms* [22] *have little protection against predators. Mollusks such as the white ramshorn snail* [23], *and the swan* [24] *and zebra mussels* [25] *benefit from tough outer shells that protect them*

Life in the Tidal Zone

How shore-dwellers find a livelihood

European tidal ranges vary dramatically, from the Baltic Sea, which has virtually no tide, to the Bristol Channel, which has spring tide variations of over 33 feet (10 meters). Temperature variants are also dramatic. In the Gulf of Bothnia, which lies off the Swedish and Finnish coasts, the summer surface temperature reaches 55°F (13°C), but in winter this area of coastal water freezes. Taking into account these variations, as well as the diverse physical environments, it is no wonder that seashores contain some of the most specialized plants and animals in the world.

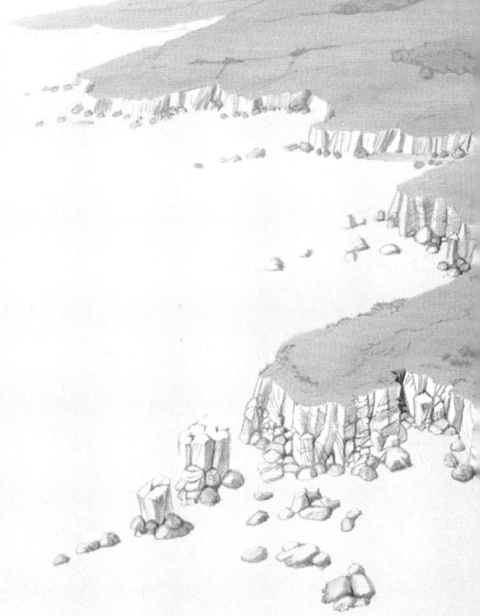

Unlike inland ecosystems, which may extend homogeneously over considerable areas, the ecosystems along a shoreline may alter frequently, suddenly and dramatically as the shoreline changes, for example, from a sandy stretch to cliffs to dingy mud. Within this diversity, the animals and plants on the shore display marked zonation, with the result that a number of typical species are identifiable at each tidal level. This zonation arises from the differences in the tolerances of animals and plants to the various physical factors, such as periodic desiccation and dramatic temperature changes, the consequent adaptations to the available ecological niches, and compromises on competition.

Rocky shores range from those that are extremely exposed to those that are comparatively sheltered. The more exposed shores have a greatly extended upper region (or "splash zone") inhabited by organisms that are essentially terrestrial, such as lichens and insects. The main area of the shore is colonized by animals and low-growing algae that are able to withstand the wind and waves.

Shelter from wind and waves increases the possibility of more massive growth forms. In addition, the ground surface around boulders provides many more micro-habitats in which specialized animals and plants can live. The gradual slope of the shore also leads to greater horizontal zonation.

The hidden hordes

Many organisms live buried in the sediment of coastal sand and mud, particularly when the tide is out. The organisms are usually filter feeders waiting for the tide to bring in organic material, on which they feed. Sandy shores tend to be the more exposed, and provide homes in typically zoned bands for species of worm, crustacean and mollusk. The fauna of muddy shores are more restricted, for they have to withstand the clogging nature of the mud.

Sand and mud without rocks cannot support many types of algae. The algae that are common are thus those that do not attach themselves to an anchorage but instead form thick mats on the water surface. On mud flats, similar mats of eelgrass flourish thanks to the ramifications of the plant's root system. Such shores present an important food resource for wildfowl and waders.

***Shoreline zonation** is most pronounced in temperate regions, where the prevailing weather comes from the same general direction. An exposed windward shore receives the full force of wind, waves and currents, and presents a much harsher environment than a sheltered leeward shore. Variations in high and low tide also affect zonation. In areas where there is little tide, the zonation is restricted. Zonation reflects the vertical movement of seawater. Sheer cliffs and outcrops of rock may present purely vertical zonation, with no apparent horizontal shift between the location of the various species.*

***A sheltered rocky shore** [**B**] presents a fairly blurred pattern of zonation with a considerable overlap. Only on the upper shore are the zones distinct. In the splash zone, the lichens* Xanthoria parietina *[1] and* Verrucaria maura *[2] grow on rocks. The upper shore is shared by the star barnacle [3], spiral wrack [4] and channel wrack [5]. The wetter middle shore supports knotted wrack [6] and bladder wrack [7]. Among these live the common limpet [8], flat periwinkle [9], some toothed wrack [10], the acorn barnacle [11] and the beadlet anemone [12]. The edible periwinkle [14] has a large range; sugar kelp [13]*

***Rockpools** [**A**] support a varied community. The common limpet [1], the white tortoiseshell limpet [2] and the edible periwinkle [3] congregate on surrounding rocks. The common mussel [4], edible crab [5] and sea oak [6] prefer the edge. The common starfish [7] and fifteen-spined stickleback [8] are active predators. Sea lettuce [9] shelters the common prawn [10]. The butterfly blenny [11] hunts on the bottom of the pool. The common hermit crab [12] attaches beadlet anemones [13] to its shell, and breadcrumb sponges [14] and snakelocks anemones [15] provide further cover.*

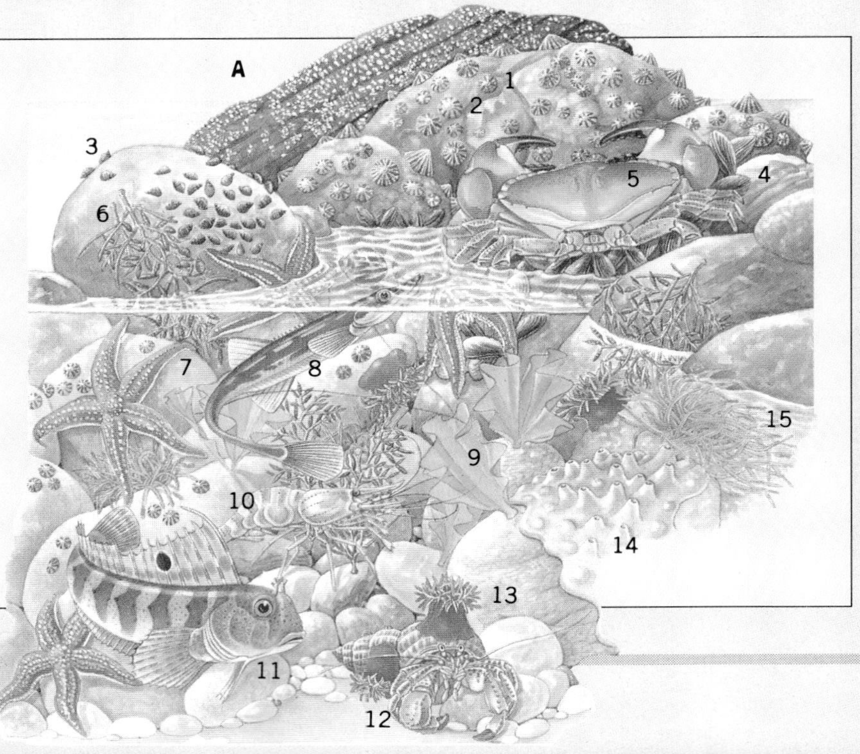

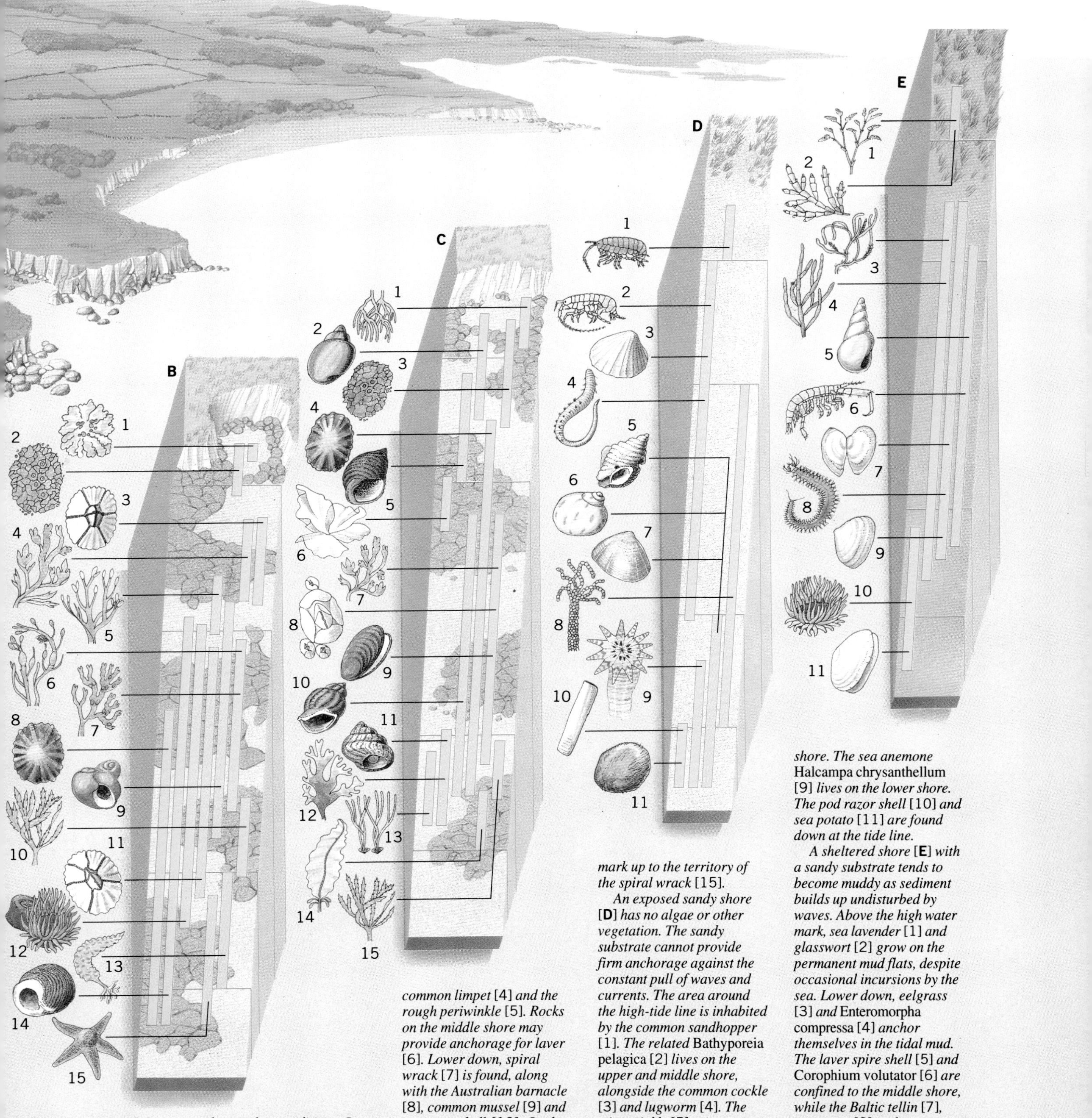

and the common starfish [15] *live around the low-water mark.*

An exposed rocky shore [**C**] *has a more sharply divided zonation owing to the rougher conditions. Sea ivory* [1] *and the hardy small periwinkle* [2] *share the splash zone with* Verrucaria maura [3]. *The upper shore is moist enough for both the common limpet* [4] *and the rough periwinkle* [5]. *Rocks on the middle shore may provide anchorage for laver* [6]. *Lower down, spiral wrack* [7] *is found, along with the Australian barnacle* [8], *common mussel* [9] *and common whelk* [10]. *On the lower shore live the gray topshell* [11], *carragheen* [12] *and thongweed* [13]. *Dabberlocks* [14] *has a range from the low-water mark up to the territory of the spiral wrack* [15].

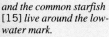

An exposed sandy shore [**D**] *has no algae or other vegetation. The sandy substrate cannot provide firm anchorage against the constant pull of waves and currents. The area around the high-tide line is inhabited by the common sandhopper* [1]. *The related* Bathyporeia pelagica [2] *lives on the upper and middle shore, alongside the common cockle* [3] *and lugworm* [4]. *The sting winkle* [5], *common necklace shell* [6] *and rayed trough shell* [7] *are confined to the lower shore, but the sand mason worm* [8] *extends up to the middle shore. The sea anemone* Halcampa chrysanthellum [9] *lives on the lower shore. The pod razor shell* [10] *and sea potato* [11] *are found down at the tide line.*

A sheltered shore [**E**] *with a sandy substrate tends to become muddy as sediment builds up undisturbed by waves. Above the high water mark, sea lavender* [1] *and glasswort* [2] *grow on the permanent mud flats, despite occasional incursions by the sea. Lower down, eelgrass* [3] *and* Enteromorpha compressa [4] *anchor themselves in the tidal mud. The laver spire shell* [5] *and* Corophium volutator [6] *are confined to the middle shore, while the Baltic tellin* [7], *rag worm* [8] *and peppery furrow shell* [9] *have a slightly wider range. The beadlet anemone* [10] *and sand gaper* [11] *are found only on the lower shore.*

Connections: **Algae** 112 **Coastlines** 56 **Estuaries** 334 **Filter Feeders** 196 **Ocean Life** 322 **Tides** 48 **Wetlands** 324

Life Beneath the Waves

How ocean ecosystems work

It is the fate of more than 90 percent of all freely mobile organisms in the ocean to be the prey of others. And this applies to all the vertical zones of the seas, from "benthic" animals that live on the seabed, to those that can swim against currents at higher levels (the "nekton"), to those organisms that live at the surface – plankton. The strange and diverse creatures that constitute the plankton, both plant and animal life-forms, represent the enormously wide base of a pyramidal food chain that supplies virtually all the dietary requirements of ocean dwellers.

There are no forests in the sea – apart from the so-called "kelp forests" off the Californian coasts. Kelp and other seaweeds are the only generally visible forms of plant life in the planet's vast oceans, and they are confined to a very narrow coastal strip. Seen from the air, seaweeds constitute a pencil line sketched in at the edges of the ocean. So what do marine animals eat? Where does the food come from to sustain great shoals of fish, turtles, seals, dolphins, whales and seabirds?

The answer is that the "forests of the ocean" are microscopic and thus not visible to the human eye. The "trees" in this invisible forest take the form of single-celled organisms collectively known as algae, although they are of diverse evolutionary origins. In their millions they live at the surface of the sea bathed in seawater and sunshine, which are almost all they need to make food. They feed by photosynthesis, using the Sun's energy to forge a molecule of sugar from six molecules of carbon dioxide. This unseen forest is known as the phytoplankton, and the microscopic animals that feed on them are called the zooplankton.

Water is far more effective than air in filtering out sunlight – only the upper waters of the oceans are well lit. Below 330 feet (100 meters) the light from above is too dim to support photosynthesis. In coastal waters, the phytoplankton layer is even thinner, for the waters are cloudy with stirred-up sediment. Sunlight is extinguished by a few yards of this turbid water.

Cloudy waters

Despite this, cloudy water is a rich source of phytoplankton. Throughout the oceans, a shortage of nutrients sets limits on the growth of the surface-dwelling algae, because gravity is working against their interests. In a terrestrial forest, leaves that fall to the ground land close to the tree's roots so that when they decompose the nutrients are readily absorbed. In the ocean, gravity has the opposite effect: it takes nutrient-rich debris – dead animals, food scraps and droppings – down through perhaps thousands of feet of water to rest on the seabed. En route it sustains scavengers such as gulper eels and deep-sea angler fish, and on the seabed, rattails and eellike hagfish compete for what is left. Bacteria rot away the remains, but the nutrients they release into the water, such as nitrates, silicates and phosphates, remain on the ocean bed far from the place they are most needed by the phytoplankton to continue photosynthesizing.

Due to local climatic conditions and ocean currents, there are huge differences in the productivity of the various oceans. Along the continental shelves, for example, waves, currents and violent storms frequently disturb the seabed, bringing both sediment and dissolved nutrients to the surface, and so producing water that is cloudy but fertile. It is for this reason that the world's richest fisheries are located on the continental shelves – which is why national fishing zones (such as those around Iceland and the United Kingdom) are so fiercely contested. In the wide ocean it is a different story: the nutrients are locked away on the seabed, some miles beneath the surface.

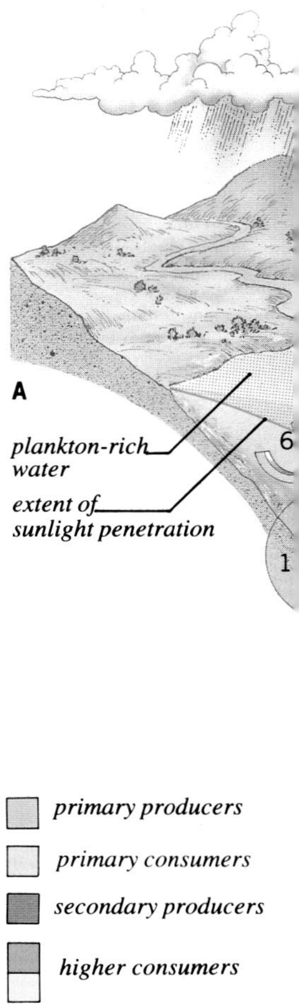

***The rich variety of life** in the seas* [**A**] *exists mostly above 660 ft (200 m), and almost all of it is sustained by plankton. Phytoplankton*[1] *thrive in areas where the water is rich in nutrients, and where there is sufficient sunlight for photosynthesis to occur. River estuaries* [2] *are a good source of nutrients, but a far more important source is the recycling of nutrients locked in the bodies of dead animals. The nutrients, released by the action of bacteria, are brought to the surface in several ways. In cold waters inversions occur, by which the cold water at the surface sinks to the bottom, while warmer water rises, bringing the nutrients with it. In some tropical and subtropical areas, strong offshore winds drive back the surface waters, sucking up nutrient-rich water from beneath. Alternatively, nutrients are brought to the surface as ocean currents run against peaks* [3] *on the ocean floor. These vertical movements of water are known as upwellings* [4]. *However, it is the coastal waters lying above continental shelves* [5] *that are richest, because the nutrients are churned up by storms and wave action* [6]. *The simple food chain* [**B**] *shows the importance of plankton as the basic source of food for ocean animals. Phytoplankton* [7] *are primary producers – they are food for zooplankton such as the filter feeders* [8], *krill* [9], *microalgal feeders* [10] *and fish fry* [11]. *Zooplankton are primary consumers, and they in turn are eaten by larger sea creatures, including codlike fish* [12], *blue whales* [13] *and arrowworms* [14]. *This selection of animals is representative of the secondary consumers, but many of these are in turn taken by larger creatures such as the Antarctic giantfish* [15] *and squid* [16] *– known as the higher consumers. The ocean food chain extends to include the efficient hunting sea creatures such as the leopard seal* [17]. *Closer to the shore, the food chain* [**C**] *is similar. The basic source of production, the phytoplankton* [18] *are taken by invertebrate filter feeders* [19]. *The primary consumers in this food chain include the invertebrate carnivores* [20], *such as starfish, as well as the predatory fish* [21].

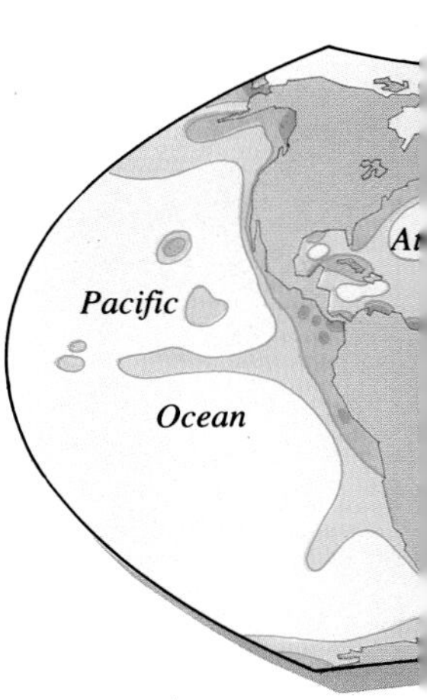

Distribution of phytoplankt… measured in ounces of carb… per acre per day

>70
35–70
20–35
<20

Connections: Algae 112 Animal Adaptation to Pressure 146 Biolunimescence 276 Estuaries 334 Filter Feeders 196 Fish 124 Jet-propelled Animals 212

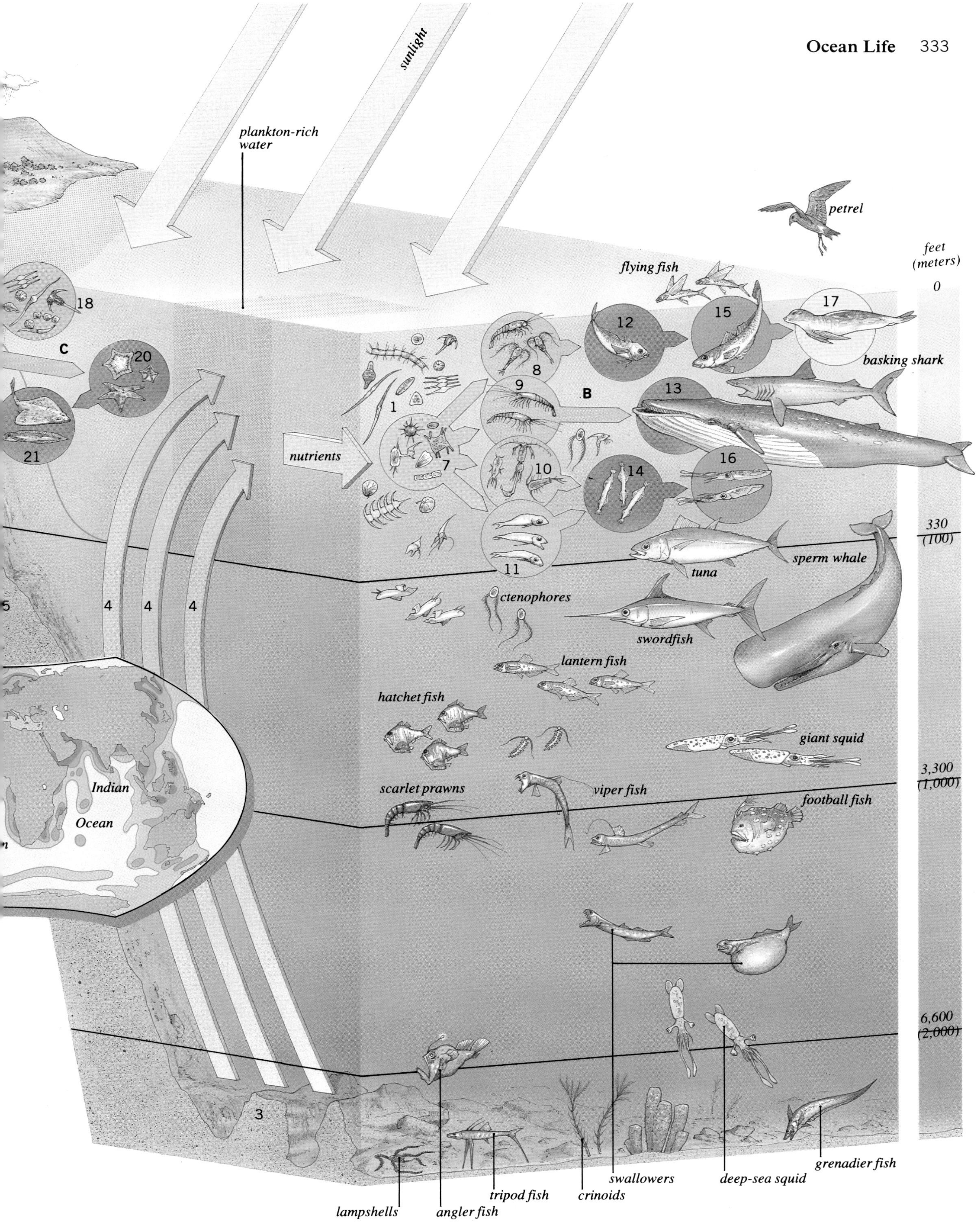
sunlight
plankton-rich water
petrel
feet (meters)
0
flying fish
18
17
15
12
C
20
basking shark
8
9
B
13
1
nutrients
7
21
10
14
16
11
330 (100)
sperm whale
tuna
5
4
4
4
ctenophores
swordfish
lantern fish
hatchet fish
giant squid
Indian
Ocean
3,300 (1,000)
scarlet prawns
viper fish
football fish
6,600 (2,000)
3
grenadier fish
swallowers
deep-sea squid
tripod fish
crinoids
lampshells
angler fish

Doors to the Sea

How life is adapted to estuaries

In order to survive, the inhabitants of estuaries must cope with huge daily salinity fluctuations – sometimes from fully saline water to completely fresh water – which would kill most creatures. In addition, the mudbanks that are home for most of an estuary's animal life are eroded and re-formed almost every day. But the survivors of this difficult and challenging environment often flourish: 35,000 surface-dwelling *Hydrobia* snails or 5,000 burrowing cockles and tellins, for example, may be found in an area as small as a square yard.

Estuaries are usually extremely sheltered habitats. But tidal fluctuations, together with variations in river flow (as a result of rain or drought), continually alter the salinity of the water. In the truly tidal areas of the estuary, salt content can vary from almost fully saline (35 parts of salt per thousand) to almost completely fresh (less than 0.5 parts of salt per thousand). The great mud expanses also experience dramatic variations in temperature between summer and winter.

The brown seaweeds that colonize estuaries do so wherever a suitable substrate is available. The various species each have their own zone for colonization, reflecting their different tolerances to the conditions. Green algae does not require a hard substance to which to attach itself, but rather grows in a large tangled mass, trapping the sediment beneath and often creating conditions of low oxygen. Eelgrass can also occur in dense growths over the mudflats, but, unlike green algae, this has a branched root system for attachment, which helps to stabilize the mudbanks.

Animal life

A number of specialized animals have become adapted to living in estuaries. Types of mollusk include mussels, vast beds of which are found on the lower reaches of many estuarine mudflats. Ragworms and lugworms are two burrowing species that can tolerate the conditions of estuarine mudflats and can occur in densities of over 850/square yard (1,000/square meter). Crustaceans, however, are much less adaptable to the harsh conditions, but one species that has succeeded is the sandhopper.

The muddy larder

Such a vast food resource inevitably attracts large numbers of predators, both fish and birds. The typical estuary fish is the flounder. It spawns in the sea, but the adult can survive in the estuary or even in fresh water for a considerable period. Other species, such as gobies, three-spined sticklebacks and gray mullet, are also typical of estuaries.

Estuaries are, however, probably best known for their vast numbers of overwintering waterfowl. Flocks of thousands of waders wheel in the winter sky above countless geese that have flown down from arctic regions to graze on the eelgrass and green algae.

Salinity is the key factor *determining the distribution of life in an estuary. Because salinity varies both with the seasons and during the day, there is a constant interchange of habitat. Upstream, and away from tidal influence and saline water* **[A]**, *there are largely freshwater species such as eelgrass* [1], Enteromorpha intestinalis [2] *and* Gammarus pulex [3]. *Lower down* **[B]**, *the estuary is usually populated by marine species that can tolerate cold water with a low salt content. These include the seaweeds: toothed wrack* [4], *knotted wrack* [5], *bladder wrack* [6] *and channel wrack* [7]. *Animals found here include the rough periwinkle* [8], *the shore crab* [9], *ragworms* [10], *common mussels* [11], *star barnacles* [12], *common eels* [13], *gray mullet* [14] *and Baltic tellin* [15]. *In mixing waters* **[C]** *live common necklace shells* [16], *common cockles* [17], *keel worms* [18], *edible periwinkles* [19], *the flounder* [20], *the sandhoppers* Corophium volutator [21] *and* Orchestia gammarella [22], *flat*

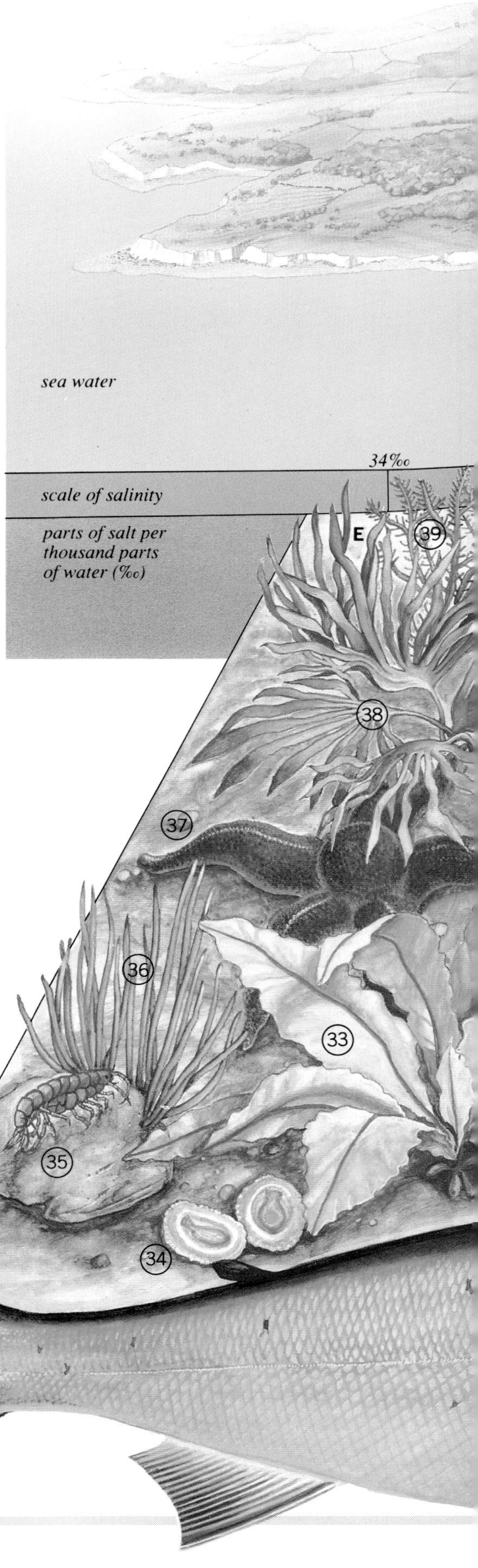

Connections: Ocean Life 332 River Life 328

periwinkles [23], *acorn barnacles* [24], *oarweed* [25] *and toothed wrack* [26]. *In areas* [**D**] *of higher salinity, spider crabs* [27], *pepper dulse* [28], *common limpets* [29], *purple topshell* [30], *beadlet anemones* [31] *and carageen* [32] *are all to be found.*

At the mouth *of the estuary* [**E**] *are found species that can tolerate little freshwater, such as dabberlocks* [33], *the limpet,* Patella aspera [34], Gammarus locusta [35], *thongweed* [36], *common starfish* [37], *cuvie* [38] *and* Corallina officinalis [39]. *The salmon* [40] *undergoes changes to adapt to the decreasing salinity as it swims upstream to spawn in the same river every year.*

Rivers 58 **Seashores** 330 **Swimming** 210 **Wetlands** 324

The Underwater Jungle

How coral reefs are such rich habitats

Coral reefs have been estimated to support one-third of all the world's fish species, and possibly as many as half a million different animal species altogether. Like the rain forests, they are truly ancient – they may well have been in existence for 500 million years. The bewildering variety of shapes and forms in a reef provide the reef animals with an almost endless supply of places to shelter, hide and feed. Most obvious of the multitude of vertebrates that live there are the fish, which teem in their millions and bring a brilliant variety of colors to the water.

The basis of a coral reef is the coral itself. Corals are simple colonial organisms formed of large numbers of polyps, tiny animals similar to sea anemones. The polyps feed on floating animal plankton, trapping the tiny creatures in their tentacles. Each polyp secretes limestone, and as the polyps die, new polyps grow on their remains, resulting in the steady expansion of the reef. Inside the coral polyps live tiny golden-brown algae (*dinoflagellates*) known as zooxanthellae. The exact relationship of the algae to the polyps is unclear, but the zooxanthellae (of which there may be one million/square inch –150,000/square centimeter) seem to benefit from the polyp's protection and in return, absorb polyp waste products, such as carbon dioxide and phosphates, that are essential for algae growth. The polyps meanwhile use the oxygen and some organic compounds photosynthesized by the algae. This highly efficient symbiosis is at the heart of the rich coral reef ecosystem. Blue-green algae also play a vital role by cementing together the coral.

Plankton thrive in the clear, well-lit reef waters. These tiny plants, such as *diatoms* and dinoflagellates, are food for herbivorous zooplankton, such as salps, which are in turn eaten by carnivorous zooplankton, including iridescent, jewellike copepods.

Night commuters

Many animal plankton species commute up and down the reef waters, feeding near the surface at night and retreating down to shelter in the reef in daylight. This is probably to avoid daytime predators, such as fish, that hunt by sight. Many corals feed at night too, for the richest supply of animal plankton food is found at night.

In the underwater jungle, larger animals abound. Sea cucumbers feed on the organic debris in the sand grains, passing the sand through their guts. Shellfish, fish and sea urchins graze the plants on the coral.

Reefs are fragile ecosystems that can be threatened by overfishing, fish dynamiting, mining the coral, silt runoff from land erosion and pollution. Fortunately, some reefs – such as Australia's Great Barrier Reef – are protected and managed. A natural enemy and prolific breeder is the crown of thorns starfish, which can eat up to 55 square feet (5 square meters) of coral a year.

***A coral reef** extracts nutrients from the sea and redistributes them through a complex food chain where every available niche is exploited. The reef front is the fastest-growing area, steeply sloping to the seabed. Plate coral* [1] *grows large plates to expose the maximum possible area to the poor light; nurse sharks* [2] *may lay eggs in protective coral crevices.*

Growth at the reef crest is hampered by dry periods at low tide. Behind the seaward side is the sheltered shallow reef flat, home for many fast-growing staghorn corals [3], *whose branches raise them above rivals, like brain coral* [4], *in the quest for sunlight. Broken-off branches may regrow, spreading the colony. Home for small fish like the black-backed butterfly fish* [5], *lower coral parts are mostly dead and algae covered, food for various invertebrates and the purple-headed parrotfish* [6], *which bites off bits of coral, ingesting it with the algae and excreting the coral as fine sand, thus building up the shallow-water area. Its gaudy colors enable it to mark out territory and scare off enemies. Like parrotfish, plankton-eating fairy basslets* [7] *change sex to suit circumstances. Among the world's most venomous animals, olive sea snakes* [8] *prey on such fish. Cleaner wrasse* [9] *eat parasites on the skin of other fish, often predators that "suspend hostilities" for grooming sessions. Giant clams* [10] *eat a mix of filtered food and photosynthetically produced nutrients from symbiotic algae living in its lips.*

Gorgonian coral [11] *flexes more than most corals, allowing it to survive close to turbulent water surfaces, unlike many others.*

Sea grass [12] *pollinates underwater, with currents carrying the pollen; providing a home for many creatures, it is food for few animals except turtles, urchins and dugongs.*

Tubular [13] *and vase* [14] *sponges have structures like chimneys that create updrafts to suck water over their feeding chambers. The male frigate bird's* [15] *huge red sack is inflated during courtship. Fish eaters, these birds often force others, like the blue-footed booby* [16], *to drop prey, catching it as it falls.*

***Coral reefs** (map) grow in waters over 68°F (20°C), thriving at about 75°F (24°C). Reefs grow beyond the tropics in sufficiently warm currents. Corals do not grow in freshwater discharges and cannot grow in water full of sediment, which smothers the fragile reef organisms. Optimum depth for growth is a few yards below the surface, where oxygen and sunlight are abundant; growth rarely occurs below about 230 ft (70 m). In ideal conditions, healthy reefs grow up to 25 mm (1 in) a year.*

Connections: Algae 112 Animal Conservation 344 Fish 124 Island Formation 52 Ocean Life 332 Seas and Oceans 44 Symbiosis 192

coral reefs
68°F
(20°C)
68°F
(20°C)
reef crest
reef flat
15
10
4
6
7
5
3
13
12
8
11
1

Fouling the Earth

How we are polluting the land

Between 1947 and 1953 nearly 22,000 tons of chemical waste were dumped at a waste disposal site in the eastern United States. The site was sold to the local board of education for just one dollar with the warning that it should not be disturbed. In fact, the warning was unheeded and a whole new neighborhood was built. In 1976, people noticed a foul-smelling liquid, and local residents fell sick. Children and pregnant women were evacuated in 1978 and Love Canal, the site in upper New York State, became one of the most infamous examples of contaminated land.

Contaminated land is a widespread problem of unknown dimensions in most industrialized countries. At Lekkerkerk, in the Netherlands, close to three hundred houses were built on an old refuse dump, which contained organic solvents and other wastes. At another site, near Utrecht, the former home of an asphalt works, tar penetrated the soil to depths of 150 feet (45 meters). Reclaiming the site cost many millions of dollars, and involved isolating the vitiated earth by sinking steel plates to the full depth of contamination. The polluted soil was then excavated and incinerated to burn off the tars.

Buried and forgotten

The millions of tons of tars, solvents, pesticides and domestic refuse created each year by modern technological society is mainly dumped into landfill sites, often just a convenient hole in the ground.

As rain seeps into these rubbish heaps, it forms a toxic liquid, or leachate, which percolates into the underlying earth. In such sites migration of the leachate was once considered to be a good thing. The idea was that the poisons, which become more diluted the farther they migrate from the site, would be broken down by bacteria, or become attached to soil particles at low concentrations. But the strategy has not worked. Western Germany has 35,000 problem sites, and Denmark estimates that 2,000 sites have seriously contaminated the surrounding groundwater.

Sealed containers

New landfill sites are designed to prevent leachate from forming and, if it forms, to keep it from migrating. These sites will eventually be practically sealed off with a low-permeability clay layer and impermeable plastic liners around the base and sides, as well as an impermeable cap.

In all landfill sites anaerobic bacteria, which thrive in the compacted air-free layers of rubbish, digest organic matter and produce a combination of methane and carbon dioxide gases. The methane that escapes out of the top of the site or the sides not only contributes to global warming, but as an explosive gas can also seep into houses, with potentially disastrous consequences.

New landfill sites are engineered to collect the methane, which can be used in the same way as natural gas – which is exactly what it is. Organic waste, including domestic sewage, is separated from other materials. Then anaerobic bacteria can be used to generate methane in a controlled environment. And the sludge that remains after the bacteria are finished can make a safe, sterile fertilizer. If the sludge is mixed with straw, or any low-grade organic waste, its high nitrogen content can feed other bacteria that will break down the organic waste and create a pleasant-smelling manure or compost, which can be used on the land. In China, this is exactly what does happen to organic waste and sewage in almost every village.

The most toxic waste, which cannot legally be dumped at landfill sites in the countries of origin, often finds its way to dumping grounds in the developing countries. Over 40,000 tons of industrial waste is shipped to

***Many elements** are naturally radioactive* [**A**]*, so there is always a background radiation, even when there is no man-made or accidentally generated radioactivity in the vicinity. Alpha radiation occurs when an unstable atomic nucleus emits an alpha* [1] *particle (which is a helium nucleus – two protons and two neutrons). Alpha (α) radiation is stopped by flesh. Beta (β) radiation occurs when an electrically neutral neutron emits a negative electron (which is the beta* [2] *particle) and becomes in the process a positive proton* [3]*. Beta particles can be stopped by a thin sheet of aluminum (Al). X rays and gamma (γ) rays are both electromagnetic radiation (types of light), created when a nucleus loses energy without undergoing structural change* [4]*. X rays can be stopped by a sheet of lead (Pb), while high-energy gamma rays will penetrate thin sheets of lead, but may be stopped by a layer of concrete.*

A

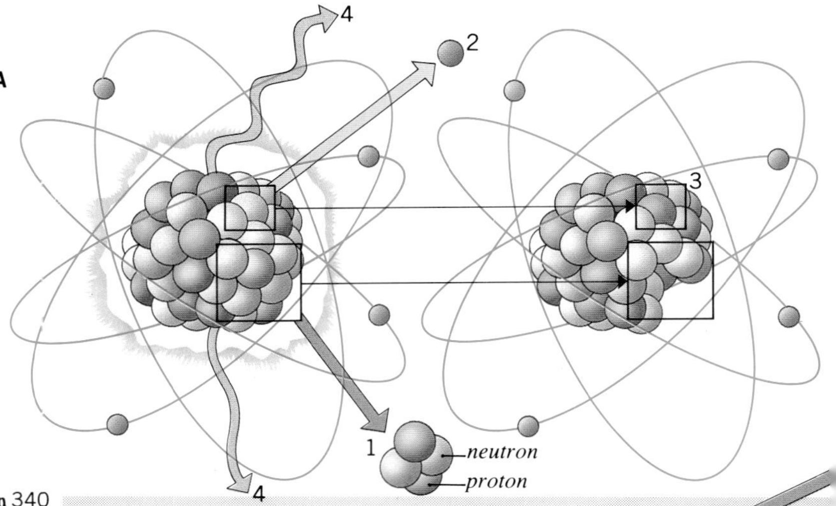

B

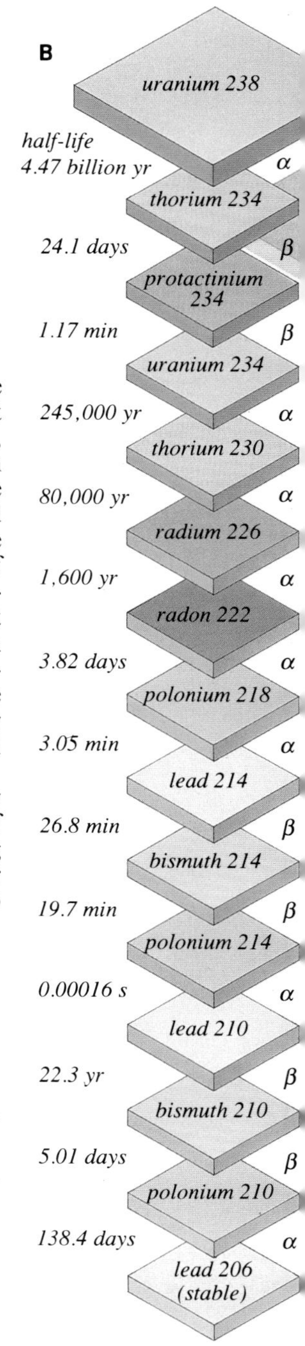

Connections: **Air Pollution** 342 **Decomposers** 204 **Water Pollution** 340

C

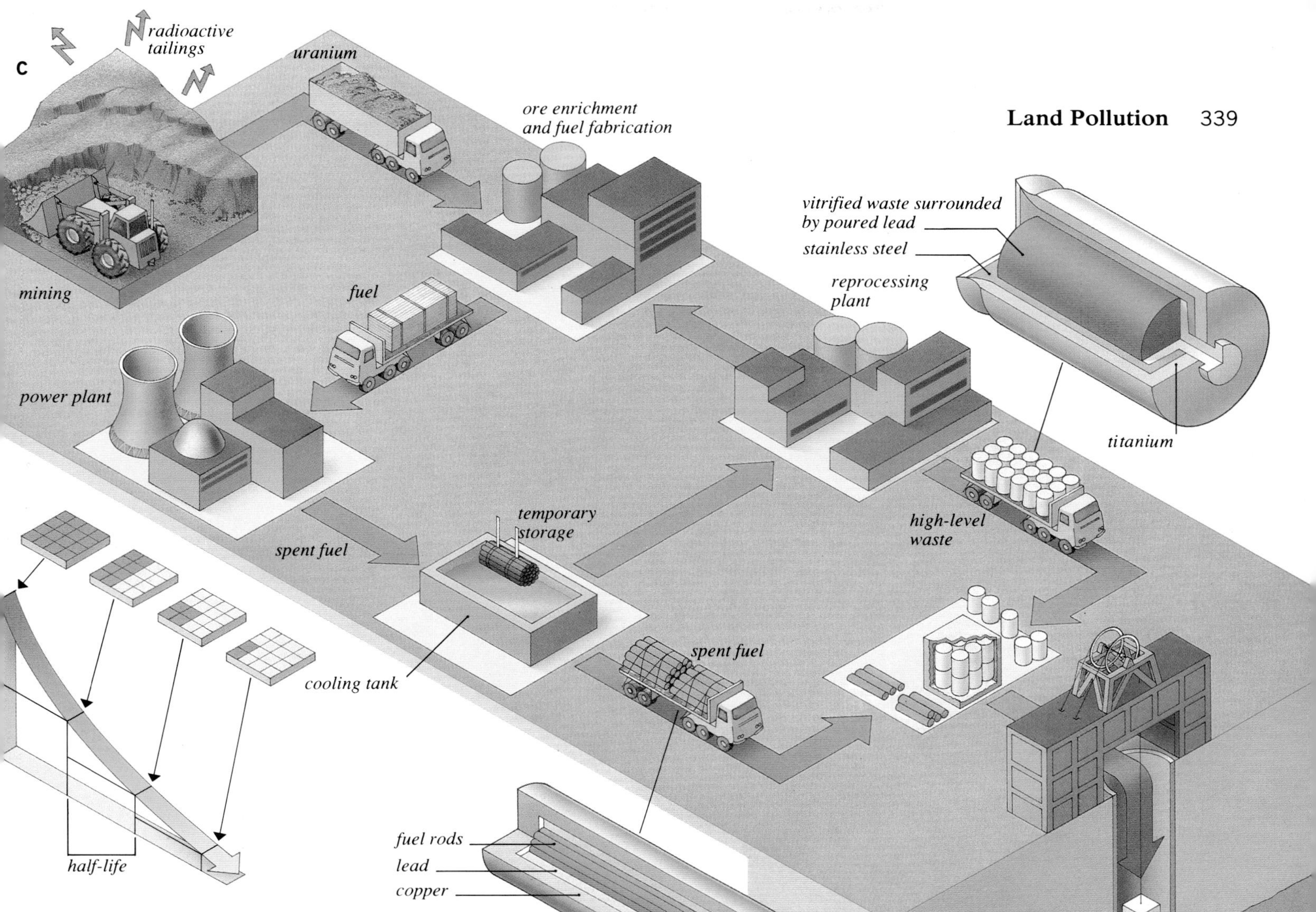

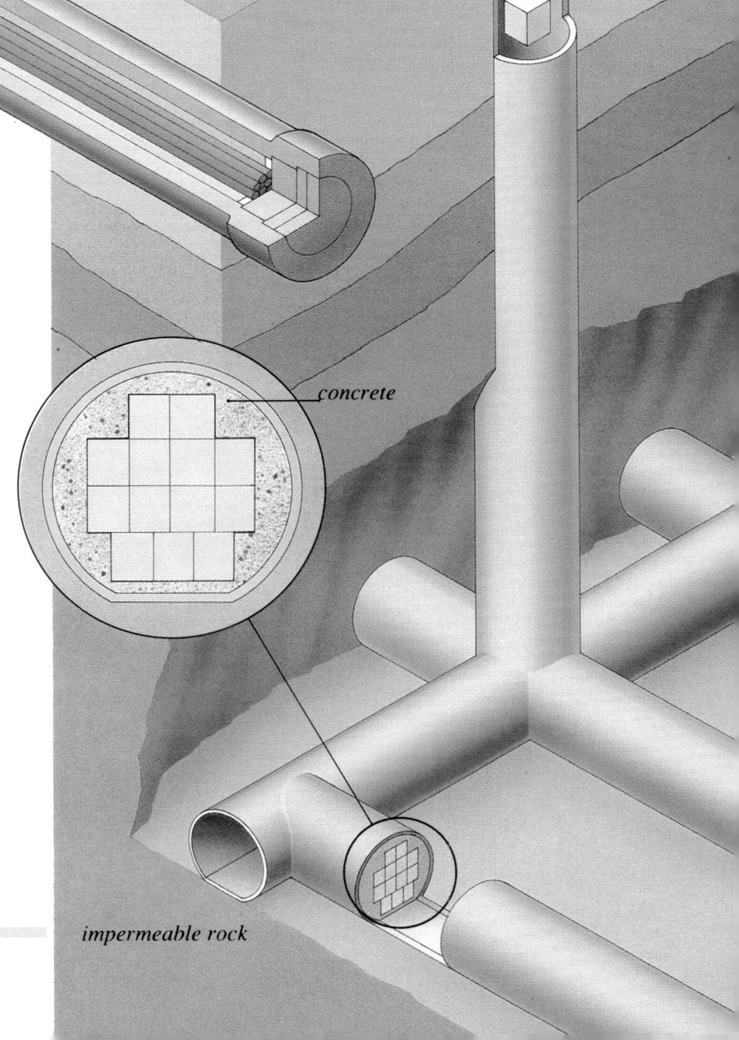

***The half-life** of a radioactive element is the time taken for half of the existing amount of the substance to decay into some other element* [**B**]*. This element may in its turn be radioactive, with its own, different, half-life. The diagram shows the half-life of uranium-238 (*the parent radionucleide*) and of each of its products (*the daughter radionucleides*). Each daughter decays into another daughter until the process reaches lead-206, which is nonradioactive or stable. The isolation required for radioactive waste depends on the lifetime of the waste concerned, and on the lifetimes of its daughters.*

The nuclear industry [**C**] *affects the environment long before any power is generated. The mining of uranium produces great quantities of tailings, the rock left over after the useful ore has been extracted. The tailings contain over 80 percent of the radioactivity that was present in the original ore, and millions upon millions of tons of this debris now lie on the Earth's surface, being leached by rain and dispersed by wind. The spent fuel that is created when power is generated may be disposed of, or could go into temporary storage for uranium to be regained, or for reprocessing into usable radioactive plutonium. Either process creates its own waste. Waste disposal is a serious problem. Initially it is kept in canisters. One Swedish scheme has two types of canister, one for the waste from reprocessing and one for spent fuel rods. These are sealed using a variety of materials to prevent leakage. A number of canisters are usually set in concrete or resin units, which are then buried in tunnels in impermeable rock before the tunnels are themselves sealed up with concrete. However, because of the heat generated by the waste from reprocessing, and because of its high concentration, extra care may need to be taken. The high-level waste from Britain's reprocessing operations is kept in special double-walled tanks at the nuclear plant in Sellafield.*

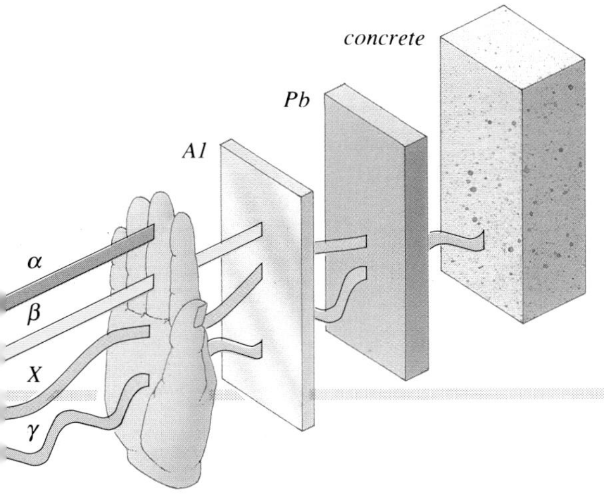

The Seeping Sickness

How we are polluting the world's water

In the 1960s, as a result of eating fish contaminated with mercury that had built up from 30 years of dumping, over 40 people died in the Japanese town of Minamata, while many more of the 50,000 inhabitants suffered pins and needles, a lack of feeling in the limbs, slurred speech and tunnel vision. The world's oceans are also threatened by oil spills – normally dreadful accidents that decimate the life along thousands of miles of shoreline – but in the Gulf they have been used deliberately as a new form of ecological warfare. Many other deadly substances find their way into the sea.

Minamata disease first alerted the world to the fact that mercury dumped in the sea was not dispersing but was building up in the bodies of marine animals. Other heavy metals that are present in industrial waste, such as lead – used in some paints and petrol – and cadmium – used for bright pigments and zinc smelting – are notorious poisons. In addition, the effect of two or more heavy metals, or other poisons, is usually greater than the sum of their individual effects.

Deadly dispersal

Not all pollutants face the same fate in the water. For instance, heavy metals easily become attached to fine-grained sediments carried in a river. When the river slows and meets the sea in a bay or an estuary, the sediments sink and are deposited on the bottom, concentrating the pollution, which is readily taken in by filter-feeders such as oysters and other shellfish. Other dangerous pollutants, such as the *chlorinated hydrocarbons,* of which DDT is an example, are insoluble and cannot be broken down by organisms. For this reason they accumulate in the fatty tissue of animals low down in the food chain, and as these animals are eaten by their natural predators, the concentration of DDT increases and continues to do so as it passes higher up the food chain. In the end, animals from shrimp to fish to carnivorous seabirds are put at risk.

PCBs (*polychlorinated biphenyls*) are another harmful pollutant. They are used in manufacturing paints, plastics, adhesive-coating compounds, hydraulic fluids and electrical equipment. They have similar effects to DDT, but are more persistent, gradually collecting in seafloor sediments, where they are then slowly released into the overlying water.

Fit for drinking?

A comparatively new water pollution problem that threatens fresh water and underground drinking supplies is nitrate. There are three main causes for the increase of nitrate in fresh water: the use of nitrate fertilizers; discharges from sewage works; and intensive livestock farming, producing waste rich in nitrates.

It is intensive farming that has in general caused the biggest recent increase in nitrate concentration. Farmers use inorganic nitrogen-rich fertilizer almost twice as much as they did ten years ago. Much of this soluble nitrate then soaks down into the water table. Nitrate put on fields is harmless in itself, but bacteria in our mouths and throats convert it to nitrite or nitrosamine, which can cause stomach cancer. Nitrite can also affect the blood's ability to carry oxygen through the body. It is responsible for "blue baby syndrome," whereby the baby turns blue for lack of oxygen and can die, strangled from the inside. Too much nitrate in water also leads to the excessive growth of algae, which when it dies, uses up oxygen as it decomposes. This lack of oxygen kills off much of the other life in the river or lake – a process called *eutrophication,* which is ruining many rural rivers and lakes.

It is easy to blame industrialists and farmers for polluted water, but our cherished, unblemished fruit and vegetables owe a lot to pesticides. Moreover, industries use chemicals to produce the automobiles, colored toilet paper and many other products that we demand.

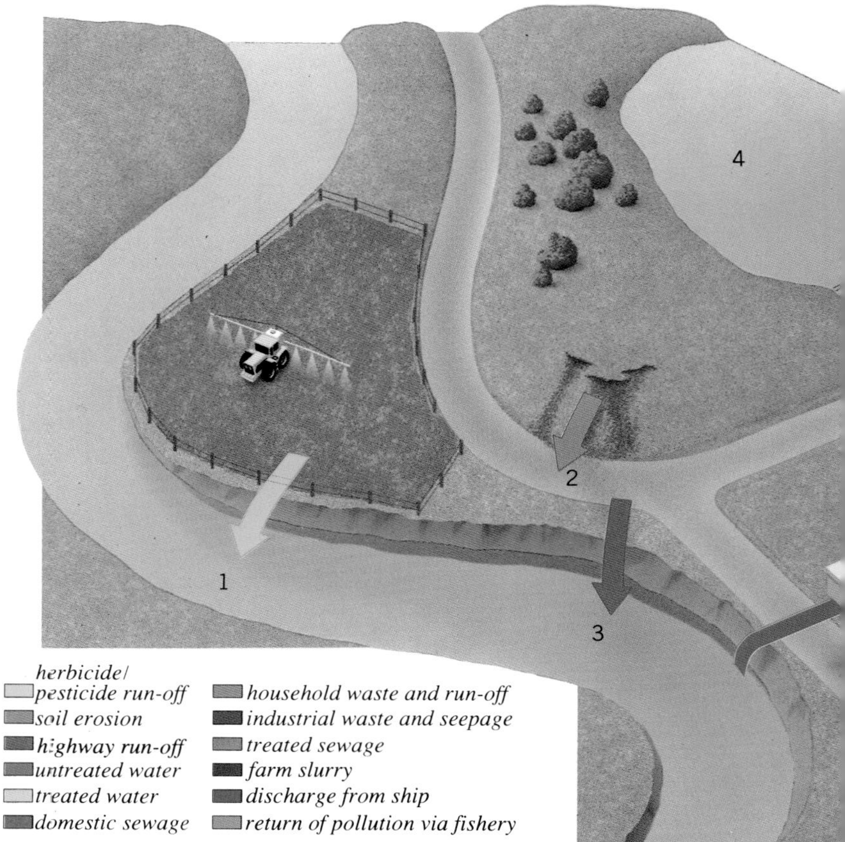

Many human activities *pollute the world's water and damage the environment. Some pesticides, for example, do not decompose after they have been sprayed on the fields, and when it rains, they may well be washed into rivers [1]. Large, hedgeless fields are now common in modern farming. If rainfall is sufficiently heavy, these fields are subject to soil erosion [2], whereby the top layer of soil is washed away. This impoverishes the land and may cause rivers to silt up. Another source of pollution is the millions of tons of salt used on icy roads every winter. These are washed into rivers [3], harming the*

wildlife. Our massive consumption of water has also resulted in the creation of more and more reservoirs [4], *often at the expense of prime agricultural land, while our homes* [5] *are full of chemicals used for cleaning purposes. These cannot always be removed, and the filter beds* [6] *used to do so can often be rendered ineffectual by household chemicals such as bleach. Moreover, a great deal of untreated domestic sewage flows straight into the sea* [7], *contaminating tidal mud.*

Oil Pollution

The environmental effect of a spill of light oil in a hot climate is different from a heavy crude spill off Alaska. How far the oil spreads depends on how quickly the volatile parts of the oil evaporate, and then how quickly the thicker, more viscous, residue breaks up. Some of the lighter oil fractions dissolve in the water under the slick, increasing the danger to life in the water. The oil forms tar balls that are either washed up on the shore or absorbed onto particles, which then sink to the bottom. Oil poisons and smothers the creatures that live in and around the sea, and the most prominent victims are seabirds. The first step in fighting an oil spill is to stop the oil from spreading. Currents and prevailing winds are checked, and booms lowered to trap the oil. It is then treated with special detergents.

People also generate *huge amounts of rubbish* [8], *which can create toxic runoff. Seepage of buried industrial waste* [9] *is more dangerous.*

Some factories even expel industrial waste directly into rivers [10]. *Pollution is also generated by the sludge farmers put on their land* [11]. *Sludge, a form of sewage, should be a good, cheap fertilizer. However, the pollutants found in sewage can make sludge dangerous. Intensive farming of livestock such as pigs* [12] *has led to increased amounts of liquid manure, which enters rivers and contaminates them. Another method of disposing of sewage is dumping at sea* [13]. *The chemicals in the sludge enter the marine food chain and many fish caught for us to eat* [14] *have traces of these pollutants.*

The Corrosive Sky

How the world's air is being polluted

The rain that falls in some areas of the Western world is as acid as lemon juice. Although rain is naturally slightly acidic anyway, for the many aquatic organisms that live in waters exposed to such extremes, the corrosive environment spells death. The acid even eats away at the surfaces of buildings and statues in towns. In the European forests, the trees are much more vulnerable still. Damage to forests by acid rain is affecting more than half of all trees in Czechoslovakia, Germany, Greece, the Netherlands, Norway, Poland and Britain.

We are daily poisoning the air we breathe. The tetraethyl lead added to gasoline to prevent engine knock is emitted by exhaust pipes and concentrates in air and dust. This is especially harmful in cities, and a study in Turin showed that 30 percent of the lead in the inhabitants' blood came from petrochemicals. Lead in the blood can cause stomach pains and headaches, irritation, coma and death. And very low levels of lead can affect the brains of growing children. There is growing international pressure to ban the use of leaded gasoline, but lead is only one of the harmful substances – many of them carcinogenic – regularly pumped into the air as a by-product of our industrialized society.

The comparatively recent rapid industrialization in Europe and North America has added tremendous quantities of acidic pollutants to the atmosphere, and taller chimneys have encouraged the pollution to spread to distant, previously unaffected, areas. The result has been lifeless lakes, dying forests and contaminated soils.

Food chains may be disrupted within forests as the leaves of damaged trees become deficient in calcium through leaching processes. Leaves are eaten by caterpillars, which, in turn, are eaten by nesting birds. These birds produce very thin and easily damaged eggshells, or even no eggs at all, and the bird population accordingly declines.

Most sulfur *[1] leaves factory chimneys as the gaseous sulfur dioxide (SO_2), and most nitrogen [2] is also emitted as one of the nitrogen oxides (NO or NO_2), both of which are gases. These gases may be dry deposited – absorbed directly by the land, by lakes or by the surface of vegetation [3]. If they are in the atmosphere for any time, the gases will oxidize (gain an oxygen atom) and go into solution as acids [4]. Sulfur dioxide will become sulfuric acid (H_2SO_4), and the nitrogen oxides will become nitric acid (HNO_3). The acids usually dissolve in cloud droplets, and may travel great distances before being precipitated as acid rain. Catalysts such as hydrogen peroxide, ozone and ammonium help promote the formation of acids in clouds [5]. More ammonium (NH_4) can be formed when some of the acids are partially neutralized [6] by airborne ammonia (NH_3). Acidification increases with the number of active hydrogen (H^+) ions dissolved in an acid [7]. Hydrocarbons emitted by – for example – car exhausts [8] will react in sunlight with nitrogen oxides to produce ozone [9]. Although it is invaluable in the stratosphere, low-level ozone causes respiratory problems, and also hastens the formation of acid rain. When acid rain falls on the ground, it dissolves and liberates heavy metals and aluminum (Al) [10]. When it is washed into lakes, aluminum irritates the outer surfaces of many fish. Eventually their gills become clogged with mucus and they die. As acid rain falls or drains into a lake, the pH of the lake falls. Perfectly neutral water would have a pH of 7, and each drop of 1 point on the pH scale means that acidity has increased tenfold. Naturally occurring water always contains dissolved substances that make it slightly acid or alkaline, even without man's intervention. Experiments in Canada indicate that minnows [11] and shrimps [12] begin to disappear when the pH reaches 6. Trout [13], which feed on them, also begin to decline. At pH 5.6 the exoskeletons of crayfish [14] soften and become overrun with infestation. All that is left is a clear lake with a lush carpet of green algae [15] and moss. Alkaline soils containing calcium carbonate can help neutralize acid rain, and some countries add lime or limestone to protect vulnerable lakes.*

Forests suffer the effects of acid rain through damage to leaves, through the loss of vital nutrients, and through the increased amounts of toxic metals liberated by acids, which damage roots and soil microorganisms.

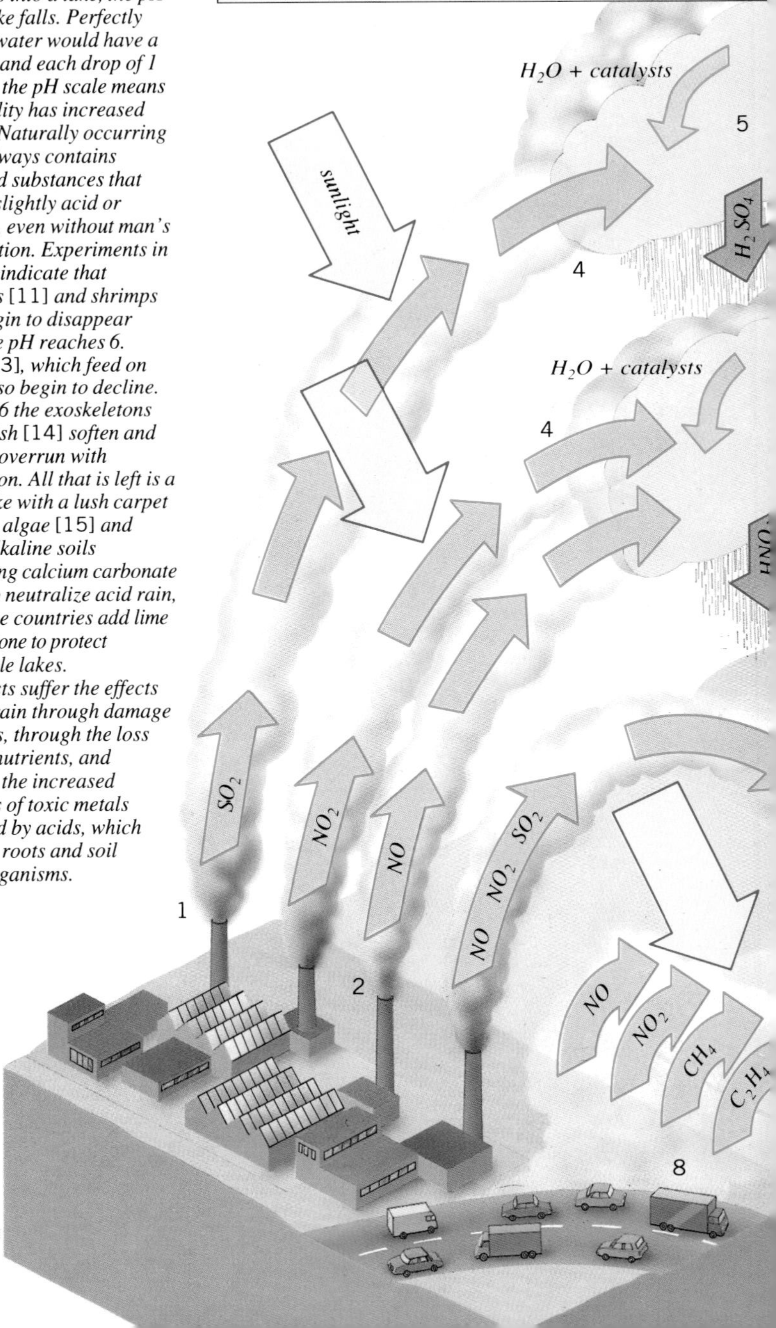

Smog

Smogs occur when pollutants accumulate in a shallow layer of cold air trapped under warmer air (a situation known as a temperature inversion). Where poor-quality coals are used extensively, as in eastern Europe, sulfurous smogs often envelop cities in cold winters. Such smogs have been largely eliminated from western Europe and North America, only to be replaced by summer photochemical smogs, caused by hydrocarbons and nitrogen oxides from vehicle exhausts reacting with sunlight to form eye-stinging ozone.

Los Angeles, enclosed between mountains and the sea, suffers severe photochemical smogs in summer. With its population increasing by 1 million every four years, smogs were expected to worsen; but tough new controls were introduced in 1989 to eliminate them by 2007.

***Some soils** are more vulnerable to acidification than others. The map (below) indicates areas of various degrees of susceptibility, based on soil type. Laterite soils of many tropical regions, for example, are highly sensitive to acidification. Tundra soils and soils that contain high levels of calcium and sodium are the least susceptible.*

low
moderate
high-moderate
high
large emissions

Acid rain may contaminate supplies of drinking water by dissolving and thus liberating toxic metals in the soil. The corrosion of underground drinking-water pipes is another source of danger. One very serious consequence is the harm caused to the liver and kidneys of young children by an accumulation of high levels of copper.

Future prospects

Since the early 1980s some progress has been made in tackling acid rain. A number of industrially based countries have signed international agreements to reduce emissions of sulfur and nitrogen oxides, especially from large coal-fired power plants. Environmentalists have strongly suggested, however, that such promised reductions do not go far enough. Soils and lakes of many regions are now so acidified that the situation will remain problematical for many decades.

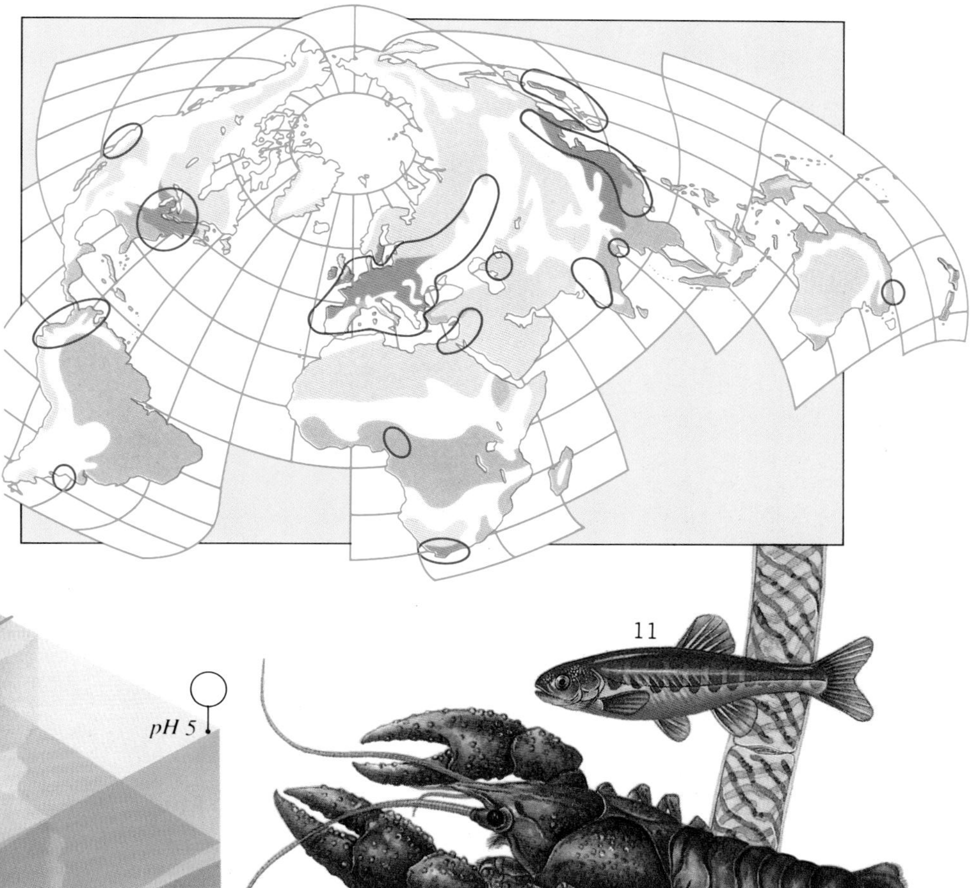

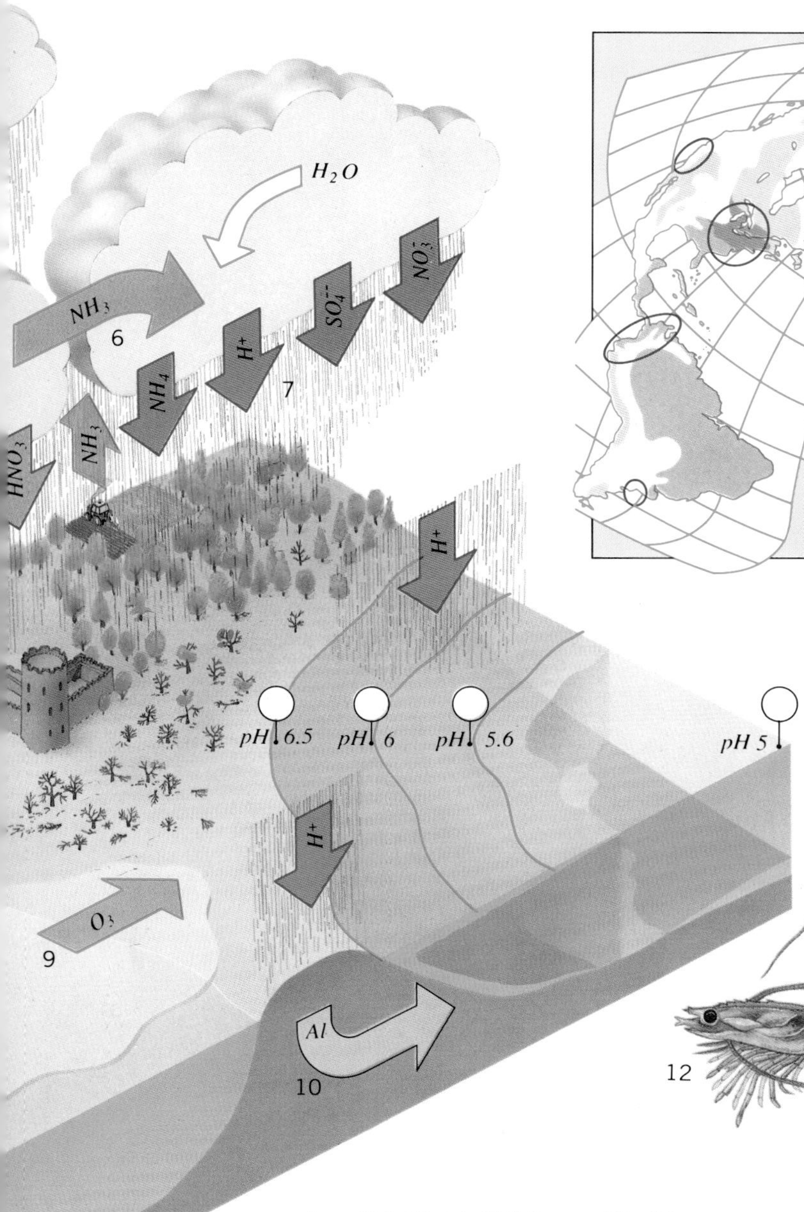

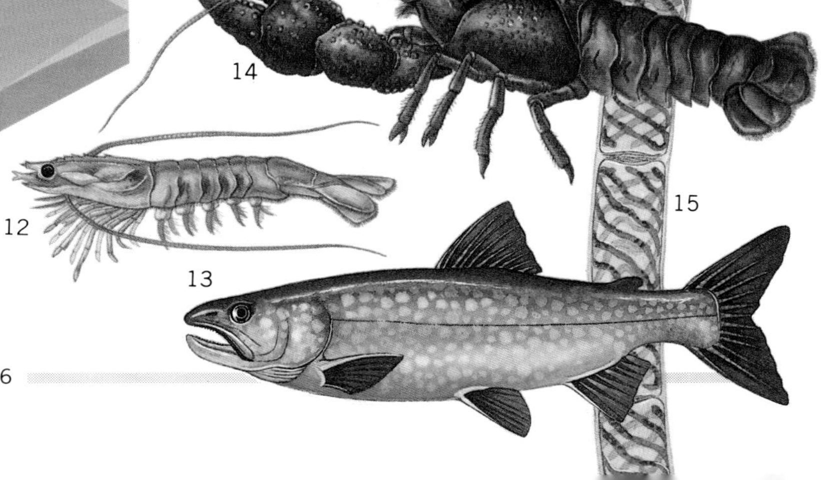

Ozone Hole 68 Rain, Hail, Snow and Fog 76

Born to be Wild

How animals are being conserved

Before the nineteenth century, roughly one animal species became extinct a year – but in the last 100 years the rate of extinction has accelerated, and we are now losing our wildlife at a rate of several species a day. Nobody knows how many species of animal live on Earth – despite over 200 years of continuous research. A current estimate puts the number at about 30 million, of which only a million are known to scientists. There is very little legal hunting of endangered species now, but as we destroy natural habitats, more and more of the Earth's animals are in danger.

A

The hawksbill turtle [**A**] *is endangered for two reasons. The hawksbill is a prize catch for fishermen; its eggs and flesh are a delicacy. But its shell plates also provide tortoiseshell, which has been highly prized since ancient times, especially in Asia and Japan. Objectors to the total conservation of the hawksbill turtle cite the "rights" of certain local groups to their traditional protein sources; and others cite the "rights" of Japanese artisans to practice their traditional craft. All of these objections could be met by the establishment of hawksbill farms that breed turtles for their meat and shells. Commercial demand would be met from farmed animals, and the wild population could be protected. However, this is not possible under current CITES (Convention on International Trade in Endangered Species) regulations, because the initial start-up stock would be taken from the wild.*

An extinct animal is lost forever. By the year 2000, human activity may well wipe out three or four species every hour. Conservation tries to save species and their habitats from such destruction. There are two main approaches. On-site conservation sets aside areas as nature reserves, or takes special care of the environment to benefit wildlife. A variation involves reintroducing locally extinct animals and plants, or even restoring entire habitats, with their vast multitudes of species and ecological interactions. Off-site conservation involves taking animals into captivity, where they are maintained as a living collection.

Zoo tales

Zoos were originally just for public entertainment. Little thought was given to the animals' welfare, or the possibility of breeding. In fact, by taking rare animals from the wild, zoos menaced the very species they wished to exhibit. Today, zoos only remove rare animals from the wild if they are in danger, and only with a view to breeding. The offspring may then be returned to the wild. By the 1960s the oryx – an antelope native to the Arabian Peninsula – had been reduced by hunting to a few scattered populations in southern Arabia. In 1962 some survivors were captured and taken to the United States for captive breeding. By the early seventies the last wild ones had been shot. From a zoo population of about 150 the species was reintroduced to protected areas in Jordan, Oman and Israel. Later, the Oman herd was released to fend for itself, and for the present this beautiful animal survives in the wild.

Breeding problems

Many animals, however, are very hard to breed in zoos. Perhaps the most famous is the giant panda. Female pandas are only sexually receptive for about three days a year, and are very fussy about choosing a mate. It is also very hard to tell if a panda is pregnant. There have been many attempts to overcome these problems with artificial insemination, but this has a 75 percent failure rate, and only one in four cubs survives.

Successful conservation requires an understanding of a species' reproductive strategy – for sea turtles, such as the green turtle [**B**], *nest location is critical. A nest on the lower shore* [1] *provides hatchlings with the best chance of crawling to the relative safety of the sea. However, wave action may expose the eggs to lethal sunlight and predators. A nest dug on the middle shore* [2] *provides greater security, but that crucial first journey to the sea is longer. The upper shore* [3] *affords best security, but is farthest from the sea. Choice of nesting site becomes a calculated risk, balancing one hazard against another. The situation is further complicated by the fact that recent research has shown that the sex of the hatchlings is determined by the temperature of the nest, which itself is dependent on exposure to the Sun. A warm nest on the exposed lower shore* [1] *produces females. An intermediate nest in the*

Connections: **Animal Growth and Development** 166 **Biological Clocks** 274

The African elephant's unique position as largest land animal has made it the focus of intense debate. As a species, the African elephant is undoubtedly endangered. However, in some parts of the continent, such as Zimbabwe, the elephant population is rising, putting pressure on the local environment. This has led to conflicting attitudes to the contentious issue of the ivory trade. Some countries see a total ban as the only way to preserve the number of elephants, while others consider that ivory produced by the culling of a healthy population is a legitimate form of national income.

B

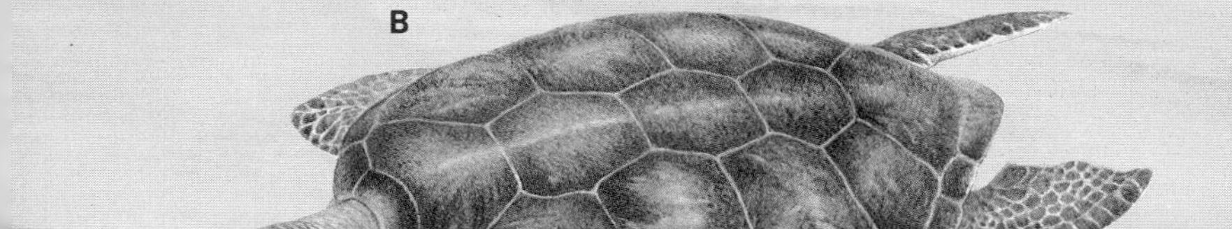

partially shaded middle shore [2] produces both males and females, and a cool nest on the upper shore [3] produces males. Somehow, while taking into account the survival of eggs and hatchlings, green turtles have also evolved the knack of locating their nests so as to produce the optimum ratio of males to females in the population. The new knowledge concerning the sex determination of hatchlings has called into question the conservation strategy of "head-starting" (taking wild eggs and rearing them until they hatch). During the 1970s, Operation Green Turtle head-started about 100,000 individuals in the Carribean Sea. This was done without any reference to the sex of the hatchlings. As green turtles take about 50 years to achieve sexual maturity, scientists will have to wait another 30 years before they can assess the effects of Operation Green Turtle.

Brave new world

Artificial reproduction involves many difficulties. Even closely related species require different conditions – such as blood hormone levels – to reproduce successfully, so procedures that work on, say, one big cat will fail with another. The semen for artificial insemination is taken by applying an electric shock when the animal is anesthetized.

Embryo transplants have been performed between different species, allowing a common antelope to be used as a surrogate mother for a much rarer one. A common problem in captive breeding is to keep the gene pool sufficiently rich, to avoid interbreeding. To this end zoos keep studbooks, and only arrange breeding swaps with other zoos after careful consultation. Despite all the problems, many animals breed readily in zoos. Some species of tiger have more members in zoos than in the wild, and lionesses breed so easily that many are fitted with contraceptive devices.

Isolation's end

Conservation started with attempts to preserve appealing animals. More recently there has been a growing concern for entire habitats, to protect a wide range of life-forms in balanced systems. At least a quarter of the Earth's species are under threat through loss of habitat. Island dwellers are especially at risk in the modern world, and more than half the world's threatened and endangered birds come from islands. Sailors and settlers have in the past brought their own livestock to islands, as well as less welcome passengers. Pigs, deer, rats, cats, stoats and ferrets have all caused havoc with native populations by competing with, or preying on them. Rats are the most deadly immigrants. Where they become endemic they are impossible to wipe out, and often the only solution is to move native creatures to another island.

Animal rights

There are now many international agreements governing the traffic in wild animals, and in the products taken from wild animals, such as ivory. Most of these spring from the meeting in Washington, D.C., in 1973, which resulted in the Convention on International Trade in Endangered Species (CITES). Policing these agreements can, however, be a conservationist's nightmare.

Plant Conservation 346 Reptile Reproduction 170

Save the Plants

Why we need to conserve our plant life

***Indifference about the** plight of wild plants has led to weak enforcement of CITES rules. Though all orchids and cacti are controlled plants, attractive specimens, like the rare Knowlton's cactus (right), are much prized by collectors and are frequently taken from the wild to be smuggled and traded.*

Around the world about 60,000 flowering plant species, a quarter of the Earth's total, are threatened with extinction. It is impossible to give a precise figure, because many areas of the tropics are, in botanical terms, relatively unknown. Expeditions record new species, but many are lost before they are known to science. On a global scale the main threat to plant species is loss through forest clearance, wetland drainage, the spread of towns and modern farming methods. Tropical hardwoods, orchids and cacti are also exploited more directly – by smugglers and collectors.

Measures to halt the decline of plant species have met with limited success. At a national or state level the first step is to document the plant species that are in need of conservation attention. Most temperate countries now have plant "red data" books, which list all the threatened species, but these have not yet been prepared for the tropics. People living traditional life-styles within the tropical rain forests use a wide variety of plants to supply many of their basic needs. As a genetic storehouse the rain forests hold great potential as a source of medicinal plants, fruits, gums, fibres, resins, oils and biochemicals for industry. Without up-to-date information on these potentially life-saving plants, no one can be sure of their continued existence.

Danger lists

Nearly all European countries have incorporated lists of threatened plant species into conservation legislation, with the result that these species are protected in the wild. The main piece of international legislation relating to plant conservation is the Convention on International Trade in Endangered Species on Wild Fauna and Flora (CITES). This agreement has been signed by over 100 countries, which are obliged to control the import and export of a range of threatened plants and animals, and their products.

Natural protection

Protection of plant species in their natural habitats is the best form of conservation. This allows populations to evolve in their native environment, and means that ecosystems can be preserved along with the species they contain. In many parts of the world, information on plant species occurring within national parks and other protected areas is still very scarce.

Certain vegetation types are in urgent need of greater protection. For example, less than 4 percent of the world's tropical rain forest is protected in national parks or reserves. Frequently, the protected areas that have been declared exist on paper only and suffer from illegal habitat clearance.

Endangered plant species can benefit from artificial propagation in nurseries and botanic gardens. There have been some cases of successful reintroduction to the wild of plants that are close to extinction. A plant conservation project on the island of St. Helena in the South Atlantic, for example, is helping to save one of the world's most endangered island floras. Islands frequently have a high proportion of endemic plant species that grow nowhere else in the world. These species are often particularly vulnerable to extinction because of their limited distribution and because they are threatened by introduced species of plants and animals.

St. Helena has 50 endemic species, seven of which have already been lost in the wild. A further 40 are threatened with extinction. Over the past 10 years, plants of 14 threatened species have been propagated and replanted in the wild. One of the main successes has been with the St. Helena ebony, *Trochetiopsis melanoxylon*. This was thought to be extinct until a solitary tree was found in 1980. Now over 3,000 cuttings have been planted on the island.

***Plants are in danger** across the planet for a variety of reasons. Here, we introduce a few of the species most at risk, arranged as follows:*

1 ***Scientific name of plant***
▲ *common name of plant*
◆ *location*
● *description*
■ *habitat*
✖ *how plant is threatened*

1 ***Serruria roxburghii***
▲ *No common name*
◆ *South Africa, northwest of Wellington*
● *Small, delicate shrub*
■ *Dry places in deep, white, sandy soil*
✖ *Extension of farmlands; restricted to two isolated patches of natural vegetation.*

2 ***Zamia floridana***
▲ *Coontie, wild sago*
◆ *USA, Florida and Georgia*
● *Fernlike plant with featherlike leaves*
■ *In dry, sandy soils, flat pinelands and coastal dunes*
✖ *The tourist industry and the resulting development of resorts throughout Florida and Georgia.*

3 ***Passiflora herbertiana* (*subspecies* *insulaehowei*)**
▲ *Passionfruit*
◆ *Lord Howe Island*
● *Climber*
■ *Lowland rain forest*
✖ *At present only known from two definite locations. One site is threatened by grazing from cattle and possible future housing; the other is adjacent to a track used by cattle and feral pigs.*

4 ***Artemisia granatensis***
▲ *No common name*
◆ *Sierra Nevada, Spain*
● *Cushion-shaped perennial*
■ *Acid schists at 10,500 ft (3,200 m) or above*
✖ *Continual collecting; the leaves and stalks were once popular as an ingredient in drinks. The plant is scarce but still sought after.*

5 ***Swallenia alexandrae***
▲ *Eureka dune grass*
◆ *United States, California*
● *Stiff, perennial grass with erect flowering shoots*
■ *Sand dunes in creosote scrub in deserts*
✖ *Off-road vehicles like trail bikes and dune buggies.*

6 ***Caryota no***
▲ *Fishtail palm*
◆ *Indonesia and Malaysia, where it usually only occurs as a single tree*
● *A spectacular palm, the largest species in the genus*
■ *Lowland tropical forest*
✖ *Mainly the destruction of its habitat, but also local people, who remove the edible apex or "cabbage" to use as a vegetable; villagers also obtain sago from the pith of the trunk, thus killing the tree.*

7 ***Phoenix theophrash***
▲ *Cretan date palm*
◆ *Crete and southwest Turkey*
● *A many-stemmed palm up to 30 ft (10 m) high*
■ *Usually on sandy alluvial sites close to the sea*
✖ *Tourists, people camping under the trees and autos being driven into the groves, all of which prevent regeneration of the species. The palms could prove invaluable for breeding new hybrid cultivars of dates.*

8 ***Lodoicea maldivica***
▲ *Coco de mer, double coconut*
◆ *Seychelles*
● *Palm bearing a black, usually 2-lobed, nut weighing 22–50 lb (10–18 kg)*
■ *Hill slopes and valleys from near sea level to about 1,000 ft (300 m)*
✖ *Mainly severe exploitation of the nuts for tourists.*

9 ***Erica chrysocodon***
▲ *No common name*
◆ *South Africa, in the Fransch Hoek Mountains*
● *Small, erect shrub up to 18 in (45 cm) high*
■ *Marshes*
✖ *The few remaining mature individuals occur on the edge of a firebelt. Other threats include illegal picking and the possible widening of a road alongside the colony.*

Connections: Algae 112 Animal Conservation 344 Boreal Forests 316 Conifers and Cone Bearers 118 Cycles of Life 206 Flowering Plants 120 Fungi 110

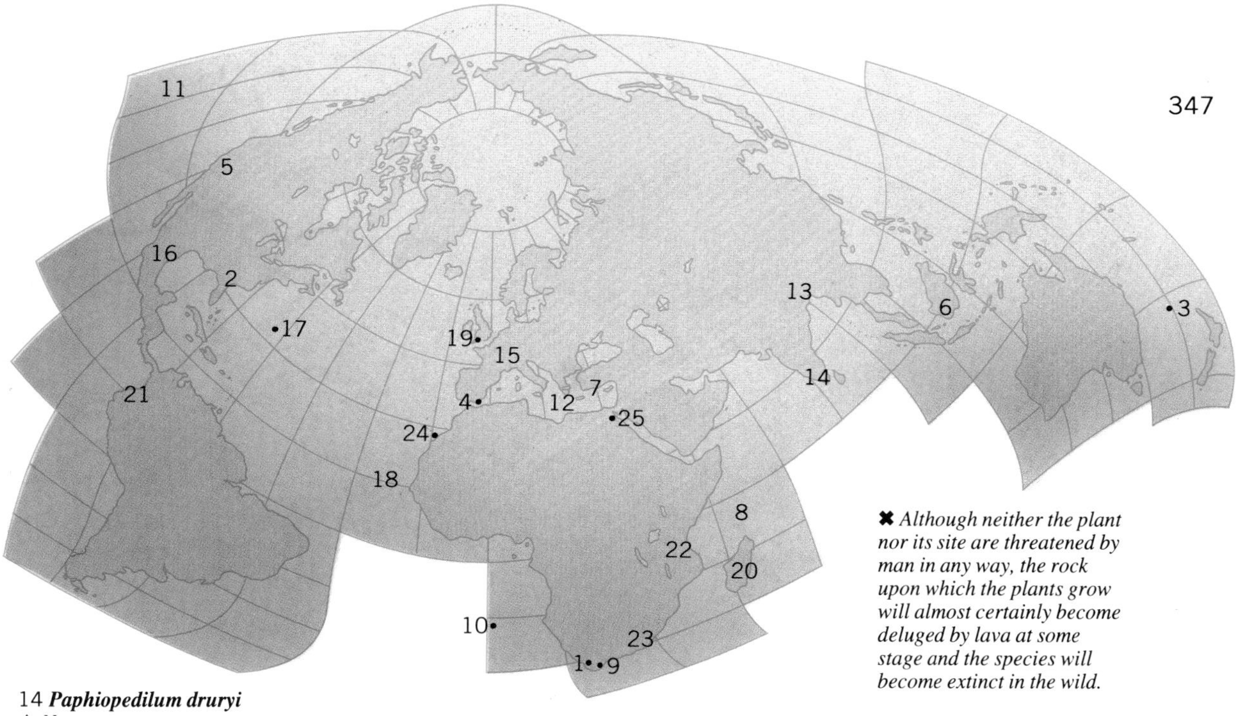

10 ***Wahlenbergia linifolia***
▲ *No common name*
◆ *St. Helena, South Atlantic*
● *Small shrub*
■ *On rocky bluffs on the windward side of the island's central ridge*
✖ *Easily swamped by flax and other tall plants. It was also heavily damaged by the introduction of goats.*

11 ***Hibiscadelphus giffardianus***
▲ *No common name*
◆ *Hawaii*
● *Tree about 25 ft (7 m) tall*
■ *Originally perched on the outer, southeastern rim of a collapsed lava tube, which discouraged browsers.*
✖ *Only one tree was ever found in the wild and this died in 1930. The species had already been brought into cultivation and could be reintroduced to the wild. By 1968 there were 10 healthy trees. The plant was probably heavily depleted by lava flows. More recently, cattle have destroyed much of the remaining habitat and disturbed the plant's pollinator, a rare endemic Hawaiian bird, which has also declined in numbers.*

12 ***Narcissus viridiflorus***
▲ *Green daffodil*
◆ *Mediterranean area*
● *Perennial herb*
■ *Coastline and islands*
✖ *The overcollecting of the underground bulbs; also the exploitation of the habitat. The bulbs are now almost unobtainable commercially.*

13 ***Vanda coerulea***
▲ *Blue vanda*
◆ *India, Myanmar, Thailand.*
● *An orchid with spectacular, skyblue flowers*
■ *Humid evergreen forests*
✖ *Massive plant collection and exportation. Also the oak trees on which these plants grow are harvested for wood and charcoal.*

14 ***Paphiopedilum druryi***
▲ *No common name*
◆ *Kerala State, south India*
● *Orchid*
■ *In sunny forest floors amidst grasses, sedges, etc., with, and sometimes growing on,* Euphorbia *species*
✖ *Possibly extinct in the wild, through forest fires and excessive collecting.*

15 ***Dianthus gratianopolitanus***
▲ *Cheddar pink*
◆ *France, Switzerland and England*
● *Pink herb*
■ *Steep, rocky slopes*
✖ *Nearly annihilated in the wild by plant collectors*

16 ***Ariocarpus trigonus***
▲ *Living rock cactus*
◆ *Mexico*
● *Stonelike cactus*
■ *Desert*
✖ *Threatened by collection, it is slow-growing and lacks the shoots necessary for propagation.*

17 ***Juniperus bermudiana***
▲ *Bermuda juniper*
◆ *Bermuda*
● *Evergreen tree up to 65 ft (20 m) high*
■ *On hillsides and along marshes*
✖ *Approximately 90% of these trees died between 1944 and 1950 due to a severe infestation of two accidentally introduced types of scale insect.*

18 ***Dracaena draco***
▲ *Drago, Canadian dragon tree*
◆ *Canary Islands, Cape Verde Islands and Madeira*
● *Umbrella-shaped tree 15–30 ft (5–10 m) in height, with a silvery-gray trunk*
■ *Stony volcanic cliffs from sea level to 1, 700 ft (500 m).*
✖ *Restricted in the wild to small, scattered populations. If the trunk is cut, it yields a dark red gum resin known as "Dragon's Blood," reputed to have medicinal and magical properties. It has been much exploited and has been destroyed or removed from all accessible sites. Fortunately this species is quite common in cultivation.*

19 ***Rhynchosinapis wrightii***
▲ *Lundy cabbage*
◆ *Lundy Island, off the coast of north Devon in Britain*
● *Perennial herb with a rosette of featherlike leaves*
■ *Mostly restricted to a quarter-mile of cliff in the southeast corner of the island*
✖ *Sheep, goats and deer on the island; the spread of bracken and possibly introduced rhododendron.*

20 ***Catharanthus coriaceus***
▲ *No common name*
◆ *Madagascar; only known from a few localities*
● *Small erect shrub up to 1 ft (40 cm) high*
■ *Rocky patches in low, deciduous forest*
✖ *The forests are intensely threatened by bush fires and are becoming further depleted in size each year.*

21 ***Persea theobromifolia***
▲ *"Caoba"*
◆ *Ecuador*
● *Canopy tree 100–130 ft (30–40 m) tall; flowers gray-green, in panicles*
■ *Canopy species restricted to mature lowland wet forest.*
✖ *The tree's habitat has almost been entirely converted to plantations of bananas and oil palms. The total known population is probably no more than 12 reproducing individuals.*

22 ***Saintpaulia ionantha***
▲ *African violet*
◆ *Isolated hills in East Africa*
● *Small herb with colorful blooms and hairy leaves*
■ *Vertical cracks at the base of cliffs*
✖ *Local people cutting down trees for firewood cause the previously shaded African violets to be exposed to the hot African sun. This bleaches the leaves and destroys the flowers.*

23 ***Aloe polyphylla***
▲ *Spiral aloe*
◆ *Lesotho, southern Africa*
● *A large rosette of triangular leaves producing coral-colored flowers*
■ *In the seepage areas of west-facing slopes*
✖ *Local farmers, who, believing the leaves to have medicinal properties, have chopped up the plants and placed them in the drinking water of their poultry.*

24 ***Centaurea junoniana***
▲ *No common name*
◆ *Canary Islands, in one location in the extreme south of La Palma*
● *Woody perennial with numerous stems 12–40 in (30–100 cm) high*
■ *Rock crevices in the coastal zone, facing the sea*
✖ *Although neither the plant nor its site are threatened by man in any way, the rock upon which the plants grow will almost certainly become deluged by lava at some stage and the species will become extinct in the wild.*

25 ***Cyperus papyrus* subspecies *hadidi***
▲ *No common name*
◆ *Egypt, in the Wadi Natroun in the Western Desert*
● *Giant perennial sedge*
■ *Among reeds in fresh-water marsh around saline soda lakes. Confined to a few scattered locations where fresh water emerges from underground sources.*
✖ *Habitat loss due to water abstraction and changes in irrigation patterns.*

The pink periwinkle *(above) is used to treat childhood leukemia. Many other endangered plants or trees may prove – too late – to be invaluable to medicine. For example, the Moreton Bay chestnut, from the Australian rain forest, has been found to contain a drug that may help in the fight against AIDS.*

Grasslands 302 Mosses and Ferns 116 Temperate Forests 318 Tropical Forest Plants 322

Glossary

acetylcholine A chemical that acts as a neurotransmitter at many synapses.

aerobic (1) Requiring oxygen. (2) An environment containing oxygen. C.f.**anaerobic**.

aestivation State of seasonal torpor (minimal activity) during the summer. C.f. **hibernation**.

alate Winged reproductive form of an insect.

albedo The percentage of the Sun's radiation that is reflected by the Earth back into space.

altricial Birds hatching in an immature and dependent state.

alveolus Tiny spherical chamber in the lungs where gaseous exchange occurs.

amino acids Small molecules from which proteins are made.

anaerobic (1) Not dependent on oxygen. (2) An environment not containing oxygen. C.f.**aerobic**.

angiosperms Group of flowering plants **(dicotyledons** and **monocotyledons)** that bear seeds within matured ovaries.

anther Fertile part of a flower **stamen** that produces **pollen**.

anthracites Coal containing very high proportion of carbon.

antheridium Male reproductive organ that bears sperm in nonseed plants.

apocrine Type of mammalian sweat gland producing viscous, odorous secretion.

archegonium Female reproductive organ that bears the egg cells in various land plants (mosses and ferns).

artery Vessel carrying blood away from the heart.

ATP (Adenosine triphosphate) High-energy molecule used to provide power for metabolic processes in most cells.

autonomic Bodily functions, such as the actions of the heart, smooth muscles and glands, which occur involuntarily.

autotrophic Able to synthesize essential organic compounds from simple inorganic molecules. C.f.**heterotrophic.**

axon Long, thin extension of a **neuron** which conducts impulses away from the cell body.

bacteriophage A virus that infects bacteria.

base Nitrogenous molecule that, along with sugar and phosphate, forms a **nucleotide**.

benthic (1) The lower zone of a body of water. (2) Animals and plants at the bottom of oceans and lakes.

bioluminescence The production of light by living organisms.

biomass Total dry mass of all the living organisms in an area.

biosphere The entire area of the Earth that contains living organisms.

bronchus One of two air passages leading to the lungs.

cambium A layer of **meristematic** tissue in plant stems which produces lateral growth, so increasing diameter.

capillary Very small, thin-walled vessel between arteries and veins, carrying blood to tissues and allowing exchange of materials with cells.

carbohydrate Compound made of carbon, hydrogen and oxygen atoms in a 1:2:1 ratio, such as sugar and starch.

carpel Female sex organs of flower, comprising **stigma**, **style** and **ovaries**.

catalyst A substance that can influence the rate of a chemical reaction but is not itself changed by that reaction.

cerebrum In vertebrates, the largest and anterior part of the brain, consisting of two hemispheres, and associated with personality, emotion and intellect in man.

chitin An insoluble, horny substance that is part of the **exoskeleton** of many invertebrates, particularly arthropods.

chlorofluorocarbons (**CFC**) Chlorine-based compounds used as aerosol propellants, as refrigerants and in foam packaging. CFCs are greenhouse gases, and are also responsible for ozone depletion.

chlorophyll Green pigments found in plants, essential for the absorption of light energy in **photosynthesis**.

chloroplast An **organelle** found in the **cytoplasm** of plant cells, containing **chlorophyll** and the other pigments necessary for **photosynthesis.**

chordate Animals that, at some stage in their development, possess a notochord – a flexible rod that acts as an internal skeleton.

chromatin Complex of **DNA** and proteins found in the **chromosomes**.

chromosome Rod-shaped structure seen in the cell **nucleus** containing **genes** and made up of **DNA.**

cirque A large armchair-shaped hollow high in a mountain, containing a lake or at the head of a glacier.

cloaca A posterior chamber in lower vertebrates which receives urine and digestive wastes, and may also serve as part of the reproductive system.

coleoptile Sheath that protects the **plumule** when the seed germinates.

conceptacle A small chamber in the fronds of some algae and fungi which bears the reproductive organs.

convection currents Movement of particles from high density (low temperature) to low density (high temperature).

Coriolis force An apparent force that deflects the path of a body moving relative to the Earth, due to the rotation of the Earth.

cornea Protective, transparent outer covering of the eye.

cortex (1) The outer layer of an organ. (2) Tissue in plant stem or root between the **epidermis** and the **vascular** tissues.

cyanobacteria Photosynthetic **prokaryotic** microorganisms, also known as blue-greens.

cytoplasm The living contents of the cell, excluding the **nucleus.**

dendrite Branched nerve fiber that conducts nerve impulses toward the nerve cell body.

diatom Unicellular alga with silicon-based cell wall in two tightly fitting halves.

dicotyledon A plant that has an embryo with two cotyledons (seed leaves).

dinoflagellate Unicellular alga, with cellulose shell, and two **flagella**.

diploid Containing two sets of paired **chromosomes**; twice the **haploid** number.

dipole (1) Two electric charges or magnetic poles of equal size but opposite signs which are separated by a small distance. (2) Molecule with separate centers of positive and negative charge.

DNA (deoxyribonucleic acid) The genetic material of all cells found in the **nucleus**, consisting of a double (or sometimes single) strand of **nucleotides** arranged into particular sequences to produce coded information in the form of **genes**.

drumlins Earth mounds formed by glacial deposition.

eccrine Type of mammalian sweat gland producing a watery secretion that is important in temperature regulation.

ecosystem A natural, stable system produced by the interaction of living organisms and the nonliving environment.

ectoparasite Parasite living on the surface of its host.

endocrine gland Ductless gland that secretes hormone directly into the bloodstream.

endocytosis Transport of material into a cell by **vesicles**, which pinch off the cell membrane into the cell and enclose materials as they do so.

endoparasite Parasite living inside its host's body.

endoplasmic reticulum A network of membranes found in the **cytoplasm.** On the rough type, protein synthesis occurs at the **ribosomes**, the smooth type produces **vesicles**.

endosperm Nutritive tissue in the embryo sac of plant seeds which provides food for the developing embryo.

enzyme A protein that catalyses reactions in living organisms. C.f. **catalyst.**

epidermis The outer layer of cells protecting the bodies of plants and animals.

equinox (1) One of the two occasions (vernal and autumnal) in the year when day and night are of equal length. (2) The point at which the Sun is over the equator.

eukaryote Organism whose cells contain a **nucleus**, which is surrounded by a nuclear membrane. C.f. **prokaryote**.

eutheria The subclass consisting of all **placental** mammals.

eutrophication Nutritional enrichment of a body of water that results in massive plant growth. Ultimately, the plants decay, which leads to oxygen depletion and subsequent death of wildlife.

exocytosis Transport of material out of cells by the fusion of **vesicles** inside the cell with the membrane, such that their contents are expelled.

exoskeleton A supportive and protective structure on the outside of the body, made from hard materials, including **chitin**.

flagellum A long, taillike extension found on the surface of many cells, used primarily for locomotion.

follicle Small cavity or sac of cells.

fovea Area of the **retina** where visual resolution is greatest.

gametes The **haploid** sex cells – in females the egg and in males the sperm – two of which fuse to form a **zygote.**

gametophyte The **haploid** or **gamete**-producing stage in a plant's life cycle.

ganglion A cluster of nerve cells, usually outside the central nervous system.

gene Unit of hereditary information specified by a particular sequence of **DNA** in a **chromosome.**

Golgi body or **Golgi apparatus** Membranous **organelle** found in **eukaryotic** cells responsible for modifying and sorting proteins made in the **endoplasmic reticulum**.

Gutenberg discontinuity The discontinuity that separates the Earth's **mantle** from the core.

guyot Submarine mountain whose top has been flattened by the erosive action of the sea.

gymnosperms Group of seed plants whose seeds are not enclosed in **ovaries.** The seeds are commonly carried in cones, like those of conifers and cycads.

haploid Containing one set of **chromosomes,** half the **diploid** number.

haustorium A threadlike extension of **hypha** which is used to penetrate a plant's tissues and absorb nutrients.

hemotoxin A poison affecting the blood.

hermaphrodite An organism that has both male and female reproductive organs, and can produce both male and female **gametes.**

heterotrophic Requiring both organic and inorganic molecules to synthesize the organic compounds essential for life. C.f. **autotrophic.**

hibernation State of seasonal torpor (minimal activity) during the winter. C.f. **aestivation**.

homoiothermic "Warm blooded" – maintaining a constant internal body temperature using metabolic heat.

hypha A thread or filament found in groups of branching fungi.

hypothalamus Part of the vertebrate brain, situated at the base of the brain, that controls hunger, thirst and related **autonomic** functions.

igneous rocks Rocks formed by cooling from a molten state. C.f. **metamorphic**, **sedimentary**.

inflorescence A cluster of flowers.

infrared Part of the light spectrum invisible to humans, with wavelengths longer than red light.

instar A stage in an insect's development between molts.

ion Atom or molecule with electrical charge.

Jacobson's organ An accessory scent organ in the roof of the mouth of some reptiles.

keratin A horny, protein-based substance found in nails and in some horns.

lateral line A line of sensory organs in fish that can detect nearby vibrations in the water.

lignin Hard, woody substance found in plant stems and roots.

lipids Fatty molecules insoluble in water which are an important component of membranes.

lithosphere The Earth's crust and part of its mantle, forming a rigid shell.

littoral (1) Zone relating to the shore of lakes, and where **photosynthesis** can take place. (2) The intertidal zone of the seashore. (3) Relating to animals that inhabit the littoral zone.

loess Clay and silt, made up of very fine particles, deposited by the wind.

lysosome An **organelle** found in many cells, containing **enzymes** that break down molecules.

mandibles The lower jaw in vertebrates, or the principal external mouthparts in many invertebrates.

mantle Generally molten part of the Earth lying between the **Moho** and **Gutenburg discontinuities**.

meiosis Cell division that reduces by half the number of **chromosomes** and thus gives rise to **haploid** cells, the male or female **gametes**.

meristem The zone of growth in a plant where cells divide by **mitosis**.

mesosphere Part of the atmosphere where temperature decreases with height, between 30 and 50 miles above the Earth.

metamorphic rocks A class of rocks that have been changed from their original form by the action of heat or pressure. C.f. **igneous, sedimentary**.

metatheria A subclass of primitive mammals that give birth to live young, which then develop in the mother's pouch. Made up of only the marsupial order.

microclimate Separate environment in a small space next to an object and influenced by that object.

microtubule Protein-based scaffold-like structure in the **cytoplasm** of many cells.

mitochondrion A spherical or thread-like organelle found in many types of cell, whose main function is to produce **ATP**.

mitosis Division of the cell **nucleus** that provides each daughter cell with a complete set of **chromosomes**.

Mohorovičić discontinuity (Moho) The boundary between the Earth's crust and **mantle**, from 6 miles deep under the ocean to 44 miles deep beneath mountain belts.

monocotyledon A plant that has an embryo with only one cotyledon (seed leaf).

monogamy Mating with only one partner.

moraine Debris carried and deposited by glaciers in various forms, terminal, lateral, medial and ground.

mycelium The mass of **hyphae** that makes up the vegetative body of fungi.

nekton Free-swimming aquatic organisms.

neoteny The persistence of juvenile characteristics in an adult form.

neuron A nerve cell, consisting of a cell body, and **dendrites** and an **axon**, which conduct nervous impulses.

neurotoxin A poison affecting the nervous system.

neurotransmitter A chemical used by nerve cells to comunicate with each other, usually across a **synapse**, so transmitting a nerve impulse.

nucleosome A beadlike subunit of **chromatin**. It is made up of a length of **DNA** wrapped around a protein core.

nucleotides Molecules made up of a nitrogen **base**, a sugar and a phosphate group, which form the building blocks of **DNA** and **RNA.**

nucleus (1) **Organelle** that is the control center of **eukaryotic** cells, containing the genetic information in the form of **DNA**. (2) The core of an atom, containing the protons and neutrons.

ommatidium One of the visual units of the compound eye of arthropods and mollusks.

oogonium An egg-bearing sex organ found in some fungi and algae.

operculum (1) A bony fold that covers and protects the gill slits of fish. (2) In some plants, an opening that allows the dispersal of seeds.

organelle Small, specialized structure within a cell that allows the cell to compartmentalize different functions.

orogeny The **tectonic** processes that produce mountains by the deformation of the Earth's crust.

osmosis The movement of a solvent through a semipermeable membrane (which allows only a solvent to cross and not dissolved particles) from a lower concentration solution to a higher concentration.

ovary (1) In plants, a structure at the base of the **carpel** containing the **ovules**. (2) In animals, one of the pair of female gonads, which produce the eggs.

ovipositor The egg-laying organ found in female insects.

ovule The part of the seed plant found in the **ovary** that contains the egg cell and develops into the seed after fertilization.

ovum The female **gamete**, or egg cell.

oxbow lake A lake created when a river meander is cut off by deposition of silt.

pelagic (1) The middle to surface zone of a body of water, usually in reference to the sea. (2) Organisms that inhabit this specific zone.

petiole A stalk that attaches a leaf to the plant stem.

pheromones Chemicals secreted by animals to affect the behavior, usually sexual, of other animals.

phloem The **vascular** tissue that transports sugars in most plants.

photon A discrete packet of light energy.

photosynthesis The production of carbohydates from carbon dioxide and water using the energy of light.

phytoplankton Tiny, **photosynthetic** algae found in seas, oceans and lakes.

pinna Outer part of the mammalian ear.

pistil The **stigma**, **style** and **ovary** of a flower.

pituitary The major vertebrate **endocrine gland**, found at the base of the brain and responsible for the production of various hormones.

placenta The organ that connects the fetus with its mother in the womb of placental mammals, and through which nutrients and gases are exchanged.

plastid A class of membrane-bound **organelles,** including **chloroplasts,** found in **eukaryotic**, **photosynthetic** cells.

plumule The first embryonic shoot poduced by a plant.

poikilothermic "Cold blooded" – having a body temperature that is reliant on external environmental factors, such as the Sun.

pollen Microscopic grains in seed plants that contain the male **gamete**.

pollinium Sticky clump of **pollen** grains found in orchids.

polyandry Females mating with more than one male.

polygyny Males mating with more than one female.

polynya An open body of water lying among sea ice.

precocial Birds hatching in a condition in which they are immediately able to move about. C.f. **altricial.**

prehensile Adapted for grasping; usually in reference to tails that can wrap around objects for support.

prokaryote Cell in which the **nucleus** is not surrounded by a nuclear membrane. C.f. **eukaryote**.

prototheria The subclass of primitive mammals now found only in Australasia, consisting of the monotremes (egg layers).

pupa Inactive stage in the development of some insects. It follows the larval stage and comes before the adult stage.

pupil Aperture in the center of the eye through which light enters.

radicle The first embryonic root produced by a plant.

refraction Change in direction of sound or light waves as they pass from one medium to another, forcing a change in velocity.

respiration (1) The breaking down of food molecules in the cell to generate energy, using oxygen and producing carbon dioxide. (2) The exchange of gases between an organism and its environment.

retina The inner layer of the eyeball, containing the light-sensitive cells.

rhizome An underground, horizontal stem that may develop new plants from buds.

ribosome A small **organelle** made of protein and **RNA** where protein synthesis takes place in the cell.

Richter scale Scale that measures the size of an earthquake.

rift valley A long, thin valley produced when land between two faults subsides.

RNA (ribonucleic acid) Single-stranded molecule of **nucleotides**, occurring in various forms – messenger **RNA**, ribosomal **RNA**, transfer **RNA** – which are involved in the synthesis of protein.

roche moutonnée Rocky mound formed by glacial erosion.

Rossby wave Vast wave occurring in an atmospheric jet stream, giving rise to depressions.

ruminant Mammal that chews the cud, ie, regurgitates its food in order to chew it more.

sebum Oily secretion produced by skin glands that protects and lubricates hair and skin.

sedimentary rocks Rocks formed from organic remains, chemical precipitates or weathered particles of preexisting rock. C.f. **igneous**, **metamorphic**.

semicircular canals Liquid-filled structures found in the inner ear of most vertebrates (usually two or three in number) vital for balance and orientation.

sepals Leaflike outermost parts of a flower that protect the rest of the flower enclosed in the bud.

solstice (1) Either the longest day of the year (summer solstice) or the shortest (winter solstice). (2) The point when the Sun is over the Tropic of Capricorn or the Tropic of Cancer.

spermatophore Packet of sperm cells enclosed in a capsule produced by various invertebrate and some vertebrate males.

sporangium Spore case found in some plants and fungi.

sporogonium Spore-bearing capsule found in mosses and liverworts.

sporophyte The **diploid** stage of a plant's life cycle that produces spores by **meiosis.**

stamen Male sex organs of flower, comprising an **anther** and a filament.

stigma Part of the **carpel** that is receptive to **pollen**.

stomata Tiny pores in the leaf **epidermis**, each surrounded by two guard cells, through which gases are exchanged.

stratosphere Part of the atmosphere containing the ozone layer, up to 30 miles above Earth. The stratosphere absorbs heat and radiation.

stratovolcano Cone-shaped, crater-topped volcano that has blown its top off after intense buildup of internal pressure.

style Part of the **carpel** connecting the **stigma** to the **ovary.**

stylet A slender, pointed probe, used by many fluid-feeding invertebrates to penetrate tissues.

subduction zone Area where one **tectonic** plate moves under another.

sublimation The vaporization of a solid, without passing through a liquid stage.

sunspot Disturbance on the surface of the Sun which appears as a dark spot.

supercooling Cooling a liquid down to a temperature at which freezing should occur but does not.

symbiosis An intimate association between two or more living organisms that are dependent upon each other.

synapse Junction between nerve cells where the nervous impulse is chemically transmitted by **neurotransmitters.**

tectonics The theory that the Earth's **lithosphere** is made of rigid, mobile plates.

teleosts The group of modern fish; bearing scales, bony skeletons and paired fins.

thermocline Boundary between two layers of water, the lower of which is much cooler.

thermosphere Part of the atmosphere between 50 and 250 miles above the Earth, where temperature increases steadily with height, up to a maximum of over 2,910°F.

thorax (1) In mammals the upper body between the head and the abdomen. (2) The middle division of the arthropod body.

thylakoids Membranous disks, arranged in stacks, found inside **chloroplasts**.

transpiration The loss of water by evaporation from the surface of leaves.

troposphere The lowest part of the atmosphere, reaching from the surface to about 6 miles above the Earth at the poles, and to about 11 miles at the equator.

tsunami Large, destructive wave caused by submarine geological events, such as earthquakes.

ultraviolet Part of the light spectrum invisible to humans, with wavelengths shorter than violet light. Some ultraviolet light is thought to cause skin cancers.

uterus Part of the female reproductive tract in which the fetus develops, also known as the womb.

vacuole A fluid-filled space within cells, bounded by a membrane.

vascular bundle Vessels made up of xylem and phloem tissue that carry sugars, water and minerals in plants.

vein (1) Vessel that carries blood to the heart. (2) The **vascular bundle** in a leaf which also acts as support for the leaf blade.

vesicle A small, spherical membranous sac or cavity found in many types of cell.

wadi A desert watercourse that is normally dry and flows only during the rainy season.

xylem The **vascular** tissue that carries water and dissolved minerals in most plants.

zooplankton Tiny, **nonphotosynthetic** organisms found in lakes and seas.

zoospore An asexually produced spore that can move by use of a whiplike **flagellum**.

zygote A fertilized egg.

Index

Page references in *italics* refer to illustration, captions or boxed text.

E

F

Acknowledgments

Picture Credits

P.14		Jack Finch
P.15		David C Fritis
P.20		Dieter and Mary Plage/Bruce Coleman
P.22		GSF Picture Library
P.25	top	Alfred Pasieka/SPL
	left	Patricia Tye
	right	Geoscience Features
P.26	left	Alain le Garsmeur/Impact
P.26	right	Dr Jeremy Burgess
P.28		By courtesy of Trustees of the British Museum
P.30	top	Bob Gibbons/Ardea
	bottom	G.A. Maclean/OSF
P.36		Sinclair Stammers
P.37	top	Haroldo Palo/nhpa
	bottom	John Mason/Ardea
P.38		Alfred Pasieha/Bruce Coleman
P.40		Franz-Dietrich Miotke
P.43	top	John Lythgoe/Planet Earth
	bottom	Joel Bennett/Survival Anglia
p.49		Ronald Toms/OSF
P.54		Alan Hutchison Library
P.56		Messerchmidt/ZEFA
P.58		Don Hadden/Ardea
P.60		Michael Fogden/OSF
P.66		Sinclair Stammers
P.71		MSA/SPL
P.73		Pillitz/Network
P.76		Claude Nuridsany & Marie Perennou/SPL
P.78		H. Binz/FLPA
P.80		A Weiner/Liaison/Gamma/Frank Spooner
P.82		SPL
P.86-7		Colorific
P.92		Ken Lucas/Planet Earth
P.104		Dr Lee Simon/SPL
P.105	left	Tektoff/CNRI/SPL
	bottom	Omikron/SPL
	center	NIBSC/SPL
P.111		David Scharf/SPL
p.116		Hans Reinhard/Bruce Coleman
P.118		Sid Roberts/Ardea
P.120	left	Laurie Campbell/NHPA
	right	Alfred Pasieka/Bruce Coleman
	center	Wanscheidt/Bruce Coleman
P.125		Planet Earth
P.139		John Mason/Ardea
P.140		C. Carvalho/FLPA
P.144		Orion/Bruce Coleman
P.153	center	Bruce Iverson/SPL
	bottom	Dr Jeremy Burgess/SPL
P.155		MPL Fogden/Bruce Coleman
p.157		Holt Studios
p.159		Premaphotos
P.163	top	Jane Burton/Bruce Coleman
	bottom	Jane Burton/Bruce Coleman
P.165		Ken Lucas/Planet Earth Pictures
P.170-1		Gerald Cubitt/Bruce Coleman
P.172		Bruce Thomson/NHPA
P.184		Nuridsany & Perenous/SPL
P.186		Bruce Coleman
P.187		Premaphotos
P.188	left	C.B. & D.W.Frith/Bruce Coleman
	right	David Scarf/SPL
P.190		Premaphotos
P.191		ABiophoto Associates
	bottom	JAL Cooke/OSF
p.192		Clem Haagner/Ardea
P.194		Adrian Warren/Ardea
P.195		K.G.Vock/Okapia/OSF
P.198		David Shale/Survival Anglia
P.199		ANT/NHPA
P.210		Herwarth Voigtmann/Planet Earth
P.213		Norbert Wu
P.220		Konrad Wolke/Bruce Coleman
P.223		Martin Dohrn/SPL
P.233		David Thompson/OSF
P.234		Premaphotos
P.237		Premaphotos
P.238		Ivan Polunin/NHPA
P.248	left	Sally Anne Thompson/Animal Photography
	top	Christian Gazimek/Okapia/OSF
	center	Clem Haagner/Ardea
	bottom	Eichhorn/Zingel/FLPA
P.252		Jane Burton/Bruce Coleman
P.254	top	Mantis Wildlife Films/OSF
	bottom	Peter Parks/OSF
P.255		Robert J Erwin/NHPA
P.256		Erwin & Peggy Bauer/Bruce Coleman
P.261		P.Permy/FLPA
P.265	top left	LP.Morris/Ardea
	top right	Bryn Campbell/Biofotos
	left	Dieter Mary Plage/Survival Anglia
	right	Peter Scones/Planet Earth
P.266		OSF
p.268		Premaphotos
P.270		Peter Ward/Bruce Coleman
P.276	left	Planet Earth Pictures
	bottom	Michael Fogden/OSF
P.290		NHPA
P.292		David Thompson/OSF
P.298	left	David Curl/OSF
	right	David Macdonald/OSF
P.340		John E. Swedberg/Ardea
P.342		J.B.O'Rourke/ZEFA
P.345		Eichhorn/Zinger/FLPA
P.346		Stephen Kraseman/NHPA

Artists and agents

Eric T. Budge
Joanne Cowne
John Davies
Bill Donohoe
Sandra Doyle
Roy Flooks
John Francis
Mick Gillah
Greensmith Asso.
Ron Haywood
Kevin Jones Asso.
Frank Kennard
Pavel Kostel
Ian Lewington
Richard Lewington
Richard Lewis
Ken Lilley
Ruth Lindsey
Sue McCormick
Malcolm McGregor
Colin Rose
Peter Sarsons
Les Smith
Malcolm Smythe
Ed Stuart
Bernard Thornton Agency
Peter Visscher